Use of sediment quality guidelines and related tools for the assessment of contaminated sediments

Other titles from the Society of Environmental Toxicology and Chemistry (SETAC)

Ecological Assessment of Aquatic Resources: Linking Science to Decision-Making
Barbour, Norton, Preston, Thornton, editors
2004

Amphibian Decline: An Integrated Analysis of Multiple Stressor Effects
Greg Linder, Sherry K. Krest, Donald W. Sparling
2003

Metals in Aquatic Systems:
A Review of Exposure, Bioaccumulation, and Toxicity Models
Paquin, Farley, Santore, Kavvadas, Mooney, Winfield, Wu, Di Toro
2003

Silver: Environmental Transport, Fate, Effects, and Models:
Papers from Environmental Toxicology and Chemistry, 1983 to 2002
Gorusch, Kramer, La Point
2003

Contaminated Soils: From Soil–Chemical Interactions to Ecosystem Management
Lanno, editor
2003

Environmental Impacts of Pulp and Paper Waste Streams
Stuthridge, van den Heuvel, Marvin, Slade, Gifford, editors
2003

Porewater Toxicity Testing: Biological, Chemical, and Ecological Considerations
Carr and Nipper, editors
2003

Reevaluation of the State of the Science for Water-Quality Criteria Development
Reiley, Stubblefield, Adams, Di Toro, Erickson, Hodson, Keating Jr, editors
2003

Ecological Risk Assessment of Contaminated Sediments
Ingersoll, Dillon, Biddinger, editors
1997

For information about SETAC publications, including SETAC's international journals, *Environmental Toxicology and Chemistry* and *Integrated Environmental Assessment and Management*, contact the SETAC Administrative Office nearest you:

SETAC Office
1010 North 12th Avenue
Pensacola, FL 32501-3367 USA
T 850 469 1500 F 850 469 9778
E setac@setac.org

SETAC Office
Avenue de la Toison d'Or 67
B-1060 Brussels, Belgium
T 32 2 772 72 81 F 32 2 770 53 86
E setac@setaceu.org

www.setac.org
Environmental Quality Through Science®

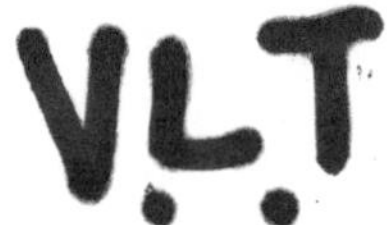

Use of sediment quality guidelines and related tools for the assessment of contaminated sediments

EDITED BY
Richard J. Wenning
Graeme E. Batley
Christopher G. Ingersoll
David W. Moore

PROCEEDINGS FROM THE PELLSTON WORKSHOP ON
USE OF SEDIMENT QUALITY GUIDELINES AND RELATED TOOLS
FOR THE ASSESSMENT OF CONTAMINATED SEDIMENTS
18–22 AUGUST 2002
FAIRMONT, MONTANA, USA

COORDINATING EDITOR OF SETAC BOOKS
Andrew Green
INTERNATIONAL LEAD ZINC RESEARCH ORGANIZATION
RALEIGH, NORTH CAROLINA, USA

PUBLISHED BY THE SOCIETY OF ENVIRONMENTAL TOXICOLOGY AND CHEMISTRY (SETAC)

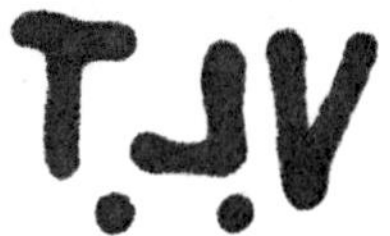

Cover by Kerri Charlton
Indexing by IRIS

Library of Congress Cataloging-in-Publication Data

Pellston Workshop on Use of Sediment Quality Guidelines and Related Tools for the Assessment of Contaminated Sediments (2002 : Fairmont Hot Springs, Mont.)
Use of sediment quality guidelines and related tools for the assessment of contaminated sediments / edited by Richard J. Wenning ... [et al.].
p. cm.
Proceedings from the Pellston Workshop on use of Sediment Quality Guidelines and Related Tools for the Assessment of Contaminated Sediments, held 18-22 August 2002 in Fairmont, Montana, USA.
Includes bibliographical references and index.
ISBN 1-880611-71-6 (alk. paper)
1. Sediments (Geology)—Analysis—Congresses. 2. Contaminated sediments—Congresses. I. Wenning, R. J. (Richard J.) II. SETAC (Society) III. Title.

QE471.2.P45 2005
551.3'04—dc22 2004058966

International Standard Book Number 1-880611-71-6

Printed in the United States of America
10 09 08 07 06 05 10 9 8 7 6 5 4 3 2 1

∞ The paper used in this publication meets the minimum requirements of the American National Standard for Information Sciences—Permanence of Paper for Printed Library Materials, ANSI Z39.48-1984.

Reference Listing: Wenning RJ, Batley GE, Ingersoll CG, Moore DW, editors. 2005. Use of sediment quality guidelines and related tools for the assessment of contaminated sediments. Pensacola (FL): Society of Environmental Toxicology and Chemistry (SETAC). 815 p.

SETAC Publications

Books published by the Society of Environmental Toxicology and Chemistry (SETAC) provide in-depth reviews and critical appraisals on scientific subjects relevant to understanding the impacts of chemicals and technology on the environment. The books explore topics reviewed and recommended by the Publications Advisory Council and approved by the SETAC North America Board of Directors, SETAC Europe Council, or the SETAC World Council for their importance, timeliness, and contribution to multidisciplinary approaches to solving environmental problems. The diversity and breadth of subjects covered in the publications reflect the wide range of disciplines encompassed by environmental toxicology, environmental chemistry, hazard and risk assessment, and life-cycle assessment. SETAC books attempt to present the reader with authoritative coverage of the literature, as well as paradigms, methodologies, and controversies; research needs; and new developments specific to the featured topics. The books are generally peer reviewed for SETAC by acknowledged experts.

SETAC publications, which include Technical Issue Papers (TIPs), workshop summaries, newsletter (SETAC Globe), and journals (*Environmental Toxicology and Chemistry* and *Integrated Environmental Assessment and Management*), are useful to environmental scientists in research, research management, chemical manufacturing and regulation, risk assessment, life-cycle assessment, and education, as well as to students considering or preparing for careers in these areas. The publications provide information for keeping abreast of recent developments in familiar subject areas and for rapid introduction to principles and approaches in new subject areas.

SETAC recognizes and thanks the past coordinating editors of SETAC books:

C.G. Ingersoll, Columbia Environmental Research Center
U.S. Geological Survey, Columbia, MO, USA

T.W. La Point, Institute of Applied Sciences
University of North Texas, Denton, TX, USA

B.T. Walton, U.S. Environmental Protection Agency
Research Triangle Park, NC, USA

C.H. Ward, Department of Environmental Sciences and Engineering
Rice University, Houston, TX, USA

Acknowledgments

This book presents the proceedings of a Pellston Workshop convened by the Society of Environmental Toxicology and Chemistry (SETAC) in Fairmont, Montana, USA, in August 2002. The 55 scientists involved in this workshop represented 8 countries and offered expertise in ecology, ecotoxicology, engineering, environmental regulation, and risk assessment. Their goals were to examine the scientific underpinnings of sediment quality guidelines (SQGs) and to provide recommendations on the appropriate use of SQGs in the assessment and management of contaminated sediments.

The workshop was made possible by the generous support of many organizations, including

- American Chemistry Council,
- American Petroleum Institute,
- Autorità Portuale di Venezia (Italy),
- CSIRO-Australia,
- E.I. duPont de Nemours & Company,
- ENVIRON International Corporation,
- Environment Canada,
- International Lead Zinc Research Organization,
- Kennecott Utah Copper Corporation,
- MEC Analytical Systems (now Weston Solutions, Inc),
- National Science Foundation,
- Nations Port,
- Port Authority of New York / New Jersey,
- Rohm and Haas Company,
- San Diego Unified Port District,
- Southern California Coastal Water and Resource Protection Agency,
- US Army Corps of Engineers,
- US Geological Survey, and
- US Navy.

The workshop also would not have been possible without the very capable management and excellent guidance provided by Taylor Mitchell, Linda Longsworth, and Greg Schiefer and the support of SETAC Executive Director Rod Parrish. In particular, the efforts of Mimi Meredith in the production of this book, despite the inconvenience of a hurricane in September 2004, are gratefully acknowledged.

Richard J. Wenning
Graeme E. Batley
Christopher G. Ingersoll
David M. Moore

Contents

7 Role of Sediment Quality Guidelines and Other Tools in Different Aquatic Habitats... 267

PETER M CHAPMAN, WESLEY J BIRGE, ROBERT M BURGESS, WILLIAM H CLEMENTS, W SCOTT DOUGLAS, MICHAEL C HARRASS, CHRISTER HOGSTRAND, DANNY D REIBLE, AMY H RINGWOOD

13 Predictive Ability of Sediment Quality Guidelines Derived using Equilibrium Partitioning 557

DOMINIC M DI TORO, WALTER J BERRY, ROBERT M BURGESS, DAVID R MOUNT, THOMAS P O'CONNOR, RICK C SWARTZ

14 Use of Sediment Quality Guidelines in Damage Assessment and Restoration at Contaminated Sites in the United States .. 589

RON GOUGUET

15 Evaluation of Sediment Quality Guidelines for Management of Contaminated Sediments in New York–New jersey Harbor 607

W SCOTT DOUGLAS, MARK REISS, JAMES LODGE

16 Influence of Confounding Factors on SQGs and Their Application to Estuarine and Marine Sediment Evaluations 633

JACK Q WORD, WILLIAM W GARDINER, DAVID W MOORE

List of figures

List of tables

About the editors

Richard J. Wenning is a Principal and leader of the sediment and ecology practice at ENVIRON. He has more than 15 years' experience using human health and ecological risk assessment and risk-driven problem-solving methods to evaluate and develop solutions to environmental contamination. He is widely regarded as an expert in several aspects of chemical risk assessment, including source identification, exposure modeling, food web modeling, and probabilistic analysis. He has considerable experience in both human health and ecological risk assessment of persistent, bioaccumulative organic chemicals. Rick has served as an independent peer-reviewer of exposure and risk studies for the US Environmental Protection Agency (USEPA), several state agencies, and agencies in Australia and Italy. He has provided expert testimony on exposure assessment and chemical fingerprinting in product liability litigation and negotiations for allocating cleanup costs at hazardous waste sites. He has managed interdisciplinary teams of engineers and scientists on multi-year contaminated sediment investigations, HSWA facility assessments, and contaminated property investigations. He has worked with clients throughout the US and in Australia, Canada, China, Italy, and New Zealand. Rick is active on the science advisory boards of several professional organizations and has published extensively on human health, ecological, and contaminated sediment issues. He is a senior editor of several scientific publications and the editor-in-chief of SETAC's journal *Integrated Environmental Assessment and Management.*

Graeme E. Batley is a Chief Research Scientist and Director of the Centre for Environmental Contaminants Research, CSIRO Energy Technology, (a division of Australia's government scientific research organization CSIRO) based in Sydney, Australia. He has been an active researcher for more than 30 years in the analytical and environmental chemistry of trace contaminants in natural water systems, in particular with respect to trace metal and organometal speciation and bioavailability in estuarine waters and sediments. Graeme played a lead role in the development of sediment quality guidelines for Australia and New Zealand and in the recent revision of their water quality guidelines. He is author of some 300 scientific publications, and editor of several books. He was recipient of the Royal Australian Chemical Institute's Analytical Medal in 1991, its Environment Medal in 1995, and shared in the CSIRO Chairman's Medal in 1996. He holds BSc (Hons1), MSc, PhD, and DSc degrees from the University of New South Wales, Sydney, Australia. Graeme was instrumental in the formation of SETAC Asia/Pacific, serving as its Foundation President for 5 years from 1996. He is joint editor of Chemical Speciation and Bioavailability and a member of the editorial board of *Integrated Environmental Assessment and Management.*

Christopher G. Ingersoll is an Aquatic Toxicologist with the United States Geological Survey (USGS) at the Columbia Environmental Research Center in Columbia, Missouri. He received his bachelor's (1978) and master's (1982) degrees from Miami University at Oxford, Ohio and his doctorate (1986) from the University of Wyoming at Laramie, Wyoming. He has conducted research to develop methods for assessing the bioavailability of contaminants in sediment. Specifically, he has coordinated the development of chronic toxicity tests with amphipods, which have been used to evaluate contaminated sediments in several areas including the Great Lakes, the upper Mississippi River, the Clark Fork River in Montana, and the Calcasieu Estuary in Louisiana. He has also conducted studies to field validate laboratory toxicity and bioaccumulation tests with invertebrates using the sediment quality triad. Chris has worked with a variety of other organizations to develop sediment quality guidelines that can be used to predict the incidence of toxicity in sediments. He has worked with the American Society for Testing and Materials (ASTM) and the USEPA in developing standard methods for conducting toxicity and bioaccumulation tests with contaminated sediments. His research interests also include evaluating the sensitivity of threatened and endangered species to contaminants, including freshwater mussels. Chris is currently the Chair of ASTM Subcommittee E47.03 on Sediment Assessment and Toxicology.

David W. Moore, Ph.D., is currently Aquatic Toxicology Practice Leader for Weston Solutions, Inc., Carlsbad, California, USA. He obtained his doctoral training in environmental toxicology at The University of South Carolina. From 1988 to 1998, David worked as aquatic toxicologist and then as research team leader at the US Army Corps of Engineers, Waterways Experiment Station, Vicksburg, Mississippi. His research focused on the development of new test methods for assessing contaminated sediments and the development of interpretive tools (e.g., Environmental Residue Effects Database) to enhance regulatory evaluations of contaminated sediments. David's current projects include a field validation study of chronic sublethal bioassays, development of sediment management plans for ports and harbors, and conducting special studies in support of environmental and human health risk assessments at contaminated sites throughout the US. David has authored or co-authored more 30 publications on topics relating to the assessment and management of contaminated sediments.

Workshop participants*

William J. Adams
Rio Tinto HSE
Murray, Utah, USA

Wolfgang Ahlf
Technical University
Department of Environmental Engineering
Hamburg, Germany

Michelle L. Anghera
University of California-Los Angeles
Los Angeles, California, USA

Marc P. Babut
Research Unit
Freshwater Ecosystems Biology
Lyon, France

Graeme E. Batley
CSIRO Energy Technology
Centre for Advanced Analytical Chemistry
Bangor, New South Wales, Australia

Renato Baudo
Italian Institute of Hydrobiology
National Research Center
Verbania Pallanza, Italy

Steven M. Bay
Southern California Coastal Water Research Project
Toxicology Department
Westminster, California, USA

Walter J. Berry
US Environmental Protection Agency (USEPA)
Atlantic Ecology Division, ORD / NHEERL
Narragansett, Rhode Island, USA

Wesley J. Birge
University of Kentucky
School of Biological Sciences
Lexington, Kentucky, USA

Thomas L. Bott
Stroud Water Research Center
Avondale, Pennsylvania, USA

Todd S. Bridges
US Army Corps of Engineers
USERDC-WES-ES-F
Vicksburg, Mississippi, USA

Steven S. Brown
Toxicology Department
Rohm and Haas Company
Spring House, Pennsylvania, USA

Robert M. Burgess
USEPA Atlantic Ecology Division / NHEERL
Narragansett, Rhode Island

G. Allen Burton
Wright State University
Institute for Environmental Quality
Dayton, Ohio, USA

Bart Chadwick
SPAWAR
Marine Environmental Quality Branch
San Diego, California, USA

Peter M. Chapman
EVS Environment Consultants
North Vancouver, British Columbia, Canada

James R. Clark
Exxon Mobil Research and Engineering Co.
Fairfax, Virginia, USA

William H. Clements
Department of Fishery and Wildlife Biology
Colorado State University
Fort Collins, Colorado, USA

Mark Crane
Crane Consultants
Faringdon, Oxfordshire, United Kingdom

Stefano Della Sala
Venice Port Authority
Safety and Environment Department
Venice, Italy

Dominic M. Di Toro
HydroQual, Inc.
Mahwah, New Jersey, USA

Philip B. Dorn
Shell Global Solutions US, Inc.
Westhollow Technology Center
Houston, Texas, USA

W. Scott Douglas
New Jersey Department of Transportation
Office of Maritime Resources
Trenton, New Jersey, USA

Stephen J. Ells
USEPA
Washington DC, USA

Robert M. Engler
US Army Corps of Engineers
Waterways Experiment Station
Vicksburg, Mississippi, USA

L. Jay Field
National Oceanic and Atmospheric Administration (NOAA)
Office of Response and Restoration
Coastal Protection and Restoration Division
Seattle, Washington, USA

Ron Gouguet
NOAA Coastal Resource Coordinator
Dallas,Texas, USA

Andrew S. Green
International Lead Zinc Research Organization, Inc.
Dept of Environment and Health
Research Triangle Park, North Carolina, USA

Thomas H. Gries
Washington Department of Ecology
Sediment Management Unit
Olympia, Washington, USA

Michael C. Harrass
BP Amoco Chemicals
Naperville, Illinois, USA

Kay T. Ho
USEPA
Atlantic Ecology Division / NHEERL
Narragansett, Rhode Island, USA

Christer Hogstrand
King's College London
Division of Life Sciences
London, United Kingdom

Jeffrey L. Hyland
NOAA National Ocean Service
Center for Coastal Environmental Health and Biomolecular Research
Charleston, South Carolina, USA

Christopher G. Ingersoll
US Geological Survey
Columbia Environmental Research Center
Columbia, Missouri, USA

D. Scott Ireland
USEPA
Office of Science and Technology
Washington DC, USA

Wayne Landis
Western Washington University
Inst. of Environmental Toxicology and Chemistry
Bellingham, Washington, USA

Peter F. Landrum
NOAA
Great Lakes Environmental Research Laboratory
Ann Arbor, Michigan, USA

Edward R. Long
ERL Environmental
Salem, Oregon, USA

Eileen M. Maher
San Diego Unified Port District
San Diego, California, USA

James P. Meador
NOAA
NMFS/Northwest Fisheries Science Center
Seattle, Washington, USA

Charles A. Menzie
Menzie-Cura & Associates, Inc.
Chelmsford, Massachusetts, USA

David W. Moore
MEC Analytical Systems, Inc.
Carlsbad, California, USA

David R. Mount
USEPA, ORD
Duluth, Minnesota, USA

Thomas P. O'Connor
NOAA
Silver Spring, Maryland, USA

Rod Parrish
Society of Environmental Toxicology and Chemistry (SETAC)
Pensacola, Florida, USA

Linda M. Porebski
Environment Canada
Marine Environment Division
Hull, Quebec, Canada

Danny D. Reible
Louisiana State University
Hazardous Substance Research Center
Baton Rouge, Louisiana, USA

Trefor B. Reynoldson
Environment Canada
National Water Research Institute
Burlington, Ontario, Canada

Amy H. Ringwood
Marine Resources Research Institute
Charleston, South Carolina, USA

Jacqueline D. Savitz
Oceana
Pollution Campaign
Washington DC, USA

Gregory E. Schiefer
SETAC
Pensacola, Florida, USA

James P. Shine
Harvard School of Public Health
Department of Environmental Health
Boston, Massachusetts, USA

Paul K. Sibley
University of Guelph
Department of Environmental Biology
Guelph, Ontario, Canada

Ralph G. Stahl, Jr.
DuPont Company
Corporate Remediation Group
Wilmington, Delaware, USA

Joost Stronkhorst
RIKZ, Ministry of Transport, Public Works and Water Management
National Institute for Coastal and Marine Management
The Hague, The Netherlands

Richard C. Swartz
USEPA (Retired)
Placida, Florida, USA

Richard J. Wenning, Chairperson
Environ International Corporation
Emeryville, California, USA

Jack Q. Word
MEC Analytical Systems, Inc.
Sequim, Washington, USA

* Affiliations were current at the time of the workshop.

Introduction to the workshop

1

RICHARD J WENNING,
GRAEME E BATLEY,
CHRISTOPHER G INGERSOLL,
DAVID W MOORE

A Pellston Workshop on Sediment Quality Guidelines

This book is the result of discussions that took place at the Pellston Workshop on Use of Sediment Quality Guidelines and Related Tools for the Assessment of Contaminated Sediments. The workshop, sponsored by the Society of Environmental Toxicology and Chemistry (SETAC), was held 18–22 August 2002 in Fairmont, Montana, USA. Three previous SETAC workshops, two that focused on sediment ecological risk assessment (ERA) (Dickson et al. 1987; Ingersoll et al. 1997) and one on porewater toxicity testing (Carr and Nipper 2003), along with a recent workshop addressing the application of weight-of-evidence (WOE) methods in ERA (Burton et al. 2002), focused on how, when, and why ERAs are needed in sediment assessments. Each of these workshops has provided guidance on performing risk assessments of chemicals in sediment and on evaluating the key chemical and biological factors.

The purpose of this workshop was to take an additional step forward on the subject of ERA of sediments and provide a forum for further discussions among scientists, environmental regulators, and environmental managers on procedures that can be used to integrate the information derived from the various ecological, biological, and chemical studies often performed in contaminated sediment assessments. The lines of evidence (LOEs) developed through the application of different field assessment tools often include the consideration of chemically based sediment quality guidelines (SQGs) to evaluate sediment contamination and to help practitioners in sediments

Use of Sediment Quality Guidelines and Related Tools for the Assessment of Contaminated Sediments
Wenning RJ, Batley GE, Ingersoll CG, Moore DW, editors.
 ISBN 1-880611-71-6

assessment and management to formulate risk management decisions. At the heart of this workshop was the question of how SQGs fit into the assessment process. The workshop focused on 3 aspects of this question:

1) the scientific foundation underlying the different available SQG methods,
2) methods to improve the integration of SQGs with information derived from multiple chemical and biological LOEs into different sediment quality assessment frameworks, and
3) identification of the limitations and uncertainties associated with different SQG approaches.

The workshop brought together 55 experts in the fields of sediment assessment and management from Australia, Canada, France, Germany, the United Kingdom, Italy, the Netherlands, and the United States (US) for 6 days of discussion on the use of SQGs and other sediment assessment tools. Participants (see Appendix 1) included representatives from regulatory and nonregulatory government agencies, academia, industry, environmental groups, and consulting firms involved in assessment, investigation, management, and basic research on contaminants in sediment. Participants were assigned to 5 different workgroups and asked to address one of the following 5 topics:

1) scientific underpinnings of the currently available SQG schemes,
2) predictive ability of SQGs for toxicity and bioaccumulation assessments,
3) application of other related sediment assessment tools,
4) role of SQGs and other tools in different sediment assessment frameworks, and
5) role and relative appropriateness of SQGs and other tools in different aquatic environments.

In each of these discussions, participants were urged to seek consensus, where possible, on specific technical issues of concern to sediment assessment and management practitioners and to identify recommendations for future research that could lead to improvements in the existing methods available to quantify the effects, or absence of effects, associated with the presence of chemical contaminants in sediment. The chapters in this book include a synopsis of the discussions in each of the workgroups, as well as a set of supporting technical papers that further explore key topics. It is our hope that this book provides a useful summary of the current state of knowledge and that it will serve as a basis for further discussions in the foreseeable future.

Introduction

Environmental scientists, engineers, and regulatory authorities throughout the world have devoted considerable resources to assessment, management, and remediation of chemical contaminants in sediments. Sediment quality has become a serious and potentially costly economical and ecological issue for navigation dredging projects, waterway restoration programs, recreational and commercial fisheries management, water quality protection, and natural resource restoration.

Because of the magnitude of the economic and natural resources at stake, sediment management has become the focus of regional, national, and international policy debate, one that engages diverse elements of society. Numerous scientific and engineering studies have shown that addressing contaminated sediments is a complex undertaking. Complexities arise from the great variability in the physical and biogeochemical characteristics of sediments; human and ecological receptors; and the cultural, social, and economic values associated with different freshwater, estuarine, and marine environments. Because of these complexities, the assessment and management of contaminated sediments in ports and harbors, rivers, lakes, and at hazardous waste sites does not easily lend itself to simple "one-size-fits-all" investigation methods and presumptive management decisions.

In response to the increasing demands from society for greater environmental protection of aquatic resources and the restoration of impaired or degraded rivers and estuaries, scientists in several countries have developed a variety of methods for evaluating the degree to which sediment-associated chemicals might adversely affect aquatic organisms. While some methods have focused on the derivation of numeric chemically based sediment concentration limits, other methods have focused on less-specific narrative statements or benchmarks that frame acceptable and potentially unacceptable sediment quality. The results of both approaches are typically referred to as "SQGs."

Increasingly, SQGs are used as informal benchmarks or aids to interpretation of sediment chemistry data. SQGs have been used to interpret historical trends, identify potential problem chemicals or reaches in a waterway, interpret or design ambient monitoring programs, classify hot spots, establish baseline conditions in non-urbanized systems, rank contaminated waterways, and help choose sites for more detailed studies (Long and MacDonald 1998). Other applications of SQGs, which have generated considerable controversy within the scientific, industrial, and environmental regulatory communities, include identifying the need for source control measures to address certain chemicals before release, triggering regulatory action as mandatory standards, and establishing target remediation objectives (USEPA 1997; Barnthouse and Stahl 2002).

Over the past few years, several concerns have been raised regarding the use of SQGs in sediment quality assessments. One of these concerns is the ability of SQGs to adequately predict the presence or absence of chronic toxicity to sediment-dwelling organisms in field-collected sediments. A second concern relates to the ability of SQGs to predict effects resulting from the bioaccumulation of sediment-associated contaminants. Another concern focuses on the ability of SQGs to establish cause-and-effect relationships. A fourth concern deals with the ability of SQGs developed for 1 endpoint (e.g., amphipod mortality measured in the laboratory) to be predictive of another (e.g., effects on organisms exposed in the field).

These concerns have not been eased by the use of SQGs in conjunction with other tools such as sediment toxicity tests, bioaccumulation, and benthic community surveys in a WOE evaluation to assess the hazards posed by the occurrence of chemicals in sediments (Ingersoll et al. 1997; Chapman et al. 2002). This is due, in large part, to the absence of guidance on what constitutes a WOE approach and the lack of guidance in the implementation of quantitative, rather than subjective or qualitative, procedures for evaluating multiple biological and chemical LOEs generated from field investigations (Burton et al. 2002).

Difficulties have become apparent in the context of efforts in the US, the European Union, and elsewhere to develop or improve sediment assessment frameworks for different management situations. According to Nord (2001), for example, the inconsistent approaches used in the US to assess or manage contaminated sediments may be due, in part, to the lack of communication or consensus by regulators and the scientific community on a number of key technical issues; this situation has resulted in inaction or ineffective implementation of regulatory policies concerned with addressing contaminated sediments. While SQGs may be appropriate in some situations, scientists generally acknowledge there are several limitations and uncertainties associated with different SQG approaches that have the potential to cause confusion and concern among sediment assessment and management practitioners.

Although there has been considerable emphasis on the use of SQGs for informal (i.e., nonregulatory) purposes and as initial screening values in enforcement and regulatory programs, there has not been broad agreement on how SQGs should be used in different regulatory and management frameworks to define cleanup actions or injuries to natural resources. In the US, the Puget Sound Dredged Material Management Program and Washington State's Sediment Management Standards Program are the among the few programs that have established, at least in part, how SQGs could be used in regulatory decision making (Washington State 2004).

Workshop Goals

Each workgroup was initiated with one or more discussion papers followed by a discussion session held under the direction of a chairperson. It was the responsibility of the chairperson and members of the workgroup to capture the main points and to identify the major issues for inclusion in the final workshop proceedings. The charge of each workgroup is summarized as follows.

Workgroup 1: Review scientific underpinnings associated with different SQG approaches

The main objective of this workgroup was to critically review the different approaches used to derive SQGs worldwide, focusing on the technical foundation, strengths and limitations, and methodological uncertainties associated with the different approaches with a view to identifying recommendations to improve on the development and application of SQGs (Chapter 3). There are a number of complex scientific issues concerning the derivation of SQGs for which clarification could aid in understanding the scientific foundations of the different approaches and their importance to sediment management decisions. Specifically, participants addressed 5 questions:

1) What are the uncertainties, deficiencies, strengths, and limitations of SQGs derived from matching toxicity and chemistry data with field-collected sediments?
2) What additional qualifiers of SQGs are required to account for bioavailability?
3) What are the strengths and limitations of SQGs derived from equilibrium partitioning (EqP)?
4) Are there issues of changing sediment chemistry, organism uptake pathways, feeding strategies, and trophic levels that might limit the acceptability of SQGs as predictors of the health of sediment ecosystems?
5) Has the scientific basis for the various uses of SQGs been field validated?

Workgroup 2: Evaluate use of SQGs to estimate potential for effects, or no effects, of sediment-associated contaminants in laboratory toxicity tests and benthic community assessments

The main objective of this workgroup was to evaluate the predictive ability of effects-based SQGs in laboratory toxicity tests and in benthic community assessments (Chapter 4). A primary concern among scientists and regulators evaluating the application of SQG approaches in the assessments of sediment quality is the ability to demonstrate that SQGs are sufficiently predictive of the presence, or absence, of toxicity to sediment-dwelling organisms or to higher trophic-level organisms. A related concern has been the ability to predict bioaccumulation of chemicals by benthic organisms and the effects of that bioaccumulation either directly on sediment-dwelling organisms or through trophic transfer of contaminants. It is clear that protection of the ecosystem beyond direct effects on benthic organisms is critical to the long-term management of contaminated sediments. The source of the current debate is the extent to which SQGs have been used in some instances without an understanding of their intended use or the quantitative extent of their predictive ability. Specifically, participants addressed the following 3 questions:

1) How well do SQGs represent the potential for effects or no effects observed in laboratory toxicity tests and in field studies of benthic communities?
2) How well do SQGs represent the potential for effects or no effects in organisms as a result of contaminant uptake and/or trophic transfer?
3) How have SQGs been applied and validated in the field as part of sediment management and risk management decision making?

Workgroup 3: Assess role of other tools available for evaluating sediment contamination

The main objective of this workgroup was to evaluate the use of other tools as either a supplement or an alternative to SQGs in the assessment of contaminated sediments (Chapter 5). An increasingly common emphasis of the different sediment frameworks that have been suggested or applied by various US and international regulatory agencies is the use of decision trees or an integrated WOE approach in sediment quality identification, assessment, and management. A common criticism of the different SQG approaches included in some of these frameworks is their inability to account for site-specific environmental conditions or unique biological characteristics, leading to either inadequate or overly conservative risk management decisions.

Improvements in 3 areas of sediment assessment have been advocated:

1) the integration of biological information,
2) the integration of routes of exposure considerations, and
3) the inclusion of biologically based tools to supplement the information provided by more traditional chemically based SQGs.

Because different SQGs do not explicitly consider routes of exposure to aquatic biota, it has been suggested that improvements to SQGs that account for different possible exposure scenarios could lead to more scientifically credible risk-based assessments. A closely related issue that has also confounded different sediment frameworks is the interpretation of information generated from multiple chemical and biological LOEs. The difficulties attributed to contrasting biological and/or chemical results with SQGs has led sediment assessment and management practitioners to ask for clearer guidance from the scientific community on the use of one or more WOE approaches (Ingersoll et al. 1997; Burton et al. 2002; Chapman et al. 2002).

In this workgroup, participants addressed 5 questions:

1) What types of data should be generated for one or more frameworks at different types of sites to determine the applicability of SQGs and reduce technical uncertainties at a particular site?
2) What can be done to improve the understanding of the uncertainties (or reduce the uncertainties) associated with more conventional assessment methods (e.g., biological testing and benthic infaunal analysis) used to derive SQGs?
3) How can we make SQGs more site specific?
4) Can the accuracy of assessing sediment quality be improved using biologically based thresholds in a WOE process (as opposed to using chemically based threshold SQGs)?
5) What are the uncertainties, deficiencies, strengths, and limitations of biologically based thresholds?

Workgroup 4: Explore role of SQGs and related chemical and biological assessment tools in sediment assessment and management frameworks

The main objective of this workgroup was to explore whether SQGs alone or as part of a broader framework are sufficient to provide technical advice on the assessment and management of contaminated sediment (Chapter 6). Several different sediment assessment and management frameworks have been proposed in the US and elsewhere. For example, in the specific area of dredged material assessment, the 1996 protocol to the London Convention developed by more than 80 countries represents one approach (GIPME 2000). In the US, the National Research Council (NRC) report on management of polychlorinated biphenyl (PCB)-contaminated sediments presents a different approach using a risk-based framework (NRC 2001). Such a framework would incorporate chemistry, ecotoxicology, bioaccumulation, effects on benthic community structure, and both temporal and spatial considerations of sediment conditions. In these cases, as well as others, the risks and tradeoffs posed by different LOEs are poorly understood and rarely explicitly evaluated in different sediment frameworks. With this in mind, participants were charged to address 4 questions:

1) What are the required elements and decision points of sediment assessment frameworks?
2) How can these elements best be assembled in an assessment or WOE decision?
3) What is the utility of different SQG schemes as part of a WOE approach to contaminated sediment assessment and management?
4) How can LOEs and WOE best be used in an overall sediment assessment framework to make sediment quality assessment decisions?

Workgroup 5: Assess use of SQGs and related tools for evaluation of sediments in aquatic environments

The main objective of this workgroup was to evaluate the use of SQGs and other tools across a range of different aquatic environments (Chapter 7). Technical concerns focused on the use of SQGs across a range of different freshwater, estuarine, and marine ecosystems, and their use in environments that involve multiple contaminants, multiple exposure pathways, multiple ecological (and human) receptors, variations in resource use and management, or differences in temporal or spatial scales. There is growing recognition among scientists that 6 factors should be considered in conjunction with SQGs when sediment quality is addressed:

1) chemical and biological changes over spatial scales (sometimes measured in meters or kilometers, and at other times measured in centimeters),
2) chemical and biological changes over temporal scales (similarly, sometimes measured in hours, while at other times measured in years),
3) physical heterogeneity of the contaminated material,
4) watershed characteristics,
5) interactions of chemical mixtures, and
6) biological diversity at different trophic levels.

For several of these factors, chronic stresses unrelated to contaminant levels may pose more important challenges to resident ecological receptors; therefore, characterizing chronic stresses of sediment origin is increasingly recognized as an important issue in sediment quality assessment. Consequently, participants addressed the following 2 questions:

1) Are SQGs alone sufficient for management decision making in different aquatic environments?
2) How applicable are SQGs and different tools for characterization of sediment quality in different aquatic environments?

References

Barnthouse L, Stahl RG. 2002. Quantifying natural resource injuries and ecological service reductions: Challenges and opportunities. *Environ Manag* 30:1–12.

Burton GA, Batley GE, Chapman PM, Forbes VE, Schlekat CE, Smith PE, den Besten PJ, Barker J, Reynoldson T, Green AS, Dwyer RL, Berti WR. 2002. A weight-of-evidence framework for assessing sediment (or other) contamination: Improving certainty in the decision-making process. *Hum Ecol Risk Assess* 8: 1675–1696.

Carr RS, Nipper MJ, editors. 2003. Porewater toxicity testing. Pensacola (FL): Society of Environmental Toxicology and Chemistry (SETAC).

Chapman PM, McDonald BG, Lawrence GS. 2002. Weight-of-evidence frameworks for sediment quality and other assessments. *Hum Ecol Risk Assess* 8:1489–1515.

Dickson KL, Maki AW, Brungs WA. 1987. Fate and effects of sediment-bound chemicals in aquatic systems. New York: Pergamon.

[GIPME] Global Investigation of Pollution in the Marine Environment. 2000. Guidance on assessment of sediment quality. London: International Maritime Organization.

Ingersoll CG, Dillon T, Biddinger RG, editors. 1997. Ecological risk assessments of contaminated sediment. Pensacola (FL): Society of Environmental Toxicology and Chemistry (SETAC). 389 p.

Long ER, MacDonald DD. 1998. Recommended uses of empirically derived sediment quality guidelines for marine and estuarine ecosystems. *Hum Ecol Risk Assess* 4:1019–1039.

[NRC] National Research Council. 2001. A risk management strategy for PCB-contaminated sediments. Washington DC: National Academy.

Nord MA. 2001. Recommendations for the implementation of a national sediment quality policy in the United States. *Hum Ecol Risk Assess* 7:641–650.

[USEPA] US Environmental Protection Agency. 1997. The incidence and severity of sediment contamination in surface waters of the United States, Volume 1: National sediment quality survey. Washington DC: USEPA. EPA 823/R-97/006.

Washington State. 2004. Chapters 173 through 204 of the Washington State Administrative Code.

Executive summary

2

RICHARD J WENNING, WILLIAM J ADAMS, GRAEME E BATLEY, WALTER J BERRY, WESLEY J BIRGE, TODD S BRIDGES, G ALLEN BURTON, PETER M CHAPMAN, W SCOTT DOUGLAS, ROBERT M ENGLER, CHRISTOPHER G INGERSOLL, DAVID W MOORE, RALPH G STAHL, JACK Q WORD

Sediment quality guidelines (SQGs) are intended to be either protective of biological resources, or predictive of adverse effects to those resources, or both. SQGs for assessing sediment quality relative to the potential for adverse effects on sediment-dwelling organisms have been derived using both mechanistic and empirical approaches. Effect and no-effect chemical concentrations in sediment have been derived using the following approaches:

- equilibrium partitioning (EqP; Di Toro, Mahony, et al. 1991; Di Toro, Zarba, et al. 1991; Ankley et al. 1996; NYSDEC 1998; Di Toro and McGrath 2000),
- screening-level concentration (SLC; Persaud et al. 1993; Von Stackelberg and Menzie 2002),
- effects range low and effects range median (ERL and ERM; Long et al. 1995; USEPA 1996),
- threshold effects level and probable effects level (TEL and PEL; MacDonald et al. 1996; Smith et al. 1996; USEPA 1996),
- apparent effects threshold (AET; Barrick et al. 1988; Ginn and Pastorok 1992; Cubbage et al. 1997),
- "consensus-based" evaluation (Swartz 1999; MacDonald, DiPinto, et al. 2000; MacDonald, Ingersoll, et al. 2000), and
- logistic regression modeling (LRM; Field et al. 1999, 2002).

The underlying supposition in the derivation of effects-based SQGs is that these guidelines can be used, at least to some extent, as a substitute for direct measures of potential adverse effects of contaminants in sediments on benthic organisms. The mechanistically based SQGs have been developed and tested using laboratory spiked sediments and compared to toxicity tests by using field-collected sediments. The empirically based SQGs have typically been developed using large databases with matching measures of sediment chemistry and toxicity from field-collected samples. These databases also have been used to evaluate the ability of SQGs to predict sediments to be either toxic or nontoxic in laboratory tests or in benthic community assessments in a variety of freshwater, estuarine, and marine environments. Generally, the results of analyses conducted with these databases in conjunction with field validation studies involving contaminated sediments suggest that some effects-based SQGs can provide reasonable predictions, with associated error rates and uncertainties, for the potential for acute (e.g., primarily 10-d laboratory toxicity) effects or no effects on benthic organisms.

The issue of causality is often contentious for empirically based SQGs because the cause of observed toxicity or biological degradation from field-collected sediments may or may not be a result of the contaminants measured in sediments but may, instead, be a response to 1 or more co-occurring chemicals or confounding factors associated with sediment geochemistry or benthic habitat characteristics. Additional uncertainties arise when the ecological receptors potentially at risk and the characteristics of the sediment are not appropriate surrogates or adequately similar to conditions used in the sediment toxicity studies that led to the development of SQGs.

Numerous studies have shown that the predictive performance of some effects-based SQGs can be described relative to their intended purposes of predicting primarily acute adverse effects, or the lack thereof, in field validation studies using laboratory tests or benthic community surveys. Nonetheless, over the past few years, several concerns have been expressed regarding the use of SQGs in sediment quality assessments. One of these concerns is the ability of SQGs to adequately predict the presence or absence of chronic toxicity to sediment-dwelling organisms in field-collected sediments. A second concern relates to the ability of SQGs to predict effects resulting from bioaccumulation of sediment-associated contaminants. Another concern focuses on the ability of SQGs to establish cause and effect relationships. A fourth concern deals with the ability of SQGs developed for 1 endpoint (e.g., amphipod mortality measured in the laboratory) to be predictive of another (e.g., effects on organisms exposed in the field).

These concerns have not been eased by the use of SQGs in conjunction with other tools such as sediment toxicity tests, bioaccumulation, and benthic community surveys in a weight of evidence (WOE) for assessing the hazards associated with contaminated sediments (Ingersoll et al. 1997; Chapman et al. 2002). This is due, in large part, to the absence of guidance on what constitutes a WOE approach and the lack of guidance on the implementation of quantitative, rather than subjective or qualitative, procedures for evaluating multiple biological and chemical lines of evidence (LOEs) generated from field investigations (Burton et al. 2002).

Scientific Underpinnings of Sediment Quality Guidelines

The different approaches used to develop SQGs can be broadly subdivided into 2 categories: 1) empirically based guidelines, and 2) mechanistically based guidelines (Chapter 3). Empirical guidelines are derived from databases of sediment chemistry (concentrations of specific sediment contaminants) and observed biological effects (e.g., those derived from sediment toxicity tests and benthic community information). These data are arrayed on a continuum of increasing chemical concentration. Various algorithms are used to define specific concentrations associated with particular levels of effect or no effect. Common examples include ERLs, ERMs, TELs, PELs, and AETs. In contrast, mechanistic guidelines are derived from a theoretical understanding of the factors that govern bioavailability of sediment contaminants, and known relationships between chemical exposure or uptake and toxicity. EqP theory forms the basis for all current mechanistic guidelines. Mechanistic guidelines account for bioavailability through normalization to sediment characteristics that affect bioavailability, primarily organic carbon for nonionic organic chemicals and simultaneously extracted metal (SEM) and acid volatile sulfide (AVS), organic carbon, or other sediment fractions for metals. Empirical guidelines may use either dry weight concentrations or normalized concentrations, depending on the guideline.

Though the scientific underpinnings of the different SQG approaches vary widely, none of the approaches appear to be intrinsically flawed. All approaches reviewed are grounded in concepts that, viewed in isolation, are sound. Potential issues of scientific appropriateness of SQGs are, therefore, more a matter of whether a particular application of an SQG is consistent with its underlying principles.

In the ideal case, SQGs would unequivocally delineate between sediments that cause biological effects and those that do not; in other words, chemicals in sediments at concentrations below their respective numeric SQG values would show no biological effects, while chemicals that occur at concentrations above their respective numeric SQG values would show biological effects. In reality, the occurrence of biological effects does not show such a clearly delineated relationship. Instead, the distribution of biological effects generally shows a relationship characterized by ranges of chemical concentration where biological effects are rare, where cases of both effects and no effects are found, and where biological effects are likely. Normalization to organic carbon or other characteristics that control bioavailability of sediment contaminants may affect the distribution of effect and no-effect data, but no normalization technique has yet come close to complete discrimination, nor is this likely. Factors that cause overlap between effect and no-effect data are many and include contributions of other chemicals to effects, unaccounted-for differences in chemical availability, differences in response among organisms, and errors in measurement of either chemical or response.

The generalized concentration–response model depicted in Figure 3-1, with the probability of biological effects increasing with increasing chemical concentration, can be used as a framework to consider different SQG approaches. The probability of effects is low until it reaches a threshold effect (TE) guideline. At the high end of the distribution is the probable effect (PE) guideline, above which effects are increasingly likely. Between TE and PE lies a transition zone where adverse biological effects may or may not occur, but within which the probability of adverse effects increases. SQGs can be thought of as vertical lines across this concentration–response curve. Different SQG derivation procedures are intended to represent TE, PE, or midrange effect (ME) guidelines. A TE guideline is a contaminant concentration below which biological effects are predicted to occur in very few cases, whereas a PE guideline is intended to identify the contaminant concentration above which effects are predicted to occur in most cases. Intermediate to these would be ME guidelines, which identify contaminant concentrations within the transition zone wherein the probability of effects is substantively above background, but below which effects are not always observed.

In this context, chemically based numeric SQGs can be effective for identifying concentration ranges where adverse biological effects are unlikely, uncertain, and likely to occur. A similar concept could be applied to the severity of the effect (e.g., reduced survival or reproductive impairment), which typically increases as the probability of toxicity increases.

While not all SQGs are derived from a concentration–response model, this paradigm can still be used to conceptualize differences among SQGs. SQGs can vary in the way that chemical concentration is expressed, such as dry weight concentration of a chemical, the organic-carbon normalized concentration, or some expression of a mixture, such as total narcotic potency (Di Toro and McGrath 2000). SQGs can also vary in the way that biological effect is expressed, such as lethality to amphipods versus benthic community response.

By definition, the existence of a sediment quality transition zone means that no SQG paradigm can unequivocally segregate all sediments showing effects from those that do not. Nonetheless, SQGs are often employed such that they are expected to separate toxic and nontoxic sediments. In such applications, the interpretation of exceeding a guideline varies according to where the guideline lies along the concentration–response continuum. If one uses a TE-type SQG to define toxic (or unacceptable) sediments, then errors will tend to be characterized as false positives; that is, some sediments will be classified as toxic when in fact they are not.

On the other hand, selecting a PE-type guideline will result in predominately false negative errors, in which some sediments will be classified as nontoxic when in fact they are toxic. As one moves from a TE guideline toward a PE guideline, the chances of false positives decrease and the chances of false negatives increase. ME guidelines can be expected to have intermediate rates of false positives and false negatives relative to TE and PE guidelines, respectively; whether this is more or less desirable depends on the management purpose. In some situations, it may be desirable to have a guideline that will identify most of the potentially toxic samples, even if many of those samples are not actually toxic (TE guideline). In all cases, application of SQGs in a toxic or nontoxic context must be cognizant of the types and rates of errors associated with each type of SQG.

Beyond the TE–ME–PE concepts, interpreting exceedances of SQGs must also recognize differences between empirically based and mechanistically derived guidelines. Unlike most empirically based SQGs, mechanistically based guidelines are intended to assess the effects of only those chemicals for which the guideline is derived; they address the question, "Is the concentration of this chemical (or chemicals) sufficient, by itself, to cause effects?" Although the likelihood of adverse effects is thought to be elevated when the guideline is exceeded, the likelihood of a sediment causing adverse effects when it exceeds an individual mechanistically based guideline cannot be predicted. This is due, in part, because discrete guideline values do not address the potential synergistic effects of other chemicals that may be present. For guidelines developed using data from field-collected sediments, which typically contain mixtures of chemicals, there is some implicit consideration of other chemicals present in the sediment, although the nature of this mixture effect is not explicitly quantified.

The issues associated with mixtures of contaminants in the environment are pervasive. The occurrence of biological effects in sediments is almost always associated with mixtures of sediment-bound contaminants. Empirically based SQGs may be used to incorrectly attribute the adverse biological effects caused by a mixture of contaminants to a single contaminant. Conversely, mechanistically derived SQGs based on causality may underestimate the adverse biological effects attributable to a mixture of contaminants in sediment.

Chemical interactions in sediments and their resulting biological effects are poorly understood. The simplest conceptual model is additivity, which predicts the effects of a contaminant mixture as the sum of the effects of the individual contaminants. Additivity is generally thought to apply to chemicals with the same mechanism of action. Thus, if concentrations of chemicals with the same mode of action are normalized to a reference concentration such as an LC50 (lethal concentration for 50% of test organisms) or an SQG, the sum of the normalized numeric SQGs may be logically used in the concentration–response analysis to predict the probability of biological effects. This additivity model is appropriate for SQG approaches that reflect adverse biological effects caused by specific contaminants (e.g., mechanistic SQGs such as those derived from EqP theory) and has been applied to polycyclic aromatic hydrocarbons (PAHs; Swartz et al. 1995), narcotic chemicals (Di Toro and McGrath 2000), and metals (Ankley et al. 1996). At present, these additivity approaches are generally restricted to chemicals with similar modes of action, because there is no theoretically rigorous method for modeling or predicting the interactions among contaminants that act with different modes of action.

Methods are available for aggregating empirically based SQGs, such as calculating the sum, maximum, or average of SQG quotients derived from the ratio of the concentration of each chemical in sediment to its respective SQG value. Although intuitively appealing, the theoretical basis for how best to do this is less clear than for mechanistically based guidelines. Individual empirically based SQGs based on the co-occurrence of effects and chemistry (e.g., ERLs and PELs) may not be causally related to specific contaminants. Instead, empirical SQGs for individual contaminants indicate effects caused by the entire mixture of sediment chemicals. For this reason, adding empirical SQG-normalized concentrations across chemicals may be, in effect, incorporating additional conservatism regarding effects attributable to contaminant mixtures and may not be as justifiable as doing so for mechanistic guidelines. The calculation of SQG quotients has the potential to provide more information on the nature of sediment contamination relative to effects, and there is evidence that these approaches improve the predictive ability of empirical SQGs

(MacDonald, Ingersoll, Berger 2000; MacDonald, Di Pinto, et al. 2000; USEPA 2000c; Ingersoll et al. 2001; Field et al. 2002; Chapter 4). Despite the technical uncertainties, in most cases the use of mixture models and SQG quotients improves the ability of SQGs to predict effects in the laboratory and in the field, compared to evaluations based on exceeding single SQGs.

Another issue affecting the interpretation of sediment quality is the definition and quantification of the bioavailable fraction of sediment contaminants. Normalizing contaminants to binding phases (e.g., PAHs to organic carbon, metals to AVS) theoretically provides a better consideration of bioavailability. Spiked sediment experiments, for both acute and chronic exposures, have demonstrated that the applications of these normalization methods improve the predictive ability of mechanistically based SQGs. However, there are additional factors known to affect bioavailability that have not yet been incorporated into SQGs, such as unusual carbon types (e.g., soot, ash, and black carbon) or geochemistry phases controlling metal bioavailability in oxic sediments. Interestingly, evaluations of empirically based SQGs have found that dry weight normalization often results in predictions of adverse biological effects that are as good or better than the normalizations thought to reflect bioavailability (e.g., Barrick et al. 1988). The reasons for this apparent discrepancy remain unclear. Nonetheless, it seems an inescapable conclusion that at some level, the accuracy of SQGs will be limited by the ability to measure and incorporate the factors that account for bioavailability in sediments.

The third issue affecting the interpretation of sediment quality lies in the effects used to define acceptable and unacceptable sediment quality. The selection of data and biological endpoints used to derive SQGs can affect their relevance to protection of benthic communities and the inferences that SQGs provide regarding contaminant conditions in sediment. Differences among SQGs often reflect differences in the sensitivity of the biological indicators used to document effects. SQGs may change with test species, life stages, response endpoints, test duration, and taxonomic composition of benthic communities. Some SQGs incorporate information from one or more aquatic toxicity databases and are intended to protect 95% of species from chronic effects (e.g., EqP using water-quality criteria endpoints). Some empirical SQGs are based on extensive databases that include a great variety of effects of sediment chemicals on benthic species (e.g., ERL, ERM, TEL, PEL, and AET). Other SQGs are based exclusively on adverse biological effects on a single taxonomic group (e.g., ΣPAH for marine amphipods). The choice of biological indicators can significantly influence the relevance of different SQGs in sediment assessments as screening tools for evaluating the impact of sediment-bound contaminants on benthic organisms or other aquatic biota.

There are inherent limitations in the ability of any SQG approach to accurately and reliably indicate adverse effects on benthic communities, especially at sediment contaminant concentrations that lie between the TE and PE on a concentration–response curve. The general expectation for SQGs is that they index the potential for effects on benthic communities. However, the connectivity between benthic community structure and function and the endpoints on which the SQGs are derived varies and is generally incomplete. A number of factors that affect organism response to contaminated sediments are not, or are not fully, captured by existing SQG approaches. For example, SQGs do not directly address avoidance of sediment contaminants by sediment-dwelling organisms or predator–prey interactions with regard to contaminant trophic transfers (Von Stackelberg and Menzie 2002). Some studies have compared SQGs with field surveys of contamination, toxicity, and benthic community structure (e.g., Swartz et al. 1994; Ingersoll et al. 1997; Chapter 4). These studies generally show a correspondence between increasing contamination and the disappearance or reduction of sensitive benthic taxa in sediments where a numeric SQG has been exceeded. However, impacts on benthic communities have also been observed at contaminant concentrations well below SQGs derived using laboratory toxicity data. Conversely, there have been some cases when adverse effects on benthic organisms were not observed in sediment where one or more SQGs were exceeded.

Despite the uncertainties involved in different SQG approaches, chemically based, empirical SQGs derived using different methods and assumptions appear to converge, suggesting important underlying relationships relative to causality. Diverse scientific and statistical methods have been used to quantify the potential for adverse biological effects at different contaminant concentrations. All of the different methods provide some information on the relationship between sediment contamination and adverse biological effects; however, none of the approaches by itself is entirely satisfactory. The recent development of a consensus-based approach (Swartz 1999; MacDonald, Ingersoll, Berger 2000; MacDonald, Di Pinto, et al. 2000) indicates that, for some contaminants, estimates of TE and ME guidelines for the most widely used empirical SQGs (e.g., ERLs, ERMs, TELs, PELs, and probable effects concentrations [PECs]) are within 1 order of magnitude or less. The mechanistic and empirical SQGs for total PAHs are also reasonably consistent, although empirically based SQGs for individual chemicals are not (Di Toro and McGrath 2000).

Research associated with the development of SQGs has made large contributions to the assessment of sediments and has identified many of the key factors that must be considered. That said, SQGs are based on a variety of assumptions and endpoints, and none can separate with 100% accuracy sediments that do and do not cause biological effects. For that reason, the application of SQGs to decision-making processes must always be consistent with the derivation and uncertainties of those SQGs. While continued development of SQGs will undoubtedly lead to further improvements, there are inherent limitations to the ability to quantify biological effects (including bioavailability) in sediment. While improvements in SQGs may narrow the range of concentrations where effects are uncertain, SQGs should be incorporated into a larger WOE framework to better evaluate the potential for adverse biological effects.

Predictive Ability of Sediment Quality Guidelines

The quantitative extent to which chemically based numeric SQGs are predictive of the presence or absence of toxicity to sediment-dwelling organisms or to higher trophic level organisms is a critical concern among scientists and agencies evaluating the application of one or more numeric SQG approaches (Chapter 4). Users of the different SQG approaches should understand how well various SQGs predict the presence or absence and the extent of toxicity in sediment samples. The predictive ability of various SQGs to represent the potential for effects or no effects of contaminants on organisms in freshwater, estuarine, and marine environments is the probability of observing the presence or absence of effects within incremental ranges of sediment contaminant concentrations as defined by SQGs based on the specific endpoints and benthic taxa evaluated. Results of these evaluations are typically expressed as the percentage of samples expected to be affected (e.g., % toxic samples); however, some evaluations have also been based on the degree of the response (e.g., % mortality). In this context, chemically based numeric SQGs are defined as the concentration of sediment-associated contaminants that is associated with a high or a low probability of observing adverse biological or bioaccumulation results.

A wide range of published laboratory and field studies in freshwater, estuarine, and marine environments encompassing more than 8000 sediment samples indicate that empirical effects-based SQGs can be used to assess the probability of observing effects with known statistical levels of confidence. Among the different empirical SQG approaches, the incidence of effects, or the degree of the response, increases with increasing sediment contamination. These comparisons are primarily based on the presence or absence of mortality in 10-d amphipod tests using whole (i.e., bulk) sediments. Among all of the sediment toxicity data sets examined, the lowest incidence of adverse biological effects (less than about 10% of the samples) was identified at contaminant concentrations less than the low range of empirically based SQGs (i.e., below a mean quotient of about 0.1); the highest incidence of toxicity (greater than about 75%) was observed at contaminant concentrations above the upper range of empirical SQGs (i.e., above a mean quotient of about 1.5 to 2.3).

For mechanistic-based EqP SQGs, the predictive ability was demonstrated initially on the results of 10-d toxicity tests using single contaminants followed by mixtures of metals and PAHs in spiked sediments (Di Toro, Zarba, et al. 1991; Di Toro, Mahony, et al. 1991; Swartz et al. 1995; Hansen et al. 1996). For organic chemicals, the range of uncertainty for predicting acute toxicity in sediment (i.e., the LC50 concentration) appears to deviate from the best estimate by about a factor of 2 in concentration (USEPA 1993, 2000b). For metals, the SEM–AVS method initially predicted only the lack of toxicity, with no exceptions. The more recent excess organic-carbon normalized SEM method predicts the presence of toxicity as well, with an uncertainty range (25% of the samples) within which both toxicity and no toxicity are found (USEPA 2000a).

Chronic toxicity tests conducted with freshwater sediments and evaluations of benthic community structure in estuaries indicate that responses occur in ranges below those where an empirical SQG would predict toxicity in 10-d toxicity tests. These differences are about 6-fold lower for the chronic responses in freshwater toxicity tests and about 10-fold lower for estuarine benthic community responses. However, the relative influence of natural factors (e.g., grain size or total organic carbon) versus chemical-induced toxicity on benthic communities in estuaries and the adequacy of the reference sites selected in these benthic community studies have not been fully quantified. In a limited number of chronic metals and metals mixture experiments, the excess organic-carbon normalized SEM method predicts the presence or absence of effects at the lower bound (100 µmol/g of organic carbon) of the acute range in virtually all of the tests conducted to date. Use of chronic laboratory toxicity tests and controlled field-colonization studies and mesocosm studies that control for confounding factors are needed to better estimate the impacts observed on benthic communities exposed to contaminated sediments.

SQGs based on partitioning theory attempt to causally relate sediment concentration to toxicity. The availability and success or failure of the different mechanistic-based SQG approaches depends on the adequacy of the partitioning model and its parameters and on the assumption that exposure is either from pore water or sediment particles. Empirical SQGs for total PAH and total PCBs are similar to comparable guidelines derived using theoretical approaches. This concordance suggests that these mixtures may be causally implicated in the toxicity observed in a substantial number of sediments. However, the results for metals are much different, with the SEM–AVS model predicting metal toxicity only in laboratory-spiked sediments either with very high metal concentrations or containing no AVS and low organic-carbon concentrations.

Importantly for both laboratory toxicity tests and benthic community studies, an incremental increase in effects has frequently been observed with an incremental increase in contamination as defined by different SQG approaches. However, direct measurement of toxicity in the laboratory and/or benthic community impacts in the field is required to determine if an individual sample with moderate contamination is toxic or nontoxic.

Many confounding factors can potentially cause spurious conclusions regarding the relationship between sediment-bound contaminants and occurrence of toxicity or the lack of toxicity. Confounding factors can be biological in that organisms are not exposed to whole sediment or pore water and, thus, may underpredict toxicity (e.g., behavior). Confounding factors can also be physical, whereby one or more sediment characteristics influence a biological response that co-varies with contaminant concentrations (e.g., grain size and total organic carbon). Additionally, the chemical state of the contaminants (e.g., the form of Pb in paint chips, lead shot, tar balls, and metal ore), the nature of the sediment matrix (e.g., black carbon, peat, and wood chips), and laboratory artifacts (e.g., water temperature and homogenization or compositing of heterogeneous sediments) can also reduce the predictive ability of various SQG approaches.

Bioaccumulation, in and of itself, is not an effect, and none of the effects-based SQGs (e.g., ERMs, EqP, and AETs) were designed or intended to be predictive of toxic residues for sediment-dwelling organisms as a result of bioaccumulation from sediment. However, there are approaches that may allow connections to be established between measures of effect and no-effect tissue residues that lead to new bioaccumulation-based SQGs. These connections can be derived from tissue-residue effects values (e.g., median lethal residue [LR50] or from regional background tissue reference values derived using empirically based biota–sediment accumulation factors [BSAFs] or measured BSAFs for nonionic organic compounds). Current evidence suggests that BSAFs for nonionic organic compounds are relatively consistent on a contaminant-specific basis; however, site-specific geochemical factors may cause deviations from the general bioavailability trends. While bioaccumulation-based SQGs have been proposed and, in some cases, implemented as chemical-specific numeric SQGs, the predictive ability of this approach has yet to be adequately validated by field experimentation. Further, establishing cause and effect from sediments through bioaccumulation and trophic transfer should be evaluated on a site-specific basis because of the unique characteristics of different ecological receptors and food web interactions in freshwater, estuarine, and marine environments. Thus, the probabilities of adverse biological effects due to bioaccumulation at higher trophic levels cannot readily be predicted using current SQG approaches in the absence of ERA.

Use of Related Sediment Assessment Tools

The primary focus of sediment assessments should be determining the potential for biological impairment (Chapter 5). The simplest of all assessments might include the use of a single LOE such as a set of toxicity tests, a benthic community survey, or the use of SQGs to make a decision regarding impairment. However, the initial screening of contaminated sediments may be insufficient for making decisions due to uncertainty. In such cases, practitioners may opt to pursue additional LOEs that improve certainty. This may involve using a WOE approach or a more formal ERA and can include developing a conceptual model, understanding organism linkages, selecting measurement and assessment endpoints, characterizing exposure, and performing a risk characterization. Regardless, the WOE approach must be clearly defined a priori with regard to how complementary and contrasting LOEs will be evaluated.

At least 4 key LOEs should be developed:

1) sediment contaminant chemistry and geochemical characteristics,
2) benthic invertebrate community structure,
3) sediment toxicity, and
4) bioaccumulation and biomagnification data (Grapentine et al. 2002).

The integration of these and other sediment assessment tools should

- improve the ability to establish impairment of aquatic biota and to establish reference or benchmark conditions,
- ensure that sediment assessments include consideration of important ecological and food web linkages,

- provide guidance on how to formulate LOEs in an overall WOE approach to assist in making a decision as to whether or not sediment contamination has resulted in biological impairment, and
- help to identify potential sources of contaminants introduced into sediments.

The United States Environmental Protection Agency (USEPA 1992) ERA framework provides the option of either making a decision with the current level of uncertainty or performing more tests to reduce uncertainty (Ingersoll et al. 1997). Such assessments may not be required in every circumstance because they may be burdensome, delay action, and increase cost. However, each of the 4 components of the current ERA paradigm—problem formulation, exposure characterization, effects characterization, and risk characterization—can help guide managers toward an appropriate level of assessment using appropriate tools (Ingersoll et al. 1997).

The problem formulation stage is arguably the most critical stage of the sediment assessment process. It is during this stage that the appropriate tools for screening and assessing the state of biological impairment should be selected. To assist in making decisions regarding the potential for biological impairment where more than a single LOE is needed, an iterative approach to problem formulation is recommended for collecting data within the sediment ERA framework. The importance of the problem formulation stage and, in particular, the conceptual model development and characterization of reference (i.e., background) conditions are frequently overlooked in sediment assessments and lead to incomplete or overly simplified ERAs that may not reflect an accurate assessment of the potential for biological impairment. To avoid situations where sediment quality is mischaracterized as either over- or underprotective, the following 8 study design elements should be considered:

1) identification of relevant exposure pathways associated with hydrodynamics, groundwater upwelling, sediment (surface and deep), and food;
2) selection of an appropriate conceptual model relating stress and response;
3) selection of appropriate reference locations, background stressor levels, and natural stressors;
4) selection of optimal measurement and assessment endpoints;
5) laboratory and field quality control procedures to ensure observations with the lowest possible variability;
6) identification of appropriate site-specific species for laboratory testing;

7) environmental sampling to ensure adequate statistical power to detect pre-specified biological changes in responses and spatial and/or temporal characterization; and
8) selection of appropriate statistical methods for LOE analyses and characterization of effects using biologically based methods that integrate resident biota and sediment toxicity into a WOE matrix.

Sediment assessments are frequently based on temporal or spatial data that do not adequately reflect co-occurrence of multiple contaminants or stressors, or the proper exposure pathway considerations. Exposures should be assessed in a manner that reflects species biological and behavioral characteristics, and the critical time periods and locations that reflect co-occurrence between stressor and receptor. With regard to characterization of ecological effects, adverse biological effects in a water body are defined as those that can impact populations, such as development, reproduction, and survival. Sediment assessments based on a single LOE or incomplete LOE often overlook the potential for long-term effects or the trophic transfer of contaminants that result in food chain effects.

A relatively simple risk characterization model, involving the comparison of sediment chemistry measurements to one or more SQGs, can be improved by incorporating probabilistic models into the integration of exposure and effects LOEs. This, in itself, does not necessarily increase the complexity of the underlying hazard quotient model, but simply adds a useful layer of quantitative interpretation. When confronted by conflicting multiple LOEs, the first task in risk characterization is to determine how to interpret apparent conflicting information between different LOEs (Grapentine et al. 2002). For example, the results of sediment toxicity tests may indicate no effects when comparison of sediment chemistry with an SQG suggests that toxicity should occur. This may be due to insufficient sensitivity on the part of the toxicity test organism or to overly conservative estimates used to derive the SQG. A transparent, and preferably quantitative, approach is required to integrate the information derived from multiple LOEs.

A tiered approach to data collection is recommended whereby screening-type tools are used in Tier 1, and tools that enable detailed characterization of either exposure or effects, or those that provide interpretation of both exposure and effects data in sediment assessments, are used in Tier 2. Tier 1 tools are typically low in cost, easy to use, and may yield information that tends to be general, rather than specific. Tier 2 tools may be more challenging or expensive to use, but often provide more information and greater specificity. Exposure and effects data form the 2 major LOEs in a Tier 2 assessment. A critical requirement for selecting the appropriate sediment assessment tools is to ensure that the appropriate tools are matched with the appropriate investigation questions. This will depend, in part, on the stage and complexity of the assessment process; earlier stages and simpler assessments may require only relatively simple tests that facilitate screening. Later stages, such as those deemed necessary following screening, may require more complex tests or tools to better delineate the extent of sediment impairment. The selection of the appropriate tools allows for development of defensible LOEs that can be used to determine sediment quality and the potential for biological impairment.

The importance of characterizing both natural and anthropogenic stressors and exposure conditions at multiple levels of the food web is often recognized in principle, though seldom accomplished in practice, during the ERA process. Traditional assessment tools such as SQGs do not always provide predictive power when multiple contaminants and stressors are present and may result in unacceptably high levels of false positive and false negative conclusions.

It is apparent that more effective ecological assessment approaches are needed to link the magnitude, frequency, and duration of exposure with biological effects and to provide better definition of when adverse ecological effects occur. A more quantitative and logical framework for using the WOE process for sediment assessment is necessary where substantial uncertainty exists. Such a framework should include the following critical elements:

- site-wide conceptual model,
- linkage of "exposure" and "effects" and conceptual model components,
- characterization of exposure profiles for key natural and anthropogenic stressors,
- appropriate reference characterization and comparison methods,
- appropriate quantification methods used to integrate LOEs,
- critique of advantages and limitations of each LOE used,
- evaluation of each LOE versus causality criteria,
- a method for combining exposure and effects LOEs into a WOE matrix for interpretation, and
- showing causality linkages in the conceptual model.

Appropriate statistical approaches (e.g., regression, analysis of variance [ANOVA], and multivariate methods) should be used within each LOE to define reference conditions and potentially impaired conditions. This approach is most useful when incorporated into the initial and final study design stages (e.g., the problem formulation and risk characterization stages of an ERA [Burton et al. 2002]).

In order to establish causality, a multiple step process is necessary, using diagnostic protocols and weighing the strength of evidence that supports each potential cause. This can be done via 7 causal considerations:

1) co-occurrence (spatial correlation),
2) temporality (temporal correlation),
3) effect magnitude (strength of link),
4) consistency of association (at multiple sites),
5) experimental confirmation (field or laboratory),
6) plausibility (likelihood of stressor–effect linkage), and
7) specificity (stressor causes unique effect [USEPA 2000d]).

The results of these expert judgments can be summarized in a tabular decision matrix, for example, by conveying judgments of different LOEs into a hierarchal scale (e.g., 1 to 3 or 4, or "+" and "–" values). Using a ranking scheme, logical comparisons can be made of multiple LOEs, comprising both exposure- and effects-based LOEs. If the WOE process adequately characterizes and links stressor exposure with biological impairment using relationships to reference (i.e., background) conditions, then uncertainty will be better understood in decisions concerning the source, occurrence, and severity of sediment-related impairment of aquatic biota.

Use of Sediment Quality Guidelines in Sediment Assessment Frameworks

Chemistry data have been used for decades by state and federal regulatory agencies in different countries to assess and manage contaminated sediments (Chapter 6). SQGs have been developed recently to better define the relationship between sediment chemistry and toxicity, providing regulatory agencies with additional insight into the importance of sediment chemistry data. Increasing interest in the development of risk-based sediment assessment frameworks to guide assessments and management decisions has led to questions concerning the role of SQGs within a sediment assessment and management process that makes use of multiple LOE to reach management decisions based on a WOE. At present, nearly 20 sediment assessment frameworks have been proposed or used by regulatory authorities in different countries. These frameworks include several key characteristics that should be preserved or refined in the future, including the use of multiple tiers, multiple LOEs (including both chemical and biological information), and an iteration process that facilitates refinement of an assessment as data are collected and analyzed. With respect to the use of SQGs in current sediment assessment frameworks, there is strong interest in having a range of SQGs that can be used to assess and classify sediments.

An assessment framework provides a structure and process for conducting a sediment assessment that leads to a management action. As such, a framework that meets regulatory or environmental objectives should define appropriate uses for SQGs as part of a risk-based evaluation. Environmental regulatory agencies have been encouraged by the scientific community to develop logical and orderly sediment assessment frameworks and to ensure that assessments are comprehensive, transparent, and consistent.

A sediment assessment framework should be structured to ensure that any evaluation that follows the steps of the framework is comprehensive and complete in its consideration and analysis of present and future exposures, effects, and human and ecological risks at the site of concern. All routes of exposure and types of effects will not occur at every site; however, a comprehensive assessment framework, if followed, should require consideration of the likelihood for all possible routes of exposure and the potential for adverse biological effects to ensure that required or important site-specific environmental factors are not omitted from the evaluation process.

Assessment frameworks should also provide a measure of transparency to sediment investigations and management, as well as facilitate meaningful participation in the assessment and decision-making process by scientists, regulatory agencies, and representatives of affected communities. Active stakeholder involvement throughout the assessment process is essential to ensuring that the results of the assessment can be successfully applied within the decision-making process. Finally, development and application of an assessment framework will facilitate consistent application of the assessment and management process at different sites.

The objectives for a sediment assessment will vary. For example, a sediment assessment may be conducted in connection with a dredging project to achieve navigable depths in a channel, to determine the need for remedial action outside a navigation channel, or as part of a more general watershed or water quality assessment. These varying objectives will dictate particular investigation tools and methods for quantifying exposures, effects, and ecological and human health risks in the assessment, including the application of SQGs.

Significant effort should be invested at the outset of a sediment investigation to develop a comprehensive conceptual model for contaminated sediments in the aquatic environment under consideration in the assessment. The conceptual model is the basis for formulating project-specific questions that drive subsequent sediment assessment activities. Programmatically defined conceptual models can be adapted and applied for some routine management applications (Cura et al. 1999).

A related fundamental requirement of sediment assessments concerns identification of meaningful sediment and site-specific questions a priori and the selection of specific LOE and assessment tools. Some assessment questions cannot be addressed with the current suite of available SQGs. For example, a common assessment question "Does the sediment contain bioaccumulative chemicals that pose an unacceptable risk to upper trophic levels?" cannot be addressed by comparing whole-sediment chemistry measurements to the most widely used effects-based SQGs. Considerable effort should be devoted to formulating and refining specific and detailed questions that must be answered to reach conclusions about the presence and magnitude of risk.

There is strong merit for using SQGs and other sources of information in a screening, or initial, assessment phase to identify sediments that pose little potential for adverse biological risks. In cases where comparison of whole-sediment chemistry data to SQGs results in ambiguous answers to the assessment questions concerning the presence of unacceptable risk, the assessment should proceed to a secondary, more detailed, sediment or site assessment after revising, as necessary, the list of contaminants of potential concern, the conceptual model, and the assessment questions.

Key activities that should be included in a detailed sediment assessment are

- defining measurement and assessment endpoints,
- selecting LOEs within 3 general categories (assessing direct exposure or effects in the water column, assessing direct exposure or effects to the benthic community, and assessing indirect exposure and effects through contaminant trophic transfer),
- selecting and applying assessment tools within the chosen LOEs,
- analyzing the collected information to reach conclusions based on a WOE, and
- revising the conceptual model to identify remaining data gaps or to communicate conclusions about risks.

Integrating information within and among LOEs will be necessary to reach conclusions about the risks posed by sediment. For example, the sediment-effects LOEs may include information from 3 different toxicity tests, 3 different metrics for describing the status of the benthic community, and critical body residue toxicity data for key contaminants. Incongruities among different LOEs must be resolved; for example, 1 toxicity test may indicate the presence of toxicity, while a different test may show a lack of toxicity. At this stage of a risk assessment, SQGs can provide substantial aid to data interpretation where apparent conflicts exist among benthic effects data. If toxicity tests provide evidence of toxicity, but sediment concentrations for all contaminants are below lower threshold guidelines (e.g., ERLs), then the conclusion that benthic effects are unlikely would be strengthened by the information provided by SQGs. More than 1 LOE should be used to resolve these challenges, including information about conditions known to interfere with toxicity tests (e.g., test species' sensitivity to sediment grain-size distribution, sediment concentrations of ammonia and hydrogen sulfide). Used in this fashion, chemistry data and SQGs elucidate cause and effect relationships and strengthen or clarify conclusions about ecological risks.

When assessment questions have been satisfactorily addressed and conclusions reached concerning the extent and magnitude of risks, the framework transitions to comparison and selection of management alternatives. This phase involves a process for

- listing the practical management alternatives,
- comparing the risks associated with implementing those alternatives,
- comparing the costs of implementing the alternatives, and
- apportioning the sediment at a site among the selected alternatives.

It is evident from the available scientific literature that there are no zero-risk options for managing contaminated sediments. Consequently, comparative risk analysis methods are needed and should be developed and used to evaluate and select the appropriate management options. Effective risk management requires comparing the risks and costs associated with the full spectrum of available management alternatives. This comparative approach may require several iterations between site assessment activities and analyses conducted as part of the evaluation of management alternatives. Effective sediment management requires developing monitoring strategies to verify the accuracy of risk predictions made during sediment assessment and the appropriateness of management decisions based on these predictions. Such a verification step also provides an opportunity for strengthening the assessment process and framework.

Apportioning contaminated sediment among more than 1 management alternative requires developing a logic or metric for using the information derived from different LOEs developed during the assessment to rank the risks associated with contaminant conditions in the sediment. Most of the controversy surrounding the use of SQGs to assess and manage risks concerns the extent to which SQGs can be used as the logical basis for distinguishing sediments that require special management from those that do not. The degree to which SQGs can contribute to decision making will depend on the nature of the risks at the site. In most cases, conclusions about the nature, extent, and magnitude of risks are developed from information supplied by multiple LOEs. The logic or metrics used to apportion sediment among management alternatives must be tied directly to the LOEs describing the potential risks. To do otherwise will likely lead to false conclusions about the level of protection being afforded by the selected course of action.

For example, if the risks associated with a particular site are caused by indirect exposures to wildlife or humans through trophic transfer of bioaccumulative chemicals, then the logic used to apportion sediment among different management alternatives must be derived from the LOEs associated with indirect routes of exposure and effects. In this specific case, the distinguishing logic could involve deriving risk-based concentrations for prey items that would be protective of upper trophic level receptors using a bioaccumulation model. Deriving such concentrations would require the use of site-specific data on the receptors of concern and their use of the site.

Because most of the widely used SQGs are limited to providing information about effects to benthic invertebrates (e.g., ERLs and ERMs, PELs and TELs, AETs, and EqP-derived methods), their relevance to apportioning sediment among different management alternatives is limited to cases where risks are limited to the benthic organisms. However, even in such cases, the uncertainties associated with SQGs will constrain the extent to which these values can be used alone to reach credible management decisions.

Successful implementation of a sediment assessment framework includes provisions for frequent communication with different stakeholder groups. It is particularly important to solicit stakeholder input when the questions that ultimately drive sediment assessment activities are being formulated and management alternatives are being evaluated.

While following a well-ordered and structured assessment and management framework will help to reduce uncertainty in risk estimates and increase the likelihood that assessments produce information useful for decision making, ultimately, the soundness of the assessment and consequent management decisions will depend on the strength of the underlying science. Strengthening the scientific foundation of sediment assessments will depend on advancements being made in the following key areas:

- assessing cumulative exposure and effects from contaminant mixtures,

- assessing bioaccumulation and trophic transfer potential for sediment-associated contaminants,
- incorporating more realistic spatial and temporal considerations into exposure assessment, and
- developing robust comparative risk assessment approaches to evaluate and select management alternatives.

Assessing Sediments in Different Aquatic Environments

There are physical, chemical, and biological factors in the environment that complicate and introduce uncertainty into the derivation and application of SQGs and other sediment assessment tools (Chapter 7). Sediments are heterogeneous and dynamic. Important physical and chemical properties, such as grain size, sulfide levels, organic carbon type, and content, may vary at small scales (millimeters) or large scales (estuaries) within even a single assessment area. Exposures to sediment-associated contaminants can occur by different routes, such as via the sediment–water interface, pore water, direct contact, or ingestion. Biological factors reflecting species-specific differences in physiology, biochemistry, and behavior result in varying tolerances, acclimation, or adaptation, which result in different levels of adverse biological effects. In addition, sediments tend to be contaminated by mixtures of chemicals whose potential interactions are not well characterized and whose bioavailability can be variable and challenging to predict. None of these factors are absolutely critical to the derivation or application of SQG approaches or to other sediment assessment tools.

In general, several LOE are needed to properly evaluate contaminated aquatic environments. The absence of information or inadequate appreciation of the variation in any 1 of the following 5 areas will detract from an adequate understanding of the aquatic system relative to the occurrence and potential effects of contamination:

1) nature and extent of contamination;
2) expected or acceptable diversity and abundance of benthic biota in the absence of contamination;
3) bioavailability, bioaccumulation, and effects of contamination (the potential for chronic as well as acute effects) on aquatic organisms;
4) stability of sediments and contaminants (fate and transport); and
5) risk of contamination to aquatic biota and associated resources.

A number of tools (not including specialized studies such as toxicity identification evaluations [TIEs], biodegradation studies, exposure surrogates, or modeling) are available to obtain some or all of this information:

- numeric SQGs,
- sediment toxicity tests (chronic as well as acute),
- resident exposed communities (not necessarily restricted to the benthic invertebrates),
- bioaccumulation, and
- biomarkers and/or histopathology.

Each tool has its own inherent strengths and weaknesses. These tools should be applied as needed to meet the objectives of the sediment assessment and as appropriate to the specific environment. In this regard, assessment frameworks and conceptual models are needed to help apply these tools appropriately and to ensure that appropriate exposure routes and site-related taxa are considered.

At present, chemical analysis of whole sediment is an adequate initial estimate of exposure; however, adjustments to account for bioavailability or chemical speciation can improve exposure estimates. The potential for adverse biological effects to aquatic organisms is reasonably measured using benthic community analysis (e.g., diversity, abundance, and presence or absence of key species), analysis of contaminant residues in tissue, and toxicity testing of appropriate, representative taxa.

SQGs alone are sometimes, but not always, sufficient for management decision making. Different tools are needed to characterize sediment quality in different environments. It is apparent that, with the possible exception of biomarkers, sediment assessment tools are not equally applicable to evaluations of different aquatic habitats. Also evident is the need for additional standardized methods and procedures to further validate SQGs in estuarine and stream habitats. There is a complex interplay between physicochemical and biological components that will dictate some level of uncertainty regarding contaminant bioavailability and effects, so sensitive diagnostic tools may be especially valuable in validating predictive models. Thus, it is critical that the right assessment tools be selected to match the ecological system being evaluated.

Path Forward

The development of a uniformly accepted set of SQGs presents the major scientific challenge that most likely will be impossible to achieve. Contemporary understanding of several fundamental scientific principles such as the definition of bioaccumulation potential and sediment toxicity indicative of unacceptable, or significant, adverse biological effects underscores the complexity of any detailed assessment of contaminated sediments. Nevertheless, it may be possible to define a protocol, based primarily on biological testing and ERA, that could be consistently used to evaluate sediment-bound contaminants and define acceptable concentrations for waterway-specific, ecosystem-specific, regional, and possibly national applications.

The path forward to achieve this goal involves additional research. One of the reasons for the transition zone between TE and PE guidelines derived from a contaminant concentration–response curve is an incomplete understanding of the factors controlling the bioavailability of contaminants in sediment and the biological and ecological factors that alter an organism's response to contaminant exposures. Factors that create the transition zone are several: some reflect intrinsic variability (e.g., measurement uncertainty, and chemical, biological, or ecological variability), some reflect errors in design or interpretation of experiments, and some reflect an incomplete scientific understanding of microenvironments, biogeochemistry, bioavailability, nonchemical stressors, and biological activity. All of these factors should be the focus of additional research. The ability to interpret the importance of these and other factors is further limited by an incomplete understanding of mixtures and the potential for unmeasured contaminants.

A related area of research is needed to improve the understanding of the variation in site-specific bioavailability of both sediment-bound organic contaminants and metals. A better understanding of the factors controlling bioaccumulation of metals and the importance of metal tissue residues is needed to confirm the relationships established by mechanistic SQG approaches such as EqP and the use of SEM–AVS analysis.

Consequently, perhaps the highest research priority in the field of sediment quality assessment is to further develop or refine the current approaches for estimating chemically based numeric SQGs from field effects of sediment-bound contaminants on benthic ecosystems. Certainly, the currently available SQGs can be improved; but, any new or revised SQGs must be demonstrated with empirical independent observations to be more predictive of effects and more protective of valued biological resources than those that are currently available. The discrimination of adverse biological effects associated with exposure to a single contaminant or mixtures of contaminants from those responses attributable to other noncontaminant stressors (e.g., grain size, organic enrichment, salinity, habitat modification, and exotic species) is essential in this research.

In addition, research is needed to support the development of new SQGs for characterizing exposure and effects from bioaccumulative contaminants. None of the currently available SQGs adequately address trophic transfer mechanisms and the potential for some contaminants to move within aquatic food webs. Progress in this area will improve sediment assessments and management decision making, given the increasingly ubiquitous occurrence of bioaccumulative contaminants in the environment.

Research is also needed to develop single-contaminant SQGs based on spiked sediments, or evaluation of field samples where there are a limited number of contaminants. Future analyses of the predictive ability of SQGs should include an evaluation of the sensitivity and efficiency of SQGs to different environmental factors. To date, sensitivity and efficiency analyses have been performed primarily for AETs. Future evaluations of the predictive ability of SQGs should also include controlled benthic community colonization studies, mesocosm studies, and chronic laboratory tests to better account for abiotic and habitat factors influencing the response of benthic invertebrates in the field. Further research is also needed to determine the conditions where site-specific SQGs may differ from SQGs derived from a range of toxicity studies reported in the literature.

Finally, training in the role of SQGs as 1 tool among several to evaluate sediment quality is highly recommended. Short courses could be given that address the development and application of SQGs and other measures of sediment quality. Once developed, these short courses should be presented at regional, national, and international scientific meetings. Aside from the various chemical and biological areas of research needed to further elucidate the significance of contaminants in sediment, there is an urgent need to provide training to professionals charged with sediment management. Among the many issues currently debated regarding the use of SQGs, the interpretation of contaminants in sediment using SQGs has not often been consistent with the intended use of the different SQG approaches.

References

Ankley GT, Di Toro DM, Hansen DJ, Berry WJ. 1996. Technical basis and proposal for deriving sediment quality criteria for metals. *Environ Toxicol Chem* 15:2053–2055.

Barnthouse L, Stahl RG. 2002. Quantifying natural resource injuries and ecological service reductions: Challenges and opportunities. *Environ Manag* 30:1–12.

Barrick R, Becker S, Pastorok R, Brown L, Beller H. 1988. Sediment quality values refinement: 1988 update and evaluation of Puget Sound AET. Bellevue (WA): Prepared by PTI Environmental Services for US Environmental Protection Agency.

Burton Jr GA, Batley GE, Chapman PM, Forbes VE, Schlekat CE, Smith PE, den Besten PJ, Barker J, Reynoldson T, Green AS, Dwyer RL, Berti WR. 2002. A weight-of-evidence framework for assessing sediment (or other) contamination: Improving certainty in the decision-making process. *Hum Ecol Risk Assess* 8: 1675–1696.

Carr RS, Nipper MJ, editors. 2003. Porewater toxicity testing. Pensacola (FL): Society of Environmental Toxicology and Chemistry (SETAC).

Chapman PM, McDonald BG, Lawrence GS. 2002. Weight-of-evidence frameworks for sediment quality and other assessments. *Hum Ecol Risk Assess* 8:1489–1515.

Cubbage J, Batts D, Briedenbach S. 1997. Creation and analysis of freshwater sediment quality values in Washington State. Olympia (WA): Environmental Investigations and Laboratory Services Program, Washington Department of Ecology.

Cura JJ, Heiger-Bernays W, Bridges TS, Moore DW. 1999. Ecological and human health risk assessment guidance for aquatic environments. Vicksburg (MS): US Army Engineer Waterways Experiment Station. Technical Report DOER-4.

Dickson KL, Maki AW, Brungs WA. 1987. Fate and effects of sediment-bound chemicals in aquatic systems. New York: Pergamon.

Di Toro DM, Mahony JD, Hansen DJ, Scott KJ, Carlson AR, Ankley GT. 1991. Acid volatile sulfide predicts the acute toxicity of cadmium and nickel in sediments. *Environ Sci Technol* 26:96–101.

Di Toro DM, McGrath JA. 2000. Technical basis for narcotic chemicals and polycyclic aromatic hydrocarbon criteria. II. Mixtures and sediments. *Environ Toxicol Chem* 19:1971–1982.

Di Toro DM, Zarba CS, Hansen DJ, Berry WJ, Swartz RC, Cowan CE, Pavlou SP, Allen HE, Thomas NA, Paquin PR. 1991. Technical basis for establishing sediment quality criteria for non-ionic organic chemicals using equilibrium partitioning. *Environ Toxicol Chem* 10:1541–1583.

Field LJ, MacDonald DD, Norton SB, Severn CG, Ingersoll CG. 1999. Evaluating sediment chemistry and toxicity data using logistic regression modeling. *Environ Toxicol Chem* 18:1311–1322.

Field LJ, MacDonald DD, Norton SB, Ingersoll CG, Severn CG, Smorong D, Lindskoog R. 2002. Predicting amphipod toxicity from sediment chemistry using logistic regression models. *Environ Toxicol Chem* 21:1993–2005.

Ginn TC, Pastorok RA. 1992. Assessment and management of contaminated sediments in Puget Sound. In: Burton Jr GA, editor. Sediment toxicity assessment. Boca Raton (FL): Lewis. p 371–397.

[GIPME] Global Investigation of Pollution in the Marine Environment. 2000. Guidance on assessment of sediment quality. London: International Maritime Organization.

Grapentine L, Anderson J, Boyd D, Burton GA, De Barros C, Johnson G, Marvin C, Milani D, Painter S, Pascoe T, Reynoldson T, Richman L, Solomon K, Chapman PM. 2002. A decision-making framework for sediment assessment developed for the Great Lakes. *Hum Ecol Risk Assess* 8:1641–1655.

Hansen DJ, Berry WJ, Mahony JD, Boothman WS, Di Toro DM, Robson DL, Ankley GT, Ma D, Yan Q, Pesch CE. 1996. Predicting the toxicity of metals-contaminated field sediments using interstitial concentrations of metal and acid volatile sulfide normalizations. *Environ Toxicol Chem* 15:2080–2094.

Ingersoll CG, Dillon T, Biddinger RG, editors. 1997. Ecological risk assessment of contaminated sediment. Pensacola (FL): Society of Environmental Toxicity and Chemistry. 389 p.

Ingersoll CG, MacDonald DD, Wang N, Crane JL, Field LJ, Haverland PS, Kemble NE, Lindskoog RA, Severn CG, Smorong DE. 2001. Predictions of sediment toxicity using consensus-based freshwater sediment quality guidelines. *Arch Environ Contam Toxicol* 41:8–21.

Kane-Driscoll S, Menzie CA. 2001. Review of toxicology of PAHs in invertebrate aquatic organisms. Palo Alto (CA): EPRI. 1006594.

Long ER, MacDonald DD, Smith SL, Calder FD. 1995. Incidence of adverse biological effects within ranges of chemical concentrations in marine and estuarine sediments. *Environ Manag* 19:81–97.

Long ER, MacDonald DD. 1998. Recommended uses of empirically based, sediment quality guidelines for marine and estuarine ecosystems. *Hum Ecol Risk Assess* 4: 1019–1039.

MacDonald DD, Carr RS, Calder FD, Long ER, Ingersoll CG. 1996. Development and evaluation of sediment quality guidelines for Florida coastal waters. *Ecotoxicology* 5:253–278.

MacDonald DD, DiPinto LM, Field J, Ingersoll CG, Long ER, Swartz RC. 2000. Development and evaluation of consensus-based sediment effect concentrations for polychlorinated biphenyls (PCBs). *Environ Toxicol Chem* 19:1403–1413.

MacDonald DD, Ingersoll CG, Berger T. 2000. Development and evaluation of consensus-based sediment quality guidelines for freshwater ecosystems. *Arch Environ Contam Toxicol* 39:20–31.

[NRC] National Research Council. 2001. A risk management strategy for PCB-contaminated sediments. Washington DC: National Academy.

Nord MA. 2001. Recommendations for the implementation of a national sediment quality policy in the United States. *Hum Ecol Risk Assess* 7:641–650.

[NYSDEC] New York State Department of Environmental Conservation. 1998. Technical guidance for screening contaminated sediments. Albany (NY): Division of Fish and Wildlife, Division of Marine Resources. 38 p.

Persaud D, Jaagumagi R, Hayton A. 1993. Guidelines for the protection and management of aquatic sediment quality in Ontario. Toronto (ON): Water Resources Branch, Ontario Ministry of the Environment. 27 p.

Smith SL, MacDonald DD, Keenleyside KA, Ingersoll CG, Field J. 1996. A preliminary evaluation of sediment quality assessment values for freshwater ecosystems. *J Gt Lakes Res* 22:624–638.

Swartz RC, Cole FA, Lamberson JO, Farraro SP, Schults DW, DeBen WA, Lee H, Ozretich JR. 1994. Sediment toxicity, contamination and amphipod abundance at a DDT- and dieldrin-contaminated site in San Francisco Bay. *Environ Toxicol Chem* 13:949–962.

Swartz RC, Schults DW, Ozretich RJ, Lamberson JO, Cole FA, DeWitt TH, Redmond MS, Ferraro SP. 1995. ΣPAH: A model to predict the toxicity of field-collected marine sediment contaminated by polynuclear aromatic hydrocarbons. *Environ Toxicol Chem* 14:1977–1987.

Swartz RC. 1999. Consensus sediment quality guidelines for PAH mixtures. *Environ Toxicol Chem* 18:780–787.

[USEPA] US Environmental Protection Agency. 1992. Framework for ecological risk assessment. Washington DC: USEPA. EPA-630/R-92/001.

[USEPA] US Environmental Protection Agency. 1993. Technical basis for deriving sediment quality criteria for nonionic organic contaminants for the protection of benthic organisms by using equilibrium partitioning. Washington DC: USEPA. EPA-822/R-93-011.

[USEPA] US Environmental Protection Agency. 1996. Calculation and evaluation of sediment effect concentrations for the amphipod *Hyalella azteca* and the midge *Chironomus riparius*. Chicago: USEPA. EPA-905/R-96/008.

[USEPA] US Environmental Protection Agency. 1997. The incidence and severity of sediment contamination in surface waters of the United States. Volume 1, National sediment quality survey. Washington DC: USEPA. EPA 823/R-97/006.

[USEPA] US Environmental Protection Agency. 2000a. Draft equilibrium partitioning sediment guidelines (ESG) for the protection of benthic organisms: Metal mixtures (cadmium, copper, lead, nickel, silver, and zinc). Washington DC: USEPA. EPA-822-R-00-005.

[USEPA] US Environmental Protection Agency. 2000b. Draft technical basis for the derivation of equilibrium partitioning sediment guidelines (ESG) for the protect of benthic organisms: Nonionic organics. Washington DC: USEPA. EPA-822-R-00-001.

[USEPA] US Environmental Protection Agency. 2000c. Prediction of sediment toxicity using consensus-based freshwater sediment quality guidelines. Chicago: USEPA. EPA-905/R-00/007.

[USEPA] US Environmental Protection Agency. 2000d. Stressor identification guidance document, Washington DC: USEPA. EPA-822-B-00-025.

Von Stackelberg K, Menzie CA. 2002. A cautionary note on the use of species presence and absence data in deriving sediment criteria. *Environ Toxicol Chem* 21: 466–472.

3

Scientific underpinnings of sediment quality guidelines

GRAEME E BATLEY, RALPH G STAHL, MARC P BABUT, THOMAS L BOTT, JAMES R CLARK, L JAY FIELD, KAY T HO, DAVID R MOUNT, RICHARD C SWARTZ, ANDRÉ TESSIER

Sediment quality guidelines (SQGs) are tools which, using a variety of methods and assumptions, relate the concentrations of contaminants in sediment to some predicted frequency or intensity of biological effects. Their development was driven by the needs to assess the kinds and degrees of sediment contamination and to evaluate mitigation efforts, and over the last fifteen years, they have become an important component of many sediment assessment and/or management programs (Chapters 6 and 9). Many SQG approaches exist, each with differing strengths and limitations that have important implications for their effective and appropriate use. This chapter critically reviews the science underpinning common SQGs, thus setting the stage for discussions on the applications and extensions of SQG approaches presented in subsequent chapters.

This discussion of the technical basis for SQGs begins by highlighting the physical, chemical, and biological complexities of sediments and the organisms associated with them. Next a synopsis is provided of the most common SQGs, followed by a discussion of key elements of sediment assessment that reflect heavily on the derivation and application of SQGs, including issues of effect endpoints, mixtures, causality, and bioavailability. This synopsis is followed by 2 sections dealing with uncertainty in the context of SQGs, and general considerations in applying SQGs in sediment assessment. The chapter closes with a series of consensus points and suggestions for future research.

Use of Sediment Quality Guidelines and Related Tools for the Assessment of Contaminated Sediments
Wenning RJ, Batley GE, Ingersoll CG, Moore DW, editors.
 ISBN 1-880611-71-6

Complexities of the Sediment Environment

Physical and chemical setting

While the substrates of lakes, rivers, and oceans can range from solid rock down to the finest particles, the term "sediment" is generally applied to depositional material with grains sizes from sand through silty-sand, to sandy-silt, to clays. Even within this range, particles vary in surface area over orders of magnitude and in chemical composition, both of which affect the nature and number of binding sites for metal and organic contaminants. Grain size often defines whether a sediment is a good habitat for biota, affecting the suitability for and stability of animal burrows, for example, and it influences benthic community structure. Some species show preferences for a particular grain-size, while others do not. Fine sediments are often those having the highest concentrations of contaminants on a dry weight basis, owing to their greater relative surface area and a correspondingly higher density of chemical sorption sites.

Decaying detrital particulate organic matter is distributed among mineral and amorphous particles in sediments and is a site for bacterial activity. These and organic coatings on inorganic particles provide binding sites for both metal and organic contaminants. Organic carbon is a particularly important binding phase for hydrophobic organic contaminants. While often lumped under generalized measures such as "total organic carbon" or "loss on ignition," it should be recognized that this particulate organic carbon represents a variety of different chemistries, owing to different sources and different stages of decomposition. In anthropogenically influenced sediment, other forms of organic carbon may be present, such as ash, soot, wood chips, oils, and tars (Chapter 4). The composition of sediment organic carbon can be very important, because the partitioning behavior of contaminants can vary substantially among these different materials. The composition of the organic carbon is also an important consideration, as discussed in later chapters, in the applicability of SQGs derived from diverse systems.

The microbial degradation of labile organic matter in whole sediments determines the redox potential (Eh) and the pH observed at various depths in whole sediment and is responsible for a variety of secondary reactions involving metals (e.g., desorption, release to pore water, formation of sulfide and associated fixation of trace metals, and precipitation of trace metals as sulfides). Sulfide, produced by sulfate-reducing bacteria, can sequester many cationic metals from pore waters through the formation of highly insoluble metal sulfides. The oxidation of iron(II) is acid forming and can facilitate metal release to pore waters.

Superficially, sediments may seem relatively homogeneous, but this is not generally the case. Because the flux of labile organic matter to the sediments is usually much faster than the diffusive flux of oxygen across the sediment–water interface, it is commonly observed that oxygen concentrations drop to zero close to the sediment–water interface. The oxic zone may vary in thickness from a few millimeters in silty sediments to several centimeters in coarser riverine and estuarine sands and is underlain by a sub-oxic and an anoxic area. Sediment is often anoxic well above the depth to which many benthic animals burrow (generally <20 cm). This oxygen gradient, along with other reactions described above, leads to vertical zonation in sediments and pore waters of pH, Eh and various chemical species, including Pb and Mn, and trace metals.

This heterogeneity of sediment can exist not only vertically, but horizontally as well. Recent studies using microsensors have revealed both vertical and horizontal heterogeneity in the concentration of solutes such as iron and manganese (Shuttleworth et al. 1999) and oxygen (Glud et al. 2000) at the millimeter scale in sediments. Digital image analysis has also been used to assess the heterogeneity of chemical binding phases in sediments (Bull and Williamson 2001). Small-scale heterogeneity is also found for bacterial activity (Stemmer, Burton, and Leibfritz-Frederick 1990; Stemmer, Burton, and Sasson-Brickson 1990). At a larger scale, particle sorting by currents results in differences in both physical and chemical composition of sediments, sometimes at the meter scale or below. While this horizontal heterogeneity may be most pronounced in rivers and estuaries, even lakes show differences between the fine-grained profundal sediments and coarser littoral sediments.

Interactions between organisms and sediment

Marine, estuarine, and freshwater sediments are inhabited by diverse communities of invertebrates. Macroinvertebrates commonly found in freshwater sediments include insect larvae, crustaceans, oligochaetes, and mollusks (Fisher 1982; Matisoff 1995). Most of these animals live in the upper 10 cm of the sediment, where they construct burrows (Fisher 1982; Matisoff and Wang 1998). The benthic community in marine sediments has great taxonomic diversity, including polychaetes, amphipods, and bivalves, some of which burrow to depths greater than 20 cm (Matisoff 1995).

With low oxygen concentrations and potentially toxic concentrations of sulfide (Wang and Chapman 1999), sediment can be a somewhat hostile environment. To exploit the sediment as habitat, benthic organisms have evolved a number of compensatory structural, physiological, and behavioral characteristics. A common strategy is to create a burrow that is irrigated with comparatively oxygen-rich overlying water. In doing so, such organisms satisfy their respiratory needs, and they evacuate toxic compounds such as sulfide (Meyers et al. 1987, 1988) and metabolic byproducts (Kristensen 1988). Bioirrigation can also be involved in animal feeding and gamete transport (Kristensen 1988).

Several observations suggest that the irrigation of burrows can create microenvironments whose chemistry differs greatly from that of anoxic sediment and pore water. Microelectrodes have been used to characterize oxygen microprofiles surrounding burrows. As the microelectrode penetrates into the surficial sediments, oxygen typically decreases steadily with depth but increases sharply where the electrode tip contacts a burrow, and then sharply decreases when it again penetrates into the adjacent sediment. Such microprofiles have been reported for inhabited burrows of marine polychaetes (Jorgensen and Revsbech 1985; Fenchel 1996) and freshwater insects (the alderfly *Sialis velata* and the mayfly *Hexagenia limbata*; Wang et al. 2001). Penetration of oxygenated water into the sediment surrounding a burrow creates oxidized layers of sediments that can be seen as color changes (Aller and Aller 1986; Matisoff 1995; Fenchel 1996) and as gradients of solid-phase iron and manganese around burrows (Aller and Yingst 1978; Aller and Aller 1986). Distinct distributions in bacteria and meiofauna occur around burrows (Aller and Aller 1986). Also, Zn (Aller and Yingst 1978) and Cd (Petersen et al. 1998) enrichment has been measured in sediments close to burrows. These observations indicate that sediment geochemistry, in and around burrows, is different from that of the whole sediment (i.e., a few millimeters away from the burrows). Water in the irrigated burrows should resemble more oxic overlying water rather than anoxic pore waters; likewise, sediment in burrow walls should resemble more oxic surficial sediment than the surrounding anoxic sediment.

While irrigation of burrows represents one way in which benthic organisms interact with the surrounding sediments, there are other strategies as well. Some organisms are primarily epibenthic, foraging primarily in the surficial layers of sediment. Organisms like oligochaetes span the oxic and anoxic layers of sediment; by locating their respiratory exchange in the posterior end of their body, they are able to feed with their mouth in anoxic layers of sediment while their tail protrudes into the oxygenated overlying water. Many marine bivalves use siphon tubes to inspire overlying water, while physically residing in deeper anoxic sediment.

Routes of chemical exposure for benthic organisms

The variety of benthic organisms and their behaviors result in diverse pathways of exposure to sediment-associated contaminants. Complexity results from various combinations of exposure to

- pore water;
- food particles, including prey organisms, from the anoxic sediment;
- oxic overlying water; and
- food particles, including prey, from the oxic sediments.

The relative importance of these various sources depends on the feeding, burrowing, and irrigating behavior of the organisms, which for most taxa are poorly understood.

Very few experiments have determined unambiguously the relative importance of food and water sources of contaminants to benthic animals. The few exceptions are the laboratory and field experiments that showed Cd accumulation by the insect larvae *Chaoborus* (Munger and Hare 1997; Munger et al. 1999) and *Sialis* (Roy and Hare 1999) and that selenium in the bivalve *Macoma balthica* (Luoma et al. 1992) came mainly from their food. Bioaccumulation of metals from the diet has been demonstrated for several invertebrates (see Luoma 1995 for a review). Landrum (1989) has shown that uptake of polycyclic aromatic hydrocarbon (PAH) from pore waters is not necessarily the dominant process for benthic fauna; the amphipod *Diporeia* sp. absorbed 20%, 48%, and 100% of tetrachlorobiphenyl, hexachlorobiphenyl, and benzo[*a*]pyrene (BaP), respectively, from ingestion of sediment rather than pore water (Landrum and Robbins 1990). Algae, detritus, and meiofauna are primary foods for benthic-feeding macrofauna. Benthic-feeding fish acquired polychlorinated biphenyls (PCBs) by ingesting either sediment-associated organisms (polychaetes, copepods) or whole sediment (Rubinstein et al. 1984; DiPinto and Coull 1997). Even for organisms as small as protozoa, grazing on bacteria accounted for 75% of the increase in bioconcentration of a hexachlorobiphenyl (Wallberg et al. 1997).

Although uptake kinetics of a contaminant from food depends on feeding rate and assimilation efficiency (Iannuzzi et al. 1996), other factors influence food selection. For example, the mouth size of an organism determines in part the size range of particles and by inference, any associated contaminants being ingested, an important consideration if a contaminant is not evenly associated with all particle sizes (Harkey et al. 1994; Rodriguez et al. 2001). Even protozoa appear to have morphological features that limit the size of particles ingested (Rassoulzadegan et al. 1988; Dolan 1991). Biochemical cues (e.g., cell wall composition) also can influence food selection by protozoa (Verity 1991; Matz et al. 2002). Because of the preferential sorption of many contaminants to organic matter, the ingestion of high organic content sediment also implies the potential for ingestion of greater amounts of organic contaminants (Landrum and Robbins 1990).

In addition to selective ingestion, selective digestion also has been reported (e.g., oligochaetes digested only some of the bacterial species they ingested; Wavre and Brinkhurst 1971). These considerations mean that organisms may not be exposed to the contaminant concentration measured in the whole sediment, an important consideration when SQGs are applied in contaminated sediment risk management situations. Animal physiological capabilities and biochemical composition (e.g., lipid contents) also influence body burdens (e.g., Leversee et al. 1982; Harkey et al. 1994; Bremle and Ewald 1995; Borchert et al. 1997; Leppänen and Kukkonen 2000), and growth of organisms during the exposure period should be taken into consideration in analyses of bioaccumulation data (Means and McElroy 1997; ASTM 2002a).

Benthic organisms that live within sediments are not necessarily exposed to pore water or to the bulk anoxic sediments. This issue has been addressed recently for Cd in field experiments (Warren et al. 1998; Hare et al. 2001). The approach entailed creating a Cd gradient in the sediments of 2 low-Cd lakes with Cd-spiked sediments maintained in containers at the bottom of these lakes; the containers were left open for about 1 year to allow colonization by indigenous benthic invertebrates. Comparison among containers of the density of the taxa, their Cd concentrations, and the Cd concentrations in pore water and sediments consistently suggests that some benthic animals that irrigate their burrow have only a limited exposure to anoxic pore water or to anoxic sediment particles. The majority of the Cd taken up by these animals appeared to come from the overlying water compartment (either the overlying water itself or food or particles whose Cd concentration is related to that in the overlying water). These results suggest that the protection of benthic animals from metal pollution should consider metals in both the water column and sediment compartments. They also suggest that conclusions about sediment toxicity from static sediment toxicity tests could be misleading. In such tests, metals are likely to exchange with the overlying water, and the response of animals taking up metal from the overlying water could be misconstrued as toxicity due to metal in sediment. It is worth noting, however, that laboratory experiments by Ingersoll et al. (2000) showed toxic effects only when amphipods (*Hyalella azteca*) were exposed in direct contact with sediments compared to exposure only to overlying water.

Overview of Procedures Used to Derive Sediment Quality Guidelines

Introduction

From the discussion above, it is clear that the sediment environment, and the interaction of benthic organisms with it, is complex. Unfortunately, in many respects we seem to know enough to recognize more potentially important complexities, but we are not always equipped to address these rigorously in sediment assessment. The need to make sediment assessment and management decisions has generally been driven by societal and legal mandates that do not provide the luxury of delaying sediment decision making until all complexities are resolved. The demand for tools to aid decision making is ongoing and intense, and has led to the development of SQGs by a variety of methods. This section provides a synopsis of the derivation methods for many of the common SQGs. We emphasize that the intent here is only to describe the derivation methods; discussion of strengths and limitations is pursued in subsequent sections in this chapter.

Existing SQGs can be categorized into 3 groups, according to their derivation:

1) empirically based (or "co-occurrence") guidelines,
2) mechanistically based guidelines (e.g., equilibrium-partitioning [EqP]) guidelines, and
3) "consensus-based" guidelines.

Empirical guidelines are derived from large data sets containing paired information on the contaminant concentrations and biological effects (or lack thereof) associated with individual sediments or study sites. Various analysis approaches are then applied to relate chemical concentrations to the frequency of biological effects. Mechanistically based guidelines are designed to predict sediment toxicity based on an understanding of the chemical and biological processes that influence toxicity. Existing mechanistic SQGs all have their roots in EqP theory, where concentrations of chemical in sediment are related to the corresponding concentrations in pore water. Specific SQGs are then established by determining the chemical concentration in water that is associated with an intended level of protection, then calculating the corresponding concentration in sediment at equilibrium. Consensus-based guidelines aggregate guidelines developed using different methods, and generate new SQGs from the central tendencies of guidelines with different intents (e.g., threshold effect, median effect).

Empirical approaches to sediment quality guidelines

The empirical guidelines are based primarily on field data, collating matching sediment chemistry and biological effects data and using various analytical approaches to relate chemical concentrations to the frequency of biological effects (Table 3-1; Chapters 4 and 12). Because they are based on environmental samples, they implicitly deal with contaminant mixtures, and the measured biological effects reflect the cumulative interactions of all chemicals in the mixture. Generally, no attempt is made to assign a biological effect to an individual chemical stressor, but rather it is assigned equally to all likely toxic chemicals that co-occur in the mixture (Table 3-2). Several different empirical approaches have been used to develop SQGs to provide interpretive tools for the evaluation of sediment chemistry in marine, estuarine, and freshwater ecosystems. The original empirical approaches used SQG values as thresholds for individual chemicals. Recently developed approaches combine individual chemical SQGs or concentration-response models to better address the probability of toxicity for mixtures of chemicals in environmental samples (Chapter 4). These approaches take into account the magnitude of individual chemical concentrations relative to their respective SQG values for predicting effects, rather than relying on individual threshold values. While most of these approaches model the probability of toxicity (i.e., toxic or nontoxic models), the magnitude of the response (e.g., decreased survival) shows a strong correlation with the predicted probability of toxicity (Fairey et al. 2001; Field et al. 2002).

Effects ranges

The effects range low (ERL) and effects range median (ERM) values were developed to provide effects-based guidelines that could be used to interpret the sediment chemical measurements from national monitoring programs (Long and Morgan 1990; Long and MacDonald 1992; Long et al. 1995; Tables 3-1 and 3-2). To ensure that the resulting guidelines were applicable in a national monitoring program, the approach involved compiling an extensive database of matching chemistry and biological effects from many areas. Thus, data were compiled from studies conducted at numerous marine and estuarine areas on the Atlantic, Pacific, and Gulf coasts of North America. Freshwater data were included in the initial data compilation (Long and Morgan 1990), but later they were excluded and replaced with additional saltwater data in revised calculations (Long et al. 1995). A broad range of effects measurements was included in the database. Biological responses were included from acute toxicity tests, benthic community assessments, and spiked sediment toxicity tests. Typically, the data used to derive the ERL or ERM values for each chemical came from toxicity tests performed with sea urchins, mysids, polychaetes,

Table 3-1 Summary of sediment quality guideline (SQG) approaches and underlying data attributes

SQG	Narrative intent	Citation	Data source	Scope	Endpoints	Route of exposure	Duration	Lab v. field
ERL	Concentration below which effects are observed infrequently	Long and Morgan 1990; Long and MacDonald 1992	Database of biological effects concentration, including other sediment SQGs, toxicity tests, and field benthic surveys	National; combined marine and freshwater (Long and Morgan 1990; Long and MacDonald 1992; marine only (Long et al. 1995)	Predominantly lethality, wide range of test organisms	Pore water, whole sediment	Variable	Mixed
ERM	Concentration above which effects are often observed							
TEL	Concentration below which effects are observed infrequently	MacDonald et al. 1996; Smith, MacDonald, et al. 1996	Database of biological effects concentration, including other sediment SQGs, toxicity tests, and field benthic surveys	National; marine (McDonald et al. 1996) and freshwater (Smith, MacDonald, et al. 1996)	Predominantly lethality, wide range of test organisms	Pore water, whole sediment	Variable	Mixed
PEL	Concentration above which effects are often observed							
AET	Concentration above which adverse effects are always observed for specified endpoint	Barrick et al. 1988; Malek 1992; Cubbage et al. 1997	Database of paired sediment chemistry and effects data for toxicity and benthic community endpoints	Puget Sound, Washington State USA; marine (Barrick et al. 1988	Marine: bivalve larvae abnormality; amphipod mortality; Microtox; benthic community structure	Varies with endpoint; elutriate, whole sediment, in situ	Variable	Mixed
			Database of paired sediment chemistry and effects data for toxicity endpoints	Washington State USA; freshwater (Cubbage et al. 1997)	Freshwater: amphipod mortality, *Chironomus* mortality and growth, Microtox			

Table 3-1, *cont'd*

SQG	Narrative intent	Citation	Data source	Scope	Endpoints	Route of exposure	Duration	Lab v. field
ERL, TEL	Concentration below which effects are observed infrequently	USEPA 1996; Ingersoll et al. 1996	Database of paired sediment chemistry and effects for several toxicity test endpoints	Regional (Great Lakes, USA) and USA and Canada; freshwater	Amphipod mortality, growth and reproduction; *Chironomus*	Whole sediment	Short and long term	Lab
ERM, PEL	Concentration above which effects are often observed							
NEC	Concentration above which adverse effects are always observed for specified endpoint							
SLC	Concentration below which effects are observed infrequently	Neff et al. 1987; Environment Canada 1992	Marine (Neff et al. 1987) or freshwater (Environment Canada 1992) benthic population abundance	New England (US) coastal waters (Neff et al. 1987); St. Lawrence River (Canada; EC 1992)	Benthic organism occurrence relative to contaminant concentration	In situ	Long term	Field
LRM	Frequency of toxicity associated with chemical concentration	Field et al. 1999, 2002	Database of paired sediment chemistry and amphipod toxicity	National; marine	Amphipod mortality	Whole sediment exposure	10-d	Lab

Table 3-1, *cont'd*

SQG	Narrative intent	Citation	Data source	Scope	Endpoints	Route of exposure	Duration	Lab v. field
NEL	No adverse effects on water quality, water uses, or benthic organisms; designed to protect against biomagnification	Persaud et al. 1992	Persaud et al. 1992	Provincial (Ontario, Canada); freshwater	Benthic community structure, fish consumption	Mixed	Variable	Mixed
LEL	Most sensitive uses may be affected							
SEL	Majority of benthic organisms will be affected							
EqP (non-ionic organics	Protection of sensitive species from chronic effects	Di Toro et al 1991; van der Kooij 1991; USEPA 1993	Acute and chronic toxicity from water column toxicity tests	National - USA (Di Toro et al. 1991; USEPA 1993); The Netherlands (van der Kooij 1991); marine and freshwater	Acute and chronic toxicity	Ingestion and pore water	Short and long term	Lab
ΣPAH	Protection of sensitive amphipods	Swartz et al. 1995; Swartz 1999	Database of paired sediment chemistry and amphipod toxicity	National; marine	Amphipod mortality	Whole sediment exposure	10-d	Lab
EqP (narc-osis)	Protection of sensitive species from chronic effects	Di Toro et al. 2000; Di Toro and McGrath 2000	Acute and chronic toxicity from water column toxicity tests	National; freshwater and marine	Acute and chronic toxicity	Ingestion and pore water	Short and long term	Lab

Table 3-1, *cont'd*

SQG	Narrative intent	Citation	Data source	Scope	Endpoints	Route of exposure	Duration	Lab v. field
EqP (metals; USA)	Protection of sensitive species from chronic effects	Di Toro et al. 1990; Ankley et al. 1996	Acute and chronic toxicity from water column toxicity tests	National; marine and freshwater	Acute and chronic toxicity	Ingestion and pore water	Short and long term	Lab
EqP (metals, The Netherlands)	Protection of sensitive species from chronic effects	van der Kooij 1991	Toxicity data from water only exposures, acute and chronic; sensitivities distribution approach applied for determining the water quality guideline. Empirical Kd values (i.e., means) obtained from a database including 800 sites	National; freshwater	Acute and chronic toxicity	Ingestion and pore water	Short and long term	Lab
Consensus of PAH SQGs	Protection of benthic ecosystems from chronic effects	Swartz 1999	SQG values derived by other methods	National; marine	Threshold, median, extreme effect concentrations (TEC, MEC, EEC) determined as arithmetic mean of multiple SQGs	Mixed	Mixed	Mixed

Table 3-1, *cont'd*

SQG	Narrative intent	Citation	Data source	Scope	Endpoints	Route of exposure	Duration	Lab v. field
Consensus of PCB SQGs	Protection of benthic ecosystems from chronic effects	MacDonald, DiPinto, et al. 2000	SQG values derived by other methods	National; freshwater and marine	TEC, MEC, EEC determined as geometric mean of multiple SQGs	Mixed	Mixed	Mixed
Consensus of SQGs	Protection of benthic ecosystems from chronic effects	MacDonald, Ingersoll, Berger 2000	SQG values derived by other methods	National; freshwater	TEC, PEC determined as geometric mean of multiple SQGs	Mixed	Mixed	Mixed

ERL = effects range low
ERM = effects range median
TEL = threshold effects level
PEL = probable effects level
AET = apparent effects threshold
NEC = no-effect concentration
SLC = screnning-level concentration
LRM = logistic regression modeling
NEL = no-effect level
LEL = lowest effect level
SEL = severe effect level
EqP = equilibrium partitioning
PAH = polycyclic aromatic hydrocarbon
PCB = polychlorinated biphenyl

Table 3-2 Summary of scientific considerations in SQG approaches

SQG	Citation	Bioavailability	Causality	Mixtures	Other stressors	Application considerations
ERL ERM	Long and Morgan 1990; Long and MacDonald 1992	Chemical concentrations are normalized to sediment dry weight with no adjustment for differences among sediments in bioavailablility	Not addressed	Considered implicitly, but not explicitly	Some nonchemical stressors are implicitly incorporated in source data; not explicitly included	Based on correlation of effects with stressors; may not know which stressor, or unknown combination of the stressors is responsible for effects associated with the SQG for individual chemical. SQG is based on aggregate effect of unknown mixture of chemicals; application to sites with single or limited contaminants, SQG may overestimate effects. Although expressed as dry wt only, organic carbon normalization has not been found to improve predictive ability.
TEL PEL	MacDonald et al. 1996; Smith, MacDonald, et al. 1996	Chemical concentrations are normalized to sediment dry weight with no adjustment for differences among sediments in bioavailability	Not addressed	Considered implicitly, but not explicitly	Some nonchemical stressors are implicitly incorporated in source data; not explicitly included	As for ERL, ERM
AET	Barrick et al. 1988; Malek 1992; Cubbage et al. 1997	Have been developed on both dry weight basis (no sediment-specific adjustment for bioavailability) and normalized to organic carbon	Attempts to address causality by focusing on concentrations where sediments are always toxic; true relationship uncertain	Based on samples that include mixtures; derivation procedure may affect degree to which mixture consideration is valid	Not considered	Misinterpretation of outliers may cause extreme values to be calculated. Interpretation of causality and mixture effects is uncertain. Region-specificity of derivation must be considered if applied outside region for source data. For a specific endpoint, additional data can only increase the value.

Table 3-2, *cont'd*

SQG	Citation	Bioavailability	Causality	Mixtures	Other stressors	Application considerations
ERL, TEL ERM, PEL NEC	Ingersoll et al. 1996; USEPA 1996	Chemical concentrations are normalized to sediment dry weight with no adjustment for differences among sediments in bioavailablility	Not addressed	Considered implicitly, but not explicitly	Some nonchemical stressors are implicitly incorporated in source data; not explicitly included	As for ERL, ERM
SLC	Neff et al. 1987; Environment Canada 1992	Some chemicals normalized to organic carbon	Derivation procedure may influence guidelines toward causality; true relationship uncertain	Considered implicitly, but not explicitly	Some nonchemical stressors are implicitly incorporated in source data; not explicitly included	As for ERL, ERM
LRM	Field et al. 1999, 2002	Chemical concentrations are normalized to sediment dry weight with no adjustment for differences among sediments in bioavailablility	Not addressed	Considered implicitly, but not explicitly	Some nonchemical stressors are implicitly incorporated in source data; not explicitly included	As for ERL, ERM
NEL LEL SEL	Persaud 1992	Derived from a combination of existing methods; specifics depend on method applied	Derived from a combination of existing methods; specifics depend on method applied	Derived from a combination of existing methods; specifics depend on method applied	Derived from a combination of existing methods; specifics depend on method applied	Derived from a combination of existing methods; specifics depend on method applied; not applicable to non-biomagnifying trace metals.

Table 3-2, *cont'd*

SQG	Citation	Bioavailability	Causality	Mixtures	Other stressors	Application considerations
EqP (nonionic organics)	Di Toro et al. 1991; van der Kooij 1991; USEPA 1993	Normalized to organic carbon	Derived to explicitly address singular effects of subject chemical (or group of chemicals)	Not considered except for guidelines developed explicitly for mixtures (e.g., PAH mixtures)	Not considered	Partitioning models for low hydrophobicity and/or ionic organic chemicals have not been established/incorporated. Dependent on assumption that actual chemical behavior is approximated by equilibrium model, regardless of the route of exposure. Assumes the actual partition coefficient in the sediment is adequately represented by literature or calculated Koc value (site-specific adjustment could be developed).
ΣPAH	Swartz et al. 1995; Swartz 1999)	Normalized to organic carbon	Derived to explicitly address effects of PAH mixtures	Derived to explicitly address effects of PAH mixtures	Some non-chemical stressors are implicitly incorporated in source data; not explicitly included	Dependent on assumption that actual chemical behavior and toxicity are approximated by EqP and other submodels. Based on 10-d lethality tests; may not address longer-term effects. Directly considers 13 PAH; relies on extrapolation to full mixture.
EqP/ Narcosis	Di Toro et al. 2001; Di Toro and McGrath 2001	Normalized to organic carbon	Derived to explicitly address singular effects of all nonionic narcotic chemicals	Derived to explicitly address effects of mixtures of nonionic narcotic chemicals	Not considered	Dependent on assumption that actual chemical behavior and toxicity are approximated by equilibrium partitioning and narcosis models.
EqP (metals; USA)	Di Toro et al. 1990; Ankley et al. 1996; USEPA 1994a	Acid volatile sulfide (AVS) and simultaneously extracted metal (SEM) approach used to account for differences in metal bioavailability	Derived to address effects of 6 metals.	Addresses mixtures of Cd, Cu, Zn, Pb, Ni, and Ag; additional chemicals not considered.	Not considered	Recognizes only AVS as binding phase; other binding phases not incorporated. AVS not primary binding phase in oxic sediments. Experimental artifacts have been identified in some AVS/SEM measurements. Metal accumulation can still occur when sulfide is in excess.

Table 3-2, *cont'd*

SQG	Citation	Bioavailability	Causality	Mixtures	Other stressors	Application considerations
EqP (metals; NL)	van der Kooij 1991	Chemical concentrations are normalized according both to grain size and organic carbon	Derived to explicitly address effects of individual metals	Each metal considered individually	Not considered	Kd values are empirical and therefore do not allow an extrapolation to other geographical areas. Kd values are means, so some aspects sediment-specific variability may be unrecognized.
Consensus of PAH SQGs	Swartz 1999	Both dry weight and organic carbon–normalized guidelines are expressed on organic carbon basis	Combines guidelines with a range of causal connection.	Derivation explicitly addresses PAH mixtures	Not explicitly addressed.	Attempts to compensate for biases of individual guidelines by looking for central tendency; however, consensus based on average may obscure conceptual differences among guidelines. McDonald, Ingersoll, and Berger (2000) apply guidelines in quotient-based framework.
Consensus of PCB SQGs	MacDonald, DiPinto, et al. 2000	Chemical concentrations are normalized to sediment dry weight with no adjustment for differences among sediments in bioavailablility		Combines guidelines with range of mixture consideration		
Consensus of SQGs	MacDonald, Ingersoll, Berger 2000					

amphipods, shrimp, bivalves, bacteria (Microtox assay with aqueous extracts), copepods, and fish. Where possible, both acute and chronic test endpoints were included, although acute tests were more common. Typically the author's designation of toxicity was used, but this was not always the case. Field data endpoints included abundances of major benthic taxa (echinoderms, bivalves, annelids, arthropods, and amphipods) and indices of species richness. The EqP-derived acute and chronic values were included along with amphipod survival data from spiked sediment toxicity tests.

Previously published guidelines based on EqP-derived acute and chronic values and results of spiked sediment toxicity tests were also entered into the database as reported, and classified as "effects" entries in the guideline derivations. Co-occurrence data, field-collected data with matching sediment chemistry and biological effect measures, represented the largest component of the effects database used to derive the effects-range guidelines. These data were subjected to a screening process prior to inclusion in the database to ensure that chemistry and effects measurements were collected from the same sample and that methods were described and consistent with standard protocols (Long and Morgan 1990; Long et al. 1995; Ingersoll et al. 1996; MacDonald et al. 1996). For example, toxicity data were excluded if sediments were frozen prior to toxicity testing or control survival was unacceptable. Trace metals data were included only if strong acid digestion was used. In addition, data were evaluated to determine the degree of concordance between chemistry and effects measurement.

The presence of multiple contaminants, many of which may be present at very low concentrations, frequently complicates evaluating the relationship between the concentration of an individual contaminant and toxicity in field-collected sediments. Consequently, the field-collected data were subjected to a second round of screening. Entries in the database from field studies for each chemical were classified as an "effects" entry if the mean concentration in the toxic samples in a study was greater than twice the mean of the nontoxic samples for that chemical in the study. This data screening step, which takes into account the no-effect data in creating the "effects" database, was intended to exclude data for a chemical from the derivation of the ERL or ERM where the effect was not associated with a chemical gradient (i.e., where the concentration associated with the effect was less than twice the concentration of the no-effect concentration).

To derive the guidelines, the dry weight–normalized database for each chemical was sorted in ascending concentration. The ERL values for a contaminant represented the lower 10th percentile concentration of the effects data, while the ERM values were the median concentration of the effects data.

Effects levels

The derivation of the threshold effects levels (TELs) and probable effects levels (PELs) for marine (MacDonald et al. 1996) and freshwater sediments (Ingersoll et al. 1996; Smith, MacDonald, et al. 1996) was similar to that used for the effects range approach. The biological effects database for sediments (BEDS), which was initially prepared for derivation of the effects range values, was expanded considerably. It included a wide variety of benthic and toxicological endpoints determined in many separate studies conducted in numerous locations. The screening process for inclusion of data in the BEDS database was the same as that for the effects range approach. As in the derivation of the effects range values, SQGs available from other approaches (e.g., EqP, AET) and results of spiked sediment toxicity tests were also included. SQGs that were originally expressed as organic carbon-normalized values were converted to dry weight assuming 1% organic carbon, which was consistent with the average organic carbon content reported in other studies included in the BEDS database.

Entries into the BEDS database from field studies included all observations of adverse biological effects that occurred at chemical concentrations at least 2-fold above nontoxic or reference conditions. They were treated as "effects" entries. The major difference between the effects range and the effects level approaches is that the latter approach used data from "no-effects" observations as well as those in which adverse effects were observed. The "no-effects" data included all observations of no adverse biological effects such as observed at reference sites or the average chemical concentrations that were less than a 2-fold elevation above those in nontoxic conditions (MacDonald et al. 1996).

As with the effects range approach, the databases for each substance were assembled in ascending order. The TEL was calculated as the geometric mean of the 15th percentile of the effects data set and the 50th percentile of the no-effects data set. The PEL was calculated as the geometric mean of the 50th percentile of the effects data set and the 85th percentile of the no-effects data set. These critical points in the ascending data tables were chosen following testing to identify the concentrations that best predicted the absence of toxicity (concentrations < TELs) and the presence of toxicity (concentrations > PELs).

Apparent effects thresholds

The apparent effects threshold (AET) approach was developed in Puget Sound, Washington, USA, to identify concentrations in the sediment above which adverse effects for a specific endpoint would always be expected to occur. The database used in the derivation of AETs consisted of synoptic sediment chemistry and effect measures from data collected in Puget Sound using consistent chemical and biological testing protocols.

The AET is determined as the chemical concentration above which statistically significant biological effects always occurred in the Puget Sound database used to create the values (Barrick et al. 1988). The principal AET values are based on whole sediment toxicity tests with marine amphipods (mostly *Rhepoxynius abronius*), tests on elutriates using oyster larvae and Microtox, and biological effects in the field as measured by the abundance of major taxa of benthic infauna. In addition, AET SQGs are expressed as the highest (HAET) and lowest (LAET) of these 4 kinds of AETs. Using the combined sets of AETs as a battery of tests, samples that had chemical concentrations below the LAET could be considered unlikely to have adverse effects from any of the test endpoints, and samples that had concentrations exceeding the HAET were likely to have adverse effects for multiple endpoints. The final set of AETs for Puget Sound were used to establish regional dredged-material guidelines (PTI 1991) and were adopted as criteria for use in sediment source control and cleanup programs in Washington State.

Freshwater AETs were developed using the same basic approach from a database including synoptic chemistry and biological effects data from the states of Washington and Oregon (Cubbage et al. 1997). Endpoints included 10- to 14-d mortality tests with *Hyalella azteca* (whole sediment) and Microtox (saline elutriate). Because the AET is based on the highest no-effect concentration, which may set the value for a chemical anomalously high due to unique substrates or random errors, Cubbage et al. (1997) proposed a probable apparent effects threshold (PAET). This was determined as the 95th percentile of the sample concentrations with no significant biological effects above the lowest concentration associated with an effect. SQGs were developed on both a dry weight– and organic carbon–normalized basis (for organic chemicals only). The dry weight values for PAH were found to be more reliable than the organic carbon-normalized values. Reliability was evaluated as sensitivity, which refers to the percentage of toxic samples that were predicted to be toxic, and efficiency, which refers to the accuracy of the predictions of effects (the percentage of the total number of samples predicted to be toxic that were observed to be toxic).

Sediment effect concentrations

Freshwater sediment effect concentrations (SECs) for specific toxicity test endpoints were determined from a database of matching field-collected sediment toxicity and sediment chemistry (Ingersoll et al 1996; USEPA 1996). SECs were developed for *Hyalella azteca* (10-d and 28-d) survival, growth, and maturation and for *Chironomus riparius* (10- to 14-d) survival and growth; the sources were a database primarily from the Great Lakes and several data sets from other parts of the United States (US). Several types of SECs were developed that generally followed the methodology of the effects range, effects level, and apparent effects threshold approaches. Unlike the effects range and effects threshold approaches, the SECs were developed from individual samples rather than summarized data from individual studies. Prior to deriving the effects range and effects level values for individual chemicals, a data screening step similar to that used by Long and Morgan (1990) was employed. Toxic samples were included in the effects database if the concentration of each toxic sample exceeded the mean of the nontoxic samples for the entire database. SECs were derived for dry weight– and organic carbon–normalized (for nonionic organics) concentrations. SEC values (analogous to ERLs, ERMs, TELs, PELs, and AETs) were derived for total metals, simultaneously extracted metals (SEMs), PCBs, and PAHs on a dry weight basis and normalized to organic carbon for organics.

Screening-level criteria

The screening-level concentration (SLC) approach (Table 3-1) was derived from field data on the co-occurrence of nonionic chemical concentrations in sediment and the presence or absence of a number of benthic species (Neff et al. 1986, 1987). A cumulative frequency distribution of stations at which a particular species was present was plotted against the organic carbon–normalized concentration of a chemical stressor to derive the species screening level concentration (SSLC). The SSLC was defined as the concentration at the 90th percentile of this frequency distribution. SSLCs for a large number of species were then plotted in another frequency distribution. The SLC was originally defined as the chemical stressor concentration above which 95% of the SSLCs were found (Neff et al. 1987). The SLC provides an estimate of the highest concentration for a specific contaminant that can be tolerated by 95% of benthic species (Neff et al. 1986). As part of the action plan for the St Lawrence River, Environment Canada used the SLC method to define 2 guidelines for organic and metal contaminants from the SSLC distribution: 1) the 15th percentile, representing the lowest effects level, and 2) the 90th percentile, representing a severe effects level (Centre Saint-Laurent 1992).

Ontario guidelines: No-effect level, lowest effect level, severe effect level

Persaud et al. (1992) defined 3 different guidelines:

1) the "no-effect level" (NEL), implying no effect on the water quality or its uses, and benthic organisms accordingly;
2) the "lowest effect level" (LEL), above which some uses may be affected; and
3) the "severe effect level" (SEL), above which the majority of benthic organisms would be impaired.

These levels were determined according to a variety of methods, including the EqP approach (NEL, LEL), the SLC approach (LEL and SEL). Background concentrations were also considered (e.g., metals NEL). Organic compound SELs are normalized to organic carbon.

Logistic regression modeling

The logistic regression modeling (LRM) approach focuses on establishing the probability of adverse effect as a function of sediment chemical concentration (Field et al. 1999, 2002). A large database of field-collected marine sediments was compiled that had matching measured concentrations of sediment contaminants and effects data from sediment toxicity tests with marine amphipods (*Ampelisca abdita* and *R. abronius*). Prior to being included in the database, all of the candidate data sets were critically evaluated to ensure that sample collection and handling procedures, toxicity testing protocols and environmental conditions, control responses, and analytical methods were consistent with established procedures (Long et al. 1995; MacDonald et al. 1996; Field et al. 1999). The classification of samples as toxic or nontoxic was standardized in the database; all test samples that were determined to be statistically different from the negative control and had less than 90% survival were classified as toxic.

A preliminary data-screening step was undertaken before the regression models for individual contaminants were developed. Data were screened to exclude toxic samples in which the selected contaminant was a poor indicator of the observed toxicity. Following the general screening approach used by USEPA (1996) and similar to that used by others (Long and Morgan 1990; MacDonald et al. 1996), individual toxic samples for each chemical were eliminated if the concentration of that contaminant was less than or equal to the average concentration in nontoxic samples. All nontoxic samples were included in these analyses. Samples from reference stations were treated the same as other samples. Some may have been designated as toxic, but this was not investigated in the study. If they were toxic and passed the screening step, they were included in the development of the individual chemical models and were also included in the development of the mixture models. The data for chemical concentrations that were less than the reported detection limits were not used to develop the logistic models.

The screened data were then used to develop LRMs that relate chemical concentration to a predicted probability of toxicity. Because the logistic regression establishes a continuous relationship between chemical concentration and probability of effect, this approach can be used to define SQGs associated with any desired probability. The individual regression models for 37 chemicals (10 metals, 22 PAHs, total PCBs, and 4 pesticides) were combined into a single model, using either a maximum or an average probability predicted from the chemicals analyzed in a sample to estimate the probability of toxicity for a sample.

SQG quotient approaches

The sediment concentrations for an individual chemical can be divided by its SQG to produce a unitless quotient value. If the quotient is less than 1.0, the SQG is not exceeded. The quotient approach can be applied to any set of SQGs, but has most commonly been applied to marine ERMs and PELs (Long et al. 1998, 2000; Long and MacDonald 1998; Fairey et al. 2001) and freshwater probable effects concentration (PECs; "Consensus approach for 28 chemicals of concern in freshwater sediments," p 70; MacDonald, DiPinto, et al. 2000; Ingersoll et al. 2001). Because the quotient is unitless, values can be readily compared among different chemicals. Quotient values for multiple chemicals can characterize overall sediment contamination.

Three methods of quotient analyses are most commonly used:

1) the number of chemicals with quotients >1.0 can be compared among sites or experimental treatments,
2) a mean quotient value can be calculated for all chemicals or all members of a particular chemical class, and
3) the quotient values can be summed or averaged for all chemicals within a chemical class (e.g., metals, PAHs, PCBs, organochlorine pesticides) and summed to provide a single value for the entire sample.

These approaches take into account the magnitude of the SQG exceedance and provide a way to compare studies or samples where different numbers of chemicals were measured. No consistent method for deriving a mean quotient has been established. A variety of mathematical frameworks have been used to calculate an overall quotient value (Long et al. 1998, 2000; Long and MacDonald 1998; Fairey et al. 2001; Ingersoll et al. 2001).

Mechanistic approaches to sediment quality guidelines

As explained previously, existing mechanistically based SQGs are based on EqP theory. While EqP can and has been used to derive SQGs, its scope is actually broader, providing an approach to interpreting chemical bioavailability and effects in sediments in general, as well as a mechanism for developing SQGs.

EqP-based sediment quality guidelines for nonionic organic chemicals

Experiments conducted by Adams et al. (1985) provide an excellent demonstration of the principles behind EqP. In these experiments, midge larvae were exposed to kepone spiked into 3 sediments with differing organic carbon content. Results from these exposures showed markedly different responses in the 3 sediments relative to the dry weight–normalized (bulk) concentration of kepone in the sediment. This shows that sediment composition is important to determining the bioavailability of a chemical in that sediment. However, if these toxicity results are expressed on the basis of kepone concentration in pore water, the data collapse into a single concentration-response curve. This suggests that the toxicological potency of kepone in sediment is proportional to its concentration in pore water. It can also be shown that the concentration of a chemical such as kepone in pore water that produces toxicity is comparable to the concentration causing similar toxicity in a water column exposure.

If the toxicity of a chemical such as kepone in sediment is proportional to its concentration in water, then the concentration of this chemical in sediment that will cause toxicity can be estimated if one understands the relationship between chemical concentration in pore water and that in sediment. For nonionic organic chemicals such as kepone, sediment organic carbon is the primary sediment phase that controls the partitioning of chemical between sediment and pore water. Thus, expressing the concentration of nonionic organic chemicals in sediment relative to organic carbon should place toxicity responses observed in different sediments on an equivalent basis. Indeed, when the kepone data of Adams et al. (1985) are expressed as kepone concentration per unit organic carbon, a single dose-response pattern is observed.

Combining the above finding with those from other experiments, Di Toro et al. (1991) proposed that this assessment approach could be used to establish SQGs for nonionic organic chemicals. In this EqP approach, Di Toro et al. (1991) proposed that the partitioning of a chemical between sediment and pore water be represented by a simple equilibrium equation:

$$C_{SOC} = C_{PW} \times K_{OC} \qquad (1),$$

where C_{SOC} is the concentration of the chemical in the sediment per unit mass of organic carbon, C_{PW} is the concentration of the chemical in pore water, and K_{OC} is the partition coefficient of the chemical to sediment organic carbon. If one then replaces C_{PW} with the chemical concentration in water associated with a biological effect in a water column exposure ($C_{effect\text{-}water}$)

$$C_{SOC} = C_{effect\text{-}water} \times K_{OC} \qquad (2),$$

then C_{SOC} becomes the chemical concentration in sediment (normalized to organic carbon) predicted to cause that same biological effect in benthic organisms, provided there is an equal sensitivity of aquatic and benthic species. This relationship was evaluated with several benthic organisms exposed to several different nonionic organic chemicals, and was found to correctly predict toxicity within a factor of 2 (USEPA 1993a, 1993d, 1993e).

Thus, to derive an SQG for a nonionic organic chemical using EqP, just 2 values are needed: 1) a $C_{effect\text{-}water}$ value that reflects the degree of biological effect appropriate to the SQG and 2) K_{OC}. With respect to $C_{effect\text{-}water}$, one of the advantages of the EqP approach is that it allows incorporation of the extensive data available on water column toxicity for a wide variety of aquatic organisms, so that substantial additional testing is not required. The K_{OC} for nonionic organic chemicals can be estimated from their octanol–water partition coefficients (K_{OW}) (e.g., Di Toro et al. 1991). While these generic K_{OC} values are typically used in SQGs, it is also possible to generate and apply site-specific K_{OC} when necessary to reflect site-specific conditions.

It is important to note that application of EqP theory to sediment assessment is not limited to any particular biological endpoint. The water-only effect concentration ($C_{effect\text{-}water}$) in Equation 2 can be established on the basis of any endpoint of importance to an assessment: lethal or sublethal effects on a specific organism, a group of organisms, or effects on large assemblages of organisms. As outlined below, specific SQGs derived using EqP rely on specific effect endpoints, but again this can be adjusted as appropriate to assessment needs.

EqP-based sediment quality guidelines for individual nonionic organic chemicals

EqP-based SQGs for individual nonionic organic chemicals have been proposed in the US (Di Toro et al. 1991; USEPA 1993a, 1993d, 1993e, 2000) and in the Netherlands (van den Kooij 1991). Both approaches derive guidelines by setting the water-only effect concentration equal to a concentration estimated to protect a large percentage of aquatic species from acute or chronic effects. The US approach uses the final chronic value (FCV) from the

USEPA ambient water quality criteria (AWQC; Stephan et al. 1985). Thus, the guideline derivation is

$$C_{guideline}\ (\mu g/g\ OC) = FCV \times K_{OC} \qquad (3).$$

The Dutch approach is completely parallel, except that instead of the FCV value, C_{effect} is derived using a different statistical approach to select a concentration estimated to protect 95% percent of aquatic species.

Using the Dutch approach, SQGs were developed for 66 individual organic chemicals, while 81 were determined through a standard approach where the water quality guidelines were derived by applying assessment factors (Sijm et al. 2001). In 1993, the USEPA issued draft sediment quality criteria (SQC) for 5 chemicals: phenanthrene, acenaphthene, fluoranthene, endrin, and dieldrin) (USEPA 1993a–e). Since that time, guidelines for individual PAHs have been abandoned in favor of a combined approach for PAH mixtures (see "Application of EqP to metals," p 66), and final values for endrin and dieldrin, now called "EqP sediment benchmarks" (ESBs), have now been published (USEPA 2003a, 2003c, 2003d). In addition, the USEPA has published a compendium of EqP-based guidelines for a number of other nonionic organic chemicals (USEPA 1993f). While published for informational purposes, none of the current USEPA SQGs have been formally incorporated into specific regulatory programs.

ΣPAH model. As described above, USEPA originally published draft SQC for 3 individual PAHs. While technically appropriate for exposures to a single PAH, these SQC were withdrawn because their application to field sediments was complicated by the reality that PAHs in the environment never occur as individual chemicals, but as complex mixtures comprised of scores of PAH compounds. Judging sediment toxicity on the basis of single PAH compounds risked underestimation of effects because sediments that did not exceed EqP-based guidelines for individual PAH could still show toxicity due to the combined effect of all PAHs present in the mixture. Thus, to effectively predict the potency of PAHs in sediments, some mechanism to aggregate the toxicity of PAH mixtures was necessary.

Swartz et al. (1995) were the first to develop such an approach with their PAH sediment toxicity model. The ΣPAH model estimates the probability that a PAH-contaminated sediment will be toxic to marine and estuarine amphipods, based on a combination of EqP, quantitative structure–activity relationships (QSARs), toxic units, additivity, and concentration–response submodels. In very brief terms, the potency of individual PAHs is estimated using a demonstrated relationship between 10-d LC50 concentration in water-only tests (conducted in the absence of sediment) and the K_{OW} of the compound. Those concentrations are related to concentrations in sediment using EqP calculations, and the potencies of individual PAHs are aggregated using concentration additivity. These components yield an estimated potency for the PAH mixture. In a final step, this estimated potency is "calibrated" against results from 10-d toxicity tests with field-collected sediments contaminated with PAHs.

Two SQGs are derived from the ΣPAH model: 1) the ΣPAH Mixture LC50, and 2) the ΣPAH Toxicity Threshold. The ΣPAH Mixture LC50 represents PAH concentrations expected to cause 50% amphipod mortality. A threshold of significant amphipod toxicity is also calculated, which is 18.6% of the ΣPAH Mixture LC50 SQG.

In the derivation of the ΣPAH model, the 10-d QSAR was extrapolated to highly hydrophobic PAHs. The model assumes that partial toxic units of these highly hydrophobic compounds are additive in the prediction of total mixture toxicity. However, Boese et al. (1999) found no toxicity in mixtures of highly hydrophobic PAHs, even when the predicted toxicity of the mixture was a factor of 2 more than the LC50. Potential explanations for this discrepancy include 1) that the QSAR may overestimate the potency of highly hydrophobic PAHs or 2) that effects of these PAHs are not additive with those of lower K_{OW} PAHs. In either case, the suggestion is that the ΣPAH model may overpredict toxicity of sediments with a predominance of highly hydrophobic PAHs.

EqP for mixtures of nonionic chemicals: Narcosis approach. The development of the ΣPAH model recognized several important aspects of accounting for mixtures of chemicals in sediment. This general approach was advanced further by Di Toro et al. (2000) and Di Toro and McGrath (2000) in the development of an SQG approach for mixtures of not just PAHs, but all nonionic organic chemicals acting through a common mode of action known as "narcosis." The details of the analysis are too complex to reproduce here, but in concept, the narcosis approach is similar to the ΣPAH model. Water-only toxicity of individual chemicals was estimated from their K_{OW}, water concentrations were related to sediment concentrations using EqP calculations, and the effects of the mixture were estimated by adding the potencies of the individual mixture components. While conceptually similar to the ΣPAH model, the narcosis approach provided several enhancements:

- It established a firmer basis for a K_{OW}-based prediction of toxicity for individual PAHs by incorporating a much larger toxicological data set;

- it allowed for explicit consideration of all PAHs present in a mixture, rather than an empirical approach;
- it provided a mechanistic basis for the assumption of additive toxicity in PAH mixtures;
- it incorporated a range of sensitivities across species; and
- it provided a mechanism to consider chronic toxicity.

The primary differences from the ΣPAH approach are that the narcosis approach

- considers all narcotic chemicals, not just the 13 PAHs in the ΣPAH model;
- is based on toxicity data for a wide variety of aquatic organisms rather than just amphipods; and
- explicitly considers chronic toxicity beyond 10-d lethality.

Based on the narcosis approach, the USEPA has developed an SQG for PAH mixtures (USEPA 2000). The USEPA PAH mixture guideline has minor differences in the way that the water-only effect concentrations are estimated, and the resulting guideline is not quite a factor of 2 lower than the guideline concentrations calculated by Di Toro and McGrath (2000). The USEPA approach also deals explicitly with only PAH compounds, although the potential for influence by other narcotic compounds is acknowledged.

Although the narcosis–EqP mixture approach (Di Toro et al. 2000; Di Toro and McGrath 2000) was described above in terms of PAH mixtures, the approach actually extends beyond PAHs to all nonionic organic chemicals that act through the narcosis mode of action. This provides coverage for additional sediment contaminants such as chlorobenzenes.

Application of EqP to metals

The description and applications of EqP described thus far have centered on nonionic organic chemicals, but the EqP concept has been applied to metals as well. In essence, it is assumed that the toxicity of a metal in sediments is proportional to its concentration in pore water, and that the concentration of a metal in pore water is related to that in sedimentary phases by appropriate equilibrium relationships (e.g., Di Toro et al. 1990; Berry et al. 1996; Hansen et al. 1996). The application of EqP to metals in sediments is, however, complicated by the existence of multiple binding phases for metals, differences in binding phases between oxic and anoxic layers of sediments, and complex chemistry.

EqP approach for trace metals in oxic sediments. An application of EqP to trace metals in oxic sediments was developed by Jenne et al. (1986). It assumes that 1) a metal is bound to a limited number of solid phases present in oxic sediment, and 2) the metal bound to these phases is in quasi-equilibrium with the free metal ion in the pore water. The major metal binding phases in oxic sediments are particulate organic matter and iron and manganese oxyhydroxides (Luoma and Bryan 1981; Lion et al. 1982; Luoma and Davis 1983; Sigg et al. 1987). The measurements needed to apply this approach are the concentrations of binding phases and those of bound metal in the oxic layers of sediments. With this information, the activity of the metal in pore water can be calculated using the appropriate binding constants. Ideally, surface complexation modeling, using intrinsic binding constants (Dzombak and Morel 1990; Stumm 1992), should be used. However, there is a lack of intrinsic binding or adsorption constants for phases other than ferrihydrite. This approach has been used successfully to relate Cd concentrations in a freshwater bivalve to sediment Cd concentrations using semi-empirical binding constants (Tessier et al. 1993).

The approach also explains the results of earlier field studies that showed that the prediction of metal concentrations in estuarine and freshwater bivalves was greatly improved when the concentrations of trace metals (As, Cu, Hg, and Pb) extracted from oxic sediment were normalized to organic carbon or iron concentrations (Luoma and Bryan 1978; Langston 1980, 1982; Tessier et al. 1983, 1984). This EqP approach was considered by the USEPA in the 1980s but has been abandoned, probably because it is difficult to sample the thin layer of oxic sediments in most natural systems.

EqP approach for trace metals in anoxic sediments. In anoxic sediments, the chemistry of trace metals is often dominated by reactions with sulfide. An application of EqP to trace metals in anoxic sediments, known as the acid volatile sulfide (AVS) model, was developed by Di Toro et al. (1990). According to this model, the iron in sedimentary iron monosulfide, FeS(s) (= AVS), is able to be exchanged by a divalent trace metal that forms a solid sulfide less soluble than FeS(s), releasing equivalent amounts of iron into pore waters. As long as the trace metal concentration added to the sediment is less than that of AVS, free-metal ion activity in the pore water is maintained at very low levels and the sediment is not predicted to be toxic because of metals. When the metal concentration added becomes greater than that of AVS, free metal ion activity increases sharply in the pore water and the sediment can become toxic as a result of metals. The measurements needed to apply this model are the concentrations of metals that are extractable by dilute hydrochloric acid at the same time as dissolution of FeS (AVS), and these are referred to as simultaneously extracted metals (SEM).

The AVS model assumes that

- the sulfide ligand in anoxic sediments competes effectively with any other dissolved or solid-phase ligand for binding divalent metal ions known to react strongly with sulfide, such as Cd, Cu, Hg, Ni, Pb, and Zn;
- AVS is a major reactive pool of solid-phase ligand in sediments;
- equilibrium between dissolved and solid phases prevails in sediments; and
- the toxicity of metals in sediments is determined by the activity of the dissolved metal in the bulk pore water.

Ankley et al. (1996) and USEPA (1994a) described approaches to develop EqP-based guidelines for Cd, Cu, Pb, Zn, and Ni in aggregate. Ag was later added to this group of AVS-responsive metals. These approaches focus on AVS as the primary factor controlling metal bioavailability in sediments. A sediment meets the guideline if either the total molar SEM concentration is less than the total AVS concentration or the summed potency of metals in pore water is below the applicable water quality criterion. The applicable water quality criterion for a metal mixture is calculated by taking the ratio of the metal concentration in pore water to the FCV for that metal (usually termed a "toxic unit" [TU]), then summing those ratios across metals; if the sum of the ratios is less than 1, the guideline is met. The pore water approach was included to recognize the existence of metal binding phases beyond AVS and to provide some applicability to oxic sediments where AVS is not a factor.

EqP metal SQGs based on empirical Kd values. The SEM–AVS approach for SQG development attempts to predict partitioning from specific binding phases. An alternative approach, used by the Netherlands to develop SQGs for metals, is to use empirically derived partition coefficients (Kd) to describe the relationship between metal concentration in sediment and in pore water (van den Kooij 1991). This Kd value is used in a partition equation equivalent to Equation 2 to specify the concentration of each metal in sediment that is associated with the target level of biological effect. The $C_{effect\text{-}water}$ is set equal to the applicable water quality criterion. This can be derived through a conventional procedure using assessment factors or a more sophisticated approach using toxicological sensitivity distribution models (Aldenberg and Slob 1993; Aldenberg and Jaworska 2000). Based on these models, SQGs can be derived for differing levels of species protection. It should be noted that this approach requires data for an appropriate number of species before the statistical rigor is acceptable. Another feature of the Dutch approach is that it considers background concentrations of metals in determining SQGs by determining a "maximum permissible addition" of metal above background concentrations (Crommentuijn et al. 2000).

Overview of consensus approaches

This "second generation" of SQGs attempts to provide a synthesis of multiple SQGs consistent with the intended application of the individual guideline approaches. The term "consensus" in this context does not mean agreement among scientists but rather that a wide variety of SQGs from various sources were combined into a single SQG or range of SQGs. The consensus method is not simply making a list of available SQGs for a particular chemical and then calculating the average. The consensus stems from the idea that if different methods for deriving SQGs result in a quantitatively similar concentration, then the validity of the result is greatly enhanced. Only then is the calculation of a consensus guideline justified. Even if consensus of different SQGs is not evident for a particular chemical, the method still serves the useful function of identifying potential variation among different SQGs.

The first consensus approach was published by Swartz (1999) and focused on PAHs. He stressed the "mixture paradox," which proposes that an SQG for a single PAH derived from laboratory tests using spiked sediments spiked will underestimate toxic effects in field sediments, because PAHs occur as mixtures in the field and their toxicity is the aggregate of all PAHs in the mixture. In contrast, an empirically derived SQG for a single PAH will over-represent the effects actually caused by that single compound because the toxic effects of the mixture are being ascribed to a single component of the mixture. Swartz argues, therefore, that guidelines for individual PAHs seem inappropriate regardless of the derivation method, and PAH guidelines should be based on an expression of the mixture of PAHs, such as total PAHs. Swartz (1999) re-calculated TPAH guidelines for several SQG approaches based on 13 compounds and found them to agree within a factor of about 4 in 3 different clusters. These clusters corresponded to 3 effects ranges, that is, a threshold, a median, and a high (extreme) effects range. Therefore, he proposed 3 consensus-based guidelines, named "threshold effect concentration" (TEC), "median effect concentration" (MEC), and an "extreme effect concentration" (EEC). The TEC and MEC are defined as the arithmetic means of the guidelines in each cluster, while the EEC was equal to the AET and was not a cluster of guidelines.

Consensus sediment quality guidelines for PCBs

A similar approach was used by MacDonald, DiPinto, et al. (2000) for PCBs. They also adopted the concept of 3 categories, namely TEC, MEC, and EEC. These consensus-based effects concentrations were developed through a stepwise approach: The existing guidelines expressed on an organic carbon-normalized basis were first converted into dry weight–normalized concentrations, assuming a 1% organic carbon level, then these guidelines were split into the 3 aforementioned categories. Only empirically derived guidelines were used. The geometric mean was used for determining each of the consensus guidelines. Other possible indicators of central tendency were considered, but the geometric mean was preferred because it was supposed to better minimize the effects of single values on the estimate of central tendency and because the distribution of the SQGs in each category was unknown.

Consensus approach for 28 chemicals of concern in freshwater sediments

The third approach, developed by MacDonald, Ingersoll, and Berger (2000), used only 2 categories of consensus-based guidelines, namely a threshold effect concentration (TEC again) and a probable effect concentration (PEC). The approach was applied to 28 chemicals of concern, comprising metals, PAHs, PCBs, and pesticides. The TECs are intended to identify contaminant concentrations below which harmful effects on sediment-dwelling organisms are not expected, while the PECs are intended to identify contaminant concentrations above which harmful effects on sediment-dwelling organisms are expected to occur frequently. TECs include the aforementioned TEL, ERL, LEL (Persaud et al. 1992), SLC-derived "minimum effect values" (Environment Canada 1992), and sediment quality advisory levels (USEPA 1997). PECs include PEL, ERM, SEL (Persaud et al. 1992), and SLC-derived "toxic effect thresholds (Environment Canada 1992). The derivation process is similar to the one used for PCBs, including an evaluation of the applicability of existing SQGs to the process. The collated SQGs were further considered for use in this study if 3 evaluation criteria were met:

1) the relevant details concerning the method of derivation were sufficient to understand and apply it,
2) the SQGs were based on empirical data relating contaminant concentrations to adverse effects on sediment-dwelling organisms (if the purpose was to protect that type of organism), and
3) originality, that is, SQGs were not simply adopted from another source.

The TECs and PECs were calculated by determining the geometric means of the SQGs for each category.

Comparing the Derivations of Sediment Quality Guidelines: Implications for Their Interpretation

Tables 3-1 and 3-2 summarize background information and the key characteristics of the SQG approaches outlined in "Overview of Procedures Used to Derive SQGs" (p 45). This section also provides a comparison of the ways in which these SQGs deal with key issues in sediment assessment, including the biological endpoints they address and the extent to which they consider bioavailability, causality, mixtures and nonchemical stressors. In this section, we discuss several of these issues in detail and emphasize how differences between guidelines affect their application and interpretation.

Biological endpoints

In general terms, all of the SQGs described in Tables 3-1 and 3-2 have the presumptive goal of protecting benthic communities from adverse effects. However, only the SLC and the benthic AET are explicitly based on benthic community effects. While data on benthic community effects are also included in the databases for the effects range and effects level approaches, these and all other SQGs are based primarily (if not exclusively) on other measures of biological effect, primarily laboratory toxicity tests.

Laboratory toxicity tests used in SQG development vary in the both duration (e.g., acute vs. chronic) and endpoints assessed (e.g., lethality, growth, development, reproduction), as well as the organisms used. For the sake of this discussion, acute tests are defined as those with a lethality endpoint and generally a short exposure (≤10 d), and chronic tests as those tests with a sublethal endpoint (e.g., growth, reproduction) and a longer exposure. This distinction is somewhat blurry, however, as some common 10-d exposures also measure sublethal effects (USEPA 1994b; ASTM 2002b). The range of species and endpoints used to develop SQGs varies widely and should be considered when selecting and applying SQGs. SQGs such as ERL or ERM and TEL or PEL (Long and Morgan 1990; McDonald et al. 1996) are based on extensive databases that include data on a variety of species and endpoints. Other SQGs (e.g., amphipod AET, LRM, ΣPAH) are based exclusively on effects on a single taxonomic group.

Compared to water column toxicity tests, the range of commonly used sediment test methods is much narrower. While test methods have been developed for chironomids (Anderson 1980; Ingersoll and Nelson 1990; Benoit et al. 1997), benthic algae (Adams and Stauber 2004), bivalves (USEPA 1994b; Roper et al. 1995; Ringwood and Charles 2002), polychaetes (Environment

Canada 2001; ASTM 2002c) nematodes (Hoss et al. 2001), and oligochaetes (Phipps et al 1993), amphipods are by far the most common organisms used in sediment toxicity testing and represent the bulk of the available data. Amphipods are generally considered to be sensitive to a variety of toxicants; *R. abronius* in particular, is thought to be one of the more sensitive marine amphipods. Although the empirical evidence suggests that amphipods are generally sensitive to toxicants, this is only a generalization. For example, freshwater amphipods are much less sensitive to dieldrin than are chironomids (USEPA 1993b); *H. azteca*, the most commonly used freshwater amphipod, is over 100-fold less sensitive to dieldrin than is the midge, *Chironomus tentans* (D Mount, unpublished data).

In addition to emphasizing the use of amphipods, many of the available sediment toxicity data are based on lethality, rather than on sublethal endpoints. While it seems intuitive that sublethal responses would constitute a more sensitive endpoint than lethality, the evidence is mixed and the results are probably species- (and perhaps toxicant-) specific. Swartz et al. (1985) saw no change in sensitivity between the 10-d and the 30-d *R. abronius* tests. In another study, however, Ingersoll et al. (2001) compared the 10-d and the 28-d *H. azteca* tests and found an acute/chronic ratio of approximately 6. In a study of 16 different sediments, Defoe and Ankley (1998) found that the 10-d growth endpoint for the midge *C. tentans* showed a more significant response to sediments than did the lethality endpoint for *C. tentans* or *H. azteca*. Further discussion of acute versus chronic responses is given in Chapter 12.

In the derivation of water quality guidelines, the issue of varying species sensitivity is addressed by using data from larger toxicity data sets containing data from taxonomically diverse organisms. For SQGs derived directly from sediment toxicity tests, this approach is not really feasible because of the limited range of available methods and the even more limited range in the available toxicity data. The degree to which this compromises the ability of these SQGs to protect benthic communities is uncertain; while the potential for underestimation of effects is clear, effects measured in specific sediment toxicity tests have been correlated to effects measured in benthic communities in the field. There is a small but growing body of literature that links many acute and sublethal endpoints to higher levels of biological organization (McGee et al. 1999; Nipper et al. 1998; Kuhn et al. 2000, 2002; Haywood 2003). More studies of this type need to be performed, particularly for community effects.

It is important to note that typical sediment toxicity testing with benthic organisms is not an appropriate means of assessing risk from chemicals whose primary effects are mediated through bioaccumulation, trophic transfer, and subsequent effects in higher-level predators. This includes chemicals such as methylmercury and dioxins and dioxin-like chemicals. Assessment of these chemicals should be based on methods directly addressing bioaccumulation pathways.

While still relying on toxicity data, SQGs derived with EqP use toxicity data gathered from water column tests. As explained in "Mechanistic approaches to SQGs" (p 62), EqP theory is based on the assumption that the response to sediment-associated chemicals is the same as would be observed in a water-column toxicity test with a water concentration equal to that in the pore water of the sediment. Guidelines proposed by USEPA (2003a, 2003c 2003d, 2004a, 2004b) are derived using the final chronic value (FCV) from the corresponding water quality criterion. Based on available water column toxicity data, the FCV is intended to protect 95% of tested species from lethal or sublethal effects (Stephan et al. 1985). The availability of toxicity data for water column exposures is much greater than for sediment exposures; consequently, an advantage of the EqP approach is that it allows the incorporation of toxicity data for a wider range of species. Some have raised concerns about the appropriateness of using toxicity data for pelagic species to predict effects on benthic organisms. Analyses by Di Toro et al. (1991) suggest that there are no systematic differences in the toxicological sensitivity of benthic or epibenthic species compared to pelagic species. However, in their guidelines for deriving EqP-based SQGs, the USEPA recommends verifying this assumption using data specific to the SQG chemical (USEPA 2003a).

Variations in toxicity test procedures and their interpretation

Guidelines for performing toxicity tests allow considerable latitude with regard to water renewal (static, bulk water changes, flow through). Choices usually depend on the organism being tested and the duration of the test. Variability in resulting guidelines can be traced in part to how tests were performed, and assumptions regarding equilibrium should be examined in light of how tests were done.

Methodological uncertainties arise from how the sediment (or pore water) is handled in toxicity testing. Is it appropriate to sieve and homogenize the sediment? Sieving is used to remove coarse particles and predators. If this is accepted, then an adequate time should be incorporated for the redox equilibria to reestablish before introducing animals. Sediment spiking is very sensitive to the procedures used and, if done incorrectly, can lead to abnormally elevated pore water contaminant concentrations (Simpson and Batley 2003). Testing sediments under abnormally high porewater concentrations can mask normally dominant exposure pathways, which were dietary rather than porewater exposure (Lee et al. 2000). Aging of sediments after spike additions is essential, but full equilibration between sediment and porewater may never be reached. At least 14 days are required for metal additions to fully equilibrate (Simpson and Batley 2003). (USEPA [2000a] and ASTM [2002d]) recommend a minimum of 1 month.) Porewater pH must be adjusted to counteract any pH changes that occur. Metal additions to sediments in the presence of AVS will give detectable porewater concentrations only when the AVS exchange capacity is exceeded. Metal spikes will displace other metals from their sulfides in the order Cu > Cd > Zn > Fe, and it is possible that an experiment to test the toxicity of added Cu, for example, may be seeing effects of displaced Cd (Simpson, Apte, and Batley 2000).Toxicity testing may be performed on whole sediments, pore waters, or elutriates. Because the focus of this workshop is on in-place sediments, elutriate tests will not be discussed in detail. The use of pore water as a test matrix has been a source of much discussion (Chapman et al. 2002; Carr and Nipper 2003). Advocates of porewater testing cite evidence that concentrations of chemicals in pore water correlate well with sediment toxicity (Adams 1987). There is the ability and ease of being able to store, handle, and test both overlying and pore waters with the same system (Carr and Chapman 1992, 1995). In addition, complete toxicity identification and evaluation (TIE) methods exist for pore waters and are more fully developed than those for whole sediments, although the latter do exist (e.g., Ankley et al. 1992; Besser et al. 1998; Burgess et al. 2000; Ho et al. 2002).There are issues with respect to porewater toxicity testing, such as porewater oxidation (which affects metal speciation and coagulation of dissolved organic matter and colloidal material with likely effects on the bioavailability of organics) that were discussed in a recent Pellston workshop (Carr and Nipper 2003). Where SQGs depend upon porewater tests, metal bioavailability may be changed and the toxicity of organic compounds may be underestimated. Organism exposure to 100% pore water may not be environmentally relevant because it lacks a dietary component of because exposure and exposure concentrations of organic chemicals may not be maintained during testing. Many benthic organisms are exposed to a mixture of overlying and porewater under field conditions (e.g., tube-dwelling amphipods may pull in large percentages of overlying water to maintain aerated burrows). The effect of exclusively using porewater tests may overemphasize exposure to many water-soluble toxicants such as ammonia or metals, while potentially underemphasizing nonsoluble toxicants. It may be debated whether organisms are actually exposed to undiluted pore water, as recovered by centrifugation. In the field, exposure through this route is frequently diluted by bioirrigation.

Bioavailability

"Bioavailability" with respect to sediment contaminants means the potential for uptake or absorption by internal organs of a sediment-dwelling organism, via pore waters or ingested particulates, including prey (NRC 2003). The bioavailability of a contaminant depends on its ability to be released from a sediment particle (desorbed or dissociated) either into pore water or in the gut of an organism, and transported across a biological membrane, where it may accumulate with potentially toxic effects. Body burden is influenced by exposure history of the organism, the physical or chemical properties of the contaminant itself and its degree of aging, as well as by sediment characteristics such as organic carbon content and composition, and grain size distribution. We measure the effects of bioavailability on an organism, namely bioaccumulation and toxicity, but these effects are species specific and depend on both the exposure pathways and the sensitivities of particular species.

We are as yet unable to chemically measure, in a sediment, the fraction of a contaminant that is bioavailable; however, measurements are available to determine concentrations that are more relevant than a total concentration for some contaminants. For example, in the case of metals, extraction of sediment particles with cold, dilute hydrochloric acid gives a better measure of metals that might be bioavailable than a measurement that includes mineralized species that are soluble only in concentrated acids. It should be noted that most SQGs are based on digestions using concentrated acids. Extractions with dilute acids have been found to typically release from 60 to 90% of metals in anthropogenically contaminated estuarine sediments (Simpson and Batley, unpublished results).

Mayer and Weston and colleagues have developed extraction procedures using gut fluids of deposit feeders such as polychaete worms to mimic what is actually extracted from sediments during gut passage (Mayer et al. 1996; Weston and Mayer 1998a, 1998b; see Chapter 11). They found that surfactant and enzymatic constituents of gut fluids could be used to mimic the biologically extractable pool of contaminants. Their method shows promise; however, many aquatic organisms of interest are too small to allow the harvest of sufficient quantities of gut fluids for extraction experiments. Additional concerns with this approach (and to some extent with all chemical extraction models of bioavailability) include the impacts of gut clearance times and other biological behavior on comparisons with results from the model extraction procedures (Penry and Weston 1998). The link between bioaccumulated metals and toxicity has not yet been fully developed.

Approaches based on EqP attempt to predict porewater concentrations based on equilibrium dissociation from sediment particles. This involves considerations of SEM–AVS for metals and the fraction of sediment organic carbon for nonionic organics. EqP theory assumes that contaminants in sediments are in equilibrium with associated pore waters, and although not all organisms are exposed to pore waters, the most sensitive species appear to be those that are exposed by this route. Recent studies (Simpson and Batley 2003) suggest that, at least for metals, thermodynamic equilibrium is rarely achieved and that dynamic changes occur constantly, driven by bioturbation, bioirrigation, tides, and other hydrodynamic disturbances that alter redox status.

The applicability of the AVS–SEM approach to wide range of heavy metals has been challenged in recent papers by Simpson et al. (1998; Simpson, Rosner, and Ellis 2000), which suggest that its application is best for Cd, Pb, and Zn. There are issues related to the behavior of non–acid-soluble sulfides, such as those of Cu, Co, and Ni, in taking into account the excess of AVS (Simpson et al. 1998). If these do not release sulfide, and do not also release metals, then there is no impact on the SEM–AVS calculation from SEM-Cu (or Ni or Co), and any excess of AVS will still react with porewater Cu, reducing its toxicity. If, however, these metal sulfides are otherwise oxidized and the metals released, then SEM–AVS will be overprotective, that is, there will be a possible prediction of an excess of SEM over AVS. Similarly, coatings of sulfides on metal oxides in some field sediments could generate the same prediction of potential toxicity, when in fact there would not be any (Simpson, Rosner, Ellis 2000). Because the uncertainties related to AVS considerations are likely to falsely show an excess of SEM over AVS in some instances, their use is still recommended in assessing metal bioavailability in sediments. There have, however, been other concerns raised over issues such as the spatial and temporal variability of AVS (Brumbaugh et al. 1994), its applicability in oxic and dynamic environments, and the fact that the AVS model refers to whole sediment and not the microenvironment of organisms that may differ, especially for organisms that irrigate burrows. That said, experimental evidence used in support of the SEM–AVS approach includes laboratory, mesocosm, and field experiments which involved benthic organisms that inhabit or establish microenvironments and/or occupy irrigated burrows (Berry et al. 1996; Hansen, Berry, et al. 1996; Hansen, Mahony, et al. 1996; Liber et al. 1996).

Although the SEM–AVS approach is a gross simplification of the overall behavior of metals in sediments, the prediction of the absence of toxicity of metals in sediments has been generally successful, including studies with burrowing organisms. This may be a mere coincidence, or it may be because of secondary relationships. For example, sediments that have SEM in excess of AVS may be more likely to have high metal concentrations in oxic layers.

Metal bioavailability in pore waters is a function of metal speciation and is thus limited by factors such as organic complexation that are well documented for surface waters. Note, however, that processes that lower pH in the pore waters, for example, oxidation of iron(II) during sediment spiking or handling, will favor increased metal release from particulates, and potentially greater bioavailability and toxicity (Simpson and Batley 2003). Porewater pH is rarely reported in sediment studies, but should be, to enable better data interpretation. A change of 1 pH unit can significantly alter the bioavailable fraction of metals and ammonia, as well as the organic compound, dieldrin (Standley 1997).

With empirical SQGs, the question of bioavailability is not overtly addressed, and only total contaminant concentrations are typically used in the databases that underpin them. It has been noted (Allen 1996) that the use of total rather than bioavailable concentrations has the potential to bias both effects and no-effects data, where the fraction of the total concentration that is bioavailable varies. For metals, this can be the case if there is a variable, inert, mineralized metal component. If dilute acids extractions were used instead of total measurements, the impact might not necessarily be significant in terms of the overall uncertainty of the measurements (see "Other stressors," p 86) or in relation to toxicity from co-occurring contaminants.

King et al. (King CK, Gale SA, Stauber JL; CSIRO, Bangor NSW Australia; unpublished data) found that using supposedly metal-sensitive marine amphipods, in tests on spiked sediments with appropriately equilibrated metal additions, and in porewater-only tests, that toxicity greatly exceeded empirical SQG values. Experiments compared the sensitivities of 3 infaunal amphipods (*Corophium* cf. *volutator*, *Chaetocorophium* cf. *lucasi*, and *Grandidierella japonica*) and 4 epibenthic species (*Melita plumulosa*, *Melita awa*, *Hyale longicornis*, and *Hyale crassicornis*) to Cu and Zn. *Melita plumulosa* was the most sensitive species, yet for Cu, the LC50 for juveniles was 520 mg/kg, and the chronic IC50 for reproduction was 320 mg/kg, compared to an SQG of 65 mg/kg.

They concluded that the apparently conservative guideline was more likely to be driven by co-occurring organic contaminants in the sediments. For organic contaminants, the EqP model (Di Toro et al. 1991) posits that the concentrations of organic contaminants dissolved in pore waters are controlled by partitioning between solid and dissolved compartments of sediments, and that this partitioning will be moderated by organic carbon content. Normalizing to organic carbon content assumes that all carbon is the same, when it is not (see Chapters 13 and 17). The use of normalization should therefore be undertaken with caution, with best applicability being over a limited range of organic carbon contents (e.g., 0.5 to 10% which is typical of marine sediments). This is true because at low TOC, non-organic carbon binding sites become more important, while at the upper end, a mostly organic carbon-dominated matrix, for example, 20% heavy oil, can drastically alter partitioning (Simpson SL, Adams MS, Stauber JL, Batley GE; CSIRO, Bangor NSW Australia; unpublished data). In addition, the presence of other, nondiagenic types of organic carbon such as wood chips, hide or hair, soot, or coal can alter partitioning significantly from what would be expected from normal, diagenic organic material (Chapter 4). Where actual partitioning differs from that predicted by typical K_{OC} values, EqP will not be predictive without correction.

As discussed earlier (in "EqP-based SQGs for nonionic organic chemicals," p 62), EqP theory requires input of a K_{OC} value. There is a source of uncertainty in this value. It is often assumed that this can be calculated from K_{OW} values, but the domain of validity for the equations used for that purpose is rarely considered, and the underlying assumptions are not always valid (Seth et al. 1999). The impact of this on derived SQGs is unknown. That said, actual partitioning can be measured in the sediment of interest and compared to literature values and/or used to derive site-specific partition coefficients. As indicated above, this may be very important where a substantial portion of the organic carbon present is not typical diagenic organic carbon. Some empirical SQGs for organics were derived without consideration of organic carbon content, but in some instances the role that organic carbon might have in moderating contaminant bioavailability has been included to permit modifying the SQG, particularly in cases where the organic content is high (see Tables 3-1 and 3-2). Organic carbon normalization is used in the AET approach in Washington State and is also explicitly considered in empirical SQGs by Barrick et al. (1988), Cubbage et al. (1997), USEPA (1996), and Field et al. (1999); however, normalization did not improve the predictive ability of the SQGs in any of these studies.It is noteworthy that with empirical SQGs, dry weight normalization often results in predictions of biological effects that are as good or better than the normalizations thought to reflect bioavailability. This may be a co-variance issue or related to the presence of other contaminants, but for single contaminants or classes of contaminants, the value of organic carbon normalization has been clearly demonstrated (Di Toro et al. 1991).

Harkey et al. (1996) correctly highlight that bioavailability also involves a temporal component. Contaminants entering a system need to be bioavailable (and a complete exposure pathway exist) on a time frame meaningful in the context of the life history of the organism in question.

Routes of exposure

Neither empirical nor mechanistic existing approaches to SQGs explicitly consider routes of exposure to contaminants. Empirical SQGs effectively encompass different exposure routes by the inclusion of toxicity data derived from organisms with differing uptake mechanisms.

There has been some debate as to the importance of exposure routes in EqP SQGs. EqP theory assumes that sediment and pore water approximate an equilibrium of chemical activity with which the organism reaches a steady state; it makes no presumption about the route of exposure (Di Toro et al.1991). If this is true, then the primary route of uptake is irrelevant. The emphasis on pore water arises from the desire to relate chemical activity in a sediment–pore water–organism system to chemical activity in water–organism systems, for which there are the most toxicological data. Because chemical activity in water and in pore water is similar to chemical concentration (at least once DOC binding has been accounted for), this approach provides a simple way to relate the 2 systems.

In practice, it is very unlikely that organisms can reach an equilibrium with pore waters (at least for metals). This issue is contentious, however, with the relative uptake kinetics from each exposure pathway being important, recognizing that organisms have active ways of accumulating, excluding, or excreting chemicals and that toxicity occurs when these mechanisms are overwhelmed.

The heterogeneity of the sedimentary environment, the diversity of organism feeding patterns and physiological capabilities, and the potential for biomagnification are factors that complicate the simple prediction of equilibrium. EqP will overestimate exposure for organisms that effectively limit their exposure, for example, through burrow irrigation for burrowing organisms, but underestimate exposure for organisms that preferentially consume particle size fractions with a higher concentration of naturally occurring organic matter and thus a higher proportion of contaminant than whole sediment. Physiological sensitivity as well as exposure will determine an organism's sensitivity to a contaminant, but from an SQG, perspective we would want to be protective against impacts on the most sensitive and maximally exposed species, and here EqP predictions are appropriate. As noted by Di Toro et al. (1991), "the representation of benthic organisms as passive encapsulations of lipid that equilibrate with external chemical concentrations is clearly only a first-order approximation." It is "an appropriate initial assumption because deviations from the first-order representation will point to necessary refinements, and for many purposes the approximation may suffice."

Mixtures

The assumptions that underpin many SQGs are that sediment toxicity and ecological effects result from a mixture of toxicants, that SQGs for the effects of individual chemicals make little sense, and that an SQG for an individual chemical should be explicitly framed in the context of the mixture. Most contaminated sediments contain contaminants from several chemical classes, and although a few of the chemicals may dominate in terms of concentrations, the toxicity of the mixture can represent the interaction of many of the components. Thus, a chemical whose concentration is not sufficient by itself to exert effects may still contribute to the toxicity of the mixture.

The effect of mixtures is further complicated by the lack of knowledge about the interaction of chemicals from different chemical classes or chemicals with different geochemical properties (e.g., PAHs, metals, ionic chemicals, high molecular weight organic compounds). There is no theoretically correct way to deal with such interactions. Similarly, the additive effects of chemicals with different modes of action (e.g., narcotics, teratogens, mutagens) cannot be dealt with by the currently available SQGs.

The simplest conceptual model of contaminant interactions is additivity, which predicts the effects of the mixture from the sum of the effects of the individual chemicals. Thus, if concentrations of individual chemicals are normalized to a reference concentration like an LC50 or an SQG, the sum of the normalized values can be used in the concentration-response model to predict the probability of biological effects. The additivity model may be appropriate for SQGs that reflect effects caused by specific chemicals (e.g., guidelines derived from EqP theory) that exert their toxicity through the same or similar mechanisms, but otherwise it may be inappropriate.

Mixtures and empirical sediment quality guidelines

Because they are based on co-occurrence, empirical SQGs for individual chemicals have little to do with the effects of individual chemicals but indicate effects caused by the entire mixture of chemicals present in a sediment (Chapter 4). Thus, the sum of SQG-normalized sediment concentrations is likely to be inappropriate for predicting the effects of the contaminant mixture. Other methods of aggregating SQG-normalized sediment concentrations (e.g., calculation of mean or maximum SQG quotients) for individual contaminants may be more appropriate to express the mixture effect for empirical SQGs. The use of a mean quotient can, however, statistically "dilute" the predicted impact of a dominant toxicant, depending on the extent that these additional compounds co-occurred in the sediments used to derive the empirical SQGs.

For a given chemical class (e.g., PAHs), the empirical SQG for each compound provides a separate estimate of the critical concentration of the mixture. Thus, in the case of PAHs, the ERM for phenanthrene is a surrogate estimate of the unknown ERM for the mixture of all PAHs (measured or unmeasured) plus any other chemicals that happen to have a high correlation with phenanthrene in the data set used to derive its ERM. The ERM for each individual PAH compound (ERMi) is, in effect, an independent surrogate for ERM of the PAH mixture. If all PAHs co-varied perfectly, SQG quotients for individual compounds (C_i/ERM_i) would be identical and equal to the unknown mixture quotient (Cmixture/ERM mixture). Also in the case of perfect co-variation, critical values of SQG quotients for individual chemicals, for example, $C_i/ERM_i = 1.0$, would be simultaneously exceeded as the level of sediment contamination increases.

These observations have several implications for the analysis of empirical SQGs:

- Empirical SQGs for individual chemicals of the same chemical class are not additive. They are not equivalent to toxic units.
- The mean C_i/ERM_i for related compounds like PAHs is the best surrogate estimate of Cmixture/ERMmixture. PAHs, PCBs, and to a lesser extent, metals and pesticides represent 4 separate classes of related, (but not necessarily) co-varying compounds. A mean C_i/ERM_i could be calculated separately as a surrogate estimate of the Cmixture/ERMmixture for each class (Ingersoll et al. 2001, 2002).
- Effects concentrations for the total measured concentration of groups of related compounds (e.g., 13 priority pollutant PAHs) provide another surrogate estimate of the Cmixture/ERMmixture for the chemical class. They may be more accurate than the C_i/ERM_i for individual compounds, but they are only a surrogate, not a direct measure of Cmixture/ERMmixture for the class.
- Estimates of the effects of complex mixtures of chemicals from different classes that do not co-vary may be approximately additive, although

it is worth noting that water-column mixture studies suggest that chemicals add within modes of action, but have little interaction where modes of action differ. In these cases, they are best represented by the sum or maximum, not average, of the surrogate estimates of Cmixture/ERMmixture for each class. In effect, SQG quotients for separate classes behave like toxic units and should be added to estimate the potential toxicity of a complex mixture. For example, at an oil-spill site where PCBs, metals, and pesticides are not contaminants of concern, the mean C_i/ERM_i for PAH compounds should not be divided by 4 to estimate the toxicity of the complex mixture. An average C_i/ERM_i for the 4 chemical classes might obfuscate the toxicity potential of just 1 class. Where more than 1 class is present, the mean C_i/ERM_i for each class should be added. Note that the sum and the average will be identical if there is an identical number of chemicals being evaluated. The average should be used if there is not a consistent number of chemicals being evaluated.

- Within a chemical class, the relative magnitude of C_i/ERM_i for individual compounds does not indicate the relative contribution of the compound to the predicted effect. For example, a PAH compound might never reach toxic concentrations in the data set used to estimate its empirical SQG. Thus, at a specific site, its SQG quotient could be higher than other PAH compounds that are actually contributing to the effects of the PAH mixture.
- In contrast, the mean C_i/ERM_i of different chemical classes should indicate the relative contribution of each class to the effects of the complex mixture.

Fairey et al. (2001) formulated an SQG quotient method that was highly correlated with both the magnitude and incidence of sediment toxicity to amphipods in 3 databases. Their SQG quotient 1 (SQGQ1) is the mean value of the SQG-normalized concentration of 9 parameters: Cd, Cu, Pb, Ag, Zn, total chlordane, dieldrin, total PAH, and total PCB. SQGQ1 was the most successful of 18 different SQGQ formulations tested by Fairey et al. (2001).

Interestingly, the predictive ability of the 18 SQG quotients was greater when SQGs developed by different methods were used in the normalizations. SQGQ1 included SQGs developed by ERM (Cu, Zn, chlordane, dieldrin), PEL (Cd, Pb, Ag), and consensus (PAH, PCB) methods. An important result of the SQGQ1 derivation was that 9 chemicals could predict toxicity in diverse sediments that certainly contain a much larger number of contaminants. Fairey et al. (2001) concluded that the low frequency of toxicity (4%) among samples with SQGQ1 < 0.1 indicates that other chemicals not specifically included in SQGQ1 infrequently occur at toxicologically significant concentrations in otherwise uncontaminated sediments. SQGQ1 is a demonstration of the potential efficacy of empirical SQGs to predict effects (in this case, mortality) of complex mixtures.

The success of SQGQ1 needs to be tested with data sets that are independent of those used in its initial derivation. Despite the empirical efficacy of SQGQ1, there are some troubling aspects in its formulation. It treats 5 individual metals as equal to both PAHs and PCBs. It is difficult to accept that Ag, for example, is as important a sediment contaminant as PAHs. Also, SQGQ1 is based on the mean rather than sum of SQG quotients. Perhaps it would be better to sum the mean quotients for metals, pesticides, PAHs, and PCBs. That general formula would allow specific chemicals within each of the 4 classes to be included on a site- or region-specific basis. It would also be consistent with the implication noted by Fairey et al. (2001) that the toxicological mechanisms of the represented chemicals are additive. Nonetheless, empiricism is not necessarily bound by theoretical considerations.The justification for the use of mean quotients is the assumption of additivity (independence of effects). A limitation is that quotients may “dilute” the relative hazard of a given sample, when a few chemicals greatly exceed their guidelines while others do not. These quotients reflect the fact that many contaminants co-vary in sediments (particularly when the sampling points are not directly influenced by specific sources). Furthermore, combining chemicals in sums or quotients of SQG units assumes a common mode of toxic action, when in reality many chemical classes are likely to have different modes of toxic action, independent of each other.

Mixtures and EqP sediment quality guidelines

Because EqP-based guidelines are designed to quantify the expected toxicity of specific components of a sediment mixture, they can be adapted to address the toxicity of a chemical mixture if the toxicity of the individual components of the mixture is known, and if their toxicological interactions can be described mathematically. Three excellent examples exist for mixtures of cationic metals, PAHs, and nonionic narcotic chemicals in general.

The EqP-based approach to metal mixtures is thoroughly described in a series of articles published in the December 1996 issue of *Environmental Toxicology and Chemistry*. These articles describe the use of AVS as a primary binding phase for several cationic metals (Cu, Cd, Zn, Ni, and Pb) in sediment. In addition to assuming a common binding phase for these metals in sediment, this approach also assumes that the toxicity of mixtures of these metals to aquatic life appears to be strictly (concentration) additive, or nearly so. Accordingly, the predicted potential for toxicity does not consider the metals individually, but in aggregate.

No toxicity from these metals is expected if

- the sum of their readily extractable equivalents is less than the equivalents of sulfide extractable under the same conditions (SEM < AVS),
- the sum of their calculated toxicities in pore water is less than that described by the USEPA AWQC, or
- both are true.

While this approach attempts to aggregate the toxicity of these 6 metals, in its current form it does not extend to contributions from other metals or other chemicals to the toxicity of a larger mixture.

The initial approach to assessing PAH toxicity using EqP was the development of SQGs for individual PAHs, specifically phenanthrene, fluoranthene, and acenaphthene (USEPA 1993a, 1993d, 1993e). While theoretically correct, this approach was not directly useful in assessing PAH-contaminated sediments because field sediments typically contain mixtures of PAHs rather than single PAHs, and there is interactive toxicity within PAH mixtures. Thus, to appropriately assess the toxicity of PAH-contaminated field sediments using EqP, some means of assessing the aggregate toxicity of PAH mixtures was necessary.

The success of the ΣPAH model (Swartz et al. 1995) is a good demonstration of the dependence of mechanistic methods on empirical correlations with the estimated, but unknown, total toxicity potential of field-collected sediments. Clearly, the 13 PAH compounds used in the model do not include all of the PAHs and other chemicals that contributed to the toxicity of the field sediments used to evaluate the model predictions. The success of the model results from the presumably good correlation between ΣTUi for the 13 compounds and the actual toxicity of the field-collected sediments (Chapter 12). ΣTUi is a surrogate estimate of the unknown sediment toxicity potential in the same sense that the ERM for phenanthrene is a surrogate estimate of the same parameter.

Mixtures of nonionic narcotic chemicals are addressed in the narcosis model of sediment toxicity, discussed in "EqP for mixtures of nonionic chemicals: Narcosis approach" (p 65).

Causality

An important consideration in the scientific underpinnings of SQGs is whether they provide either an explicit or an implicit basis for assigning causality. When one or more of the chemical stressors in a sediment sample is found to exceed an SQG, the question is to what degree is that stressor or stressors likely to cause impairment or toxicity? With empirical SQGs, the narrative intent was not to determine causality, and hence the linkage between observed toxicity or impairment and a single, specific chemical stressor may generally be weak. This is a direct consequence of their derivation from tests of samples in which mixtures of chemical stressors were present. In the case of EqP-based SQGs, there may be a greater likelihood for toxicity or impairment from a specific chemical stressor because the approach itself is based on measurements of individual chemical stressors (see Tables 3-1 and 3-2, and Chapters 4, 12, and 13).

The relationship between a metal concentration and actual toxicity in the field may be poor, perhaps because the metals (or measured forms) are not responsible for the observed toxicity but simply co-vary with the chemicals that are toxic (e.g., PAHs, DDT).

It may also be true that metals are good indicators of toxicity in the field, whether or not they cause or contribute to toxicity (Field et al. 2002).

Assigning the cause of observed toxicity or impairment in field-derived samples is difficult if the assignment is based solely on comparing a chemical stressor concentration to its particular SQG. Numerous workers have used Hill's criteria or similar approaches (Fox 1991) to help identify the cause of a particular observation (Suter 1993; USEPA 2000b). As used by the USEPA (2000b) in strength of evidence analysis of causality, these criteria require consideration of strength, co-occurrence, temporality, biological gradient, and complete exposure pathway for the case being studied, together with plausibility, specificity, analogy, and predictive performance, based on this or other cases or test situations. However, in other instances, applying Hill's criteria failed to establish the cause of water quality impairment in a series of Colorado streams (Canton and van der Veer 1997). Multivariate statistical techniques also have been employed to establish which among a suite of chemical and physical stressors might have produced the observed toxicity and benthic community impairment (Watzin et al. 1997). As these authors noted, no single stressor was solely responsible, and in some cases, both chemical and physical stressors combine to produce the observed effects (Preston 2002).

Technical advances in TIEs with sediments and spiked sediments may offer avenues of investigation for establishing causality, although these are not without limitations (Hunt et al. 2001; Chapter 7). Other approaches and considerations for establishing causality in multiply stressed systems can also be employed for sediment evaluations (Harris et al. 1994; Foran and Ferenc 1999). Harris et al. (1994), for example, assigned risk values to particular ecosystem stressors and coupled that with impairment criteria to evaluate which of several stressors were most responsible for degradation of Green Bay, Lake Michigan, USA. Similarly, Parkhurst et al. (1997) detailed the use of habitat models in watershed-scale aquatic ecological risk assessments to make comparisons among physical and chemical stressors in multiply stressed systems. Both of these approaches could be helpful in evaluating causes of toxicity or impairment at contaminated sediment sites.

Reasonable certainty about causal relations between contaminants and biological effects is considered an attribute of mechanistic methods and a detriment of empirical methods for deriving SQGs. Mechanistic approaches are based on controlled laboratory experiments in which causality is not an issue. Empirical methods are based principally on correlations that do not necessarily reflect causality between sediment chemistry and toxicity or biological conditions. Both approaches provide only a surrogate toxicity estimate. In the case of empirical methods, the correlative surrogate is expected to co-vary with the toxicity of the entire contaminant mixture. In the case of the mechanistic approach, the surrogate estimate reflects toxicity that would be caused by the chemicals used in the model. It cannot account for toxicity of the majority of sediment contaminants that are not part of the model. Thus theoretically derived toxicity estimates are also surrogates that are expected to co-vary with the actual toxicity of complex contaminant mixtures. Uncertainty about causality and dealing with the complexity of mixtures are ultimately a common problem for both empirical and mechanistic methods (Chapters 4, 12, and 13).

Other stressors

Whole sediment testing is not free from matrix effects due to nonchemical stressors such as sediment grain size or salinity (Chapters 16 and 17). These natural variables may have a great influence on benthic communities in nature and may affect biological responses in laboratory tests (Chapter 12). For example, DeWitt et al. (1988) demonstrated that high levels of percent fines in the sediment may result in false positive results for the marine amphipod *R. abronius*. If interpretations of the results do not account for grain size effects, decisions or guidelines based on these tests may be overprotective. Such effects can sometimes be minimized through experimental or survey design, but they still represent a source of error in quantifying the chemical stressor effects.

In amphipod tests on estuarine sediments, the test species needs to be carefully acclimated from the culture salinity to the salinity of the test sample (Chapter 16). Altered salinity can be a stressor on such species, inducing a more sensitive response from an obviously stressed species. Temperature can also be an important stressor. Matched controls are essential to account for such differences.

Relationships between Sediment Quality Guidelines, Toxicity Tests, and Field Responses

Toxicity tests have been used to predict the effects of contaminants in sediments for many years (Environment Canada 1992; USEPA 1992b, 1994b, 2001). A measure of the success of SQGs is their ability to predict both laboratory and field toxicity. This is somewhat ironic in that toxicity tests are themselves imperfect predictors of real-world results. However, it should be recognized that the ultimate goal is to protect against or predict effects on benthic community structure in the field. For this reason, one must consider not only the ability of SQGs to predict results of toxicity tests but also the relationship between toxicity test results and actual effects on benthic communities.

With respect to toxicity, it has been suggested that tests do not generally address field issues such as sediment heterogeneity and contaminant microniches, changing sedimentary conditions, and natural variation of sedimentary conditions, because in many cases the field sample is sieved and homogenized before laboratory testing. Some of these issues may be addressed by collecting relevant sediment samples (e.g., collecting sediment at the depth that is biologically meaningful or meaningful to the particular study, testing dredged sediment at the depth to be dredged, or collecting the top 2 cm or the zone of biological activity for field assessments). However, laboratory toxicity testing will never be able to capture all of the environmental conditions of field exposures (Chapter 5), thus making a strong case for in situ toxicity testing (e.g., Chappie and Burton 1997) in combination with laboratory toxicity tests. In many instances, however, in situ tests are subject to issues that are seen as weaknesses in laboratory tests, for example, where sediment is collected and sorted before use in field chambers. Artifacts associated with chambers used for in situ testing may also influence the response of organisms exposed in the field (Chapter 5). Nevertheless, there is evidence for the concordance between toxicity test results and field effects (Swartz et al. 1982, 1994; Canfield et al. 1994, 1996; McGee et al. 1999; Long et al. 2001; Ferraro and Cole 2002; Ingersoll et al. 2004; Chapter 4 and 12).

Although contaminant exposure can be a highly dynamic process in ever-changing water column sources, issues of temporal changes in contaminant concentrations in sediments usually are of less a concern because of the relative stability of the sediment environment and the low mobility of benthic species. It is important to characterize benthic exposures and responses over time frames associated with expressions of acute and chronic toxicity in order to better understand potential impacts of contaminants on benthic species. This is valuable because many aquatic populations and communities tend to demonstrate short-term and long-term cycles in abundance and ranges. Decisions about what might be an acceptable or unacceptable level of contaminant stress on a local population of valued species or a benthic community should include consideration of natural, dynamic cycles.

Over chronic exposure timeframes, it is also important to consider the changing baseline conditions (physical, chemical, biological). The dynamic physical and chemical changes in aquatic habitats associated with transient and seasonal events contribute a baseline or background level of stress to which benthic communities must adapt. For example, daily temperature excursions affect metabolic rates. Over the longer term, changes in silt loading affect the characteristics of sediment habitat. Occasionally, strong habitat disruptions such as enhanced runoff following severe storms, limited flow due to drought, and ice scouring punctuate changes in local environmental quality. These changes generate their own set of background stresses, on which contaminant stresses are superimposed. In the absence of chemical contamination, the short-term status of benthic populations and communities can vary significantly because of these natural stresses, although local populations usually have adaptive life history strategies to quickly repopulate a temporarily stressed habitat. Issues with sediment-associated contaminants are separating chemical impacts from these baseline stresses and understanding when contaminant stresses compromise the resiliency of these populations (Baird and Burton 2001). Thus, investigators need to consider the multiplicity of stresses as they extrapolate laboratory test data and apply SQGs to field settings.

Ringwood and Keppler (2002) have demonstrated an approach that offers significant progress in separating effects of contaminant stresses from natural stresses on survival and growth under field conditions. By using a combination of laboratory and field toxicity tests and intensive physicochemical monitoring, they were able to account for effects of natural changes in salinity, temperature, dissolved oxygen, and pH on the survival and growth of *Mercenaria mercenaria*, while simultaneously accounting for toxicity from sediment-associated contaminants. Such approaches, while intensive in data collection and monitoring effort, should prove useful in separating natural stress and ecological changes from contaminant impacts. With regard to understanding the science and relevancy of SQGs, such comprehensive and effective field approaches demonstrate that knowing and applying an effects-level concentration should be part of a more complex assessment and analysis of benthic community health and sustainability at a site.

Using sediment quality guidelines to protect benthic communities

The ultimate objective of SQGs might be stated as the protection of the natural structure and function of benthic ecosystems. Although some empirical SQGs (e.g., SLCs, benthic AET) include data on benthic community abundance in the field, most are based in large part on laboratory toxicity tests. The extent to which these tests reflect potential impacts on benthic ecosystems is a matter of much debate and is discussed in Chapters 4 and 12. There have been some attempts to compare empirical SQGs with field surveys of contamination, toxicity, and benthic community structure. These studies generally show a correspondence between increasing contamination and the absence or reduction of sensitive benthic taxa at sites where SQG values are exceeded (Canfield 1996; Hyland et al. 1999). However, benthic degradation is often observed at chemical concentrations substantially below SQG values (Hyland et al. 1999). There are inherent limitations in the ability of any SQG to accurately and reliably indicate effects on benthic communitics.

It is important to understand the extent to which the combination of contaminant exposure data and supporting laboratory toxicity information can be extrapolated beyond their origin as single species, laboratory-controlled exposure-response data, to provide a basis for contaminated sediment management decisions that will be protective of benthic populations and communities at field sites. For both laboratory-based and field-based SQGs, making ecologically protective decisions requires an understanding of the degree of contaminant stress that will lead to ecologically significant changes. This may be difficult to discern for benthic communities, or indeed most aquatic ecosystems, because organisms in situ are exposed to multiple stresses, some of which are background or "natural" and all of which affect community composition. These issues are pertinent to the application of SQGs in the context of ecological risk assessments, where the goal is often protection of community and ecosystem properties (Bartell et al. 1992; USEPA 1992a; Suter 1993), rather than maintenance of individual species, and this makes the selection of local reference sites especially important.

A few studies have evaluated the degree of correspondence between empirical SQGs, results of laboratory toxicity tests conducted with contaminated sediments, and the structure of benthic communities at field sites (see Chapter 12). The strongest correspondence appears to be between sediments found toxic in laboratory studies using sensitive species, and reduced abundance and diversity in benthic communities. Ferraro and Cole (2002) conducted laboratory toxicity tests with 2 amphipods (*R. abronius* and *Leptocheirus plumulosus*) and evaluated the macrofauna community at the collection sites. Increasing toxicity correlated with reductions in numbers of species, abundance of animals, benthic biomass, and selected diversity indices. Long et al. (2001) reviewed data from several study sites to examine the concordance between sediment samples that were acutely toxic in laboratory tests and the abundance and diversity of benthic infauna at the sites where the sediments were collected. Results were fairly similar among all study sites; relatively high abundance and diversity were observed in areas where samples either were nontoxic in laboratory tests or displayed a low level of toxicity. For samples showing toxicity in the middle and higher ranges of species mortality, infaunal abundance and diversity were greatly reduced. Laboratory toxicity test results were particularly good indicators of abundance of amphipods in the benthos. These results indicate that laboratory toxicity tests with sensitive species, conducted with appropriate quality assurance and quality control (QA/QC) can generate results that will be protective of benthic communities for a particular sediment.

Correspondence between the specific chemical concentration guidelines used as an SQG and protection of benthic communities has been evaluated as well. The point of validation is whether the specific chemical concentration guidelines provide a reliable means to make decisions that protect benthic communities. Hyland et al. (1999) reviewed a significant number of studies for southeastern US estuaries and reported strong predictive relationships between SQG quotients and impacts on estuarine benthic communities. Their review showed that increasing bulk chemical concentrations in sediment were associated with increases in the degree of observed benthic community impacts. Similar results were reported by Long et al. (2002), who studied a gradient of sediment contamination in the lower Miami River, Ohio, USA, and in Biscayne Bay, Florida, USA.

However, unlike the findings of Long et al. (2002), the analyses of Hyland et al. (1999) showed that benthic indices were adversely affected at contaminant concentrations an order of magnitude lower than amphipod mortality in laboratory toxicity tests, which suggests that toxicity tests were not always predictive of field effects (see Chapter 12 for a more detailed discussion). However, Hyland et al. (1999) also observed benthic communities to respond to abiotic factors such as TOC and salinity that were correlated to concentrations of contaminants in sediment.

In another study, Paine et al. (1996) reported PAH sediment concentrations as high as 10 000 mg/kg at a site near an aluminum smelter, but sediments did not demonstrate adverse effects on benthic organisms. The observed presence of coke or pitch in the sediment was the likely reason for limited bioavailability. Similar "exceptions" have been the topic of discussions during various technical meetings or exchanges such as those found in the *SETAC Globe* newsletter. However, the extent or frequency of these kinds of inconsistencies is unclear because the information may be embedded in site remediation studies, may be reported as anecdotal information associated with other types of studies, or may simply go undocumented because results were considered negative findings. In any case, examples in which SQGs may be overprotective of benthic community effects have created a controversy that undermines confidence in their application and field relevancy.

While there are data that demonstrate that it is reasonable to expect some degree of correspondence, there have been a number of attempts to improve the precision and predictive ability of various SQG approaches (Swartz 1999; Fairey et al. 2001; Field et al. 2002). One goal has been to reduce the frequency of finding specific contaminated sites or sediment samples with high bulk chemical contamination and no apparent sediment toxicity or benthic community impacts. Because of the broad range of sites and contaminants that the SQGs have been developed from and are applied to, some managers of contaminated sites attempt to use the SQGs at a level of precision or independence beyond the scope for which they are intended. While SQGs have some acceptance in identifying contaminant levels below which no impacts might be expected and for identifying the higher levels of contamination at which benthic community effects may be obvious, the sites or samples that fall in the middle of this range require additional study in order to evaluate potential for population- and community-level impacts.

Empirically derived SQGs would seem to provide implicit field relevance because the data (at least in part) come directly from field samples. However, consideration of temporal duration and spatial extent of contaminated habitat is needed to more accurately assess the potential for population- and community-level impact. SQGs that reflect chronic toxicity effects are closer to those goals because they reflect sediment conditions required for baseline growth rates and uncompromised reproductive capacity among individuals in a test population. However, without an understanding of the population dynamics for species of concern within the contaminated habitat, application of toxic-effect SQGs, whether empirically or theoretically derived, provides for protection of the maximum rate of reproduction and growth among individuals. It has been argued that what is needed for field relevance is a basis to estimate the extent of survival, growth, and reproductive capacity necessary to sustain a local population and community when multiple stresses impinge on the individuals at a site (Baird and Burton 2001). Investigators make do with relatively simple comparisons of benthos and contaminant distributions in the field, at least for the first cut.

Studies of population dynamics and development of appropriately formatted population models have provided considerable insight into rates of survival, growth, and reproduction needed to sustain aquatic populations with various life history strategies when stressed by contaminants (Akcakaya et al. 1999; Caswell 2001). Studies by Kuhn et al. (2000, 2002) and McGee et al. (1999) have demonstrated means to connect laboratory toxicity data to input to population dynamics models for benthic amphipods and have applied this information to assess sediment-associated contaminants. These approaches provide a tool to account for and analyze the impact of incremental stresses of contaminant effects on survival, growth, and reproduction as one of the components controlling local populations. Population and community modeling and population field studies provide insight into determining the level of change that becomes significant to the sustainability of local populations. Such approaches also are a means of determining the level of incremental impact at which contaminants might change the sustainable dynamics of a population, as well as estimating, in more complicated and sophisticated evaluations, sustainability of system-level dynamics (Bartell et al. 1992).

Understanding the relation between SQGs and the "sustainability of system-level dynamics" is certainly a worthwhile goal and could be pursued in the development of better SQGs. SQGs should, however, be part of an assessment process and not an end in themselves.

Consideration of Uncertainty

Methodological uncertainty

A number of methodological issues impinge upon the chemical, physical, and biological data that underpin SQGs used to evaluate sediment quality. These issues can stem from an inadequate consideration of chemistry in biological studies, and vice versa. They will affect how well measurements of contaminant concentrations reflect what organisms are exposed to in the field, and how well laboratory toxicity tests are representative of field toxicity and ultimately of benthic community effects. It is important to be aware of these uncertainties and how best to undertake sediment sampling, handling, and testing to minimize their effects. Many uncertainties were unknown when many of the sediment databases were compiled, and awareness of them continues to increase (e.g., Batley et al. 2002; Simpson and Batley 2003).

For example, numerous studies have dealt with the artifacts that may result from differences in sediment sampling and handling (e.g., Burton 1991; Environment Canada 1994; USEPA 2001; Diamond et al. 2002), especially as they affect redox status and metal chemistry, with resultant alteration of metal availability. The sensitivity of metal chemistry is largely related to the effects of iron(II) and sulfide oxidation on partitioning of metals into pore waters, with significant effects on measured porewater concentrations (Simpson and Batley 2003). Oxidation processes can result in pH lowering of as much as 1 pH unit, which mobilizes metals into pore waters, thus overestimating their potential toxicity (Simpson and Batley 2003).

To avoid such changes that result from sampling, mixing, compositing, or homogenizing, re-equilibration for periods in excess of 14 d is required to re-establish a redox gradient before any sampling of pore waters or toxicity testing. As already discussed, there are issues of sediment heterogeneity and the importance that localized contaminant concentrations can have in determining a biological response in a system compared to an averaged response from a homogenized sample. The depth to which sediments should be sampled will also be important. In evaluating sediment quality, this depth should be relevant to the receptor organisms of concern. Consideration should also be given to the ability of the organism itself to modify pore water and sediment chemistry, for example, through burrow irrigation and the introduction of oxygenated waters at depth.

Sediment chemistry at a site is dependent on depositional, physical, and biological (bioturbation) mixing; bioirrigation; remobilization; and erosion events. Temporal changes may have a significant effect on sediment composition. Natural deposition is typically slow (on the order of mm/y), but this can be event driven. Organism impacts on sediments occur on a much shorter time scale (d).

The uncertainty issues with respect to sediment sampling, storage, and toxicity testing were considered in more detail at an earlier Pellston Workshop on Ecological Risk Assessment of Contaminated Sediment (Ingersoll et al. 1997), and as part of a workshop on Weight of Evidence Approaches to Sediment Quality Assessment (Batley et al. 2002).

The sampling, handling, and testing of pore waters which also represents a major source of uncertainties, has already been mentioned (in "Variations in toxicity test procedures and their interpretation," p 73). Selecting the appropriate methods for sampling pore waters for the purpose of porewater toxicity testing was the subject of a Pellston workshop (Carr and Nipper 2003). Issues include the stability and representativeness of porewater samples, the possibility for chemical changes, and particularly the difficulty of maintaining strictly anoxic conditions. As much as 50% of porewater metals can be lost from solution over a 24-h period because of coprecipitation with oxidized iron (Simpson and Batley 2002). More fundamental issues regarding the meaning of the porewater test itself have been raised by Chapman et al. (2002). In nature, pore waters often contain no oxygen but appreciable sulfide and ammonia; these chemicals (or their absence in the case of oxygen) can be the cause of observed toxicity in porewater tests. To avoid such undesirable toxic effects, removal of these toxic agents by oxidation or volatilization is encouraged (Carr and Nipper 2003). In doing so, pore water is transformed and is no longer representative of the original natural system that was to be tested. Furthermore, in nature, many benthic organisms irrigate their burrows and, by doing so, are not necessarily exposed to pore water. Many such organisms do not continually irrigate their burrows, and thus experience a diffusion gradient of contaminants from the surrounding pore waters. This process is rapid if strong gradients and short path lengths are involved. Consideration of these issues challenges the utility of pore water tests or at least points out to their limitation as an assessment tool for sediment quality.

Uncertainties in derived sediment quality guidelines

Uncertainty associated with experimental methodologies, and the associated precision (or lack thereof) in the experimental data, is often overlooked in both the derivation and the application of SQGs. The derived numbers depend on the number of data points and the distribution of data between high- and low-sensitivity endpoints. Associated with each endpoint is a further level of experimental uncertainty common to most toxicity tests (greater than ±20%), together with errors associated with the chemical analyses, which could also be considerable (2% to 10%), depending on the particular analyte and the concentrations being determined. Evaluations of SQG performance often consider agreement within a factor of 2 or 3 to indicate similarity or successful prediction.

Even if these uncertainties are acknowledged by the developers of SQGs, they do not always accompany the SQGs when they are incorporated into assessments or decision frameworks. The potential for inherent uncertainties to be lost can be further aggravated by the common practice of reporting SQGs to 3 or 4 significant figures; from this, some may infer that very small differences in sediment chemistry have great significance, when in fact they may not. As an illustration, the logistic regression models developed by Field et al. (1999) yield a 50% effects concentration (T50) for Zn of 245 mg/kg dry wt.; therefore, within the data set modeled, a sediment with that Zn concentration would have a 50.0% chance of being toxic. However, even though this T50 value can be calculated to 3 (or more) significant figures, small changes in Zn concentration do not substantially affect the predicted probability of toxicity. Altering this Zn concentration by ±1 mg Zn/kg yields modeled probabilities of 49.9% and 50.1% for 244 and 246 mg Zn/kg dry wt., respectively. For ±10 mg Zn/kg, the modeled probabilities are 48.5% and 51.5% (235 and 255 mg Zn/kg dry wt.). Even for changes of ±100 mg Zn/kg, the modeled probabilities only change to 31.9% and 62.2% (145 and 345 mg Zn/kg dry wt.). From these examples, it is clear that it would be inappropriate to treat SQGs as a razor edge separating "safe" from "unsafe" or "toxic" from "nontoxic." They are useful as landmarks within the continuum of sediment contamination, but they are not black and white decision points because of their inherent variability and uncertainty. It is important that this reality is recognized in regulatory applications of SQGs.

In determining the acceptability of EqP approaches, the concentration axis of the concentration-toxicity plots using a log scale more realistically reflects the acknowledged uncertainties. In EqP research, agreement within a factor of 2 between predicted and observed toxicological effects was considered good supporting evidence for the validity of prediction methods. In the development of consensus guidelines for PAHs, Swartz (1999) found that values for a particular effect level (e.g., threshold effects) from different SQGs clustered within factors of 2 to 4, and concluded that the "clusters appear to represent independent estimates of the same PAH concentration."

Uncertainties in data use and analysis

Retrospective analysis of data included in various SQG data sets with respect to data quality is often difficult because of the absence of important supporting information. Compared to chemical analyses, the application of rigorous QA/QC to ecotoxicology has gained broad acceptance only in the past decade. Adherence to accepted or quality assured protocols is essential in all aspects of biological, chemical, or physical measurements. Other issues relate to the refinement of data, and the statistical analyses applied by researchers deriving the SQG, in order to assess whether it is over- or underprotective. This has been discussed previously in the context of understanding the evolution of scientific understanding of endocrine disruption in wildlife populations (Stahl and Clark 1998).

Data screening and classification forms an important part of empirical guideline development. For example, data-screening criteria in the past have included

- similarities in protocols used to collect sediment samples (Smith, Rathbun et al. 1996),
- protocols for conducting sediment toxicity tests in the laboratory using the field-collected samples, and
- chemical analysis techniques (Long et al. 1995, 1998).

A number of these criteria appear to be met in most cases where the work has been conducted under the auspices of a single governmental agency (e.g., NOAA's Status and Trends Program; USEPA's Environmental Monitoring and Assessment Program Estuaries). The BEDS database and the Field et al. (1999) database were developed using ASTM standards. Even so, there were instances in which screening criteria were not met fully, and some segments of the data set were excluded from consideration and thus not reflected in the final SQG. Where this occurs, the authors make clear which data were not considered, the reason for the exclusion, and any influence that exclusion might have had on the final SQG.

Similarly, once screened, the data can be subject to classification or groupings (toxic, not toxic, marginally toxic) depending on the author. Uncertainties arise when different authors choose to define "toxic" in different ways. In one case, toxic may be defined using a statistical comparison, in another as a percentage difference in survival or growth. More importantly, what is the basis for a particular classification: comparison to the laboratory control or to a reference sample? Making a comparison to the laboratory control in the case of toxicity testing is likely to smooth variability among the various laboratories where the testing was conducted. Making the comparison to a reference sample (assuming it is from the same area, yet unimpacted by the chemical stressors in question) may tend to smooth differences among regions. Or is the classification based on comparison to a reference location or on an SQG (Long et al. 1998, 2000; O'Connor et al. 1998)?

A recent study by Von Stackelberg and Menzie (2002) has cautioned against the exclusive reliance on statistical and mathematical relationships between invertebrate data and sediment concentrations to derive SQGs, as was done with the Ontario SQGs (Persaud et al. 1992). In their analysis of Canada's Provincial Sediment Quality Guidelines, Von Stackelberg and Menzie (2002) showed that the normal patchiness (presence or absence) of benthic invertebrates in a sediment sample can become a misleading correlate with the concentration of a particular chemical stressor. For example, low densities of those invertebrates in a sediment sample could result from physical, chemical, or biological parameters. In small sample sizes or small data sets, these species can appear to be absent much more frequently than will other species. When these same species' presence or absence is then compared with the concentration of a particular chemical stressor, a statistically significant correlation between the two may result. Without an underlying knowledge of the patchiness of the particular species in the sediment samples, these correlations have little meaning biologically, because they are simply statistical artifacts. Hence, an additional level of uncertainty is introduced into those SQGs that are built upon species' presence or absence and upon their co-location with a particular chemical stressor.

State of the Science of Sediment Quality Guidelines as It Relates to Environmental Decision Making

In the ideal case, SQGs would be able to unequivocally delineate between sediments that cause biological effects and those that do not, that is, all sediments below the SQG value would not show effects, while all those above would show effects (Figure 3-1a). In reality, the occurrence of biological effects does not show such a clearly delineated relationship. Instead, the distribution of biological effects generally shows a relationship characterized by ranges of chemical concentrations where biological effects are rare, where cases of both effects and non-effects are found, and where biological effects essentially always occur (Figure 3-1b). Normalization of chemical stressor concentrations to organic carbon or other characteristics that control bioavailability of sediment contaminants may affect the distribution of effects and no-effects data, but no normalization technique has yet come close to complete discrimination (i.e., Figure 3-1a), nor is one likely to do so. Factors that cause overlap between effect and no-effect data are many, including contributions of other chemicals to effects, unaccounted-for differences in chemical availability, differences in response among organisms, and errors in measurement of either chemical concentrations or biological responses.

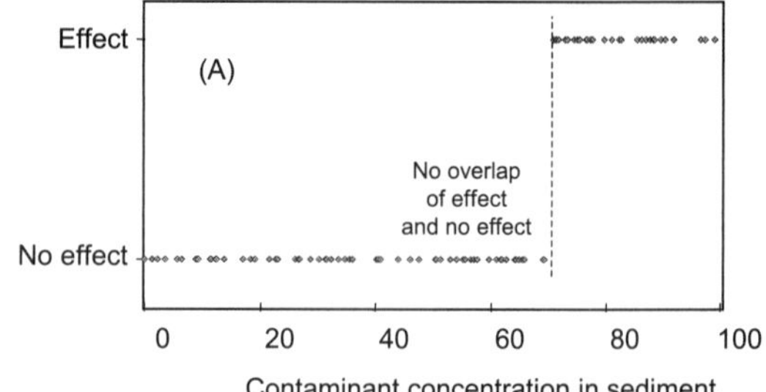

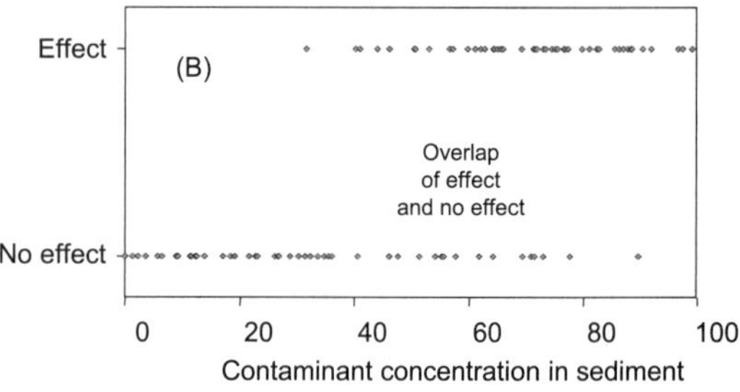

Figure 3-1 Examples of distributions of biological response data over a range of sediment contaminant concentrations: (A) Ideal case where there is no overlap between concentration ranges that do and do not show biological effects. (B) Typical actual distribution with overlap between these concentration ranges.

The various approaches to SQGs can be evaluated in relation to a generalized concentration-response model presented in Figure 3-2. In the case of contaminated sediments, this generalized model predicts the relation between the probability of biological effects and the concentration of chemical stressors in sediments. The approaches differ in

- the type of response displayed on the *y* axis,
- the chemical or mixture of chemicals whose concentration is shown on the *x* axis, and

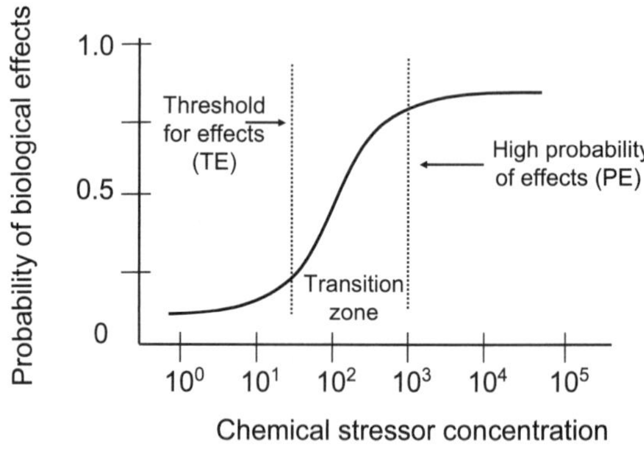

Figure 3-2 A generalized concentration–response model for contaminated sediments

- the concentration value (vertical line) that is chosen to indicate a level or probability of effects.

The selection of an SQG value reflects the intended use of the guideline and the method of its derivation.

The biological response to exposure to chemical stressors in sediments will vary with the organisms chosen and the endpoints measured. Results may represent an aggregated data set over a range of organisms, responses, and sites. The observations thus encountered may be due to a mixture of contaminants, possibly exacerbated by the effect of other stressors, or to a single chemical or class of chemicals. Generally speaking, it is not known which of these is accurate, that is, due to a single chemical stressor or mixture, without further evaluation. The concentration of this chemical stressor may be expressed as the total concentration in the sediment or some measure of the bioavailable fraction. All of these issues pose challenges for the further development of SQGs and need to be understood when SQGs are used in decision making.

By definition, biological effects are rare at concentrations below the threshold of effects (TE) (Figure 3-2). The probability of biological effects increases along the chemical gradient above the TE until it reaches a concentration above which effects have a high probability of occurring (PE). Additionally, the severity of the effect (e.g., reduced survival) increases as the probability of toxicity increases. The concentration range between the TE and PE represents a transition zone in which effects may or may not occur. The relationship between chemical concentration and biological effects always includes some gray area in which effects may or may not occur. Further discussions on the predictability of particular SQGs is provided in Chapters 4 and 12.

Although this illustration is an oversimplification of a very complex interaction between chemical stressors and benthic organisms, several critical issues in the derivation of SQGs are evident in the structure of the concentration-response model. First, all SQGs for individual chemicals can be represented as a concentration value (vertical line) on the model. The selection of the SQG value reflects the intended use of the guideline and the method of its derivation. The different sets of SQGs are often intended to represent TE, PE, or median effect (ME) levels (note that ME is not strictly a statistical median, but a qualitative expression of "mid-range"). A TE guideline is a concentration below which biological effects are expected to occur in very few cases. A PE guideline is proposed to identify the concentration above which effects are expected to occur in almost all cases. ME guidelines identify concentrations within the transition zone, with effects predicted to occur about half of the time. TE and PE guidelines cannot reliably predict effects within the transition zone for specific sediments. If exceedance of a TE guideline is used to predict effects, there will be a substantial number of false positives (i.e., biological effects will be predicted but not observed). This is a common source of error and frustration if TE guidelines are misunderstood or misapplied. Similarly, if failure to exceed a PE guideline is used to predict the absence of effects, there will be a substantial number of false negatives (i.e., biological effects observed when not predicted) within the transition zone. ME guidelines have been derived to indicate the position of a particular sediment within the transition zone; however, an ME guideline can only indicate the probability of effects. It cannot reliably predict the presence or absence of biological effects caused by chemical stressors in specific sediments.

The transition zone, as we have defined it, is a region of uncertainty. Unfortunately, it encompasses a substantial concentration range, 1 to 2 orders of magnitude for many chemical stressors (Di Toro et al. 1991). In addition, it tends to be the area where a great many scientists and decision makers find themselves during investigations of contaminated sediments. The size of the transition zone may be reduced by careful attention to sources of error, but it cannot be eliminated. It is an inherent property of the concentration–response model. Thus, it will be important to identify methods and approaches that will be employed to resolve the question of "effect or no effect" (e.g., toxicity testing) when potential exposures fall within this transition zone, particularly where risk management decisions may entail substantial expenditures. Ultimately, scientists and decision makers will have to decide how much certainty is needed to achieve an appropriate ecologically based risk management action (Stahl et al. 2001).

In summary, there are numerous sources of uncertainty and error in the derivation of the concentration–response relations that form the scientific basis of all SQGs. Diverse methods have been implemented to quantify biological effects and chemical concentrations. None of them can comprehensively and unequivocally delineate "safe" from "unsafe." The uncertainty of the transition zone (Figure 3-2) is inherent in all concentration-response relations and cannot be avoided. Despite these limitations, there may be strength in the diversity of SQGs that have been proposed. A weight-of-evidence (WOE) framework that considers multiple SQGs may be effective (Chapter 6). Several TE and PE SQGs could be used to define the limits of the transition zone. Indeed, the recent development of consensus SQGs indicates that many guidelines have generated similar estimates for TE and ME values. The absence and presence of adverse biological effects could be reliably predicted below the TE and above the PE, respectively, with the appropriate caveats expressed by Von Stackelberg and Menzie (2002). ME SQGs could indicate the presence of specific sediments within the transition zone and predict the probability of biological effects. This may be the limit of the effectiveness of SQGs. Site-specific assessment tools may be required in combination with SQGs to document the actual presence or absence of adverse effects at sites that fall within the transition zone of the concentration-response model (Chapters 4 and 5).

It is important to reinforce that SQGs generally predict ranges of sediment contamination where effects occur, but that does not mean these effects will always happen, and this prediction must be consistent with the application of guidelines. SQGs can place sediments in ranges of likely effect, but these categories will always have exceptions, and the types of "errors" that are more or less likely vary with the guideline. If it is essential that an application have a functionally low error rate, then site-specific evaluation will be necessary.

Conclusions and Recommendations

The preceding discussions have established that, generally, sound science underpins SQGs; however, there may be uncertainties in the particular numerical values that are revealed with increasing understanding of methodological issues and the effects that these might have on the results of toxicity testing. Many of the problems with SQGs have indeed arisen in their application, when the limitations associated with their derivation, and the fact that they may not always be applicable to single contaminants have been overlooked or poorly understood by scientists and decision makers. Most notable has been the application of empirical guidelines, derived on the basis of co-occurrence, to single or limited numbers of contaminants.

There are differing views as to whether SQGs should be protective of aquatic ecosystems or predictive of toxicity. The predictability of SQGs has been discussed in detail in Chapter 4. Given the uncertainties in their derivation and the limited range of organisms used in toxicity tests in these derivations, there may be questionable comfort in the assurance of ecosystem protection based on these predictions of toxicity or nontoxicity. Using SQGs as screening tools is far more defensible. Incorporating them into tiered frameworks or WOE approaches is discussed in Chapter 5.

Expectations of the reliability of SQGs often seem unreasonable. Guidance should be expected from SQGs, not unequivocal certainty. These are all imperfect methods, but they do have some utility. At a minimum, a combination of toxicological and chemical assessments should allow a more reliable classification of the vast majority of sediments, which do not pose serious environmental risks, while also correctly classifying the worst cases, for which there is little question about the severity of unacceptable contamination. Inevitably, there will be a range of uncertainty for which SQGs alone cannot adequately assess the sediment risk. Site-specific investigations of ecology, chemistry, and toxicology will be needed in such cases.

Consensus points from the workgroup

- SQGs may be most effective for identifying ranges in which effects are unlikely, uncertain, and highly likely (see Figure 3-2).
- A number of factors that affect organism response to contaminated sediments are not, or not fully, captured by existing SQG derivations.
- Sediments causing biological effects almost always contain mixtures of contaminants; to be effective, any SQG method must account for this reality. Applying empirical SQGs for a single chemical may falsely attribute effects of a mixture to that chemical. Conversely, applying SQGs based on causality may greatly underestimate effects of mixtures in field sediments.
- All SQGs have uncertainty, which must be recognized in their interpretation or application. Despite the uncertainties involved in the derivation of individual SQGs, those with independent derivation methods and assumptions often converge, suggesting important underlying relationships.
- A high research priority is to develop methods to derive SQGs from the in situ effects of sediment contaminants on benthic ecosystems. Discriminating the effects of toxic chemicals from those of natural stressors (e.g., grain size, organic enrichment, salinity, natural concentrations of chemicals) is essential in this research. Research is also needed to develop single-contaminant SQGs based on spiked sediments, or evaluation of field samples in which there are a limited number of contaminants. The development of endpoint-specific SQGs based on chronic toxicity endpoints for sensitive species is also needed.
- Despite the above limitations, there may be strength in the diversity of SQGs that have been proposed, particularly if they are derived using different databases.
- In addition to SQGs, site-specific assessment tools may be required to document the actual presence or absence of adverse effects at sites that fall within the transition zone of the concentration-response model.

- In summary, there are 3 elements to the path forward:
  1) Understand the basis and limitations of SQGs, and make their applications coherent with both.
  2) Enhance our understanding of the factors that contribute to uncertainty in the transition zone, and work to incorporate those factors into better SQG approaches.
  3) Acknowledge the reality of the transition zone, and develop complementary assessment approaches to better define the actual biological effects occurring in the field.

The way forward

The above consensus points highlight the focus needed from future research. With respect to the appropriate uses of SQGs, it is apparent that the misapplication of SQGs relates to a regulatory need. One of the biggest issues with respect to SQGs is that of defining "safe" concentration of contaminants, and this requires causality of biological effect to be able to be ascribed to a particular sediment contaminant. There is a need to deal equally with typical mixtures and single contaminants or less typical combinations of contaminants. Relating single contaminant SQGs to those of mixtures will be important. Developing SQGs for single contaminants will require spiked sediment tests, possibly using a wider range of species than is typically applied, covering different exposure pathways and sensitivities, to generate an SQG that is protective of a given percentage of species (e.g., 95%).

Investigators should ensure that future field data that underpin SQGs be qualified by a set of desirable additional data that can be used to better refine the existing empirical SQGs. The same might apply to spiked sediment tests. Regrettably, these supporting data were not often included in studies for existing databases, so that retrospective reevaluation of the data is, in many cases, not possible. Just what these parameters might be needs to be defined: total contaminants in sediment, total concentrations in pore water, AVS/SEM, total organic carbon, grain size, porewater pH (hardness, salinity), or spatial heterogeneity of sediments. Methodological issues such as deciding whether to homogenize sediments, remove large particles, include or remove predators and competitors, or sample pore waters are critical for comparing data. Site-specific assessments will always be necessary.

It has been demonstrated that life history attributes of benthic animals (e.g., feeding habits, microhabitat) can be a major determinant of their exposure and response to sediment contaminants. We know little about the tube-building and irrigation behavior of benthic animals and about their feeding behavior. Such information is needed to determine the appropriate exposure of these animals in nature and to identify those species at maximum risk. The degree to which these factors affect the accuracy or applicability of SQGs needs to be quantified and, where practical, incorporated into laboratory tests, field measures, and SQGs.

References

Adams MS, Stauber JL. 2004. Development of a whole sediment bioassay using a marine benthic microalga. *Environ Toxicol Chem* 23:1957–1968.

Adams WJ. 1987. Bioavailability of neutral, lipophilic organic chemicals contained on sediments: A review. In: Dickson KL, Maki AW, Brungs WA, editors. Fate and effects of sediment-bound chemicals in aquatic systems. New York: Pergamon. p 219–245.

Adams WJ, Kimerle RA, Mosher RG. 1985. Aquatic safety assessment of chemicals sorbed to sediments. In: Cardwell RD, Purdy R, Bahner RC, editors. Aquatic toxicology and hazard assessment: Seventh Symposium. Philadelphia: American Soc for Testing and Materials. STP 854. p 429–453.

Akcakaya HA, Burgman MA, Ginzburg LR. 1999. Applied population ecology: Principles and computer exercises using RAMAS EcoLab. Sunderland (MA): Sinauer Assoc.

Aldenberg T, Jaworska JS. 2000. Uncertainty of the hazardous concentration and fraction affected by normal species sensitivity distributions. *Ecotoxicol Environ Saf* 46:1–18.

Aldenberg T, Slob W. 1993. Confidence limits for hazardous concentrations based on logistically distributed NOEC toxicity data. *Ecotoxicol Environ Saf* 25:48–63.

Allen HE. 1996. Standards for metals should not be based on total concentration. *SETAC News* 16(6):18–19.

Aller JY, Aller RC. 1986. Evidence for localized enhancement of biological activity associated with tube and burrow structures in deep-sea sediments at the HEBBLE site, western North Atlantic. *Deep-Sea Res* 33:755–790.

Aller RC, Yingst JY. 1978. Biogeochemistry of tube-dwellings: A study of the sedentary polychaete *Ampritrite ornata* (Leidy). *J Mar Res* 36:201–254.

Anderson RL. 1980. Chironomidae toxicity tests - biological background and procedures. In: Buikema Jr AL, Cairns Jr J, editors. Aquatic invertebrate toxicity tests. Philadelphia: American Soc for Testing and Materials. ASTM STP 715. p 70–80.

Ankley GT, Di Toro DM, Hansen D, Berry W. 1996. Technical basis and proposal for deriving sediment quality criteria for metals. *Environ Toxicol Chem* 15:2056–2066.

Ankley GT, Schubauer-Berigan M, Dierkes J, Lukasewycz M. 1992. Sediment toxicity identification evaluation: Phase I (characterization), Phase II (identification) and Phase III (confirmation) modifications of effluent procedures. Duluth (MN): US Environmental Protection Agency, Environmental Research Laboratory. Draft technical report. EPA 08-91.

[ASTM] American Society for Testing and Materials. 2002a. Standard guide for determination of bioaccumulation of sediment-associated contaminants by benthic invertebrates. ASTM annual book of standards, Volume 11.05. West Conshohocken (PA): ASTM. ASTM E1688-00a.

[ASTM] American Society for Testing and Materials. 2002b. Standard guide for conducting 10-day static sediment toxicity tests with marine and estuarine amphipods. ASTM annual book of standards, Volume 11.05. West Conshohocken (PA): ASTM. ASTM E1367-99.

[ASTM] American Society for Testing and Materials. 2002c. Standard guide for conducting sediment toxicity tests with polychaetous annelids. ASTM annual book of standards, Volume 11.05. West Conshohocken (PA): ASTM. ASTM E1611-00.

[ASTM] American Society for Testing and Materials. 2002d. Standard test methods for measuring the toxicity of sediment-associated contaminants with freshwater invertebrates. ASTM annual book of standards, Volume 11.05, West Conshohocken (PA): ASTM. ASTM E1706-00.

Baird DJ, Burton GA. 2001. Ecological variability: Separating natural from anthropogenic causes of ecosystem impairment. Pensacola (FL): Society of Environmental Toxicology and Chemistry (SETAC). 336 p.

Barrick R, Becker S, Brown L, Beller H, Pastorok R. 1988. Sediment quality values refinement: 1988 Update and evaluation of Puget Sound AET, Volume 1. Bellevue (WA): PTI Environmental Services. 74 p.

Bartell SM, Gardner RH, O'Neill, RV. 1992. Ecological risk estimation. Chelsea (MI): Lewis. 252 p.

Batley GE, Burton, GA, Chapman, PM, Forbes VE. 2002. Uncertainties in sediment quality weight of evidence assessments. *Hum Ecol Risk Assess* 8:1517–1547.

Benoit DA, Sibley PK, Juenemann JL, Ankley GT. 1997. *Chironomus tentans* life-cycle test: Design and evaluation for use in assessing toxicity of contaminated sediments. *Environ Toxicol Chem* 16:1165–1176.

Berry WJ, Hansen DH, Mahony JD, Robson DL, Di Toro DM, Shipley BP, Rogers B, Corbin JM, Boothman WS. 1996. Predicting the toxicity of metal-spiked laboratory sediments using acid-volatile sulfide and pore water normalizations. *Environ Toxicol Chem* 15:2067–2079.

Besser JM, Ingersoll, CG, Leonard E, Mount DR. 1998. Effect of zeolite on toxicity of ammonia in freshwater sediments: Implications for sediment toxicity identification evaluation procedures. *Environ Toxicol Chem* 17:2310–2317.

Boese BL, Winsor M, Lee H, Echols S, Pelletier J, Randall R. 1995. PCB congeners and hexachlorobenzene biota sediment accumulation factors for *Macoma nasuta* exposed to sediments with different total organic carbon contents. *Environ Toxicol Chem* 14:303–310.

Borchert J, Karbe L, Westendorf J. 1997. Uptake and metabolism of benzo(*a*)pyrene absorbed to sediment by the freshwater invertebrate species *Chironomus riparius* and *Sphaerium corneum. Bull Environ Contam Toxicol* 58:158–165.

Bremle G, Ewald G. 1995. Bioconcentration of polychlorinated biphenyls (PCBs) in chironomid larvae, oligochaete worms and fish from contaminated lake sediment. *Mar Freshw Res* 46:267–273.

Brumbaugh WG, Ingersoll CG, Kemble NE, May TW, Zajicek JL. 1994. Chemical characterization of sediments and pore water from the upper Clark Fork River and Milltown Reservoir, Montana. *Environ Toxicol Chem* 13:1971–1994.

Bull DC, Williamson RB. 2001. Prediction of principal metal binding solid phases in sediments from color image analysis. *Environ Sci Technol* 35:1658–1662.

Burgess RM, Cantwell MG, Pelletier MC, Ho KT, Serbst JR, Cook HF, Kuhn A. 2000. Development of a toxicity identification evaluation (TIE) procedure for characterizing metal toxicity in marine sediments. *Environ Toxicol Chem* 19: 982–991.

Burton Jr GA. 1991. Assessing freshwater sediment toxicity. *Environ Toxicol Chem* 10: 1585–1627.

Canfield TJ, Dwyer FJ, Fairchild JF, Haverland PS, Ingersoll CG, Kemble NE, Mount DR, La Point TW, Burton GA, Swift MC. 1996. Assessing contamination in Great Lakes sediments using benthic invertebrate communities and the sediment quality triad approach. *J Gt Lakes Res* 22:565–583.

Canfield TJ, Kemble NE, Brumbaugh WG, Dwyer FJ, Ingersoll CG, Fairchild JF. 1994. Use of benthic invertebrate community structure and the sediment quality triad to evaluate metal contaminated sediment in the Upper Clark Fork River. *Environ Toxicol Chem* 13:1999–2012.

Canton SP, Van Derveer WD. 1997. Selenium toxicity to aquatic life: An argument for sediment-based water quality criteria. *Environ Toxicol Chem* 16:1255–1259.

Carr RS, Chapman DC. 1992. Comparison of solid-phase and pore-water approaches for assessing the quality of marine and estuarine sediments. *Chem Ecol* 7:19–30.

Carr RS, Chapman DC. 1995. Comparison of methods for conducting marine and estuarine sediment pore water toxicity tests-extraction, storage and handling techniques. *Arch Environ Contam Toxicol* 28:69–77.

Carr RS, Nipper M, editors. 2003. Porewater toxicity testing: Biological, chemical, and ecological considerations. Pensacola (FL): Society of Environmental Toxicology and Chemistry (SETAC).

Caswell H. 2001. Population matrix population models: Construction, analysis and interpretation. Sunderland (MA): Sinauer Associates Inc.

Centre Saint-Laurent. 1992. Critères intérimaires pour l'évaluation de la qualité des sédiments du Saint-Laurent. Quebec City (PQ): Environment Canada and Québec Ministry of Environment. 28 p.

Chapman PM, Wang F, Germano DD, Batley GE. 2002. Pore water testing and analysis: The good, the bad and the ugly. *Mar Pollut Bull* 16:969–972.

Chappie DJ, Burton GA. 1997. Optimization of in situ toxicity tests with Hyalella azteca and Chironomus tentans. *Environ Toxicol Chem* 16:559–564.

Crommentuijn T, Sijm D, de Bruijn J, van den Hoop M, van Leeuwen K, van de Plassche E. 2000. Maximum permissible and negligible concentrations for metals in the Netherlands, taking into account background concentrations. *J Environ Manag* 60:121–143.

Cubbage J, Batts D, Briedenbach S. 1997. Creation and analysis of freshwater sediment quality values in Washington State. Olympia (WA): Environmental Investigations and Laboratory Services Program, Washington Department of Ecology.

Defoe DL, Ankley GT. 1998. Effect of storage time on toxicity of freshwater sediments to benthic macroinvertebrates. *Environ Pollut* 99:123–131.

DeWitt TH, Ditswoth GR, Swartz RC. 1988. Effects of natural sediment features on survival of the phoxocephalid amphipod, *Rhepoxynius abronius. Mar Environ Res* 25:99–124.

Diamond J, Burton Jr GA, Scott J. 2002. Procedures for collection, storing, and manipulating sediments and interstitial waters for chemical and toxicological analyses. In: Whittemore R, editor. Handbook on sediment quality. Fairfax (VA): Water Environment Federation. p 139–197.

DiPinto LM, Coull BC. 1997. Trophic transfer of sediment-associated polychlorinated biphenyls from meiobenthos to bottom-feeding fish. *Environ Toxicol Chem* 16:2568–2575.

Di Toro DM, McGrath JA. 2000. Technical basis for narcotic chemicals and polycyclic aromatic hydrocarbon criteria. II. Mixtures and sediments. *Environ Toxicol Chem* 19:1971–1982.

Di Toro DM, Mahony JD, Hansen DJ, Scott KJ, Hicks MB, Mayr SM, Redmond MS. 1990. Toxicity of cadmium in sediments: The role of acid volatile sulfide. *Environ Toxicol Chem* 9:1487–1502.

Di Toro DM, Zarba CS, Hansen DJ, Berry WJ, Swartz RC, Cowan CE, Pavlou SP, Allen HE, Thomas NA, Paquin PR. 1991. Technical basis for establishing sediment quality criteria for nonionic organic chemicals using equilibrium partitioning. *Environ Toxicol Chem* 10:1541–1583.

Di Toro DM, McGrath JA, Hansen JD. 2000. Technical basis for narcotic chemicals and polycyclic aromatic hydrocarbon criteria. I. Water and tissue. *Environ Toxicol Chem* 19:1951–1970.

Dolan JR. 1991. Guilds of ciliate microzooplankton in the Chesapeake Bay. *Estuar Coast Shelf Sci* 33:137–152.

Dzombak DA, Morel FMM. 1990. Surface complexation modeling. New York: Wiley.

Environment Canada. 1992. Biological test method: Acute test for sediment toxicity using marine or estuarine amphipods. Ottawa (ON): Environment Canada, Conservation and Protection. EPS 1/RM/26. 83 p.

Environment Canada. 1994. Guidance document for preparation of sediments for physicochemical characterization and biological testing. Ottawa (ON): Environment Canada, Conservation and Protection. EPS 1/RM/29.

Environment Canada. 2001. Biological test method: Test for survival and growth in sediment using spionid polychaete worms (*Polydora cornuta*). Ottawa (ON): Environment Canada, Conservation and Protection. EPS 1/RM/41.

Fairey R, Long ER, Roberts CA, Anderson BS, Phillips BM, Hunt JW, Puckett HR, Wilson CJ. 2001. An evaluation of methods for calculating mean sediment quality guideline quotients as indicators of contamination and acute toxicity to amphipods by chemical mixtures. *Environ Toxicol Chem* 20:2276–2286.

Fenchel T. 1996. Worm burrows and oxic microniches in marine sediments. I. Spatial and temporal scales. *Mar Biol* 127:289–295.

Ferraro SP, Cole FA. 2002. A field validation of two sediment-amphipod toxicity tests. *Environ Toxicol Chem* 21:1423–1437.

Field LJ, MacDonald DD, Norton SB, Ingersoll CG, Severn CG, Smorong D, Lindskoog R. 2002. Predicting amphipod toxicity from sediment chemistry using logistic regression models. *Environ Toxicol Chem* 21:1993–2005.

Field LJ, MacDonald DD, Norton SB, Severn CG, Ingersoll CG. 1999. Evaluating sediment chemistry and toxicity data using logistic regression modeling. *Environ Toxicol Chem* 18:1311–1322.

Fisher JB. 1982. Effects of macrobenthos on the chemical diagenesis of freshwater sediments. In: McCall PL, Tevesz MJS, editors. Animal-sediment relations. New York: Plenum Pr. p 177–218.

Foran JA, Ferenc SA. 1999. Multiple stressors in ecological risk and impact assessment. Pensacola (FL): Society of Environmental Toxicology and Chemistry (SETAC).

Fox GA. 1991. Practical causal inference for ecoepidemiologists. *J Toxicol Environ Health* 33:359–373.

Glud RN, Gundersen JK Ramsing NB. 2000. Electrochemical and optical oxygen microsensors for in situ measurements. In: Buffle J, Horvai G, editors. In situ monitoring of aquatic systems. New York: Wiley.

Hansen DJ, Berry WJ, Mahony JD, Boothman WS, Di Toro DM, Robson DL, Ankley GT, Ma D, Yan Q, Pesch CE. 1996. Predicting the toxicity of metal contaminated field sediments using pore water concentrations of metals and acid volatile sulfide normalizations. *Environ Toxicol Chem* 15:2080–2094.

Hansen DJ, Mahony JD, Berry WJ, Benyi SJ, Corbin JM, Pratt DM, Di Toro DM, Able MB. 1996. Chronic effect of cadmium in sediments on colonization by benthic marine organisms: An evaluation of the role of pore cadmium and acid-volatile sulfide in biological availability. *Environ Toxicol Chem* 15:2126–2137.

Hare L, Tessier A, Warren L. 2001. Cadmium accumulation by invertebrates living at the sediment-water interface. *Environ Toxicol Chem* 20:880–889.

Harkey GA, Lydy MJ, Kukkonen J, Landrum PF. 1994. Feeding selectivity and assimilation of PAH and PCB in Diporeia spp. *Environ Toxicol Chem* 13:1445–1455.

Harkey GA, Van Hoof PL, Landrum PF. 1996. Bioavailability of polycyclic aromatic hydrocarbons from a historically contaminated sediment core. *Environ Toxicol Chem* 14:1551–1560.

Harris HJ, Wenger RB, Harris VA, Devault DS. 1994. A method for assessing environmental risk: A case study of Green Bay, Lake Michigan, USA. *Environ Manag* 18:295–306.

Hayward JMR. 2003. Field validation of long-term sediment toxicity tests [PhD thesis]. Columbia (MO): Univ Missouri.

Ho KT, Burgess RM, Pelletier MC, Serbst JR, Ryba SA, Cantwell MG, Kuhn A, Raczelowski P. 2002. An overview of toxicant identification in sediments and dredged materials. *Mar Pollut Bull* 44:286–293.

Hoss S, Henschel T, Haitzer M, Traunspurger W, Steinberg CEW. 2001. Toxicity of cadmium to *Caenorhabditis elegans* (Nematoda) in whole sediment and pore water: The ambiguous role of organic matter. *Environ Toxicol Chem* 20:2794–2801.

Hunt JW, Anderson B, Phillips BM, Tjeerdema R, Taberski KM, Wilson C, Puckett HM, Stephenson M, Fairey R, Oakden JM. 2001. A large-scale categorization of sites in San Francisco Bay, USA, based on the sediment quality triad, toxicity identification evaluations, and gradient studies. *Environ Toxicol Chem* 20:1252–1265.

Hyland JL, Van Dolah RF, Snoots TR. 1999. Predicting stress in benthic communities of southeastern U.S. estuaries in relation to chemical contamination of sediments. *Environ Toxicol Chem* 18:2557–2564.

Iannuzzi TJ, Harrington NW, Shear NM, Curry CL, Carlson-Lynch H, Henning MH, Su SH, Rabbe DE. 1996. Distributions of key exposure factors controlling the uptake of xenobiotic chemicals in an estuarine food web. *Environ Toxicol Chem* 15:1979–1992.

Ingersoll CG, Dillon T, Biddinger GR. 1997. Ecological risk assessment of contaminated sediments. SETAC Pellston Workshop on Sediment Ecological Risk Assessment; 1995 Apr 23–28; Pacific Grove, CA. Pensacola (FL): Society of Environmental Toxicology and Chemistry (SETAC). 390 p.

Ingersoll CG, Haverland PS, Brunson EL, Canfield TJ, Dwyer FJ, Henke CE, Kemble NE. 1996. Calculation and evaluation of sediment effect concentrations for the amphipod *Hyalella azteca* and the midge *Chironomus riparius*. *J Gt Lakes Res* 22:602–623.

Ingersoll CG, Hayward JMR, Jones JR, Wang N. 2004. A field assessment of long-term laboratory sediment toxicity tests with the amphipod *Hyalella azteca* and the midge *Chironomus tentans*. Washington DC: USEPA Office of Science and Technology. In review.

Ingersoll CG, Ivey CD, Brunson EL, Hardesty DK, Kemble NE. 2000. An evaluation of the toxicity: Whole-sediment vs. overlying-water exposures with the amphipod *Hyalella azteca. Environ Toxicol Chem* 19:2906–2910.

Ingersoll CG, MacDonald DD, Brumbaugh WG, Johnson BT, Kemble NE, Kunz JL, May TW, Wang N, Smith JR, Sparks DW, Ireland SD. 2002. Toxicity assessment of sediments from the Grand Calumet River and Indiana Harbor Canal in northwestern Indiana. *Arch Environ Contam Toxicol* 43:153–167.

Ingersoll CG, MacDonald DD, Wang N, Crane JL, Field LJ, Haverland PS, Kemble NE, Lindskkog RA, Severn C, Smorong DE. 2001. Predictions of sediment toxicity using consensus-based freshwater sediment quality guidelines. *Arch Environ Contam Toxicol* 41:8–21.

Ingersoll CG, Nelson MK. 1990. Testing sediments with *Hyalella azteca* (Amphipoda) and *Chironomus riparius* (Diptera). In: Landis WG, Van der Schalie WH, editors. Aquatic toxicology and risk assessment, 13th Volume. Philadelphia: American Soc for Testing and Materials. ASTM STP 1096. p 93–109.

Jenne EA, Di Toro DM, Allen HE, Zarba CS. 1986. An activity-based model for developing sediment criteria for metals. In: Lester JN, Perry R, Sterritt RM. editors. Proceedings International Conference on Metals in the Environment. London: Selper Ltd. p 560–568.

Jorgensen BB, Revsbech NP. 1985. Diffusive boundary layers and the oxygen uptake of sediments and detritus. *Limnol Oceanogr* 30:111–122.

Kristensen E. 1981. Direct measurement of ventilation and oxygen uptake in three species of *Tubicolous polychaetes* (*Nereis* spp.). *J Comp Physiol* 145:45–50.

Kristensen E. 1988. Benthic fauna and biogeochemical processes in marine sediments: Microbial activities and fluxes. In: Blackburn TH, Sorensen J. editors. Nitrogen cycling in coastal marine environments. New York: Wiley.

Kuhn A, Munns WRJ, Poucher S, Champlain D, Lussier S. 2000. Prediction of population-level response from mysid toxicity test data using population modeling techniques. *Environ Toxicol Chem* 19:2364–2371.

Kuhn A, Munns WRJ, Serbst, Edwards P, Cantwell MG, Gleason T, Pelletier MC, Berry W. 2002. Evaluating the ecological significance of laboratory response data to predict population-level effects for the estuarine amphipod *Ampelisca abdita. Environ Toxicol Chem* 21:865–874.

Landrum PF. 1989. Bioavailability and toxicokinetics of polycyclic aromatic hydrocarbons sorbed to sediments for the amphipod *Pontoporeia hoyi. Environ Sci Technol* 23:588–595.

Landrum PF, Robbins JA. 1990. Bioavailability of sediment-associated contaminants to benthic invertebrates. In: Baudo R, Enesy J, Muntau H, editors. Sediments: Chemistry and toxicology of in-place pollutants. Ann Arbor (MI): Lewis. p 237–236.

Langston WJ. 1980. Arsenic in U.K. estuarine sediments and its bioavailability to deposit-feeding bivalves. *J Mar Biol Assoc UK* 60:869–881.

Langston WJ. 1982. Distribution of mercury in British estuarine sediments and its bioavailability to deposit-feeding bivalves. *J Mar Biol Assoc UK* 62:667–684.

Lee BG, Griscom SB, Lee JS, Choi HJ, Koh CH, Luoma SN, Fisher NS. 2000. Influences of dietary uptake and reactive sulfides on metal bioavailability from aquatic sediments. *Science* 287:282–284.

Leppänen MT, Kukkonen JVK. 2000. Fate of sediment-associated pyrene and benzo[*a*]pyrene in the freshwater oligochaete Lumbriculus variegatus (Muller). *Aquat Toxicol* 49:199–212.

Leversee GJ, Giesy JP, Landrum PF, Gerould S, Bowling JW, Fannin TE, Haddock JD, Bartell SM. 1982. Kinetics and biotransformation of benzo(*a*)pyrene in *Chironomus riparius. Arch Environ Contam Toxicol* 11:25–31.

Liber K, Call DJ, Markee TP, Schmude KL, Balcer MD, Whiteman FW, Ankley GT. 1996. Effects of acid-volatile sulfide on zinc bioavailability and toxicity to benthic macroinvertebrates: A spiked-sediment field experiment. *Environ Toxicol Chem* 15:2113–2125.

Lion LW, Altmann RS, Leckie JO. 1982. Trace-metal adsorption characteristics of estuarine particulate matter: Evaluation of contribution of Fe/Mn oxide and organic surface coatings. *Environ Sci Technol* 16:660–666.

Long EE, Hameedi MJ, Sloane GM, Read LB. 2002. Chemical contamination, toxicity, and benthic community indices in sediments of the lower Miami River and adjoining portions of Biscayne Bay, Florida. *Estuaries* 25:622–637.

Long ER, Field LJ, MacDonald DD. 1998. Predicting toxicity in marine sediments with numerical sediment quality guidelines. *Environ Toxicol Chem* 17:714–727.

Long ER, Hong CB, Severn CE. 2001. Relationships between acute sediment toxicity in laboratory tests and the abundance and diversity of benthic infauna in marine sediments: A review. *Environ Toxicol Chem* 20:46–60.

Long ER, MacDonald DD. 1992. National status and trends program approach. In: Sediment classification methods compendium, Chapter 14. Washington DC: US Environmental Protection Agency, EPA 823-R-92-006.

Long ER, MacDonald DD. 1998. Perspective: Recommended uses of empirically derived, sediment quality guidelines for marine and estuarine ecosystems. *Hum Ecol Risk Assess* 4:1019–1039.

Long ER, MacDonald DD, Severn CG, Hong CB. 2000. Classifying the probabilities of acute toxicity in marine sediments with empirically-derived sediment quality guidelines. *Environ Toxicol Chem* 19:2598–2601.

Long ER, MacDonald DD, Smith SL, Calder FD. 1995. Incidence of adverse effects within ranges of chemical concentrations in marine and estuarine sediments. *Environ Manag* 19:81–97.

Long ER, Morgan LG. 1990. The potential for biological effects of sediment-sorbed contaminants tested in the National Status and Trends Program. Rockville (MD):

National Oceanic and Atmospheric Agency. Technical Memorandum NOS OMA 5.

Luoma SN. 1995. Prediction of metal toxicity in nature from toxicity tests: Limitations and research needs. In: Tessier A, Turner DR, editors. Metal speciation and bioavailability in aquatic systems. New York: Wiley.

Luoma SN, Bryan GW. 1978. Factors controlling the availability of sediment-bound lead to the estuarine bivalve Scrobicularia plana. *J Mar Biol Assoc UK* 58:793–802.

Luoma SN, Bryan GW. 1981. A statistical assessment of the form of trace metals in oxidized estuarine sediments employing chemical extractants. *Sci Total Environ* 17:165–196.

Luoma SN, Davis JA. 1983. Requirements for modelling trace metal partitioning in oxidized estuarine sediments. *Mar Chem* 12:159–181.

Luoma SN, Johns C, Fisher NS, Steinberg NA, Oremland RS, Reinfelder JR. 1992. Determination of selenium bioavailability to a benthic bivalve from particulate and solute pathways. *Environ Sci Technol* 26:485–491.

MacDonald DD, Carr RS, Calder FD, Long ER, Ingersoll CG. 1996. Development and evaluation of sediment quality guidelines for Florida coastal waters. *Ecotoxicology* 5:253–278.

MacDonald DD, DiPinto LM, Field J, Ingersoll CG, Long ER, Swartz RC. 2000. Development and evaluation of consensus-based sediment effect concentrations for polychlorinated biphenyls. *Environ Toxicol Chem* 19:1403–1413.

MacDonald DD, Ingersoll CG, Berger TA. 2000. Development and evaluation of consensus-based sediment quality guidelines for freshwater ecosystems. *Archiv Environ Contam Toxicol* 39:20–31.

Malek J. 1992. Apparent effects threshold approach. In: Sediment classification methods compendium, Chapter 11. Washington DC: US Environmental Protection Agency. EPA 823-R-92-006.

Matisoff G. 1995. Effects of bioturbation on solute and particle transport in sediments. In: Allen HE, editor. Metal contaminated aquatic sediments. Ann Arbor (MI): Ann Arbor Pr. p 201–272.

Matisoff G, Wang X. 1998. Solute transport in sediments by freshwater infaunal bioirrigators. *Limnol Oceanogr* 43:1487–1499.

Mayer LM, Chen Z, Findlay RH, Fang J, Sampson S, Self RFL, Jumars PA, Quetel C, Donard OFX. 1996. Bioavailability of sedimentary contaminants subject to deposit-feeder digestion. *Environ Sci Technol* 30:2641–2645.

McGee BL, Fisher DJ, Yonkos LT, Ziegler GP, Turley S. 1999. Assessment of sediment contamination, acute toxicity and population viability of the estuarine amphipod *Leptocheirus plumulosus* in Baltimore Harbor, Maryland, USA. *Environ Toxicol Chem* 18:2151–2160.

Means JC, McElroy AE. 1997. Bioaccumulation of tetrachlorobiphenyl and hexachlorobiphenyl congeners by *Yoldia limatula* and *Nephtys incisa* from bedded sediments: Effects of sediment and animal-related parameters. *Environ Toxicol Chem* 16:1277–1286.

Meyers MB, Fossing H, Powell EN. 1987. Microdistribution of pore meiofauna, oxygen and sulfide gradients, and the tubes of macro-infauna. *Mar Ecol Prog Ser* 35:223–241.

Meyers MB, Powell EN, Fossing H. 1988. Movement of oxybiotic and thiobiotic meiofauna in response to changes in pore-water oxygen and sulfide gradients around macro-infaunal tubes. *Mar Biol* 98:395–414.

Munger C, Hare L. 1997. Relative importance of water and food as cadmium sources to an aquatic insect (*Chaoborus punctipennis*): Implications for predicting Cd bioaccumulation in nature. *Environ Sci Technol* 31:891–895.

Munger C, Hare L, Tessier A. 1999. Cadmium sources and exchange rates for *Chaoborus* larvae in nature. *Limnol Oceanogr* 44:1763–1771.

Neff JM, Bean DJ, Cornaby BW, Vaga RM, Gulbransen TC, Scanlon JA. 1986. Sediment quality criteria methodology validation: Calculation of screening level concentrations from field data. Washington DC: Report to US Environmental Protection Agency, Office of Water Regulation and Standards. 60 p.

Neff JM, Word JQ, Gulbransen TC. 1987. Recalculation of screening level concentrations for nonionic organic contaminants in marine sediments. Final Report. Washington DC: US Environmental Protection Agency, Office of Water.

Nipper MG, Roper DS, Williams EK, Martin ML, Van Dam LF, Mills GN. 1998. Sediment toxicity and benthic communities in mildly contaminated mudflats. *Environ Toxicol Chem* 17:502–510.

[NRC] National Research Council. 2003. Bioavailability of contaminants in soils and sediments: processes, tools and applications. Committee on Bioavailability of Contaminants in Soils and Sediments, National Research Council, Washington DC: National Academies Pr. 420 p.

O'Conner TP, Daskalakis KD, Hyland JL, Paul JF, Summers JK. 1998. Comparisons of sediment toxicity with predictions based on chemical guidelines. *Environ Toxicol Chem* 17:468–471.

Paine MD, Chapman PM, Allard PJ, Murdoch MH, Minifie DJ. 1996. Limited bioavailability of sediment PAH near an aluminum smelter: Contamination does not equal effects. *Environ Toxicol Chem* 15:2003–2018.

Parkhurst BR, Warren Hicks WJ, Creager CS. 1997. Methods for assessing watershed scale aquatic risks for multiple stressors. In Dwyer FJ, Doane TR, Hinman ML, editors. Environmental toxicology and risk assesment: Modelling and risk assessment, Volume 6. Philadelphia: Amer Soc Testing and Materials. ASTM STP 1317.

Penry DL, Weston DP. 1998. Digestive determinants of benzo(*a*)pyrene and phenathrene bioaccumulation by a deposit-feeding polychaete. *Environ Toxicol Chem* 17:2254–2265.

Persaud D, Jaagumagi R, Hayton A. 1992. Guidelines for the protection and management of aquatic sediment quality in Ontario. Toronto (ON): Ontario Ministry of the Environment, Water Resources Branch. 27 p.

Petersen K, Kristensen E, Bjerregaard P. 1998. Influence of bioturbating animals on flux of cadmium into estuarine sediment. *Mar Environ Res* 45:403–415.

Phipps GL, Ankley GT, Benoit DA, Mattson VR. 1993. Use of the aquatic oligochaete *Lumbriculus variegatus* for assessing the toxicity and bioaccumulation of sediment-associated contaminants. *Environ Toxicol Chem* 12:269–274.

Pourriot R. 1977. Food and feeding habits of rotifera. Archiv für Hydrobiologie, *Beiheft Ergebnissse Limnologie* 8:243–260.

Preston BL. 2002. Spatial patterns in benthic biodiversity of Chesapeake Bay, USA (1984–1999): Association with water quality and sediment toxicity. *Environ Toxicol Chem* 21:151–162.

PTI Environmental Services. 1991. Pollutants of concern in Puget Sound. Seattle (WA): US Environmental Protection Agency, Region 10. EPA 910/9-91-003.

Rassoulzadegan F, Laval-Peuto M, Sheldon RW. 1988. Partitioning of the food ration of marine ciliates between pico- and nanoplankton. *Hydrobiologia* 159:75–88.

Ringwood AHK, Charles J. 2002. Comparative in situ and laboratory sediment bioassays with juvenile *Mercenaria mercenaria. Environ Toxicol Chem* 21:1651–1657.

Ringwood AH, Keppler CJ. 2002. Comparative in situ and laboratory sediment toxicity tests with juvenile *Mercenaria mercenaria. Environ Toxicol Chem* 21: 1651–1657.

Rodriguez P, Martinez-Madrid M, Angel J, Navarro E. 2001. Selective feeding by the aquatic oligochaete *Tubifex tubifex* (Tubificidae, Clitellata). *Hydrobiologia* 463: 133–140.

Roper DS, Nipper MG, Hickey CW, Martin ML, Weatherhead MA. 1995. Burial, crawling and drifting behaviour of the bivalve *Macomona liliana* in response to common sediment contaminants. *Mar Pollut Bull* 31:4–12.

Roy I, Hare L. 1999. Relative importance of water and food as cadmium sources to the predatory insect *Sialis velata* (Megaloptera). *Can J Fish Aquat Sci* 56:1143–1149.

Rubinstein NI, Gilliam WT, Gregory NR. 1984. Dietary accumulation of PCBs from a contaminated sediment source by a demersal fish (*Leiostomus xanthurus*). *Aquat Toxicol* 3:331–342.

Schönborn W. 1984. The annual energy transfer from the communities of Ciliata to the population of *Chaetogaster diastrophus* (Gruithuisen) in the River Saale. *Limnologica* (Berlin) 16:15–23.

Seth R, Mackay D, Muncke J. 1999. Estimating the organic carbon partition cooefficient and its variability for hydrophobic chemicals. *Environ Sci Technol* 33: 2390–2394.

Shuttleworth SM, Davison W, Hamilton-Taylor J. 1999. Two-dimensional and fine structure in the concentrations of iron and manganese in sediment pore-waters. *Environ Sci Technol* 33:4169–4175.

Sigg L, Sturm M, Kistler D. 1987. Vertical transport of heavy metals by settling particles in Lake Zurich. *Limnol Oceanogr* 32:112–130.

Sijm D, De Bruijn J, Crommentuijin T, Van Leeuwen K. 2001. Environmental quality standards: Endpoints or triggers for tiered ecological effect assessment approach? *Environ Toxicol Chem* 20:2644–2648.

Simpson SL, Apte SC, Batley GE. 1998. Effect of short term resuspension events of trace metals speciation in polluted anoxic sediments. *Environ Sci Technol* 32: 620–625.

Simpson SL, Apte SC, Batley GE. 2000. Effect of short term resuspension events on the oxidation of cadmium, lead and zinc sulfide phases in anoxic sediments. *Environ Sci Technol* 34:4533–4537.

Simpson SL, Batley GE. 2003. Disturbances to metal partitioning during toxicity testing of iron(II)-rich pore waters and whole sediments. *Environ Toxicol Chem* 22:424–432.

Simpson SL, Rosner J, Ellis J. 2000. The competitive displacement reactions of Cd, Cu and Zn added to a polluted sulfidic estuarine sediment. *Environ Toxicol Chem* 19:1992–1999.

Smith SL, MacDonald DD, Keenleyside KA, Ingersoll CG, Field LJ. 1996. A preliminary evaluation of sediment quality assessment values for freshwater ecosystems. *J Gt Lakes Res* 22:624–638.

Smith VE, Rathbun JE, Rood SG, Huellmantel LL. 1996. Technical considerations in sediment quality surveys. *J Gt Lakes Res* 22:512–522.

Stahl Jr RG, Bachman R, Barton AL, Clark JR, de Fur PL, Ells SJ, Piitinger CA, Slimak MW, Wentsel RS. 2001. Risk management: Ecological risk based decision-making. Pensacola (FL): Society of Environmental Toxicology and Chemistry (SETAC). 192 p.

Stahl Jr RG, Clark JR. 1998. Uncertainties in the risk assessment of endocrine modulating substances in wildlife. In: Kendall R, Dickerson R, Giesy J, Suk W, editors. Principles and processes for evaluating endocrine disruption in wildlife. Pensacola (FL): Society of Environmental Toxicology and Chemistry (SETAC). p 431–448.

Standley, LJ. 1997. Effect of sedimentary organic matter composition on the partitioning and bioavailability of dieldrin to the oligochaete *Lumbriculus variegatus*. *Environ Sci Tech* 31:2577–2583.

Stange K, Swackhamer DL. 1994. Factors affecting phytoplankton species-specific differences in accumulation of 40 polychlorinated biphenyls (PCBs). *Environ Toxicol Chem* 13:1849–1860.

Stemmer BL, Burton Jr GA, Leibfritz-Frederick S. 1990. Effect of sediment test variables on selenium toxicity to Daphnia magna. *Environ Toxicol Chem* 9: 381–389.

Stemmer BL, Burton Jr GA, Sasson-Brickson G. 1990. Effect of sediment spatial variance and collection method on cladoceran toxicity and indigenous microbial activity determinations. *Environ Toxicol Chem* 9:1035–1044.

Stephan CE, Mount DI, Hansen DJ, Gentile JH, Chapman GA, Brungs WA. 1985. Guidelines for deriving numerical national water quality criteria for the protection of aquatic organisms and their uses. Springfield (VA): National Technical Information System. NTIS #PB85-227049.

Stumm W. 1992. Chemistry of the solid-water interface. New York: Wiley.

Suter GW. 1993. Ecological risk assessment. Chelsea (MI): Lewis.

Swartz RC. 1999. Consensus sediment quality guidelines for polycyclic aromatic hydrocarbon mixtures. *Environ Toxicol Chem* 18:780–787.

Swartz RC, Cole FA, Lamberson JO, Ferraro SP, Schults DW, DeBen WA, Lee HI, Ozretich JH. 1994. Sediment toxicity, contamination and amphipod abundance at a DDT and dieldrin contaminated site in San Francisco Bay. *Environ Toxicol Chem* 13:949–962.

Swartz RC, DeBen WA, Sercu KA, Lamberson JO. 1982. Sediment toxicity and the distribution of amphipods in Commencement Bay, Washington, USA. *Mar Poll Bull* 13:359–364.

Swartz RC, Schults DW, Ozretich RJ, Lamberson JO, Cole FA, DeWitt TH, Redmond MS, Ferraro SP. 1995. ?PAH: A model to predict the toxicity of field-collected marine sediment contaminated by polynuclear aromatic hydrocarbons. *Environ Toxicol Chem* 14:1977–1987.

Tessier A, Campbell PGC, Auclair JC. 1983. Relationship between trace metal partitioning in sediments and their bioaccumulation in freshwater pelecypods. In: Proceedings 4th International Conference on Heavy Metals in the Environment. Edinburgh (UK): CEP Consultants. p 1086–1089.

Tessier A, Campbell PGC, Auclair JC, Bisson M. 1984. Relationships between the partitioning of trace metals in sediments and their accumulation in the tissues of the freshwater mollusc *Elliptio complanata* in a mining area. *Can J Fish Aquat Sci* 41:1463–1472.

Tessier A, Couillard Y, Campbell PGC, Auclair JC. 1993. Modeling Cd partitioning in oxic lake sediments and Cd concentrations in the freshwater bivalve *Anodonta grandis. Limnol Oceanogr* 38:1–17.

[USEPA] US Environmental Protection Agency. 1992a. Framework for ecological risk assessment. Washington DC: USEPA. EPA 630/R-92/001.

[USEPA] US Environmental Protection Agency. 1992b. Sediment classification methods compendium. Washington DC: USEPA. Sediment Oversight Technical Committee. Final Report EPA 823/R-92-006.

[USEPA] US Environmental Protection Agency. 1993a. Sediment quality criteria for the protection of benthic organisms. Acenaphthene. Washington DC: USEPA, Office of Water. Report EPA 822/R-93-013.

[USEPA] US Environmental Protection Agency. 1993b. Sediment quality criteria for the protection of benthic organisms. Dieldrin. Washington DC: USEPA, Office of Water. Report EPA 822/R-93-015.

[USEPA] US Environmental Protection Agency. 1993c. Sediment quality criteria for the protection of benthic organisms. Endrin. Washington DC: USEPA, Office of Water. Report EPA 822/R-93-016.

[USEPA] US Environmental Protection Agency. 1993d. Sediment quality criteria for the protection of benthic organisms. Fluoranthene. Washington DC: USEPA, Office of Water. Report EPA 822/R-93-012.

[USEPA] US Environmental Protection Agency. 1993e. Sediment quality criteria for the protection of benthic organisms. Phenanthrene. Washington DC: USEPA, Office of Water. Report EPA 822/R-93-014.

[USEPA] US Environmental Protection Agency. 1993f. Technical basis for deriving sediment quality criteria for nonionic contaminants for the protection of benthic organisms by using equilibrium partitioning. Washington DC: USEPA, Office of Water. Report EPA 822/R-93-011.

[USEPA] US Environmental Protection Agency. 1994a. Briefing report to the EPA Science Advisory Board on the equilibrium partitioning approach to predicting metal bioavailability in sediments and the derivation of sediment quality criteria for metals. Vol. 1. Washington DC: USEPA, Office of Water. Report EPA 822/D-94-002.

[USEPA] US Environmental Protection Agency. 1994b. Methods for asssessing the toxicity of sediment-associated contaminants with estuarine and marine amphipods. Washington DC: USEPA, Office of Research and Development. Final Report EPA 600-R-94-025.

[USEPA] US Environmental Protection Agency. 1994c. Short-term methods for estimating the chronic toxicity of effluents and receiving waters to marine and estuarine organisms. Washington DC: USEPA Office of Research and Development, Report EPA-600-4-91-003.

[USEPA] US Environmental Protection Agency. 1996. Calculation and evaluation of sediment effect concentrations for the amphipod Hyalella azteca and the midge Chironomus riparius. Chicago: USEPA. Report EPA 905-R-96-008.

[USEPA] US Environmental Protection Agency. 1997. The incidence and severity of sediment contamination in surface waters of the United States. Volume 1: National sediment quality survey. Washington DC: USEPA. Report EPA 823-R-97-006.

[USEPA] US Environmental Protection Agency. 2000a. Methods for measuring the toxicity and bioaccumulation of sediment-associated contaminants with freshwater invertebrates, second edition, Washington DC: USEPA. Report EPA–600-R-99-064.

[USEPA] US Environmental Protection Agency. 2000b. Stressor identification guidance document Washington DC: USEPA. Report EPA 822-B-00-025.

[USEPA] US Environmental Protection Agency. 2001. Methods for collection, storage and manipulation of sediments for chemical and toxicological analyses; technical manual. Washington DC: USEPA, Office of Water. Final Report EPA 823-B-01-002.

[USEPA] US Environmental Protection Agency. 2003a. Procedures for the derivation of equilibrium partitioning sediment benchmarks (ESBs) for the protection of benthic organisms: Dieldrin. Washington DC: USEPA, Office of Research and Development. EPA-600-R-02-010.

[USEPA] US Environmental Protection Agency. 2003b. Procedures for the derivation of equilibrium partitioning sediment benchmarks (ESBs) for the protection of benthic organisms: Endrin. Washington DC: USEPA, Office of Research and Development. EPA-600-R-02-009.

[USEPA] US Environmental Protection Agency. 2003c. Procedures for the derivation of equilibrium partitioning sediment benchmarks (ESBs) for the protection of benthic organisms: PAH mixtures. Washington DC: Office of Research and Development. EPA-600-R-02-013.

[USEPA] US Environmental Protection Agency. 2003d. Technical basis for the derivation of equilibrium partitioning sediment benchmarks (ESBs) for the protection of benthic organisms: Nonionic organics. Washington DC: USEPA, Office of Water. Final Report EPA-600-R-02-014.

[USEPA] US Environmental Protection Agency. 2004a. Procedures for the derivation of equilibrium partitioning sediment benchmarks (ESBs) for the protection of benthic organisms: Metal mixtures (cadmium, copper, lead, nickel, silver, and zinc). Washington DC: Office of Research and Development. EPA-600-R-02-011.

[USEPA] US Environmental Protection Agency. 2004b. Procedures for the derivation of equilibrium partitioning sediment benchmarks (ESBs) for the protection of benthic organisms: Nonionics compendium. Washington DC: Office of Research and Development. EPA-600-R-02-016.

Van Der Kooij LA, Van De Meent D, Van Leeuwen CJ, Bruggeman WA. 1991. Deriving quality criteria for water and sediment from the results of aquatic toxicity tests and product standards: Application of the equilibrium partitioning method. *Water Res* 6:697–705.

Verity PG. 1991. Feeding in planktonic protozoans: Evidence for non-random acquisition of prey. *J Protozoo* 38:69–76.

Von Stackelberg K, Menzie CA. 2002. A cautionary note on the use of species presence and absence data in deriving sediment criteria. *Environ Toxicol Chem* 21: 466–472.

Wallberg R, Bergqvist PA, Andersson A. 1997. Potential importance of protozoan grazing on the accumulation of polychlorinated biphenyls (PCBs) in the pelagic food web. *Hydrobiologia* 357:53–62.

Wang F, Chapman PM. 1999. Biological implications of sulfide in sediment: A review focusing on sediment toxicity. *Environ Toxicol Chem* 18:2526–2532.

Wang F, Tessier A, Hare L. 2001. Oxygen measurements in the burrows of freshwater insects. *Freshw Biol* 46:317–327.

Wang H, Kostel JA, St. Amand AL, Gray KA. 1999. 2. The response of a laboratory stream system to PCB exposure: Study of periphytic and sediment accumulation patterns. *Water Res* 33:3749–3761.

Warren L, Tessier A, Hare L. 1998. Modelling cadmium accumulation by benthic invertebrates in situ: The relative contributions of sediment and overlying water reservoirs to organism cadmium concentrations. *Limnol Oceanogr* 43:1442–1454.

Watzin MC, McIntosh AW, Brown EA, Lacey R, Lester DC, Newbrough KL, Williams AR. 1997. Assessing sediment quality in heterogeneous environments: A case study of a small urban harbor in Lake Champlain, Vermont, USA. *Environ Toxicol Chem* 16:2125–2135.

Wavre M, Brinkhurst RO. 1971. Interactions between some tubificid oligochaetes and bacteria found in the sediments of Toronto Harbor, Ontario. *J Fish Res Board Can* 28:335–341.

Weston DP, Mayer LM. 1998a. Comparison of in vitro digestive fluid extraction and traditional in vivo approaches as measures of polycyclic aromatic hydrocarbon bioavailability from sediments. *Environ Toxicol Chem* 17:830–840.

Weston DP, Mayer LM. 1998b. In vitro digestive fluid extraction as a measure of the bioavailability of sediment-associated polycyclic aromatic hydrocarbons: Sources of variation and implications for partitioning models. *Environ Toxicol Chem* 17: 820–829.

Predictive ability of sediment quality guidelines

4

JACK Q WORD, BARBARA B ALBRECHT, MICHELE L ANGHERA, RENATO BAUDO, STEVEN M BAY, DOMINIC M DI TORO, JEFFERY L HYLAND, CHRISTOPHER G INGERSOLL, PETER F LANDRUM, EDWARD R LONG, JAMES P MEADOR, DAVID W MOORE, THOMAS P O'CONNOR, JAMES P SHINE

The quantitative extent to which sediment quality guidelines (SQGs) are predictive of the presence, or absence, of toxicity of contaminated sediment to sediment-dwelling organisms or to higher trophic-level organisms is a critical concern among scientists and regulators evaluating the application of one or more numeric SQG approaches in assessments of sediment quality. Users of these guidelines should understand how well various SQGs predict the presence or absence and extent of toxicity in sediment samples. The ability of various SQGs to represent the potential for effects or no effects of contaminants on organisms in freshwater, estuarine, and marine environments was examined through a review of the published literature focused on 3 specific questions:

1) How well do SQGs represent the potential for effects or no effects observed in laboratory toxicity tests and in field studies of benthic communities?
2) How well do SQGs represent the potential for effects or no effects in organisms as a result of contaminant uptake and/or trophic transfer?
3) How have SQGs been applied and validated in the field as part of sediment management and risk management decision making?

The predictive ability of various SQGs to represent the potential for effects or no effects of contaminants on organisms in freshwater, estuarine, and marine environments was defined for this chapter as the probability of observing the presence or absence of effects within incremental ranges of sedi-

Use of Sediment Quality Guidelines and Related Tools for the Assessment of Contaminated Sediments
Wenning RJ, Batley GE, Ingersoll CG, Moore DW, editors.
©2005 Society of Environmental Toxicology and Chemistry (SETAC). ISBN 1-880611-71-6

ment contaminant concentrations as defined by SQGs based on the specific endpoints and benthic taxa evaluated. Results of these evaluations are typically expressed as the percentage of samples expected to be affected (e.g., % toxic samples) or are based on the degree of the response (e.g., % mortality). In this context, a chemically based numeric SQG is defined as the concentration of sediment-associated contaminants that is associated with a high or a low probability of observing adverse biological or bioaccumulation results, depending on its purpose and narrative intent (i.e., is the SQG designed to be predictive of toxic or nontoxic conditions in sediment?). Alternative approaches for evaluating predictive ability of SQGs are also discussed (e.g., efficiency and sensitivity; Shine et al. 2003). For purposes of discussion, the SQGs were classified into 2 categories: 1) those derived using empirical approaches (such as effects range low [ERL], effects range median [ERM], probable effects concentration [PEC], apparent effects threshold [AET]) or 2) those derived using mechanistic approaches (such as equilibrium partitioning [EqP] for nonionic organic compounds, simultaneously extracted metal–acid volatile sulfide [SEM–AVS], organic carbon normalizations, and critical body residues). An important outcome of this chapter and of other chapters in this book is a clear delineation of the uses and limitations of both types of SQGs. The uses of approaches to compare sediment chemistry to background, reference, or baseline concentrations were not included in this evaluation because these approaches are not based on effects of contaminants in sediment.

The following sections provide a summary of discussions on 10 topics dealing with the predictive ability of SQGs:

1) use of empirical SQGs to predict the presence or absence of toxic effects;
2) use of mechanistic SQGs to predict the presence or absence of toxic effects;
3) use of SQGs based on an understanding of how they were derived, their narrative intent, and their predictive ability;
4) use of controlled benthic community colonization studies and controlled mesocosm studies;
5) potential confounding factors;
6) influence of the chemical state of the contaminants on the predictive ability of SQGs;
7) influence of an unusual sediment matrix on the predictive ability of SQGs;
8) single exceedances of empirical SQGs;

9) SQGs for total polycyclic aromatic hydrocarbon (PAH) and total polychlorinated biphenyls (PCBs); and
10) bioaccumulation-based SQGs.

Use of Empirical Sediment Quality Guidelines to Predict Presence or Absence of Toxic Effects

Empirically derived SQGs can be used to predict the probability of the presence or absence of toxic effects with a known level of statistical confidence based on the results of the analyses of the data sets evaluated to date. This conclusion is based on considerable effort that has been expended to quantify the predictive ability of various SQGs (Chapter 12). Thus far, the majority of studies have been focused on SQGs derived using empirical approaches (Chapter 12). Numerous studies have been published in which paired sediment chemistry and either laboratory toxicity or benthic community data were compared for freshwater, estuarine, or marine habitats. Conditions along all 3 coastlines of the United States (US) and in Hawaii have been represented in estuarine and marine studies. Conditions in the Great Lakes, midwestern lakes, and many rivers across North America have been represented in freshwater studies. The importance of this issue is apparent by the level of effort put forth to quantify the predictive ability of SQGs (Table 12-24 in Chapter 12).

The studies summarized in Chapter 12 were seldom designed specifically to quantify the predictive ability of SQGs (e.g., Table 12-2 in Chapter 12). These analyses were conducted with data acquired from monitoring or research programs conducted for other purposes. However, data from these various studies are representative of conditions in the field where SQGs would be used. Therefore, these data can be used retrospectively to evaluate the ability of SQGs to correctly classify samples as being toxic in the case of laboratory tests or having an impaired benthos in the case of field studies. Most of the data used to examine the predictive ability of SQGs for toxicity were derived from acute 10-d amphipod mortality tests. A limited number of studies have examined the predictive ability of SQGs for sublethal responses or chronic exposures of estuarine or marine species exposed to whole sediment in the laboratory. A greater amount of chronic or sublethal data is available for freshwater samples, but most of these studies used a single species of amphipod.

The correspondence between SQGs and toxicity in other matrices (e.g., pore water [Table 12-2 in Chapter 12], elutriates [AETs for oysters or Microtox], or organic extracts) has been documented in some regional studies, but sufficient data are not available to summarize their overall predictive ability of SQGs in these other matrices on a larger scale. Further, there is little confidence that toxicity tests conducted with these other matrices reflect the bioavailable fractions of sediment-associated contaminants or real-world conditions in bedded sediments. In particular, the chemical changes that invariably occur during extraction of pore water (e.g., the precipitation of iron hydroxide) have unknown effects on the bioavailability of contaminants. Therefore, the discussions at the workshop focused on evaluating the predictive ability of SQGs using data from whole-sediment samples.

The underlying supposition in most studies of predictive ability is that the percentages of samples indicating toxicity or benthic community impacts would be similar to the narrative intent of the SQG being used in an evaluation. That is, the incidence of effects (e.g., % toxic) and the degree of the response (e.g., % mortality) would be expected to increase with an incremental increase in chemical concentration, as defined by SQGs (Chapter 12). To date, most evaluations of predictive ability of SQGs have focused on the incidence of effects, but not on the degree of the effects (Table 12-24 in Chapter 12).

A wide range of laboratory and field studies in freshwater, estuarine, and marine sediments indicate that empirical SQGs can be used to assess the probability of observing effects with known levels of statistical confidence (Chapter 12). The incidence of effects frequently increases with increasing sediment contamination based on the empirical SQGs. For example, the lowest incidence of effects was measured at concentrations less than the low-range SQGs (e.g., mean quotient <0.1), and the highest incidence of toxicity is observed at chemical concentrations above the upper-range SQGs (e.g., mean quotient >1.5 to 2.3; Table 12-24 in Chapter 12; mean SQG quotients are calculated by dividing the concentration of each contaminant by its respective SQG and then calculating the average of these individual quotients; see Chapter 12 for additional detail). Evaluations of predictive ability were not attempted to identify chemical concentrations below which no effects were observed and above which such effects were always observed (e.g., toxicity relative to a negative control or reference sediment; USEPA 2000b; ASTM 2003). Rather, most evaluations were conducted to quantify the incidence of toxicity (predictive efficiency) over ranges in chemical concentrations defined either by the numbers of empirical SQGs exceeded or mean quotients. These low-range SQGs are intended to be indicative of chemical concentrations below which effects would be infrequent; however, these low-range SQGs were not intended to be highly predictive of adverse effects (Chapter 12). The lowest incidence of effects occurs at concentrations less than low-range SQGs, and the incidence increases incrementally in chemical concentration ranges defined by midrange and upper-range SQGs (Table 12-24 in Chapter 12). The incidence of toxicity is relatively low in sediments in which only 1 midrange SQG is exceeded. For example, 14% and 23% of samples were classified as highly toxic in amphipod survival tests with only 1 probable effects level (PEL) or 1 effects range median (ERM) concentration exceeded, respectively (Long, Field, MacDonald 1998). Therefore, it cannot be assumed that there would be a high probability of toxicity in other sediments with similar chemical characteristics.

Data from these evaluations showed that the incidence of effects in acute laboratory tests was generally 10% or less when all chemical concentrations were less than all effects-based empirical SQGs or mean quotients were < 0.1. Similarly, low incidences in degraded benthic conditions in estuaries occurred when mean quotients were about 0.01 to 0.02 (Table 12-24 in Chapter 12). In contrast, greater than about 75% of the samples were either toxic in the laboratory tests at mean quotients exceeding 1.5 or indicated degraded benthic conditions when mean quotients exceeded about 0.1 to 0.8 (depending on the type empirical SQG evaluated; Table 12-24 in Chapter 12). Observations of a <10% incidence of effects in the lower contaminant range and a >70% incidence of effects in the upper contaminant range are consistent with the original narrative intent of the empirical SQGs. In addition, the same level of predictive ability for laboratory toxicity tests has been shown to apply for samples with extremely high concentrations of single chemicals. The percentages of samples indicating effects were intermediate (about 50%) in sediments with moderate degrees of contamination as gauged by the empirical SQGs (Table 12-24 in Chapter 12). Importantly, an incremental increase in effects has been observed with an incremental increase in contamination as gauged by the empirical SQGs. However, direct measurement of toxicity or benthic community impacts in the field are required to determine if an individual sample with moderate contamination is toxic or not toxic.

The type of test conducted or endpoint measured influences the incidence of toxicity observed among various SQGs. For example, in freshwater toxicity tests, a higher incidence of toxicity is observed in longer-term tests when growth is evaluated (Table 12-24 and Figure 12-3 in Chapter 12). In addition, a growing body of evidence (Swartz et al. 1994; Hyland et al. 1999, 2003; Brown et al. 2000; Long et al. 2002; MacDonald, Ingersoll, Smorong, et al. 2002) suggests that impacts on estuarine benthic communities may occur at chemical concentrations below those that are identified in 10-d laboratory toxicity tests (e.g., mean quotient of >0.1 to 0.4 for benthic community assessments versus mean quotients >1.5 for laboratory tests). It has been suggested in these studies that the impacts observed on the benthic communities at lower chemical concentration ranges may reflect the varying sensitivities of multiple species and life stages to longer-term exposures that may be persisting over several generations. Such differences should not be misinterpreted to imply that laboratory toxicity data are invalid or of lesser value than the benthic assessment data. Rather, these comparisons indicate that sensitivity to contaminants may vary with different receptors, endpoints, and exposure regimes. The approximately 10-fold difference between the responses observed with the benthic communities and the responses observed in 10-d amphipod tests is consistent with acute-to-chronic ratios often observed in water-only toxicity tests for many chemicals (Chapter 12).

The relative influence of natural controlling factors (e.g., salinity, grain size, TOC, erosional effects) versus chemical-induced toxicity on the condition of benthic communities is often difficult to separate and adds complexity to the interpretation of data. Both types of factors may act together to influence the changes in benthic communities observed in contaminated sediments. Such factors need to be taken into account in efforts to assess benthic community-level effects as well as laboratory toxicity. The use of chronic laboratory toxicity tests also may be needed to reflect more accurately subtle effects observed in the field. Additionally, the evaluation of the extent of bioaccumulated residue taken synoptically with measures of effects could help to better establish cause–effect relationships (See "Bioaccumulation-Based SQGs," p 149).

Recommendations

- Consistent procedures should be used in the assessment of predictive ability in order to facilitate comparisons among different data sets (e.g., efficiency, sensitivity, specificity).
- Additional analyses should be performed with existing data to evaluate the predictive ability of SQGs for which less information on predictive ability is available (e.g., AETs, screening-level concentrations [SLCs]).
- Additional data should be generated to examine the relationship between SQGs and chronic effects in laboratory tests or in benthic community assessments.
- Additional evaluations are needed on the influence of confounding factors and other co-occurring stressors on the ability of SQGs to predict effects, particularly in benthic community assessments.

Use of Mechanistic Sediment Quality Guidelines to Predict Presence or Absence of Toxic Effects

Mechanistic SQGs based on equilibrium partitioning (EqP) theory attempt to causally relate sediment concentration to toxicity (Chapter 3). The availability and success or failure of these mechanistic SQGs depends on the adequacy of the partitioning model and its parameters and on the assumption that exposure is from either pore water or sediment particles or both. A fundamental component of the development of SQGs based on EqP is a careful and comprehensive comparison of EqP predictions initially using laboratory-spiked sediments and then field contaminated sediments. As a consequence, quite a lot is known about the predictive ability of these mechanistically based SQGs (Chapter 13). Assessing the predictive ability of EqP-based SQGs is in some ways a more straightforward exercise than for empirical SQGs, which are derived from the field observations for which they make predictions. Mechanistically based SQGs are derived from effects concentrations measured in water-only exposures. In sediment exposures, the effect is predicted to occur when the same concentration occurs in the pore water of the sediment. The equivalent sediment concentrations are derived from partitioning theory applied to the pore water–sediment particles system, which is assumed to be at equilibrium. Hence both EqP predictions in pore water or in whole sediment can be tested directly.

The data demonstrating the ability of measured porewater concentrations to predict toxicity based on independently measured water-only LC50s are presented in Chapter 13. For organic chemicals (Figure 13-2 in Chapter 13), the predicted sediment LC50s are within a factor of about 2 of observed LC50s. For metals (Figure 13-3 in Chapter 13), the predictions for lack of toxicity are essentially always correct. However, there is very little predictive power until the porewater concentration exceeds 10 to 100 times the water-only LC50. The reason is that metal complexation in pore water affects bioavailability, and this is not taken into account when total dissolved metal is used to represent bioavailable metal. The problem is that the chemical state of the metal in the water-only test is not the same as in the pore water. However, when the porewater concentration is large enough, the issue of bioavailability becomes less important (the complexing ligands are saturated and toxicity results).

The second component of EqP predictions is the partitioning model that predicts porewater concentrations from solid-phase chemistry. For neutral hydrophobic organic chemicals (e.g., PAHs, chlorinated hydrocarbons) the partitioning model is based on sediment organic carbon because for neutral hydrophobic chemicals it is the sediment phase in which they reside (Karickhoff et al. 1979; Di Toro 1985; USEPA 2000c). The data for single chemical laboratory-spiked sediments suggests a factor of 2 or 3 error in predicting LC50s (Figures 13-4 and 13-5 in Chapter 13). For mixtures of PAHs, the same level of predictive ability exists (Figures 13-6 and 13-7 in Chapter 13). For metals, the AVS is used because metal–sulfide complexes are very insoluble and act to sequester the metal (Di Toro et al. 1990, 1992). The metal concentration used to compare to AVS is the SEM with the sulfide. The sum of all the metals that form sulfides more insoluble than iron sulfide (ΣSEM) is used. The prediction that for ΣSEM/AVS <1 no toxicity is observed, is virtually always the case (Figures 13-9 to 13-11 in Chapter 13). More recently, sediment organic carbon is used in addition to AVS to predict toxicity (USEPA 2000a; Besser et al. 2003). The SQG is based on organic carbon–normalized excess SEM, that is, (ΣSEM – AVS) / f_{OC}. If Σ(SEM – AVS) / f_{OC} < 130 μmol/gOC, no toxicity is predicted. If Σ(SEM – AVS) / f_{OC} > 3400 μmol/gOC, then toxicity is predicted. About 25% of the data are in the ambiguous range where no prediction is possible (Figure 13-12 in Chapter 13).

The predictive ability of EqP-based SQGs for field collected sediments is more difficult to assess because it is necessary to know the concentrations of all chemicals in the sediment that are contributing to the toxicity. For these chemicals, it is necessary to have an SQG for each chemical and a method to predict the toxicity of the mixture. In practice, it is usually known what class of chemical is causing the toxicity. The data for DDT (Figure 13-13 in Chapter 13) suggests a factor of 2 with some outliers. For PAH mixtures (Figures 13-14 and 13-15 in Chapter 13) and fuel oil (Figure 13-16 in Chapter 13), again a factor of 2 is suggested. Observations of the reduction of amphipod populations for sediments exceeding the predicted total PAH SQG support the applicability of EqP-based SQGs for PAHs to field-contaminated sediments.

For metals, the situation for field-contaminated sediments is similar to the laboratory-spiked sediments. The prediction that for ΣSEM/AVS < 1 no toxicity is observed, is virtually always the case (Figure 13-18 in Chapter 13). It is remarkable that the excess carbon-normalized SEM method also predicts the results of chronic sediment toxicity tests. The chronic data indicate that the lower bound for excess ΣSEM derived from the 10-d toxicity tests also is predictive for chronic exposure because all but 1 sediment exhibited no effects below (ΣSEM – AVS) /f_{OC} < 130 μmol/gOC. However, there are too few data for even a rough statistical evaluation.

The EqP method is based on a partitioning model that is used to determine the sediment concentration that is in equilibrium with the pore water. If this calculation is incorrect, then the predictions are inaccurate ("Influence of Chemical State of Contaminants on Predictive Ability of SQGs," p 142, and "Influence of an Unusual Sediment Matrix on the Predictive Ability of SQGs," p 143). A small source of difficulty is the assumption that the organism is exposed via either pore water or sediment particle ingestion. For some epibenthic organisms, the exposure can be primarily from the overlying water. Also, pore waters are often anoxic, and most benthic organisms will avoid immersion in such water by respiring and eating in burrows ventilated with overlying water or by inhabiting the sediment–water interface. These organisms are then exposed to water that may be intermediate between pore water and overlying water. Nonetheless, so long as overlying water is not toxic, sediments with nontoxic pore water should not be toxic to benthic organisms.

The original EqP values can be derived for any chemical for which a water-only effects concentration is available for the test organism and for which a sediment–porewater partitioning model is available. These conditions are fulfilled for Type I narcotic chemicals: aliphatic, aromatic, and halogenated hydrocarbons; PAHs, alcohols, ethers, and ketones (Di Toro and McGrath 2000); and some pesticides (e.g., DDT and metabolites, endrin, dieldrin). For metals, single EqP values and mixture criteria exist for Cd, Cu, Ni, Pb, and Zn (USEPA 2000a). EqP values that apply to a single test organism (e.g., 10-d amphipod mortality) can also be derived. The LC50 or EC50 in water-only exposures for that organism is required as well as a partitioning model. Usually the partitioning model is the limiting factor. The water-only LC50 can be determined by toxicity testing. However, the partitioning model should apply to all the sediment types for which EqP values are to be applied. Such general partitioning models are not yet available for certain classes of chemicals (e.g., organic acids and bases that are ionized at sediment pHs, metals that do not form insoluble sulfide complexes or that partition to other phases in the sediments [e.g., arsenite, arsenate, and other oxyanions], and more complex molecules with unknown partitioning behavior such as pharmaceuticals).

EqP values can also be derived for mixtures of chemicals if a model of toxicity of the mixture in water-only exposures is available. For total PAHs, narcotic additivity is used (Chapter 13). If other chemicals are present and no toxicity interaction model is available, then EqP cannot account for these interaction effects.

It has often been pointed out that EqP values do not apply if the porewater–sediment system is not at equilibrium. There are 2 scenarios to consider: 1) whether equilibrium is presented in stable sediments and 2) the situation for dynamic situations where sediments are being disturbed. For stable sediments, the contact time is much longer (years) than the usual time for complete sorption equilibrium to occur (days to months). In fact, the aging phenomena in sediments and soils actually result in more resistant sorption and higher apparent partition coefficients. For dynamic situations, if the concentrations are varying, then any of the SQGs would not be appropriate because the data sets from which the values have been derived were mainly from sediment samples in stable situations.

It should be pointed out that the EqP model itself has no parameters that are fit to the sediment toxicity data. The model attempts to predict toxicity or lack of toxicity in sediment exposures from toxic concentrations measured in water-only exposures and using either porewater or sediment chemistry. Therefore, the model can be falsified by data that do not conform to the predictions. This is actually an important feature of the model because failures point to areas that need further investigation.

Recommendations

- EqP values should be derived for other important classes of chemicals by developing the appropriate partitioning and mixture toxicity models.

Use of Sediment Quality Guidelines Based on Their Derivation, Narrative Intent, and Predictive Ability

SQGs should only be used with an understanding of how they were derived, their narrative intent, and their predictive ability. Effects-based SQGs have been demonstrated to be predictive of effects or no effects over a range of freshwater, estuarine, and marine habitats for the endpoints used to develop the SQGs (e.g., primarily for 10-d sediment toxicity or for benthic community responses in estuaries; Table 12-24 in Chapter 12 and "Use of Empirical SQGs to Predict the Presence or Absence of Toxic Effects," p 123). For example, an increase in the incidence of toxicity with increasing mean quotients has been reported for organisms exposed to freshwater, estuarine, or marine sediments collected from several regions, basins, and coastal areas across various habitats in North America (e.g., Tables 12-2, 12-10, and 12-11 in Chapter 12). The results of these analyses indicate that SQGs can be used to predict toxicity of sediments on both a regional and national basis across a variety of freshwater, estuarine, and marine habitats using the same methods used to establish the SQGs. Although these analyses indicate the extent to which

SQGs can be used to predict toxicity and benthic community impairment in a variety of habitats, SQGs should not be applied to make decisions without an understanding of their derivation method, narrative intent, predictive ability, and limitations (Long and MacDonald 1998; Chapter 3). An understanding of the capabilities and limitations is needed to minimize the inappropriate use or misinterpretation of SQGs. Knowledge of the factors controlling bioavailability and tolerance of organisms are also needed to more fully interpret the predictive ability of SQGs at a site of interest ("Use of Mechanistic SQGs to Predict the Presence or Absence of Toxic Effects," p 127; "Influence of an Unusual Sediment Matrix on the Predictive Ability of SQGs," p 143; "Single Exceedances of Empirical SQGs," p 145; "Sediment Quality Guidelines for Total PAH and Total PCBs," p 147). Site-specific studies of the predictive ability of SQGs elevates confidence in their appropriate use at a location of interest and validates the use of SQGs in management decisions at the site. Regional variation in contaminant mixtures, chemical state ("Influence of Chemical State of Contaminants on Predictive Ability of SQGs," p 142), or matrix type ("Influence of an Unusual Sediment Matrix on the Predictive Ability of SQGs," p 143) can reduce the performance of SQGs at a location of interest.

In addition to understanding the limits to the predictive ability of SQGs, the users of the guidelines should understand the methods that were used to derive the SQGs and the narrative intent of the SQGs. The users of SQGs should take time to learn how SQGs were derived (Chapter 3) and how to use SQGs or any other tool used to assess sediments (Chapters 5 and 6). Information on the derivation and application of SQGs is available in the primary literature and in the papers developed in this book (Chapters 3, 6, 12, 13). For example, low-range SQGs were established to be conservative estimates of toxicity and typically are not predictive of a high incidence of effects.

Currently available SQGs were derived with data generated to evaluate effects on sediment-dwelling organisms. Therefore, these guidelines should not be applied to evaluate the direct effects of contaminated sediment on other biota such as fish, birds, or other wildlife. Because SQGs were generated with data from sediments, these guidelines should not be applied to materials such as upland soils, de-watered dredge materials, metal ores, or mine tailings.

Site-specific studies of the predictive ability of SQGs may not be necessary for small-scale projects or when the concentrations of contaminants in sediments are well below low-range SQGs at a site of interest (Table 12-24 in Chapter 12). However, for larger-scale projects or when concentrations are elevated above SQGs, it is desirable to generate site-specific information on the predictive ability of SQGs using a variety of assessment endpoints (e.g., MacDonald, Ingersoll, Moore, et al. 2002).

Studies intended to evaluate the predictive ability of SQGs in laboratory toxicity tests or in benthic community surveys should be designed to include the collection of samples that represent the ranges of chemicals of concern in the sediments at the site of interest. Bioaccumulation measures should also be made with field-collected benthos to demonstrate exposure and for comparison, where possible, to known toxic tissue residues. This will assist in establishing cause–effect relationships ("Bioaccumulation-Based SQGs," p 149). Historical data should be reviewed to help identify areas to sample and the numbers of samples to be collected from the site of interest. The measures of sediment chemistry and measures of effects in laboratory tests or benthic surveys should be performed on splits of the same samples. Additionally, detection limits for each chemical of concern should be established below the low-range SQGs in order to apply SQGs to evaluate predictive ability (Table 12-24 in Chapter 12). A battery of whole-sediment toxicity tests should be evaluated with organisms known to be responsive to the chemicals of concern in the sediment matrices at the site of interest. Infaunal community structure analyses should also accompany the laboratory testing to complete the sediment quality triad information (Chapters 5 and 6).

Assessments are often conducted in areas where there is little or no site-specific information on the predictive ability of SQGs. Guidance can be provided on the sampling effort required to confirm that the presence or absence of toxicity at a site of interest conforms to predictive expectations from national SQGs (Table 12-24 in Chapter 12). These models rely on our current understanding of the predictive ability of SQGs. For example, concentration–response curves (Figures 12-2 and 12-3 in Chapter 12) or logistic models (Figure 12-5 in Chapter 12) have been derived to estimate the probability of toxicity associated with a given level of contamination. The goodness-of-fit of these models could be used to construct statistical tools indicating the number of samples required to reasonably demonstrate that the observed presence or absence of effects at a site does not substantially deviate from existing models based on national SQGs. The number of samples evaluated should also be determined based on the size of the site of interest, the heterogeneity of the sediments, an understanding of the hydrologic and geophysical characteristics of the site, and the degree of confidence needed for the assessment (USEPA 2000b; ASTM 2003).

Three criteria have been used to establish confidence in the use of SQGs (Long and MacDonald 1998): reliability, comparability, and predictive ability. Reliability is evaluated as the agreement between the narrative intent of the SQGs and distribution of effects data within the derivation database. Comparability is the degree of concordance among SQGs derived with different methods, but with the same narrative intent. Finally, Long and MacDonald (1998) recommend analyses of the predictive ability of SQGs using data from the site of interest to determine how the data compare to data from similar environments elsewhere.

Three aspects of SQGs should be considered in an evaluation of predictive ability: efficiency, sensitivity, and specificity (Ginn and Pastorok 1992; Shine et al. 2003). Efficiency describes the ability of the SQG to correctly predict the occurrence of effects (e.g., toxicity or benthic community impairment) and is usually expressed as the incidence of effects (% toxic) among the total number of samples with chemical concentrations above or below various SQGs (Table 12-24 in Chapter 12). Efficiency describes the extent to which a decision for further testing is correct. Sensitivity is a measure of the ability of the SQG to detect effects and is usually expressed as the percentage of affected samples that were correctly predicted to be affected (Table 12-7 in Chapter 12). Sensitivity describes the extent to which an SQG is protective of the environment. That is, it describes how well an SQG correctly classifies a toxic sample as toxic. Specificity describes the ability of the SQG to correctly classify samples exhibiting no effect. The predictive ability of SQGs can be described using 2 metrics: efficiency and sensitivity. However, an evaluation of specificity may be valuable in assessing the appropriateness of a specific SQG for some applications (e.g., screening for further testing). Traditionally, most analyses of predictive ability of SQGs have focused on evaluating the incidence of toxicity (e.g., efficiency) across incremental ranges of SQGs (Table 12-24 in Chapter 12). A useful SQG will be both efficient and sensitive. However, efficiency and sensitivity are not necessarily related. In certain situations, an SQG providing high efficiency may have low sensitivity and vice versa. Revisions of SQGs may be needed as a result of these evaluations (Shine et al. 2003). Additionally, the presence of unique chemicals for which no SQGs are available also may be identified at the site of interest.

The efficiency and sensitivity of different SQGs can vary substantially (Long, Field, MacDonald 1998) and is often an intended characteristic of the narrative intent of the guideline. For example, the efficiency and sensitivity of Puget Sound AET values varied from 37% to 67% and from 57% to >90%, respectively (Barrick et al. 1988). Low-range SQGs are intended to be protective of the environment and thus may be very sensitive (i.e., identify most of the nontoxic samples), yet may be relatively inefficient (incorrectly classify a number of nontoxic samples as toxic). SQG efficiency and sensitivity for the site of interest should be calculated over a wide range of contamination (e.g., MacDonald and Ingersoll 2002; MacDonald, Ingersoll, Smorong, et al. 2002; MacDonald, Ingersoll, Moore, et al. 2002) and compared to a national or regionally robust database using a statistical method such as logistic regression (Field et al. 2002) or receiver operating characteristic curve analysis (Shine et al. 2003). Differences in sensitivity or efficiency, especially for low or high contamination situations, may indicate the presence of confounding factors ("Potential Confounding Factors," p 137; Chapter 16) or variations in chemical and matrix type that would constrain the use of national SQGs ("Influence of Chemical State of Contaminants on Predictive Ability of SQGs," p 142, and "Influence of an Unusual Sediment Matrix on the Predictive Ability of SQGs," p 143). Analysis of the magnitude of the response (e.g., % mortality) in addition to the incidence of effects (% toxic) provides additional information regarding the predictive ability of SQGs (Table 12-2 in Chapter 12, MacDonald, Ingersoll, Moore, et al. 2002). Use of test response data facilitates comparisons among studies where different statistical methods were used to classify the toxicity of samples. Because the test response describes the magnitude of the effect, its use may aid in interpreting variation in the predictive ability of SQGs (MacDonald, Ingersoll, Moore, et al. 2002). Evaluating the magnitude of the response allows for a comparison of critical values such as the minimum significant difference from control (Thursby et al. 1997; Long, MacDonald, et al. 1998).

Recommendations

- There is a need to better educate the users of SQGs. Short courses could be developed that deal with the development and application of SQGs and other measures of sediment quality. Once developed, these short courses should be presented at annual or regional meetings of the Society of Environmental Toxicology and Chemistry (SETAC) or at other regional, national, or international meetings.
- It is useful to develop site-specific data to determine if SQGs are predictive of effects or no effects at a site of interest. The number of samples needed to conduct a site-specific evaluation of the utility of SQGs can be estimated from the variance associated with established concentration–response relationships for SQGs (e.g., Figures 12-2, 12-3, and 12-5 in Chapter 12).
- Future analyses of predictive ability should include an evaluation of both sensitivity and efficiency of SQGs. To date, these analyses have

been performed primarily for AETs or have been used to compare the predictive ability of empirical SQGs and SEM–AVS (Long, Field, and MacDonald 1998; Chapter 12).

Use of Controlled Benthic Community Colonization Studies and Controlled Mesocosm Studies

Future evaluations of the predictive ability of SQGs should include controlled benthic community colonization studies and controlled mesocosm studies. Historically, assessments of impacts of contaminated sediment on benthic communities have involved the sampling of organisms across a gradient of multiple chemicals at a series of sampling sites (Chapter 12). These measures of benthic community structure and condition can then be evaluated as part of a sediment quality triad assessment of sediment chemistry and sediment toxicity (Chapters 5 and 6). The benefit of conducting these assessments is that direct measures of the effects of contaminants in sediment can be evaluated in situ under natural and realistic exposure conditions. However, the tradeoff is that there are numerous confounding factors associated with these types of studies ("Potential Confounding Factors," p 137). For example, it is difficult to establish appropriate reference sites that encompass the sediment conditions observed across the gradient sampled. In addition, interpretation of observed benthic community effects is confounded by variation in the physicochemical characteristics of the sediments and variation in the habitats sampled (e.g., grain size, total organic carbon [TOC], salinity, depth, currents, hydrological conditions; "Potential Confounding Factors," p 137; Chapters 5 and 16). Therefore, assessments of impacts of contaminated sediments on benthic communities would benefit from the use of additional controlled colonization and mesocosm studies that account for potential confounding factors.

Colonization studies have been used by ecologists for decades to evaluate the factors controlling distributions and population dynamics of sediment-dwelling organisms (Vanderhorst 1980, 1981; Cowell 1984; Hyland et al. 1985; Ruth et al. 1994; Hare 1995). Such procedures typically involve deployment of either replicated hard substrate samples (e.g., horizontal multi-plate Hester-Dendy samplers [Klemm et al. 1990] or vertical single-plate hard substrates [Schoener 1983]) or sediment trays deployed in specific locations of interest. The colonization of these substrates by benthic invertebrates is then monitored over a set period of time to evaluate processes such as recruitment or recovery from anthropogenic impacts. The assumption is made in these studies that reduced abundance or diversity within a treatment better reflects what happens in situ compared to measures of sediment toxicity in the laboratory or measures of benthic communities in the field that may be influenced by abiotic factors or by habitat differences.

These methods for conducting colonization studies have been adapted to evaluate the impacts of contaminated sediments in the field (e.g., Tagatz et al. 1983, 1986, 1987; Mattsson and Notini 1985; Blackman et al. 1988; Spies et al. 1988; Matthiessen and Thain 1989; Parrish et al. 1989; Berge 1990; Plante-Cuny et al. 1993; Hare et al. 1994; Watzin et al. 1994; Christie and Berge 1995; Morrisey et al. 1995; Liber et al. 1996; Flemer et al. 1997; Watzin and Roscigno 1997; Warren et al. 1998; Olsgard 1999; Roach et al. 2001; Ingersoll et al. 2004). These studies typically have involved the deployment of a dilution series of contaminated sediments in small trays (e.g., 0.5 to 2.0 L) placed in a variety of habitat types (e.g., ponds, lakes, estuaries). In addition to measuring changes in the benthic fauna, concurrent measures of sediment toxicity in the laboratory and measures of sediment chemistry can be performed on splits of the samples placed in the trays (Parrish et al. 1989; Liber et al. 1996; Ingersoll et al. 2004). Sources of the contaminants have included sediments spiked with materials such as cadmium (Hare et al. 1994), zinc (Watzin et al. 1994; Liber et al. 1996; Watzin and Roscigno 1997), copper (Morrisey et al. 1995; Olsgard 1999), creosote (Tagatz et al. 1983), dibutyl phthalate (Tagatz et al. 1986), fenvalerate (Tagatz et al. 1987), drilling mud (Blackman et al. 1988; Matthiessen and Thain 1989), antifouling paint (Matthiessen and Thain 1989), oil (Spies et al. 1988; Berge 1990; Plante-Cuny et al. 1993), or chlorpyrifos (Flemer et al. 1997). In contrast, Parrish et al. (1989) and Roach et al. (2001) evaluated contaminated sediment collected from the field placed into colonization trays. Importantly, these studies have been designed to control or account for sediment physical characteristics (e.g., grain size, TOC) and habitat characteristics (e.g., depth, lighting, current). It is critical to follow standard procedures for preparing spike sediments or diluted sediment for testing in these studies (e.g., ASTM 2003). These types of colonization studies with benthic invertebrates have been used to field validate mechanistic SQGs (Hare et al. 1994; Liber et al. 1996; Ingersoll et al. 2004) and empirical SQGs (Ingersoll et al. 2004).

Controlled mesocosm studies also are recommended as an approach for linking laboratory and field results. Such an approach brings important aspects of field-exposure conditions into the laboratory. For example, at the National Oceanic and Atmospheric Administration (NOAA) laboratory in Charleston SC, USA, mesocosm facilities have been used to mimic major features of tidal estuaries (Lauth et al. 1996). This facility includes benthic and pelagic phases exposed to varying stages of a tidal cycle. Strengths of this and similar mesocosm approaches (e.g., Hansen and Tagatz 1980; Clark and Noles 1994; Graney et al. 1994) include the ability to maintain replicated exposures over a range of treatment levels and to control for natural confounding factors.

Recommendations

- Field-colonization and mesocosm studies should be used as a means to further field validate the ability of SQGs to predict impacts on benthic communities in situ under controlled conditions.
- Reciprocal transplant experiments should be included as a type of field-colonization study. For example, a combination of spiked and reference-site sediments could be deployed across a range of reference and contaminated sites to evaluate colonization of benthic invertebrates.
- When conducting field-colonization studies, both reference and treatment samples should be compared to samples of the surrounding ambient benthic community. The study design and sampling parameters should be selected to provide a basis to account for confounding factors that could otherwise influence the response of the benthic community (e.g., salinity, grain size, TOC, depth).

Potential Confounding Factors

Efforts to estimate sediment toxicity and benthic community effects in relation to SQGs need to account for potentially confounding factors. The response of organisms in laboratory toxicity tests or the benthic community assessment may potentially be influenced by a variety of sediment conditions beyond contaminants. Importantly, many of these factors can co-vary with gradients in sediment chemistry. These factors may include variables such as ammonia, sulfide, salinity, grain size, sediment stability, organic carbon content of sediment, food quantity and quality. Additionally, the response of benthic communities in the field also may be influenced by factors such as water depth, hydrologic condition, season, or other characteristics of the habitat. Additional biological factors that can influence the response of organisms to contaminants include sensitivity to contaminants, behavior or feeding type (e.g., epibenthic, tube dwelling, or burrowing; direct ingestion of sediment versus filter feeding), presence of indigenous organisms (e.g., predators, congenerics), and seasonal changes in the sensitivity or abundance of organisms. The consistency and accuracy of taxonomic identifications among laboratories can lead to differences in results. Mishandling of sediments for toxicity tests or for chemical analyses may also lead to spurious results.

Seasonal or temporal variation in the bioavailability of contaminants in sediment and the method used to collect or manipulate the sediments or organisms from the field also can influence the outcome of sediment assessments. Finally, the selection of the organisms evaluated and the endpoints measured in the laboratory or in the field can influence the relationships between the responses of the organisms and sediment chemistry.

It is important to understand the contribution of these various factors on the outcome of either laboratory or field studies. In the laboratory, these various factors should be accounted for in order to determine the contribution of contaminants to the observed response of the test organisms. In the field, benthic communities must be able to cope with these factors to survive in the environment. The relative contribution of these various factors needs to be accounted for in field studies to determine the contribution of contaminants to the observed response of benthic communities. The challenge is to determine if contaminants in sediment are contributing to impacts on benthic communities beyond the influence of these naturally occurring factors.

The influence of many of the factors listed above can be accounted for by proper experimental design. For laboratory toxicity tests, USEPA (1994, 2000c) and ASTM (2003) report tolerance ranges for tests organisms to factors such as ammonia, grain size, and organic carbon. These physicochemical characteristics of test sediment should be within the tolerance limits of the test organisms. Ideally, the limits of a test organism should be determined in advance and controls for factors including grain size and organic carbon should be included to determine the potential influence of these features. The effects of sediment characteristics on the results of laboratory sediment tests can also be addressed with regression equations to account for the potential influence of these factors (DeWitt et al. 1988; USEPA 1994, 2000b; SEA 1996; ASTM 2003). However, unless confounding factors co-occur with chemical gradients, the observation of <10% toxic samples at concentrations below the low empirical SQGs suggests that confounding factors may not be dominating the prediction by empirical SQGs ("Use of Empirical SQGs to Predict the Presence or Absence of Toxic Effects," p 123).

Chapter 16 describes a process to address the potential influence of confounding factors on estimates of sediment toxicity (i.e., selection, testing, accounting, removal, replacement [STARR]). This process is designed to help in planning and implementing experiments that address the potential influence of confounding factors (CFs) in sediment assessments. The acronym represents a process inclusive of the following steps: the Selection process for determining appropriate test organisms or benthic endpoints based on a series of questions of concern, the Testing procedures or the design of the study that are implemented to separate the effects of chemical contaminant concentrations from CFs, an Accounting of the influence of CFs, Removal of the CF to determine if the observed adverse affect can be attributed to the removed CF, and Replacement of the removed CF to demonstrate that the degree of adverse effect is confirmed.

A growing body of evidence suggests that impacts on estuarine benthic communities may be occurring at chemical concentrations below those that are identified in 10-d laboratory toxicity tests with single species and mortality as an endpoint (Table 12-24 in Chapter 12). However, the relative influence of natural factors (i.e., grain size or TOC) versus chemical-induced toxicity on benthic communities has not been fully quantified (Chapter 12).

Studies of the relationships between the responses of benthic communities to contaminants in sediment should focus on reducing uncertainty due to CFs. One approach used recently and that has been suggested here is to identify the physical, chemical, and biological factors in reference conditions that influence key structural or functional attributes of benthic communities (e.g., numbers of species and individuals, diversity, biomass, dominance, presence of indicator species, relative percentages of various feeding types). With such an approach, one can then judge a sample as having a degraded benthos based on a comparison of select biological attributes to reference ranges of those same attributes expected under reference or relatively uncontaminated conditions for similar types of habitats (Reynoldson et al. 1995; SEA 1996; Hyland et al. 1999, 2003; Chapters 5 and 6). An important prerequisite for any efforts along this line is to collect data on appropriate CFs in order to support the statistical analyses (e.g., multivariate data analysis).

Ammonia is an example of a CF that has been demonstrated to influence the response of organisms in some laboratory toxicity tests with sediments. Moore et al. (1997) reported that the probability of toxicity due to ammonia alone in sediment toxicity tests conducted with the marine amphipod *Leptocheirus plumulosus* may run as high as 18% in dredged material evaluations. Additionally, ammonia concentrations were observed to influence the response of marine amphipods in more than 60% of the dredged materials evaluated at a specific site. Moore et al. (1997) also reported that porewater ammonia concentrations in dredged material were higher in dredged materials than in surface sediment by a factor of 4, with the porewater ammonia concentrations in dredged material averaging 40 mg N/L (total ammonia as nitrogen). This average concentration in dredged material exceeds the threshold of 20 mg ammonia N/L established by USEPA Region II and USACE NY District dredging projects by a factor of 2 (Ferretti et al. 2000). Frazier et al. (1996) also reported that porewater ammonia concentration increase with sediment depth at all study sites and as such deeper cores would be expected to have more toxicity associated with this factor. Additionally, ammonia concentrations were observed to influence the response of marine amphipods in all of the dredged-materials evaluated at a single site (Barrows et al. 1996).

This site-specific identification of ammonia toxicity led to the development of modified testing procedures to address the influence of increased toxicity due to ammonia in assessments of dredged materials using marine sediment toxicity tests with amphipods (USEPA 1994). Studies conducted with marine sediments collected from the upper 2 cm have not found ammonia to routinely contribute to the toxicity observed in sediment tests with marine amphipods (RC Swartz, personal communication). Similarly, whole-sediment tests with marine amphipods and porewater tests with sea urchins indicate that only 2% of the surficial samples tested nationwide in the NOAA National Status and Trends Program had concentrations of unionized ammonia that exceeded low effect concentrations in these tests (ER Long, KJ Scott, RS Carr; personal communications). However, testing of dredged material from marine environments, which typically included deeper sediments, has routinely encountered toxicity associated with ammonia (Frazier et al. 1996; Moore et al. 1997). There is a need to determine whether ammonia acts as a confounding factor (e.g., a laboratory artifact) or whether ammonia is actually a secondary toxicant associated with a temporary disruption of the microbial community passing from freshwater into an estuarine environment (Rysgaard et al. 1999) or a more permanent disruption associated with sediment contamination (Sverdrup et al. 2002).

Other sediment attributes that can result in increased mortality include the presence of adequate quantities of quality food. Older buried sediments with little to no chemical contamination have been shown to be harmful to test organisms as a result of food that is of insufficient quality (Pinza et al. 1996). In this case, sediment with low TOC (<0.1%) had low survival of the polychaetes (*Nephtys caecoides*). The lack of sediment contamination and the low organic carbon content implicated the potential toxicity as being the result of low food availability. Addition of quality organic detritus (dried *Enteromorpha* spp.) that increased TOC concentration to 0.2% reduced the observed mortality and resulted in acceptable survival of test organisms.

The absence of an effect associated with elevated chemical contaminants can indicate that the chemicals present in the sediment are not available to the test organisms. For example, the lack of an effect associated with concentrations of SQGs that would predict effects could occur when organisms are isolated from exposure to elevated concentrations of contaminants (e.g., species that live in burrows, in biological mats, or in tubes such as *Corophium* spp.), the biological endpoint has low sensitivity to the mixture of contaminants, or the bioavailable fraction of the contaminants may be less than the total measured contaminant concentration ("Influence of Chemical State of Contaminants on Predictive Ability of SQGs," p 142). This circumstance can also arise when the field-collected sample is not thoroughly homogenized, resulting in an inconsistent relationship between the chemical analyses and the actual exposure concentrations.

The influence of CFs on the predictive ability of SQGs is not always known. Many of the studies summarized in Table 12-24 in Chapter 12 did not sufficiently report measures of potential confounding factors. Future assessments of sediments should document and report the influence of CFs in both laboratory and field studies.

Recommendations

- Existing guidance should be followed for addressing potential CFs, such as analyses for ammonia, grain size, and hydrogen sulfide in sediment toxicity tests (e.g., ASTM 2003 and USEPA 1994, 2000b).
- The database used to develop and evaluate SQGs should be reexamined to search for potential evidence of confounding factors (e.g., STARR process outlined in Chapter 16).
- Future studies of benthic community responses should better account for the influence of CFs. The application of approaches that provide a basis for comparing benthic condition in samples to the optimum reference conditions for similar ecological types will permit a better assessment of the influence of CFs and contaminants on structural and functional characteristics of benthic communities.
- Mesocosm studies and manipulative field studies ("Use of Controlled Benthic Community Colonization Studies and Controlled Mesocosm Studies," p 135) should also be used to further validate the ability of SQGs to predict impacts on benthic communities under controlled conditions.

Influence of Chemical State of Contaminants on Predictive Ability of Sediment Quality Guidelines

The chemical state of the contaminants can reduce the predictive ability of the SQGs. For example, chemicals may exist as solid phases such as zero valence metals (e.g., lead shot), as insoluble mineral ores (e.g., smelter slag or $CuFeS_2$), or as components of complex mixtures such as tar balls. Insoluble manmade substances also can contain potentially toxic chemicals (e.g., chips of antifouling paint). The questions are "How do these various physical states for chemicals of concern affect the predictions of SQGs?" and "How prevalent are these conditions in sediment?"

The state of a chemical in the sediment is known to affect the bioavailability of that chemical. In the most extreme case, a chemical in the form of an inert insoluble compound, impervious to attack by any biological agent, will not have an effect associated with its chemical components. Nevertheless, the chemical will be determined to be present in the sediment if the analytical method is designed to liberate the chemical. For example, the method for total recoverable metal will dissolve zero valence metals, and the rigorous solvent extractions will dissolve tar balls. The other extreme is a chemical with a very low partition coefficient to sediment solids (e.g., ammonia; K_p about 1). In this instance, the chemical is essentially all in the pore water and would probably exhibit the same bioavailability in any sediment.

If an SQG does not explicitly consider bioavailability (e.g., an SQG based on total concentration), then the "bioavailable" fraction implicit in this SQG is determined by the data set used to establish the relationship between total chemical concentration and toxic effect. The usual case for organic chemicals is that the chemical either exists in the pore water or is sorbed to sediment particles. While empirical SQGs are based on total concentrations of chemicals rather than on a bioavailable fraction, it is recognized that if a chemical cannot be transferred to an organism, it cannot exert an effect. The field samples that were used to develop and evaluate empirical SQGs probably have differences in the bioavailable fraction of contaminants; however, there is probably an average range of bioavailability inherent in the samples used to develop the empirical SQGs. The empirical SQGs may not be applicable to unusual situations that cause bioavailability to be outside that range (e.g., "Influence of an Unusual Sediment Matrix on the Predictive Ability of SQGs," p 143).

Mechanistic SQGs that are explicitly intended to account for bioavailability use partitioning models to predict the concentration of chemicals in pore water. If the partitioning or chemical speciation model does not explicitly consider phases such as paint chips or tar balls, then the prediction of porewater concentrations, and therefore the toxicity, will be incorrect. For metals, an example would be including the mineral phase in the partitioning model. For organic chemicals, examples would be inclusion of nonaqueous-phase liquids (NAPLs) or tar balls as part of the calculation of porewater concentration.

Recommendations

- If it is expected that unusual chemical speciation is present in a sediment sample or is observed when collecting the samples, then it would be expected that SQGs would not accurately predict toxicity. In these instances, a direct determination of toxicity or effects is needed.
- If the concentrations of particular chemicals are well above the upper-range SQGs (e.g., Table 12-24 in Chapter 12) and toxicity is not observed, then the presence of these unusual phases should be suspected. For these cases, a measurement of the chemical concentration in pore water of the sediment can be used to evaluate whether the chemical in sediment is elevated to a concentration of concern.

Influence of an Unusual Sediment Matrix on Predictive Ability of Sediment Quality Guidelines

The presence of an unusual sediment matrix (e.g., black carbon, peat, wood chips) can reduce the predictive ability of the SQGs. SQGs, whether empirically derived or mechanistically derived, have been determined using sediments with “typical” matrices; that is, the geochemical make-up of the sediments used to generate or evaluate the SQGs generally fell within the bounds commonly observed in natural sediments. However, there may be instances where certain atypical constituents of the sediment matrix may alter the expected bioavailable form of a contaminant. This would in turn alter the expected bioavailability of that contaminant, compromising the ability of an SQG to predict toxicity in that atypical sediment matrix. These errors can be in both directions, resulting in either an overprediction or an underprediction of effects. There is a need to be aware of the likelihood that these binding phases may be present in a given sediment under study.

Only a portion of the contaminants present in sediments is available for uptake into aquatic organisms. For example, “natural” organic carbon such as fulvic and humic acids can bind both metal and organic contaminants, thereby reducing bioavailability. The predictive ability of SQGs can be modified by the presence of these binding phases. This is explicitly addressed with mechanistic SQGs and implicitly accounted for through empirical SQGs. However, there can be other forms of organic carbon that behave differently than "natural" organic carbon, leading to either an overestimate or an underestimate of sorptive capacity. SQGs may not be designed to account for these variations and thus may not adequately predict the presence or absence of adverse effects.

An example of this type of a sediment constituent altering bioavailability is the presence of black carbon (e.g., soot carbon). Black carbon, which under most analytical schemes will be detected as part of the TOC fraction, has an exceptionally high sorptive affinity for planar organic compounds such as PAHs (Gustafsson et al. 1997a, 1997b). Partition coefficients for PAH sorption to black carbon can be several orders of magnitude higher than partition coefficients to "natural" organic carbon (Acardi-Dey and Gschwend 2002). By incorrectly assuming that all the measured organic carbon in sediment is "natural" organic carbon, sediments with unusually high levels of black carbon may have a greater sorption of PAHs than expected. The application of a mechanistic SQG to these types of sediments would result in an overprediction of bioavailability by underpredicting the sorption of contaminants to the sediment matrix. When empirical SQGs were applied, an overprediction of bioavailability also would occur because the unusually high sorptive capacity of the sediment is greater compared to the sorptive capacity of the sediments used to generate the SQG. Black carbon is ubiquitous but can be particularly elevated when the source of contaminants is of pyrogenic origin, such as near smelters or bodies of water receiving urban street runoff.

A second example of a sediment constituent that may alter bioavailability is the presence of wood fragments, plant fibers, or peat. Wood or peat residues might be measured as contributing to the TOC under some analytical methods. However, unlike black carbon, the sorptive affinity of contaminants for wood fragments and peat are generally less than for "natural" organic carbon (Mackay and Gschwend 2000). In cases where peat or wood fragments are present, an overestimate of the sorption of contaminants might occur, resulting in an underestimate of bioavailability. Visual inspection of the sample may indicate whether peat or wood fragments are present in a sediment sample.

Other confounding factors include the inability to adequately remove shell fragments before analyzing for TOC. As above, shell fragments do not adsorb contaminants to an appreciable extent and might be measured as organic carbon under most analytical methods. Typically, shell fragments can be removed before organic carbon analysis (either by a gentle acid treatment or by physical removal). Large amounts of limestone in the geological matrix will cause a similar problem. However, in sediments with a large amount of shell material, more aggressive treatments may be necessary to adequately remove the shell material. As with the wood fragments or peat, failure to remove shell fragments can result in an overestimate of the sorption of contaminants. Finally, SQGs and sediment toxicity tests should not be applied to geological matrices other than aquatic sediments (e.g., terrestrial soil or metal ore). The geochemical characteristics of these matrices are different from aquatic sediments, and application of SQGs to these matrices other than aquatic sediments is either uncertain or is unjustified.

Recommendations

- If these types of confounding factors are expected in the sediment matrix (e.g., visual inspection, proximity to a smelter or wood mill), the SQGs would not expected to be predictive of either toxicity or nontoxicity. Large shells or wood chips should be removed from the sample before chemical analysis. If these materials cannot be removed, a direct measurement of effects is recommended using toxicity tests or benthic community surveys. Similarly, if the result of a direct measurement of effects is inconsistent with contaminant concentrations, investigations should be conducted to determine if these confounding factors are present in the sediment matrix.
- Further testing of the constituents of the sediment matrix may be warranted to explain why observed effects (or lack of effects) are inconsistent with available SQGs. Protocols for analysis of these matrix constituents are described (e.g., Gustafsson et al. 1997a, 1997b) and may need to become routine analytical tools to aid in the interpretation of sediment data.

Single Exceedances of Empirical Sediment Quality Guidelines

Toxic sediment in which only a single empirical SQG is exceeded should not be assumed to be toxic as a result of the presence of that substance. It is important to accurately predict the presence or absence of toxicity in field-collected sediments. It is also important to identify the factors that are causing or substantially contributing to the observed sediment toxicity. Such information enables limited resources to be focused on the highest priority sediment issues. Empirical SQGs have been developed and evaluated primarily using field-collected sediments that contain complex mixtures of contaminants. As such, the relationships between toxicity and chemistry are based on association rather than cause (e.g., the co-occurrence of other PAHs likely contributes to the observed toxicity). While a sample with a single empirical SQG exceeded may be toxic, this chemical may not be related to the factors causing the toxicity. For example, exposures conducted with metal-spiked sediments indicate that toxicity is overestimated when individual empirical SQGs are used to predict the toxicity of metals in freshwater, estuarine, or marine sediments (Berry et al. 1996; Besser et al. 2001, 2003). Additionally, Di Toro and McGrath (2000) reported that empirical SQGs for individual PAHs were 1 to 2 orders of magnitude lower than the narcotic concentrations that are observed or predicted to cause toxic effects in aquatic organisms in sediment. However, when SQGs are compared for total PAHs developed using empirical and mechanistic approaches, there is much better agreement because more of the components that contribute to the toxic response are considered in mixtures. See "Sediment Quality Guidelines for Total PAH and Total PCBs" (p 147) for a discussion on consensus-based SQGs that build upon the agreement among the various empirical and mechanistic SQGs. See Chapter 3 for a discussion of approaches that have been used to estimate total PAHs and total PCBs in sediment samples.

There are methods that can be used to determine the chemicals causing toxicity. The mechanistic SQGs, for example, are specifically designed to identify concentrations of contaminants in sediment that cause toxicity. Thus, the first step should be to compare the sediment chemistry to the available mechanistic SQGs. Because these SQGs predict the porewater concentrations, the predictions can be compared to direct measurements. In fact, examining porewater chemistry is the most direct method currently available to determine the nature of the toxic chemical. Other available experimental methods identify chemicals in sediment that cause toxicity, including spiked-sediment toxicity tests or toxicity identification evaluations (TIEs). These approaches can be used to provide a basis for identifying the concentrations of sediment-associated contaminants that are causing or substantially contributing to sediment toxicity (Ingersoll et al. 1997; MacDonald, DiPinto, et al. 2000, MacDonald, Ingersoll, and Berger 2000). Unfortunately, only limited relevant data are available to assess effects of spiked sediments (MacDonald, Ingersoll, and Berger 2000). Differences in spiking procedures, equilibration time, and lighting conditions during exposures confound the interpretation of the results of sediment spiking studies, especially for PAHs. Moreover, many sediment spiking studies have been conducted to evaluate bioaccumulation using relatively insensitive test organisms or in sediments containing mixtures of chemical substances (MacDonald, Ingersoll, and Berger 2000). TIEs might also be useful in helping to identify major groups of chemicals that contribute to the observed toxicity. However, TIE methods may lack the specificity needed to identify specific chemicals causing toxicity (e.g., Besser et al. 1998). Evaluation of bioaccumulated residue can also add evidence for causality, particularly if the residue concentration is above a recognized tissue effect level ("Bioaccumulation-Based SQGs," p 149).

Recommendations

- Sediments in which only a single empirical SQG is exceeded should not be assumed to be toxic on the basis of the presence of that substance alone.
- Spiked-sediment tests and TIE methods should be used to help identify chemicals that may cause the observed toxicity in a sediment sample (see also "Sediment Quality Guidelines for Total PAH and Total PCBs," below).

Sediment Quality Guidelines for Total PAH and Total PCBs

SQGs for total PAHs and total PCBs derived using empirical approaches are similar to guidelines derived using mechanistic approaches with a similar narrative intent. This concordance suggests that these mixtures are causally implicated in the toxicity observed in a substantial number of sediments. Empirical SQGs for mixtures of contaminants in sediments have been reported for total PCBs or total PAHs (Swartz 1999; MacDonald, DiPinto, et al. 2000, MacDonald and Ingersoll 2001; Table 12-21 in Chapter 12). These SQGs for estimating chronic effects of total PAHs or total PCBs have also been calculated using equilibrium partitioning (mechanistic SQGs for total PAHs based on narcosis [Swartz 1999; Di Toro and McGrath 2000] and mechanistic SQGs for total PCBs calculated using the results of water-only toxicity tests [MacDonald, Ingersoll, Smorong, et al. 2002]). Unlike the case for individual chemicals described in "Single Exceedances of Empirical SQGs" (p 145), these mechanistic and empirical SQGs are in reasonable agreement. Because the 2 methods used to derive SQGs for total PAHs or total PCBs are very different and depend on entirely different sets of data, this concordance presumably tells us something important about the causes of toxicity in sediments.

Swartz (1999) reported that SQGs for total PAHs with a consistent narrative intent were similar among values derived using mechanistic or empirical approaches and that this similarity was probably not coincidental. Furthermore, Swartz (1999) concluded that these empirical SQGs reflect causal rather than correlative effects when expressed as total PAHs rather than when expressed as empirical SQGs for individual PAHs. Similarly, MacDonald, DiPinto, et al. (2000) and MacDonald, Ingersoll, and Berger (2000) compared midrange or upper-range empirical SQGs and mechanistic SQGs derived for total PAHs (summarized in Table 12-21 in Chapter 12) or for total PCBs (0.2 to 0.8 µg/L no-effect concentrations determined in freshwater or marine water-only chronic toxicity tests with individual aroclors). The results of these evaluations also indicated that these empirical SQGs were generally comparable to the mechanistic SQGs for total PAHs or total PCBs (i.e., within a factor of 3; MacDonald et al. 1996; Smith et al. 1996, MacDonald, DiPinto, et al. 2000; MacDonald, Ingersoll, and Berger 2000). The fact that multiple applications of mechanistic and empirical approaches yield similar results, with the total PAH results being very similar, suggests that for many of the sediments, it is not the individual compounds, but the mixtures of PAHs (and to a lesser extent) PCBs that may cause or substantially contribute to toxicity observed in many field-collected sediments.

There is one difference between the empirical SQGs and the mechanistic SQGs: the mechanistic SQGs are normalized to organic carbon concentrations (e.g., µmol/g organic carbon for mechanistic SQGs, versus µg/kg dry weight for empirical SQGs). In order to make the 2 values equivalent, it is necessary to use an average molecular weight and an average sediment organic carbon (Di Toro and McGrath 2000). The data sets used for the development of empirical SQGs often have a relatively narrow range of organic carbon concentrations (e.g., mean of 4.3% [standard deviation 4.8, $n = 865$] for a freshwater sediment database and mean of 1.8% [standard deviation of 1.9, $n = 2599$] for a marine sediment database; DD MacDonald, personal communication), which potentially explains why organic carbon–normalized empirical SQGs are no more predictive than the dry weight–normalized empirical SQGs (Barrick et al. 1988; Long et al. 1995, 1998; Ingersoll et al. 1996). However, if the sediment that is being evaluated has an organic carbon concentration substantially above these mean concentrations (e.g., >8% for freshwater sediment or >4% for marine sediment), then the organic carbon–normalized empirical SQGs should be considered. Additional analyses should be performed using these databases to determine if normalization to organic carbon improves the predictive ability of empirical SQGs for total PAHs or total PCBs. It should also be noted that mechanistic SQGs do not apply to sediments with an organic carbon content < 0.1 to 0.2% because of the breakdown of the organic carbon partitioning model (Karickhoff et al.1979; Di Toro 1985).

Recommendations

- It is important to calculate total PAHs or total PCBs using a consistent number of compounds (e.g., Di Toro and McGrath 2000).
- Mechanistic and empirical SQGs could be used in combination to identify concentrations of total PAHs or total PCBs in sediment that may be causing or substantially contributing to toxicity.
- Additional analyses of freshwater and marine databases should be conducted to further evaluate the influence of organic carbon normalization of empirical SQGs for total PAHs and total PCBs.

Bioaccumulation-Based Sediment Quality Guidelines

Existing effects-based SQGs (e.g., ERMs, AETs) are not designed or intended to predict bioaccumulation-based effects. However, biota–sediment accumulation factors (BSAFs) for nonionic organic compounds may be used to generate SQGs predictive of effects in sediment-dwelling organisms. Bioaccumulation is only the first step and should be linked to predicted tissue residue effects concentrations. The desire for cost-effective screening tools for contaminated sediments has resulted in the development of a variety of numerical SQGs. Nearly all of the approaches developed to date are neither designed nor intended to be protective of effects through bioaccumulation Thus, there is a need under certain regulatory programs in the US (e.g., the Marine Protection, Research and Sanctuaries Act [MPRSA] and the Clean Water Act [CWA]) to assess the potential of sediment-associated contaminants to bioaccumulate and to evaluate potential effects resulting from bioaccumulation of chemicals from sediment. Although there have been limited efforts to establish such effects-based SQGs for bioaccumulation on a regional basis for use in screening-level risk assessments, the predictive ability of these methods has not been evaluated, and the uncertainty associated with their application is unknown (Chapter 11). One such effort developed for the New York State Department of Environmental Conservation used equilibrium partitioning to back-calculate sediment concentrations for nonionic organics from consumption limits used in the derivation of water quality criteria (WQC) for the protection of human health and wildlife (NYDEC 1998). While bioaccumulation-based SQGs are proposed and have been proposed, and in some cases have been implemented as SQGs (e.g., NYDEC 1998), the predictive ability of these guidelines has not been evaluated. The State of Washington also uses equilibrium partitioning and trophic transfer factors to derive sediment quality values for nonpolar organics protective of human health (Chapter 11). This section focuses on the science of predicting toxic response from sediment exposure using bioaccumulation and tissue concentrations associated with adverse effects.

While bioaccumulation of a chemical into an organism does not in itself represent an adverse effect, bioaccumulation can be used to generate SQGs representing adverse effects. Establishing an SQG based on bioaccumulation can be accomplished through either a mechanistically or empirically based tissue residue-effects framework (Chapter 11). One approach proposed by Shepard (1998) generates a protective tissue concentration from a waterborne exposure by use of WQC and bioconcentration factors (BCFs) for a compound. This approach estimates the tissue concentration associated with toxic effects from acute WQC and applies a 10-fold acute-to-chronic ratio to obtain an estimated tissue residue value for chronic responses. WQC are established to be protective of 95% of the species tested, and therefore the tissue concentration should be protective as well. Mixtures can be addressed in a toxic unit approach in an additive manner. Issues of biotransformation should be specifically considered when using this approach because some of the WQC include readily biotransformed compounds. Where biotransformation is an issue, the approach may not work particularly well. These tissue concentrations could be linked directly to sediment concentrations through an equilibrium partitioning calculation that assumes that the calculated porewater concentration represents the effective exposure.

The critical body residue (CBR) approach can be used in a fashion similar to the method described above. However, in this case, the tissue concentrations associated with adverse effects are used. These can be the lethal residue (e.g., LR50), effective residue (e.g., ER25) or lowest observed effect residues (LOERs)) determined in dose–response studies (Chapter 11). The connection to sediments can then be made based on class-specific BSAFs. This approach has been developed for tributyltin (Meador et al. 2002a) to determine 3 different response metrics (e.g., mortality, growth, and imposex) in snails. A similar approach was used for PCB tissue concentrations in fish and site-specific BSAFs (Meador et al. 2002b). This method is generally not recommended for higher trophic levels unless reliable BSAFs can be determined. The CBR SQG approach may also be applicable as a screening tool for those contaminants where tissue residues and effects can be modeled (e.g., those exhibiting narcosis) and generalized BSAFs from equilibrium partitioning theory can be applied. These adverse tissue concentrations could be related directly to sediment, based on the toxicant class-specific BSAF relationships (Chapter 11) or based on site-specific BSAFs for sediment-dwelling organisms.

Alternatively, it is possible to establish a bioaccumulation-based SQGs through the development of regional background tissue concentrations for toxicants (Metcalf and Eddy 1995). The regional background numbers were established by collecting the tissue concentration data from monitoring studies for reference sites in the region and then ranking these values for a particular contaminant. It is then possible to select a rank (e.g., 90th percentile) to establish the tissue concentrations that represent the appropriate regional background for a particular compound. Unlike the calculation above, these tissue concentrations have no definitive relationship to effects. However, they are presumed to be protective because they represent the regional reference background. Further, because a background tissue concentration is selected for each compound of interest, the issue of mixtures of additivity does not have to be considered. In this case, biotransformation is not a complicating factor because the concentrations in the database are based on measured concentrations in sediment-dwelling organisms. These tissue concentrations can then be related to the sediment concentration through the BSAFs on a compound class-specific basis as suggested for the other approaches described above (BCF and CBR). Both of these approaches should provide estimates that would establish the sediment concentrations that would be protective of organisms from toxicity. Both of the above approaches attempt to address direct and indirect effects on sediment-dwelling organisms but are not specifically designed to protect against trophic transfer.

It is important to note that attempts have been made to develop bioaccumulation-based tissue residue values for metals (Shephard 1998) which could be linked to sediment concentrations in a manner similar to that described above for organics. Additional research is required before such approaches can be more fully developed and applied for bioaccumulative metals.

Finally, while there are approaches for generating bioaccumulation-based SQGs for predicting potential effects of organics in sediment-dwelling organisms, we are not as advanced in our ability to estimate potential effects to higher trophic levels through food web transfer. Any assessment of potential effects to higher trophic levels will still be site specific and should rely on ecological risk assessment techniques. The accuracy of such assessments will be largely dependent on the knowledge of the system, the availability of appropriate input parameters (trophic transfer coefficients), and the exposure scenarios being evaluated.

Recommendations

- All the approaches outlined above should be evaluated and assessed against existing databases to verify their predictive ability prior to application.
- Regionally developed BSAFs also should be verified periodically.
- In general, BSAFs used to develop bioaccumulation-based SQGs should be derived from steady-state tissue concentrations.
- Predicting the availability, uptake, and potential effects of bioaccumulating metals (e.g., Hg, Se, and possibly As) requires additional research before bioaccumulation-based SQGs can be developed and applied.
- Even though the CBR-SQG approach works well for tributyltin and exhibits low variability, it has not been explored for other toxicants. Several existing databases for tissue residue effects (e.g., the USACE Environmental Residue Effects Database at www.wes.army.mil/el/ered and Jarvinen and Ankley 1999) could be used to examine this approach for other contaminants of concern, in addition to values produced by theoretical applications.

Summary

The following consensus statements were developed regarding the ability of SQGs to predict effects or no effects in laboratory toxicity tests, benthic community assessments, or bioaccumulation assessments of sediment-dwelling organisms:

- Empirically derived SQGs can be used to predict the probability of the presence or absence of toxic effects with a known level of statistical confidence based on the results of the analyses of the data sets evaluated to date.
- Mechanistic SQGs based on partitioning theory attempt to causally relate sediment concentration to toxicity. The availability and success or failure of mechanistic SQGs depends on the adequacy of the partitioning model and its parameters and on the assumption that exposure is either from pore water or sediment particles or both.
- SQGs should only be used with an understanding of their derivation, their narrative intent, and their predictive ability.
- Future evaluations of the predictive ability of SQGs should include controlled benthic community colonization studies and controlled mesocosm studies.

- Efforts to estimate sediment toxicity and benthic community effects in relation to SQGs need to account for potentially confounding factors.
- The chemical state of the contaminants (e.g., paint chips, lead shot, tar balls, metal ore) can reduce the predictive ability of the SQGs.
- The presence of an unusual sediment matrix (e.g., black carbon, peat, wood chips) can also reduce the predictive ability of SQGs.
- Toxic sediment in which only a single empirical SQG is exceeded should not be assumed to be toxic because of the presence of that substance.
- SQGs for total PAHs and total PCBs derived using empirical approaches are similar to guidelines derived using mechanistic approaches with a similar narrative intent. This concordance suggests that these mixtures are causally implicated in the toxicity observed in a substantial number of sediments.
- Existing effects-based SQGs (e.g., ERMs, AETs) are not designed or intended to predict bioaccumulation-based effects. However, BSAFs for nonionic organic compounds may be used to generate SQGs predictive of effects in sediment-dwelling organisms. Bioaccumulation is only the first step and should be linked to predicted tissue residue effects concentrations.

The following research needs were identified for improving our understanding of the predictive ability of SQGs:

- Develop a better understanding of the reasons why benthic communities in estuaries appear to be more sensitive to contaminants in sediment, compared to 10-d laboratory tests (e.g., conduct additional data analyses or conduct controlled laboratory and field studies).
- Develop procedures to better understand and account for the potential influence of confounding factors in toxicity, bioaccumulation, and benthic community studies.
- Further validate the predictive ability of SQGs in the laboratory and in the field for single chemicals and for complex mixtures.
- Develop a better understanding of additive, antagonistic, and synergistic effects of chemical mixtures in sediments.
- Further evaluate the ability of SQGs to predict chronic and sublethal endpoints. Additional research is needed to develop methods for assessing effects of contaminated sediment using under represented groups of organisms (e.g., beyond amphipods, midge, and polychaetes).

- Further evaluate the predictive ability of SQGs using controlled benthic community colonization studies, controlled mesocosm studies, or in situ toxicity testing to better account for abiotic and habitat factors influencing the response of benthic invertebrates in the field.
- Further evaluate the predictive ability of SQGs using measures of efficiency and sensitivity (e.g., Shine et al. 2003).
- Improve our understanding of the relationship between chemical residues and toxic response.
- Improve our ability to account for variation in site-specific bioavailability to improve the potential for developing bioaccumulation-based SQGs.
- Develop a better understanding of the factors controlling bioaccumulation of metals and the importance of metal tissue residues.

References

Accardi-Dey A, Gschwend PM. 2002. Assessing the combined roles of natural organic matter and black carbon as sorbents in sediments. *Environ Sci Technol* 36:21–29.

[ASTM] American Society for Testing and Materials. 2003. Standard test methods for measuring the toxicity of sediment-associated contaminants with freshwater invertebrates. ASTM annual book of standards, Volume 11.05. West Conshohocken (PA): ASTM. E1706-00.

Barrick R, Becker S, Brown L, Beller H, Pastorok, R. 1988. Sediment quality values refinement: 1988 update and evaluation of Puget Sound AET, Vol. I. Bellevue (WA): PTI Environmental Services. PTI Contract C717-01.

Barrows ES, Antrim LD, Pinza MR, Gardiner WW, Kohn NP, Gruendell BD, Mayhew HL, Word JQ, Rosman LB. 1996. Evaluation of dredged material proposed for ocean disposal from federal projects in New York and New Jersey and the Military Ocean Terminal (MOTBY). PNL-11280. Prepared for the US Army Corps of Engineers-New York District by the Pacific Northwest National Laboratory operated by Battelle Memorial Institute for the US Department of Energy under Contract DE-AC06-76RLO 1830. Sequim (WA): Battelle Memorial Inst.

Berge JA. 1990. Macrofauna recolonization of subtidal sediments. Experimental studies on defaunated sediment contaminated with crude oil in 2 Norwegian fjords with unequal eutrophication status. Community responses. *Mar Ecol Prog Serv* 66:103–115.

Berry WJ, Hansen DJ, Mahony JD, Robson DM, Di Toro DM, Shipley BP, Rogers B, Corbin JM, Boothman WS. 1996. Predicting the toxicity of metal-spike laboratory sediments using acid-volatile sulfide and interstitial water normalizations. *Environ Toxicol Chem* 15:2067–2079.

Besser JM, Allert AL, Hardesty DK, Ingersoll CG, May TW, Wang N. 2001. Evaluation of metal toxicity in streams of the Upper Animas River watershed, Colorado. Columbia (MO): USGS. Biological Science Report BSR 2001-0001.

Besser JM, Brumbaugh WG, May TW, Ingersoll CG. 2003. Effects of organic amendments on the toxicity and bioavailability of cadmium and copper in spiked formulated sediments. *Environ Toxicol Chem* 22:805–815.

Besser JM, Ingersoll CG, Leonard E, Mount DR. 1998. Effect of zeolite on toxicity of ammonia in freshwater sediments: Implications for sediment toxicity identification evaluation procedures. *Environ Toxicol Chem* 17:2310–2317.

Blackman RAA, Fileman TW, Law RJ, Thain JE. 1988. The effects of oil-based drill-muds in sediments on the settlement and development of biota in a 200-day tank test. *Oil Chem Pollut* 4:1–19.

Brown SS, Gaston GR, Rakocinski CF, Heard RW. 2000. Effects of sediment contaminants and environmental gradients on macrobenthic community trophic structure in Gulf of Mexico estuaries. *Estuaries* 23:411–424.

Christie H, Berge JA. 1995. In situ experiments on recolonization of intertidal mudflat fauna to sediment contaminated with different concentrations of oil. *Sarsia* 80:175–185.

Clark JR, Noles JL. 1994. Contaminant effects in marine/estuarinesystems: Field studies and scaled simulations. In: Graney RL, Kennedy JH, Rodgers JH, editors. Aquatic mesocosm studies in ecological risk assessment. Boca Raton (FL): Lewis. p 47–60.

Cowell BC. 1984. Benthic invertebrate recolonization of small-scale disturbances in the littoral zone of a subtropical Florida lake. *Hydrobiologia* 109:193–205.

DeWitt TH, Ditsworth GR Swartz RC. 1988. Effects of natural sediment features on the phoxocephalid amphipod, *Rhepoxynius abronius*: Implications for sediment toxicity bioassays. *Mar Environ Res* 25:99–124.

Di Toro DM 1985. A particle interaction model of reversible organic chemical sorption. *Chemosphere* 14:1503–1538.

Di Toro DM, Mahony JD, Hansen DJ, Scott KJ, Carlson AR. 1992. Acid volatile sulfide predicts the acute toxicity of cadmium and nickel in sediments. *Environ Sci Technol* 26:96–101.

Di Toro DM, Mahony JD, Hansen DJ, Scott KJ, Hicks MB, Mayr SM, Redmond MS. 1990. Toxicity of cadmium in sediments: The role of acid volatile sulfide. *Environ Toxicol Chem* 9:1487–1502.

Di Toro DM, McGrath JA 2000. Technical basis for narcotic chemicals and polycyclic aromatic hydrocarbon criteria. II. Mixtures and sediments. *Environ Toxicol Chem* 19:1971–1982.

Field LJ, MacDonald DD, Norton SB, Ingersoll CG, Severn CG, Smorong D, Lindskoog R. 2002. Predicting amphipod toxicity from sediment chemistry using logistic regression models. *Environ Toxicol Chem* 21:1993–2005.

Ferretti JA, Calesso DF Hermon TR. 2000. Evaluation of methods to remove ammonia interference in marine sediment toxicity tests. *Environ Toxicol Chem* 19:1935–1941.

Flemer DA, Ruth BF, Bundrick CM, Moore JC. 1997. Laboratory effects of microcosm size and the pesticide chlorpyrifos on benthic macroinvertebrate colonization of soft estuarine sediments. *Mar Environ Res* 43:243–263.

Frazier BE, Naimo TJ, Sandheinrich MB. 1996. Temporal and vertical distribution of total ammonia nitrogen and unionized ammonia nitrogen in sediment pore water from the upper Mississippi River. *Environ Toxicol Chem* 15:92–99.

Ginn T C, Pastorok RA. 1992. Assessment and management of contaminated sediments in Puget Sound. In: Burton GA Jr, editor. Sediment Toxicity Assessment. Boca Raton (FL): Lewis. Chapter 16. p 371–397.

Graney RL, Kennedy JH, Rodgers JH, editors. 1994. Aquatic mesocosm studies in ecological risk assessment. Boca Raton (FL): Lewis. 736 p.

Gustafsson O, Gschwend PM. 1997. Soot as a strong partition medium for polycyclic aromatic hydrocarbons in aquatic systems. Molecular markers in environmental Geochemistry ACS symposium series. *Environ Sci Technol* 671:365–381.

Gustafsson O, Haghseta F, Chan C, MacFarlane J, Gschwend PM. 1997. Quantification of the dilute sedimentary soot phase: Implications for PAH speciation and bioavailability. *Environ Sci Technol* 31:203–209.

Hansen DJ, Tagatz ME. 1980. A laboratory test for assessing impacts of substances on developing communities of benthic estuarine organisms. In: Eaton JG, Parrish PR, Hendricks AC, editors. Aquatic toxicology, 3rd symposium. Philadelphia: American Soc for Testing and Materials. STP 707. p 40–57.

Hare L. 1995. Sediment colonization by littoral and profundal insects. *J North Am Benthol Soc* 14:315–322.

Hare L, Carignan R, Huerta-Diaz JM. 1994. A field study of metal toxicity and accumulation by benthic invertebrates; implications for the acid-volatile sulfide (AVS) model. *Limnol Oceanogr* 39:1653–1668.

Hyland JL, Balthis WL, Engle VD, Long ER, Paul JF, Summers JK, Van Dolah RF. 2003. Incidence of stress in benthic communities along the U.S. Atlantic and Gulf of Mexico coasts within different ranges of sediment contamination from chemical mixtures. *Environ Monitor Assess* 81:149–161.

Hyland JL, Hoffman EJ, Phelps DK. 1985. Differential responses of two nearshore infaunal assemblages to experimental petroleum additions. *J Mar Res* 43:365–394.

Hyland JL, Van Dolah, Snoots TR. 1999. Predicting stress in benthic communities of southeastern U. S. estuaries in relation to chemical contamination of sediments. *Environ Toxicol Chem* 18:2557–2564.

Ingersoll CG, Dillon T, Biddinger RG, editors. 1997. Ecological risk assessment of contaminated sediment. Pensacola (FL): Society of Environmental Toxicology and Chemistry (SETAC).

Ingersoll CG, Haverland PS, Brunson EL, Canfield TJ, Dwyer FJ, Henke CE, Kemble NE, Mount DR, Fox RG. 1996. Calculation and evaluation of sediment effect concentrations for the amphipod *Hyalella azteca* and the midge *Chironomus riparius*. *J Great Lakes Res* 22:602–623.

Ingersoll CG, Wang N, Hayward JMR, Jones JR, Jones SA. 2004. A field assessment of long-term laboratory sediment toxicity tests with the amphipod *Hyalella azteca* and the midge *Chironomus tentans*. Report prepared for the USEPA Office of Science and Technology, Washington DC. Forthcoming.

Jarvinen AW, Ankley GT. 1999. Linkage of effects to tissue residues: Development of a comprehensive database for aquatic organisms exposed to inorganic and organic chemicals. Pensacola (FL): Society of Environmental Toxicology and Chemistry (SETAC).

Karickhoff SW, Brown DS, Scott T. 1979. Sorption of hydrophobic pollutants on natural sediments. *Water Res* 13:241–248.

Klemm DJ, Lewis PA, Fulk F, Lazorchak JM. 1990. Macroinvertebrate field and laboratory methods for evaluating the biological integrity of surface waters. Washington DC: US Environmental Protection Agency. EPA 600/4-90-030.

Lauth JR, Scott GI, Cherry DS, Buikema AL. 1996. A modular estuarine mesocosm. *Environ Toxicol Chem* 15:630–637.

Liber K, Call DJ, Markee TP, Schmude KL, Balcer DM, Whiteman FW, Ankley GT. 1996. Effects of acid-volatile sulfide on zinc bioavailability and toxicity to benthic macroinvertebrates: A spiked-sediment field experiment. *Environ Toxicol Chem* 15:2113–2125.

Long ER, Field LJ, MacDonald DD. 1998. Predicting toxicity in marine sediments with numerical sediment quality guidelines. *Environ Toxicol Chem* 17:714–727.

Long ER, Hameedi MJ, Sloane GM, Read LB. 2002. Chemical contamination, toxicity, and benthic community indices in sediments of the lower Miami River and adjoining portions of Biscayne Bay, Florida. *Estuaries* 25:622–637.

Long ER, MacDonald DD. 1998. Recommended uses of empirically derived, sediment quality guidelines for marine and estuarine ecosystems. *Human Ecol Risk Assess* 4:1019–1039.

Long ER, MacDonald DD, Smith SL, Calder FD. 1995. Incidence of adverse biological effects within ranges of chemical concentrations in marine and estuarine sediments. *Environ Manag* 19:81–97.

Long ER, MacDonald DD, Cubbage JC, Ingersoll CG. 1998. Predicting the toxicity of sediment-associated trace metals with simultaneously-extracted trace metal: acid-volatile sulfide concentrations and dry weight-normalized concentrations: a critical comparison. *Environ Toxicol Chem* 17:972–974.

Mackay AA, Gschwend PM. 2000. Sorption of monoaromatic hydrocarbons to wood. Sorption of monoaromatic hydrocarbons to wood. *Environ Sci Technol* 34:839–845.

MacDonald DD, Carr RS, Calder FD, Long ER, Ingersoll CG. 1996. Development and evaluation of sediment quality guidelines for Florida coastal waters. *Ecotoxicology* 5:253–278.

MacDonald DD, DiPinto LM, Field J, Ingersoll CG, Long ER, Swartz RC. 2000. Development and evaluation of consensus-based sediment effect concentrations for polychlorinated biphenyls. *Environ Toxicol Chem* 19:1403–1413.

MacDonald DD, Ingersoll CG. 2001. An overview of the toxic effects of sediment-associated polycyclic aromatic hydrocarbons (PAHs), with special reference to the 8335 Meadow Avenue site in Burnaby, British Columbia. Vancouver (BC): Prepared for Legal Services Branch, Ministry of the Attorney General.

MacDonald DD, Ingersoll CG. 2002. Development and evaluation of sediment quality standards for the waters of the Colville Indian Reservation including Lake Roosevelt and Okanogan River. Nespelem (WA): Confederated Tribes of the Colville Reservation.

MacDonald DD, Ingersoll CG, Berger TA. 2000. Development and evaluation of consensus-based sediment quality guidelines for freshwater ecosystems. *Arch Environ Contam Toxicol* 39:20–31.

MacDonald DD, Ingersoll CG, Moore DRJ, Bonnell M, Brenton RL, Lindskoog RA, MacDonald DB, Muirhead YK, Pawlitz AV, Sims DE, Smorong DE, Teed RS, Thompson RP, Wang N (MacDonald Environmental Sciences). 2002. Calcasieu estuary remedial investigation/feasibility study (RI/FS): Baseline ecological risk assessment (BERA). Dallas (TX): US Environmental Protection Agency, Region 6. Document control nr. 3282-941-RTZ-RISKZ-14858.

MacDonald DD, Ingersoll CG, Smorong DE, Lindskoog RA, Sloane G. 2002. Development and evaluation of numerical sediment quality assessment guidelines for Florida inland waters. Tallahassee (FL): Florida Dept Environmental Protection.

Matthiessen P, Thain JE. 1989. A method for studying the impacts of polluted marine sediments on intertidal colonising organisms; tests with diesel-based drilling mud and tributyltin antifouling paint. *Hydrobiologia* 188/189:477–485.

Mattsson J, Notini M. 1985. Experimental recolonization by macrozoobenthos in a bay polluted by municipal sewage, oil and heavy metals compared to an unpolluted Baltic bay. *Ophelia* 24:111–124.

Meador JP, Collier TK, Stein JE. 2002a. Determination of a tissue and sediment threshold for tributyltin to protect prey species for juvenile salmonids listed by the U.S. Endangered Species Act. *Aquat Conserv: Mar Freshw Ecosyst* 12: 539–551.

Meador JP, Collier TK, Stein JE. 2002b. Use of tissue and sediment based threshold concentrations of polychlorinated biphenyls (PCBs) to protect juvenile salmonids listed under the U.S. Endangered Species Act. *Aquat Conserv: Mar Freshw Ecosyst* 12:493–516.

Metcalf and Eddy 1995. Background concentrations of contaminants in benthic invertebrate tissue. Boston: US Environmental Protection Agency (USEPA) Region 1.

Moore DW, Bridges TS, Gray BR, Duke BM. 1997. Risk of ammonia toxicity during sediment bioassays with the estuarine amphipod *Leptocheirus plumulosus. Environ Toxicol Chem* 5:1020–1027.

Morrisey DJ, Underwood AJ, Howitt L. 1995. Development of sediment quality criteria. A proposal from experimental field studies of effects of copper on benthic organisms. *Mar Pollut Bull* 31:372–377.

[NYDEC] New York State Department of Environmental Conservation. 1998. Technical Guidance for Screening Contaminated Sediments. Albany: Division of Fish, Wildlife and Marine Resources, NYDEC. 38 p.

Olsgard F. 1999. Effects of copper contamination on recolonization of subtital marine soft sediment - an experimental field study. *Mar Pollut Bull* 38:448–462.

Parrish PR, Moore JC, Clark JR. 1989. Dredged material effects assessment: Single species toxicity/bioaccumulation and macrobenthos colonization tests. Oceans '89: The global ocean. Volume 2, Ocean pollution. New York: Institute of Electrical and Electronics Engineers. p 611-616. EPA/600/D-89/134.

Plante-Cuny MR, Salen-Picard C, Grenz C, Plante R, Alliot E, Barranguet C. 1993. Experimental field study of the effects of crude oil, drill cuttings, and natural biodeposits on microphyto- and macrozoobenthic communties in the Mediterranean area. *Mar Biol* 117:355–366.

Pinza MR, Mayhew HL, Word JQ. 1996. Evaluation of Older Bay mud sediment from Richmond Harbor, California. Sequim (WA): Pacific Northwest National Laboratory. PNNL-111318.

Reynoldson TB, Day KE, Bailey RC, Norris RH. 1995. Methods for establishing biologically based sediment guidelines for freshwater quality management using benthic assessment of sediment. *Austral J Ecol* 20:198–219.

Roach AC, Jones AR, Murray A. 2001. Using benthic recruitment to assess the significance of contaminated sediments: The influence of taxonomic resolution. *Environ Pollut* 112:131–143.

Ruth BF, Flemer DA, Brendrick CM. 1994. Recolonization of estuarine sediments by macroinvertebrates: Does microcosm size matter? *Estuaries* 17:606–613.

Rysgaard S, Thastum P, Dalsgaard T, Christensen PB, Sloth NP. 1999. Effects of salinity on NH4+ adsorption capacity, nitrification, and denitrification in Danish estuarine sediments. *Estuaries* 22:21–30.

Schoener A. 1983. Colonization rates and processes as an index of pollution severity. Rockville (MD): National Oceanic and Atmospheric Administration. NOAA Technical Memorandum OMPA 27.

[SEA] Striplin Environmental Associates. 1996. Development of reference value ranges for benthic infauna assessment endpoints in Puget Sound. Prepared for the Washington Department of Ecology by Striplin Environmental Associates. Olympia (WA). 45 p plus appendices.

Shephard B. 1998. Quantification of ecological risks to aquatic biota from bioaccumulated chemicals. International Sediment Bioaccumulation Conference Proceedings. Washington DC: US Environmental Protection Agency, Office of Water. EPA 823-R-98-002.

Shine JP, Trapp CJ, Coull BA. 2003. Use of receiver operating characteristic curves to evaluate sediment quality guidelines for metals. *Environ Toxicol Chem* 22: 1642–1648.

Smith SL, MacDonald DD, Keenleyside KA, Ingersoll CG, and Field J. 1996. A preliminary evaluation of sediment quality assessment values for freshwater ecosystems. *J Great Lakes Res* 22:624–638.

Spies RB, Hardin DD, Toal JP. 1988. Organic enrichment or toxicity? A comparison of the effects of kelp and crude oil in sediments on the colonization and growth of benthic infauna. *J Exp Mar Biol Ecol* 124:261–282.

Sverdrup LE, Ekelund F, Krogh PH, Nielsen T, Johnsen K. 2002. Soil microbial toxicity of eight polycyclic aromatic compounds: Effects on nitrification, the genetic diversity of bacteria, and the total number of protozoans *Environ Toxicol Chem* 21:1644–1650.

Swartz RC. 1999. Consensus sediment quality guidelines for polycyclic aromatic hydrocarbon mixtures. *Environ Toxicol Chem* 18:780–787.

Swartz RC, Cole FA, Lamberson JO, Ferraro SP, Schults DW, DeBen WA, Lee H, Ozretich JR. 1994. Sediment toxicity, contamination and amphipod abundance at a DDT- and dieldrin-contaminated site in San Francisco Bay. *Environ Toxicol Chem* 13:949–962.

Tagatz ME, Plaia GR, Deans CH. 1986. Toxicity of dibutyl phthalate-contaminated sediment to laboratory- and field-colonized estuarine benthic communities. *Bull Environ Contam Toxicol* 37:141–150.

Tagatz ME, Plaia GR, Deans CH, Lores EM. 1983. Toxicity of cresote-contaminated sediment to field- and laboratory-colonized estuarine benthic communities. *Environ Toxicol Chem* 2:441–450.

Tagatz ME, Stanley RS, Plaia GR, Deans CH. 1987. Responses of estuarine macrofauna colonizing sediments contaminated with fenvalerate. *Environ Toxicol Chem* 6:21–25.

Thursby GB, Heltshe J, Scott KJ. 1997. Revised approach to toxicity test acceptability criteria using a statistical performance assessment. *Environ Toxicol Chem* 16:1322–1329.

[USEPA] US Environmental Protection Agency. 1994. Methods for measuring the toxicity of sediment-associated contaminants with estuarine and marine invertebrates. Washington DC: USEPA. EPA 600/R-94/025.

[USEPA] US Environmental Protection Agency 2000a. Equilibrium partitioning based sediment guidelines (ESGs) for the protection of benthic organisms: Metals mixtures (cadmium, copper, lead, nickel, silver, and zinc). Washington DC: USEPA, Office of Water Regulations and Standards, Criteria and Standards Division.

[USEPA] US Environmental Protection Agency. 2000b. Methods for measuring the toxicity and bioaccumulation of sediment-associated contaminants with freshwater invertebrates, 2nd ed. Washington DC: USEPA. EPA/600/R-99/064.

[USEPA] US Environmental Protection Agency 2000c. Technical basis for the derivation of equilibrium partitioning based sediment guidelines (ESGs) for the protection of benthic organisms: Nonionic organics. Washington DC: USEPA, Office of Water Regulations and Standards, Criteria and Standards Division.

Vanderhorst JR, Blaylock JW, Wilkinson P, Wilkinson M, Fellingham GW. 1980. Recovery of Strait of Juan de Fuca intertidal habitat following experimental contamination with oil. Second Annual Report. Washington DC: US Environmental Protection Agency. EPA 600-17-80-140.

Vanderhorst JR, Blaylock JW, Wilkinson P, Wilkinson M, Fellingham GW. 1981. Effects of experimental oiling on recovery of Strait of Juan de Fuca intertidal habitats. Washington DC: US Environmental Protection Agency. EPA 600/7-81-008.

Warren LA, Tessier A, Hare L. 1998. Modelling cadmium accumulation by benthic invertebrates in situ: The relative contribution of sediment and overlying water reservoirs to organism cadmium concentrations. *Limnol Oceanogr* 43:1442–1454.

Watzin MC, Roscigno PR. 1997. The effects of zinc contamination on recruitment and early survival of benthic invertebrates in an estuary. *Mar Pollut Bull* 34:443–455.

Watzin MC, Roscigno FP, Burke WD. 1994. Community-level field method for testing the toxicity of contaminated sediments in estuaries. *Environ Toxicol Chem* 13:1187–1193.

Using sediment assessment tools and a weight-of-evidence approach 5

WILLIAM J ADAMS, ANDREW S GREEN, WOLFGANG AHLF, STEVEN S BROWN, G ALLEN BURTON, BART CHADWICK, MARK CRANE, RON GOUGUET, KAY T HO, CHRISTER HOGSTRAND, TREFOR B REYNOLDSON, AMY H RINGWOOD, JACQUELINE D SAVITZ , PAUL K SIBLEY

Sediment assessments may be performed for a variety of purposes, including dredging and dredged sediment disposal, evaluating sediments as a capping material, determining sediment quality, and assessing biological impairment and the status of environment monitoring trends. The simplest of all assessments might include the use of single lines of evidence (LOEs) such as a small set of toxicity tests (bioassays) or the use of sediment quality guidelines (SQGs) to make a decision regarding impairment. In many cases, however, a more detailed assessment will be required to increase certainty and better characterize the degree and extent of contamination. Moving beyond the simplest assessment requires careful consideration of the types of data that will be needed to make a determination of sediment quality. This consideration includes developing a conceptual model, understanding organism linkages, selecting measurement and assessment endpoints, characterizing exposure, and performing a risk characterization. Hence, in many cases, the ecological risk assessment (ERA) paradigm provides a useful context for sediment assessment.

It is recognized that there are cases in which very minimal data sets or the use of one of the LOEs may be sufficient to make a decision. However, initial screening data from contaminated sediments often are not sufficient for making decisions because of the uncertainties associated with the minimal information typically provided by screening LOEs. In such cases, practitioners may choose to pursue additional LOEs that improve certainty. This choice may involve undertaking a full weight-of-evidence (WOE) assessment

Use of Sediment Quality Guidelines and Related Tools for the Assessment of Contaminated Sediments
Wenning RJ, Batley GE, Ingersoll CG, Moore DW, editors.

or a more formal risk assessment. WOE is used in this chapter to mean that multiple LOEs are integrated and evaluated in a framework that allows for a decision to be made even when all of the data are not consistent. Risk assessments provide the option of either making a decision with the current level of uncertainty or performing additional tests to reduce the uncertainty. Such assessments may not be required in every circumstance because they may be unnecessarily burdensome, delay action, and increase cost.

There are many reasons why single LOEs may not be adequate for assessing sediment biological impairment, including failure to sufficiently characterize spatial and temporal heterogeneity (thus, exposure and effects), sampling and sample manipulation artifacts, inadequacy of surrogate species or measurement endpoints, stressor interactions, indirect (unmeasured) stressor effects, and unmeasured stressors and receptors. These can result in either over- or underestimates of toxicity. Because of such uncertainties, there is growing recognition that multiple LOEs used in a strategic, WOE (also known as strength- or burden-of-evidence) fashion can improve the accuracy of the risk assessment process and aid in the decision-making process (Burton, Batley, et al. 2002; Burton, Chapman, Smith 2002; Chapman et al. 2002; Forbes and Calow 2002; Grapentine et al. 2002; Reynoldson et al. 2002; Smith et al. 2002). Where this approach is applied, it is essential that a logical group of complementary LOEs be linked and that they be used in a systematic study design that adequately defines key spatial and/or temporal components or trends, keeping in mind the need to link exposure and effects endpoints. It is also important that this approach be undertaken in a transparent process with meaningful public participation and that it neither be used in a way to increase financial burden on government or industry nor be misapplied to avoid or delay cleanup decisions.

The intent of this chapter is to provide critical evaluation of, and recommendations for, the application and integration of sediment assessment tools to

- improve the ability to distinguish impairment of biological systems and establish reference or benchmark conditions,
- ensure that sediment assessments include careful consideration of important ecological and food-web linkages, and
- provide guidance on how to formulate LOEs in an overall WOE approach to assist in making a decision as to whether or not sediment contamination has resulted in biological impairment.

This chapter is built upon the premise that there are at least 4 key LOEs that can be developed:

1) sediment chemistry (including contaminant concentrations) and sediment physical properties (e.g., granulometry, organic carbon content),
2) benthic invertebrate community structure,
3) sediment toxicity, and
4) bioaccumulation and biomagnification data (Grapentine et al. 2002).

Most sediment assessment tools that allow for LOEs and WOE to be developed fit within these categories. In specialized cases, tools could be used to provide additional LOEs.

Sediment assessment tools are defined as any of the current scientific techniques that are available for collecting data of known quality that can be used in an integrated fashion, leading to a decision on biological impairment. In this chapter, tools currently available are discussed together with ways that they can be used to develop multiple LOEs in an integrated WOE approach useful for making decisions regarding extent and degree of impairment or lack thereof. The use of these tools is considered in the context of a framework (see Chapter 6) for sediment assessment and with a view of providing practical guidance regarding the application of a WOE in the decision-making process. The decision process may be used in assessing the extent and nature of contamination as well as the need for sediment management (see Chapter 6).

The following questions are fundamental to discussions on the use of WOE approaches and associated tools:

- What types of data should be generated for one or more frameworks at different types of sites to determine the applicability of SQGs and to reduce technical uncertainties at a particular site?
- What can be done to improve our understanding of the uncertainties (or reduce the uncertainties) associated with more conventional assessment methods (biological testing, benthic infaunal analysis, etc.) used to derive SQGs?
- How can we use our understanding of the methods used to derive SQGs to make them more site specific?
- Can the accuracy of assessing sediment quality be improved using biologically based thresholds in a WOE process (as opposed to using chemically based threshold SQGs)?
- What are the uncertainties, deficiencies, strengths, and limitations of biologically based thresholds?

This chapter provides information to help the sediment assessment practitioner select appropriate tools to characterize exposure and effects and to integrate the data into a WOE approach allowing for a decision to be made as to whether or not biological impairment from contaminants has occurred.

Sediment Assessment Approach

The tools and techniques that are used to perform sediment assessments are used throughout the assessment process, starting with the problem formulation and carried through the exposure, effects, and risk characterizations. Key aspects of each part of the sediment assessment process are discussed below.

Problem formulation

The problem formulation tier of the assessment consists of 6 components:

1) site description,
2) a conceptual model that describes the critical linkages in the biological community as well as linkages to potential sources and routes of exposure,
3) discussion of the valued resources to be protected,
4) selection of measurement and corresponding assessment endpoints,
5) reference condition characterization, and
6) study design (USEPA 1992).

The importance of the problem formulation tier and, in particular, the development of a conceptual model and characterization of the reference condition are frequently overlooked in sediment assessments. This oversight often leads to incomplete or oversimplified assessments that may not provide an accurate basis upon which to judge the potential for biological impairment. To avoid situations in which sediment quality is mischaracterized (either over- or underprotection), careful consideration should be given to the following study design elements during the problem formulation tier: identification of key exposure pathways in the conceptual model (e.g., pore water, flow regime (low or high), groundwater upwelling and surfacewater down welling, sediment stratigraphy [surface versus depth], and food sources), selection of appropriate number of samples and statistical methods for LOE analyses, characterization of effects using biologically based methods (including resident biota and toxicity assays integrated into a WOE matrix), and guidelines to support decision making in the absence of sufficient data. Additional details on developing a conceptual model are provided in Chapter 6.

Exposure characterization

The purpose of the sediment exposure analysis is to develop an exposure profile. The exposure profile quantifies the nature and extent of stressors (e.g., which stressor does a receptor interact with), including the magnitude and the spatial and temporal patterns of contact for the pathways identified in the problem formulation or conceptual model. This exposure profile is used in the risk characterization tier of the overall assessment.

The first step in developing the exposure profile requires an initial analysis of known or suspected contaminants, receptors, and pathways leading to the development of the conceptual model. From the conceptual model, a working hypothesis is developed for receptors of interest that are potentially exposed to contaminated media via complete exposure pathways (i.e., evaluation of co-occurrence of contaminant and organism). Exposure of benthic receptors to chemical contaminants released to the environment via complete exposure pathways is generally expressed in terms of concentration in the exposure medium and should include sediment, food, water, and pore water. All complete pathways determined in the conceptual model should be evaluated to ensure a thorough assessment (Chapter 6). The assessment would include evaluation of the magnitude, duration, and frequency of exposure.

Ideally, these exposure indicators would be tightly coupled to effects indicators. For plants and animals to be at risk, chemicals must be present in the environment at concentrations above their respective toxicity reference values, and the plants or animals must come in contact with the contaminated media. Coupling of exposure and response in sediment assessment is often inferred from co-occurrence or correlative analysis (see Chapter 3) because few toxicity data are available to interpret exposure estimates expressed as contaminant concentrations in tissue. In many cases, evaluation of exposure to nonchemical stressors must also be conducted to understand the context of chemical exposure. Environmental and physical factors can modify exposure, including bioavailability, home range, mobility, and life-cycle attributes.

Effects characterization

In this phase of the analysis, the relationship between the stressor and the measurement endpoints is evaluated using the logical structure and linkages provided by the site conceptual model. The characterization is built upon an evaluation of effects that are tied to the potential stressors and are relevant to the environmental protection of the site. The exposure–response analysis for a site describes the relationship between the magnitude, frequency, or duration of a contaminant stressor in an experimental or observational setting and the magnitude of response. Any extrapolations that are required to relate measurement to assessment endpoints (e.g., between species, between response levels, from laboratory to field) are explained. In the characterization phase, a stressor–response profile (e.g., dose response for chemical stressors) is developed to quantify the relationship of the stressor to the assessment endpoints. In the effects assessment, there is an effort to evaluate the effect of stressors on measurement endpoints so they can be related to the assessment endpoints using the logical structure provided by the conceptual model. This step can be thought of as interpreting biological effects (measurement endpoints) along an exposure concentration gradient (i.e., development of the dose–response curve). This stressor–response profile is used in the risk characterization tier of the overall assessment.

Techniques for ecological effects assessment may include the following: observational field studies, including benthic community structure analysis; laboratory or field toxicity tests; interspecies extrapolation of effects; interchemical extrapolation based on knowledge of their modes of action (e.g., quantitative structure-activity relationships and biological or ecological modeling to extrapolate from measurement endpoints to assessment endpoints); and use of biomarkers.

Benthic bioassessment, which relies on the analysis of infaunal and epibenthic communities to evaluate contaminated sediments, has gained increasing acceptance in both the scientific and regulatory communities. Protecting biodiversity and maintaining a robust ecosystem spatially and temporally will require attention to subtle degrees of changes that fall within the limits of natural variability observed in indigenous communities. To achieve this, careful selection of reference sites is required. Reynoldson et al. (1997) have described an approach to link habitat characteristics, toxicity testing, and analysis of benthic structures. Confounding effects of physical stressors such as currents or sediment grain size must be addressed by using models, reference-area measurements, or experimental designs to separate the effects of physical factors from those of chemicals (see Chapters 3 and 6).

Biological effects testing and characterization of contaminated sediments is a major component of environmental quality assessment. Frequently, this testing is confined to acute and chronic laboratory bioassay protocols. Reviews of sediment testing methods and approaches have been presented by Burton and MacPherson (1995), Chapman (1995), and Chapman and Wang (2001). It is generally agreed that a suite of bioassays is necessary to adequately describe a general toxicity because of different organism sensitivities and exposure routes and because of increased discriminatory power gained by including multiple toxicological endpoints. It is important to select bioassays in a manner that is consistent with the exposure pathways identified in the conceptual model and that are relevant to the protection of the defined assessment endpoints.

If additional information is needed to characterize toxicity, other tools such as in situ testing, inclusion of an ecological epidemiological approach to take into account factors other than source information, and nonstandard laboratory bioassays tests for endocrine disruptors, genotoxicity, or biomarkers may be considered (see Chapter 3). Additional details on tools currently available for use are contained in the section titled " Selecting Appropriate Tools for Sediment Assessment" (p 170).

Risk characterization

During the risk characterization phase of a site assessment, information on organism exposure to contaminated sediments is combined with information on biological effects to characterize overall risk to the environment. At its simplest, this step may involve comparing chemical measurements on sediments from the site with an SQG. This is similar to deriving a hazard quotient by comparing a predicted environmental concentration (PEC) with a predicted no-effect concentration (PNEC), but with the SQG replacing the PNEC. The PEC can be determined either by modeling (e.g., using equilibrium partitioning [EqP] approaches to predict sediment concentrations) or by direct measurement of sediment contaminants. The SQG or PNEC may be estimated in a variety of ways, with some approaches leading to more conservative values than others. We caution against the use of SQGs as stand-alone numbers for decision-making purposes, particularly when the cause-and-effect relationship is important. The manner in which the values are derived does not exclude the possibility that co-occurring contaminants may be responsible for observed or predicted effects.

In some regulatory frameworks (e.g., that of the EU), an SQG or PNEC is determined by applying an assessment factor to a suitably sensitive single chemical toxicity test result or by using a low percentile of effect to summarize a species sensitivity distribution for such results. Assessment factors, when used, vary according to the amount of toxicity information that is available and to the regulatory system within which they are applied. They are designed to account for uncertainties in toxicity that result from the limited amount of toxicity information, for differences in interspecies tolerances, and for problems associated with extrapolating from laboratory to field and from short-term to long-term effects. As further information becomes available, assessment factors tend to be reduced as a reflection of reduced uncertainty.

The simple risk characterization model described above can be extended by incorporating probabilistic integration of exposure and effects, as described later in this chapter. This approach, in itself, does not necessarily increase the complexity of the underlying hazard quotient model, but simply adds a useful layer of quantitative interpretation. Needed complexity is introduced to risk characterization when several LOEs are available. These may include elements such as results from benthic surveys and direct toxicity testing of sediments, rather than reliance upon surrogates for site toxicity, such as SQGs and PNECs.

When confronted by multiple LOEs, the first task in risk characterization is to determine how to interpret and apply differences in the LOEs (Grapentine et al. 2002). For example, a more sensitive sublethal toxicity test result should normally be used in preference to less sensitive lethal results for the same species, and results from comprehensive chemical analysis at a site should normally be preferred to modeled estimates of sediment contamination. Some uncertainty in risk characterization is introduced when non-redundant multiple LOEs do not agree. For example, direct assessment of sediment toxicity through toxicity testing may not show effects when comparison of sediment chemistry with an SQG suggests that toxicity may occur. This may be due to insufficient sensitivity on the part of the toxicity test organism or to overly conservative estimates used to derive the SQG. Which value should be used to characterize risk? A transparent approach, preferably quantitative, is clearly required to integrate multiple LOEs, and this approach is a main focus for the section titled "Developing the WOE for Decision Making" (p 196).

Selecting Appropriate Tools for Sediment Assessment

Sediment assessments require that appropriate tools for screening and/or assessing the state of biological impairment be selected. In this section, we describe classes of tools that can be used to assess sediment quality. Because SQGs can be chemical or biological in origin, the selection of tools should be conducted in a manner that reflects the unique characteristics of each category. Many of these tools, particularly those associated with toxicity testing, have been described in detail elsewhere (Burton 1991; Ingersoll et al. 1997). The primary focus of this section is to provide detailed descriptions for alternative techniques and for techniques considered to be new and emerging. However, to ensure a holistic presentation of the various existing tools, a brief discussion has been included on each of the classes of techniques with the goal of providing readers with some guidance regarding their application in a WOE framework, which is presented in "Tools that facilitate interpretation" (p 182). In previous reviews, sediment toxicity tests were assessed using the following criteria (Burton 1991; Ingersoll et al. 1997):

- precision,
- ecological relevance,
- causality,
- sensitivity,
- interferences,
- standardization,
- discrimination,
- bioavailability, and
- field validation.

The relative importance of these characteristics varied, depending on the project and situation. For this reason, it was often difficult for the authors of the reviews to offer practical recommendations regarding which tools are most appropriate at different tiers of a sediment risk assessment (Baird and Burton 2001).

Description of existing tools

Table 5-1 summarizes existing categories of tools used for sediment assessment. The table is organized to identify 2 tiers of tools: Tier I contains screening-type tools, while Tier II contains tools that provide more diagnostic or specific answers. Tier I tools are typically lower in cost, are easier to use, and may yield information that tends to be general, or integrative, rather than specific and diagnostic. Tier II tools are typically more challenging or expensive to use but often provide more information and greater specificity. To aid the reader in choosing tools that encompass both exposure and effect LOEs, the categories of tools listed in Table 5-1 have also been identified as either exposure or effects or those that integrate information from both exposure and effects. It should be recognized that tools measuring exposure do not, by themselves, indicate effects.

A critical part of selecting the appropriate tools is making sure that the tool matches or addresses the questions being posed. The selection of a given tool must be conducted in a manner that relates to the problem formulation (see

Table 5-1 Summary of available sediment assessment methods for direct measurement of exposure and adverse biological effects used in the assessment of contaminated sediments

Method	Measurement endpoint: Exposure	Measurement endpoint: Effects	Applications or strengths	Limitations
Methods typically used in Tier I sediment assessments				
Toxicity tests, lethal	✓	—	• Offer direct measure of short-term toxicity • Have relatively low cost • Have standardized methods • Are applicable in different matrices	• Lab–field extrapolation can be difficult • Does not predict long-term effects • May not be the most sensitive measure
SQGs	✓	✓	• Are generic • Are widely available • Identify effects or no effects of contaminants in sediments	• Predicting of effects may be imprecise • Methods are not standardized • Consistency across regulatory agencies is lacking • Site-specific data are lacking
Physicochemical profiling	✓	—	• May identify co-stressors • Helps assess bioavailability	• Cost may be high
Exposure surrogates	✓	—	• Provides an indirect measure of in situ exposure/uptake • Easy to misuse	• Exposure in active organisms may not be • Dietary exposure is not measured
Contaminant profiling	✓	—	• Direct measure of contaminants • For routine chemicals, cost relatively low • Critical for source identification • Methods well established	• Bioavailable fraction may not be reflected • Measuring and identifying all chemicals in most systems is prohibitive • Methods for new chemicals are lacking • Limited to chemicals measured
Methods typically used in both Tier I and Tier II sediment assessments				
Toxicity tests, sublethal	—	✓	• Provide direct measure of sublethal toxicity • Are relatively more sensitive • Offer some standardized methods • Can be used in different matrices	• Lab–field extrapolation can be difficult • Level of difficulty and cost are increased • May not be the most sensitive measure

Table 5-1 *cont'd*

Method	Measurement endpoint		Applications or strengths	Limitations
	Exposure	Effects		
Methods typically used in both Tier I and Tire II sediment assessments, *cont'd*				
Biomarkers	✓	✓	• Can be diagnostic (chemical specificity) • Can provide early indication of effects	• Association with organism- or community-level effects is unclear • Interpretation is difficult • Noncontaminant effects may be indicated
Tissue residue	✓	—	• Provides direct evidence of exposure • Provides direct evidence of bioaccumulation or bioavailability • Useful in considering trophic transfer	• Linkage to effects needs better validation • Normalization to lipids for organic chemicals may be needed • Methods are not widely standardized • Cost may be high • Whole organism versus target tissue needs consideration
Methods typically used in Tier II sediment assessments				
Microcosm test, multispecies	✓	—	• Has increased relevancy to field conditions • Can allow discrimination of confounding factors • May allow better assessment of long-term effects	• Level of difficulty and cost are high • Interpretation may be difficult • Reproducibility may be difficult
Toxicity identification evaluation	✓	✓	• Links specific chemical classes with effects (causality) • Can be used with different matrices	• Field validation is needed • Sensitivity is limited to toxicity test or organisms used • Precision and accuracy of chemical analyses are limiting • Level of difficulty and cost are increased
In situ toxicity testing	—	✓	• Directly measures effects in the field • Reflects realistic exposure conditions • Can identify co-stressors	• Cage effects may confound interpretation of results • Logistical challenges exist in deep or high energy systems • Control over variables is lacking • Starvation of test organisms may result

Table 5-1 *cont'd*

Method	Measurement endpoint		Applications or strengths	Limitations
	Exposure	Effects		
Methods typically used in Tier II sediment assessments, *cont'd*				
Infaunal and epifaunal assessment	✓	—	• Directly measures effects in the field • Assesses community level effects • Can indicate spatial extent of contamination	• Significant effort and cost are required • Reference sites must be carefully selected • Logistical challenges may be great • Ability to identify causality is limited
Mapping (GIS)	✓	—	• Helps visualize spatial (pattern) and co-occurrence • Can indicate spatial extent of contamination • Can be used with management decisions	• Indicates correlation, not causation • Large amount of data may be required
Probabilistic models	✓	✓	• Use all available data • Provide quantifiable interpretation of risk and uncertainty associated with assessment	• Large amount of data may be required • Results may be difficult to communicate • Robustness of model depends upon quantity and quality of data • Selection of distributions can be difficult, subject to best professional judgment
Environmental fate models	✓	—	• Provide prediction of exposure in system assessed • Can provide understanding of complex processes controlling exposure (bioavailability)	• Model must be calibrated to system being assessed • Model assumptions must be recognized and adhered to • Predictions need validation with empirical data • Robustness of model depends upon quantity and quality of data
Food web models	✓	✓	• Provide linkage of sediments to higher trophic level effects • Help define direct and indirect ecosystem effects	• Model must be calibrated to system being assessed • Model assumptions must be recognized and adhered to • Predictions need validation with empirical data • Robustness of model depends upon quantity and quality of data

also "Developing the Weight of Evidence for Decision Making," p 196). This selection will depend, in part, on the tier of the assessment process; earlier tiers may require only relatively simple tests that facilitate screening (Tier I tests in Table 5-1) whereas later tiers, such as those deemed necessary following screening, may require more complex tests or tools to better delineate the extent and nature of sediment impairment (Tier II tests in Table 5-1).

Tools to assess exposure

Chemical, geochemical, physical, and related tools

Sediment investigations involve a range of physical, chemical, and geochemical analysis tools. These tools are generally focused on delineation of the spatial extent of contamination and the evaluation of potential levels of exposure for ecological receptors. Tools to characterize chemistry, geochemistry, and the physical components of a site are generally separated into 2 categories, including 1) chemical measurements in various media (solid phase, pore water, overlying water, elutriate) that are used to evaluate the magnitude, extent, and frequency of exposure and 2) supporting physical and geochemical measurements (e.g., pH, salinity, particle size distribution) that are used to help assess bioavailability or identify co-stressors that may be present and to aid in the interpretation of habitat quality. Applications, strengths, and limitations of these general categories are summarized in Table 5-1 and have been discussed at length in several review papers (Ingersoll et al. 1997). The most common exposure measures include

- contaminant concentrations in whole sediment, pore water, overlying water, and elutriates;
- sediment grain-size distribution;
- sediment total organic carbon (TOC) content;
- acid volatile sulfides (AVS) and simultaneously extracted metals (SEM);
- porewater pH;
- salinity (marine and estuarine sites) or alkalinity;
- hardness and conductivity (freshwater sites);
- ammonia concentration; and
- sulfide concentration.

The physicochemical parameters to be characterized will depend upon the site conceptual model and the data quality requirements for the particular site. Integration of these measurements with other LOEs should be carefully considered during parameter selection to ensure that they are relevant to the evaluation of exposure for assessment endpoints.

Bioaccumulation

Accumulation of sediment contaminants such as PCBs in benthic organisms as well as organisms higher in food chains, including marine mammals and humans, has been well documented (Dewailly et al. 1993; DiPinto and Coull 1997; Maruya and Lee 1998). Despite the major role that sediments may play in bioaccumulation of many organic compounds and metals in aquatic biota and in water-dependent terrestrial biota at higher trophic levels, sediment quality assessments frequently do not include evaluation of bioaccumulation. Importantly, neither the EqP nor co-occurrence approaches address the transfer of contaminants through food chains and the resulting effects. Direct measurement or modeling of bioaccumulation may therefore provide additional key information in assessing potential risks associated with contaminated sediments (see Chapters 3 and 4).

Bioaccumulation may be assessed directly in the laboratory (Ankley et al. 1993; USEPA 2000a), via in situ studies (Burton 1999; Burton et al. 2003; Pauwel and Sibley 2000) and/or through the use of predictive models. Ingersoll et al. (1997) evaluated uncertainties associated with bioaccumulation using 4 categories:

1) laboratory-based approaches under controlled exposure conditions,
2) field collection of potentially exposed organisms,
3) development of predictive bioaccumulation models, and
4) assessment of food web transfer.

Laboratory-based approaches were considered to be the most reliable across all criteria, followed by field-based methods, bioaccumulation models, and assessments of food web transfer. However, all of these approaches were considered to have poor relevance for the protection of ecology because they generally fail to connect bioaccumulation as an exposure phenomenon to ecologically relevant adverse effects.

Several food chain models have been proposed to evaluate the bioaccumulation of contaminants in fish that feed on lower trophic levels (e.g., Thomann et al. 1991, 1992a, 1992b, 1995, 1997; Gobas 1993; Suarez and Barber 1994; Gobas et al. 1995; Froese et al. 1998; Sharpe and Mackay 2000). Of these, the Thomann and Gobas models have gained widest scientific acceptance and use. These models have largely been developed in relation to lipophilic compounds and are best applied to chemicals where uptake is by passive diffusion.

Thomann et al. (1995) and others have shown that uptake of metals can be modeled using a simple relationship that is a function of the sediment–water partitioning, the uptake and elimination rate constants, the metal assimilation efficiency from food, the bivalve feeding rate, and the organism growth rate. Indeed, recent advances in bioaccumulation models, especially those incorporating relationships (predictions) with adverse biological effects, offer opportunities for reducing uncertainties around the ecological significance of bioaccumulation (see "Sediment Assessment Approach," p 166, and "Selecting Appropriate Tools for Sediment Assessment," p 170).

Surrogate tools to assess exposure

A wide variety of diffusion-type samplers have been developed as surrogate methods for determining exposure and uptake of chemicals in aquatic organisms. These samplers include vapor-diffusion samplers, water-to-water samplers, solvent or lipid-to-water samplers (semipermeable membrane devices [SPMDs], solid-phase micro-extraction [SPME]), and diffusive gradients in thin films (DGT). The vapor and water-to-water diffusion samplers consist of either air or deionized water, respectively, inside a polyethylene membrane (Vroblesky 1997). Among the most common devices are the water-to-solvent samplers (SPMDs), which are passive samplers for monitoring and assessing trace levels of organic compounds (Huckins et al. 1993). In side-by-side comparisons with *Salmo trutta*, the uptake constants from SPMDs were 1 to 1.5× greater than fish tissue (Meadows et al. 1998).

For the in situ measurement of trace chemicals in waters, DGT shows some promise. The recently developed technique allows direct assessment of exposure in various media (e.g., sediment or water column) for short (i.e., hours) or long (i.e., days or months) time periods. The quantification of exposure in sediments with DGT has been correlated with metal concentrations in sediment pore water, providing a surrogate measure for the bioavailable fraction of metals in sediments (Zhang et al. 1995). Further assessment of the methodology is needed, but such surrogate devices are generally easy to use, can be applied under laboratory or in situ conditions, and are relatively inexpensive.

Fate and exposure models

Numerous models are available for evaluating the fate and exposure of chemicals in the environment. Models developed for assessing exposure vary from simple, 1-dimensional, continuous, analytical solution models to complex, 3-dimensional, fate and transport models. A review of fate and transport and other exposure models is provided by Paquin et al. (2003). Some relatively simple models that have proven very useful for screening-type assessments of chemical exposure are the EqP approach using carbon normalization for nonionic organic chemicals (Di Toro et al. 1991) and the EqP-based SEM–AVS model for metals (Ankley et al. 1991, 1996). These 2 models have been directly used in developing mechanistic-based SQGs for organic chemicals and metals (see Chapter 3). More advanced models such as the "Mackay Model" provide a tool for predicting the mass gain and loss of chemicals in multiple compartments such as water and sediment based upon the concept of fugacity (Mackay 1991).

The most complex class of exposure models, fate and transport models, can take into account processes such as particulate transport, hydrodynamics, scouring, spatial and temporal aspects, and other key aspects to predict the fate and transport of chemicals in complex systems (e.g., lake, river, estuary). These models have proven very useful in management-type decisions regarding cleanup and remediation of complex systems where contamination may exist.

The ability of all of these models to predict exposure concentrations correctly depends on the extent that the model is calibrated to a site-specific situation. If a model has been validated for a number of chemicals, and the chemical-specific parameters are available, then it can be used with confidence. However, predictions using uncalibrated models, or using data from inappropriate systems, are not reliable in almost all but the simplest cases. Further, one must be aware of the assumptions that each of these models uses to understand their limitations and correctly apply their predictions.

Tools to assess effects

Community surveys and related tools

Benthic community surveys. Benthic invertebrate community surveys are widely used to assess the potential biological effects of sediment contamination. Resh and McElvary (1993) provide a useful review of this methodology. An array of measurement endpoints are available in this approach, from simple structural measures such as taxa richness, abundance and biomass, to more descriptive interpretive tools (e.g., species diversity; evenness; Ephemeroptera, Plecoptera, and Trichoptera taxa; species–area–biomass curves; Pearson's similarity index; Index of Biotic Integrity; trophic structure), to more sophisticated approaches such as the predictive modeling and multivariate approaches used in freshwater (Wright et al. 2001) and marine systems (Warwick and Clarke 1991). In their review of approaches for sediment quality assessment, Ingersoll et al. (1997) concluded that surveys of benthic communities using these types of structural metrics were reliable and considered to have high ecological relevance. Methods recently developed in Canada and Australia, based on predictive modeling (Resh and McElvary 1993), and in the United States (US), based on multimetrics (Barbour et al. 1999), have been used to generate numeric criteria (SQGs) based on invertebrate communities.

Benthic surveys are highly integrative and useful as general screening tools, but they are limited with respect to diagnosing causality. Recent studies suggest, however, that the types of differences or changes observed in invertebrate assemblages can be used effectively to provide evidence of cause or stressors (Yoder and Rankin 1995; Walley and Fontama 1997; Fletcher et al. 2001). The development of these benthic community patterns is generally limited to freshwater streams and lakes. Benthic community surveys also can be combined with studies conducted at lower levels of biological organization (e.g., mechanistic and single-species toxicity studies) to reveal specific causal relationships (Brown 1995; Culp et al. 2000; Clements et al. 2002; Schmidt et al. 2002).

Benthic community distributions are strongly influenced by factors other than contaminants. Benthic distributions in streams, rivers, lakes, estuaries, and other systems are, in fact, largely organized on the basis of natural environmental gradients such as hydrologic regimes, substrate or sediment types, salinity, and temperature (Gaston and Nasci 1988; Diaz and Schaffner 1990; Resh and McElvary 1993). Multivariate statistical techniques have been used with some success to identify the relative influence of sediment contaminants and natural environmental gradients on benthic community structure in freshwater and estuarine systems (Rakocinski et al. 1997; Brown et al. 2000; Culp et al. 2000; Diamond et al. 2002; Reynoldson et al. 2002). Sorting out these influences is a significant challenge in the context of sediment quality assessments. However, approaches using multiple reference sites and predictive models developed in the Great Lakes can help to resolve these influences (Reynoldson and Wright 2000), when the situation allows for the data collection.

Other community surveys. While benthic invertebrates arguably represent the most appropriate group for assessing sediment quality, it is frequently necessary to use other taxonomic groups. Benthic invertebrates are a major source of food for many fish, waterfowl, and other aquatic and semi-aquatic animals, which, in turn, may be linked to human diets. Trophic transfer of sediment-associated contaminants (bioaccumulation) is a common phenomenon; therefore, benthic invertebrates are frequently not the ecological receptors of primary concern in sediment risk assessments. For this reason, other aquatic species such as attached algae, phytoplankton, zooplankton, emergent and submergent plants, and, more commonly, fishes, reptiles, amphibians, water birds, and mammals are often surveyed when assessing contaminated sites (Karr 1991). Wading birds and other animals that feed in the littoral zones of lakes, marshes, and intertidal zones and mudflats of coastal estuaries may be counted and observed or sampled (e.g., nesting success, population assessment, tissue analysis) in order to assess the potential indirect effects of sediment contaminants (USEPA 1997; Adams et al. 1998). Assessment endpoints focused on non-macroinvertebrate species may, therefore, be extremely important in assessing risks from sediment contaminants.

Toxicity tests

Sediment toxicity tests, alone or in concert with other tools, represent the most commonly applied biological approach to assess sediment quality. Toxicity tests must be carefully chosen and, where possible, include the most sensitive life stage of the most sensitive organism present at the site (or a surrogate species). However, acute toxicity tests might be limited in their ability to predict long-term effects on species and communities. The use and application of toxicity tests in marine and freshwater environments have been reviewed extensively (Burton 1991; Ingersoll et al. 1997), so we will not consider them in detail here. However, to aid the reader in understanding the various classes of toxicity tests and the uncertainty associated with them, the results of an uncertainty analysis from a previous workshop is referenced (Chapter 18 in Ingersoll et al. 1997). The main findings are summarized below.

Ingersoll et al. (1997) identified 6 sediment phases, individually or in combination with a particular type of test species:

1) whole sediment using benthic invertebrates,
2) whole sediment using water column organisms,
3) organic extracts of whole sediment,
4) suspended solids,
5) elutriates, and
6) pore water isolated from whole sediment.

Overall, across all of the criteria, tests with whole sediments were considered to be better than, or at least as good as, tests using all other exposure phases. Importantly, interferences (biotic or abiotic factors, such as grain size or salinity, that can influence the response endpoints beyond the direct effects of specific contaminants) existed in tests with all phases. A lack of adequate information upon which to form a definitive judgment was identified for whole-sediment tests under the criteria of discrimination, bioavailability, and field validation.

The reliability of several endpoints typically used in laboratory toxicity tests was considered. These endpoints were

- survival,
- growth,
- reproduction,
- behavior,
- life tables,
- development, and
- biomarkers.

Of these, survival scored highest, or equal to the highest, across all criteria, with the exception of discrimination. This exception is because the "all or nothing" binary response obtained with this endpoint may not distinguish between marginally contaminated sediments. Interestingly, the point was made that survival tests in 1 species may occasionally be more sensitive than sublethal tests in another. Growth, reproduction, behavior, and development endpoints scored relatively high but were considered to be less reliable than survival with a substantially greater lack of knowledge for several criteria. Use of life tables and biomarkers as endpoints scored least well. Both were considered to suffer from poor precision, causality, interference, lack of standardization, and inadequate field validation. In addition, their ecological relevance was considered poor. However, there was a substantial lack of knowledge across many criteria for most endpoints with the exception of survival.

It is important to point out that in situations of low-level contamination, effects such as reproduction may only become manifest after exposure encompasses one or more generations. There are a number of chronic toxicity tests that may help elucidate toxicity in these situations, and these would most often be applied in a second phase (Tier II) of the sediment assessment process (Benoit et al. 1997; Ingersoll et al. 1998, 2000). While most currently available chronic tests typically encompass a single generation, tests that incorporate one or more generations may also provide important insights, particularly from the standpoint of making quantitative predictions of effects at the population level. Such tests may be used in conjunction with life-table analyses, although it is important to realize that such analyses have not been widely applied in sediment quality assessments (Ingersoll et al. 1997). Moreover, they can be costly compared to the more traditional short-term toxicity tests. A sediment algal chronic test recently developed by Franklin et al. (2002) shows promise in terms of sensitivity and cost and can measure algal enzyme activity as well. Alternatively, if sufficient information exists on the life history and demographic characteristics of an organism, potential long-term effects could be modeled using population models (Sibley et al. 1997; Jorgensen and Bendoricchio 2002; Pastorok et al. 2002).

Toxicity identification and evaluation

Toxicity identification evaluation (TIE) methods are a diagnostic tool using both toxicity testing and chemical analysis in an iterative, logical approach to arrive at the identification of a toxicants and link cause and effect in sediments and waters (Mount and Anderson-Carnahan 1988, 1989; Norberg-King et al. 1991; Mount et al. 1993; Burgess et al. 1996). TIEs may be performed on pore waters (Schubauer-Berigan and Ankley 1991; Ho et al. 1997; Doe et al. 2003), whole sediments (Ho et al. 2002), and in situ (Nordstrom and Burton 2000). Bioassay-directed fractionation is a more directed form of TIE, combining toxicity tests and fractionation of organic solvent extracts that is generally limited to organic toxicants. TIEs are one of the few tools designed to identify specific chemical toxicants (classes) that cause effects. TIEs are not generally used in a screening level because of the effort (cost) necessary to perform the studies. Because TIEs use both chemical analyses and toxicity testing, the limitations of both of these methods are inherent in the TIE process. For example, if the toxicity test used within the TIE approach is not sensitive to the toxicant, then the approach will fail to identify the toxicant. Likewise, if analytical methods have not been developed for a particular toxicant or class of toxicants, the resulting conclusions may be uncertain or misleading. In situ TIEs have been shown to be a very sensitive approach for detecting sediment contamination from freshwater pore waters (Nordstrom and Burton 2000).

Tools that facilitate interpretation

A number of tools enable characterization of both exposure and effects, either simultaneously or in stages, and thus are considered to be interpretive rather than specific measures of exposure or effects. These tools are discussed briefly below.

Sediment quality guidelines

While numerical SQGs can play an important role in assessing the significance of sediment contamination, their derivations have limitations that do not lend themselves to be stand-alone tools in many situations. The limitations and strengths of SQGs are described in detail in Chapter 3, and in Ingersoll et al. (1997, Chapter 18, Table 18-7). Many of the limitations can lead to an under- or overprediction of toxicity. While the accuracy of the predictions can be improved through careful choice of decision points (i.e., SEM / AVS = 1 vs. 1.5), SQGs, along with other interpretative tools, are best used in a WOE approach (Burton, Batley et al. 2002; Burton, Chapman, Smith 2002; Chapman et al. 2002; Grapentine et al. 2002; Reynoldson et al. 2002).

Community surveys

The 2 approaches frequently used for interpreting community survey data are multimetrics and multivariate predictive modeling. Extensive literature describes and compares these 2 approaches and appropriate study designs and interpretation methods (Barbour et al. 1999; Wright et al. 2001). In both cases, the approaches require data from multiple reference sites as opposed to a single reference site. Multimetrics use derived descriptors of the assemblage that describe richness, diversity, sensitivity, and function (e.g., number of taxa, dominance, biotic indices, functional feeding groups). These individual metrics are enumerated and integrated into a single score that describes the condition of the community and is compared to values derived from a matched set of reference sites, the matching usually being based on geographic or ecoregional characteristics. Multivariate predictive modeling uses derived multivariate descriptors (e.g., ordination axes, similarity matrices) that use the basic taxa information. In this approach, test sites are matched to a subset of reference sites using a probabilistic model that matches the habitat to the community. Thus, the issue of scale of the site assessment (i.e., small site versus geographic reference data) is appropriately considered. Traditional assessments, which used contaminant concentration gradient or upstream–downstream type assessments, are discussed below.

The alternatives to these 2 approaches are the traditional before and after, control and impact (BACI) and control and impact (CI) studies. These studies require careful design to avoid issues related to pseudoreplication and assumptions about the suitability of the control sites (Hurlbert 1984). These latter designs largely rely on univariate analyses (e.g., analysis of variance [ANOVA]) to identify differences on either direct species counts or derived descriptors (richness, abundance, diversity, etc.). These approaches are well documented in Norris and Georges (1993). Additionally, contaminant assessment using a concentration gradient analysis has been frequently used, and the data are typically analyzed using regression analysis (Hickey and Clements 1998).

Biocriteria

Biocriteria (biological criteria) are based on biological assessments such as community surveys as opposed to chemical measurements in order to classify sediment quality. In the US, biocriteria are considered to be field-based assessments of benthic communities and are not necessarily linked to any management guidelines. In Europe, biocriteria are considered to be bioassays that are linked to national regulations and are limited to a few examples for the evaluation of dredged material.

In Canada, numerical biological guidelines have been developed for both invertebrate community composition and sediment toxicity for 10 test endpoints using 4 invertebrate species. In both cases, the guidelines were developed from information on the 3 LOEs (chemistry, toxicity, and community structure) from 233 reference sites in the Great Lakes. The use of such a large reference site database allowed the guidelines to incorporate effects of natural variation in sediment quality (Reynoldson et al. 1995, 1997).

Mapping techniques

Because aquatic ecosystems are affected by so many different characteristics of the watershed, effects assessment and management can be significantly improved by using tools that show the geographic relationship between measurements taken at specific places and the characteristics specific to that region (i.e., issues of scale). The challenge of evaluating multiple environmental issues can be made easier by combining scientific data and watershed characteristics into a geographic information system (GIS) (Diamond et al. 2002).

Sediment quality investigations are generally conducted in a spatial context; therefore, their designs usually involve investigation along spatial stressor gradients. If the investigations capture appropriate spatial coordinates (*x,y,z*), it can be useful to plot responses on maps. GISs allow scientists to analyze a variety of data (such as sediment contaminant concentrations, tissue data, aquatic species occurrence, and habitat characteristics) in combination with a watershed's features and land uses. Sediment contaminant concentrations, toxicity and tissue data, natural resources, and potential habitat restoration projects can be overlaid on a watershed's features and land uses and displayed on maps at flexible spatial scales. Comparing these data using overlays can help elucidate co-occurrences and quantify spatial correlation.

Alternative techniques

Many of the traditional tools discussed above and presented in Table 5-2 represent the foundation of sediment quality assessment and will continue to do so in years to come. However, numerous alternative approaches have been introduced in recent years that either have not been widely applied or have not been standardized for application in a sediment assessment context. These include in situ assessments, probabilistic risk assessment approaches and species sensitivity distributions, the use of bioassays incorporating functional endpoints (e.g., processes such as leaf decay), and biomarkers that can be used to assess both exposure and effect.

In situ evaluations

In situ testing using caged organisms (Chappie and Burton 2000) or recolonization studies (Hansen et al. 1996) have been increasing around the world for the past several years. In situ testing may improve the accuracy and assessment of exposure by removing laboratory-related artifacts and allowing for natural fluctuations and interactions of suspended solids, light, temperature, flow, etc. (Burton et al. 1996, 2003; Burton 1999; Burton, Gallagher, et al. 2001; Burton, Scott, et al. 2001). In addition, in situ testing reduces the possible artifacts and altered bioavailability resulting from the collection, transport, storage, and manipulation of sediments. The hydrodynamics of surface waters and groundwater–surface water transition zones, and rapidly changing physicochemical conditions, can now be monitored in situ, thereby better defining organism exposures and stressor profiles.

In situ testing, as with all assessment approaches, has limitations (Burton 1999). Exposure from within a cage is not a true exposure because of reduced flow, accumulation of suspended solids, reduced ability to avoid predation, and possible wall effects. Other limitations include logistical challenges such as deployment in very deep or high current systems, food and nutrient variation, and vandalism.

A key aspect of in situ testing is the need for stressor source identification. Organisms identified as water column or benthic are often exposed to other ecosystem compartments (water, sediment, food) that can harbor stressors. For example, many benthic invertebrates are primarily exposed to surficial waters (e.g., Hare et al. 2001), and daphnids and fish often forage on the sediment surface or on intact resuspended sediments. All exposure routes must be characterized to identify key receptors. In aquatic ERAs, exposures from various pathways are often only crudely characterized, and there is often no understanding of how much each exposure compartment contributes to the overall exposure of the receptor. All benthic organisms are treated equally as if they are exposed in the same manner, but this approach fails to recognize that different benthic organisms may be exposed to different ecosystem com-

Table 5-2 Summary of sediment in situ exposure and effects methods for assessment of contaminated sediments

Method	Applications or strengths	Limitations
Caged organisms	• Offer direct, accurate exposure assessment • Include exposures with natural fluctuations of light, temperature, contaminants • Eliminate laboratory artifacts	• Chambers have reduced flow compared to natural conditions • Logistical deployment can be difficult • Vandalism can destroy equipment
Recolonization studies	• Allow for evaluation of indigenous species • Incorporate multiple organism life stages • Allow for chronic evaluation of stressors	• Extended periods of exposure are required • Storm events can wash out chambers • Deployment or retrieval can be difficult
Flux chambers	• Allow for transport of contaminants from solid to liquid phase • Expose organism to site conditions • Facilitate analyses of oxygen, nutrients, ammonia, etc.	• Membrane fouling can occur • Sedimentation can bury chambers • Deployment or retrieval can be difficult
Mini piezometers	• Quantify hydraulic gradients and flow patterns, including upwellings and downwellings • Allow for analysis of water samples	• Deployment can be difficult • Direct exposure of in situ organisms is not allowed
Field TIE chambers	• Provide for collection of water samples for chemical analyses • Allow for on-site toxicity assessment and identification of causative agents of effects • Require no manipulation or storage of samples	• Deployment or retrieval can be difficult • Specialized equipment is required for field deployment
Seepage meters	• Allow for collection of integrated water samples for contaminant and effects assessment • Provide for estimates of flow rates	• Meters are subject to vandalism • Organisms are tested independent from sediment

partments such as overlying waters, suspended solids, surficial sediments or organic debris, deeper sediments, pore waters, or groundwaters, or may be differentially exposed in time, depending upon life history characteristics such as stage of life cycle development.

A key aspect of in situ testing, therefore, should be to assess the dominant exposure pathways associated with the dominant benthic populations at a site. The surrogate test organisms commonly used for toxicity and bioaccumulation testing may not be adequate indicators of important receptors at the test site of concern. Ideally, indigenous organisms can be used for in situ toxicity and bioaccumulation testing with an understanding of the life history and trophic characteristics of the benthic community. This understanding allows for better evaluations not only of key exposure pathways, but also of ecosystem functioning (such as plant/animal diet ratio, feeding mechanism, food size, food acquisition behavior, energy and substance transfer) (Pavluk et al. 2000). However, if appropriate indigenous species are not available, then surrogate species (e.g., *Corbicula* sp., *Lumbriculus* sp., *Hyalella* sp., *Hexagenia* sp., *Chironomus* sp., *Ampelisca* sp., *Rhepoxynius* sp., *Eohaustorius* sp., *Leptocheirus* sp., *Neanthes* sp. [Adams and Rowland 2002]) appropriate for major exposure pathways should be selected for in situ testing. Contaminant exposure from the various exposure compartments should be linked to receptors and effects via targeted chemical sampling of that compartment and organism (i.e., tissue and effects endpoints).

To properly link exposure and effects, the assessor needs to identify the dominant receptor habits such as where the organisms reside, population dynamics, and the nature of their habitat (e.g., surface water, surficial sediment, size fraction of organic matter, pore water, deep sediments). Organisms that are in contact with sediments for some or all of their life cycle may come into contact with contaminants via several exposure compartments, including

- surface water (low and high flows),
- organic matter at the sediment–water interface (plant or animal, small to large size),
- surficial sediment particles (inorganic or organic),
- nonsurficial sediments with surface water irrigation,
- nonsurficial sediments with no irrigation,
- pore water, and
- groundwater advection.

These exposure compartments may vary in importance through the life of an organism, vary through space and time, and, obviously, vary with species. Within each of these compartments, chemical stressors may be partitioned into dissolved, colloidal, or particulate fractions. For each major exposure pathway, an appropriate surrogate species should be selected for in situ testing.

A number of emerging in situ methods for the evaluation of chemical exposure via both diffusive (gradient or biologically mediated) and advective fluxes show promise for providing quantitative exposure data in systems where these pathways are suspected of contributing to or controlling biological effects (Tengberg et al. 1996; Chadwick et al. 1999; USEPA 2000a). A potentially important exposure pathway for contaminated sediments is via benthic fluxes driven by the desorption or dissolution of contaminants from the solid phase into a more mobile and bioavailable dissolved phase (Burgess and Scott 1992). These fluxes are driven by small-scale gradients and are very difficult to preserve when sediment samples are collected for laboratory evaluations. Thus, studies have focused on the development of in situ chambers for the quantification of these fluxes (Tengberg et al. 1996). These benthic flux chambers were originally developed for the evaluation of oxygen and nutrient fluxes (Bott et al. 1978; Berelson et al. 1987; Brown and King 1987; Sayles 1992) but have evolved to include chambers and techniques for assessing the fluxes of some classes of inorganic and organic contaminants (Hampton and Chadwick 2000).

A similar but often overlooked exposure pathway in sediments is via the advection of contaminants to sediments and waterways from connecting hydraulic zones (such as groundwater) (Valiela and D'Elia 1990; Chadwick and Largier 1999; Moore 1999; Montlucon and Sanudo-Wilhelmy 2001). At inland stream and lake systems, these fluxes generally occur in association with groundwater hydraulic gradients that may fluctuate significantly depending on season, water management processes, rainfall, and other factors (Lee and Cherry 1978). In estuarine and coastal systems, the effects of upland hydraulic gradients are often complicated by shorter time-scale events such as waves, tides, and storm surges (Lewis 1987; Chadwick et al. 1999, 2002).

A number of techniques for evaluating exposure at these sites have been developed and evaluated and were recently summarized in USEPA (2000b). A brief summary of in situ methods, advantages, and limitations are provided in Table 5-2. In stream systems, effective exposure assessment has been demonstrated using arrays of nested mini-piezometers (Dean et al. 1999; Greenberg et al. 2002). These techniques provide a means of quantifying the hydraulic gradients and, in association with measurements or estimates of hydraulic conductivity, a means detecting areas of upwelling groundwaters or downwelling surface waters (Fetter 1994). Concurrent collection of water samples can be used to quantify chemical concentrations at potential exposure points (Lendvay et al. 1998). Seepage meters provide another means of quantifying exposure at these sites (Belanger and Mikutel 1985; Cherkauer and McBride 1988). Recent advances in seepage meter technology have led to the capability to monitor flow continuously (Taniguchi and Fukuo 1993) and to sample water samples simultaneously for chemical characterization (Chadwick et al. 2002). Direct integration of these exposure measurements with appropriate measurement endpoints (e.g., in situ sediment water interface tests, bioaccumulation tests) provides the potential for improvements in the assessment of these pathways.

After key exposure routes have been identified and adverse biological effects documented, it is often necessary to identify the dominant stressors (Greenberg et al. 2002). While this may be done via laboratory TIE procedures, or a series of spiking experiments, an alternative cost-effective approach that has been used is in situ TIEs and modified caged exposures. These exposures can optimize stressor class exposure into the following categories: suspended solids, photo-induced toxicity from polycyclic aromatic hydrocarbons (PAHs), ammonia, metals, and nonionic organics. For field TIE exposures, a modified chamber design was used that pumped pore water through resins (zeolite, Chelex, Ambersorb) that selected for particular chemical types (ammonia, metals, nonionic organic chemicals) and then passed into an organism exposure chamber (Nordstrom and Burton 2000). Site comparisons between the field and laboratory TIEs have shown that the field approach generally is more sensitive, primarily because acute toxicity is often lost in the laboratory during sample manipulations. Field TIEs are somewhat more complicated to perform than laboratory TIEs and require rapid testing and careful analysis to avoid contamination.

At sites where PAHs are present, it is essential to consider the toxic interactions of photo-induced toxicity. In all cases during low flow conditions, chambers showed increased mortality of test organisms. However, at high flow conditions, toxicity was reduced, likely because of high turbidity blocking penetration of UV light and PAH binding to suspended particles (Ireland et al. 1996).

Suspended solids and siltation are dominant stressors of aquatic ecosystems (USEPA 1995; Waters 1995), yet they are rarely characterized in an ERA. While contaminants may be associated with suspended solids and siltation, impairment is often partially a physical phenomenon. How can biological impairment be attributed to contaminated sediments if suspended solids from upstream are significant stressors? To ascertain the effects of suspended solids (elevated turbidity occurred during high flows), in situ chambers are fitted with varying sizes of mesh with both single and double layers. The effect of suspended solids has been shown within cages where solids exposure is highest and may be the greatest stressor even in contaminated urban waterways (Burton and Moore 1999; Burton and Pitt 2001).

Species sensitivity distributions and probabilistic approaches

Probabilistic approaches to ERA have been applied with increasing frequency in recent years. Probabilistic approaches involve the comparison of distributions of biological effects with distributions of exposure concentrations as a basis for determining the likelihood that effects thresholds will be exceeded and, thus, the risk of adverse effects incurred. Biological effects data may include species sensitivity distributions (SSDs) (Posthuma et al. 2001) or information on the behavioral or feeding ecology of single species (Hart 2001). If exposure data are collected over time at a particular site, the degree of overlap of the exposure distribution with the effects distribution can be used to estimate the joint probability of exposure and toxicity, allowing for either a full overview of the probability of adverse effects or the calculation of exceedance probabilities for response at a fixed assessment criterion (Solomon 2001). Selection of the criterion is somewhat subjective and may vary according to jurisdiction, but it is most often set at a concentration equivalent to the 10th percentile of the SSD in North America (SETAC 1994) or to the 5th percentile in Europe (Aldenberg and Jaworska 2000); these terms equate to 90% and 95% species protection respectively (see Chapter 9). Although methods of probabilistic ecological risk assessment (PERA) are relatively new and continue to evolve, they have already been applied widely in the context of waterborne contaminants, particularly pesticides (Giesy et al. 1999; Giddings et al. 2000; Hart 2001). Only recently, however, has the application of PERA in the context of contaminated sediments been proposed. The potential application of PERA in sediment quality assessments, including a consideration of its advantages and disadvantages, was recently discussed by Solomon and Sibley (2002).

The primary advantage of PERAs is that they use all relevant biological effects data and, when combined with exposure distributions, enable quantitative estimations of risk (Solomon and Sibley 2002). Moreover, the data can be revised with additional toxicity and exposure information, providing an opportunity to make the decision criteria more robust (Solomon 2001). However,

PERA also has a number of disadvantages linked to the limitations of SSDs on both ecological and statistical grounds (Forbes and Forbes 1993; Newman et al. 2000; Grist et al. 2002). In addition to such general criticisms, distributional analyses are most effective in the presence of either abundant species toxicity data or information on individual behavioral or feeding ecology plus exposure data, but these are often lacking in sediment toxicology. Where sufficient data do exist, PERA represents a significant improvement over the sole use of single-species toxicity data such as the LC50 in standard deterministic risk assessments. The approach allows for the incorporation of regression approaches for measuring toxicity responses. However, PERA often relies on laboratory-based information such as LC/EC50s or, less frequently, on no-observed-effect concentrations (NOEC) or low toxicity values (e.g., EC10) to establish distributions, which may not have great relevance to field situations. It is possible that effects and exposure information from in situ sediment assessments (see "In situ evaluations," p 185) could be used to formulate site-specific distributions, but this formulation has not been attempted to date.

Other factors also affect the application of PERA in sediment quality assessments, most of which relate to difficulties in providing an accurate exposure distribution profile. Some of these represent the same issues as those currently faced by sediment toxicologists, including

- how to account for or incorporate the different exposure routes in sediments (to date, PERA methods have largely been developed for use with waterborne contaminants where exposure generally originates from a single compartment);
- how to incorporate or account for bioavailability (and spatiotemporal variation in bioavailability) of sediment stressors in the exposure distribution;
- how to evaluate bioaccumulative substances, which may or may not behave similarly under laboratory conditions when compared to contaminated field sites (a method has been proposed [Balk et al. 1995] in which effect concentrations in organisms are compared to exposure concentrations in organisms from the exposed site);
- how to incorporate exposure resulting from the presence of mixtures of contaminants, especially when modes of toxic action differ; and
- how to account for spatiotemporal variability in exposure to contaminants.

On the latter point, Solomon and Sibley (2002) suggest that, when comparing exposures to species sensitivity distributions, it may be most appropriate to consider organisms in classes related to their relative mobility. That is, if the area of contamination is small relative to the range of the organism, the probability of exposure is reduced. If the organism is relatively sedentary, the probability of exposure is much higher if the organism occurs within the contaminated area. Such an approach is similar to that developed for PERA of vertebrate animal species for which information on behavioral and feeding ecology is available (Hart 2001).

Realistic concerns of many practitioners are the costs and resources required to undertake comprehensive, WOE assessments of sediment quality. Indeed, these concerns are one reason for the widespread appeal of the use of chemical-based SQGs. The application of PERA in the assessment of sediment quality, once refined, could represent a cost-effective approach that falls between the WOE approaches discussed in this chapter and the more rapid, but possibly over- or underprotective, chemical-based approaches discussed in other chapters.

Use of microbial communities and application of functional endpoints

A number of recent developments in the use and application of functional endpoints, particularly in relation to microbially mediated processes, may prove to be useful in higher tiers of the sediment assessment process. Benthic microbial communities represent the foundation of numerous ecosystem processes (e.g., nutrient cycling, mineralization), and sediment-associated stressors that affect the diversity and function of this community can have a profound effect on ecosystem function as a whole. To date, functional endpoints have received limited attention with respect to their use in sediment toxicity assessments because of a perceived lack of sensitivity, poor discrimination of biological impairment, and poor early warning diagnostic potential (Cairns et al. 1992, 1995; Cairns and Pratt 1995). This section briefly reviews several promising techniques for assessing microbial-related endpoints.

The measurement of dimethyl sulfoxide (DMSO) reduction could be chosen as an indicator of bacterial activity. DMSO is reduced to dimethylsulfide by enzymes within the electron transport chain under oxic as well under anoxic conditions, although anaerobic reduction rates are higher (Ahlf and Gratzer 1999). Thus, only sediments of a similar redox potential can be compared. Considering the short experimental duration and high sensitivity, this method could easily be included in a sediment testing procedure to provide unique information about effects on bacterial communities. A survey in the river Rhine demonstrated an inverse relationship between Micotox toxicity and DMSO reduction, which suggests that organic pollutants decrease the activity of the indigenous bacteria (Ahlf, Braunbeck, et al. 2002; Ahlf, Hollert, et al. 2002).

Several microbial-based, community-fingerprinting techniques have been developed in recent years that may be useful in the assessment of chemical stressors on benthic microbial communities. One of these approaches is sole carbon source use (Garland and Mills 1991), often referred to as "community-level physiological profiling" (CLPP). This metabolic fingerprint approach provides a qualitative indication of changes in the diversity and metabolic activity of bacterial communities by assessing changes in carbon utilization patterns in relation to exposure to stressors. Although largely developed for use in soils, bacterial assemblages derived from contaminated sediments could be grown on substrates containing different carbon sources and their development assessed spectrophotometrically over a 7-d growth period. This information is compared to bacterial assemblages, derived from appropriate reference sites, using a multivariate approach (e.g., principal component analysis [PCA]) or diversity indices. Although CLPP has been widely applied in ecological studies, it is important to recognize that this technique is cultivation dependent and, as such, may account only for a small proportion of the microbial community.

Measurement of ribonucleic acid (RNA), deoxyribonucleic acid (DNA), or fatty acids can overcome cultivation-dependent concerns with microbial responses to environmental stressors. Using this approach, whole community fatty acid profiles can be derived from phospholipid components of cellular membranes of microorganisms directly extracted from environmental samples (Peterson and Klug 1994). This approach has been used in a number of applications, primarily with soils (Haack et al. 1994; Peterson and Klug 1994; Heipieper et al. 1996), but may be hindered by the fact that knowledge of the qualitative and quantitative distributions of fatty acids in environmental microorganisms and the effects of growth conditions on their distribution is limited (Heuer and Smalla 1997).

A second cultivation-independent approach is denaturing gradient gel electrophoresis (DGGE). This approach uses the 16S rDNA subunit, which is highly conserved among bacteria, as an indicator of microbial diversity (Heuer and Smalla 1997). In DGGE, a soil or sediment sample is extracted to isolate the bacteria and subjected to polymerase chain reaction (PCR) to amplify and facilitate sequencing of the 16S rDNA subunit. The resulting banding patterns for each sample are then compared as a basis for assessing differences in structural diversity between samples. Sequencing of selected bands can be used to provide additional insights into the structural composition of the microbial community.

While DGGE largely provides information on the structure of bacterial or fungal communities, the fact that it fingerprints the most dominant metabolically active species means that it could be used to make inferences about the functional capacity of the microbial community (Santegoeds et al. 1998). The primary advantages of DGGE are that it is culture independent, unlike many of the traditional techniques (e.g., direct counts) and more recent techniques (e.g., CLPP), comparatively easy to use, and relatively inexpensive. Those advantages that a much higher proportion of the microbial community can be assessed with respect to its response to environmental stressors than is the case with many traditional approaches, including CLPP.

A third microbial approach that has been useful for assessing contamination is indigenous microbial enzyme analysis (Burton 1988). Four effluent-impacted streams across the US were assessed using conventional aquatic test organisms and a battery of microbial activity assays as toxicity indicators. Microbial assays of sediment enzyme activity included alkaline phosphatase, amylase, arylsulfatase, electron transport (dehydrogenase), galactosidase, glucosidase, and protease. Significant correlations were observed between sediment enzyme activity and stream biota or chemical measurements of water quality. These relationships, at diverse test sites, suggest these short-term microbial assays are useful assessment tools.

Biomarkers

Both biomarkers of exposure and effect may be useful in detecting bioavailability or diagnosing possible causes of toxicity. Biomarkers of effect are likely to carry more weight in a risk assessment because they simultaneously demonstrate both exposure and adverse effects that fall outside natural homeostasis.

Biomarkers of effect show both that an organism has been exposed to a toxicant and that harm to the organism has occurred. Like other direct measures of environmental toxicity, the main advantages of biomarkers are that they can potentially detect both measured and unmeasured contaminants and provide an integrated biological response. Changes in biochemistry and cell performance can be rapid after exposure to toxic compounds and indicate that the compound is bioavailable and has reached a site of toxic action (Timofeeva and Stom 1990). Biochemical and cellular biomarkers that have been used to investigate contaminated sediments include metallothionein and mixed function oxidase induction, and acetylcholinesterase inhibition and lysosomal stability (e.g., Ringwood et al. 1998, 1999). Additionally, the effects of sediment-bound organophosphorus and organochlorine insecticides on acetylcholinesterase activity of chironomid larvae have been related to adult fitness (Callaghan et al. 2001; Hirthe et al. 2001; Crane et al. 2002).

Biomarkers have been recommended as potential early warning systems that allow the detection of reversible departures from normal individual health (Depledge and Fossi 1994). In the context of contaminated sediments, these biomarkers are likely to mean the detection of bioavailability from apparently low levels of exposure rather than a temporal early warning to indicate that organisms in an assemblage are at an early stage of exposure. The reason is that contaminated sediment problems are usually historical in nature, so organisms will already have been exposed over a prolonged period. Exceptions include, but are not limited to, "new" substances. While useful as an early warning indicator, the diagnostic potential of biomarkers is perhaps a more robust characteristic of their use. Some biomarkers are specific to particular chemicals or groups of chemicals, while others are more general responses to stress. For example, inhibition of aminolevulinic acid dehydratase in an organism indicates exposure to lead (Johnson et al. 1999) and, more broadly, induction of mixed function oxidases can be caused by many different organic compounds (Peakall 1994).

The very rapid technical development within the field of molecular genetics has provided new tools that are extremely powerful and potentially useful in assessing effects caused by environmental toxicants, including those present in sediments. There have also been significant refinements in some existing molecular techniques that may render them suitable for assessment of sediment toxicity. This field of study is rapidly emerging.

Analysis of gene transcripts (i.e., mRNA) from key biomarker genes (i.e., CYP1A1, metallothionein, glutathione-*S*-transferase) is one emerging tool. For downstream analysis of specific mRNA, real-time quantitative PCR (Q-PCR) has been a significant development. This method can be made fully quantitative and has a sensitivity of only a few molecules of the mRNA of interest. Furthermore, analysis of up to 300 or more samples and standards can be conducted in a few hours, allowing large-scale analyses at limited cost.

Through DNA array technology, it is theoretically possible to simultaneously quantify the expression of all genes in an organism, tissue, or cell (transcriptomics). The array is used to measure gene expression in exposed and control animals to produce global analyses of genes that are differentially expressed in response to the treatment. Because each toxicant is likely to produce a unique gene expression profile, this technique could be used to identify significant toxicants in complex chemical mixtures. The analyses can also provide insights into mechanisms of toxicity and the compensatory responses used by the organism in an exposure scenario (Hogstrand et al. 2002). DNA array analysis holds exceptional promise as a diagnostic tool, but the resources required for each new species are still elaborate and expensive to develop.

Recently, transgenic organisms, such as yeast, *C. elegans,* and zebra fish, have been produced that carry reporters that are sensitive to toxicants (Power et al. 1998). Typically, gene regulatory sequences that are stimulated by toxicants are fused to the coding region of a gene that can be easily assayed (e.g., ß-galactosidase or fluorescent proteins). Such transgenic organisms have been produced for the rapid assay of responses to metals, planar aromatic hydrocarbons, and xenoestrogens. The great advantages of these transgenic biomonitoring organisms are that the response can be made exclusive to a narrow group of environmental toxicants and that the read-out endpoint is trivial to measure. Disadvantages include legislative limitations for use of genetically modified organisms and the limited ecological relevance of the response.

Summary of tool selection

This section has described the main biological tools available for the assessment of contaminated sediments. Analytical chemistry, toxicity tests in the laboratory and the field, biomarkers, and ecological community surveys are among the major tools available for use. Not all of these techniques need to be used under all circumstances, but in complex situations it is likely that many will be needed to reduce uncertainty to a suitable level for reliable decision making. Data from these tools can be analyzed and summarized by a variety of quantitative techniques to provide several LOEs for input to risk characterization. The way in which these LOEs should be combined in decision making is discussed in the next section.

Developing the Weight of Evidence for Decision Making

Approaches to sediment quality assessment

It is critical that practitioners approach sediment assessment in a meaningful way, recognizing there may be unique management goals in each case, carefully choosing the endpoints and environmental variables necessary to assess whether those goals are achieved, and providing the necessary amount of objective information to decision makers. This assessment should occur in a framework that is adequately robust yet sensitive to both the cost and the timeframe in which the data are needed. With a diverse set of sediment contamination scenarios, as well as source control and sediment management demands, sediment quality assessment must be an adaptable process. For relatively simple applications, such as regulatory permits for small private docks or boat launches, a sediment assessment using only 1 or 2 LOEs might suffice. Comparison of sediment chemical concentrations with effects thresholds might be adequate to reach a determination of no concern. Such an assessment is fairly straightforward. However, in more complex situations (PCBs in a large river system), such an approach would leave an unacceptable amount of uncertainty and would call for a more robust WOE approach. The remainder of this section focuses on the integration and interpretation of sediment assessment tools and LOEs in the context of a more robust WOE approach to decision making.

Concepts and rationale

Use of single assessment methods or single LOEs frequently produces results that are equivocal about the significance of field contamination. Commonly used, single LOE approaches include laboratory toxicity tests, bioaccumulation testing, chemical analyses for comparison to chemical-specific SQGs, biological field assessments, and/or modeling. When data for various LOEs were compared for their predictability in describing biological responses they were similar only 40% to 70% of the time (Burton, Gallagher, et al. 2001). This may be an unacceptable for each, when used alone. It is apparent that additional LOEs and, in many cases, more effective ecological assessment approaches are needed to link the magnitude, frequency, and duration of exposure with biological effects. This more comprehensive approach can increase certainty concerning significant adverse ecological effects (Baird and Burton 2001). These issues can be addressed in greater detail with the addition of alternative assessment tools (e.g., in situ biologically based LOEs), improved linkage of LOEs, and a more comprehensive WOE assessment strategy. These additional LOEs help to remove major uncertainties in risk characterization.

Traditional assessments of surface water, soil, and sediment quality tend to emphasize a chemical-specific approach. This approach has numerous uncertainties as described in Chapter 3 and in Ingersoll et al. (1997) and requires extrapolation and comparison of a chemical concentration to acceptable or adverse biological responses. Using biologically based approaches with appropriate numeric biological targets (i.e., acceptance criteria) provides a direct assessment of site-specific adverse effects (Burton 1999, 2001; Reynoldson and Wright 2000; Ahlf, Braunbeck, et al. 2002; Burton et al. 2003). If adverse responses are observed or results are equivocal, then additional LOEs, such as chemical guideline exceedances, are useful in implicating causality. This approach can also help interpret indigenous biological indices data, rank stations of primary concern, and guide remediation strategies.

A quantitative and logical framework for using the WOE process for sediment assessment is described in Figure 5-1. It clearly defines essential elements in the assessment and decision-making process, thereby improving the certainty of conclusions about whether or not impairment exists because of sediment contamination and, if so, which stressors and biological species (or ecological responses) are of greatest concern. "Certainty" is used to describe the steps of the WOE framework by properly characterizing and implementing each step; thus, the uncertainty in the conclusion and decision-making process is reduced. Key elements of the WOE process include the conceptual model and its components, linkage of exposure and effect LOEs, characterization of key natural and anthropogenic stressor exposure profiles, appropriate reference comparison methods, appropriate quantification methods used to integrate LOEs, a critique of the advantages and limitations of each LOE used, evaluation of each LOE versus causality criteria, and combination of the exposure and effects LOE into a WOE matrix for interpretation to show causality linkages in the conceptual model. Although establishing causality is a major objective of a WOE approach, the inability to do so does not indicate that the system is not impaired (see "Lines of evidence evaluation," p 205).

Ecosystem quality and reference condition

Defining ecosystem quality is closely linked to the WOE step of defining a reference condition. This process is extremely important because it forms the basis of all decisions regarding impairment. These decisions are made by comparison to reference sites, benchmarks (i.e., standardized conditions) and/or control values and provide the basis for minimizing uncertainty. While the definition of ecosystem quality is a subjective, there should be a clear separation between the subjective stage (e.g., defining the amount of acceptable change) and the quantitative stage of actually detecting that change. Defining the appropriate reference condition includes selecting the environmental parameters to compare, their concentrations or levels, and their location (Reynoldson et al. 1995, 1997). Because many of the areas being assessed for environmental impairment are located in human-dominated watersheds, it is often difficult to find an appropriate reference condition.

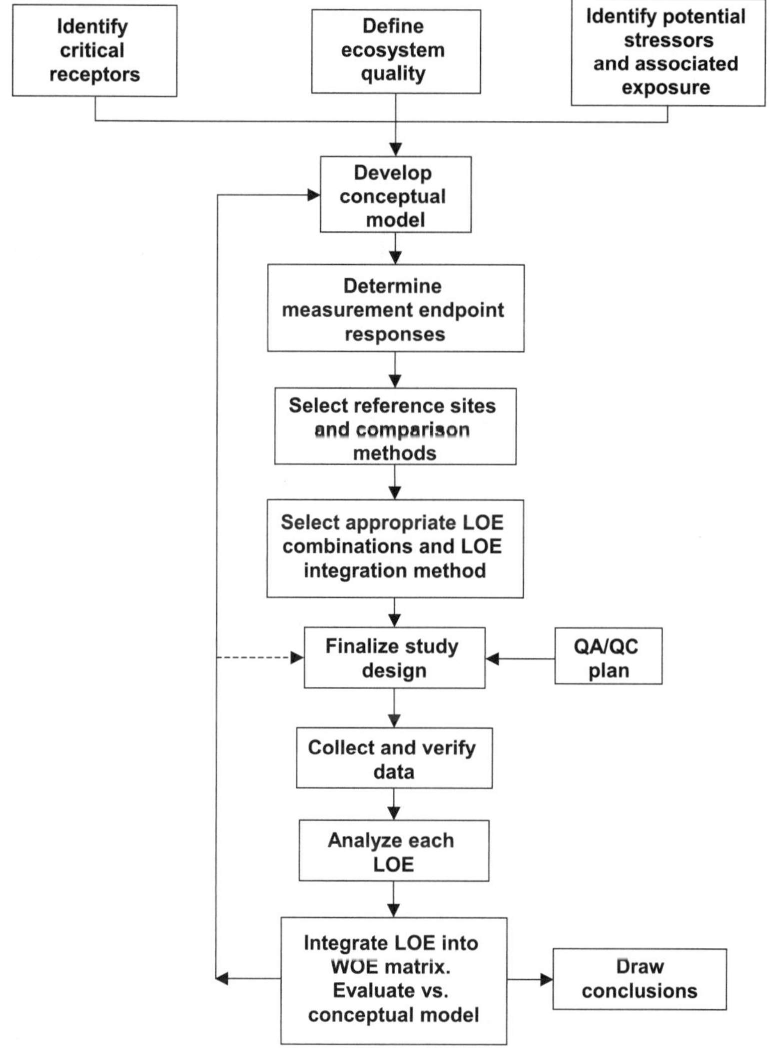

Figure 5-1 Critical elements (CEs) of the weight-of-evidence (WOE) framework (Reprinted with permission from Burton Jr GA, Batley GE, Chapman PM, Forbes VE, Smith EP, Reynoldson T, Schlekat CE, den Besten PJ, Bailer AJ, Green AS, Dwyer RL. 2002. A weight-of-evidence framework for assessing sediment (or other) contamination: improving certainty in the decision-making process. *Hum Ecol Risk Assess* 8:1675–1696. © CRC Press.)

Many of the reference classification systems are based on physical and chemical standards that are linked to designated beneficial uses and may not have an ecological basis. Studies of many major urban and agricultural watersheds across the US identified elevated levels of nutrients and pesticides in sediments and fish tissues (USGS 1999). Virtually every "reference" site surveyed in Ohio had agricultural or urban inputs and detectable toxicity in sediments and/or storm waters (GA Burton, unpublished data), suggesting that these sites may be inappropriate references. However, it is important to note that reference sites do not have to be pristine and may represent the best sites available or even a theoretical or historical condition (Hughes et al. 2000; Bailey et al. 2003). These authors use the following definition of reference condition: "Where sites that are minimally exposed to the stressors of interest." Specific operational criteria for reference sites that are strictly reliant on lack of exposure to the stressors of interest must also be defined. Such sites will encompass the range of variability in biological conditions that are minimally exposed to the stressors. Thus, in watersheds dominated by human activities, where a pristine reference site is not available, the appropriate reference may not be pristine but rather a minimally affected site possessing desirable species diversity. However, it should be noted that the definition of minimally impaired must be completely transparent and well defined so that it does not compromise the ability to demonstrate impairment in problem formulation.

A confounding factor in the selection of a reference condition has been that each water body may be very unique, thus making it difficult to mimic and certainly difficult to characterize. For example, while stream ecologists generally acknowledge the importance of landscape attributes (Poff and Ward 1990; Townsend et al. 1997), the traditional focus in stream ecology has been on small-scale studies conducted within a single watershed (Wiley et al. 1997). Because land use, hydrology, geomorphology, and vegetation directly or indirectly control physical, chemical, and biological processes in streams, an understanding of the complex interactions among these landscape features is necessary to

- define reference conditions,
- accurately measure conditions, and/or
- predict effects of anthropogenic stressors.

In estuaries and harbors, reference conditions are often confounded by variations in salinity, sediment properties (e.g., grain size, TOC, consolidation), temperature, and other habitat conditions such as turbidity or light regimes and water depth. Because contamination often co-associates with fine-grained, high-TOC sediments, it may be particularly difficult to find reference areas with these sediment characteristics that are not contaminated.

Stream ecologists have traditionally employed a hierarchical approach in which stream reaches are nested within broader ecological levels (segments, catchments, basins) to classify stream ecosystems. One important assumption of this approach is that stream ecosystems within the same ecoregion are more similar to each other than to those in different ecoregions. Although this assumption may be valid across broad spatial scales (e.g., ephemeral streams in the southwestern North America or southern Europe are quite different from Rocky Mountain streams and those in northern Europe, respectively), some studies have shown little correspondence between ecoregion and community composition (Corkum 1990; Richards et al. 1993; Reynoldson et al. 1997; Hawkins et al. 2000). Thus, it appears that classification of stream ecosystems by ecoregion alone is not sufficient to account for significant variation in aquatic communities.

A useful, ecoregion-based approach for assessing stream quality in Europe using benthic macroinvertebrates has been published recently (AQEM 2002a). This approach defines typical reference conditions across Europe based on land use in the catchment area, local hydromorphology, and chemical and biological conditions. The authors use multimetric approaches that define expected biological conditions for the typical types of streams, establishing variability and significant differences based on multivariate techniques (see multiple publications and software at www.aqem.de). A multiple reference site approach has also been used in studies of the effects of sediment contaminants and natural environmental gradients on benthic communities in Gulf of Mexico estuaries (Brown et al. 2000; Rakocinski et al. 2000). A comprehensive discussion of the strengths and limitations of the multiple reference site approach and predictive modeling, including appropriate numbers of reference sites and spatial and temporal scale issues, is available in Wright et al. (2001).

In situations where natural variation is more stochastic or unidentified, it may be difficult to distinguish all but the most extreme examples of perturbation. Regardless, some understanding of the natural spatiotemporal variation of community structure is essential for any biomonitoring program (Hildrew and Giller 1994; La Point et al. 2000; Baird and Burton 2001). Annual changes in faunal composition are not always predictable, being influenced by floods and droughts. The validity of models constructed from reference sites collected from a narrow time window is a concern, and repeated sampling of a subset of reference sites to examine temporal change is necessary (Burton et al. 2003). Assessors should beware, however, of pseudoreplication through time and should consider use of multiple analysis of variance (MANOVA) approaches, rather than analysis of variance (ANOVA) to account for interdependence of sequentially sampled variables (Scheiner 1993; von Ende 1993).

As bioassessment approaches have evolved over the past decade (concomitant with advances in basic benthic ecology), the ability to address natural variation has evolved (Maltby 1999; Baird and Burton 2001). Spatial variation has been addressed by classifying water bodies into homogeneous units of physico-chemical and hydrological units (Abell et al. 2000; AQEM 2002b) and in the reference condition approach by developing predictive models for probabilistic matching of multiple reference sites to test sites (Wright et al. 2001).

The appropriate reference condition will vary depending on study objectives and site characteristics. Just as measurement endpoints have associated variance, so do the descriptors of reference condition, which must be considered in quantitative comparisons. Thus, reference site selection and estimation of natural variability are absolutely critical to the success of the process. These are best accomplished through the use of appropriate statistical methods. If a sufficiently diverse and inclusive reference site database has been established (e.g., Great Lakes, Reynoldson and Wright 2000; UK, Wright et al. 1996; Australia, Norris and Hawkins 2000; coastal Gulf of Mexico, Brown et al. 2000; Germany, Heise and Ahlf 2002), then these reference sites can be repeatedly used for site assessment. The choice of procedures depends on the number of sites, the amount and type of data, and the statistical characteristics of the data. Some guidance for consideration in comparing habitat and geographical features, which should be similar between reference and exposure areas, are presented in Table 5-3. Some modification may be required on a case-by-case basis. However, the approach presented is one that has been demonstrated to be robust (Reynoldson and Wright 2000).

In summary, definition of ecosystem quality and characterization of reference condition provide the foundation for determining which sites have acceptable environmental quality and which are impaired. All sites vary in their physico-chemical and biotic characteristics, so it is naïve to assume that one reference site will adequately reflect conditions at the test site. Therefore, it is critical that the full range of possible reference conditions is considered in the assessment process by having available information from multiple reference sites. From these multiple reference sites, a method is required to select a subset that have typical physical, chemical, and biological conditions of least impaired reference conditions but are otherwise representative of the site in question (Wiegers et al. 1998; Brown et al. 2000; Reynoldson et al. 2002). Thus, the optimal reference condition can be better refined as additional information is provided that links key physical and chemical conditions (e.g., hydrology, temperature, altitude, organic matter, nutrients, hardness, and river or harbor versus near-shore lake sediments) with biological population and community indices.

Selection of lines of evidence

Various traditional and alternative LOEs are used to characterize exposure and effects, and each has unique advantages and limitations. These are discussed at length at the beginning of this section and in Table 5-1. Issues to consider in the selection of measurement endpoints include data quality and reliability, as well as the linkage between the endpoint and ecological stress (Chapman et al. 2002).

The exposure LOEs (Table 5-2 tools) help to characterize stressor exposure to biota and consist of physical (e.g., habitat descriptors, flow, sediment grain size, salinity, pH, temperature, surface and groundwater flow), chemical (e.g., inorganic and organic), and biological (tissue residues and some biomarker responses) endpoints. Some of the key issues that raise uncertainty in the linkage of exposure to effects with these LOEs include biological relevance of parameters measured, characterization of spatiotemporal profiles, characterization of key exposure compartments (e.g., whole sediment, pore water, groundwater upwellings, downwellings, low and high flows), the role of environmental variables (e.g., grain size, organic carbon, carbonates, AVS), relevance of tissue concentrations, and sampling of appropriate home range.

The effects LOEs (Table 5-1 tools) characterize responses of site biota and tend to focus on laboratory toxicity testing and characterizations of indigenous biological community structure. Key uncertainty issues in the risk assessment process for these LOEs include level of acceptable variance and statistical power, known Type I and II error rates appropriately sensitive and ecologically relevant test species, response characterization of benthos representative of dominant exposure pathways and indigenous populations appropriate controls or references, appropriate biotic indices, and appropriate spatial and temporal representation.

In a WOE approach, the various LOEs described should not be redundant, and each should provide unique and useful information to the assessment process. LOEs should include both exposure indicators and effects responses. For example, habitat suitability profiles, bioaccumulation, and physicochemical characterizations provide information for exposure characterization, while toxicity, in situ testing, and indigenous community responses characterize the effects component of the assessment. The optimal combination of LOEs to use on an assessment of sediment quality depends on the situation and questions

Table 5-3 Habitat and geographical features that should be similar between reference and exposure areas[a]

Features	Estuarine marine environments	Rivers and streams	Lakes
Ecoregion	***	*	*
Drainage basin	*	***	***
Basin area if different basins	*	*	*
Land use	***	***	***
Other inputs	*	***	***
Dominant habitat types	***	***	***
Substratum	***	***	***
Shoreline structure		***	***
Riparian vegetation		***	***
Stream order		*	
Channel gradient		*	
Depth	***	***	
Velocity		***	
Bankfull width		***	
Channel pattern		***	
Elevation		*	*
Salinity	***		
Temperature	***		
Tidal regime	***		
Geological origin			*
Morphometry			*
Slope from shoreline			*
Water clarity or total suspended solids (TSS)	*		*
Trophic state	***	***	***
Stratification characteristics	***		***

[a] Reynoldson and Wright 2000.
* These features should be similar.
*** These features must be similar.

being addressed but usually will include a minimum of 3 observational LOEs, 2 of which are biologically based (e.g., biosurveys and toxicity) (Burton, Batley, et al. 2002; Grapentine et al. 2002). Because 3 LOEs are recommended, single LOEs such as sediment or water chemistry and tissue chemical concentrations should not be used alone in an assessment. When sediment or water chemistry measures are used, extent of contaminant bioavailability should be addressed when possible (Meyer 2002). Optimally, the LOEs will be chosen to work together in a way that allows quantitative assessment of specific exposure and effects pathways that are identified in the site conceptual model. The quantitative relationships between 2 groups of LOEs should be linked in the risk characterization phase of the ERA. The linkage process can occur via multiple methods, as described by Burton, Batley, et al. (2002); Burton, Chapman, and Smith (2002); Grapentine et al. (2002); Reynoldson et al. (2002); and Burton et al. (2003) (Table 5-1 and Figure 5-1).

Lines of evidence evaluation

Each LOE should be analyzed carefully using quantitative methods (Bailer et al. 2002; Reynoldson et al. 2002; Smith et al. 2002). A variety of approaches have been used in detecting change between reference and test conditions. The major types of study design are listed below:

- control/impact design (C/I),
- multiple control/impact design (MC/I),
- simple gradient design (SG),
- radial or multiple gradient design (RMG), and
- reference condition approach (RCA).

These designs fall into 3 basic categories with different philosophical approaches:

1) the C/I or MC/I designs which are ANOVA-type designs used to determine the magnitude of difference between unexposed (control) areas and exposed (impact) areas, in which approach reference sites are selected on the basis of best professional judgment that are most similar to the test site in all aspects but the stressor;
2) the gradient (simple, radial, or multiple) designs, which are intended to examine changes in community structure along a physical and/or effluent gradient which are better suited to regression analyses, but may not include an appropriate reference site; and
3) the multivariate approach of the reference condition, which compares potential impaired or test stations to matched multiple reference sta-

tions with a statistical evaluation demonstrating the extent to which the reference and test sites are matched.

Whichever approach is used, the data within an LOE should not be integrated or pooled in a manner (such as condensing down to 1 number) that causes useful information to be lost (such as condensing down to 1 number). This recommendation applies both within an LOE and in the LOE integration process. However, within an LOE, multiple measures may be integrated if approached correctly. For example, the Benthic Assessment of Sediment Toxicity (BEAST) approach uses multivariate statistics to combine both multiple endpoints within LOEs and measures of sediment physicochemistry, with benthic communities and sediment toxicity to identify reference versus impaired conditions (Reynoldson et al. 1995, 2002; Grapentine et al. 2002). An important consideration in the approach is how information is consolidated within a single LOE into 1 summary number. An acceptable example would be consolidating benthic invertebrate data into various indices (e.g., diversity, taxa richness, multivariate descriptors) and then combining the indices into an overall site (assessment area) metric (e.g., similarity index). Often this is necessary, informative, and useful in minimizing uncertainty associated with multiple testing when comparisons are made to the reference condition. When available information is integrated into a single LOE (Smith et al. 2002), linking the method of data integration should be linked to the study objectives.

It is pointed out that when multiple metrics are integrated into a single metric (LOE), some information is lost. It is possible with this approach to use independent measures and/or the composite index to evaluate a site for effects as compared to a similar analysis of reference site data.

An important component of the WOE assessment process is identifying which stressors are the primary cause of any impairment. This evaluation of causality should be conducted, as far as possible, with individual LOE analyses that are integrated to establish how well each LOE links stressors with adverse biological responses. An objective of the WOE assessment process is the determination of which stressors are the primary cause of impairment (Burton, Gallagher, et al. 2001; Burton, Scott, et al. 2001; Clements et al. 2002; Norton et al. 2002; Schmidt et al. 2002; Suter et al. 2002). This evaluation of causality should be conducted both with individual LOEs (see tools in "Selecting Appropriate Tools for Sediment Assessment," p 170) and within the framework for WOE matrix (see next section on how to integrate LOEs into a WOE matrix). The approach uses results of both laboratory and field measurements to develop quantitative assessments of stressor–response relationships but can be followed by the use of more diagnostic tools to identify causality. This approach is likely to be successful in relatively simple systems, but in complex systems, multiple stressor–multiple source situations may not always be entirely successful. Establishing causality is critical for a number of reasons: it allows managers to identify stressors and sources and develop tracking and cleanup actions; it is also critical in understanding the action and effects of stressors in an ecosystem. However, causality and irrefutable linkages of source and stressor and can be one of the most difficult connections to develop. Because of the complexity of ecosystems, the best that can often be developed are plausible connections between stressors and effects. Use of a WOE approach can be used to develop logical linkages in the absences of hard evidence for causality. Even when causality cannot be proven, management decisions may still need to be made under this uncertainty. Lack of causality should not preclude ecosystem protection.

The WOE approach uses results of quantitative assessments of stressor-effect relationships followed by a combination of quantitative findings and best professional judgment in the causality decision. A consensus-based process that should occur during the problem formulation tier is the determination of acceptable effect limits (i.e., critical effect size). These will involve both ecological and statistical criteria. Various processes have been described for establishing these limits, ranging from best professional judgment to more quantitative approaches (e.g., Harris et al. 1994; Bailer et al. 2002; Burton, Batley, et al. 2002; Grapentine et al. 2002; Reynoldson et al. 2002; Smith et al. 2002). These limits may be study specific because they are affected by study design (e.g., power, characterization accuracy), societal values, and understanding of ecosystem components, dynamics and inter-relationships.

Appropriate statistical approaches (e.g., regression, ANOVA, and multivariate methods) should be used within each LOE to define reference and impaired conditions. This approach is most useful when incorporated into the initial and final study design tiers (e.g., the problem formulation and risk characterization tiers of a risk assessment) (Burton et al. 2003). This WOE framework does not, however, focus on societal or economic concerns that are also important components of an overall project but which may be addressed in the problem formulation tier of the risk assessment process through involvement of key stakeholders.

In order to establish causality, a multi-step process may be necessary using diagnostic protocols and weighing the strength of evidence that supports each potential cause. This establishment can be done via 7 causal considerations:

1) co-occurrence (spatial correlation),

2) temporality (temporal correlation),
3) effect magnitude (strength of link),
4) consistency of association (at multiple sites),
5) experimental confirmation (field or laboratory),
6) plausibility (likelihood of stressor-effect linkage), and
7) specificity (stressor causes unique effect) (modified from USEPA 2000a).

If the WOE process adequately links stressor exposure with biological impairment using statistically valid relationships and the results are related to the appropriate reference condition, then uncertainty will be minimized in decisions concerning the source, occurrence, and severity of sediment-related impairment of aquatic biota. It is important to note that the level of certainty required in the decision must be addressed beforehand and incorporated within the framework, and there must be an appropriate mechanism for further increasing certainty if required.

Integration across lines of evidence into a weight-of-evidence matrix

The integration and interpretation required to determine whether significant impairment exists and establish the link between exposure and effects (i.e., causality) is often a multi-step process involving eliminating alternatives, using diagnostic protocols, and weighing the strength of evidence that supports each potential cause. In the final step, strength of evidence can be established using the following criteria (Suter 1993):

- the adverse effect should be regularly associated with exposure to the stressor,
- the stressor (or indicator of exposure) should be found in the affected receptor (e.g., organism, population, community),
- and the adverse effect should be manifest in unimpaired species (i.e., healthy, unexposed prior) following exposure under controlled experimental conditions.

After such evaluations are conducted, individual expert qualitative judgments on causality should be integrated. The results of these expert judgments can be summarized in a tabular decision matrix, for example, by converting to ranks (e.g., 1 to 4 or "+" and "–" values), as shown by Chapman (1990, 1996), Ingersoll and MacDonald (2002), Grapentine et al. (2002), and USEPA (2000b). This approach can be useful to both experts and stakeholders but requires that the WOE critical elements be conducted in a thorough, high-quality manner and the conclusions from the individual LOEs be quantitatively based with an adequate level of power. This then allows an interdisciplinary team to combine the LOEs into a WOE matrix table for the decision-making process, and such matrix tables may have a fixed set of outcomes into which site specific cases will fit (e.g., Grapentine et al. 2002). The combination of these LOEs should be pursued in a manner that is consistent with the stressors, exposure pathways, and receptors that were identified in the conceptual model, rather than simply evaluating all possible combinations of LOEs. By using criteria derived through consensus for individual LOEs, quantitative assessment within LOEs, and a sensible integration of LOEs into the WOE, subjectivity in decision making can be restricted to where it is appropriate or unavoidable.

Uncertainty must be considered with respect to effect size in evaluating evidence. Effect size is defined as the amount of ecological change that is important enough to signal concern. Effect size is a critical component of any assessment study design using power analysis with hypothesis testing approaches. Effect size should be defined a priori. A simple approach to defining effect size is through statistical rareness (reduction in frequency) relative to a reference (Smith 2002). Another critical component associated with evaluating evidence is the role of error rates in the decision process. Such errors can be of 2 kinds, Type I and Type II. Type I errors refer to deciding that there is an effect when there is not an effect, while Type II errors refer to not recognizing an effect when there is one. A typical statistical approach is to specify the Type I error rate at a low level then use sample size to control the Type II error rate. This approach is precautionary toward the null hypothesis, which is often stated as "there is no effect." As a result, the hypothesis-based assessment process is frequently not protective of the environment (Peterman 1990). A priori establishment of these error rates and the power required to detect them can be used to set the appropriate risk to the environment and the potential stressor or stressee on a case-by-case basis. When regression analyses are used, uncertainty has to be addressed in the experimental design through selection of an appropriate number of samples or sites that are utilized.

Individual LOEs convey different types of information (e.g., exposure or effects categories [Table 5-2]), and effects that are evident from one "effects" line may reflect only certain types of effects. Large effects in different lines may indicate different types of problems and concerns. For example, if there is no biological effect but the chemical concentrations exceed guidelines or standards, this may indicate a potential rather than a realized problem. Alternatively, if the biological information demonstrates effect but the chemical does not, this may indicate a non-sediment associated problem or an unmeasured stressor, and further resolution is required. Thus, integration of all the lines into a single number is likely to over-simplify the evidence. Examples of WOE integration of various LOEs are presented by Bailer et al. (2002), Forbes and Calow (2002), Grapentine et al. (2002), Reynoldson et al. (2002), and Smith et al. (2002).

Example of a weight-of-evidence approach

In the Great Lakes, progress in remediating contaminated sediment has been slow because of the difficulty in reaching consensus on scientific, social, and economic decisions (Zarull et al. 1999). Scientific assessments have been hampered because of complex interactions between sediment contaminants, biota, and contaminants in the overlying water, and the potential for contaminant movement through the aquatic food web. Grapentine et al. (2002) summarized a process that allows the manager or user to understand both the information required and the rationale for determining potential problems associated with contaminants in sediments. While the method and logic used were not intended to be prescriptive, they suggested that if any elements of this system were excluded, or alternative methods were used, then an equal degree of explanation and the associated rationale should be provided. The underlying philosophy of the framework was that observations of elevated concentrations of contaminants in sediments alone are not indications of ecological degradation. Rather, it is the biological responses, both short and long term, to those contaminants that are the concern. A recommendation on remedial or other management activity should be supported by evidence of an adverse biological effect either on the biota resident in the sediment or on biota that are affected by contaminants originating from the sediment. Effects can be direct or indirect and may be physical, chemical, and/or biological. Grapentine et al. (2002) identified 4 lines of evidence (in addition to knowledge on the stability of sediments) required for their weight of evidence approach:

1) sediment chemistry and grain size,
2) benthic invertebrate community structure,
3) sediment toxicity, and
4) invertebrate body burdens or biomagnification.

Results for the LOEs in this approach are integrated in 2 stages: 1) synthesizing the data within each category of attributes and 2) combining the conclusions for the categories to determine the overall status of sites. In the first stage, multivariate ordination is used for characterizing sites in terms of 2 or 3 "composite" variables and assessing differences between test and reference sites (see Reynoldson et al. 1997 for a full description). However, other methods can be used to make decisions within LOEs and to provide the information necessary to use the decision framework (Chapman 1996). Overall assessment of a site is achieved by integrating the information obtained both within and among the above 4 LOEs and modified from Long and Chapman (1985), Chapman (1996), and the BEAST model (Reynoldson et al. 1995) frameworks. The integration approach, while resulting in 16 combinations for the individual elements and current status, interpretation and management recommendations, produces 4 basic interpretations:

1) that the sediments do not present a risk,
2) that there are adverse effects that require risk management evaluation (i.e., need for risk reduction),
3) that there is a need for both risk management evaluation and further investigation because of equivocal results, and
4) that there is no immediate need for risk management evaluation, but further investigation is required.

References

Abell J, Emerson S, Renaud P. 2000. Distributions of TOP, TON and TOC in the North Pacific subtropical gyre: Implications for nutrient supply in the surface ocean and remineralization in the upper thermocline. *J Mar Res* 58:203–222.

Adams WJ, Brix KV, Cothern KA, Tear LM, Cardwell RD, Fairbrother A, Toll JE. 1998. Assessment of selenium food chain transfer and critical exposure factors for avian wildlife species: need for site-specific data. In: Little EE, Delonay AJ, Greenberg BM. editors. Environmental toxicology and risk assessment: Seventh Volume. Philadelphia: American Soc for Testing and Materials. p 312–342.

Adams WJ, Rowland CD. 2002. Aquatic toxicology test methods. In: Environmental handbook of ecotoxicology. 2nd edition. Ann Arbor (MI). Lewis. p 19–43.

Ahlf W, Braunbeck T, Heise S, Hollert H. 2002. Sediment and soil quality criteria. In: Burden, FR, McKelvie I, Förstner U, Guenther A. editors. Environmental monitoring handbook. New York: McGraw-Hill. p 17.1–17.18.

Ahlf W, Gratzer H. 1999. Erarbeitung von Kriterien zur Ableitung von Qualitätszielen für Sedimente und Schwebstoffe - Entwicklung methodischer Ansätze. *UBA Texte* 41/99:1–171.

Ahlf W, Hollert H, Neumann-Hensel H, Ricking M. 2002. A guidance for the assessment and evaluation of sediment quality: A German approach based on ecotoxicological and chemical measurements. *J Soils Sed* 1:1–6.

Aldenberg T, Jaworska JS. 2000. Uncertainty of the hazardous concentration and fraction affected for normal species sensitivity distributions. *Ecotoxicol Environ Saf* 46:1–18.

Ankley GT, Cook PM, Carlson, AR, Call DJ, Swenson JA, Corcoran F, Hoke RA. 1993. Bioaccumulation of PCB's from sediments by oligochaetes and fishes: comparison of laboratory and field studies. *Can J Fish Aquat Sci* 49:2080–2085.

Ankley GT, Di Toro DM, Hansen DJ. 1996. Technical basis and proposal for deriving sediment quality criteria for metals. *Environ Toxicol Chem* 15:2056–2066.

Ankley GT, Phipps GL, Leonard EN, Benoit DA, Mattson VR, Kosian PA, Cotter AM, Dierkes JR, Hansen DJ, Mahoney JD. 1991. Acid-volatile sulfide as a factor mediating cadmium and nickel bioavailability in contaminated sediments. *Environ Toxicol Chem* 10:1299–1307.

[AQEM] Assessment System for the Ecological Quality of Streams and Rivers throughout Europe using Benthic Macroinvertebrates. 2002a. Experiences with different stream assessment methods and outlines of an integrated method for assessing streams using benthic macroinvertebrates. Available from: www.aqem.de. Accessed 1 Jul 2004.

[AQEM] Assessment System for the Ecological Quality of Streams and Rivers throughout Europe using Benthic Macroinvertebrates. 2002b. Manual for the application of the AQEM system. Available from: www.aqem.de. Accessed 1 Jul 2004.

Bailer AJ, Hughes MR, Kyoungah S, Noble R, Schaefer R. 2002. A pooled response strategy for analyzing multiple responses to develop relative risk rankings in a weight of evidence evaluation of sediment contamination. *Hum Ecol Risk Assess* 8: 1597–1612.

Bailey RC, Norris RH, Reynoldson TB. 2003. Bioassessment of freshwater ecosystems: the reference condition approach. New York: Springer. 184 p.

Baird D, Burton Jr GA. 2001. Ecological variability: Separating natural from anthropogenic causes of ecosystem impairment. Pensacola (FL): Society of Environmental Toxicology and Chemistry (SETAC).

Balk F, Okkerman PC, Dogger JW. 1995. Guidance document for aquatic effects assessment. Paris: Organization for Economic Cooperation and Development (OECD).

Barbour MT, Gerritsen J, Snyder BD, Stribling JB. 1999. Rapid bioassessment protocols for use in streams and wadeable rivers: periphyton, benthic macroinvertebrates, and fish. 2nd ed. Washington DC: US Environmental Protection Agency, Office of Water. EPA 841-B-99-002.

Belanger TV, Mikutel DF. 1985. On the use of seepage meters to estimate groundwater nutrient loading to lakes. *Water Res Bull Am Water Res Assoc* 21:265–272.

Benoit DA, Sibley PK, Jeunemann JJ, Ankley GT. 1997. *Chironomus tentans* life-cycle test: Design and evaluation for use in assessing toxicity of contaminated sediments. *Environ Toxicol Chem* 16:1165–1176.

Berelson WM, Hammond DE, Smith Jr KL, Jahnke RA, Devol AH, Hinga KR, Rowe GT, Sayles F. 1987. In-situ benthic flux measurement devices: bottom lander technology. *Mar Technol Soc J* 21:26–32.

Bott TL, Brock JT, Cushing CE, Gregory SV, King D, Petersen RC. 1978. A comparison of methods for measuring primary productivity and community respiration in streams. *Hydrobiologia* 60:3–12.

Brown SS. 1995. Bioavailability and toxicity of sediment-associated contaminants in freshwater and estuarine sediments [PhD dissertation]. Oxford (MS): Biology Dept, Univ Mississippi. 229 p.

Brown SS, Gaston GR, Rakocinski CF, Heard RW. 2000. Effects of sediment contaminants and environmental gradients on macrobenthic community trophic structure in Gulf of Mexico estuaries. *Estuaries* 23:411–424.

Brown SS, King DK. 1987. Community metabolism in natural and agriculturally disturbed riffle sections of the Chippewa River, Isabella County, Michigan. *J Freshw Ecol* 4:39–51.

Burgess RM, Ho KT, Morrison GE, Chapman G, Denton DL.1996. Marine toxicity identification evaluation (TIE) procedures manual: Phase I Guidance Document. Washington DC: USEPA Office of Research and Development. EPA 600/R-96/054.

Burgess RM, Scott KJ. 1992. The significance of in-place contaminated marine sediments on the water column: Processes and effects. In: Burton Jr GA, editor. Sediment toxicity assessment. Boca Raton (FL): Lewis. p 129–165.

Burton Jr GA. 1988. Sediment impact assessments using microbial activity tests, In: Lichtenberg J, Winter J, Weber C, Fradkin L, editors. Chemical and biological characterization of municipal sludges, sediments, dredge spoils and drilling muds. Philadelphia: American Soc for Testing and Materials (ASTM). STP 976. p 300–310.

Burton Jr GA. 1991. Assessing the toxicity of fresh-water sediments. *Environ Toxicol Chem* 10:1585–1627.

Burton Jr GA. 1999. Realistic assessments of ecotoxicity using traditional and novel approaches. *J Aquat Ecosyst Health Manag* 2:1–8.

Burton Jr GA, Batley GE, Chapman PM, Forbes VE, Smith EP, Reynoldson T, Schlekat CE, den Besten PJ, Bailer AJ, Green AS, Dwyer RL. 2002. A weight-of-evidence framework for assessing sediment (or other) contamination: improving certainty in the decision-making process. *Hum Ecol Risk Assess* 8:1675–1696.

Burton Jr GA, Chapman P, Smith E. 2002. Weight of evidence approaches for assessing ecosystem impairment. *Hum Ecol Risk Assess* 8:1657–1673.

Burton Jr GA, Gallagher J, Schwab B, Rowland C, Greenberg M, Irvine C, Johnson J, McElroy M, Leppanen C, Lavoie D, Nordstrom N. 2001. Sediment contamination assessment methods vs. biological concern values. Abstr Annu Meet Soc Environ Toxicol Chem; 2001 Nov 11–15; Baltimore, Maryland, USA. Pensacola (FL): Society of Environmental Toxicology and Chemistry (SETAC).

Burton Jr GA, Ingersoll CG, Burnett L, Henry M, Hinman M, Klaine S, Landrum P, Ross P, Tuchman M. 1996. A comparison of sediment toxicity test methods at three Great Lake areas of concern. *J Gt Lakes Res* 22:495–511.

Burton Jr GA, MacPherson C. 1995. Test methods for measuring sediment toxicity. In: Hoffman D, Hoffman DJ, Rattner BA, Burton Jr GA, Cairns Jr J, editors. Handbook of ecotoxicology. Boca Raton (FL): Lewis. p 70–103.

Burton Jr GA, Moore L. 1999. An assessment of storm water runoff effects in Wolf Creek, Dayton, OH. Final report. Dayton (OH): City of Dayton.

Burton Jr GA, Nelson MK, Ingersoll C. 1992. Freshwater benthic toxicity assays. In: Sediment toxicity assessments. Boca Raton (FL): Lewis. p 213–240.

Burton Jr GA, Pitt R. 2001. Stormwater effects handbook: A tool box for watershed managers, scientists and engineers. Boca Raton (FL): CRC/Lewis. 924 p.

Burton Jr GA, Pitt R, Clark S. 2000. The role of whole effluent toxicity test methods in assessing stormwater and sediment contamination. *CRC Crit Rev Environ Sci Technol* 30:413–447.

Burton Jr GA, Rowland CD, Greenberg MS, Lavoie DR, Nordstrom JF, Eggert LM. 2003. A tiered, weight-of-evidence approach for evaluating aquatic ecosystems. In: Munawar M, editor. Sediment quality assessment and management: Insight and progress. Ecovision World Monograph Series. Hamilton (ON): Aquatic Ecosystem and Management Society Publ. p 3–21.

Burton Jr GA, Scott DD, Cormier SM, Suter GW, Dorward-King EJ. 2001. Identifying watershed stressors using database evaluations linked with field and laboratory studies: a case example. In: Baird D, Burton Jr GA, editors. Ecosystem variability: separating natural from anthropogenic causes of ecosystem impairment. Pensacola (FL): Society of Environmental Toxicology and Chemistry (SETAC). p 233–254.

Cairns Jr JJ, Neiderlehner BR, Smith EP. 1992. The emergence of functional attributes as endpoints in ecotoxicoogy. In: Burton Jr GA, editor. Sediment toxicity assessment. Boca Raton (FL): Lewis. p 111–128.

Cairns Jr JJ, Niederlehner BR, Smith EP. 1995. Ecosystem effects: Functional endpoints. In: Rand GM, editor. Fundamentals of aquatic toxicology. 2nd ed. Washington DC: Taylor & Francis. p 589–607.

Cairns J, Pratt JR. 1995. Ecological restoration through behavioral change. *Restor Ecol* 3:51–53.

Callaghan A, Hirthe G, Fisher T, Crane M. 2001. Effect of short-term exposure to the organophosphorus pesticide chlorpyrifos on developmental parameters and biochemical biomarkers in *Chironomus riparius* Meigen. *Ecotoxicol Environ Saf* 50:19–24.

Chadwick B, Davidson B, Hampton T, Groves J, Guerrero J, and Stang P. 1999. Offshore porewater and flux chamber sampling of San Diego Bay sediments at Site 9, Naval Air Station, North Island. San Diego: SPAWAR Systems Center. Technical Report 1799. 39 p.

Chadwick DB, Largier JL. 1999. The influence of tidal range on the exchange between San Diego Bay and the ocean. *J Geochem Res-Oceans* 104:29885–29899.

Chadwick DB, Groves JG, He L, Smith C, Paulsen R, Harre B. 2002. New techniques for evaluating water and contaminant exchange at the groundwater-surface water interface. *Oceans Conference Record (IEEE)* 4:2098–2104.

Chapman PM. 1990. The Sediment Quality Triad approach to determining pollution induced degradation. *Sci Tot Environ* 97:815–825.

Chapman PM. 1995. Ecotoxicology and pollution: key issues. *Mar Pollut Bull* 31:167–177.

Chapman PM. 1996. Presentation and interpretation of sediment quality triad data. *Ecotoxicology* 5:327–339.

Chapman, PM, McDonald BG, Lawrence GS. 2002. Weight of evidence issues and frameworks for sediment quality (and other) assessments. *Hum Ecol Risk Assess* 8:1489–1516.

Chapman PM, Wang F. 2001. Assessing sediment contamination in estuaries. *Environ Toxicol Chem* 20:3–22.

Chappie DJ, Burton Jr GA. 2000. Applications of aquatic and sediment toxicity testing in situ. *J Soil Sed Contam* 9:219–246.

Cherkauer DA, McBride JM. 1988. A remotely operated seepage meter for use in large lakes and rivers. *Ground Water* 26:165–171.

Clements WH, Carlisle DM, Courtney LA, Harrahy EA. 2002. Integrating observational and experimental approaches to demonstrate causation in stream biomonitoring studies. *Environ Toxicol Chem* 21:1147–1155.

Corkum LD. 1990. Intrabiome distributional patterns of lotic macroinvertebrate assemblages. *Can J Fish Aquat Sci* 47): 2147–2157.

Crane M, Sildanchandra W, Kheir R, Callaghan A. 2002. Relationship between biomarker activity and developmental endpoints in *Chironomus riparius* Meigen exposed to an organophosphate insecticide. *Ecotoxicol Environ Saf* 53:361–369.

Culp JM, Lowell RB, Cash KJ. 2000. Integrating mesocosm experiments with field and laboratory studies to generate weight-of-evidence risk assessments for large rivers. *Environ Toxicol Chem* 19:1167–1173.

Dean SM, Lendvay JM, Barcelona, MJ, Adrians P, Katopodes ND. 1999. Installing multilevel sampling arrays to monitor ground water and contaminant discharge to a surface water body. *Ground Water Model Remed* 19: 90–96.

Diamond JM, Bressler DW, Serveiss VB. 2002. Assessing relationships between human land uses and the decline of native mussels, fish and macroinvertebrates in the Clinch and Powell River watershed, USA. *Environ Toxicol Chem* 21:1147–1155.

Diaz RJ, Schaffner LC. 1990. The functional role of estuarine benthos. In: Haire M, Krome EC, editors. Perspectives on the Chesapeake Bay: advances in estuarine science. Gloucester (VA): Chesapeake Research Consortium. Report No. CBP/TRS41/90. p 25–56.

Depledge MH, Fossi MC. 1994. The role of biomarkers in environmental assessment. 2. Invertebrates. *Ecotoxicology* 3(3):161–172.

Dewailly E, Ayotte P, Bruneau S, Laliberte C, Muir DCG, Norstrom RJ. 1993. Inuit exposure to organochlorines throug the aquatic food chain in Arctic Quebec. *Environ Health Persp* 101:618–625.

DiPinto LM, Coull BC. 1997. Trophic transfer of sediment-associated polychlorinated biphenyls from meiobenthos to bottom-feeding fish. *Environ Toxicol Chem* 16:2568–2575.

Di Toro DM, Zarba CS, Hansen DJ, Berry WJ, Swartz RC, Cowan CC, Pavlou, SP, Allen HE, Thomas NA, Paquin PR. 1991. Technical basis for establishing sediment quality criteria for nonionic organic chemicals using equilibrium partitioning, *Environ Toxicol Chem* 10:1541–1583.

Doe KG, Burton Jr GA, Ho KT. 2003. Pore water toxicity testing: An overview. In: Carr RS, Nipper M, editors. Porewater toxicity testing: biological, chemical and methodological considerations. Pensacola (FL): Society of Environmental Toxicology and Chemistry (SETAC). p 125–142.

Fetter CW. 1994. Applied hydrogeology, 3rd ed. Upper Saddle River (NJ): Prentice Hall. p 308–314, 368–370.

Fletcher R, Reynoldson TB, Taylor WD 2001. The use of benthic mesocosms for the assessment of sediment contamination. *Environ Pollut* 115:173–182.

Forbes TL, Forbes VE. 1993. A critique of the use of distribution-based extrapolation models in ecotoxicology. *Function Ecol* 7:249–254.

Forbes VA, Calow P. 2002. Applying weight of evidence in a retrospective ecological risk assessment when quantitative data are limited. *Hum Ecol Risk Assess* 8:1625–1640.

Franklin MF, Stauber JL, Lim RP, Petocz P. 2002. Toxicity of metal mixtures to a tropical freshwater alga (*Chlorella* sp): The effect of interactions between copper, cadmium, and zinc on metal cell binding and uptake. *Environ Toxicol Chem* 21:2412–2422.

Froese KL, Verbrugge DA, Ankley GT, Niemi GJ, Larsen CP, Giesy JP. 1998. Bioaccumulation of polychlorinated biphenyls from sediments to aquatic insects and tree swallow egges and nestlings in Saginaw Bay, Michigan, USA. *Environ Toxicol Chem* 17:484–492.

Garland JL, Mills AL. 1991. Classification and characterization of heterotrophic microbial communities on the basis of patterns of community-sole-carbon-source utilization. *Appl Environ Microbiol* 57:2351–2359.

Gaston GR, Nasci JC. 1988. Trophic structure of macrobethic communities in the Calcasieu Estuary, Louisiana. *Bull Environ Contam Toxicol* 49:922–928.

Giddings JM, Hall LWJ, Solomon KR. 2000. An ecological risk assessment of diazinon from agricultural use in the Sacramento–San Joaquin River basins, California. *Risk Anal* 20:545–572.

Giesy JP, Solomon KR, Coates JR, Dixon KR, Giddings JM, Kenaga EE. 1999. Chlorpyrifos: Ecological risk assessment in North American aquatic environments. *Rev Environ Contam Toxicol* 160:1–129.

Gobas FAPC. 1993. A model for predicting the bioaccumulation of hydrophobic organic chemicals in aquatic food webs: Application to Lake Ontario. *Ecol Model* 69:1–17.

Gobas FAPC, Zhang X, Wells R. 1995. Time response of the Lake Ontario ecosystem to virtual elimination of PCBs. *Environ Sci Technol* 29:2038–2046.

Grapentine L, Anderson J, Boyd D, Burton Jr. GA, DeBarros C, Johnson G, Marvin C, Milani D, Painter S, Pascoe T, Reynoldson T, Richman L, Solomon K Chapman PM. 2002. A decision making framework for sediment assessment developed for the Great Lakes. *Hum Ecol Risk Assess* 8:1641–1655.

Greenberg M, GA Burton Jr., Rowland CD. 2002. Optimizing interpretation of in situ effects: impact of upwelling and downwelling. *Environ Toxicol Chem* 21:289–297.

Grist EPM, Leung KMY, Wheeler JR, Crane M. 2002. Better bootstrap estimation of hazardous concentration thresholds for aquatic assemblages. *Environ Toxicol Chem* 21:1515–1524.

Haack SK, Garchow H, Odelson, DA, Forney LJ, Klug MJ. 1994. Accuracy, reproducibility, and interpretation of fatty acid methyl ester profiles of model bacterial communities. *Appl Environ Microbiol* 60:2483–2493.

Hampton T, Chadwick B. 2000. Quantifying in situ metal contaminant mobility in marine sediments. San Diego: SPAWAR Systems Center San Diego. Technical Report 1826.

Hansen D, Mahony J, Berry W, Benyi S, Corgin J, Pratt S, Di Toro D, Able M. 1996. Chronic effect of cadmium in sediments on colonization by benthic marine organisms: An evaluation of the role of interstitial cadmium and acid volatile sulfide in biological availability. *Environ Toxicol Chem* 15:2126–2137.

Hare L, Tessier A, Warren L. 2001. Cadmium accumulation by invertebrates living at the sediment–water interface. *Environ Toxicol Chem* 20:880–889.

Harris HJ, Wenger RB, Harris VA, Devault DS. 1994. A method of assessing environmental risks: a case study of Green Bay, Lake Michigan, USA. *Environ Manag* 18:259–306.

Hart ADM, editor. 2001. Probabilistic risk assessment for pesticides in Europe. implementation and needs. Report from the European Workshop on Probabilistic Risk Assessment for the Environmental Impacts of Plant Protection Products; 2001 June; The Netherlands. York (UK): Central Science Laboratory.

Hawkins CP, Norris RH, Gerritsen J, Hughes RM, Jackson SK, Johnson RK Stevenson RJ. 2000. Evaluation of the use of landscape classifications for the prediction of freshwater biota: synthesis and recommendations. *J North Am Benth Soc* 19:541–556.

Heipieper HJ, Meulenbeld G, von Oirschot Q, de Bont JAM. 1996. Effect of environmental factors on the *trans/cis* ratio of unsaturated fatty acids in *Pseudomonas putida* S12. *Appl Environ Microbiol* 62:2773–2777.

Heise S, Ahlf W. 2002. The need for new concepts in risk management of sediments. *J Soils Sed* 2:4–8.

Heuer H, Smalla K. 1997. Application of denaturing gradient gel electrophoresis and temperature gradient gel electrophoresis for studying soil microbial communities. In: Van Elsas JD, Trevors JT, Wellington EMH, editors. Modern soil microbiology. New York: Marcel Dekker.

Hickey CW, Clements WH. 1998. Effects of heavy metals on benthic macroinvertebrate 745 communities in New Zealand streams. *Environ Toxicol Chem* 17:2338–2346.

Hildrew AG, Giller PS. 1994. Patchiness, species interactions and disturbance in stream communities. In: Giller P, Hildrew A, Rafaelli D, editors. Aquatic ecology: Scale pattern and process. Oxford (UK): Blackwell Scientific. p 21–62.

Hirthe G, Fisher T, Crane M, Callaghan A. 2001. Short-term exposure to sublethal doses of lindane affects developmental parameters in *Chironomus riparius* Meigen, but has no effect on larval glutathione-s-transferase activity. *Chemosphere* 44: 583–589.

Ho KT, Burgess RM, Pelletier MC, Serbst JR, Ryba SA, Cantwell MG, Kuhn A, Raczelowski P. 2002. An overview of toxicant identification in sediments and dredged materials. *Mar Pollut Bull* 44:286–293.

Ho KT, McKinney RA, Kuhn A, Pelletier MC, Burgess RM. 1997. Identification of acute toxicants in New Bedford Harbor sediments. *Environ Toxicol Chem* 16:551–558.

Hogstrand C, Balesaria S, Glover CN. 2002. Application of genomics and proteomics for study of the integrated response to zinc exposure in a non-model fish species, the rainbow trout. *Comp Biochem Physiol* 133B:523–535.

Hogstrand C, Wood CM, Bury NR, Wilson RW, Rankin JC, Grosell M. 2002. Binding and movement of silver in the intestinal epithelium of a marine teleost fish, the European flounder (*Platichthys flesus*). *Comp Biochem Physiol C-Toxicol Pharmacol* 133:125–135.

Huckins JN, Manuweera, GK Petty JD, Mackay D, Lebo JA. 1993. Lipid containing semipermeable membrane devices for monitoring organic contaminants in water. *Environ Sci Technol* 27:2489–2496.

Hughes RM, Paulsen SG, Stoddard JL. 2000. EMAP-surface waters: a multiassemblage, probability survey of ecological integrity in the U.S.A. *Hydrobiologia* 422/423:429–443.

Hurlbert SH. 1984. Pseudoreplication and the design of ecological field experiments. *Ecol Monogr* 54:187–211.

Ingersoll CG, Ankley GT, Baudo R, Burton GA Jr., Lick W, Luoma SN, MacDonald DD, Reynoldson TB, Solomon KR, Swartz RC, Warren-Hicks WJ. 1997. Workgroup summary report on uncertainty evaluation of measurement endpoints used in sediment ecological risk assessment. In: Ingersoll CG, Dillon T, Biddinger GR, editors. Ecological risk assessment of contaminated sediments. Pensacola (FL): Society of Environmental Toxicology and Chemistry (SETAC). p 297–352.

Ingersoll CG, Brunson FJ, Dwyer FJ, Hardesty DK, Kemble NE. 1998. Use of sublethal endpoints in sediment toxicity tests with the amphipod *Hyalella azteca*. *Environ Toxicol Chem* 17:1508–1523.

Ingersoll CG, MacDonald DD. 2002. Guidance manual to support the assessment of contaminated sediments in freshwater ecosystems. Volume III: Interpretation of the results of sediment quality investigations. Chicago: USEPA Great Lakes National Program Office. EPA-905-B02-001-C.

Ireland DS, Burton Jr GA, Hess GG. 1996. In situ toxicity evaluations of turbidity and photoinduction of polycyclic aromatic hydrocarbons. *Environ Toxicol Chem* 15:574–581.

Johnson GD, Audet DJ, Kern JW, LeCaptain LJ, Strickland MD, Hoffman DJ, McDonald LL. 1999. Lead exposure in passerines inhabiting lead-contaminated floodplains in the Coeur d'Alene River Basin, Idaho, USA. *Environ Toxicol Chem* 18:1190–1194.

Jorgensen SE, Bendoricchio G. 2002. Fundamentals of ecological modelling. 3rd ed. Amsterdam: Elsevier.

Karr JR. 1991. Biological integrity: a long-neglected aspect of water resource management. *Ecol Appl* 1:66–84.

La Point TW, Waller WT. 2000. Field assessments in conjunction with WET testing. *Environ Toxicol Chem* 19:14–24.

Lee DR, Cherry JA. 1978. A field exercise on groundwater flow using seepage meters and mini-piezometers. *J Geol Educ* 27:6–10.

Lendvay JM, Sauck WA, McCormick ML, Barcelona MJ, Kampbell DH, Wilson JT, Adriaens P. 1998. Geophysical characterization, redox zonation, and contaminant distribution at a groundwater/surface water interface. *Water Resour Res* 43:3545–3559.

Lewis JB. 1987. Measurements of groundwater seepage flux onto a coral reef: spatial and temporal variations. *Limnol Oceanogr* 32:1165–1169.

Long ER, Chapman PM. 1985. A sediment quality triad: measures of sediment contamination, toxicity and infaunal community composition in Puget Sound. *Mar Pollut Bull* 16:405–415.

Mackay D. 1991. Multimedia environmental models: the fugacity approach. Chelsea (MI): Lewis.

Maltby L. 1999. Studying stress: The importance of organism-level responses. *Ecol Appl* 9:431–440.

Maruya KA, Lee RE. 1998. Biota sediment accumulation and trophic transfer factors for extremely hydrophobic polychlorinated biphenyls. *Environ Toxicol Chem* 17:2463–2469.

Meadows JC, Echols KR, Huckins JN, Borsuk FA, Carline RF, Tillitt DE. 1998. Estimation of uptake rate constants for PCB congeners accumulated by semipermeable membrane devices and brown trout (*Salmo trutta*). *Environ Sci Technol* 32:1847–1852.

Meyer JS. 2002. The utility of the terms "bioavailability" and "bioavailable fraction" for metals. *Mar Environ Res* 53:417–423.

Montlucon D, Sanudo-Wilhelmy SA. 2001. Influence of net groundwater discharge on metal and nutrient concentrations in a coastal environment: Flanders Bay, Long Island, New York. *Environ Sci Technol* 35:480–486.

Moore WS. 1999. The subterranean estuary: A reaction zone of ground water and sea water. *Mar Chem* 65:111–125.

Mount DI, Anderson-Carnahan L. 1988. Methods for aquatic toxicity identification evaluations: Phase I toxicity characterization procedures. Duluth (MN): US Environmental Protection Agency. EPA/600-3-88/034.

Mount DI, Anderson-Carnahan L. 1989. Methods for aquatic toxicity identification evaluations: Phase II toxicity identification procedures. Duluth (MN): US Environmental Protection Agency. EPA/600-3-88/03.

Mount DI, Norberg-King T, Ankley G, Burkhard LP, Durhan EJ, Schubauer-Berigan MK, Lukasewycz M. 1993. Methods for aquatic toxicity identification evaluations: Phase III toxicity confirmation procedures for samples exhibiting acute and chronic toxicity. Duluth (MN): US Environmental Protection Agency. EPA/600/R-92/08.

Newman MC, Ownby DR, Mézin LCA, Powell DC, Christensen TL, Lerberg SB, Anderson BA. 2000. Applying species sensitivity distributions in ecological risk assessment: assumptions of distribution type and sufficient numbers of species. *Environ Toxicol Chem* 19:508–515.

Norberg-King T, Mount DI, Amato JR, Jensen DA, Thompson JA.. 1991. Methods for aquatic toxicity identification evaluations: Phase I Toxicity Characterization

Procedures. 2nd ed. Duluth (MN): US Environmental Protection Agency. EPA-600/6-91/005.

Nordstrom JF, Burton Jr GA. 2000. In situ vs. laboratory toxicity identification evaluations. Abstr World Congress Meet. Society of Environmental Toxicology and Chemistry (SETAC) Europe; Brighton, England. Pensacola (FL): SETAC.

Norris RH, Georges A. 1993. Analysis and interpretation of benthic macroinvertebrate surveys. In: Rosenberg DM, Resh VH editors. Freshwater biomonitoring and benthic macroinvertebrates. New York: Chapman & Hall. p 234–286.

Norton SB, Cormier SM, Suter II GW, Subramanian B, Lin E, Altfater D, Counts B. 2002. Determining probable causes of ecological impairment in the Little Scioto River, Ohio, USA: Part 1. Listing candidate causes and analyzing evidence. *Environ Toxicol Chem* 21:1112–1124.

Paquin P, Santore RC, Farley KJ, Kavvadas C, Wu KB, Mooney K, Di Toro DM. 2003. A review: Exposure, bioaccumulation, and toxicity. Metals in aquatic systems. Pensacola (FL): Society of Environmental Toxicology and Chemistry (SETAC).

Pastorok RA, Bartell SM, Ferson S, Ginzburg LR. 2002. Ecological modelling in risk assessment: Chemical effects on populations, ecosystems, and landscapes. Boca Raton (FL): Lewis.

Pauwel S, Sibley PK. 2002. Testing for toxicity and bioaccumulation in freshwater sediments. In: A handbook on sediment quality. Alexandria (VA): Water Environment Federation (WEF). p 199–258.

Pavluk TI, bij de Vaate A, Leslie HA. 2000. Development of an index of trophic completeness for benthic macroinvertebrate communities in flowing waters. *Hydrobiologia* 427:135–141.

Peakall DB. 1994. The role of biomarkers in environmental assessment: introduction. *Ecotoxicology* 3:157–160.

Peterman R. 1990. Statistical power analysis can improve fisheries research and management. *Can J Fish Aquat Sci* 47:2–15.

Peterson SO, Klug MJ. 1994. Effects of sieving, storage, and incubation temperature on the phospholipid fatty acid profile of a soil microbial community. *Appl Environ Microbiol* 60:2421–2430.

Poff NL, Ward JV. 1990. Physical habitat template of lotic systems - recovery in the context of historical pattern of spatio-temporal heterogeneity. *Environ Manag* 14:629–645.

Posthuma L, Suter II GW, Traas TP. 2001. Species sensitivity distributions in ecotoxicology. Boca Raton (FL): Lewis.

Power RS, David HE, Mutwakil MHAZ, Fletcher K, Daniells C, Nowell MA, Dennis JL, Martinelli A, Wiseman R, Wharf E, De Pomerai DI. 1998. Stress-inducible transgenic nematodes as biomonitors of soil and water pollution. *J Biosci* 23:513–526.

Rakocinski CF, Brown SS, Gaston GR, Heard RW, Walker WW, Summers JK. 1997. Macrobenthic responses to natural and contaminant-related gradients in northern Gulf of Mexico estuaries. *Ecol Appl* 7:1278–1298.

Rakocinski CF, Comyns BH, Peterson MS. 2000. Relating environmental fluctuation and the early growth of estuarine fishes: Ontogenetic standardization. *Trans Am Fish Soc* 129:210–221.

Resh VH, McElvary EP. 1993. Contemporary quantitative approaches to biomonitoring using benthic macroinvertebrates. In: Rosenberg DM, Resh VH, editors. Freshwater biomonitoring and benthic macroinvertebrates. New York: Chapman & Hall. p 159–194.

Reynoldson TB, Day KE, Bailey RC 1995. Biological guidelines for freshwater sediment based on benthic assessment of sediment (the BEAST) using a multivariate approach for predicting biological state. *Aust J Ecol* 20:198–219.

Reynoldson TB, Norris RH, Resh VH, Day KE, Rosenberg DM 1997. The reference condition: a comparison of multi-metric and multivariate approaches to assess water-quality impairment using benthic macroinvertebrates. *J North Am Benth Soc* 16:833–852.

Reynoldson TB, Thompson SP, Milani D. 2002. Integrating multiple toxicological endpoints in a decision-making framework for contaminated sediments. *Hum Ecol Risk Assess* 8:1659–1585.

Reynoldson TB, Wright JF. 2000. The reference condition: problems and solutions. In: Wright, JF, Sutcliffe DW, Furse MT, editors. Assessing the biological quality of freshwaters. RIVPACS and other techniques. Ambleside (UK): Freshwater Biological Assn. p 293–303.

Rhoads DC. 1974. Organism-sediment relations on the muddy sea floor. *Oceanogr Mar Biol Annu Rev* 12:263–300.

Richards C, Host GE, Arthur JW. 1993. Identification of predominant environmental-factors structuring stream macroinvertebrate communities within a large agricultural catchment. *Freshw Biol* 29:285–294.

Ringwood AH, Conners DE, Hoguet J. 1998. Effects of natural and anthropogenic stressors on lysosomal destabilization in oysters *Crassostrea virginica. Mar Ecol Progr Ser* 166:163–171.

Ringwood AH, Conners De, Keppler CJ. 1999. Cellular responses of oysters, *Crassostrea virginica*, to metal-contaminated sediments. *Mar Environ Res* 48: 427–437.

Santegoeds CM, Ferdelman TG, Muyzer G, de Beer D. 1998. Structural and functional dynamics of sulfate-reducing populations in bacterial biofilms. *Appl Environ Microbiol* 64:3731–3739.

Sayles FL. 1992. Benthic landers: taking the laboratory to the seafloor. *Oceanus* 35:8–10.

Scheiner SM. 1993. MANOVA: Multiple response variables and mutispecies interactions. In: Scheiner SM, Gurevitch J, editors. Design and analysis of ecological experiments. New York: Chapman & Hall. p 94–112.

Schmidt TS, Soucek DJ, Cherry DS. 2002. Integrative assessment of benthic macroinvertebrate community impairment from metal-contaminated waters in tributaries of the upper Powell River, Virginia, USA. *Environ Toxicol Chem* 21:2233–2241.

Schubauer-Berigan MK, Ankley GT. 1991. The contribution of ammonia, metals and nonpolar organic compounds to the toxicity of sediment interstitial water from an Illinois River tributary. *Environ Toxicol Chem* 10:925–939.

[SETAC] Society of Environmental Toxicology and Chemistry. 1994. Aquatic Dialogue Group: Pesticide risk assessment and mitigation. Pensacola (FL): SETAC. p 65.

Sharpe S, Mackay D. 2000. A framework for evaluating bioaccumulation in food webs. *Environ Sci Technol* 34:2373–2379.

Sibley PK, Benoit DA, Ankley GT. 1997. The significance of growth in Chironomus tentans sediment toxicity tests: Relationship to reproduction and demographic endpoints. *Environ Toxicol Chem* 16:336–345.

Smith EP. 2002. BACI design. In: El-Shaarawi AH, Piegorsch WW,editors. Encyclopedia of environmetrics. Chichester (UK): J Wiley. p 141–148.

Smith EP, Lipkovich I, Ye K. 2002. Weight of evidence (WOE): quantitative estimation of probability of impact. *Hum Ecol Risk Assess* 8:1585–1596.

Solomon KR. 2001. Ecotoxicological risk assessment of pesticides in the environment. 2nd ed. In: Krieger RI, Doull J, Gammon D, Hodgson E, Reiter L, Ecobichon D, Ross J, editors. Handbook of pesticide toxicology. Volume 1, Principles. New York: Academic Pr.

Solomon KR, Sibley PK. 2002. New concepts in ecological risk assessment: Where do we go from here? *Mar Pollut Bull* 44:279–285.

Suarez LA, Barber MC. 1994. FGETS Version 3.0.18 user's manual. Athens (GA): US Environmental Protection Agency.

Suter II GW, Norton SB, Cormier SM. 2002. A methodology for inferring the causes of observed impairments in aquatic ecosystems. *Environ Toxicol Chem* 21:1101–1111.

Suter II GW. 1993. Ecological risk assessment. Boca Raton (FL): Lewis.

Taniguchi M, Fukuo Y. 1993. Continuous measurements of ground-water seepage using an automatic seepage meter. *Ground Water* 31:675–679.

Tengberg A, De Bovee F, Hall P, Berelson W, Chadwick B, Ciceri G, Crassous P, Devol A, Emerson S, Gage J, Glud R, Graziottin F, Gundersen J, Hammond D, Helder W, Hinga K, Holby O, Jahnke R, Khripounoff A, Lieberman S,

Nuppenau V, Pfannkuche O, Reimers C, Rowe G, Sahami A, Sayles F, Schurter M, Smallman D, Wehrli B, De Wilde P. 1996. Benthic chamber and profile landers in oceanography: a review of design, technical solutions and functioning. *Progr Oceanogr* 35:253–294.

Thomann RV, Connolly JP, Parkerton TF. 1992a. Modeling accumulation of organic chemicals in aquatic food webs. In: Gobas FAPC, McCorquodale JA, editors. Chemical dynamics in fresh water ecosystems. Boca Raton (FL): Lewis.

Thomann RV, Connolly JP, Parkerton TF. 1992b. An equilibrium model of organic chemical accumulation in aquatic food webs with sediment interactions. *Environ Toxicol Chem* 11:615–629.

Thomann RV, Mahony JD, Mueller R. 1995. Steady-state model of biota sediment accumulation factor for metals in two marine bivalves. *Environ Toxicol Chem* 14:1989–1998.

Thomann RV, Mueller JA, Winfield RP, Huang CR. 1991. Model of the fate and accumulation of PCB homologues in Hudson Estuary. *ASCE J Environ Enr* 117:161–177.

Thomann RV, ShkreliF, Harrison S. 1997. A pharmacokinetic model of cadmium in rainbow trout. *Environ Toxicol Chem* 16:2268–2274.

Timofeeva SS, Stom DJ. 1990. Perspecitives of employing enzymoindication methods in ecological monitoring. *Acta Hydrochim Hydrobiol* 18:639–648.

Townsend CR, Scarsbrook MR, Doledec S. 1997. Quantifying disturbance in streams: alternative measures of disturbance in relation to macroinvertebrate species traits and species richness. *J North Am Benthol Soc* 16:531–44.

[USEPA] US Environmental Protection Agency. 1992. Framework for ecological risk assessment. Washington DC: USEPA Risk Assessment Forum. EPA/630/R-92/001.

[USEPA] US Environmental Protection Agency. 1995. National water quality inventory. 1994 Report to Congress. Washington DC: USEPA, Office of Water. EPA841-R-95-005.

[USEPA] US Environmental Protection Agency. 1997. Mercury study report to Congress. Volume III: Fate and transport of mercury in the environment. Washington DC: USEPA Office of Air Quality Planning and Standards and Office of Research and Development. EPA-452/R-97-005.

[USEPA] US Environmental Protection Agency. 2000a. Methods for measuring the toxicity and bioaccumulation of sediment-associated contaminants with freshwater invertebrates. 2nd ed. Washington DC: USEPA. EPA/600/R-99/064.

[USEPA] US Environmental Protection Agency. 2000b. Stressor identification guidance document. Washington DC: USEPA, Office of Water and Office of Research and Development. EPA/822/B-00-025.

[USGS] US Geological Survey. 1999. The quality of our nation's waters: Nutrients and pesticides. Reston (VA): USGS. Circular 1225. Fact sheets 116–99. Available from: http://water.usgs.gov/pubs/circ/circ1225/pdf/index.html. Accessed 8 Sep 2004.

Valiela I, D'Elia C. 1990. Groundwater inputs to coastal waters: Introduction. *Biogeochemistry* 10(3):175.

von Ende CN. 1993. Repeated-measures analysis: growth and other time-dependent measures. In: Scheiner SM, Gurevitch J, editors. Design and analysis of ecological experiments. New York: Chapman & Hall. p 113–137.

Vroblesky DA, Hyde WT. 1997. Diffusion samplers as an inexpensive approach to monitoring vocs in ground water: *Ground Water Monitor Remed* 17:177–184.

Walley WJ, Fontama VN. 1997. New approaches to river quality classification based upon artificial intelligence. In: Wright JF, Sutcliffe DW, Furse MT, editors. Assessing the biological quality of fresh waters. Ambleside (UK): Freshwater Biological Assn. p 63–280

Warwick RM, Clarke KR. 1991. A comparison of some methods for analysing changes in benthic community structure. *J Mar Biol Assess* 71:225–244.

Waters TF. 1995. Sediments in streams: sources, biological effects, and control. Bethesda (MD): American Fisheries Society. Monograph #7.

Wiegers JK, Feder HM, Mortensen LS, Shaw DG, Wilson VJ, Landis WG. 1998. A regional multiple stressor rank-based ecological risk assessment for the fjord of Port Valdez, AK. *Hum Ecol Risk Assess* 4:1125–1173.

Wiley MJ, Kohler SL, Seelbach PW. 1997. Reconciling landscape and local views of aquatic communities: lessons from Michigan trout streams. *Freshw Biol* 37:133–48.

Wright JF, Blackburn JH, Gunn RJM, Furse MT, Armitage PD, Winder JM, Symes KL, Moss D. 1996. Macroinvertebrate frequency data for the RIVPACS III sites in Great Britain and their use in conservation evaluation. *Aquat Conserv* 6:141–167.

Wright JF, Sutcliffe DW, Furse MT, editors. 2001. Assessing the biological quality of freshwaters. RIVPACS and other techniques. Ambleside (UK): Freshwater Biological Assn.

Yoder CO, Rankin ET. 1995. Biological response signatures and areas of degradation value: new tools for interpreting multimetric data. In: Biological assessment and criteria: tools for water resource and decision-making. Boca Raton (FL): Lewis. p 263–288.

Zarull MA, Hartig JH, Krantzber G, Burch K, Cowgill D, Hill G, Miller J, Sherbin IG. 1999. Contaminated sediment management in the Great Lakes basin ecosystem. *J Gt Lakes Res* 25:412–422.

Zhang H, Davison W, Miller S, Tyck W. 1995. In situ high resolution measurements of fluxes of Ni, Cu, Fe, and Mn and concentration of Zn and Cd in porewaters by DGT. *Geochim Cosmochim Acta* 59: 4181–4192.

A framework for assessing and managing risks from contaminated sediments 6

TODD S BRIDGES, WALTER J BERRY, STEFANO DELLA SALA, PHILIP B DORN, STEPHEN J ELLS, THOMAS H GRIES, D SCOTT IRELAND, EILEEN M MAHER, CHARLES A MENZIE, LINDA M POREBSKI, JOOST STRONKHORST

Introduction

Chemistry data, including sediment quality guidelines (SQGs), have been used for a number of years by regulatory agencies in different countries to assess and manage contaminated sediments (see extensive reviews in Chapters 9 and 10). Increasing interest in the use of risk-based sediment assessment frameworks to guide assessments and management decisions has led to questions concerning the role of SQGs within an assessment and management process that makes use of multiple lines of evidence (LOEs) to reach management decisions (e.g., Sijm et al. 2001). An assessment framework guides the assessment–management process by providing structure, organization, and flow for the actions to be taken in assessing and managing risks. As such, a framework that meets a specific program's objectives will delimit appropriate uses for SQGs.

Environmental regulatory agencies should be encouraged to develop logical and orderly sediment assessment frameworks to ensure that assessments are comprehensive, transparent, and consistent. A sediment assessment framework should be structured to ensure that any evaluation that follows the steps of the framework is comprehensive and complete in its consideration and analysis of present and future exposures, effects, and human and ecological risks at the site of concern. All routes of exposure and types of effects will not occur at every site; however, a comprehensive framework, if followed, should require consideration of the likelihood for all possible routes of exposure and effects to ensure that required or important site-

Use of Sediment Quality Guidelines and Related Tools for the Assessment of Contaminated Sediments
Wenning RJ, Batley GE, Ingersoll CG, Moore DW, editors.

specific environmental factors are not omitted from the evaluation process. Assessment frameworks should also provide a measure of transparency to sediment investigation and management to facilitate meaningful participation in the assessment and decision-making process by scientists, regulatory agencies, and representatives of affected communities. Active stakeholder involvement throughout the assessment process is essential to ensure that the results of the assessment can be successfully applied within the decision-making process. Finally, development and application of an assessment framework will help ensure consistent application of the assessment and management process at different sites.

Points of emphasis for designing or applying sediment assessment frameworks

At present, nearly twenty sediment assessment frameworks have been proposed or used by regulatory authorities in different countries. Following our review of these frameworks (Chapter 10), we reached several conclusions regarding the design and use of assessment frameworks to characterize and manage risks posed by sediments. These conclusions are presented in the remainder of this section; this discussion is followed by a description of a generic risk-based framework that incorporates our sense of the key elements of the sediment assessment–management process and the role of SQGs within that process.

Need for considering program objectives

Frameworks are developed and used to address specific programmatic goals. The objectives motivating a specific sediment assessment will, in part, be determined by the specific regulatory programs relevant to the assessment. Knowledge of these drivers must be factored into the assessment process to ensure that a complete set of information is collected and analyzed to aid decision making. Some sponsoring programs may place emphasis on limiting chemical inputs to the environment without a specific linkage to resulting risks, whereas other programs will give greater emphasis to quantifying exposures, effects, and risks. A regulatory program charged with linking risks to contaminant sources (with the purpose of implementing source controls) will likely differ from an assessment sponsored by a program with the simpler charge of identifying only sediments that pose a risk. A sediment assessment framework used in a regulatory program charged with monitoring broad spatial and temporal patterns in sediment quality in coastal environments will in all likelihood differ from one that governs navigation dredging operations. The existence of such differences is expected and reasonable. The degree of success achieved in using a specific sediment assessment framework within the context of a regulatory program will be determined, in large part, by the extent to which program-specific objectives are acknowledged and accounted for when the assessment framework is designed and applied.

Need for stakeholder involvement

To be effective, assessment and management processes must allow for the involvement of stakeholders. Successful implementation of a sediment assessment framework will include frequent opportunity for stakeholders to provide input to the process. It is particularly important to get stakeholder input when the goals and questions that will drive assessment activities are being formulated and management alternatives are being evaluated. However, these are not the only instances in which stakeholder involvement in the assessment process will be of value; for example, stakeholders can provide valuable information about the history of a site and its current use by the public.

Multiple tiers and lines of evidence

Effective evaluation frameworks will make use of multiple assessment tiers and multiple LOEs. The types and amount of information necessary to reach management decisions will vary from site to site or project to project. Most sediment assessments will involve the use of a variety of physical, chemical, and biological information in order to reach decisions about the presence or absence of risk and how best to manage evident risks. No single LOE alone will generally provide sufficient information for effective decision making.

To conserve time and other resources, assessment frameworks can be designed with tiers to encourage investigations that optimize the level of effort expended in the assessment with respect to the complexity of both the site and assessment questions that must be answered to reach management decisions. A tier in this sense is a phase or stage in the assessment process that concludes with a decision to either 1) exit the assessment process because sufficient information has been collected to answer the assessment questions about the need for management, or 2) continue the assessment because insufficient information exists to reach a management decision. In many cases, management decisions may be possible during the initial stages (tiers) of an assessment when convincing evidence for or against the presence of risk is evident. In more ambiguous circumstances, or where the complexity of the site requires, more comprehensive assessments and data collection may be required in subsequent tiers before credible management decisions can be made. One strength of an assessment framework structured in such a tiered fashion is that the framework includes clear decision points that address the need to continue the evaluation.

Assessment frameworks also should be structured to allow for iteration. As information is collected and analyzed during an evaluation, the assessment process must allow for making additions and refinements to the conceptual model and assessment questions that are formulated during the initial stages of assessment. Such iteration allows the assessment to become more focused as the evaluation proceeds.

Comparative decision making

Selecting the management alternatives to implement at a site will involve analyzing tradeoffs. For this reason, a comparative analysis should inform decision making. The use of a comparative analysis is necessitated by the fact that there are no zero-risk options for managing contaminated sediment. Making effective risk management decisions requires comparing the risks and costs associated with the full spectrum of available management alternatives. Such a comparative approach will involve close exchange between sediment assessment activities and the process used to select management alternatives.

Conceptual models and assessment questions

Effective sediment assessments will be based on thoroughly developed conceptual models and assessment questions (risk hypotheses). Sediment assessments are inherently complex, owing to the dynamics and heterogeneity apparent at most sites and the nature of the assessment questions commonly posed. For these and other reasons, significant effort should be invested in developing a comprehensive conceptual model for the specific contaminated sediment or site being evaluated. A conceptual model is a written description and visual representation of predicted relationships between ecological entities and the stressors to which they may be exposed (USEPA 1998). During the conduct of an assessment, a conceptual model will be the basis for formulating project-specific questions that will drive subsequent assessment activities. Sediment assessments should be driven by site-specific questions rather than proclivities for applying specific assessment tools. Considerable time and energy should be dedicated to formulating and refining specific and detailed questions that must be answered to reach conclusions about the presence and magnitude of risk. Assessment questions lead to the selection of LOEs and specific assessment tools to apply during the assessment.

Monitoring

An assessment and management framework applied to sediments should include a phase in which risk estimates (predictions) are verified and monitored. Risk assessments produce descriptions or predictions about risks and the stressors driving those risks. Management decisions will be made based on these descriptions and predictions. Effective sediment management requires developing a monitoring strategy to check the accuracy of risk descriptions and to modify, as necessary, management strategies or decisions. The results of monitoring activities also can be used as feedback on the performance of the assessment framework.

Role of sediment quality guidelines

It is evident from the structure and organization of many of the sediment assessment and management frameworks in use today that regulatory authorities see a benefit to using SQGs during the assessment–management process (Chapters 9 and 10). One of the chief benefits to using SQGs, and other sources of information, during the initial phases of an evaluation is to identify sediments that require no further evaluation because they pose little potential for risk, thereby conserving time and other resources. Sediment assessments are expensive and time consuming. Some efficiency can be achieved if sufficient evidence can be assembled early in the process to reach a credible conclusion that the sediments pose little potential for risk. For some regulatory programs, there may also be merit in incorporating approaches within the initial stages of an assessment for collecting information from multiple LOEs, including the use of SQGs, that can be used to identify sediments that pose some high potential for risk. Such approaches can allow pragmatic decisions to bypass further evaluation and accept more restrictive management when the costs of further analysis are likely to exceed the costs of a restrictive management option. In cases where a comparison to SQGs results in ambiguous or incomplete answers to the assessment questions concerning the presence and nature of unacceptable risks, further assessment and analysis will be required.

A Generic Risk-Based Sediment Assessment Framework

The assessment framework outlined in Figure 6-1 incorporates the key elements of existing sediment assessment frameworks and provides a structured, defined role for SQGs within a risk-based evaluation. The framework is composed of 6 major phases of activity:

1) pre-assessment,
2) initial assessment,
3) sediment or site assessment,
4) evaluation and selection of management alternatives,

5) verification and monitoring, and
6) process adaptation.

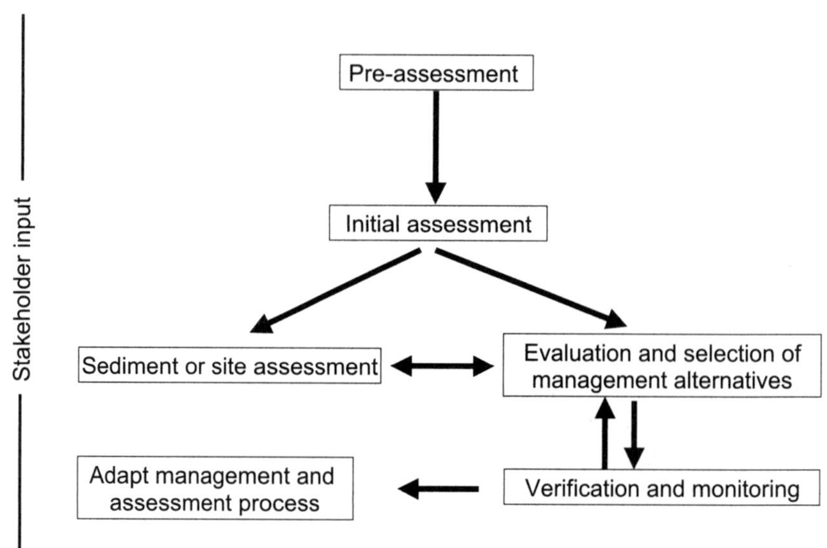

Figure 6-1 Generic assessment and management framework for contaminated sediments

As with other published assessment frameworks, this generic framework is structured with tiers. While the assessment frameworks described in Chapter 10 differ in their breadth, scope, structure, and inclusion of specific assessment tools, the framework described in the remainder of this chapter was developed as a generic model framework that could be adapted to meet a range of specific programmatic objectives.

Pre-assessment

During the pre-assessment phase of an evaluation, the reasons why a sediment assessment is being conducted should be defined. Assessments may be contemplated to address any number of programmatic and/or regulatory goals or objectives. The reasons motivating the assessment will impact nearly every aspect of how the assessment is conducted, from the structure and complexity of the conceptual model to the need for specific expertise on the assessment team, as well as the projected costs of the assessment.

An assessor must be aware of any programmatic requirements and/or constraints to ensure that the evaluation produces an assessment sufficient for decision making. Such program mandates may even include directives regarding the use of SQGs. International conventions, national laws and regulations, and programmatic guidance have been written that govern dredged material assessment (e.g., USEPA/USACE 1998; Stronkhorst et al. 2001), integrated watershed assessment and management (e.g., de Deckere 2000; Martin and Kennedy 2000), and contaminated site assessment (e.g., USEPA 2000). Therefore, sediment assessments can be conducted in support of a navigation dredging project to achieve navigable depths in a channel, as part of a more general watershed or water quality assessment, or to determine the need for remedial action at a contaminated site. Assessments conducted in these various contexts would likely vary in fundamental ways that are shaped by the specific program and regulatory objectives involved, requirements regarding the use of specific assessment tools, or the presence of set decision points. The generic framework described in this chapter assumes no particular regulatory constraints but rather identifies best practice.

Assessments for navigation dredging

The most routine sediment assessments occur in support of navigation dredging operations. Most of the sediments assessed in navigation dredging programs present minimal risk to the environment. For example, in the United States (US) dredging program, only 5% to 10% of the 300 million cubic yards of sediment dredged annually to support navigation activities requires special management or handling due to the presence of risk from contaminants (NRC 1997). However, compliance with international conventions (e.g., the London Convention) and national laws and regulations requires the conduct of sediment assessments to determine how the material should be managed. The objective is to move the sediment out of the navigation channel. The results of the sediment assessment will be used to determine how the material, once removed from the channel, will be managed. The management alternatives range broadly from unrestricted, open water disposal or beneficial uses (e.g., beach nourishment or habitat creation) for materials posing no or minimal risk, to confined aquatic (i.e., capping) or upland disposal for sediments where risks are evident (USEPA/USACE 1992).

In navigation dredging programs, the presumption is that risks will be relatively low in most cases and that a principal function of an assessment framework will be to provide a consistent, reliable, and understandable process for making compliance decisions. Because such assessments will be conducted on a routine basis, the conceptual model used in such assessments may be standardized for repeated use and adaptation. In addition, the LOEs and contaminants of concern may be "hardwired" within the framework, based on past experience within the program. Indeed, some standardization is often helpful in reducing the costs of assessments and for allowing decisions to be made in a consistent manner by professionals with different technical backgrounds. In such cases, assessment frameworks may be more prescriptive in nature. However, any assessment framework should retain sufficient flexibility for assessors to respond to changes in conditions over time and to accommodate unique project conditions.

To conserve resources, given the large number of routine assessments that will be conducted in navigation dredging programs, attention should be given to developing initial assessment procedures that allow for early identification of minimal risk sediments and sediments presenting obvious high risks. In most cases, making such distinctions involves using a range of chemical (e.g., SQGs), physical (e.g., grain size, proximity to contaminant sources), and biological (e.g., toxicity tests) LOEs. Screening approaches used during the initial assessment to identify low-risk sediments should be conservative and seek to establish conditions in which effects on the environment or human health are unlikely to occur. Reaching such distinctions during the initial assessment will appropriately focus assessment resources on the minority of cases where ambiguities exist regarding risk and the need for special management. Several jurisdictions (e.g., Canada, Hong Kong, Australia, The Netherlands) require the use of SQGs in which a list of chemicals is identified for routine use (Chapters 9 and 10). In the US dredging program, the use of SQGs is more constrained by existing regulatory language that requires the use of bioassays. The International Convention on the Prevention of Marine Pollution by Dumping of Wastes and Other Matter (1972; London Protocol on ocean disposal) requires party nations to set action levels using chemical or biological measures in order to identify sediments that require special management.

Assessments for monitoring and watershed management

Sediment assessments are also conducted as part of large-scale monitoring efforts. Examples of such programs include the National Oceanic and Atmospheric Administration's (NOAA's) National Status and Trends program (Long et al. 1995), the US Environmental Protection Agency's Environmental Monitoring and Assessment Program (USEPA 1998), and the US National Sediment Inventory (USEPA 2001). Such programs strive to provide information about patterns and trends in contamination and potential risk over very large spatial and temporal scales to provide information to the public on the status of natural resources and to provide feedback to specific regulatory programs. The challenges of developing or implementing a sediment assessment framework for such purposes includes the need to apply a consistent set of assessment tools in a repeatable fashion over time to help ensure that data types are produced in a manner that permits patterns and trends to be recognized. Depending on the program objectives, such efforts may also require development and use of relatively simple metrics (quantitative and/or qualitative) to convey information to the public or other constituencies of the program.

Over the last several years, sediments have been increasingly identified as playing an important role in limiting designated uses of waterbodies. The total maximum daily load (TMDL) program in the US is concerned with identifying the loading limits of contaminants to waterbodies to ensure that relevant standards or criteria are met for the waterbodies' designated use. In such a program, sediments are recognized as both a sink and a source of contaminants. Developing load allocations and attainment goals and strategies may require conducting a sediment assessment to determine the extent to which the sediment source contributes to the degraded condition of the waterbody. In most cases, such assessments will be conducted over regional or local spatial scales. The principal challenge for such efforts concerns how to quantify the extent to which sediments contribute to degraded conditions in comparison to other sources (e.g., land-based point and nonpoint discharges).

Contaminated site assessments

Sediment assessments also will be performed to inform management decisions at hazardous waste sites. Within the context of a cleanup program (e.g., the US Superfund program), some risk is presumed to exist at the site at the beginning of the assessment; in most cases, a site is nominated to such a program because evidence supports the presence of some risk. The information collected during the assessment will be used to determine the nature, extent, and magnitude of that risk and to aid in the selection of the best set of management technologies to apply at the site.

A program with the charge to oversee the assessment and management of such sites will encounter sites with a broad range of environmental conditions and receptors. This diversity will require development and use of a flexible assessment framework that emphasizes the need to develop a site-specific conceptual model to guide the design and conduct of the assessment. Less prescription with regard to choice of methods and tools will be necessary in such a program. Allowing for such site specificity generally results in more complex assessments involving more LOEs.

Making management decisions at hazardous waste sites, each possessing a unique set of environmental conditions and receptors, will require development of site-specific cleanup targets or goals in order to apportion sediment among selected management options. The development of such site-specific goals will require close coordination and interaction with local stakeholders. The range of available management options is larger for cleanup programs compared to navigation dredging programs because sediments can be managed in place, under certain conditions, while still meeting the overall objectives of the program, that is, to prevent further degradation, to remediate risks, and to restore use.

Initial assessment

The Initial Assessment phase of the framework includes these 4 activities:

1) collecting and analyzing existing and preliminary data,
2) developing a conceptual model for the sediment or site,
3) developing sediment or site-specific assessment questions, and
4) comparing initial data to appropriate SQGs (Figure 6-2).

An Initial Assessment will begin with collecting, reviewing, and analyzing any existing or preliminary data for the sediment or site. Such data would include information about contaminant sources, receptors of concern, and biological or chemical data from the sediment or site. This data collection effort should involve a review of data quality and would culminate in identifying preliminary lists of contaminants and receptors of concern.

The initial data collection and review should address the following questions:

- Are there local sources of contamination, either past or present (e.g., industrial or municipal discharges; shipping; inputs from industrial, municipal, or agricultural sources; spills and urban surface runoff)?
- What chemicals may have been released from these sources, that is, what are the contaminants of concern (COCs)?
- What data on physical, chemical, toxicological, and biological characteristics are available, and are these data of sufficient quality?
- What are the key receptors of concern (ROCs)?

Following the initial data collection and analysis, a conceptual model for the site is developed. Such a model identifies and describes contaminant sources; the processes linking those sources to the sediment in question; the physical, chemical, and biological processes occurring within the sediment that affect exposure; and how receptors of concern are exposed to the contaminants as-

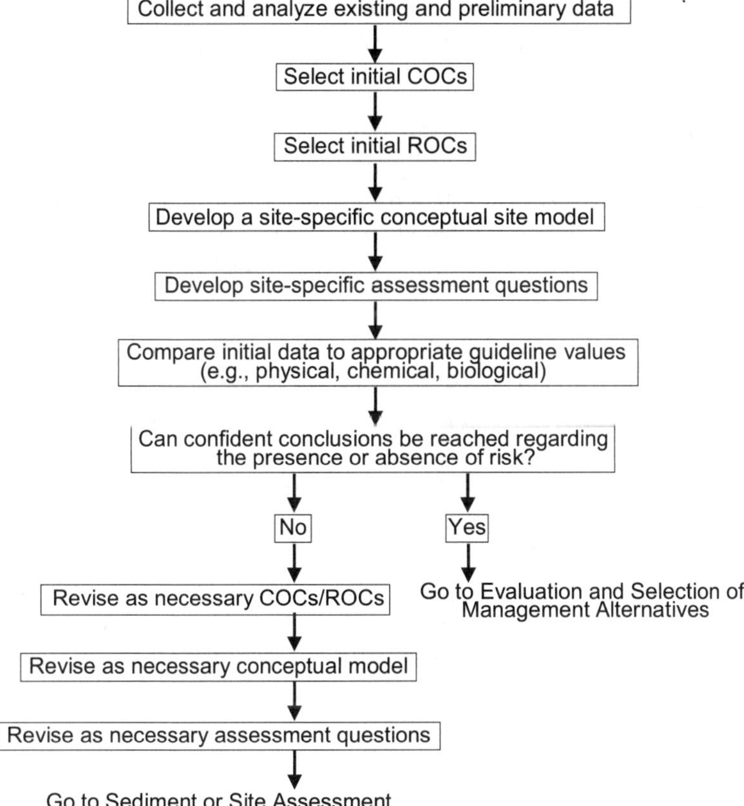

Figure 6-2 Initial assessment activities (COCs = contaminants of concern; ROCs = receptors of concern)

sociated with the sediment. Significant effort should be invested in developing a comprehensive conceptual model for the specific contaminated sediment or site under consideration because this model will be the basis for formulating the project-specific questions that will drive subsequent assessment activities.

Value of developing a conceptual model

A conceptual model is a written description and visual representation of predicted relationships between ecological entities and the stressors to which they may be exposed (USEPA 1998). While the previous statement is taken from the USEPA's Guidelines for Ecological Risk Assessment, conceptual models are also used to reflect exposures to humans for health risk assessments.

Further, conceptual models can be expanded to include other processes and endpoints of concern including economics, aesthetics, and cultural issues.

The conceptual model is critically important for defining an environmental problem and for guiding the technical and managerial approach for addressing that problem. This fact is broadly recognized by existing guidance (e.g., ANZECC/ARMCANZ 2000). It is recommended that conceptual models be developed and used in any risk-based approach involving contaminated sediment issues. The principal uses of a conceptual model are as follows (adapted from Menzie et al. 2002):

- Promote communication and understanding: Conceptual models provide a powerful tool for communicating ecological or human health (and other) issues among assessors, managers, and interested parties. For this purpose, it can be helpful to have hierarchical models at various levels of complexity so that presentations can be tailored to the level of detail appropriate for different audiences. Conceptual models are useful for generating ideas and visualizing and organizing the collective thinking of the risk management team and other interested parties who participate in developing the model. As a result, the conceptual model process can lead to a greater shared understanding of the problem by scientists, managers, and the public.
- Support specific regulatory programs and/or actions: Conceptual models can also be used to guide and scope specific regulatory actions such as the management of dredged materials, the implementation of TMDLs or the cleanup of a hazardous waste site.
- Serve as a quality assurance tool: When dealing with broad arrays of potential receptors, stressors, and relationships that are typical of aquatic environments, the risk assessor can use the conceptual model to help ensure that the right things are examined and as a "checklist" for what should be considered in addressing the problem. In general, the focus and scope of the assessment should be determined by the level of effort appropriate to decision needs. To this end, the conceptual model can be a precursor to aid in the development of a sampling and analysis plan or a quality assurance project plan.
- Construct site-specific, scenario-specific, or watershed simulation models: Conceptual models can serve as a starting point for constructing simulation models that capture important system processes (e.g., mass, equilibrium, and rate-based processes). This would support the analysis phase of an assessment. Ideally, they capture and represent the problems at hand in a way that lays a foundation for assessment and,

ultimately, action that helps preserve and protect the ecosystems in question.

- Support decisions when data are limited: The relationships depicted by a carefully thought-out conceptual model can be helpful for decision making when the time and resources to complete an assessment are not available. For example, predictions made from a conceptual model may be used to prioritize problems by illustrating which stressors, pathways, and assessment endpoints are most critical. In such cases, decisions would be based primarily on best professional judgment. These judgments can then be articulated and documented through use of a conceptual model.

Conceptual models can be used to guide the development and implementation of dredging programs, watershed and TMDL programs, and cleanup programs. As described in "Pre-Assessment" (p 232), these programs present varied types of problems and challenges for which the conceptual model serves as an important analytical and communication tool. Some environmental assessments focus specifically on a chemical contamination problem, while others such as watershed management programs and TMDLs may involve a variety of chemical, physical, and biological stressors. Conceptual models are especially valuable tools for these types of environmental assessments because they can display and describe the multiple physical, biological, and chemical stressors in a system and the sources and pathways by which they are likely to impact multiple ecological resources (Suter 1999).

Conceptual models consist of 2 principal components (USEPA 1998):

1) A set of risk hypotheses that describe predicted relationships among stressor, exposure, and assessment endpoint responses, along with the rationale for their selection. Hypotheses are assumptions made in order to evaluate logical or empirical consequences, or suppositions tentatively accepted to provide a basis for evaluation. Risk hypotheses are specific assumptions about potential risk to assessment endpoints and may be based on theory and logic, empirical data, mathematical models, or probability models. They are formulated using a combination of professional judgment and available information on the ecosystem at risk, potential sources of stressors, stressor characteristics, and observed or predicted ecological effects on selected or potential assessment endpoints.
2) A diagram that illustrates the relationships presented in the risk hypotheses. These are used to illustrate important pathways clearly and concisely and can be used to generate new questions about relationships

that help formulate plausible risk hypotheses. Typical conceptual model diagrams are flow diagrams containing boxes and arrows to illustrate relationships.

In this chapter, we focus on graphical conceptual models to illustrate the features of such models for addressing sediment contaminant problems. We emphasize 2 points: 1) the importance of recognizing the potential range of pathways associated with contaminated sediments and 2) the commonality among many contaminated sediment assessments. We do not illustrate or discuss risk hypotheses although these are "the questions" and associated "expected relationships and potential effects" that are reflected in the development of assessment endpoints (in the case of ecological assessments) or exposure scenarios (in the case of human health risk assessments.)

A basic conceptual model for sediment risk assessment. A basic, generic conceptual model was developed for sediment risk assessment (Figure 6-3). This generic model provides a starting point for guiding risk assessors, risk managers, and interested parties to recognize relationships among contaminant sources, sediment processes, and aquatic, wildlife, and human receptors. The generic model illustrates how chemical inputs to sediment can interact via different processes to impact sensitive receptors over time. The model can serve as a checklist to guide the definition of a problem and ensure that all the appropriate sources, processes, and receptors are included. It may also guide decision making with respect to focusing the analysis on those elements that are important. Because this is a basic model, not every contaminant source, sediment process, or sensitive receptor is included. It is possible that more than 1 chemical source is present or that the source has been eliminated and only historical sources are left. Also, it is possible that more than 1 sediment process could be impacting the sensitive receptors.

Sources. The generic model (Figure 6-3) includes many of the possible sources that can result in sediment contamination. These include historical as well as ongoing upstream sources. The surficial and deep sediment under "sediment processes" can be used to represent in-place sediment sources. Ongoing sources may include air emissions, permitted discharges from municipal treatment facilities, power-generating facilities, and industrial manufacturing and distribution facilities. Nonpoint sources could include soil erosion, agricultural runoff, urban runoff, oil, hazardous waste, and sewage spills. The conceptual model also shows that spills, groundwater infiltration, and subsurface nonaqueous-phase liquid (NAPL) flows can be sources.

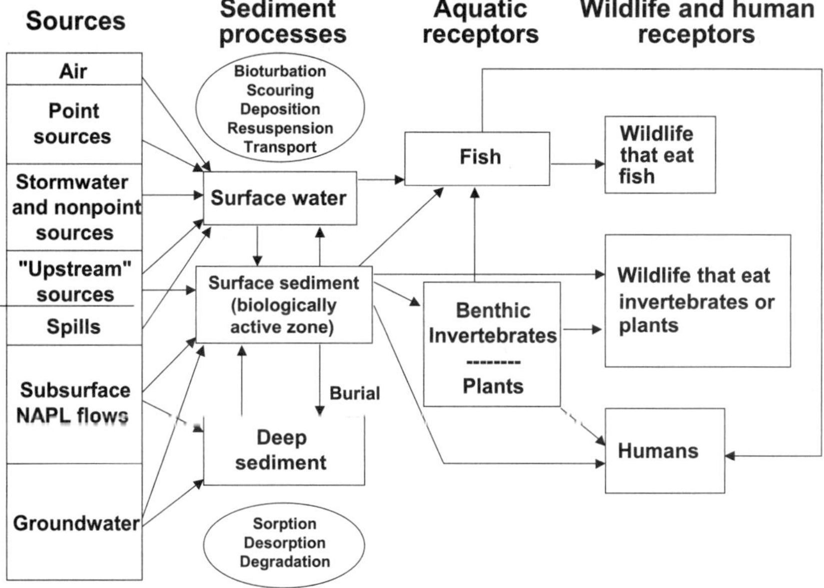

Figure 6-3 Basic conceptual model for sediment risk assessment (NAPL = nonaqueous-phase liquids. Reminder: The above processes operate at specific spatial and temporal scales. Adapted from Driscoll et al. 2002.)

Depending on the purpose of the assessment, project managers and regulators would need to identify and possibly discriminate among the various sources.

Sediment processes. The generic conceptual model identifies some of the major sediment processes that influence exposure. Here again, only a limited set of processes is mentioned to provide a starting point and to remind the reader that sediments and the surrounding environment are dynamic systems for which time will be an important factor. Sediment processes include the physical movement of sediments caused by storm events, from wave action and currents, as well as ship and tug movements. Also important are the geochemical processes that influence bioavailability.

The biologically active zone within a sediment column is where plants and benthic invertebrates will absorb and bioaccumulate contaminants. Delineating this zone will contribute to defining the relevant exposure pathways at the site. Are contaminants entering the water column and dissolving, being moved around by sediment transport, or being buried within the sediment column? Describing these pathways is followed by representing how this contaminant movement impacts sensitive receptors. Deep sediment is differentiated from shallow sediment to account for burial of historical contamination and thus its removal from the biologically active zone. The model also shows that despite these processes, the deep zone could still be a contaminant source to surficial sediments under certain conditions, for example, if the sediment column is physically disturbed.

Receptors. The basic generic model identifies a number of the receptors considered in various sediment assessments. Receptors include, but are not limited to, benthic organisms, shellfish, plants, fish, birds, and humans. Similarly, there may be various exposure pathways to each of these groups. These will vary depending on the chemicals present in the sediments. Some chemicals may be toxic but not bioaccumulate (Figure 6-4). Other chemicals may accumulate into invertebrates and/or fish (Figure 6-5), while still others may accumulate into plants (Figure 6-6). Project managers and regulators can use the generic model to consider contamination potential for these receptors via the various pathways. The generic model provides a checklist for considering what receptors may need to be considered for a specific type of assessment. Questions might include: Are benthic invertebrates feeding on contaminated sediment and then being eaten by fish or birds? Are plants absorbing the contaminants and passing the contaminants on to other

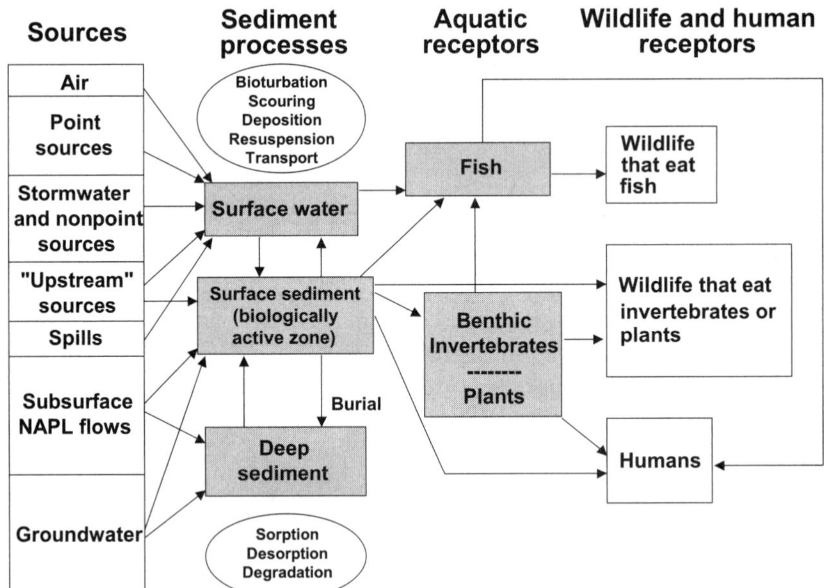

Figure 6-4 Basic conceptual model for nonbioaccumulative, water soluble, toxic chemical (Reminder: The above processes operate at specific spatial and temporal scales. Adapted from Driscoll et al. 2002.)

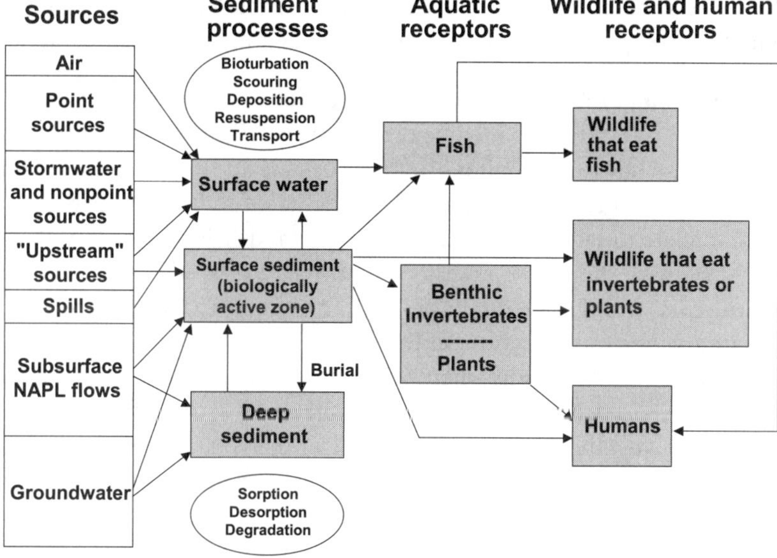

Figure 6-5 Basic conceptual model for compound that bioaccumulates in animal tissue (Reminder: The above processes operate at specific spatial and temporal scales. Adapted from Driscoll et al. 2002.)

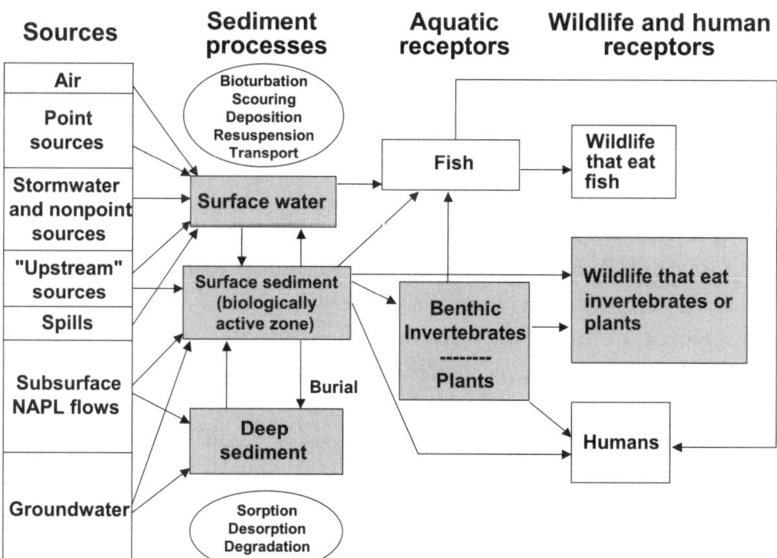

Figure 6-6 Basic conceptual model for compound that accumulates in plants (Reminder: The above processes operate at specific spatial and temporal scales. Adapted from Driscoll et al. 2002.)

wildlife? Are subsistence fishermen eating the impacted fish or shellfish? Every potential source, pathway, and receptor is included in the conceptual model.

Spatial and temporal scales. The simple conceptual models provided in Figures 6-3 to 6-6 do not include details on spatial or temporal scales. However, these obviously are important for understanding and representing relationships. Sediment contaminant assessments typically have spatial scales that relate to sediment strata and horizontal distributions, as well as relationships with the overlying water column. In real-world situations, many compartments within the sediment may be important to recognize and reflect in a conceptual model if they are important to the problem at hand. For example, there may be depths where sediments are easily mobilized and other depths that may be susceptible to scouring only during extreme storm events, if at all. These, together with representations of the vertical reach of biological zones and zones of contamination, may all be important to the analyst and manager. Maps are often the best way to depict the spatial aspects of a problem. Geographical information system (GIS) tools have proven especially helpful for capturing and displaying such information (Clifford et al. 1995). Another key spatial scale for contaminated sediment assessments is water depth, as it influences zones of exposure. For example, people, wading birds, and most emergent and submerged vegetation tend to be confined to shallow-water areas. Evaluation of human exposure via direct contact may be necessary in such shallow areas at a site. Water depth also influences many physical and geochemical processes that influence the nature and movement of sediments and indirectly the bioavailability of chemicals within the sediment matrix. Simple (or complex) picture diagrams can be helpful for illustrating depth-related aspects of a conceptual model.

Temporal scales are also important when contaminated sediments are assessed but they are not easy to reflect in a conceptual model (e.g., information about process rates). Nevertheless, an effort should be made to communicate these temporal aspects of the problem. Each of the components reflected in Figure 6-3 has a temporal aspect associated with it. Sources may be continuous (e.g., point sources from a treatment plant), periodic (e.g., stormwater discharge, dredged material disposal), or discrete (e.g., a single spill). Processes influencing sediment may occur at short (e.g., daily), periodic (e.g., erosional events due to storms), seasonal (e.g., changes in geochemistry that might influence desorption processes and bioavailability), or decadal scales (e.g., long-term deposition of sediments and burial). Biological timescales important to sediment assessments also can range on the order of days to decades and reflect physiological processes, reproduction, survival, growth, behavior, and population and ecosystem processes. Clearly, this introduces a level of detail that is not easily incorporated into a single conceptual model. However, if these processes are particularly important to the problem at hand, an effort should be made to capture them in narrative, graphical (including submodels), and/or tabular form.

The conceptual model illustrated in Figure 6-3 is a simple representation. On a site-specific basis, it may be valuable to capture the greater detail known to be important to the problem. One way to do this is to use multiple or nested hierarchical conceptual models. These diagrammatic representations also could be organized in a way that allows the viewer to zoom into and out of the various scales. Suter (1999) describes an approach for creating modular component models for activities, sites, and receptors that can be reused in different combinations in different assessments. With respect to future directions in this field, tools or software that could be used to develop and explore conceptual models at various process levels and scales could provide benefits in terms of assessment efficiency and risk communication.

Examples for various types of programs

To illustrate the commonalities among assessment problems and to show how a generic conceptual model can be used as a starting point, we have illustrated its use in 3 types of assessments: maintenance dredging with aquatic disposal, development of TMDLs, and cleanup of sediments at a hazardous waste site.

Navigation dredging. Navigation dredging is required to construct and maintain navigation channels and other port infrastructure (e.g., berthing areas). Dredged material evaluations generally begin by investigating possible sources of contamination to the sediments to be dredged. Local sources (e.g., spills, outfalls) as well as more remote upstream sources are possible.

A generic conceptual model could be used within a focused regulatory program like navigation dredging to inform what data should be collected as part of dredged material evaluation following the Ocean or Inland Testing Manuals (USEPA/USACE 1991, 1998). Such a generic conceptual model could be used as a starting point that would be amended based on project-specific data related to identification of known contaminant sources and previous data identifying COCs and ROCs. A modified, dredged material conceptual model is presented in Figure 6-7 where polychlorinated biphenyls (PCBs) and Hg are identified as the COCs. Because PCBs and Hg bioaccumulate and biomagnify in aquatic food chains, exposure pathways terminating in upper trophic-level ROCs are represented in the conceptual model.

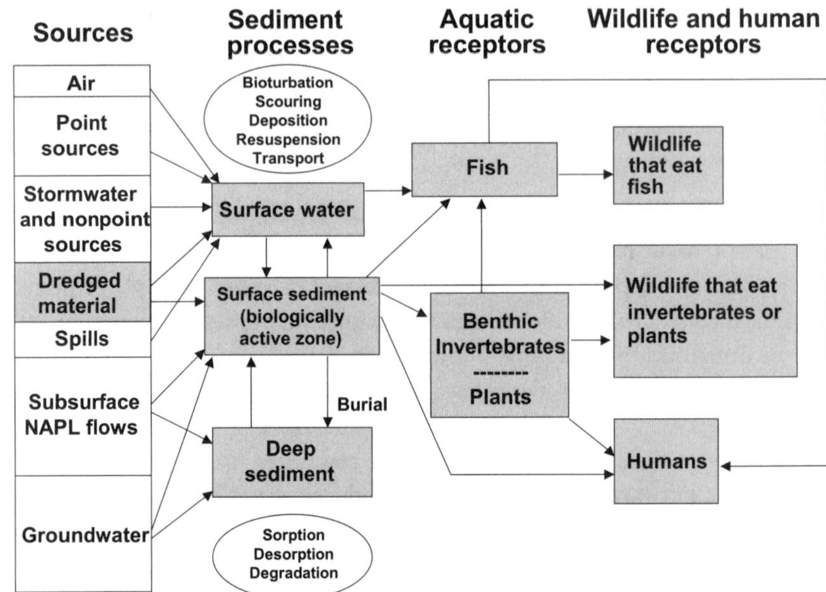

Figure 6-7 Basic conceptual model for dredged material evaluation involving PCBs and Hg (Reminder: The above processes operate at specific spatial and temporal scales. Adapted from Driscoll et al. 2002.)

TMDL development. The generic conceptual site model was also modified to illustrate a TMDL (Martin and Kennedy 2000) application involving polycyclic aromatic hydrocarbon (PAH; Figure 6-8). PAHs are introduced into the environment through 5 generalized pathways:

1) low temperature diagenesis of organic matter (part of the changes undergone by a sediment after its initial deposition),
2) the formation of petroleum and coal,
3) incomplete or inefficient combustion at moderate to high temperatures (pyrolysis),
4) biosynthesis by plants and animals, and
5) anthropogenic sources including the combustion of fossil fuels and biomass (e.g., wood) as well as chemical production that results in the formation of PAHs.

In the TMDL process, the occurrence of PAH could be identified through monitoring data for the waterway or watershed using chemical or biological tools. The identified co-occurrence of PAH and degraded biological conditions could itself lead to a conclusion that the water quality standard (attain-

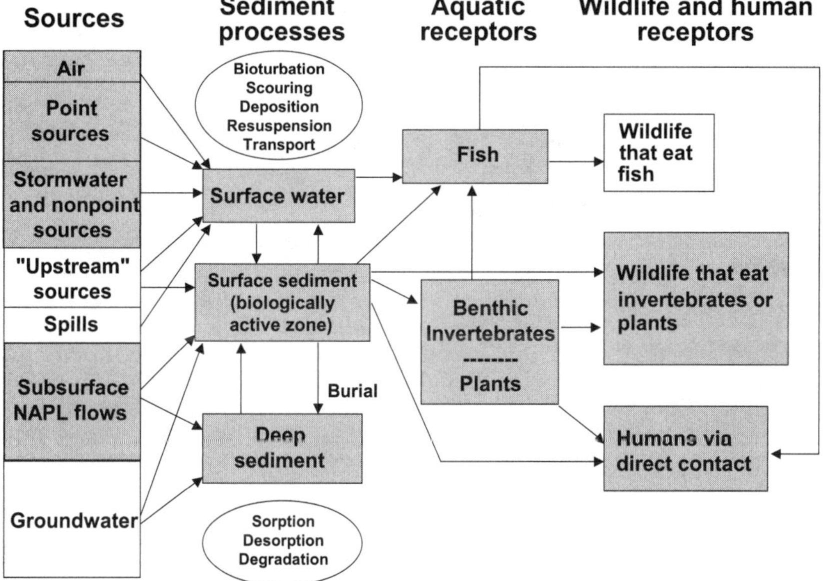

Figure 6-8 Basic conceptual model for TMDL involving PAHs in sediment (Reminder: The above processes operate at specific spatial and temporal scales. Adapted from Driscoll et al. 2002.)

able use) is not being met. Developing a plan to bring the waterbody into compliance with existing use goals will involve the development of a conceptual model. In this example (Figure 6-8), we identified potential sources as release from permitted discharges, stormwater and other nonpoint sources, subsurface flows of NAPLs achieving a dissolved phase and entering the waterway, and air deposition from relatively remote sources. The "delivery" vehicle through which the PAH sources may reach receptors is limited in the surface water due to solubility constraints; however, the pathway is included for completeness. Bedded sediments represent the most important secondary source to receptors in this system. Contaminated sediments may be isolated over time due to deposition. Exposure pathways linking the primary and secondary sources of PAH to receptors of concern include surface water to fish; sediment to benthic invertebrate, fish, and recreational swimmers; and trophic transfer to benthic invertebrates, fish, and wildlife.

Cleanup at hazardous waste site. The generic conceptual site model for a sediment risk assessment is modified to illustrate the major pathways for a Pb-contaminated hazardous waste site in Figure 6-9. Pb may originate from smelter operations, industrial processes, shooting ranges, etc. The source in

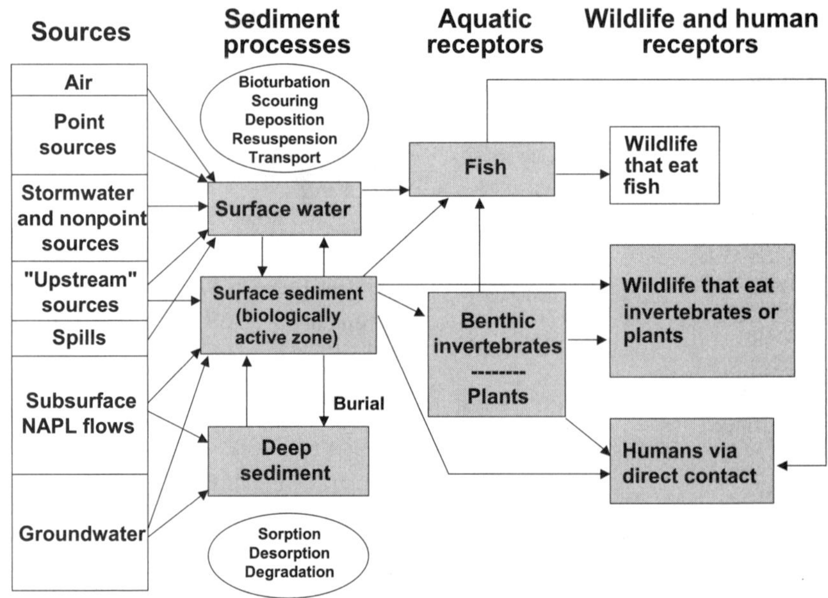

Figure 6-9 Basic conceptual model for hazardous waste site contaminated with Pb (Reminder: The above processes operate at specific spatial and temporal scales. Adapted from Driscoll et al. 2002.)

this example has been remediated, and remaining sediment concentrations are elevated above background. Contamination is heterogeneously distributed within the sediments at the site. Concern exists regarding potential flux to overlying surface waters. Contaminated sediment may be isolated through sediment deposition and burial. Potential exposure pathways for the site include mobilization of Pb into the surface water with exposures to resident fish and contact between bedded sediment and benthic invertebrates, aquatic plants, and recreational swimmers using this shallow-water site. Potential exposure through the food chain to upper trophic-level invertebrates and aquatic and terrestrial wildlife could be identified as an additional exposure pathway of concern, though the potential for Pb exposure through trophic transfer is relatively low compared to other metals (e.g., Hg, Se).

Developing assessment questions

The conceptual model will provide a basis for developing sediment- and/or site-specific assessment questions that must be answered to reach conclusions about whether the sediment or site poses risk. The design and conduct of sediment assessments should be driven by these sediment- or site-specific questions. Developing these questions will lead to selecting the LOEs and tools that will be used in the assessment. This progression of actions is important because assessment tools vary in terms of their relevance to questions. For example, the most widely used SQGs do not address one of the most common assessment questions: Do the bioaccumulative chemicals in the sediment pose an unacceptable risk to upper trophic levels (see Chapters 4 and 11)? Considerable effort should be devoted to formulating and refining specific and detailed questions that must be answered to reach conclusions about the nature and extent of risks. Examples of such questions include the following:

- Will zooplankton be adversely affected by sediment-associated contaminants at the site?
- Will oyster reproduction be adversely affected by sediment-associated contaminants during disposal operations at the management site?
- Will the mix of contaminants present in this sediment be toxic to sediment-dwelling fauna?
- Will the neogastropod molluscs inhabiting the site be adversely affected by the tributyl tin (TBT) in this dredged material?
- Are dioxin-like compounds present in this sediment at levels of concern?
- Will the fish that use the disposal site as a feeding ground be adversely affected by this sediment?
- Will fish-eating birds be adversely affected by the sediment at this site?
- Will the health of recreational anglers fishing at this site be endangered?

Use of guidelines

The Initial Assessment phase concludes with a comparison of existing or preliminary data to appropriate and relevant physical, chemical, or biological guidelines. These guidelines are used at this stage of the assessment to assist in reaching conclusions about the need for additional analysis and, if necessary, how that analysis should be focused.

At the initial assessment phase, there is merit in using SQGs in combination with other sources of information to identify sediments that require no further evaluation because they pose little potential for risk. For some programs, there also may be merit in incorporating approaches within the initial assessment for using SQGs along with other LOEs to identify sediments that pose some high potential for risk.

The kinds of information and/or guidelines that can be used in combination at this stage in the assessment to reach decisions about the need for further analysis include the following:

- Proximity to contaminant sources. Sediments that are far removed from sources of pollution are less likely to be a carrier of contaminants.
- Grain size distribution of the sediment. If the sediment is largely composed of coarse-grained material, the sediment is unlikely to be a significant carrier of contaminants.
- Tissue chemistry. Recent chemistry data from organisms collected at the site can be compared to existing health advisory levels or other tissue-based standards to reach conclusions about potential risk.
- Sediment toxicity data. Recent sediment toxicity test data can be used in some cases to reach conclusions about the need for further testing or analysis.
- Sediment chemistry. The potential for direct sediment toxicity to benthos may be assessed through the use of SQGs. Whether these SQGs are empirically or mechanistically derived (Chapter 3), their use is intended to provide insight to whether or not toxicity is expected. Lower threshold SQGs, that is, chemical levels associated with a low probability of toxicity, can be used, within the constraints of a program and logical intent of SQGs, to reach conclusions about the need for further assessment. The SQGs that are intended to identify sediments with a greater likelihood for producing effects can be used to focus assessments or accelerate consideration and selection of management alternatives. The manner in which SQGs are used within an assessment framework will be determined in large measure by the objectives and constraints of the relevant regulatory programs involved as well as by the nature of the assessment questions developed during the initial assessment. Use of SQGs must also be guided by a clear understanding of how the SQGs to be used were derived, what type and level of effects they address, their predictive ability, and their appropriate or recommended uses (Chapters 3, 4, 7, 11, 12, and 13).

In some cases, the use of guidelines, including SQGs, will be used to reach conclusions that no further assessment is required because the assessment questions could be satisfactorily addressed using information available at this stage of the evaluation. In cases where such a comparison with guideline values results in ambiguous answers to the assessment questions concerning the presence of unacceptable risk, the assessment would proceed to Sediment or Site Assessment after revising, as necessary, the list of contaminants of potential concern, the conceptual model, and the assessment questions. In cases where the assessment questions were confidently addressed through the use of guidelines, the investigation proceeds to an Evaluation and Selection of Management Alternatives.

Sediment assessment

During the sediment assessment phase of the evaluation, more comprehensive and site-specific information will be collected and analyzed for the purpose of clarifying the nature, extent, and magnitude of risks posed by a sediment or site (Figure 6-10).

Developing assessment endpoints

The first series of actions to be taken by an assessor conducting a risk assessment for sediment will broadly define the scope and focus of the assessment based on information collected during the Initial Assessment. During the Initial Assessment, the list of contaminants and receptors of concern and the conceptual model are used to develop specific assessment questions that must be answered before decisions can be reached about how to manage the sediment. These outputs of the Initial Assessment will be the basis for developing sediment or site-specific assessment endpoints. Assessment endpoints are descriptions of the environmental values that are to be protected (USEPA 1997). Some regulatory programs may use predefined assessments endpoints; however, assessment endpoints are generally expected to vary among sites due to differences in environmental conditions and receptors at aquatic sites. Example assessment endpoints for a sediment assessment could include these:

- a diverse benthic community whose structure is characteristic of non-chemically impacted areas in the watershed,
- a viable oyster fishery,
- a sustainable population of English sole,
- a sustainable local population of brown pelican, and
- a recreational fishery free of health advisories or ingestion limits.

Selecting lines of evidence

The LOEs deemed necessary to address the assessment questions are selected on the basis of the chosen COCs, ROCs, conceptual model, and assessment endpoints. An LOE is composed of distinct but related points of information. The term "line of evidence" is commonly used to refer to broadly defined categories of information, for example, sediment chemistry, toxicity test data, or benthic community survey results. Here, we refer to 6 primary LOEs as they

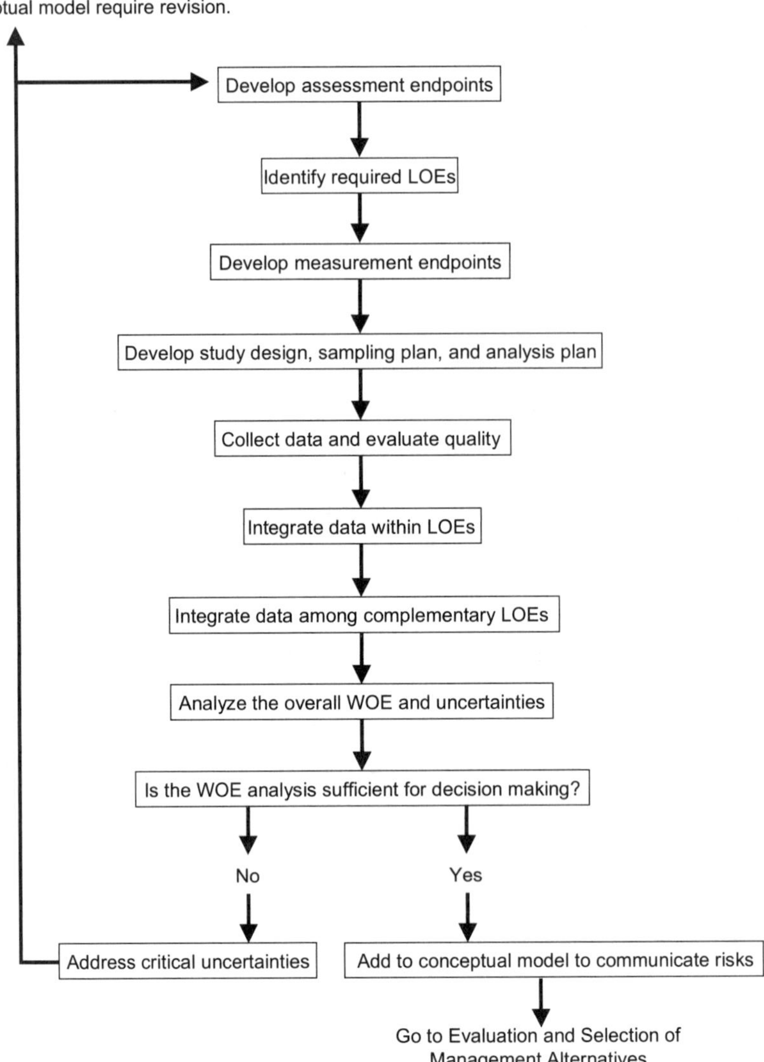

Figure 6-10 Sediment or site assessment activities

are related to the fundamental processes investigated in risk assessment, that is, exposure and effects processes (Figure 6-11). These LOEs are categorized according to the principal component of the ecosystem where the exposures and effects occur. Receptors living in close association with bedded sediments can be adversely affected by sediment-associated contaminants through direct contact with those sediments, as in the case of benthic invertebrates, plants,

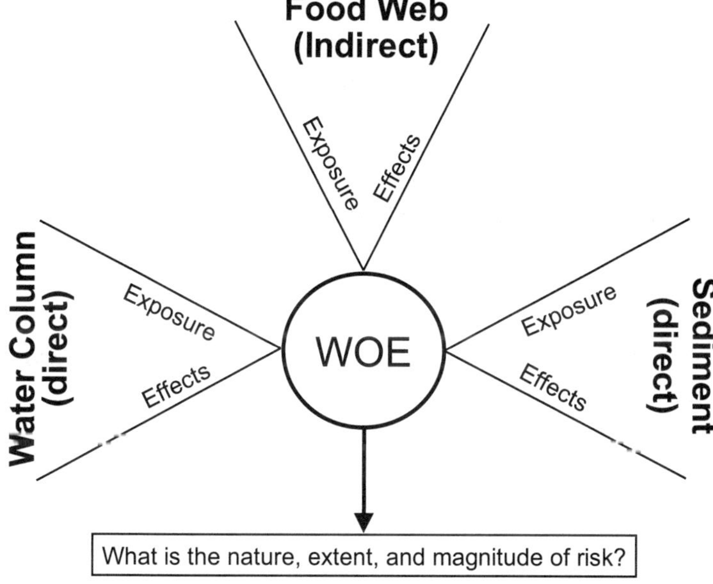

Figure 6-11 Lines of evidence for risk-based sediment assessment. Water Column, Food Web, and Sediment represent components of the ecosystem within which exposures and effects are occurring.

and demersal fish. Receptors residing in the water, for example, planktonic invertebrates and fish, can be exposed and affected by contaminated sediments suspended in the water column or through flux of contaminants into the dissolved phase. Other receptors can be adversely affected by sediment-associated contaminants in a more indirect manner through the movement of contaminants within food chains, as would be the case for fish-eating birds and mammals, including humans. For each of the 3 ecosystem components in Figure 6-11, separate LOEs are provided for exposure and effects information. This categorization of 6 principal LOEs captures all potential receptors that could be affected by contaminated sediments and the pathways through which they are exposed. The chief benefit of this categorization is that it establishes a strong linkage between the conceptual model, assessment questions, and the WOE process used to reach conclusions about the nature, extent, and magnitude of risks. Answering assessment questions concerning risks to brown pelican will require assembling points of information related to brown pelican exposure to contaminants and the potential for effects in that receptor. The points of information assembled to characterize exposure and effects to brown pelican will differ significantly from the points of information used to characterize exposure and effects to the benthic community. While sediment toxicity tests can provide useful information for characterizing the potential for benthic effects, they are not directly relevant to characterizing effects to fish-eating birds. Which LOEs are developed within an assessment will vary from site to site, depending on the combination of contaminants and receptors of concern included in the assessment. For example, a site with no bioaccumulative COCs would not include development of food web LOEs.

Selecting measures

Once the specific LOEs to be developed in the assessment are identified, specific measures of exposure and effect are selected. These measures provide a description of the information that will be collected to make inferences about exposure, effects, and ultimately risk to ROCs. For instance, in the case of an assessment endpoint concerned with the status of the benthic community, measures of effect might include a description of how sediment toxicity tests would be used to assess the potential for effects (e.g., which toxicity tests and endpoints would be used). In the case of assessment endpoints concerned with risks to fish-eating birds and mammals, relevant measures of exposure would include a description of how fish tissue concentrations will be estimated to assess exposure potential. For many sediment assessments, multiple measures of exposure and effect would be selected for each assessment endpoint. Choices about which and how many measures will be used in a given assessment will depend on the characteristics of the sites and the assessment questions and endpoints chosen. The approaches or tools that may be employed as measures of exposure and effect might include a combination of laboratory tests or analyses, field measurements, and modeling efforts, depending on the objectives and needs of the assessment. In selecting measures to be used in the assessment, attention must be given to the ecological relevance of the measures and the strength of their association with the selected assessment endpoints (Burton, Batley et al. 2002).

Study design

A study design is developed after decisions have been made as to how the potential for exposure and effects will be measured. This design will serve to guide what and how many samples are taken at specific locations within the site. The data quality objectives process provides an orderly approach for designing data collection efforts to ensure that the data are of sufficient quality and quantity to meet the objectives of the assessment (USEPA 2000). This design process will include how quality assurance and quality control procedures will be implemented and how statistical robustness and data sufficiency are to be achieved. The statistical procedures that will be used in analyzing the data will be selected, and the decision criteria used in drawing inferences from specific tests will be established. A key element in this regard will be the selection of reference conditions or sediments that will be used as a basis of comparison when exposure and effects are evaluated. The need for reference data is greatest for sediment LOEs and measures of effect. Adverse effects in toxicity tests and benthic community assessments are generally defined through comparison to a reference sediment or condition. After the data are collected and the quality of the data is established, the assessment proceeds to integrating the gathered information.

Integrating information

At least 3 levels of information integration are necessary to reach conclusions about the risks posed by a sediment. First, points of information within a given LOE must be integrated to characterize exposure or effects within the 3 ecosystem components represented in Figure 6-11. For example, the sediment–effects LOE may include information from 3 different toxicity tests, 3 different metrics for describing the status of the benthic community, and critical body residue toxicity data for key COCs. Incongruities among related points of information (segments of information, if you will) must be resolved, for example, 1 toxicity test indicates the presence of toxicity and 2 show lack of toxicity. At this stage of a risk assessment, SQGs can provide substantial aid in addressing such data interpretation challenges where apparent conflicts exist among points of information concerned with benthic effects. If 1 of 3 toxicity tests provides evidence indicating the presence of toxicity, but sediment concentrations for all COCs are below lower threshold guidelines (e.g., effects range low [ERL]), then the conclusion that benthic effects are unlikely would be strengthened by the information provided by SQGs (Chapters 4 and 12). If, on the other hand, one or more COCs exceed equilibrium partitioning (EqP)-based guidelines for contaminants that the responding test organism is known to be sensitive to, then the conclusion that effects are likely is supported. Clearly, other information could and should be used to resolve these challenges, including information about conditions known to interfere with toxicity tests (e.g., test species' sensitivity to sediment grain size distribution, sediment concentrations of ammonia and hydrogen sulfide, etc.; Chapter 16). Used in this fashion, chemistry data and SQGs can be used to develop cause and effect and other supporting arguments to strengthen or clarify conclusions.

A second level of integration occurs when the exposure and effects LOEs are integrated. For example, contaminant tissue concentrations in field-collected invertebrates and bioaccumulation test organisms (sediment–exposure LOE) could be integrated with the critical body residue data (sediment–effects LOE) to reach conclusions about risk to the benthic community. This second level of integration supports the third level of integration, where an overall weight of evidence (WOE) is assembled to address the assessment questions and reach conclusions about the nature, extent, and magnitude of risks (Figure 6-11). At each level of integration, care should be taken to recognize that there are spatial and temporal dimensions to exposure and effects processes. Stated in simple terms, 10 square meters of contaminated sediment presents less opportunity for exposure and effect than 10,000 square meters.

Consideration should also be given to controlling the loss of information during the integration process. Summarizing and combining results from multiple sources of information is the principal challenge associated with using WOE approaches in risk assessment (Chapter 5). Critical information can be lost if the dataset is oversimplified for consideration by stakeholders, which can result in underestimating the uncertainties associated with conclusions drawn from the LOEs (Burton, Chapman, et al. 2002; Burton, Batley, et al. 2002). The dual purpose of sediment risk assessment is to establish whether potential risks exist and to provide sufficient process-level information to identify where and how the risks are occurring so that effective management strategies can be designed to control the key exposures and reduce the evident risks.

The third level of integration occurs when all of the LOEs are considered to characterize the nature, extent, and magnitude of risk. Weight of evidence has been defined as "the process of combining information from multiple lines of evidence to reach a conclusion about an environmental system or stressor" (Burton, Chapman, et al. 2002). Several approaches have been used to reach conclusions about risk using a WOE (Chapter 5); however, most of these are more developed in regard to assessing risks in bedded sediment (i.e., risks to benthos) than the other 2 ecosystem components where exposures and effects can occur.

Evaluating uncertainties

Conclusions about risk should be supported by a thorough uncertainty analysis that identifies where the sources of uncertainty are in the analysis, and to the extent possible, quantifies those uncertainties. Much has recently been written on the subject of uncertainty in sediment risk assessment (Ingersoll et al. 1997; Luoma and Fisher 1997; Solomon et al. 1997; Vorhees et al. 2001;

Batley et al. 2002; von Stackelberg et al. 2002) (also Chapters 3, 4, 17). While the approaches taken to address uncertainties in sediment evaluations will vary from site to site, the amount of effort devoted to this element of the assessment should be scaled to the consequences (in terms of environmental damage or monetary costs) of the management decisions. The uncertainty analysis should provide the means of assigning a measure (qualitative or quantitative) of reliability to particular LOEs and confidence in overall conclusions.

Decision point

At this juncture of the evaluation, judgments are reached, with the input of stakeholders, regarding whether the WOE analysis is sufficient for decision making. Such judgments would be based on the extent to which each of the assessment questions is addressed by the evidence collected and the results of the uncertainty analysis. If the assessment is judged to be sufficient for decision making, the risks are then summarized using the conceptual model and other means, as appropriate, and a transition is made to evaluating and selecting management alternatives. In cases where the WOE is judged to be insufficient for decision making, critical uncertainties are addressed by iterating back into the assessment.

Evaluation and selection of management alternatives

Following the framework depicted in Figure 6-1, the management alternative selection phase of an assessment can be reached either directly following an Initial Assessment or at the conclusion of a Sediment/Site Assessment. A 2-headed arrow connects Sediment/Site Assessment with Evaluation and Selection of Management Alternatives to indicate that iteration between these 2 activities will be necessary. In cases where an early determination is reached during the Initial Assessment that risks are present, additional site or process data may be needed to guide the selection of the most appropriate management alternatives. Likewise, it may be necessary at times to conduct additional evaluations or reanalyze information collected during the Sediment/Site Assessment in order to inform the process of selecting management alternatives.

Identify feasible or available risk management alternatives

The first action to be taken in selecting a management alternative is to assemble a list of feasible or available management options (Figure 6-12). At this stage of the selection process, the list should be inclusive of the range of possible options available, that is, there should be no presumptive management alternative. Premature culling of alternatives before collecting and analyzing

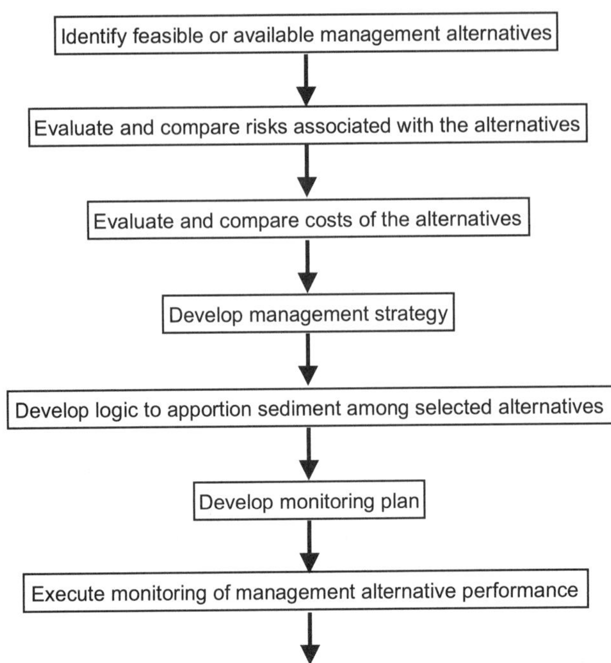

Figure 6-12 Activities related to evaluation and selection of management alternatives

sufficient information to support the selection process will invite criticism of the process and reduce credibility in the assessment (NRC 2001). The broad range of risk management alternatives for sediment sites can be grouped into the following categories and subcategories (NRC 1997, 2001):

- Controls
  - Source control
  - Constraints on site use, for example, fishing limits
- In situ management
  - Monitored natural recovery (MNR)
  - In-place capping with clean sediment
  - Treatment, for example, chemical or biological

- Ex situ management
  - Dredging followed by isolation in
    - Hazardous waste landfill
    - Confined disposal facility (CDF)
  - Dredging followed by treatment
    - Physical
    - Chemical
    - Thermal
    - Biological

General information about each of the available or feasible alternatives should be collected. Such information would include the basic logistical and engineering elements of the alternatives, distance and routes to management sites, and a listing of site features or characteristics with the potential to impact the effectiveness of the alternative to remediate risks (e.g., hydrodynamic characteristics affecting sediment stability or mobility, geotechnical properties of the sediment). Collecting this information will help identify the need for additional data collection or analysis that may be necessary before a definitive comparison can be made of the risks and benefits associated with each of the management alternatives.

Compare the risks associated with the alternatives

All management alternatives for contaminated sediments carry their own specific set of strengths and weaknesses, advantages and disadvantages, and risks. Efforts have been made to rank management alternatives for contaminated sediments in terms of their overall feasibility, effectiveness, and practicality (e.g., NRC 1997). What has emerged from these efforts is the conclusion that there is no universally superior technology for managing contaminated sediments. One of the consequences of this commonly accepted conclusion is that the decision process used to select management alternatives must include a comparison of the alternatives with respect to the characteristics of the site. Because management decisions will involve reconciling tradeoffs, most sites will require using a combination of management alternatives (NRC 1997, 2001).

One of the comparisons that should be made among the alternatives under consideration is the environmental risks that may be created or remain as a residue following implementation of the alternative (e.g., Driscoll et al. 2002). There are no zero-risk alternatives for managing contaminated sediments. Because the environmental risks associated with alternatives will vary in their nature and magnitude, making effective management decisions at contaminated sediment sites will require developing a comprehensive understanding of these risks (NRC 2001).

A comparative risk assessment of the alternatives will need to consider both short- and long-term risks. Short-term risks associated with the initial implementation of an alternative can be evaluated and described. One of the primary sources of short-term risk is exposure to contaminants or sediment released during implementation of the alternative (e.g., dredging, effluent discharges during filling of a CDF, side-streams of treatment processes). Long-term risks are generally associated with the maintenance of the selected management alternatives (e.g., structural reliability of lined or unlined containment facilities, sediment stability in the case of MNR, maintenance of cap integrity). Variation in the temporal and spatial dimensions of these short- and long-term risks should be evaluated and considered in the comparison and selection process.

Compare the costs of the alternatives

Costs must be considered when selecting management alternatives for sediments for 2 reasons: 1) money is a commodity with limits, and 2) risk management alternatives can vary in cost by a factor of more than 1000 (NRC 1997). Effective stewardship of the public and private resources that are used to execute management strategies at a contaminated site requires having information about both the risks and the costs associated with each of the management alternatives under consideration. A choice between 2 management alternatives that are equally effective at controlling the exposures resulting in unacceptable risks will be aided by information regarding differences in cost. In cases where the management alternatives differ in terms of risks and costs, decision makers must be prepared to justify the selection of more costly alternatives in terms of the additional risk reduction gained by the selection. A comprehensive analysis of costs would also include lost opportunities or access to environmental resources incurred by the public due to risks consequent to implementing an alternative. A number of approaches for accomplishing this type of analysis have been proposed and discussed within the context of contaminated sediment management, including cost–benefit or cost–effectiveness analyses, risk ranking, risk tradeoff analysis, and decision analysis (NRC 1997, 2001).

Apportioning sediment among the selected alternatives

The risks posed by sediments at a contaminated site will vary spatially across the site. Some portion of the sediments at a spatially heterogeneous site will pose minimal risk and will not, as a consequence, require any direct management action. Other portions of the site may require varying degrees of management action, depending on the nature and magnitude of the risks posed. Distinguishing the sediments at a site that present risk and require management from those that do not necessitates having a logic to make the distinction. The development and application of this logic generally results in the derivation of a regulatory standard (e.g., in the case of navigation dredging programs) or a cleanup target or remediation goal (e.g., in the US Superfund Program).

Most of the controversy surrounding the use of SQGs to assess and manage risks concerns the extent to which SQGs can be used as the logical basis for distinguishing sediments that require special management from those that do not. The degree to which SQGs can contribute to such a logic will depend on the nature of the risks at the site. As described in Figure 6-11, conclusions about the nature, extent, and magnitude of risks are developed from information supplied by as many as 6 LOEs. The logic or metrics used to apportion sediment among management alternatives must be tied directly to the LOEs describing the evident risks. To do otherwise will lead to false conclusions about the level of protection being afforded by the management decisions. If the risks associated with a particular site are caused by indirect exposures to wildlife or humans through trophic transfer of bioaccumulative chemicals, then the logic used to apportion sediment among the management alternatives must be derived from the LOEs associated with that indirect route of exposure and effects. In this specific case, the distinguishing logic could involve deriving risk-based concentrations for prey items that would be protective of upper trophic-level receptors using a bioaccumulation model. Deriving such concentrations would require the use of site-specific data on the receptors of concern and their use of the site. If direct exposures and effects in the water column were causing risks, then another logic would be required. Because most of the widely used SQGs are limited to providing information about effects to benthic invertebrates (e.g., ERL and effects range median [ERM], probable effects level [PEL] and threshold effects level [TEL], apparent effects threshold [AET], EqP-derived methods) their use in deriving a logic for apportioning sediment among management alternatives must be limited to cases in which risks are limited to the benthos. However, even in such cases, the uncertainties associated with SQGs (see Chapters 3, 4, 5, 11, 12, 13, 17) will constrain the extent to which these values can be used alone to reach credible management decisions.

The interest in using SQGs to make management decisions appears to be closely tied to the desire to have a "number", that is, a chemical concentration that delimits sediments that pose no risk from those that pose some risk. However, in most cases where risks to benthos are evident, more direct means of delimiting portions of the sediment or site posing a risk will be available (e.g., sediment toxicity test data and/or benthic community data). One of the advantages afforded by using biological data in defining the areas at a site requiring management because of risks to benthos is that these measures provide an integrated view of effect caused by all of the contaminants present in a site-specific mixture. When used in combination with biological data, SQGs will provide opportunities for clarifying and strengthening inferences about risks that lead to management decisions.

One of dangers of condensing the information about risks in a Sediment/Site Assessment into a series of chemical concentrations in sediment is that it becomes more difficult for risk assessors, risk managers, and stakeholders to access and use information about where the exposures, effects, and risks are located within the ecosystem. Recognition that the predominant risks are limited to the benthos may lead risk managers to select a different set of management alternatives than instances in which exposures, effects, and risks are more broadly distributed within the ecosystem.

Monitoring and adapting

Following the selection of management alternatives that will be implemented and the development of a logic for apportioning sediment among those alternatives, a monitoring plan should be developed. Monitoring at the site will accomplish 2 objectives: 1) track the performance of the management strategy in reducing risks and 2) provide feedback to the assessment and management process. Sediment cleanup is a complex enterprise. Monitoring conducted in a timely manner provides for adjustments to be made, if necessary, in the strategy for managing the site to achieve the objective of risk reduction. Monitoring also provides information that can be used to adjust and revise the assessment and management processes in cases where weaknesses in the process can be identified. Decisions to be made in developing a monitoring plan include the following:

- What hypotheses will be tested?
- What measurements are required to accomplish the test?
- How will the data be managed and interpreted?

The level of effort invested in monitoring should be scaled to the complexity of the site and the management strategy.

The Way Forward

Numerous frameworks are in use today to guide how risks posed by contaminated sediments are assessed and managed (Chapters 9 and 10). We have attempted in this chapter to define some of the critical elements and decision points in an assessment and management framework concerned with contaminated sediments. While following a well-ordered and structured assessment and management framework will help to reduce uncertainty in risk estimates and increase the likelihood that assessments produce information useful for decision making, ultimately, the soundness of the assessment and consequent management decisions will depend on the strength of the underlying science. Strengthening the scientific foundation of sediment assessments will depend on advancements being made in the following key areas:

- assessing cumulative exposure and effects from contaminant mixtures,
- assessing bioaccumulation and trophic transfer potential for sediment-associated contaminants,
- incorporating more realistic spatial and temporal considerations into exposure assessment, and
- developing robust comparative risk assessment approaches to evaluate and select management alternatives.

Given that research in these key areas is being actively pursued by scientists and engineers in numerous countries, we can expect that the results of future risk-based sediment assessments will play an expanded role in the management of contaminated sites.

References

Adams DA, O'Connor JS, Weisberg SB. 1998. Sediment quality of the NY/NJ harbor system: an investigation under the Regional Environmental Monitoring and Assessment Program (R-EMAP). Edison (NJ): US Environmental Protection Agency. EPA/902-R-98-001.

[ANZECC/ARMCANZ] Australian and New Zealand Environment and Conservation Council/Agriculture and Resource Management Council of Australia and New Zealand. 2000. Australian and New Zealand guidelines for fresh and marine water quality. Canberra (AU): ANZECC/ARMCANZ.

Batley GE, Burton GA, Chapman PM, Forbes VE. 2002. Uncertainties in sediment quality weight-of-evidence (WOE) assessments. *Hum Ecol Risk Assess* 8:1657–1673.

Burton Jr GA, Chapman PM, Smith EP. 2002. Weight-of-evidence approaches for assessing ecosystem impairment. *Hum Ecol Risk Assess* 8:1657–1673.

Burton Jr GA, Batley GE, Chapman PM, Forbes VE, Smith EP, Reynoldson T, Schlekat CE, den Besten PJ, Bailer AJ, Green AS, Dwyer RL. 2002. A weight-of-evidence framework for assessing sediment (or other) contamination: improving certainty in the decision making process. *Hum Ecol Risk Assess* 8:1675–1696.

Clifford PA, Barchers DE, Ludwig DF, Sielken RL, Klingensmith JS, Graham RV, Banton MI. 1995. An approach to quantifying spatial components of exposure for ecological risk assessment. *Environ Toxicol Chem* 14:895–906.

de Deckere E, de Cooman W, Florus M, Devroede-Vanderlinden M-P. 2000. A manual for the characterisation of sediments in Flemish watercourses: a TRIAD approach. Brussels: Ministry of the Flemish Community – AMINAL.

Driscoll, Kane SB , Wickwire WT, Cura JJ, Vorhees DJ, Butler CL, Moore DW, Bridges TS. 2002. A comparative screening-level ecological and human health risk assessment for dredged material management alternatives in New York/New Jersey Harbor. *Hum Ecol Risk Assess* 8: 603–626.

Ingersoll CG, Ankley GT, Baudo R, Burton GA, Lick W, Luoma SN, MacDonald DD, Reynoldson TB, Solomon KR, Swartz RC, Warren-Hicks WJ. 1997. Workgroup summary report on uncertainty evaluation of measurement endpoints used in sediment ecological risk assessments. In: Ingersoll CG, Dillon T, Biddinger GR, editors. Ecological risk assessment of contaminated sediments. Pensacola (FL): Society of Environmental Toxicology and Chemistry (SETAC). p 297–352.

Long ER, Wolfe DA, Carr RS, Thursby GB, Stern EA, Peven C, Schwartz T. 1995. Magnitude and extent of sediment toxicity in the Hudson-Raritan Estuary. Silver Spring (MD): National Oceanic and Atmospheric Administration. NOAA Technical Memorandum 88. NOAA/NOS/ORCA. 230 p.

Luoma SN, Fisher N. 1997. Uncertainties in assessing contaminant exposure from sediments. In: Ingersoll CG, Dillon T, Biddinger GR, editors. Ecological risk assessment of contaminated sediments. Pensacola (FL): Society of Environmental Toxicology and Chemistry (SETAC). p 211–237.

Martin JL, Kennedy RH. 2000. Total maximum daily loads: A perspective. Vicksburg (MS): US Army Engineer and Development Center. Environmental Effects of Dredging Technical Notes Collection (ERDC/TN EEDP-01-46).

Menzie CA, Serveiss V, Connery J, Norton S. 2002. Development of conceptual models for watershed and regional-scale environmental assessments. Washington DC: Report prepared by National Center for Environmental Assessment, USEPA.

[NRC] National Research Council. 1997. Contaminated sediments in ports and waterways: cleanup strategies and technologies. Washington DC: National Academy Pr. 295 p.

[NRC] National Research Council. 2001. A risk-management strategy for PCB-contaminated sediments. Washington DC: National Academy Pr. 432 p.

Sijm D, De Bruijn J, Crommentuijn T, van Leeuwen K. 2001. Environmental quality standards: endpoints or triggers for a tiered ecological effect assessment approach? *Environ Toxicol Chem* 20:2644–2648.

Solomon KR, Ankley GT, Baudo R, Burton GA, Ingersoll CG, Lick W, Luoma SN, MacDonald DD, Reynoldson TB, Swartz RC, Warren-Hicks WJ. 1997. Workgroup summary report on methodological uncertainty. In: Ingersoll CG, Dillon T, Biddinger GR, editors. Ecological risk assessment of contaminated sediments. Pensacola (FL): Society of Environmental Toxicology and Chemistry (SETAC). p 271–296.

Stronkhorst J, Schipper CA, Honkoop J, van Essen K. 2001. Disposal of dredged material in Dutch coastal waters. A new effect-oriented assessment framework. The Hague: National Institute for Coastal and Marine Management/RIKZ. Report RIKZ 2001.030.

Suter II GW. 1999. Developing conceptual models for complex ecological risk assessments. *Hum Ecol Risk Assess* 5:375–396.

[USEPA] US Environmental Protection Agency. 1997. Ecological risk assessment guidance for Superfund: process for designing and conducting ecological risk assessments. Washington DC: USEPA, Office of Solid Waste and Emergency Response. EPA540-R-97-006.

[USEPA] US Environmental Protection Agency. 1998. Guidelines for ecological risk assessment. Washington DC: Risk Assessment Forum. EPA/630/R-95/002F.

[USEPA] US Environmental Protection Agency. 2000. Guidance for the data quality objectives process. Washington DC: USEPA, Office of Environmental Information. EPA/600/R-96/055.

[USEPA] US Environmental Protection Agency. 2001. The incidence and severity of sediment contamination in surface waters of the United States. National sediment quality survey: second edition. Washington DC: USEPA, Office of Science and Technology. EPA 823-R-01–01. (Draft).

[USEPA/USACE] US Environmental Protection Agency, US Army Corps of Engineers. 1991. Evaluation of dredged material proposed for ocean disposal: testing manual. Washington DC: USEPA/USACE. EPA-503/8-91/001.

[USEPA/USACE] US Environmental Protection Agency and US Army Corps of Engineers. 1992. Evaluating environmental effects of dredged material management alternatives: A technical framework. Washington DC: USEPA/USACE. EPA 842-B-92-008.

[USEPA/USACE] US Environmental Protection Agency and US Army Corps of Engineers. 1998. Evaluation of dredged material proposed discharge in waters of the U.S. testing manual: Inland testing manual. Washington DC: USEPA/ USACE. EPA 823-B-98-004.

von Stackelberg KD, Burmistrov D, Vorhees DJ, Bridges TS, Linkov I. 2002. Importance of uncertainty and variability to predicted risks from trophic transfer of PCBs in dredged sediments. *Risk Anal* 22:499–512.

Vorhees DJ, Kane Driscoll SB, von Stackelberg K, Cura JJ, Bridges TS. 2002. An evaluation of sources of uncertainty in dredged material assessment. *Hum Ecol Risk Assess* 8:369–389.

7

Role of sediment quality guidelines and other tools in different aquatic habitats

PETER M CHAPMAN, WESLEY J BIRGE, ROBERT M BURGESS, WILLIAM H CLEMENTS, W SCOTT DOUGLAS, MICHAEL C HARRASS, CHRISTER HOGSTRAND, DANNY D REIBLE, AMY H RINGWOOD

In general, several lines of evidence (LOEs) are needed to properly evaluate contaminated aquatic environments (adapted from Chapman et al. 2002 and Chapters 5 and 6):

- nature and extent of contamination;
- expected or acceptable diversity and abundance of benthic biota in the absence of contamination;
- bioavailability, bioaccumulation, and effects of contamination (the potential for chronic as well as acute effects) on aquatic organisms;
- stability of sediments and contaminants (fate and transport); and
- risk of contamination to aquatic biota and associated resources.

A number of tools are available to obtain some or all of this information, each of which has its own inherent strengths and weaknesses. These tools range in complexity from numeric sediment quality guidelines (SQGs) and short-term single-species toxicity tests to integrated ecological risk assessments (ERAs). Detailed discussions of available tools and how they can be placed into a weight-of-evidence (WOE) decision framework are presented elsewhere and include the need to modify these conventional approaches to reflect local habitat conditions (Chapman et al. 1992, 2002; Burton, Batley, et al. 2002; Burton, Chapman, Smith 2002; Chapter 6). The choice of, for instance, a particular set of SQGs or a particular sediment toxicity test cannot be prescribed for every possible habitat or situation. In this chapter, the following basic tools (i.e., based on Chapter 5 but not including specialized

Use of Sediment Quality Guidelines and Related Tools for the Assessment of Contaminated Sediments
Wenning RJ, Batley GE, Ingersoll CG, Moore DW, editors.
 ISBN 1-880611-71-6

studies such as toxicity identification evaluations [TIEs], biodegradation studies, exposure surrogates, or modeling) are considered for use in different habitats or situations:

- numeric SQGs,
- sediment toxicity tests (chronic as well as acute),
- resident exposed populations and communities (i.e., not necessarily restricted to benthic organisms),
- bioaccumulation, and
- biomarkers or histopathology.

As recommended at the previous Society of Environmental Toxicology and Chemistry (SETAC) Pellston Workshop dealing with contaminated sediments (Ingersoll et al. 1997), tiered approaches, WOE, and/or ERA can be useful approaches to integration and evaluation.

The purpose of this chapter is to build on previous chapters and provide guidance to practitioners and other stakeholders regarding the use of SQGs and other tools for the assessment of situations in which benthic contamination is an issue: depositional marine and freshwater environments, estuaries, and streams and other erosional environments. The chapter begins by briefly discussing real-world physical and chemical issues that may result in uncertainty in the use of SQGs, biological and habitat issues, and regulatory issues and then proceeds to examine the situations described above, followed by a discussion and summary. Detailed discussion of uncertainties inherent in sediment assessments related to this chapter is provided in Chapter 17.

The ultimate purpose of this chapter is to answer 2 specific questions:

- Are SQGs alone sufficient for management decision-making?
- Are different tools needed to characterize sediment quality in different aquatic environments?

Physical and Chemical Issues

Three major physical and chemical sources of uncertainty are associated with current SQGs:

1) effects of dynamic conditions leading to disequilibrium,
2) variability in critical sediment normalizing constituents (e.g., acid volatile sulfide [AVS], total organic carbon [TOC]), and
3) desorption resistance and limited availability of organic contaminants that are partitioned to organic carbon phases (Table 7-1).

Table 7-1 Principal physical and chemical uncertainties in real-world application of current SQGs and suggested resolutions (See Chapter 17 for additional details.)

Source of uncertainty	Proposed resolution
Dynamic conditions	Measure magnitude of disequilibria to compare and relate to equilibrium and near-equilibrium conditions
Field variations in concentrations of sediment-normalizing constituents	Control variability via compositing or conduct power analysis to determine optimum number of replicates to quantify uncertainties
Desorption resistance and limited availability of organic contaminants	Assess field partitioning or conduct bioavailability studies

Other issues introduce uncertainty into the current SQGs, but the three listed above are the most significant. Based on the current state of the science, only potential resolutions can be proposed. Each of these areas of uncertainty represents fertile topics for future research. Each source of uncertainty is discussed in detail in Chapter 17.

The primary effect of the variability introduced by the above 3 factors is the uncertainty imparted to any calculated SQGs or other assessment tools. For example, determination of mechanistic SQGs as well as empirical SQGs (i.e., effects range low [ERL] and effects range median [ERM]) for a given suite of metals may show substantial variability with depth based on the vertical variation in AVS, though AVS does not universally influence metal bioavailability (Chapter 3). The magnitude of seasonal or other variation in AVS measurements requires a decision for the best time or conditions to sample in order to obtain representative estimates of metal bioavailability. Similarly, variability in organic carbon content or quality may affect organic bioavailability. In addition, dynamic processes such as erosion may affect these quantities or the very foundation of equilibrium partitioning (EqP)-based SQGs.

Dealing with this variability in any sediment assessment requires controlling or characterizing variability to the extent possible. Given that most assessments are simply "snapshots in time" and that variations in physical, chemical, and biological processes can occur over time, the key issue is to design the assessment to account for temporal variability. Conventionally, the most common approach for controlling variability, especially vertically and horizontally, is by compositing samples. In this approach, replicate samples are homogenized into a single sample. The greatest disadvantage to this approach is that, in the process of compositing, the uniqueness of the replicates is lost.

If the investigation supporting the sampling desires to maintain the unique qualities of the collection site replicates (e.g., for statistical hypothesis testing), compositing is not appropriate and characterizing variability may be required (Chapter 4). Conversely, if the objectives of the sampling are to make an overall assessment of sediment quality, compositing may be appropriate. Some level of compositing is often required to obtain sufficient sample and/or organisms for analyses.

The most frequently prescribed approach to characterize variability is to increase the number of replicates collected within a site. While this approach may not reduce variability such as that documented by Holland et al. (1993), it will allow for a much better quantification of how much variability is present. This solution brings with it the increased cost of collecting and analyzing more samples, which eventually becomes prohibitive. Several documents describe techniques for performing statistical power analyses that provide objective approaches for determining the needed number of replicates given acceptable levels of variability (USEPA/USACE 1991; Environment Canada 1994; USEPA 2001a). By weighing what is acceptable based on cost (i.e., how much expense is acceptable for a given project) and statistical design (i.e., how much variability is acceptable for a given project), the determination of SQGs and other tools can be accomplished with a known degree of confidence.

Finally, all practitioners need to have a firm understanding of the uncertainties inherent in the SQGs and other tools they are using (Batley et al. 2002; Chapters 3 and 17). Uncertainty is a function of both the variability in the system being assessed (e.g., physical and chemical) and the precision of the measure being performed (e.g., AVS, TOC). The practitioner must balance any assessment by choosing tools that are appropriate for the degree of variability in the system and the acceptable level of uncertainty in the results obtained. A lower level of precision may be more acceptable when used, for example, for screening purposes, than for the higher level of precision needed for a human health risk assessment or a remedial action.

Real-world physical and chemical factors that complicate and introduce uncertainty into the derivation and application of SQGs and other sediment assessment tools, including dynamic conditions, variability in sediment normalizing constituents, and desorption resistance or limited bioavailability of organic contaminants, are not "fatal" to SQGs and other sediment assessments tools. However, they do represent serious challenges and present areas in which further research and/or cost may be required to fully and accurately understand their appropriate role in assessment and management of contaminated sediments.

Biological Issues

General considerations

Biological issues that must be considered in sediment assessments that involve SQGs or other tools include biological interactions, natural variability, and exposure routes (additional details provided in Chapter 17). Biological interactions include not only competition and predation, which directly affect the structure and composition of resident communities, but also interactions between multiple stressors. Such stressors can include contaminants but can also include physical factors such as habitat changes as well as the aforementioned competition and predation, which may be particularly pronounced in the case of introduced species (e.g., zebra mussels in the Great Lakes).

Species-specific differences in physiology, biochemistry, behavior, or tolerance can mitigate adverse effects of contaminants and other stressors. Bioavailability of sediment contaminants will reflect the complex interplay between biological components (e.g., species-specific and tissue-specific differences in bioaccumulation, contaminant interactions with receptors and within tissues) and physical factors associated with sediments (e.g., redox changes in overlying waters and pore waters, bioturbation or bioirrigation).

Exposure routes

Exposure routes can include contact and ingestion of sediment, sediment pore water, and/or overlying water (Luoma et al. 2001). Overlying water has been shown, in some cases, to be the most important exposure route (Warren et al. 1998). The alimentary tract contributes considerably to, for instance, whole body metal accumulation (Hogstrand and Haux 1991; Luoma et al. 2001; Campbell et al. 2004). Intestinal metal uptake can be important in marine environments where anion complexation reduces the influence of the branchial pathway and, in many osmoregulators, drinking opens the intestinal route of uptake for waterborne metals. Because of the digestive functions of the intestine, the imbibed water becomes heavily modified in the gut, and water constituents that ameliorate gill toxicity may no longer be protective. There are differences in bioavailability of ingested sediment between animal species depending on the function of their alimentary tracts (Chapter 17).

Dietary uptake of contaminants from sediments may be direct in deposit or suspension feeders or indirect through consumption of prey from contaminated sediments (including sediment-exposed benthic diatoms or bacteria by filter feeders) or through accumulation from suspended sediments or pore water. For instance, bivalves absorb metals bound to anoxic sediments or particles coated with sulfides (Griscom et al. 2000; Lee et al. 2000).

Even if metals (or other contaminants) are firmly bound to sediments in the environment they might be bioavailable and bioreactive during feeding and digestion processes. Lee et al. (2000) found that the filter-feeding bivalve *Pomatocorbula amurensis* and the deposit-feeding bivalve *Macoma balthica* take up Ni, Zn, and Cd in reducing sediments as a function of the total extractable content of the respective metal. There were no correlations between tissue levels and porewater metal concentrations, and AVS did not influence assimilation; similar results were also observed with the deposit-feeding polychaete worm *Neanthes arenaceodentata*. Thus, no single species is a surrogate for all other species in all exposure scenarios. Consequently, several species should be tested; the choice of test species should reflect the nature and origin of the particular sediment to be assessed.

Chronic effects and biomarkers

An increasing consideration in sediment or other contaminant assessments is the possibility of chronic effects that can lead to long-term effects on population dynamics, including reproduction and recruitment (Chapters 3 and 12). Adverse effects on reproduction may be associated with reduced gamete production or gamete viability because of parental effects or toxicity via direct exposures to sensitive embryo and larval stages. For example, in studies on *M. balthica* from mudflats in the San Francisco Bay area from 1977 to 1989, the number of fully developed gametes was related to the body burden of Cu and Ag, which in turn were correlated to sediment metal concentrations (Hornberger et al. 1999).

Although resident community structure analyses implicitly include consideration of chronic effects and laboratory toxicity tests can include chronic endpoints, they may not be the best tools for such assessments of relatively subtle effects. For instance, community structure analyses are affected by a relatively large "signal to noise" ratio in comparison to single-species chronic endpoints measured either in the laboratory or by in situ assays. Laboratory toxicity tests do not typically include environmental realism such as the potential for multiple stressors. In situ sediment toxicity tests with caged organisms provide more realistic exposure regimes but do not necessarily account for natural physical, chemical, and biological factors that could also affect toxicity (Tucker and Burton 1999; Beckvar et al. 2000; Ringwood and Keppler 2002).

Tissue concentrations (of parent compounds or metabolites), when they can be measured, provide the most direct evidence of bioavailability. However, bioaccumulation or evidence of the presence of metabolites does not necessarily translate into adverse effects because toxicity depends on pharmacokinetics and because organisms may respond with various cellular mechanisms that may effectively sequester or detoxify contaminants (Mason and Jenkins 1995; Rainbow 1997). In contaminated environments, which are typically characterized by multiple stressors, synergistic or antagonistic interactions can also affect bioavailability. Similarly, it is possible that stimulation of cellular detoxification responses (e.g., heat shock proteins) by certain contaminants may serve to protect against adverse effects from other contaminants. Therefore, the use of selected physiological and cellular biomarkers can facilitate an appreciation of when homeostatic or detoxification mechanisms have been overwhelmed and can assist in the identification of stress or adverse effects (Ringwood et al. 1999; Maltby et al. 2001).

Cellular biomarker responses to contaminant exposures are thought to be among the most sensitive detectable responses and are also expected to provide the most detailed diagnostic capability (Ringwood et al. 1999; Livingstone et al. 2000). Furthermore, an important strength of sublethal organism and biomarker responses is their likelihood to provide an appreciation of the potential for chronic effects associated with long-term or episodic exposures. They are also potentially valuable for validating whether or not effects on other indicators (e.g., benthic community condition) are associated with contaminants. For example, if contaminant-specific biomarker responses of resident or deployed organisms do not indicate evidence of stress but benthic community indices do, then this indication would suggest that benthic community stress effects may not be a result of contaminants. Likewise, concordant benthic community and biomarker responses would reinforce contaminants as causative factors. At present, we can identify a handful of biomarkers that are highly sensitive and are symptomatic of impaired organismal health; these include CYP1A protein, metallothionein, lipid peroxidation, lysosomal stability, and the Comet Assay for deoxyribonucleic acid (DNA) damage. Of these, CYP1A and metallothionein also provide information on contaminant-specific effects.

Regulatory Issues

Regulations regarding the assessment and management of contaminated sediments vary widely (Nord 2001), and while intended to facilitate the process, they can sometimes lead to the development of unique challenges in particular situations. In the cases where these guidelines are used as "bright line" (i.e., definitive) criteria, the opportunity to generate data that are appropriate for cost-effective management decisions may be diminished. The regulator or manager needs to answer 2 questions regarding sediments: 1) What is the potential contribution of any given contaminant in sediment to the continued degradation of water quality? and 2) What is the potential contribution of any given contaminant in sediment to observed effects in the field or laboratory (i.e., to the health of the sediment ecosystem)? Answering these questions will require information on trends in sediment contamination.

Multijurisdictional systems have the highest level of management challenges. Rivers and estuaries make convenient political boundaries. Many of the most contaminated (and valued) systems occur at boundaries between states or nations, for example, the North Sea (bordered by several European nations), Puget Sound and the Great Lakes (bordered by the USA and Canada), Chesapeake Bay (bordered by Maryland, Virginia, and Delaware, USA) and New York Harbor (bordered by New York and New Jersey, USA). Even when contained within a given political entity, challenged sediment environments are often large, affecting large, diverse areas. Systems such as these are highly valued both economically and ecologically, generating a large number of stakeholder groups, often with disparate goals. In these situations, consensus building is required and should include discussions of appropriate goals or targets for sediment quality, resource use, economics, and public perception.

SQGs can provide a relatively easy way to communicate sediment quality to the public, but care must be taken to ensure that the SQG or set of SQGs chosen is technically appropriate for the management goal or target and the region affected. In some cases where there are sufficient field and/or laboratory data available, it may be reasonable to expend the effort to develop site- or region-specific SQGs (Chapter 6). This effort can be performed either as a database exercise (e.g., focusing on specific data from the original SQG derivation data set) or by incorporating data generated specifically for this purpose.

The initiative in watershed-based management in the United States (US; NRC 1999) has also resulted in the need to develop multijurisdictional assessments and management plans. Watershed-based planning has the added challenge of including stakeholders who can be many miles away from contaminated sediments. A recent panel on polychlorinated biphenyl (PCB)-contaminated sediments (NRC 2001) has also embraced the need for recognizing and incorporating the needs and desires of all stakeholders in the process of evaluating risks and options and making decisions. Conflicting laws, management goals, and/or scientific opinion will require negotiation on assessment tools that have been shown to have relevance to the particular system in question. Negotiation can result in setting remediation targets that reflect economic interests, public perception, and political reality, not just the scientific perceptions used to make assessments. That is not to say that the scientist does not have an important role; quite the contrary. It is in these situations that it is imperative that scientists both understand and effectively communicate the strengths and weaknesses of the available tools so that these are fully understood by the decision makers and by other stakeholders.

SQGs might be used by regulators in 4 ways:

1) They can be applied in the same regulatory frameworks as for water quality criteria (WQC; USEPA 2002).
2) SQGs might be used as chemical surrogates of the ecological status and performance of the benthic component of aquatic systems.
3) SQGs might be applied as criteria for sediment management (i.e., to classify in-place or dredged sediments as "acceptable" or "unacceptable" for specific decisions or to define remedial action or no-action zones).
4) SQGs have been proposed as indicators of natural resource damage.

Application of SQGs and other tools to various habitat types is discussed in the following sections. It is the responsibility of the assessor and/or regulatory agency involved to ensure that the correct tools are used and that the information obtained is interpreted in a scientifically defensible manner.

Assessments of Depositional Areas, Freshwater, and Marine Habitats

Waterbodies with soft bottom habitats represent areas of sediment deposition. Contaminants that are associated with sediments are likely to be found in such areas. Areas adjacent to such waterbodies are frequently developed and subject to a large number of human activities that complicate the use of SQGs and other sediment assessment tools. For discussion, 7 categories of soft bottom habitats are identified: lakes, low-gradient rivers, wetlands, ports and harbors, open ocean disposal sites, open ocean oil and gas platforms, and highly modified or constructed waterways. For each of these categories, a general narrative is provided. Specific examples are used to illustrate some aspects of sediment assessment in such habitats along with issues that may complicate sediment assessment, particularly the empirical-based SQG approaches.

Lakes

Fine-grained sediments in sheltered lakes and ponds represent one of the best opportunities to use the national freshwater numerical guidelines, particularly the empirical-based SQG approaches, because the national databases from which they were derived consist of data from depositional areas such as the Great Lakes (Persaud et al. 1993; MacDonald et al. 2000; Ingersoll et al. 2001). Many of the areas where sediment assessments are performed are low-energy environments with relatively consistent physical and chemical characteristics. Sediment management in lakes will be largely dictated by sediment loading, eutrophication, and lake morphometry. Physical and chemical issues affecting bioavailability, except where factors such as stratification and deoxygenation result in contaminant remobilization from sediments, are likely to be constant and, with minor exceptions, similar to those samples represented in the generic guidelines.

However, a few issues that may influence the accuracy of numerical guidelines should be noted. Bioavailability has been shown to be profoundly altered in sediments high in certain matrices such as wood chips, paint chips, carbon soot, or high organic carbon from sewage and paper mill discharge (Chapter 4). These matrices reduce bioavailability such that current numerical guidelines (based on standard bulk analyses) would tend to overpredict effects (Jonker and Koelmans 2002). Lakes that have enough depth to thermally stratify may have significant seasonal variability in bioavailability (Watzin et al. 1997). This same phenomenon has been noted at the base of large dams and reservoirs. Because lakes often have high sediment input or experience deposition of planktonic materials, sediment burial rates may be relatively great. High nutrient inputs can result in increased concentrations of sulfides that in turn will modulate metal toxicity to some organisms. Also, the presence of ammonia can result in toxicity. Exotic species may result in modifications to the natural benthic community.

Assessment and management approaches will vary depending on whether the site is deepwater or littoral. The discussion of wetlands later in this chapter can be applied to the littoral zone.

Onondaga Lake, a 12 km^2 lake near Syracuse in central New York, has been described as "the most polluted lake in the United States" (UFI 2002). Environmental problems include excessive nutrient loading from municipal discharge and combined sewer overflows (causing eutrophic and low-oxygen conditions), bacterial pollution, Hg and other hazardous materials from past manufacturing operations, and high sediment loading (USEPA 1995). Major initiatives have been undertaken to clean up the lake, including a 15-year program to improve sewage treatment, with initial costs of $380 million, or $1000 per person in the watershed. Progress of this program is to be monitored to evaluate success in achieving WQC for phosphorus, ammonia, nitrite, oxygen concentration, and water clarity. Elevated Hg levels in sediments resulting from former chloralkali operations in the watershed are among the contaminants of concern. Numerous investigations have been conducted and status reviews published (Effler 1996). Major modeling efforts have been used to evaluate management options, including mechanistic and mass balance models (UFI 2002).

Precision of site-specific empirical-based SQGs was evaluated in a study in Onondaga Lake (Becker et al. 2002). Toxicity tests using 10-d *Hyalella azteca* and *Chironomus tentans* (survival and biomass endpoints) and chemical analyses involved 79 samples from Onondaga Lake and 5 from a reference lake. The study concluded that SQG methods had variable levels of predictive accuracy. Overall accuracy was greatest for site-specific, freshwater apparent effects thresholds (AETs) (84%), compared to probable effects levels (PELs) (78%), ERMs (77%), ERLs (75%), and threshold effects levels (TELs) (53%). AETs also had the lowest false positive rate (8%). ERLs had the lowest false negative rate (10%). The freshwater AETs had the added benefit of helping diagnose which contaminants were likely most influential in observed toxicity. The relatively similar overall accuracy of various empirical-based SQGs illustrates both the fact that no individual approach is clearly better than any other, as well as their general applicability in a lake habitat.

In summary, because some empirically derived SQGs for freshwater were developed with large amounts of data from lake habitats, they are probably applicable without modification as screening tools. They may also help in communicating trends in sediment condition. Because SQGs do not distinguish between sources of contaminants, other tools are necessary when loadings are modeled or potential sources identified. Some sediments may contain matrices that affect chemical availability, so SQGs may lose accuracy in high organic carbon or specific matrix situations (Chapters 4 and 17).

Low-gradient rivers

As with freshwater lakes, low-gradient freshwater rivers are likely to present opportunities for the use of SQGs. This type of system is the archetype for development and application of SQGs, making their use in these habitats most straightforward. However, issues associated with hydrology and sediments are similar to those found in estuarine systems (see "Assessment of Streams and

Other Erosional Environments," p 294, i.e., sediment movement is generally down gradient, but with exceptions). Substrate in these systems can also vary considerably, and there can be substantial movement of substrate during storm events, resulting in dynamic conditions and changing receptor pathways, especially over large distances.

Low-gradient systems not only are the repositories for sediments from throughout the watershed but also tend to be highly urbanized and heavily used by a diverse stakeholder community. In some systems, or even in adjacent parts of the same system, uses can range from recreation and fisheries to industry and commercial shipping. Consequently, the appropriate use of SQGs must be determined in light of reasonable and achievable goals that are in keeping with the designated use of the waterbody. In urban systems, many of the same challenges presented later for ports and harbors may hold true and must be considered prior to application of numerical guidelines.

The Ashtabula River, Ohio, USA, is an excellent example of this type of system. The lower part of the Ashtabula River was designated as an area of concern (AOC) by the International Joint Commission (IJC) because of industrial pollution, primarily PCBs transported into river sediments from the Fields Brook Superfund site. Contamination levels in river sediments led to cessation of navigational dredging and to biological impacts and affected recreational uses in the river and the harbor in Lake Erie about 3 km downstream. The site was included as a priority for the Assessment and Remediation of Contaminated Sediments (ARCS) program, administered by the Great Lakes National Program Office of USEPA. Studies included sediment assessment, evaluation of both human health and ecological risks (Crane 1992; GLNPO 1994), and demonstration of a treatment technology (IJC 1997). Following evaluation, remedial actions were identified under the Superfund program that required the removal and treatment of approximately 30 000 m^3 of sediment and soils at the former industrial site and approximately 400 000 m^3 of sediment from the river (GLNPO 2000).

A number of issues may complicate sediment assessments in low-gradient river habitats. Physical habitat can be altered because of hardened shorelines or flood control structures. Storm events can significantly modify sediment load, influx of contaminants, movement of sediments within the system, and volume of water in the river. Systems flowing through or draining urban areas are likely to be high in organic carbon, altering contaminant bioavailability. Historical contaminants are likely to be present. Multiple contaminants are typically present, and the relative ecological significance of any particular contaminant may be unclear. Multiple sources of contaminants are likely, and identification of sources can be complicated. Large watersheds influence conditions in low-gradient rivers, increasing the likely significance of distant sources and of nonpoint sources, which may also increase the natural variability of the system, making it more difficult to distinguish anthropogenic impacts. Additionally, recreational uses and subsistence fishing increase the importance of considering human health in the assessment and management process.

In summary, because some freshwater SQGs were developed with data from low-gradient river habitats, they are probably applicable without modification as screening tools. They may also help in communicating trends regarding sediment condition or as an initial point for negotiations for resource damages or remedial actions. Where point sources are involved, upstream and downstream comparisons may be useful, and SQGs may be helpful in determining the extent of contamination but not its relative economic, social, or even ecological importance to the particular system in question.

Wetlands

Wetlands have been defined as lands that are transitional between terrestrial and aquatic systems, where the water table is at or near the surface, the vegetation is primarily hydrophytic, and the soils are hydric (Mitsch and Gosselink 1986). Primarily because of these characteristics, both freshwater and tidal wetlands are characterized by fine sediments, high levels of organic matter and sulfides, and relatively dense vegetation. They are often areas of high biological use and significance, meaning that, if contaminated, they may present an "attractive nuisance." Because wetlands serve as cover for migrating waterfowl and mammals, which provide a transport mechanism for contaminants, the potential pool of receptors is often much larger than those organisms that directly visit the site. Many countries specifically address wetlands with special regulatory frameworks to protect their critical functions such as habitat, flood control, and tidal buffering. Benthic populations are often patchy in distribution, particularly in freshwater systems, making bioassessments difficult to interpret. By their nature, wetlands are depositional, and in estuarine systems, they serve as zones of accretion. Because of this deposition, contaminants often concentrate in wetlands from diffuse sources; these sources can often be far from the wetland itself. However, high concentrations of organic matter of all types as well as AVS can result in low bioavailability of contaminants.

A good example of a wetland system is the Hackensack Meadowlands located in northern New Jersey, between the Hackensack and Passaic Rivers. It is part of the greater NY–NJ Harbor estuary (see "Ports and harbors," p 281). This system is estuarine, with salinities ranging from meso- to oligohaline. Much of the system has been highly modified by human activities, including landfilling, refuse disposal, canalization for mosquito control, and upstream damming of freshwater flows. Because of its connection with Newark Bay, the system continues to receive contaminated suspended sediment in addition to the contaminated material already in the Hackensack River from sewage effluent and industrial discharges. Solid waste and uncontrolled historical industrial landfills contribute more than 2.7 ML/hectare per year of contaminated leachate to many parts of the Meadowlands, complicating the contaminant source picture. Unfortunately, hydrodynamic alterations to the system over the last century (e.g., damming, diking, and channelization) have resulted in a steady increase in sediment accretion rates (via tidal flux) from the highly contaminated Newark Bay system downstream. Because of the surrounding development of metropolitan New York and New Jersey, the area is a wildlife refuge, containing a significant number of rare and endangered species of birds and plants, and is a breeding area for Diamondback Terrapin turtles (*Malaclemys terrapin*). The system is also a critical stopover for migratory waterfowl. Point-source control measures over the past several decades have resulted in a dramatic improvement in water quality in the area. Unfortunately, as water quality improves in the surrounding area, there is an increasing probability of exposure to wildlife as ecological use increases and some level of historically contaminated sediments persists.

Numerical guidelines such as empirical-based SQGs may not be applicable to wetland sites because of the probability of highly altered bioavailability. Conversely, EqP-based SQGs may have some applicability. Consequently, benthic and phytotoxicity tests should be used to assess the toxicity of wetland sediments. The high productivity of wetland systems makes the assessment of bioaccumulation potential especially important (Word et al. 2001). SQGs are not accurate predictors of bioaccumulation potential at this time (Chapter 11). In addition to conventional laboratory or field residue analysis to assess bioaccumulation, the practitioner may need to perform residue analysis of higher trophic levels to evaluate the potential for trophic transfer and biomagnification. Other conventional tools such as bioassessment (e.g., assessments of benthic communities) are not encouraged in wetlands because of the natural patchiness of organism distribution. Management of wetland contamination is often highly challenging because of the difficulty in removing sediments without destroying the habitat, increasing the importance of correctly identifying bioavailable fractions.

Ports and harbors

Ports and harbors typically have highly modified habitats because of navigational dredging and hardened shorelines. In many cases, the presence of piers and docks results in unnatural accretion of fine sediments. Ports are usually located in urban areas with substantial industry, resulting in the potential for high numbers of point and nonpoint pollution sources, spills, and historical mismanagement of human waste resulting in unnaturally high amounts of organic carbon. For these reasons, and perhaps simply because of longevity, many ports suffer from a legacy of pollution that results in highly variable and often severe sediment contamination. Therefore, biological diversity may or may not be natural, and reference sites may be rare or nonexistent. In addition, high socioeconomic values can be placed on urban waterways, resulting in competing interests and goals regarding environmental quality (compensation for lost habitat, including sediment pollution, may be part of the socioeconomic value placed on waterways). Many harbors are also multi-jurisdictional, with little or no cooperation among those agencies responsible for environmental regulation.

An excellent example of this type of environment is the NY–NJ Harbor (Chapter 15). The estuarine system in which it lies encompasses almost 780 km^2, drains more than 33 000 km^2, has more than 1 600 km of shoreline (75% of which is riprap or bulkhead), and is home to 100 species of fish, more than 300 species of birds, and more than 20 million people. The harbor is home to the Port of New York and New Jersey, the largest container port on the East Coast of the US and the largest petroleum port in North America. The harbor boasts more than 400 km of engineered waterways, resulting in a need to dredge about 5 million m^3 of sediment annually. The oldest industrialized watershed in North America (Passaic River), the largest wastewater treatment plant in the US (Passaic Valley Sewerage Commission), and the second largest urban center in the US (New York City) are all contained in or drain to the NY–NJ Harbor.

A large array of chemicals has been measured in the sediments in the system, producing a complex picture of highly variable concentrations and combinations of contaminants. Not surprisingly, the surface sediments in many areas are toxic. Because of historical industrial use, contamination often not only increases but also changes in composition with depth. Industrial companies located in the system have been numerous and have changed dramatically over the last 150 years, making identifying and tracking sources highly problematic. Despite these insults, there is still a viable but highly modified benthic and pelagic wildlife population that has enjoyed significant revitalization in recent decades, responding to improvements in surface water quality. However, improved water quality has often increased the number of potential receptor organisms and altered the bioavailable fraction of the complex mixture of historical contaminants.

In addition, the system is hydrodynamically complex, influenced heavily by stormwater flows, tidal cycles, and wind and contains multiple riverine inputs and outlets to the ocean. As a result, the sediment–water interface is highly dynamic, and large volumes of contaminated sediments move continually around the system (Chant et al. 2001; Rankin et al. 2001; Styles et al. 2001; Harrington et al. 2002). Complicating this picture is the fact that the system is shared by the States of New York and New Jersey, each having vastly different legal and regulatory frameworks for resource management decision-making. The NY–NJ Harbor is in the heart of a large population center, making it important to evaluate the potential for human exposure. Water quality has improved in the harbor over the decades since modern sewage treatment was implemented, increasing the use of the system for recreational boating and fishing, as well as encouraging an arguably persistent subsistence fishery. These facts make it necessary to estimate human exposure to contaminants in the sediments via ingestion and dermal contact. The potential for human health risk may, in fact, be more important in these situations than the potential ecological risk, given the highly modified benthic environment.

A variety of SQGs was recently evaluated for their applicability to NY–NJ Harbor sediments (Chapter 15). It was found that, in many cases, SQGs developed from national databases overestimated toxicity to the amphipod *Ampelisca abdita* in standard 10-d toxicity tests. The best correlation with observed toxicity was obtained when consensus-based quotients were used (Fairey et al. 2001). Field effects were not readily correlated with SQG predictions in NY–NJ Harbor either. Long et al. (2001) failed to find adequate correlation between ERL and ERM evaluations of NY–NJ Harbor sediments and synoptic bioassessment (benthic community) data. While this may be due to the presence of uncharacterized contaminants or altered bioavailability, it may also simply be a result of episodic anoxia caused by the high nutrient inputs to the system. Because of these confounding factors, regional SQGs are currently being calculated for NY–NJ Harbor using a large regional database (Chapter 15). This approach has successfully been employed in other ecosystems where regional bioavailability concerns raised questions regarding the applicability of national guidelines (MacDonald 1994).

In addition, work is ongoing in NY–NJ Harbor to evaluate the potential for chemicals not included in typical contaminant scans to influence toxicity (e.g., alkylated polycyclic aromatic hydrocarbons [PAHs], alkyl phenols, and tributyl tin [TBT]). TIEs will be used as well to attempt to link chemical concentrations to effects. Monitoring of success in restoration activities will likely include toxicity tests, bioassessments, and surficial sediment chemistry. These will be used to refine an SQG approach that can be used in the planning and potential implementation of a long-term monitoring program for the harbor. Another use for numerical guidelines will be to provide endpoints for a contaminant fate and transport model currently being developed for the harbor estuary. This model will be used in total maximum daily load (TMDL) development and as a predictive tool for estimating potential outcomes of alternative management strategies.

Ports and harbors are significant economic entities for the regions in which they are located. It is reasonable to weight socioeconomic factors when evaluating management scenarios for these areas, though there may need to be compensation for lost habitat (e.g., sediment pollution is part of the socioeconomic value). Ports and harbors, because of their urban settings, are not likely to reduce or eliminate all source inputs in a timescale concurrent with any specific restoration or remediation effort. Therefore, a tiered approach to both assessment and remediation in these areas needs to be developed and needs to be developed consistent with the fact that all or portions of these areas have been designated primarily for human use as transportation networks (much like highways and parking lots in upland areas). In very large ports and harbors, such as NY–NJ Harbor, costly remediation efforts should be scaled or targeted to those areas where the greatest environmental benefit can be achieved and sustained.

SQGs may be useful for screening and monitoring trends in ports and harbors with fine-grained depositional sediments. However, they should be validated against field-collected data to ensure that bioavailability has been properly assessed. If necessary, regional SQGs can be developed to increase predictability of effects. In cases of historical contamination, use of SQGs may be confounded by toxicity caused by the presence of chemicals for which there are no guidelines. In these cases, site-specific dose–response curves may help to define proper effects limits for the particular chemical of interest (i.e., spiked sediments or sediment dilutions). Systems that have been highly modified by human activities are likely to have altered the benthic community even in areas without contamination; therefore, SQGs are not likely to be predictive of field effects (Long et al. 2001).

For the same reason, actual reference sites for conventional bioassessment tools may be difficult to find, and bioassessment results may be as much reflective of varying salinity, grain size, temperature, and dissolved oxygen as contaminants. Bioaccumulation tests, TIEs, and site-specific dose–response tests should be used in a WOE approach for higher-level evaluations such as risk assessment and remedial design. The presence of many and diffuse contamination sources will complicate remedial efforts, and care should be taken to ensure that remedial goals are not unrealistic. Long-range planning and restoration efforts in these areas will require cost–benefit analysis to ensure that resources are used efficiently.

Open Ocean

The open ocean is typically a soft-bottom (depositional) habitat that reflects the major physical factors of water velocity and direction, bottom topography, and particulate properties. Anthropogenic contaminants may be introduced from terrestrial activities (such as dissolved or particulate materials carried by rivers and estuaries), from direct discharge of wastewaters or solid wastes (such as dredge spoils), from airborne materials deposited to surface waters, or by activities in the marine environment. Two relatively high-profile activities are considered here: open ocean disposal sites and oil and gas platforms.

Open ocean sites

Assessment and management of contaminated sediments in open ocean disposal sites is not as commonplace as near shore and riverine sites. However, there are documented areas in the open ocean that serve as sediment sinks, resulting in a fine-grained depositional mat that may concentrate contaminants from diffuse nonpoint sources such as atmospheric deposition or shoreline runoff. More commonly, point sources result from spills, ocean outfalls, or open water sites for sediments dredged from navigational waterways. Because these sites do not frequently represent areas of unique habitat, effects evaluation is usually not problematic.

Reference sites are often relatively easy to locate and sample. The US Army Corps of Engineers (USACE) has provided guidance for the evaluation of sediments proposed for ocean disposal, which includes the potential use of numerical guidelines as a screening tool (USEPA/USACE 1991). However, appropriate estimates of exposure are more difficult to obtain because the sheer size of surrounding ocean relative to the site itself makes it difficult to predict the frequency and extent of site use by any given receptor organism of interest, especially for migratory species. This difficulty may be further complicated if relatively nearby areas are actually more contaminated than the site of interest.

In the case of dredged material disposal sites, it is important to consider the potential for attraction of pelagic species resulting from topography changes and nutrient input. In addition, benthic populations in a disposal site may be altered by the addition of shallow water species along with the dredged material. Organisms associated with and that colonize the dredged material may serve as either short-term or long-term pathways for contaminants in dredged materials to be transferred to the attracted populations of pelagic organisms. This same concern is relevant for spill sites or ocean outfall locations. Transport concerns are also relevant and require a thorough evaluation of the stability of the site relative to the energy of the particular system in terms of both normal conditions and storm events.

To illustrate this category, one open water site is the Historic Area Remediation Site (HARS) off Sandy Hook, NJ. The HARS is a historical, dredged material disposal site containing fine-grained sediments from navigational dredging in NY–NJ Harbor. The site is currently being capped with clean fine-grained and sandy sediments from ongoing navigational dredging activities in the harbor. Recently, considerable attention has been paid to appropriate tools for the evaluation of exposure of historical disposed materials and the clean dredged materials being used for remediation. Of particular interest is the information necessary for human health and ecological risk assessment, including ingestion rates, site use factors, and a determination of regional biota-to-sediment accumulation factors (BSAFs) to define exposure pathways. The overall evaluation of risk presented by the presence and/or continued use of ocean disposal sites needs to be made in conjunction with an evaluation of the economic value of a site for disposal of dredged materials.

Empirically based SQGs may not be relevant in these open ocean sites. The numerical guidelines are predictors of effect, not exposure, and if exposure is reduced, then risk is also substantially reduced. Furthermore, in deep ocean disposal sites (e.g., the San Francisco Deep Ocean Site), SQGs may not be appropriate indicators of effect because they were developed from nearshore databases and nearshore organisms. These deep sites contain substantially different populations of organisms with unknown sensitivities to anthropogenic contaminants. In all cases, the practitioner needs to consider the species of relevance to these unique sites as well as the potential for exposure and transfer. A site-specific numerical guideline might be developed, but it would probably be more cost effective to determine effects using SQGs as part of a Sediment Quality Triad approach.

Oil and gas platforms

Oil and gas platforms may be fixed to the ocean floor or floating in deeper water, tethered in place. Drill cuttings, sometimes associated with petroleum hydrocarbons and drilling fluids, constitute the major materials released. The issues that complicate a sediment assessment of such sites include altered physical habitat (e.g., burial of original sediments by drill cuttings, a fixed structure to serve as a surface for colonization by immobile organisms and a habitat for mobile organisms), increased biological activity from microbial metabolism of hydrocarbons, modified benthic community reflecting increased nutrient supply, oxygen depletion of subsurface sediments, and potential ecotoxicity from hydrocarbons or other chemicals.

Sediment assessments of oil and gas platforms have included the Sediment Quality Triad approach and sediment porewater toxicity assessment. Chapman et al. (1991), evaluating a wellsite in the Gulf of Mexico, found high chemical enrichment and high laboratory toxicity in sediments within 25 m of the central platform but found little discernible change in the benthic infauna in the area. Toxicity tests used the whole sediment 10-d amphipod test with *Rhepoxynius abronius*, together with elutriate tests using the 48-h bivalve larvae test with the oyster *Crassostrea gigas* and the Microtox test of bacterial luminescence.

Carr et al. (1996) evaluated laboratory toxicity of sediment pore waters from several platforms in the Gulf of Mexico, also finding chemical enrichment and toxicity within 100 m of the platform. Toxicity tests used the sea urchin *Arbacia punctulata* fertilization and embryological development test, a 7-d polychaete test using *Dinophilus gyrociliatus*, and a 96-h copepod nauplii test using *Longipedia americana*. Not all platforms evidenced toxicity; the platform most fully evaluated was atypical because its discharge was placed within 10 m of the bottom (depth 130 m), unlike most platforms where discharges occur at or near the surface. Combined with low energy conditions at the site, the chemical concentrations were highest, and gradients were most apparent at this platform (High Island A389). Porewater toxicity (sea urchin development) was observed at 35% to 75% of stations where probable effects levels (PELs) for metals (MacDonald 1994) were exceeded. No information about benthic infauna in the area was reported.

Other approaches to sediment assessment include baseline site characterization (allowing before and after comparisons), usually as part of a permitting process. Monitoring studies, such as those established by Norwegian authorities for petroleum activity on the Norwegian Continental Shelf (SFT 2002), typically include reference sites for comparison. Data obtained included tissue levels of selected contaminants in fish, biota distribution and abundance, and chemical analyses of sediments and water. Correlations between biota, sediment characteristics, and contaminant concentrations were calculated. Regional data are collected every 3 years. Special investigations have been undertaken to evaluate impacts of past activities and potential actions such as cuttings pile removal. Data used in the assessments include use of an indicator species, the sea urchin *Echinocardium cordatum* (TNO 1999). A more recent study (UKOOA 2002) evaluated toxicity tests using the amphipod *Corophium volutator*, *E. cordatum*, the copepod *Acartia tonsa*, and the phytoplankton *Skeletonema costatum*. Bioaccumulation patterns were evaluated in the amphipod and in mesocosms containing mussels (*Mytilus edulis*), polychaetes (*Nereis virens*), and fish (*Scophthalmus maxiumus*). Cell assay techniques were used to assess endocrine disruption but were not conclusive, finding effects to be weak and variable.

These successful approaches rely on before and after comparisons, trends over time, and the use of reference sites. Chemical analyses do not rely primarily on comparison to an external numerical value (i.e., an SQG) but rather to historical and spatial data. Benthic community analyses are a common feature, providing data on abundance and taxa. Tissue measurements address bioaccumulation concerns. Monitoring on a multi-year timescale provides trend data. Changes in drilling practices (from oil-based muds to water-based muds) in the past decade mean that these chemicals are less likely to be associated with cuttings piles, reducing toxicity of these materials. One important aspect is the ability to distinguish between petrogenic hydrocarbons and PAHs (which are most likely to originate from the platform) and pyrogenic hydrocarbons and PAHs (which are likely to originate from marine transport activities).

Highly modified systems

Human activities have significantly modified or even created aquatic systems that are depositional areas for sediments and associated contaminants. Canals are an example where there was no aquatic system, or perhaps a small system, before human activity. Navigational channels are highly modified by construction and continually disturbed by ship traffic and maintenance dredging. In addition, some waterways in urban areas are dominated by wastewater discharges where perhaps 90% or more of the water in the system originates from permitted discharges such as municipal wastewater treatment works. Such systems may be viewed as the aquatic equivalent to a terrestrial highway, essentially a societal determination that the area is committed to nonecological services (though there may be a need for compensation for lost habitat). Such areas often attract development of industrial, commercial, and residential users, adding to the complexity of the evaluation by increasing the number of potential sources and types of contaminants.

The issues that complicate a sediment assessment of such sites are 6-fold:

1) Physical habitat has been altered by dredging or construction of the waterway with its hardened shorelines.
2) Defining a reference condition or expected benthic community is difficult because of the anthropogenic nature of the system. Comparison with a pre-existing community usually cannot be done or would be inappropriate because of the extensive changes in physical habitat. Similarly, application of empirical SQGs derived from naturally developed communities is problematic, given that these are not naturally developed communities.
3) These systems tend to be high in organic carbon.
4) Historical contaminants, whose origins predate current environmental laws, are likely to be present.
5) Multiple contaminants are typically present, so the relative ecological significance of any particular contaminant may be unclear and elusive.
6) Multiple sources of contaminants are likely so that identification of sources will be complicated. Allocating responsibility among sources is a necessary aspect of implementing management options.

The Houston Ship Channel is an 80 km–long constructed waterway that allows ocean-going vessels to traverse the shallow Galveston Bay to Houston, Texas, USA. Construction of the channel began in the 1870s. Patrick Bayou is a 5 km urbanized stream connected to the Houston Ship Channel, receiving permitted treated municipal and industrial wastewater discharges and stormwater runoff from adjacent industrial facilities and nearby urban and residential areas. Elevated sediment concentrations of pesticides, PAHs, metals, and PCBs have been reported (Parsons Engineering Science 2002).

Biota colonize such highly modified systems, and these systems are subject to environmental regulation and management. Under state regulations, sediment samples taken in Patrick Bayou exceeded sediment benchmarks, and sediment elutriate tests showed reduced fish survival at several stations. These data triggered requirements for further investigation under the TMDL process to determine the causes of benthic toxicity. A Sediment Quality Triad approach was adopted (Parsons Engineering Science 2002). Because SQGs were not available for each chemical being evaluated, quantitative structure–activity relationship (QSAR)-based estimates were obtained using USEPA Estimations Programs Interface for Windows (EPIWIN) software to permit calculating an estimated toxicity for the chemical suite at each station. Summed toxicity unit values were, however, less than 1, which would predict no toxicity to fish or invertebrates. Toxicity tests with sediments were conducted using amphipods (*Leptocheirus* sp.) and polychaetes (*Neanthes* sp.), but results varied over time, species, or location. Porewater toxicity tests showed no effects, indicating that the contaminants were largely sorbed to solids. TIE evaluations established that toxicity was not affected by treatment to reduce metals. Manipulations were consistent with organic contaminants but were unable to identify specific chemicals. Benthic community surveys found the biota to be equivalent to those outside the bayou. Multivariate statistical analysis failed to relate data on chemicals, toxicity, and the benthic community. However, salinity and temperature did correlate with the benthic community. Overall, the extensive and expensive ($600 000) assessment was not able to identify consistent LOEs. This experience suggests a need for improved frameworks and decision rules in using sediment assessment tools in a regulatory scheme (Chapter 6).

Determining what benthic community to expect in a highly modified location is the first challenge in sediment assessment; there is no way to make an a priori determination of the "healthy" or desired benthic community. Initial assessment of benthic communities will therefore provide the most insight into the sediment condition in highly modified sites. Identification of comparison sites provides a context for setting an expectation of what benthic communities occur under similar conditions. Selection of comparison sites is difficult but should be guided by trying to match the physical habitat conditions that constrain benthic community development, such as salinity, temperature, hydraulics, granularity, or organic carbon. Because a range of benthic communities may develop under modified conditions, it may be most appropriate to compare the site with a range of comparison sites because the assessor cannot know which comparison site provides the best match of habitat characteristics.

Sediment toxicity testing and concurrent chemical analysis are useful to establish a direct linkage between contaminants, toxicity, and the benthic community. Patterns of chemical distribution are useful to help distinguish among multiple sources, a necessary requirement to identifying management options. Developing a consensus about an effective conceptual ecological site model is a critical part of evaluating such highly modified systems.

Assessment of Estuaries

Estuaries are treated in detail here because of their highly dynamic nature and the fact that there are unique components to bioavailability of contaminants in this environment. Bioavailability is due to particle-related desorption release limitations and dynamics that influence the biologically active zone (e.g., resuspension events); as such, it differs from processes such as erosion that simply expose contaminated sediment.

General comments

Estuaries are among the most productive marine ecosystems in the world and are the primary nursery habitats for a variety of commercially and ecologically important fish and shellfish. These coastal areas are subject to freshwater inputs, generally from rivers but also from groundwater, typified by a horizontal gradient of salinity, increasing from the headwaters to the mouth. Salt wedge estuaries provide the largest extremes in salinity, both spatially and temporally. In salt wedge estuaries, river flow is dominant. Saltwater moves up the estuary in the form of a defined "wedge" whose upstream penetration is greatest during periods of low river flow and least during periods of high river flow. As a result, overlying and porewater salinities in areas affected by the salt wedge show substantial seasonal and spatial variations with concomitant changes in the benthic infauna (saltwater species move up-river with the salt wedge, and are replaced by freshwater species as the extent of salt wedge penetration declines). Similar conditions arise in fjords and estuaries that have large tidal ranges with significant fresh water inputs. In other estuaries, such as shallow, partially mixed estuaries, the area of the bottom that is transitional between marine and freshwater environments may not be as extensive, and they may not be particularly stratified. Vertically homogeneous estuaries show the smallest salinity gradients as tidal currents predominate and salinity differences are minimal because of intense vertical mixing.

Salinity, stratification, and circulation patterns can differ seasonally and spatially. It is not unusual for tens of square kilometers of bottom sediments in salt wedge estuaries to show seasonal porewater salinity differences related to seasonal differences in the extent and duration of the salt wedge (Chapman and Wang 2001); for instance, estuaries with salt wedges during periods of high river flow can become homogeneous during periods of minimal or no river flow. Circulation patterns can vary from relatively simple for a river draining into the sea to complex for a variety of rivers draining into a large embayment. Estuaries are frequently composed of a sequential series of different types of systems, ranging from deeper portions, including bays and harbors, which are connected to watersheds by dendritic tidal river and tidal creek habitats. Tidal creeks and tidal rivers are the major recipients of contaminants and nutrients associated with anthropogenic activities and function as bi-directional pipelines or conduits to associated water bodies (Sanger et al. 1999).

Consequently, estuaries should not be treated as steady-state systems, but they should not be regarded as unpredictable. Estuaries, by virtue of their circulation and resulting salinity variations, may be very variable environments. The important point is that estuarine dynamics in more variable systems must be sufficiently characterized to facilitate predictions regarding the bioavailability of contaminants and the types of benthic communities present. Additional details on such variability are provided in Chapter 17.

Categorization of estuarine sediments

The first step in any estuarine sediment assessment should be to characterize the estuary in terms of major physical or chemical gradients. Based on considerations outlined by Chapman and Wang (2001), an estuarine system should first be subdivided on the basis of salinity regimes, then on grain size, and then on other water quality parameters by measuring overlying water temperature, pH, salinity, dissolved oxygen (DO); porewater salinity and pH; sediment grain size; and TOC.

Because these parameters can be measured with a high degree of precision at a relatively nominal cost, sufficient sampling should be conducted to characterize the areas to be assessed. Porewater and sediment sampling can be done synoptically; overlying water data should be collected using in situ data loggers capable of taking measurements at 15 to 30 min intervals for >24 h. These data can then be used to determine the necessary sampling frequency and spatial intensity within different salinity regimes for more comprehensive (and expensive) sediment quality assessments. Where there is greater variability in grain size and TOC within a salinity regime, intensive sampling will be required; in more homogeneous sediments, sampling intensity can be reduced. The above initial categorization provides the basis for developing a sampling design that will ensure appropriate spatial analyses in a cost-effective manner.

After the initial characterization studies described above, sites can be characterized into various ecotomes (e.g., muddy freshwater, sandy freshwater, muddy marine, sandy marine, muddy estuarine, sandy estuarine) for evaluating sediment quality and biological effects. Note that because contaminant bioavailability and toxicity as well as benthic community structure can be affected by salinity changes, both overlying and porewater salinities should be measured synoptically when any of the assessment tools discussed for estuaries are used. Similarly, grain size and TOC will also affect contaminant bioavailability and toxicity and benthic community structure and should also be measured synoptically. Estuarine condition, particularly in the intermediate or variable salinity regions, requires site-specific consideration of the assessment tools discussed below and, possibly, the use of biomarkers (Chapters 5 and 17).

Sediment quality guidelines

SQGs are often derived from and compared to total contaminant concentrations (i.e., correlative SQGs), thus providing no ability to correct for bioavailability. Bioavailability in estuaries is affected by salinity and redox conditions (DO and pH) and by factors such as river flow, tidal flushing, and sediment resuspension events. Therefore, bioavailability and toxicity are less readily predictable than in more stable systems and can increase the potential for false positives or false negatives.

EqP approaches to deriving SQGs are not readily applicable to all estuarine sediments. The dynamic physical and biogeochemical nature of some estuaries invalidates the fundamental assumption of EqP, specifically, that a quasi-equilibrium state is achieved between contaminants in sediments and in water. Furthermore, AVS measurements, used as part of the EqP approach to assess metal contamination, vary seasonally (i.e., AVS levels tend to be lower during cooler winter months than during summer seasons), spatially (vertically within a sediment column as well as over different locales within a system, even over very short distances; see Chapter 17), and temporally (i.e., over the daily cycles of anoxia/hypoxia and oxygenation that occur in shallow estuaries) (Chapman and Wang 2001).

SQGs have not been evaluated extensively under all estuarine conditions, in particular low or varying salinities, though they have been validated relatively robustly under standard or relatively static saline conditions (e.g., 30 ppt; Chapter 12). There is a need to develop estuarine-specific SQGs that more appropriately account for low and variable salinities. Until such values are developed, existing SQGs should be used with caution for assessments in these areas. For example, normalized sediment contamination concentrations should be compared to background concentrations of sites from similar salinity regimes as recommended by Chapman and Wang (2001). Other appropriate normalizations would further improve our ability to identify contaminated sediments (Windom et al. 1989; Hanson et al. 1993; Ringwood et al. 1997).

Benthic surveys

Benthic estuarine assemblages have many attributes that make them reliable and sensitive indicators of ecological condition and pollution stress (Boesch and Rosenberg 1981; Bilyard 1987; Van Dolah et al. 1999). Benthic species composition and abundance have been used to measure the status and trends of estuaries for several decades (Sanders 1956; Rosenberg 1976; Boesch and Rosenberg 1981; Holland et al. 1987; Summers et al. 1993; Hyland et al. 1996, 1999; Weisberg et al. 1997; Balthis et al. 2002). While benthic community assessments are regarded as one of the most important assessment tools, many confounding issues make them more difficult to evaluate than is sometimes realized. Soft-bottom benthic populations can be patchy and naturally subject to dramatic seasonal differences, predation effects, physical–chemical (e.g., temperature, dessication, hypoxia) and other stresses that can make it difficult to distinguish contaminant effects from natural factors. In some cases, estuarine areas tend to have lower species diversity than areas with low or high salinity (Chapman and Wang 2001). Measures such as biodiversity and species richness, abundance, and relative abundances of key species or functional groups have been used as indicators of the integrity of benthic communities. More recent approaches have used integration of a suite of such measures into a single multimetric index. Because benthic community composition is dramatically affected by salinity regimes and sediment characteristics, it is important to compare potentially impacted communities with reference communities from similar habitats (i.e., similar salinities: low, moderate, or high; similar grain size; and TOC) or develop approaches for normalizing for these factors.

Sediment toxicity tests

Laboratory sediment toxicity tests are typically conducted using either fresh or salt waters (e.g., added to whole sediments either to provide overlying water or to produce elutriates), which may affect both contaminant bioavailability and sediment toxicity. There are at least 2 approaches to toxicity testing of contaminated sediments. One approach dictates that the toxicity of estuarine sediments should be tested as received, using organisms capable of tolerating the natural salinity and grain-size conditions. The other approach advocates standardizing testing conditions. Both approaches have valid advantages and disadvantages that must be weighed when a sediment assessment is planned. Recommendations in this regard are provided by USEPA (1994, 2001b), Environment Canada (1998), Chapman and Wang (2001), USEPA/USACE (2000), and ASTM (2004a, 2004b) and include measurements of overlying salinities, ammonia, and pH during testing. Ammonia toxicity can be an important issue for estuarine sediments, as can that of H_2S (Wang and Chapman

1999). Estuarine conditions favor the release of ammonia from sediments (Joye and Hollibaugh 1995; Rysgaard et al. 1999). Ammonia can be highly toxic in sediment toxicity tests (Kohn et al. 1994) and should be measured at least at the beginning and end of all estuarine sediment toxicity tests. In situ sediment toxicity tests with caged organisms provide the most direct evidence of potential toxicity, but it is important to be able to account for natural factors that could also affect toxicity in order to appropriately identify contaminant effects (Ringwood and Keppler 2002).

Bioaccumulation

In situ bioaccumulation assays in estuaries are conducted under actual conditions, but similar to laboratory sediment toxicity tests, laboratory bioaccumulation assays are typically conducted using either fresh or salt waters (e.g., added to whole sediments either to provide overlying water or to produce elutriates), which may affect contaminant bioavailability. In situ bioaccumulation assays are preferable to laboratory tests, but both approaches have limitations (Ingersoll et al. 1997). Estuarine sediments should be tested in the laboratory as received, with ambient salinities, using organisms capable of tolerating the natural salinity and grain-size conditions. In situ bioaccumulation studies with caged or native organisms provide more direct evidence of bioavailability.

Assessment of Streams and Other Erosional Environments

Unique features of high-gradient streams and other erosional environments are described in Chapter 17, in particular the unidirectional water flow in high-gradient streams and the lack of sediment both there and in rocky intertidal habitats.

Assessment tools

As in studies conducted in depositional habitats, benthic surveys in erosional habitats should be conducted in the same locations where sediments are collected for toxicity testing and chemical analyses. Because fine sediments comprise a relatively small portion of the benthic habitat in these systems, novel methods for assessing contaminant exposure and ecological integrity should be developed. These methods should focus on characterizing contamination in the dominant habitats of these systems. Selectively sampling uncommon depositional zones in streams and conducting sediment toxicity tests with these materials may provide misleading or variable results.

Biofilm, defined as the combination of attached algae, diatoms, and abiotic materials that accumulate on the surface of cobble and other hard substrates, may be a more relevant exposure matrix in some high-energy, erosional environments. Similar to depositional sediments, contaminant concentrations in biofilm are generally orders of magnitude higher than in overlying water. Because biofilm is widely distributed in many high-gradient streams and other erosional habitats, this material is an important route of exposure to benthic organisms. Measures of sediment chemistry, toxicity tests, and assessments of contaminant bioaccumulation for erosional environments should be conducted using biofilm. To improve integration of these measures, in situ assessments of benthic community structure should occur in the same erosional habitats where biofilm is collected. Appropriate test species must be selected accordingly.

Neptun (2001) developed a novel approach using biofilm to assess the ecological integrity of the Clark Fork River, a cobble-bottom stream in southwestern Montana, USA. Previous attempts to characterize effects of metals in this system using conventional approaches were limited (e.g., Canfield et al. 1994) because of the difficulty in locating depositional areas and the relative unimportance of fine sediments (Neptun 2001). This situation is common in many high-gradient streams. Neptun (2001) modified the Sediment Quality Triad (Chapman 1990, 1996) and conducted chemical analyses of contaminants in biofilm, toxicity tests with grazing macroinvertebrates, and in situ assessments of community structure in riffle habitats to provide an integrated measure of biological integrity. Results showed that mayflies grazing on biofilm collected from the Clark Fork River accumulated significantly greater levels of metals compared to controls. Growth of mayflies feeding on contaminated biofilm was also significantly reduced, suggesting that the low abundance of mayflies in the Clark Fork River was a result of metal exposure.

Courtney and Clements (2002) performed similar experiments to demonstrate the influence of metals in biofilm on the grazing mayfly *Baetis tricaudatus* (Ephemeroptera: Baetidae) in the Animas River, Colorado, USA. Because fine-grain sediments represented a relatively small portion of the benthic community in the Animas River, biofilm was considered a more important route of metal exposure to benthic macroinvertebrates. Organisms were exposed to clean and metal contaminated biofilm in experimental streams. Results showed significant metal bioaccumulation and reduced growth in mayflies grazing on contaminated biofilm compared to those grazing on clean substrate (Figure 7-1). These findings have important implications for assessment of contaminant effects in environments where fine sediments are uncommon.

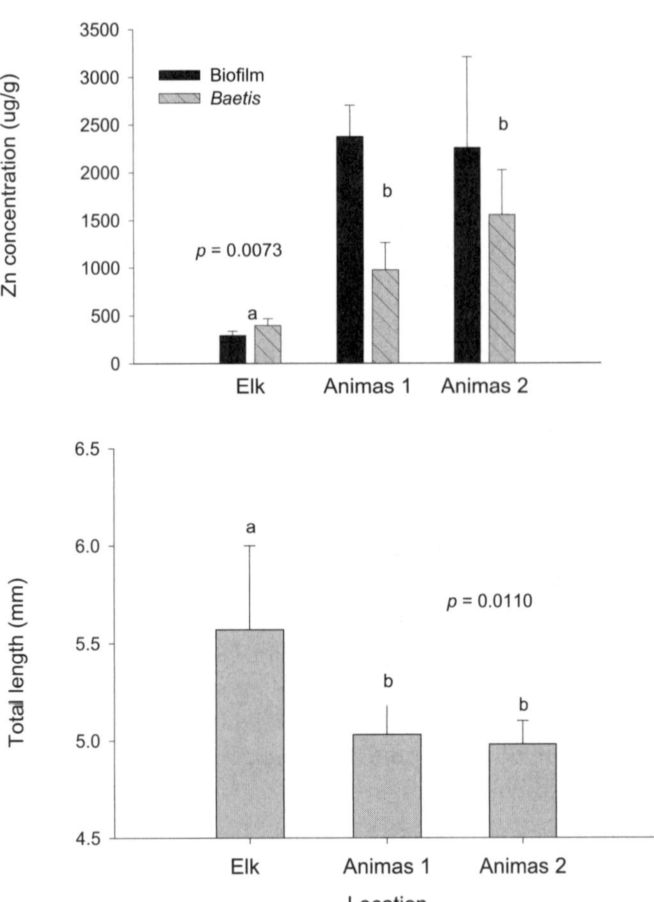

Figure 7-1 Zn concentrations (means and 95% confidence limits) in biofilm and grazing mayflies (*Baetis tricaudatus*) after 7-d exposure to reference (Elk) and metal-contaminated substrate (Animas 1 and Animas 2) in experimental streams (upper panel). Growth of *B. tricaudatus* after 7-d exposure to heavy metals in biofilm (lower panel). Results of 1-way ANOVA testing for differences among locations also shown. Data from Courtney and Clements (2002).

Because biofilm is ubiquitous in many cobble-bottom streams and forms the base of the food chain in these environments, elevated levels of contaminants in biofilm may pose significant risk to benthic organisms.

In situ experimental approaches

One of the key challenges in relating changes in community composition to levels of sediment contamination is to quantify the direct effects of sediment characteristics on measures of community structure. Many of the same features that influence bioavailability and toxicity of contaminants in sediments (e.g., grain size, organic carbon) also influence community composition. Experimental approaches in which these sediment characteristics are manipulated can provide information on their relative importance. In situ experimental approaches can also be employed to quantify the relative importance of water quality and sediment quality in environments subjected to multiple stressors. Measuring colonization of contaminated substrate or conducting community-level experiments with macroinvertebrates will allow researchers to separate the direct effects of sediment contaminants from those associated with poor habitat or degraded water quality (Chapters 4, 12, and 16). Courtney and Clements (2002) measured colonization of contaminated substrate and conducted community-level experiments in the Animas River to quantify the relative importance of substrate quality on benthic communities. Not surprisingly, these experiments demonstrated that patterns of community structure observed in the field were a result of both contaminated substrate and elevated metals in the water column. However, individual species showed variable responses to these sources of exposure. For example, heptageniid mayflies, organisms known to be highly sensitive to metals in water, were not affected by contaminated sediments.

In summary, experimental approaches are important tools for evaluating contaminant effects in environments where natural longitudinal variation confounds routine biomonitoring approaches and sediment toxicity tests. Similar field experiments may be necessary in other environments to quantify the influence of sediments relative to biotic and other factors.

Discussion and Summary

Sediments can be heterogeneous and dynamic. Physical, chemical, and biological factors complicate and introduce uncertainty into the derivation and application of SQGs and other sediment assessment tools. Physical and chemical factors include dynamic conditions, spatial and temporal variability in sediment-normalizing constituents (e.g., AVS), and desorption resistance or limited availability of some organic contaminants. Grain size, sulfide levels, and organic carbon type and content may vary significantly at small scales (millimeters) or large (estuaries) within an assessment area. Organisms' exposures to sediment-associated contaminants occur by different routes, such as via the sediment–water interface, pore water, direct contact, or ingestion. Biological factors reflecting species-specific differences in physiology, biochemistry, and behavior result in varying tolerances, acclimation, or adaptation, which result in different levels of adverse biological effects of contaminants. None of these factors are "fatal" to the derivation or application of SQG approaches nor to other sediment assessment tools; however, these factors do represent uncertainties, challenges to their further development, and areas where further research is required to better understand their appropriate and scientifically defensible use.

At present, while chemical analysis of bulk sediment can provide an adequate estimate of exposure, there are relatively easy and inexpensive adjustments to account for bioavailability or chemical speciation, which can improve exposure estimates. Such analyses should be restricted to surficial sediments except where dredging, erosion, resuspension, or other processes are likely to incorporate buried contaminated sediment into the water column or the surficial layer (Chapman et al. 2002). Adverse effects to the sediment community are reasonably measured by benthic community analysis (e.g., diversity, abundance, key species); toxicity testing of appropriate, representative taxa; and tissue residues. Some biomarkers can provide quantitative information on contaminant-specific exposure and be predictive of deleterious effects at the organism level (Chapter 17).

Within the context of the above physical, chemical, and biological realities, the 2 key findings are as follows:

- SQGs alone are sometimes, but not always, sufficient for management decision-making. As detailed in Table 7-2, developed on the basis of information contained in Chapter 17, there are 10 management reasons for sediment assessments. In 5 cases, SQGs should be used in a WOE approach with other tools; in 2 cases, they can be used alone; in 2 cases, they should not be used at all; and in 1 case, they can be used either alone or in a WOE approach with other tools, depending on the circumstances.
- Different tools are needed to characterize sediment in different environments. Table 7-3 compares the characteristics of the 5 basic sediment assessment tools described above with 3 general types of aquatic situations or habitats that are typically encountered. It is apparent that, with the possible exception of biomarkers, these tools are not equally applicable to all situations or habitats. Also evident is the need for additional standardized methods and procedures to validate SQGs in certain habitats (e.g., estuaries). Methods that incorporate dynamic aspects of certain habitats would be particularly useful (e.g., varying salinity).

Table 7-2 Management reasons for performing a sediment assessment and the role of SQGs in different sediment management scenarios

Reason for sediment assessment[a]	Role for SQGs[b]	Specific role and comments
Mapping spatial patterns	Primary	SQGs can be used to address relative patterns of contamination, including probable no effect and possible effect concentrations in sediments
Measuring temporal trends	Primary	
Determining condition of populations and communities	Secondary	As part of an ERA and/or in a tiered assessment scheme, SQGs are appropriate when used in conjunction with other tools
Estimating ecological risks, including bioaccumulation	Secondary	
Screening the suitability of proposed use or development	Secondary	
Assessing impacts of sediment dredging and / or management	Secondary	
Estimating human health risks and evaluation of biomagnification	None	SQGs have not been developed for this purpose
Determining sediment stability and transport	None	SQGs are not relevant
Remediation and restoration objectives	Primary	In cases of simple contamination where adverse effects are likely, SQGs can be used alone when the costs of further investigation outweigh costs of remediation and there is agreement to act instead of conducting further investigation
	Secondary	As part of an ERA and/or in a tiered assessment scheme, SQGs are appropriate when used in conjunction with other tools
Long-term monitoring of system status post-remediation	Secondary	As part of an ERA and/or in a tiered assessment scheme, SQGs are appropriate when used in conjunction with other tools. However, SQGs alone do not address all possible monitoring needs such as human health, exposure routes, or functional aspects of the ecosystem

[a] Adapted and modified from GIPME (2000)

[b] "Primary" can be used alone for management purposes; "Secondary" should be used with other assessment tools. In most cases, site-specific information should be generated to supplement the use of SQGs in sediment assessments.

Table 7-3 Characteristics of different sediment assessment tools[a] for describing sediment quality in depositional marine, freshwater, and estuarine environments and in stream and other erosional aquatic environments

Tools	Depositional marine and freshwater (7 types)	Estuarine	Streams and other erosional environments
Sediment chemistry including numeric SQGs	Contaminant analyses based on sediments; current SQGs usually relevant	Contaminant analyses based on sediments; current SQGs should be further validated for estuarine conditions, e.g., variable salinity	Contaminant analyses should include biofilm; current SQGs may not be relevant for riffle and erosional environments
Toxicity tests	Generally conducted in containers with sediments; standardized methods are available	Generally conducted in containers with sediments; standardized methods are available but may not be applicable to all situations; additional tests are required	Conducted with biofilm; may be necessary to provide water current in test chambers during exposure
Bioaccumulation tests	Organisms exposed to contaminants in sediments, tissue levels estimated or measured in higher trophic levels		Organisms exposed to contaminants in biofilm
Biomarkers	Dependent on the exposure; can involve field or laboratory measurements; similar tests can be applied to all habitats		
Resident aquatic community structure	Samples collected directly from benthic habitats		Samples collected from riffles

[a] Does not include specialized tools such as TIE, biodegradation studies, exposure surrogates or modeling, nor does it include frameworks such as ERA.

Table 7-4 summarizes the issues that differentiate various aquatic environments and provides general suggestions for conducting sediment assessments in 7 types of depositional habitats in addition to estuarine and stream or other highly erosional systems. Cases in which there is overlap between situations or habitats (e.g., wetlands in estuaries, watershed systems combining erosional and depositional ecosystems) will require consideration of combined issues with consequent possible modification of the suggestions for the different environments. A complex interplay between physical, chemical, and biological components will dictate some level of uncertainty regarding contaminant analysis, bioavailability, and effects, so sensitive diagnostic tools may be especially valuable in validating predictive models.

Table 7-4 Major issues and suggested approaches for sediment assessments and for using SQGs in different aquatic habitats

Situation or habitat	Issues	Suggestions
Depositional: Lakes and ponds	SQGs are generally appropriate. Thermal stratification, dynamic bioavailability, exotic species, seasonal anoxia	• Design sampling based on seasonal differences • Do not exclude influence of exotic species if present
Depositional: Low-gradient rivers and streams	SQGs are generally appropriate. Highly modified habitat, dynamic sediment–water interface, high total organic carbon (TOC), complex mixtures in urban systems, multiple diffuse sources in watershed, high recreational and commercial use	• Further verify SQGs with site-specific data; develop site-specific SQGs if needed and if practical • Define different ecoregions if present • Design sampling based on upstream and downstream comparisons if appropriate
Depositional: Wetlands	SQGs may not be specific for this environment. High organic matter and sulfides, high benthos patchiness	• Focus on WOE evaluation not individual LOE because of uncertain bioavailability • Assess bioaccumulation and biomagnification
Depositional: Ports and harbors	SQGs generally appropriate. Assessments typically address complex mixtures, high TOC, highly modified habitat, multijurisdictional issues, and high resource demand and involve multiple sources	• Further verify SQGs with site-specific data, develop site-specific SQGs if needed and if practical • Rely heavily on ground-truthing using WOE • Relate reference or background comparisons to assessment goals • Account for physical disturbances that confound benthic assessments and use of SQGs
Depositional: Open ocean	SQGs may not be appropriate. Highly modified sites; can attract biota; exposure difficult to quantify; bioavailability likely to change between dredging and placement	• Evaluate exposure based on residence time in affected area • Develop appropriate site-specific SQGs (coastally derived SQGs may not be appropriate) • Use WOE to assess effects
Depositional: Oil and gas production environments	SQGs may not be appropriate. Altered physical habitat, increased biological activity (hydrocarbon food source), modified benthos (nutrient, oxygen changes), potential ecotoxicity	• Use before and after comparisons, temporal trends • Distinguish between petrogenic and pyrogenic PAHs and hydrocarbons for source identification • Use a WOE approach

Table 7-4 *contd*

Situation or habitat	Issues	Suggestions
Depositional: Highly modified systems	SQGs may not be appropriate. Altered physical habitat, multiple sources, and no obvious reference condition. Elevated TOC, legacy contamination, multiple contaminants make ecotoxicological significance of single contaminants unclear.	• Identify comparison sites based on physical similarities (no true reference sites) • Conduct initial benthos surveys • Conduct pattern analyses to find sources and develop mass loadings
Estuaries	May require site-specific SQGs and additional bioassays. Dynamic areas dominated by salinity, grain size and other gradients; contaminants; sediments and benthic populations move bidirectionally	• Define salinity and grain size zones both temporally and spatially • Use salinity appropriate bioassays and site-specific SQGs and/or normalized background comparisons • Design sampling based on salinity and grain size zones
Nondepositional streams and other erosional ecosystems	SQGs are not appropriate. Lack of fine grain sediments; longitudinal variation in particle size, OC, benthos	• Measure other sources of exposure (e.g., biofilm) • Quantify longitudinal variation • Use sampling designs that quantify longitudinal variation

Site-specific (e.g., regional) SQGs might need to be developed for certain different aquatic situations, as noted in Table 7-3. However, SQGs would not need to be modified if recalculated for site-specific conditions. Further research is needed to determine the conditions in which site-specific SQGs would differ significantly from generic SQGs.

References

[ASTM] American Society for Testing and Materials. 2004a. Standard test method for measuring the toxicity of sediment-associated contaminants with estuarine and marine invertebrates. ASTM International annual book of standards, Volume 11.05, E1367-03. West Conshohocken (PA): ASTM.

[ASTM] American Society for Testing and Materials. 2004b. Standard test method for measuring the toxicity of sediment-associated contaminants with freshwater invertebrates. ASTM International annual book of standards, Volume 11.05, E1706-00. West Conshohocken (PA): ASTM.

Balthis WL, Hyland JL, Scott GI, Fulton MH, Bearden DW, Greene MD. 2002. Sediment quality of the Neuse River estuary, North Carolina: an integrated assessment of sediment contamination, toxicity, and condition of benthic fauna. *J Aquat Ecosyst Stress Recov* 9:213–225.

Batley GE, Burton Jr GA, Chapman PM, Forbes VE. 2002. Uncertainties in sediment quality weight-of-evidence (WOE) assessments. *Hum Ecol Risk Assess* 8: 1517–1547.

Becker DS, Bigham GN, Nielsen RD. 2002. Comparison of site-specific sediment quality values for contaminated sediment management. In: Porta A, Hinchee RE, Pellie M, editors. Management of contaminated sediments. Columbus (OH): Battelle. p 147–154.

Beckvar N, Salazar S, Salazar M, Finkelstein K. 2000. An in situ assessment of mercury contamination in the Sudbury River, Massachusetts, using transplanted freshwater mussels (*Elliptio complanata*). *Can J Fish Aquat Sci* 57:1103–1112.

Bilyard GR. 1987. The value of benthic infauna in marine pollution monitoring studies. *Mar Pollut Bull* 18:581–585.

Boesch DF, Rosenberg R. 1981. Response to stress in marine benthic communities. In: Barrett GM, Rosenberg R, editors. Stress effects on natural ecosystems. New York: Wiley. p 179–200.

Burton Jr GA, Batley GE, Chapman PM, Forbes VE, Schlekat CE, Smith PE, den Besten PJ, Bailer J, Reynoldson T, Green AS, Dwyer RL, Berti WR. 2002. A weight-of-evidence framework for assessing sediment (or other) contamination: Improving certainty in the decision-making process. *Hum Ecol Risk Assess* 8: 1675–1696.

Burton Jr GA, Chapman PM, Smith EP. 2002. Weight of evidence approaches for assessing ecosystem impairment. *Hum Ecol Risk Assess* 8:1657–1673.

Canfield TJ, Kemble NE, Grumbaugh WG, Dwyer FJ, Ingersoll CG, Fairchild JF. 1994. Use of benthic community structure and the sediment quality triad to evaluate metal-contaminated sediment in the Upper Clark Fork River, Montana. *Environ Toxicol Chem* 13:1999–2012.

Carr RS, Chapman DC, Presley BJ, Biedenbach JM, Robertson L, Boothe P, Kilada R, Wade T, Montagna P. 1996. Sediment porewater toxicity assessment studies in the vicinity of offshore oil and gas production platforms in the Gulf of Mexico. *Can J Fish Aquat Sci* 53:2618–2628.

Chant R, Glenn S, Hunter E, Rankin K, Bruno MS, Styles R, Hires R. 2001. Circulation and mixing in a complex estuarine environment: Effects on the transport and fate of suspended matter. Presented at the SETAC 22nd Annual Meeting; 2001 Nov 11–15; Baltimore, MD, USA.

Chapman PM. 1990. The Sediment Quality Triad approach to determining pollution-induced degradation. *Sci Total Environ* 97/98:815–825.

Chapman PM. 1996. Presentation and interpretation of Sediment Quality Triad data. *Ecotoxicology* 5:327–339.

Chapman PM, McDonald B, Lawrence GS. 2002. Weight of evidence issues and frameworks for sediment quality (and other) assessments. *Hum Ecol Risk Assess* 8: 1489–1515.

Chapman PM, Power EA, Burton Jr GA. 1992. Integrative assessments in aquatic ecosystems. In: Burton Jr GA, editor. Contaminated sediment toxicity assessment. Chelsea (MI): Lewis. p 313–340.

Chapman PM, Power EA, Dexter RN, Andersen HB. 1991. Evaluation of effects associated with an oil platform using the Sediment Quality Triad. *Environ Toxicol Chem* 10:407–424.

Chapman PM, Wang F. 2001. Assessing sediment contamination in estuaries. *Environ Toxicol Chem* 20:3–22.

Courtney LA, Clements WH. 2002. Assessing the influence of water quality and substratum quality on benthic macroinvertebrate communities in a metal-polluted stream: An experimental approach. *Freshw Biol* 47:1766–1778.

Crane JL. 1992. Baseline human health risk assessment: Ashtabula River, Ohio, Area of Concern. [Abstract and table of contents.] Athens (GA): US Environmental Protection Agency, Environmental Research Laboratory. EPA-905-R92-007. Available from: http://www.epa.gov/glnpo/arcs/EPA-905-R92-007. Accessed 3 Aug 2004.

Effler SW. 1996. Limnological and engineering analysis of a polluted urban lake: Prelude to environmental management of Onondaga Lake, New York. New York: Springer. 856 p.

Environment Canada. 1994. Guidance document on collection and preparation of sediments for physiochemical characterization and biological testing. Ottawa (ON):Environment Canada. Report EPS 1/RM/29.

Environment Canada. 1998. Biological test method: Reference method for determining acute lethality of sediment to marine or estuarine amphipods. Ottawa (ON): Environment Canada. Report 1/RM/35.

Fairey R, Long ER, Roberts CA, Anderson BS, Phillips BM, Hunt FW, Puckett HR, Wilson CJ. 2001. An evaluation of methods for calculating mean sediment quality guideline quotients as indicators of contamination and acute toxicity to amphipods by chemical mixtures. *Environ Toxicol Chem* 20:2276–2286.

[GIPME] Global Investigation of Pollution in the Marine Environment. 2000 Guidance on assessment of sediment quality. London: GIPME. IOC-UNEP-IMO.

[GLNPO] Great Lakes National Program Office. 1994. Assessment and remediation of contaminated sediments (ARCS) program, assessment guidance document: 8. Fish tumors and abnormalities. Available from: http://www.epa.gov/glnpo/arcs/EPA-905-B94-002/B94002-ch8.html. Accessed 3 Aug 2004.

[GLNPO] Great Lakes National Program Office. 2000. Realizing remediation II: An updated summary of contaminated sediment remediation activities at Great

Lakes Areas of Concern. Available from: http://www.epa.gov/glnpo. Accessed 3 Aug 2004.

Griscom SB, Fisher NS, Aller RC, Lee BG. 2002. Effect of gut chemistry in marine bivalves on the assimilation of metals from ingested sediment particles. *J Mar Res* 60:101–120.

Hanson PJ, Evans DW, Colby DR. 1993. Assessment of elemental contamination in estuarine and coastal environments based on geochemical and statistical modeling of sediments. *Mar Environ Res* 36:237–266.

Harrington TO, Rankin KL, Bruno MS. 2002. Frequency of sediment suspension events in Newark Bay. Presented at Protection and Restoration of the Environment IV; 2002 July 1–5; Skiathos Island, Greece.

Hogstrand C, Haux C. 1991. Binding and detoxification of heavy metals in lower vertebrates with reference to metallothionein. *Comp Biochem Physiol C Pharmacol Toxicol Endocrinol* 100:137–141.

Holland AF, Shaughnessy A, Hiegel MH. 1987. Long-term variation in mesohaline Chesapeake Bay macrobenthos: spatial and temporal patterns. *Estuaries* 10: 227–245.

Holland PT, Hickey CW, Roper DS, Trower TM. 1993. Variability of organic contaminants in intertidal sandflat sediments from Manikau Harbour, New Zealand. *Arch Environ Contam Toxicol* 25:456–463.

Hornberger MI, Luoma SN, van Geen A, Fuller C, Anima R. 1999. Historical trends of metals in the sediments of San Francisco Bay, California. *Mar Chem* 64:39–55.

Hyland JL, Herrlinger TJ, Snoots TR, Ringwood AH, Van Dolah RF, Hackney CT, Nelson GA, Rosen JS, Kokkinakis SA. 1996. Environmental quality of estuaries of the Carolinian Province 1994. Silver Spring (MD): National Oceanic and Atmospheric Administration. NOAA Technical Memorandum NOS ORCA 97.

Hyland JL, Van Dolah RF, Snoots TR. 1999. Predicting stress in benthic communities of southeastern U.S. estuaries in relation to chemical contamination of sediments. *Environ Toxicol Chem* 18:2557–2564.

[IJC] International Joint Commission. 1997. Overcoming obstacles to sediment remediation in the Great Lakes Basin. Available from: http://www.ijc.org/php/publications/html/sedrem.html. Accessed 8 Sep 2004.

Ingersoll CG, Dillon T, Biddinger RG, editors. 1997. Ecological risk assessment of contaminated sediments. Pensacola (FL): Society of Environmental Toxicology and Chemistry (SETAC). 389 p.

Ingersoll CG, MacDonald DD, Wang N, Crane JL, Field LJ, Haverland PS, Kemble NE, Lindskoog, Severn C, Smorong DE. 2001. Predictions of sediment toxicity using consensus based freshwater sediment quality guidelines. *Arch Environ Contam Toxicol* 41:8–21.

Jonker MTO, Koelmans AA. 2002. Sorption of polycyclic aromatic hydrocarbons and polychlorinated biphenyls to soot and soot-like materials in the aqueous environment: Mechanistic considerations. *Environ Sci Technol* 36:3725–3734.

Joye SB, Hollibaugh JT. 1995. Influence of sulfide inhibition of nitrification on nitrogen regeneration in sediments. *Science* 270:623–624.

Kohn NP, Word JQ, Niyogi DK, Ross LT, Dillon T, Moore D. 1994. Acute toxicity of ammonia to four species of marine amphipods. *Mar Environ Res* 38:1–15.

Lee BG, Griscom SB, Lee JS, Choi HJ, Koh CH, Luoma SN, Fisher NS. 2000. Influences of dietary uptake and reactive sulfides on metal bioavailability from aquatic sediments. *Science* 287:282–284.

Livingstone DR, Chipman JK, Lowe DM, Minier C, Mitchelmore CL, Moore MN, Peters LD, Pipe RK. 2000. Development of biomarkers to detect the effects of organic pollution on aquatic invertebrates: Recent molecular, genotoxic, cellular and immunological studies on the common mussel (*Mytilus edulis* L.) and other mytilids. *Int J Environ Pollut* 13:57–91.

Long ER, Hong CB, Severn CG. 2001. Relationships between acute sediment toxicity in laboratory tests and abundance and diversity of benthic infauna in marine sediments: A review. *Environ Toxicol Chem* 20:46–60.

Luoma SN, Hogstrand C, Bell RA, Bielmyer GK, Galvez F, LeBlanc GA, Lee BG, Purcell TW, Santore RC, Santshi PH, Shaw JR. 2001. Biological processes. In: Andren AW, Bober TW, editors. Transport, fate and effects of silver in the environment. Pensacola (FL): Society of Environmental Toxicology and Chemistry (SETAC). p 65–96.

MacDonald DD. 1994. Approach to the assessment of sediment quality in Florida coastal waters. Volume 1, Development and evaluation of sediment quality guidelines. Final Report. Tallahassee (FL): Florida Department of Environmental Protection.

MacDonald DD, Ingersoll CG, Berger TA. 2000. Development and evaluation of consensus-based sediment quality guidelines for freshwater ecosystems. *Arch Environ Contam Toxicol* 39:20–31.

Maltby L, Kedwards TH, Forbes VE, Grasman K, Kammenga JE, Munns WR, Ringwood AH, Weis JS, Wood SN. 2001. Linking individual-level responses and population-level consequences. In: Baird DJ, Burton Jr GA, editors. Ecological variability: Separating natural from anthropogenic causes of ecosystem impairment. Pensacola (FL): Society of Environmental Toxicology and Chemistry (SETAC). p 27–82.

Mason AZ, Jenkins KD. 1995. Metal detoxification in aquatic organisms. In: Metal speciation and bioavailability in aquatic systems. New York: J Wiley. p 479–579.

Mitsch WJ, Gosselink JG. 1986. Wetlands. New York: Van Nostrand Reinhold.

Neptun KM. 2001. The riffle quality triad: a novel approach to assessing pollution-induced degradation in cobble-bottom streams [MSc thesis]. Fort Collins: Colorado State Univ.

Nord MA. 2001. Recommendations for the implementation of a national sediment quality policy in the United States. *Hum Ecol Risk Assess* 7:641–650.

[NRC] U.S. National Research Council. 1999. New strategies for America's watersheds. Washington DC: National Academy Pr.

[NRC] U.S. National Research Council. 2001. A risk management strategy for PCB contaminated sediments. Washington DC: National Academy Pr.

Parsons Engineering Science. 2002. Assessment of sediment toxicity and quality in Patrick Bayou, Segment 1006, Harris County, Texas. Final report. Howard CL, La Point TW, Horne J, editors. Austin (TX): prepared for Patrick Bayou TMDL Lead Organization, submitted to Texas Commission on Environmental Quality.

Persaud D, Jaagumagi R, Hayton A. 1993. Guidelines for the protection and management of aquatic sediment quality in Ontario. Toronto: Ontario Ministry of the Environment and Energy.

Rainbow PS. 1997. Trace metal accumulation in marine invertebrates: Marine biology or marine chemistry? *J Mar Biol Assoc UK* 77:195–210.

Rankin KL, Chant R, Glenn S, Bruno MS, Styles R, Fullerton B, Harrington TO, Hires RI. 2001. Integrated collection of hydrodynamics and surface/water quality data in NY Harbor. Presented at the SETAC 22nd Annual Meeting; 2001 Nov 11–15; Baltimore, MD, USA.

Ringwood AH, Conners DE, Keppler CJ, DiNovo AA. 1999. Biomarker studies with juvenile oysters (*Crassostrea virginica*) deployed in-situ. *Biomarkers* 4:400–414.

Ringwood AH, Keppler CJ. 2002. Water quality variation and clam growth: is pH really a non-issue in estuaries? *Estuaries* 25:901–907.

Ringwood AH, DeLorenzo ME, Ross PE, Holland AF. 1997. Interpretation of Microtox solid-phase toxicity tests: The effects of sediment composition. *Environ Toxicol Chem* 16:1135–1140.

Rosenberg R. 1976. Benthic faunal dynamics during succession following abatement in a Swedish estuary. *Oikos* 27:414–427.

Rysgaard S, Thastum P, Dalsgaard T, Christensen PB, Sloth NP. 1999. Effects of salinity on NH^{4+} adsorption capacity, nitrification, and denitrification in Danish estuarine sediments. *Estuaries* 22:21–30.

Sanders HL. 1956. Oceanography of Long Island Sound, 1952-1954. X. The biology of marine bottom communities. *Bull Bingham Oceanogr Coll* 15:345–414.

Sanger DM, Holland AF, Scott GI. 1999. Tidal creek and salt marsh sediments in South Carolina coastal estuaries: I. Distribution of trace metals. *Arch Environ Contam Toxicol* 37:445–457.

[SFT] Norwegian Pollution Control Authority. 2002. Appendix 1 to the Activity Regulations: Requirements for environmental monitoring of the petroleum activity on the Norwegian Continental Shelf. Available from: http://www.npd.no/regelverk/12002/Aktivitetsfoiskriften_e.pdf. Accessed 9 Sep 2004.

Styles R, Chant R, Glenn S, Rankin K, Pence A, Burke P, Herrington T, Bruno MS, Haldeman C. 2001. Particle concentration and size distributions: implications for contaminant transport in NY-NJ Harbor New York Bight. Presented at the SETAC 22nd Annual Meeting; 2001 Nov 11-15; Baltimore, MD, USA.

Summers JK, Macauley JM, Engle VD, Brooks GT, Heitmuller PT, Adams AM. 1993. Lousianian province demonstration report EMAP - estuaries - 1991. Washington DC: US Environmental Protection Agency, Office of Research and Development. EPA/629/R-93/XXX.

[TNO] TNO Institute of Environmental Sciences, Energy Research and Process Innovation. 1999. Assessment of sediment contamination and biological effects around former OBM drilling locations on the Dutch Continental Shelf. Apeldoorn (NL): TNO-MEP. Report R-98/515.

Tucker KA, Burton Jr GA. 1999. Assessment of nonpoint-source runoff in a stream using in situ and laboratory approaches. *Environ Toxicol Chem* 18:2797–2803.

[UFI] Upstate Freshwater Institute. 2002. Onondaga Lake. Available from: http://www.upstatefreshwater.org/html/onondaga_lake.html. Accessed 11 August 2004.

[UKOOA] United Kingdom Offshore Operators Association. 2002. UKOOA Drill Cuttings Initiative final report. Available from: http://www.ukooa.co.uk/issues/drillcuttings/rdintro.htm. Accessed 3 Aug 2004.

[USEPA] US Environmental Protection Agency. 1994. Methods for assessing the toxicity of sediment-associated contaminants with estuarine and marine amphipods. Washington DC: USEPA Office of Research and Development. EPA/600/R-94/025.

[USEPA] US Environmental Protection Agency. 1995. Phase I inventory of current EPA efforts to protect ecosystems: Onondaga Lake. Available from: http://www.epa.gov/ecoplaces/part2/region2/site20.html. Accessed 11 Aug 2004.

[USEPA] US Environmental Protection Agency. 2001a. Methods for collection, storage and manipulation of sediments for chemical and toxicological analyses: Technical manual. Washington DC: USEPA Office of Science and Technology, Office of Water. EPA-823-B-01-002.

[USEPA] US Environmental Protection Agency. 2001b. Method for assessing the chronic toxicity of marine and estuarine sediment-associated contaminants with the amphipod *Leptocheirus plumulosus*. Washington DC: USEPA Office of Science and Technology, Office of Water. EPA 600/R-01/020.

[USEPA] US Environmental Protection Agency. 2002. Draft strategy for water quality standards and criteria: Strengthening the foundation of programs to protect and restore the nation's waters.Washington DC: USEPA Office of Science and Technology, Office of Water. EPA-825-R-02-001.

[USEPA/USACE] US Environmental Protection Agency/US Army Corps of Engineers. 1991. Evaluation of dredged material proposed for ocean disposal: Testing manual. Washington DC: USEPA Office of Water. EPA-503/8-91/001.

[USEPA/USACE] US Environmental Protection Agency/US Army Corps of Engineers. 2001. Methods for assessing the chronic toxicity of marine and estuarine sediment-associated contaminants with the amphipod *Leptocheirus plumulosus*. Washington DC: USEPA Office of Research and Development. EPA/600/R-01/020.

Van Dolah RF, Hyland JL, Holland AF, Rosen JS, Snoots TR. 1999. A benthic index of biological integrity for assessing habitat quality in estuaries of the southeastern USA. *Mar Environ Res* 48:269–283.

Wang F, Chapman PM. 1999. The biological implications of sulfide in sediment: A review focusing on sediment toxicity. *Environ Toxicol Chem* 18:2526–2532.

Warren LA, Tessier A, Hare L. 1998. Modelling cadmium accumulation by benthic invertebrates in situ: The relative contributions of sediment and overlying water reservoirs to organism cadmium concentrations. *Limnol Oceanogr* 43:1442–1454.

Watzin MC, McIntosh AW, Brown EA, Lacey R, Lester DC, Newbrough KL, Williams AR. 1997. Assessing sediment quality in heterogeneous environments: A case study of a small urban harbor in Lake Champlain, Vermont, USA. *Environ Toxicol Chem* 16:2125–2135.

Weisberg SB, Ranasinghe JA, Dauer DM, Schaffner LC, Diaz RJ, Frithsen JB. 1997. An estuarine benthic index of biotic integrity (B-IBI) for Chesapeake Bay. *Estuaries* 20:149–158.

Windom HL, Schropp SJ, Calder FD, Ryan JD, Smith RG Jr, Burney LC, Lewis FG, Rawlinson CH. 1989. Natural trace metal concentrations in estuarine and coastal marine sediments of the southeastern United States. *Environ Sci Technol* 23:314–320.

Word JQ, Bodensteiner S, Word LS. 2001. Results from biological and chemical evaluation of testing in Bahia Lagoon, CA. Unpublished report. Sequim (WA): MEC Analytical Systems.

Chronology of the development of sediment quality assessment methods in North America 8

ROBERT M ENGLER, EDWARD R LONG, RICHARD C SWARTZ, DOMINIC M DI TORO, CHRISTOPHER G INGERSOLL, ROBERT M BURGESS, THOMAS H GRIES, WALTER J BERRY, G ALLEN BURTON, THOMAS P O'CONNOR, PETER M CHAPMAN, L JAY FIELD, LINDA M PORERSKI

During the last several decades, sediment contamination has generated concerns that sediment management activities such as dredging and disposal may adversely affect water quality and important biological resources such as benthic organisms. Because many of the nation's waterways are located in agricultural, industrial, and urban areas, sediments are often contaminated with wastes and runoff from these sources. A number of tools have been developed for the assessment of these sediments. This chronology briefly summarizes the development of two of these tools, sediment toxicity tests and numerical chemical sediment quality guidelines (SQGs). Any chronological account of the development of these tools is complicated by their evolution along several simultaneous paths. Toxicity tests have involved assays of both whole sediments and extracts (e.g., elutriates, pore waters, and solvent extracts). SQGs have involved background concentrations, concentrations derived from sewerage assessments, biological–chemical empirical and mechanistic (or theoretical) approaches, and most recently, consensus-based approaches. This discussion documents the chronology of the development of sediment quality assessment in North America by first setting the regulatory context for sediment assessment and then following the different paths pursued in the development of assessment tools. Significant events in this chronology are outlined in Table 8-1.

Wenning RJ, Batley GE, Ingersoll CG, Moore DW, editors.
 ISBN 1-880611-71-6

Table 8-1 Significant events in chronology of development of sediment quality assessment methods and tools

Year	Event
1863	1st aquatic toxicity test by Penny and Adams
1890	Section 10 of Rivers and Harbors Act enacted
1937	Publication of culture methods for invertebrate animals by Galtsoff et al.
1960	First marine sediment bottle toxicity test using *Capitella capitata* under field exposure conditions by Reish and Barnard
1967	DeFalco's prediction of New York Harbors future sediment and water quality problems: *The Septic Tank of the Megalopolis*
1971	1st freshwater whole-sediment toxicity test by Gannon and Beeton
1971	Jensen criteria based upon background approach adopted for evaluations of dredge materials (US Army Office, Chief of Engineers)
1970s	Dredged Material Research Program (DMRP) by US Army Office Corps of Engineers (USACE)
1972	Marine Protection, Research, and Sanctuaries Act (MPRSA) and Federal Water Pollution Control Act (FWPCA) enacted by Congress
1973	Final regulations for ocean disposal of dredged material issued in *Federal Register*
1977	Ocean disposal regulations and criteria issued by USEPA/USACE
1977	Green book marine dredging by USEPA and USACE
1977	Freshwater sediment toxicity test methods published by Wentsel et al.
1979	First marine whole-sediment toxicity tests by Swartz et al (1979)
1981	First multigeneration study of the influence of Cr species on reproductive potential of polychaetes by Oshida et al.
1982	Relative performance in several candidate toxicity tests compared in Puget Sound by Chapman et al.
1983	Planning meeting held in Seattle at USEPA Region 10 to evaluate several candidate approaches to development of national sediment criteria
1983	EqP-approach for guidelines for nonpolar organics to develop national sediment criteria published by Pavlou and Weston
1984	Freshwater sediment methods by Nebeker et al.
1984	1st SETAC Pellston workshop by Dickson et al. (1987)
1985	1st toxicity-based equilibrium partitioning model by Adams et al.
1985	Sediment Quality Triad approach in Puget Sound by Long and Chapman
1985	Echinoderm embryo test adapted for use in testing sediments by Dinnel and Stober

Table 8-1, *contd.*

Year	Event
1985	Microtox tests of sediment extracts developed in Puget Sound by Schiewe et al.
1986	Screening-level concentrations (SLCs) using benthic community and sediment chemistry data published by Neff et al.
1986	AET values first developed for Puget Sound published by Beller et al., first estimation of predictive ability
1987	ASTM Subcommittee E47.03 on Sediment Toxicology established
1988	Revised AET values for Puget Sound published by Barrick et al.
1988	Puget Sound Dredged Disposal Analysis Program (PSDDA) begins using AET-based SQGs for dredged material evaluations
1989	NRC report on contaminated sediments is published
1989	USEPA Science Advisory Board (SAB) reviews AET approach or regional regulatory use
1990	First ASTM standards published dealing with marine and freshwater toxicity testing and methods for collection and manipulation of sediments
1990	SAB report issued by USEPA on the evaluation of the EqP approach
1990	Initial set of effects-range SQGs published by Long and Morgan for NOAA
1990	SEM/AVS approach for Cd published by Di Toro et al.
1990	First SETAC short course describing freshwater sediment toxicity testing methods
1991	First SETAC short course comparing various approaches for developing SQGs including EqP, ERL, ERM, and AET
1991	Draft sediment quality criteria issued by USEPA
1991	Washington (USA) State adopts Sediment Management Standards rule that includes chemical (AET-based) and biological criteria and publishes guidance manuals for both sediment site cleanups and source control activities
1991	Technical basis for criteria for nonionic organic chemicals published by Di Toro et al.
1991	Panel of experts convened in Richmond to guide the state of California on development of objectives by Lorenzato et al.
1991	Draft TIE methods guidance manual for freshwater sediments published by USEPA
1991	Development and implementation of a chronic sediment toxicity test using juvenile polychaetes by Johns et al.
1992	Relative sensitivities of marine porewater and whole-sediment tests compared by Carr and Chapman
1992	Marine toxicity methods manual issued by Environment Canada

Table 8-1, *contd.*

Year	Event
1992	Long and Morgan report reprinted
1992	SLC values for Ontario published by Persaud et al.
1992	SAB report issued by USEPA on EqP approach for nonionic organic contaminants
1993	SETAC short course describing approaches for sediment assessments as part of the USEPA Great Lakes National Program Office
1993	Draft criteria for 3 PAHs and 2 pesticides issued by USEPA; later rescinded
1994	Toxicity and bioaccumulation methods manual for freshwater and marine sediment by USEPA
1994	Final set of saltwater effects-level values prepared as SQAGs for the state of Florida by MacDonald 1994
1995	Human Reporter Gene System (HRGS) assay of sediment extracts developed for application in monitoring by NOAA by Anderson et al. (1995)
1995	SETAC short course describing USEPA freshwater sediment toxicity and bioaccumulation testing methods
1995	SAB report issued by USEPA on EqP approach for 5 metals
1995	Revised set of saltwater, effects-range SQGs published by Long et al.
1995	Sigma PAH approach to derivation of SQGs published by Swartz et al.
1995	Canadian protocol published for freshwater and marine SQGs based on modified NSTP and SSTT approaches
1996	Saltwater effects-level TELs and PELs published for Florida by MacDonald et al.
1996	Freshwater SQGs by Ingersoll et al.
1996	National freshwater and marine SQGs for Canada published by Smith et al.
1996	TIE methods guidance manual for marine toxicity published by Burgess et al.
1997	SETAC short course on use of SQGs
1997	Freshwater toxicity testing methods manuals issued by Environment Canada
1997	Second SETAC Pellston workshop on sediment quality by Ingersoll et al.
1997	First National Sediment Quality Inventory issued by USEPA
1998	Comparison of predictive ability of ERL/ERM and TEL/PEL values and mean SQG quotients by Long et al.
1998	Predictive ability of SEM:AVS and empirically derived SQGs by Long et al.
1998	Inland testing manual for sediments issued by USEPA/USACE
1999	Logistic regression model of exposure–response relationships published by Field et al., updated in 2002

Table 8-1, *contd.*

Year	Event
1999	Consensus-based PAH guidelines published by Swartz
1999	Exposure–response relationships for revised SQG quotients and benthic impacts published by Hyland et al.
2000	Toxicity and bioaccumulation methods manual for freshwater sediment by USEPA updated to describe chronic methods for amphipods and midge
2000	USEPA Office of Science and Technology prepares draft manuals on technical basis and procedures for derivation of EqP-derived sediment guidelines
2000	Consensus-based SQGs for freshwater sediments published by MacDonald et al.
2000	Consensus-based SQGs for PCBs published by MacDonald et al.
2000	Probabilities of toxicity predicted by SQGs published by Long et al.
2001	Chronic toxicity testing manual for *Leptocheirus plumulosus* by USEPA and USACE
2001	Exposure–response relationships for revised SQG quotients and amphipod survival published by Fairey et al.
2001	Predictive ability of freshwater SQGs by Ingersoll et al.
2002	Freshwater, effects-based SQGs for Florida by MacDonald et al.
2002	Use of SQGs and a WOE approach for assessing sediment injury relative to Department of the Interior regulations by MacDonald et al.
2002	Third SETAC Pellston workshop on sediment quality assessments by Wenning and Ingersoll
2002	USEPA Office of Research and Development prepares manuals on procedures for deriving EqP-based sediment benchmarks
2002	Expanded exposure–response relationships in benthic estuarine communities by Hyland et al. (2003)

Background

In the late 19th century, litigation prompted the United States (US) Congress to enact Section 10 of the Rivers and Harbors Act of 1890, prohibiting any obstruction to the navigable capacity of US waters (Ablord and O'Neill 1976). The jurisdiction of the US Army Corps of Engineers (USACE) was limited to the protection of navigation (Ablord and O'Neill 1976). The authority to implement this act through a regulatory permit program was vested in the Secretary of the Army acting through the Chief of Engineers. As discussed by Ablord and O'Neill (1976), the USACE limited its jurisdiction to the protection of navigation and limited its review of proposed activities to only those effects. Not until the late 1960s did the USACE expand the scope of its review of permit applications to include fish and wildlife, conservation, pollution, esthetics, ecology, and the general public interest (CFR 1968). At about the same time, the National Environmental Policy Act of 1969 (NEPA) required public interest review and preparation of an environmental impact statement (EIS) where activities had the potential to significantly affect the quality of the human environment (Ablord and O'Neill 1976).

Defalco (1967) predicted that the harbors and estuaries associated with large cities such as New York, New York; Los Angeles–Long Beach, California; and San Diego, California, USA would become increasingly contaminated as a result of transporting water quality problems from streams and rivers to coastal environments. As this prediction came to fruition and because of the need to maintain navigable depths in the nation's waterways by dredging, most of the initial concern regarding contaminated sediments centered on aquatic disposal of contaminated sediments. The responsibility for development of ecological criteria and guidelines regulating dredging and dredged material disposal was legislatively assigned to the US Environmental Protection Agency (USEPA) in consultation and conjunction with the USACE. Public Laws 92-532 (Marine Protection, Research, and Sanctuaries Act [MPRSA] of 1972) and 92-500 (Federal Water Pollution Control Act [FWPCA] Amendments of 1972) required the USEPA and USACE to participate in developing guidelines and criteria for regulating dredged and fill material discharges (USEPA/USACE 1973; US House of Representatives 1973; Brannon 1978). Other environmental controls included ratification of international treaties involving the control of pollution of the Great Lakes and the oceans by incorporating international concerns into environmental legislation and subsequent sediment assessment and management guidelines.

Before about 1970, the only quality assessment control of dredging, construction, and related activities was under the Rivers and Harbors Act of 1899. In the late 1960s, concern increased over possible environmental problems related to dredging. Concerns over dredged material disposal in the Great Lakes region resulted in the Federal Water Quality Administration (FWQA, predecessor of USEPA) and the USACE Buffalo District requesting the initiation of studies on the chemical characteristics of selected Great Lakes harbors. These harbor sediments were analyzed by using methods developed to characterize municipal and industrial wastes rather than sediments. Consequently, many harbors were erroneously characterized. Inadequacy of data led to revisions in the Rivers and Harbors Act of 1970 (Jones and Lee 1978), and special provisions were made for dredged material disposal activities in the Great Lakes.

The Rivers and Harbors Act also authorized the USACE to initiate a comprehensive evaluation of the environmental effects of dredged material disposal (Boyd et al. 1972) through the Dredged Material Research Program (DMRP). The DMRP was the first program to address national issues regarding the assessment and management of the disposal of dredged sediments in compliance with the newly enacted Clean Water and Ocean Dumping Acts. The DMRP was a 5-year program that was conducted from 1973 to 1978. It assessed aquatic, upland, and wetland beneficial uses and regulatory aspects of dredged material management. The DMRP formed the technology foundation for dredged material testing and assessment required by domestic legislation and by international treaty (The London Convention) that regulated ocean disposal. The DMRP also established the technology necessary to develop techniques for identifying, assessing, and managing contaminated dredged material. It took the national lead in wetlands management and restoration. The key outcome of the DMRP was that there was no single a priori solution to dredged material management; instead, there were a range of management alternatives that should be considered and evaluated for any given project. As a product of the regulatory research and legislative requirements, the USACE and USEPA initiated development of a multiple lines of evidence (LOEs), effects-based assessment approach that emphasized toxicity and bioavailability coupled with chemical and physical components in their regulatory program.

In the early 1980s, USEPA began to address the potential impact of contaminated sediments to benthic aquatic organisms in all water body and sediment types, not just harbors and dredged material. Contaminated sediments were identified as posing major and multiple risks to the beneficial uses of aquatic ecosystems throughout the Great Lakes basin, including 42 of 43 areas of concern (AOCs) identified by the International Joint Commission (IJC 1988). Contaminated sediments were linked to high levels of tumors and other abnormalities in fish and to degradation of benthos that influenced populations of fish and wildlife. The imposition of fish consumption advisories adversely affected commercial, sport, and food fisheries in many areas. Contaminated sediments also threatened the viability of many commercial ports through the imposition of restrictions on dredging of navigational channels and disposal of dredged materials. In response to the concerns that were raised regarding contaminated sediments, the USEPA launched the Assessment and Remediation of Contaminated Sediments (ARCS) Program in 1987 to support the assessment and management of contaminated sediments in the Great Lakes basin. The information that was generated under the ARCS program provided subsequent guidance for designing and implementing investigations at sites with contaminated sediments (USEPA 1994c; MacDonald and Ingersoll 2002a, 2002b).

Initial Sediment Quality Guidelines

The earliest SQGs (initially proposed as "criteria") for dredged sediment were promulgated in 1971 by the FWQA (USEPA) for the Great Lakes and were referred to as "the Jensen Criteria." These guidelines were derived from sewerage assessments, were not based upon measures of biological effects, and were named after the USEPA Headquarters manager who signed the criteria memorandum that resulted in their promulgation. They initially were applied to assess sediment contamination in the belief that they could be extrapolated to harbor sediments that had traditionally received effluents from sewage discharges. Lacking any other guidance, the FWQA (USEPA) in 1971 promulgated these criteria in a memorandum. That same year, USACE issued a circular (US Army Office, Chief of Engineers 1971) stating that the Jensen Criteria should be applied to sediments dredged from all US waters. Seven chemical constituents with concentration limits were specifically mentioned in the total-sediment (Jensen) criteria, including chemical oxygen demand (COD), total Kjeldahl nitrogen (TKN), volatile solids, oil and grease, mercury, lead, and zinc (Table 8-2). The limits were total concentrations based on the dry weight of sediment. If the concentration of any constituent exceeded the numerical limit, the material was classified as polluted and unacceptable for open-water disposal. Although the criteria were not specifically limited to the 7 constituents for which limits had been established, implementation was almost exclusively based on these criteria.

Table 8-2 Dredged material disposal criteria developed for FWQA 1971[a]

Parameter	Maximum % dry wt
Volatile solids	6.0
Chemical oxygen demand (COD)	5.0
Total Kjeldahl nitrogen (TKN)	0.10
Oil and grease	0.15
Mercury	0.0001[b]
Lead	0.005
Zinc	0.005

[a] US Army Office, Chief of Engineers 1971.
[b] A value of 0.001, which was a typographical error, appeared in the original report. These criteria are presented because of their historical importance and are not recommended for current use as SQGs.

General opposition to the Jensen Criteria developed as technical flaws became apparent. Specifically, the procedures did not account for the geochemical location of contaminants in the dredged material, did not address the potential availability of contaminants to organisms, and did not consider background concentrations of the same constituents. The procedures prescribed for use with the criteria provided only an inventory of the total amount of each constituent contained in the sediment. Table 8-3 presents an updated approach to "bulk" or Jensen Criteria later used by USEPA Region V (Bowden 1977). An application column lists the average earth's crustal abundance for comparison to the "polluted" categories. Eight parameters classified as "heavily polluted" were actually less than natural abundance. Because of these deficiencies, the MPRSA

Table 8-3 Region V sediment criteria 1977[a,b]

	Nonpolluted	Moderately polluted	Heavily polluted	Average earth's crustal abundance[c]
Volatile solids	<5%	5–8%	>8%	
COD	<40,000	40,000–80,000	>80,000	
TKN	<1000	1000–2000	>2000	
Oil and grease (hexane solubles)	<1000	1000–2000	>2000	
Lead	<40	40–60	>60	16
Zinc	<90	90–200	>200	80
Mercury	<1.0	NA	>1.0	0.5
Ammonia	<75	75–200	>200	
Cyanide	<0.10	0.10–0.25	>0.25	
Phosphorus	<420	420–650	>650	1200
Iron	<17,000	17,000–25,000	>25,000	50,000
Nickel	<20	20–50	>50	100
Manganese	<300	300–500	>500	1000
Arsenic	<3	3–8	>8	5
Cadmium	[d]	[d]	>6	0.2
Chromium	<25	25–75	>75	200
Barium	<20	20–60	>60	430
Copper	<25	25–50	>50	70

[a] Bowden 1977

[b] All ranges in mg/kg dry weight unless otherwise noted.

[c] After Goldschmidt 1954.

[d] Lower limits not established.

These criteria are presented because of their historical importance and are not recommended for current use as SQGs.

and FWPCA directed USEPA to develop regulatory criteria and guidelines in consultation and conjunction, respectively, with the USACE.

Final regulations and criteria controlling ocean disposal of dredged sediments were published by the USEPA in 1973 (*Federal Register* 1973). Chemical procedures for assessing the suitability of dredged sediments for ocean disposal consisted primarily of elutriate tests (USACE 1976), which replaced whole sediment analyses. This procedure was used to address short-term water quality impacts but not longer-term benthic impacts. Tests of toxicity and bioaccumulation were recommended only in general terms.

The MPRSA further required that the criteria for ocean disposal be updated at least every 3 years. The first updated criteria published in 1977 (*Federal Register* 1977) accounted for provisions of the Convention on the Prevention of Marine Pollution by Dumping of Wastes and Other Matter and reflected legal challenges by the National Wildlife Federation as to the adequacy of the 1973 criteria (USEPA 1977; *Federal Register* 1977). The MPRSA 1977 regulations led to the first dual agency implementation manual (USEPA/USACE 1977) to assess dredged material for ocean disposal. USEPA/USACE (1977) describes methods for conducting whole sediment toxicity tests and bioaccumulation tests and describes methods for estimating mixing at the disposal site. Sediment quality was assessed by means of liquid-phase, suspended-particulate, and whole-sediment tests, along with chemical analyses of the liquid phase. These first whole-sediment toxicity tests were based on exposure of test organisms to layered sediments (surface layer of dredged material overlaying a subsurface layer of disposed or disposal site sediments). The impact of chemical constituents to the water column was addressed by comparing their elutriate concentrations with appropriate water quality criteria (WQC) after taking initial mixing into account or through use of a liquid-phase toxicity test. These protocols were further revised to reflect the advances in science and experience with regard to toxicity and bioaccumulation assessment and the use of highly developed disposal, mixing, and transport models. This manual contains critical guides to organism selection, experimental design, statistics, model input, chemistry, and quality assurance–quality control (USEPA/USACE 1991).

Interim and final guidelines for implementation of the Public Law 92-500 (Clean Water Act) were published in the *Federal Register* Vol. 45, No. 249 on 5 September 1975 (Engler 1980). The guidelines required the proposed discharger to consider physical effects (especially impacts on wetlands) and chemical–biological interactive effects and to conduct a thorough site-selection assessment. Assessment of chemical water-column effects was by means of the elutriate test. The permitting authority also could specify that the applicant conduct elutriate and whole-sediment toxicity tests on a case-by-case basis. The authority could also require total sediment–chemical analyses and/or benthic community structure analyses when reviewing alternative sites. This regulation was revised in 1981 using the same testing protocol, and an implementation manual using the ocean manual as a format was published for inland waters in 1998 (USEPA/USACE 1998; see further description on p 326). Regulations on dredged material always have and continue to require laboratory tests of sediment toxicity and bioaccumulation.

In 1981, sediment quality indicators were proposed by the Region 6 office of USEPA (Phillips 1981) whereby interstitial (or pore) water contaminant levels were compared to the USEPA WQC. This approach suggested that interstitial pore water was an important route of uptake by aquatic organisms and used the extensive toxicological database compiled by USEPA and incorporated into derivation of the WQC. Although the database was comprised primarily of information from tests with water column organisms, the comparison was suggested as valid on the basis of water–sediment interactions. The proposed approach, however, was rejected by Hernandez (1981). Although this concept was not adopted, in 1990 the USEPA Criteria and Standards Division began a sediment criteria development program that focused on equilibrium partitioning (EqP) of contaminants between particulate and interstitial water phases of sediments. This approach also used the ambient WQC as a comparison, but sediment concentrations were normalized with sediment–water partitioning coefficients (USEPA 1990a). The relative bioavailability of nonionic organic compounds (e.g., DDT or Kepone) in sediments was shown to be driven by total organic carbon (TOC) concentrations (USEPA 1992). More recently, it has been demonstrated that acid volatile sulfide (AVS) is one of a several sediment components (TOC is another) that controls metal bioavailability (USEPA 1995; Chapter 13). These refinements of the original interstitial water concept significantly improved the reproducibility and effects validation of the approach. In the early 1990s, the first set of draft EqP-based sediment criteria were issued (USEPA 1991a–c).

During this same period, the reference element approach was developed to provide an approach for assessing metal contamination in estuarine sediments (e.g., Schropp et al. 1990). This procedure relies on normalization of metal concentrations to a reference element. Normalization of metal concentrations to concentrations of aluminum in estuarine sediments provided the most useful method of comparing metal levels on a regional basis in Florida estuaries.

However, normalization using lithium, iron, or other reference elements has been used in other estuarine regions (Loring 1991).

Sediment Toxicity Tests

Derivation of effects-based SQGs was not possible until toxicity tests were developed in the laboratory and field for testing the toxicity of sediments. Historically, the evaluation of contaminant effects in receiving systems emphasized analyses of surface waters, not sediments. Most assessments of water quality focused on water-soluble substances, and sediments were considered to be a safe repository of sorbed contaminants. This approach emphasized testing organisms in the water column without considering the fate of chemicals in sediment.

Various methods have been developed to evaluate sediment toxicity. These procedures range in complexity from short-term lethality tests that measure effects of individual contaminants on single species to long-term tests that determine the effects of chemical mixtures on the structure and function of communities. The evaluated sediment phase may include whole sediment, suspended sediment, elutriates, or sediment extracts. The amount of sediment tested can range from a few milliliters to more than 800 liters. The test organisms can include algae, macrophytes, and fishes, and benthic, epibenthic, and pelagic invertebrates (Ingersoll 1995). Sediment toxicity tests were developed to determine if sediment is harmful to benthic organisms under controlled laboratory conductions using either field-collected sediment or chemicals spiked into sediment. Toxicity testing of sediment has been used to

- determine the relationship between toxic effects and bioavailability,
- investigate interactions among contaminants,
- determine spatial and temporal distribution of contamination,
- evaluate hazards of dredge material,
- rank areas for cleanup, and
- estimate the effectiveness of remediation and management.

Sediments spiked with known concentrations of contaminants can be used to establish cause-and-effect relationships between chemicals and biological responses.

Some initial attempts were made to adapt tests previously developed for effluents and ambient water (e.g., daphnids, mollusk embryos, urchin gametes) or pure chemicals (Microtox) to tests of sediment. In 1967, Reish and Barnard conducted laboratory tests with various aqueous phases or extracts of sediments and compared results with evidence of degradation from community structures of benthic communities. Reish and Lemay (1988) suggested that these associations might be tighter if whole (whole-sediment) sediment toxicity tests were performed.

During the late 1970s and early 1980s, research began to focus on measures of adverse biological effects associated with and predicted by the contamination of in-place sediments. Most methods developed in the 1970s and 1980s and applied over the past 30 years have relied on relatively short-term ($\leq$10 d) exposures. The first whole-sediment toxicity tests were developed for freshwater sediments by Gannon and Beeton (1971) and for marine sediments by Swartz et al. (1979; Table 8-1). Gannon and Beeton (1971) assessed 24- to 48-h survival of the benthic amphipods *Pontoporeia* and *Gammarus* and the midge *Chironomus* in exposures to sediments from Great Lakes harbors. Subsequently, Prater and Anderson (1977a) conducted sediment toxicity testing with a combination of benthic and water-column organisms, under the premise that sediment reworking, particularly by hexagenid mayflies, would maximize exposure of the water column organisms to sediment contaminants. This approach, using both *Hexagenia limbata* and *Daphnia magna*, was used by Malueg et al. (1984) together with sediment chemistry and benthic community structure, to characterize contaminated sediments in Michigan but was later replaced by static tests using benthic species such as *Chironomus tentans* (Wentsel, McIntosh, Anderson 1977; Wentsel, McIntosh, Atchison 1977; Wentsel et al. 1978).

In 1980, the now commonly used freshwater benthic aquatic oligochaete *Lumbriculus variegates* was first used in sediment toxicity tests (Bailey and Liu 1980). In 1984, Nebeker et al. published a methodology for conducting acute sediment toxicity tests with *Hyalella azteca* and *C. tentans*. Milbrink (1987) first recommended the aquatic oligochaete *Tubifex tubifex* as a test species.

A major impetus for developing toxicity tests for sediments was the necessity to test prospective dredged materials before permits were issued. A sequence of manuals was published for implementation of tests for ocean or freshwater disposal. The first inland manual was issued in 1976 and was required by the 1975 CWA Dredged and Fill Material Guidelines as a USACE product. It was rudimentary and presented simple mixing models, sedimentary chemistry, algal assays for the water column, and a discussion of the developing nature of benthic assays. Because it was not a joint USEPA/USACE product and not terribly informative, it was seldom used. The 1981 CWA Guidelines prescribed joint approaches, but none were initiated until early in the1990s. The initial ocean disposal manual (the so-called "Green Book" issued in 1977 by the USEPA/USACE) mandated "effects-based" testing and included a number of toxicity and bioaccumulation tests.

One of the most important developments in sediment toxicity testing was the use of infaunal amphipods in 10-d static tests. These tests were developed primarily in response to the US regulatory requirements for evaluating the suitability of dredged material for in-water disposal. Specifically, the 1977 Ocean Dumping Regulations and Criteria (USEPA 1977a, 1977b) required testing with at least 3 species: a filter-feeder, a deposit-feeder, and a burrower. Species recommended for testing included a crustacean, an infaunal bivalve, and an infaunal polychaete. Among these three, "infaunal amphipods seem to be among the most sensitive crustaceans and, for this reason, are among the preferred organisms for solid phase bioassays" (USEPA/USACE 1977). Two amphipod taxa were recommended in 1977: *Ampelisca* spp. and *Paraphoxus* spp. (the latter genus was later renamed *Rhepoxynius*). Initially, testing was to be conducted in aquaria under flow-through conditions, with test material deposited on top of the organisms.

Swartz et al. (1979) published additional details on testing methods for amphipods with *Paraphoxus epistomus* the sole crustacean in a suite of 5 different species tested in 25-L aquaria. Three years later, Swartz et al. (1982) published a key paper that described broad-scale testing of sediments at a contaminated marine site (Commencement Bay, Washington, USA) with this crustacean, which had now been taxonomically reclassified as *Rhepoxynius abronius*. The objective of this study was not to classify dredged material but rather to test large numbers of samples quickly; thus this testing was carried out in 1-L beakers under static conditions. The 10-d sediment toxicity test with *R. abronius* (Swartz et al. 1985) was subjected to interlaboratory calibration by Mearns et al. (1986). This work provided the basis for other amphipod toxicity tests now conducted routinely using similar methods: *H. azteca* in freshwater (Nebeker et al. 1984), *Ampelisca abdita* (Scott and Redmond 1989) and *Eohaustorius estuarius* in estuarine waters (DeWitt et al. 1989), and *Leptocheirus plumulosus* also in marine waters (Schlekat et al. 1992; McGee et al. 1993). Significantly, acute toxicity tests performed with benthic amphipods were incorporated into and required in testing of prospective dredge materials and therefore became a universal tool in evaluations of sediment quality in 1991 (USEPA/USACE 1991).

In the 1970s and 1980s, investigations of sediment toxicity and bioaccumulation were limited by a lack of available standardized methods that limited the use of sediment tests in contamination assessments. The American Society for Testing and Materials (ASTM) International established Subcommittee E47.03 on Sediment Toxicology in May of 1987 with the goal of developing standard guides for assessing the bioavailability of contaminants associated with sediments. The intent of developing these standards was to provide methods that could be used to evaluate the toxicological hazard of contaminated sediment, soil, sludge, drilling fluids, and similar materials. The subcommittee began meeting biannually in the spring with ASTM International Committee E47 on Biological Effect and Environmental Fate and in the fall at the annual meeting of the Society of Environmental Toxicology and Chemistry (SETAC). Importantly, 30 to 50 individuals interested in assessment and management of contaminated sediments attending these subcommittee meetings were provided a forum to discuss development of standard methods through a consensus based process that was independent of their organizations.

The Sediment Quality Triad concept, first applied in Puget Sound (Long and Chapman 1985), was further defined in a study conducted in San Francisco Bay (Chapman et al. 1987). Methods for reporting results of the triad of analyses were later formalized (Chapman 1990, 1996). The National Research Council (NRC 1989) convened a panel of scientists in Tampa, Florida, USA, to compile information on the assessment and remediation of contaminated sediments. Following development of initial standardized protocols for toxicity tests in Puget Sound, matching chemistry and toxicity data were compiled during the late 1980s for derivation of criteria applicable to that region. The amphipod survival, larval development, and Microtox (saline extract) tests, together with a measure of benthic community effects, were used to derive Puget Sound apparent effects thresholds (AETs). The AETs formed the basis upon which regional dredging guidelines and Washington sediment cleanup and source control criteria were based (Beller et al. 1986; Barrick et al. 1988; WAC 1991).

After 5 years of work and review during the 1990s, the first official Inland Testing Manual was released in February 1998. The structure was similar to the Ocean Manual, in which a tiered approach to assessment was used. However, it presented several technical advances over the 1991 Ocean Manual. The manual addressed the discharge of navigation dredging to waters of the US and runoff and effluent from confined disposal facilities. The Inland Manual offered improved sampling, analytical, statistical, and exposure modeling assessments, compared to the 1991 Green Book. The Inland Manual included a comprehensive discussion of test conditions and test acceptability criteria for all recommended toxicity tests that cover freshwater, brackish, and marine organisms as well as the standard approaches to implement these tests. The manual listed test organisms, covering a range of sensitivities, and recommended benchmark organisms.

In 1990, ASTM International published standards on whole-sediment toxicity tests with freshwater invertebrates (acute and chronic tests with amphipods, midge, oligochaetes, and cladocerans; ASTM 2004a) and with estuarine and marine amphipods (acute and chronic tests; ASTM 2004b). Standards were also published in 1990 for collection, storage, and manipulation of contaminated sediments (ASTM 2004c) and in 1994 on guidance for designing sediment toxicity studies (ASTM 2004d). Subsequently, ASTM International standards were developed for conducting toxicity tests with polychaete (ASTM 2004e), echinoderms (ASTM 2004f), and mollusks (ASTM 2004g). In 1996, ASTM International published the first standard for evaluating bioaccumulation of contaminants from sediment with freshwater and marine invertebrates (ASTM 2004h) based on guidance provided in USEPA (1989a). Concurrently, Environment Canada and USEPA have developed a series of freshwater and marine toxicity tests that were harmonized with these ASTM standards. The Organisation for Economic Co-operation and Development (OECD) is currently developing freshwater sediment toxicity tests with midge and bioaccumulation tests with oligochaetes that are based on these ASTM, Environment Canada, and USEPA methods (Table 8-1).

Also during the 1990s, emphasis was placed on developing methods for conducting long-term toxicity tests (28- to 60-d exposures) that assessed effects on the survival, growth, or reproduction of freshwater and marine test organisms (Dillon et al. 1993; Benoit et al. 1997; Bridges and Farrar 1997; Emery et al. 1997; Gray et al. 1998; Ingersoll et al. 1998; ASTM 2002a). Also, beginning in the 1990s, toxicity tests were used by USEPA, National Oceanic and Atmospheric Administration (NOAA), and US Geological Survey (USGS) as survey tools of ambient conditions nationwide in estuarine and freshwater monitoring programs. Information from these large-scale surveys led to the development of large databases used subsequently in the derivation and field validation of SQGs (Chapter 12). Marine sediment toxicity tests with sublethal or chronic (life-cycle) endpoints were developed and applied in research and surveys conducted by NOAA and USGS. These included the Microtox tests on solvent extracts (Schiewe et al. 1985) and a variety of tests with fish tissues and invertebrates (Chapman et al. 1982, 1983) done in Puget Sound. A juvenile polychaete growth test was developed for marine sediments (Johns et al. 1991) and was subsequently adopted for use by all Washington sediment management programs. Several statistical criteria were used to evaluate the relative performance of 5 toxicity tests in San Francisco Bay (Long et al. 1990). Methods were developed for the *Neanthes arenaceodentata* growth test (Johns et al. 1991) to provide a sublethal endpoint (ASTM 2004e). The relative sensitivities of porewater tests performed with invertebrates were compared to those of adult amphipod tests of whole-sediment sediments (Carr and Chapman 1992). Methods were developed for the Human Reporter Gene System (HRGS) assays of mixed-function oxygenase induction in exposures of liver cells to organic solvent extracts (Anderson et al. 1995). The Microtox, sea urchin, and HRGS tests were incorporated along with amphipod survival tests into NOAA's sediment quality surveys of US estuaries (Long 2000). The first estimates of the spatial scales of estuarine sediment toxicity, based upon data from these tests, were published by NOAA and collaborators (Long et al. 1996; Turgeon et al. 1998). The first comprehensive nationwide inventory of contaminated sediments, in both freshwater and saltwater, was compiled by USEPA (1997) on the basis of data from chemical analyses and toxicity tests.

Effects-Based Sediment Quality Guidelines

With new sets of toxicity testing protocols being developed and applied, attention began to turn to using effects-based chemical guidelines to predict toxicity or other adverse biological effects. Several candidate methods for the development of effects-based SQGs were evaluated, including the EqP approach (Pavlou and Weston 1983; Adams et al. 1985; Di Toro et al. 1991). Separately from the approaches based on EqP, others examined empirical relationships between biological characteristics of sediment and whole chemical concentrations in field-collected samples. With the advent of the consensus approach in the late 1990s, 3 basic families of approaches to the development of numerical SQGs were pursued by different groups to satisfy a variety of objectives. The approaches consisted of theoretical (mechanistic or EqP), empirical (or correlative), and consensus approaches (Chapters 3 and 4).

In the late 1980s, the screening level-concentration (SLC) approach (an empirical approach) was used to derive values that were protective of 90% of benthic species, using data from several US estuaries (Neff et al. 1986). Based on matching sediment chemistry, toxicity, and benthic data, the AET approach (also an empirical method) was used to derive the first set of guidelines for use in the Puget Sound (Beller et al. 1986). The guidelines were later finalized with an expanded database (Barrick et al. 1988). The final set of AETs for Puget Sound were reviewed by the Science Advisory Board (SAB) convened by USEPA (1989b). AETs were used to establish regional dredged material guidelines (PSDDA 1989; WDOE 1995), and adopted as criteria for use in sediment source control and cleanup prognosis in Washington State (WAC 1991).

The development of effects-based SQGs was started within the USEPA in the mid-1980s by a workgroup consisting of Office of Research and Development (ORD) and Office of Water (OW) scientists and OW contractors. Initially these values were called "sediment quality criteria" (SQC). The SAB of USEPA reviewed the EqP, AET, and SLC approaches as the basis for developing criteria in 1988 (USEPA 1989b, 1990a, 1990b). The bioavailability of metals in sediment was evaluated using the acid volatile/simultaneously extracted metals (AVS/SEM) approach (Di Toro et al. 1990, 1991; Chapter 13). In the early 1990s, the first set of draft, EqP-based, sediment criteria were made available for public comment (USEPA 1991a–c). The SAB reviewed the EqP approach for nonionic organics in 1992 (Di Toro et al. 1991; USEPA 1992) and the AVS/SEM approach for 5 metals in 1995 (USEPA 1995). On the basis of these reviews, the EqP approach was adopted as the preferred method for developing SQC. Also, the term "criteria" was subsequently replaced by the term "guidelines" because of the legal and regulatory context of the term "criteria" as suggested in the critical review by the SAB and comments from the public. Parallel to the SAB review process, the draft SQCs were reviewed by ORD and OW scientists and by OW contractors, and all of the documents were sent out for public comment. The benchmarks for polycyclic aromatic hydrocarbon (PAH) mixtures recently (in 2002) have been drafted and are currently undergoing external peer review. Also during the late 1980s and early 1990s, the predictive ability of the Puget Sound AETs were evaluated with estimates of their efficiency, sensitivity, and overall reliability (Barrick et al. 1988; Ginn and Pastorok 1992).

The first set of effects range low and effects range median (ERL and ERM) values based upon empirical methods were published by NOAA (Long and Morgan 1990) as aids in the interpretation of sediment chemistry data generated in the National Status and Trends Program of NOAA. These values were based on analyses of matching sediment chemistry and biological data compiled from many saltwater and freshwater studies nationwide. A panel of scientists was convened to provide scientific guidance on development of sediment quality objectives for the state of California (Lorenzato et al. 1991). SLC values, based on analyses of benthic community data, were issued for Ontario (Persaud et al. 1992). The first ERLs and ERMs were criticized for merging freshwater and saltwater data and for ignoring information in which no effects were observed. In response, revised ERL and ERM values published in 1995 were based upon a significantly expanded database in which freshwater data were excluded (Long et al. 1995). Observations of both effects and no effects were included in the calculations of the threshold effects level (TEL) and probable effects level (PEL) guidelines that were issued for the marine sediments of Florida (MacDonald 1994; MacDonald et al. 1996). In Canada, interim SQGs were developed (Smith et al. 1996) for both freshwater and marine sediments according to a national protocol issued by the Canadian Council of Ministers of the Environment (CCME 1995, 1999). Initial methods guidance manuals for TIE studies, based on analyses of pore water, were published by USEPA (USEPA 1991d; Ankley and Schubauer-Berigan 1995; Burgess et al. 1996) to aid in determining which chemicals caused toxicity in sediments.

In 1994, USEPA made available for public comment draft criteria for 3 individual PAHs and 2 pesticides (USEPA 1993a–c), reflecting the state of the science in 1994. With the addition of new data, modeling, and analysis, the 3 individual PAH SQC were combined into a PAH mixtures SQC to more appropriately capture their coexistence in nature and their additive nature as toxicants. Subsequently, methods to evaluate the toxicological significance of mixtures of PAHs were developed in the 1990s (Swartz et al. 1995). The draft criteria for the 3 PAHs issued earlier were, therefore, rescinded and included in the procedures used to develop the values for the combined PAH mixtures. Methods for deriving consensus-based values for PAHs that combined the individual SQGs from different approaches subsequently were derived (Swartz 1999). The consensus approach, therefore, became the third major family of approaches to the derivation of effects-based SQGs.

The SQGs derived with empirical approaches were criticized for not being derived with causal methods. The consensus approach combined the strengths of the theoretical (mechanistic) approaches and the real-world conditions embraced in the empirical approaches. In response to concerns regarding the predictive ability of the ERL, ERM, TEL, and PEL values, analyses were conducted to quantify the predictive ability of those particular SQGs (Long, Field, MacDonald 1998; Long, MacDonald, et al. 1998; O'Connor et al. 1998). These evaluations included the first estimates of the predictive ability of mean SQG quotients (Long, MacDonald, et al. 1998). These exposure–response relationships were recalculated with an expanded database and described as probability tables (Long et al. 2000). Subsequently, Fairey et al. (2001) reexamined the list of chemicals to be used in calculations of mean SQG quotients and identified those that provided the best match with toxicity data.

The consensus-based approach was applied to guidelines developed for total polychlorinated biphenyls (PCBs) and for freshwater SQGs (MacDonald, Di Pinto, et al. 2000; MacDonald, Ingersoll, Berger 2000). The consensus-based SQGs for PCBs were derived to identify the concentrations of these substances that constituted injury to the sediments of PCB-contaminated sites. The state of Florida issued informal guidelines for freshwater sediments based upon the procedures previously used to derive their marine SQGs (MacDonald, Ingersoll, Smorong, Lindskoog, Sloane 2002). The predictive abilities of the consensus-based SQGs were evaluated in a freshwater sediment toxicity database (Ingersoll et al. 2001). The logistic regression model approach was first applied to the prediction of marine amphipod toxicity from sediment chemistry (Field et al. 1999, 2002). By the late 1990s, the overall relationships were becoming clearer between exposure as gauged with either mean SQG quotients or numbers of SQG values exceeded and the response in laboratory toxicity test with amphipods (Long et al. 1998a; Fairey et al. 2001; Ingersoll et al. 2001). However, because these studies relied upon only data from laboratory toxicity tests, little was known of such relationships using information from benthic community analyses. Relationships between mean SQG quotients and benthic community impacts in southeastern US estuaries were examined (Hyland et al. 1999). These initial analyses were followed by a reevaluation with an expanded database that included information from Atlantic and Gulf of Mexico estuaries and Puget Sound (Hyland et al. 2003). Additional data from Biscayne Bay, Florida, USA, also suggested that benthic community responses in the lower Miami River exceeded those observed in the amphipod survival tests along a strong pollution gradient (Long et al. 2002). Collectively, the outcome of these analyses suggested that benthic community indices were affected by chemical concentrations well below those in which amphipod survival in laboratory tests was affected. However, the relative influence of natural factors (i.e., salinity, grain size, or TOC) versus chemical-induced toxicity on benthic communities has not been fully quantified (Chapter 12).

In early 2002, USEPA published the collection of sediment documents as ORD technical procedure publications to provide the user community with the latest and best information known to the agency for deriving values to assess sediment contamination (USEPA 2002a–g). Also in 2002, SQGs were used as a weight-of-evidence (WOE) approach to evaluate injury to sediments in Natural Resource Damage Assessment and Restoration cases (MacDonald, Ingersoll, Smorong, Lindskog, Sparks, et al. 2002). Current research on development of TIE procedures is focusing on whole sediments (Ho et al. 2002).

During 2002, the Water Environment Federation (WEF 2002) published a handbook on sediment quality assessments that included compilations of sampling methods, analytical procedures, and the then-current understanding of the predictive abilities of SQGs in saltwater.

The role of SETAC in these activities cannot be overstated. Relevant meetings on contaminated sediments were sponsored by SETAC in Florissant, Colorado, USA (1984; Dickson et al. 1987); Monterey, California (1995; Ingersoll et al. 1997); Pensacola, Florida (2001; Carr et al. 2003); and Butte, Montana, USA (2002; Wenning and Ingersoll 2002; this volume). The SETAC annual conferences, technical short courses, and Pellston workshops have been instrumental in furthering the development of assessment methods, in fostering open debate and discussion, and in ensuring publication of scientific findings in a peer-reviewed environment. The SETAC meetings allowed global experts from academia, industry, and government to reach points of consensus and provide recommendations on SQGs. The outcome of the Pellston workshop in 2002 undoubtedly will lead to further research on the development and field validation of effects-based SQGs. The proceedings of the workshop, contained in this publication, will serve as an important benchmark in time of the development of these and related assessment tools. Thus, this book will serve as the basis for evaluating progress made in future years.

References

Ablord CD, O'Neill BB. 1976. Wetland protection and Section 404 of the Federal Water Pollution Control Act Amendments of 1972: A Corps of Engineers renaissance. *Vermont Law Rev* 1:51-115.

Adams WJ, Kimerle RA, Mosher RC. 1985. Aquatic safety assessment if chemicals sorbed to sediments. In: Cardwell RD, Purdy R, Bahner RC, editors. Aquatic Toxicology and Hazard Assessment: 7th Symposium. Philadelphia: American Soc for Testing and Materials. ASTM STP 854. p 429–453.

Anderson JW, Rossi SS, Tukey RH, Vu T, Quattrochi LC. 1995. A biomarker, P450 RGS, for assessing the induction potential of environmental samples. *Environ Toxicol Chem* 14:1159–1169.

Ankley GT, Schubauer-Berigan MK. 1995. Background and overview of current sediment toxicity identification procedures. *J Aquat Ecosyst Health* 4:133–149.

[ASTM] American Society for Testing and Materials. 2004a. Standard test methods for measuring the toxicity of sediment-associated contaminants with freshwater invertebrates: E1706-00. In: Annual book of ASTM standards, Vol. 11.05. West Conshohocken (PA): ASTM.

[ASTM] American Society for Testing and Materials International. 2004b. Standard test method for measuring the toxicity of sediment-associated contaminants with estuarine and marine invertebrates, E-1367-03. In: Annual book of ASTM International standards, Vol. 11.05. West Conshohocken (PA): ASTM.

[ASTM] American Society for Testing and Materials. 2004c. Standard guide for collection, storage, characterization, and manipulation of sediments for toxicological testing and for selection of samplers used to collect benthic invertebrates, E1391-03. In: Annual book of ASTM standards, Vol. 11.05. West Conshohocken (PA): ASTM.

[ASTM] American Society for Testing and Materials. 2004d. Standard guide for designing biological tests with sediments: E1525-02. In: Annual book of ASTM standards, Vol. 11.05. West Conshohocken (PA): ASTM.

[ASTM] American Society for Testing and Materials International. 2004e. Standard guide for conducting sediment toxicity tests with polychaetous annelids, E1611-00. In: Annual book of ASTM International standards, Volume 11.05. West Conshohocken (PA): ASTM.

[ASTM] American Society for Testing and Materials International. 2004f. Standard guide for conducting static acute toxicity tests with echinoid embryos, E1563-98 (s00s). In: Annual book of ASTM International standards, Vol. 11.05. West Conshohocken (PA): ASTM.

[ASTM] American Society for Testing and Materials International. 2004g. Standard guide for conducting static acute toxicity tests starting with embryos of four species of saltwater bivalve molluscs, E724-98 (2002). In: Annual book of ASTM International standards, Vol. 11.05. West Conshohocken (PA): ASTM.

[ASTM] American Society for Testing and Materials. 2004h. Standard guide for the determination of bioaccumulation of sediment-associated contaminants by benthic invertebrates: E1688-00a. In: Annual book of ASTM standards, Vol. 11.05. West Conshohocken (PA): ASTM.

Bailey HC, Liu DHW. 1980. *Lumbriculus variegates*, a benthic oligochaete as a bioassay organism. In: Eaton JC, Parrish RC, Hendricks AC, editors. Aquatic toxicology. West Conshohocken (PA): American Soc Testing and Materials. ASTM STP 707. p 205–215.

Barrick R, Becker S, Brown L, Beller H, Pastorok R. 1988. Volume 1, Sediment quality values refinement: 1988 update and evaluation of Puget Sound AET. Bellevue (WA): PTI Environmental Services. EPA Contract No. 68-01-4341. PTI Contract No. C717-01.144 p.

Beller H, Barrick R, Becker S. 1986. Development of sediment quality values for Puget Sound. (Tetra Tech, Inc.) Bellevue (WA): Resource Planning Associates/ U.S. Army Corps of Engineers, Seattle District for the Puget Sound Dredged Disposal Analysis Program.

Benoit DA, Sibley PK, Juenemann JL, Ankley GT. 1997. *Chironomus tentans* life-cycle test: Design and evaluation for use in assessing toxicity of contaminated sediments. *Environ Toxicol Chem* 16:1165–1176.

Boyd MB, Saucier RT, Keeley JW, Montgomery RL, Brown RD, Mathis DB, Guice CJ. 1972. Disposal of dredge spoil; problem identification and assessment and research program development. Vicksburg (MS): US Army Engineer Waterways Experiment Station, CE. Technical report H-72-8.

Bowden RJ. 1977. Guidelines for the pollutional classification of Great Lakes harbor sediments. Chicago: USEPA Great Lakes National Program Office.

Brannon JM. 1978. Evaluation of dredged material pollution potential. Vicksburg (MS): US Army Engineer Waterways Experiment Station, CE. Synthesis report DS-78-6.

Bridges TS, Farrar JD. 1997. The influence of worm age, duration of exposure, and endpoint selection on bioassay sensitivity for *Neanthes arenaceodentata* (Annelida: Polychaeta). *Environ Toxicol Chem* 16:1650–1658.

Burgess R, Ho K, Morrison G, Chapman G, Denton D. 1996. Marine toxicity identification evaluation (TIE) guidance document: Phase I. Washington DC: Office of Research and Development USEPA. EPA/600/R-96/054.

Carr RS, Chapman CD. 1992. Comparison of solid-phase and pore-water approaches for assessing the quality of marine and estuarine sediments. *Chem Ecol* 7:19–30.

Carr RS, Nipper M, editors. Porewater toxicity testing: Biological, chemical, and ecological considerations. Pensacola (FL): Society of Environmental Toxicology and Chemistry.

[CCME] Canadian Council of Ministers of the Environment. 1995. Protocol for the derivation of Canadian sediment quality guidelines for the protection of aquatic life. Ottawa: Prepared by the Technical Secretariat of the CCME Task Group on Water Quality Guidelines. Report CCME EPC-98E. 38 p.

[CCME] Canadian Council of Ministers of the Environment. 1999. Canadian environmental quality guidelines. Winnipeg (MB): Canadian Council of Ministers of the Environment.

[CFR] Code of Federal Regulations. 1968. 33 CFR Section 209.120(d). Washington DC: US Government Printing Office.

Chapman PM. 1990. The Sediment Quality Triad approach to determining pollution-induced degradation. *Sci Tot Environ* 97-8:815–825.

Chapman PM. 1996. Presentation and interpretation of Sediment Quality Triad data. *Ecotoxicology* 5:327–339.

Chapman PM, Long ER, Dexter RN. 1987. Synoptic measures of sediment contamination, toxicity and infaunal community composition (the Sediment Quality Triad) in San Francisco Bay. *Mar Ecol Progr Ser* 37.

Chapman PM, Munday DR, Morgan J, Fink R, Kocan RM, Landolt ML, Dexter RN. 1983. Survey of biological effects of toxicants upon Puget Sound biota.

II. Tests of reproductive impairment. Rockville (MD): National Oceanic and Atmospheric Administration. NOAA Technical Report NOS 102 OMS 1.

Chapman PM, Vigers GA, Farrell MA, Dexter RN, Quinlan EA, Kocan RM, Landolt M. 1982. Survey of biological effects of toxicants upon Puget Sound biota. I. Broad-scale toxicity survey. Boulder (CO): National Oceanic and Atmospheric Administration. NOAA Technical Memorandum OMPA-25.

Defalco P. 1967. The estuary-septic tank of the megalopolis. In: Lauff GH, editor. Estuaries. Washington DC: American Assn for Advancement of Science. Publication Nr. 83. p 701–703.

DeWitt T, Swartz RC, Lamberson JO. 1989. Measuring the acute toxicity of estuarine sediment. *Environ Toxicol Chem* 9:1487–1502.

Dickson KL, Maki AW, Brungs WA. 1987. Fate and effects of sediment-bound chemicals in aquatic systems. New York: Pergamon.

Dillon TM, Moore DW, Gibson AB. 1993. Development of a chronic sublethal bioassay for evaluating contaminated sediment with the marine polychaete worm *Nereis (Neanthes) arenaceodentata. Environ Toxicol Chem* 12:589–605.

Dinnel PA, Stober QF. 1985. Methodology and analysis of sea urchin embryo bioassays. Seattle (WA): Univ Washington, College of Fisheries. Fisheries Research Institute Circular No. 85-3. 19 p.

Di Toro DM, Mahony JD, Hansen DJ, Scott KJ, Hicks MB, Mayr SM, Redmond MS. 1990. Toxicity of cadmium in sediments: The role of acid volatile sulfide. *Environ Toxicol Chem* 9:1487–1502.

Di Toro DM, Zarba CS, Hansen DJ, Berry WJ, Swartz RC, Cowan CE, Pavlou SP, Allen HE, Thomas NA, Paquin PR. 1991. Technical basis for establishing sediment quality criteria for nonionic organic chemicals using equilibrium partitioning. *Environ Toxicol Chem* 10:1–43.

Emery Jr VL, Moore DW, Gray BR, Duke BM, Gibson AB, Wright RB, Farrar JD. 1997. Development of a chronic sublethal sediment bioassay using the estuarine amphipod *Leptocheirus plumulosus* (Shoemaker). *Environ Toxicol Chem* 16:1912–1920.

Engler RM. 1980. Prediction of pollutions potential through geochemical and biological procedures: Development of regulations guidelines and criteria for the discharge of dredged and fill material. In: Babu RH, editor. Contaminants and sediments. Ann Arbor (MI): Ann Arbor. 522 p.

Environment Canada. 1992. Biological test methods: Acute test for sediment toxicity using marine or estuarine amphipods. Ottawa (ON): Conservation and Protection. Report EPS 1/RM/26. 83 p.

Fairey R, Long ER, Roberts CA, Anderson BS, Phillips BM, Hunt JW, Puckett HR, Wilson CJ. 2001. An evaluation of methods for calculating mean sediment quality guideline quotients as indicators of contamination and acute toxicity to amphipods by chemical mixtures. *Environ Toxicol Chem* 20:2236–2286.

Federal Register. 1973. Ocean dumping: Final regulations and criteria. *Federal Register* 38(198). October 1973.

Federal Register. 1977. Ocean dumping: Final regulations and criteria. *Federal Register* Part VI, 42(7). January 1977.

Field LJ, MacDonald DD, Norton SB, Ingersoll CG, Severn CG, Smorong D, Lindskoog R. 2002. Predicting amphipod toxicity from sediment chemistry using logistic regression models. *Environ Toxicol Chem* 21:1993–2005.

Field LJ, MacDonald DD, Norton SB, Severn CG, Ingersoll CG. 1999. Evaluating sediment chemistry and toxicity data using logistic regression modeling. *Environ Toxicol Chem* 18:1311–1322.

Galtsoff PS, Lutz RE, Needham JC. 1937. Culture methods for invertebrate animals. Ithaca (NY): Comstock.

Gannon JE, Beeton AM. 1969. Studies on the effects of dredged materials from selected Great Lakes harbors on plankton and benthos. Milwaukee (WI): Center for Great Lakes Studies, Univ Wisconsin. Special Report Nr. 8.

Gannon JE, Beeton AM. 1971. Procedures for determining the effects of dredged sediments on biota-benthos viability and sediment selection tests. *J Water Pollut Control Fed* 43:392–398.

Ginn TC, Pastorok RA. 1992. Assessment and management of contaminated sediments in Puget Sound. In: Burton Jr GA, editor. Sediment toxicity assessment. Boca Raton (FL): Lewis. p 371–401.

Goldschmidt VM. 1954. Geochemistry. Muir A, editor. London: Clarendon Pr. 730 p.

Gray BR, Emery VL, Brandon DL, Wright RB, Duke BM, Farrar DJ, Moore DW. 1998. Selection of optimal measures of growth and reproduction for the sublethal Leptocheirus plumulosus sediment bioassay. *Environ Toxicol Chem* 17:2288–2297.

Hernandez Jr JW. 1981. JW Hernandez Jr, Deputy Administrator; Region VI Sediment Quality Indicators Memorandum of October 16, 1981, to F.E. Phillips, Deputy Regional Administrator, USEPA, Washington DC.

Ho KT, Burgess RM, Pelletier MC, Serbst JR, Ryba SA, Cantwell MG, Kuhn A, Raczelowski P. 2002. An overview of toxicant identification in sediments and dredged materials. *Mar Pollut Bull* 44:286–293.

Hyland JL, Van Dolah, Snoots TR. 1999. Predicting stress in benthic communities of southeastern U.S. estuaries in relation to chemical contamination of sediments. *Environ Toxicol Chem* 18:2557–2564.

Hyland JL, Balthis WL, Engle VD, Long ER, Paul JF, Summers JK, Van Dolah RF. 2003. Incidence of stress in benthic communities along the U.S. Atlantic and Gulf of Mexico coasts within different ranges of sediment contamination from chemical mixtures. Environmental Monitoring and Assessment, Special Issue on EMAP Symposium 2001: Coastal Monitoring Through Partnerships, April 24–27, 2001, Pensacola, FL. *Environ Monitor Assess* 81(1–3)149–161.

[IJC] International Joint Commission. 1988. Procedures for the assessment of contaminated sediment problems in the Great Lakes.. Windsor (ON): Prepared by the Sediment Subcommittee and its Assessment Work Group, Great Lakes Regional Office. 140 p.

Ingersoll CG. 1995. Sediment toxicity tests. In: Rand GM, editor. Fundamentals of aquatic toxicology. 2nd ed. Washington DC: Taylor & Francis. p 231–255.

Ingersoll CG, Brunson EL, Dwyer FJ, Hardesty DK, Kemble NE. 1998. Use of sublethal endpoints in sediment toxicity tests with the amphipod *Hyalella azteca*. *Environ Toxicol Chem* 17:1508–1523.

Ingersoll CG, Dillon T, Biddinger RG, editors. 1997. Ecological risk assessment of contaminated sediments. Pensacola (FL): Society of Environmental Toxicology and Chemistry (SETAC).

Ingersoll CG, Haverland PS, Brunson EL, Canfield TJ, Dwyer FJ, Henke CE, Kemble NE, Mount DR, Fox RG. 1996. Calculation and evaluation of sediment effect concentrations for the amphipod *Hyalella azteca* and the midge *Chironomus riparius*. *J Gt Lakes Res* 22:602–623.

Ingersoll CG, MacDonald DD, Wang N, Crane JL, Field LJ, Haverland PS, Kemble NE, Lindskoog RA, Severn C, Smorong DE. 2001. Predictions of sediment toxicity using consensus-based freshwater sediment quality guidelines. *Arch Environ Contam Toxicol* 41:8–21.

Jensen A. 1987. Criteria for sediments. In: Kay SH, Marquenie JM, editors. Application and interpretation bioassay and biomonitoring: A planning document. London: European Research Office of the United States Army. Report no. R 87/266. p 4-48–4-52.

Johns DM. Pastorok RA, Ginn TC. 1991. A sublethal sediment toxicity test using juvenile *Neanthes* sp. (Polychaeta : Nereidae). In: Mayes MA, Barron MG, editors. Aquatic toxicology and risk assessment: Fourteenth volume. Philadelphia: American Soc Testing and Materials. p 280–293.

Jones RA, Lee GF. 1978. Evaluation of the elutriate test as a method of predicting contaminant release during open-water disposal of dredged sediments and environmental impact of open water dredged material disposal. Vicksburg (MS): US Army Engineer Waterways Experiment Station, CE. Technical report D-78-45.

Long ER. 2000. Spatial extent of sediment toxicity in U.S. estuaries and marine bays. *Environ Monitor Assess* 64:391–407.

Long ER, Buchman MF, Bay SM, Breteler RJ, Carr RS, Chapman PM, Hose JE, Lissner AL, Scott J, Wolfe DA. 1990. A comparative evaluation of five toxicity tests with sediments from San Francisco Bay and Tomales Bay, California. *Environ Toxicol Chem* 9:1193–1214.

Long ER, Chapman PM. 1985. A sediment quality triad: Measures of sediment contamination, toxicity and infaunal community composition in Puget Sound. *Mar Pollut Bull* 16:405–415.

Long ER, Field LJ, MacDonald DD. 1998. Predicting toxicity in marine sediments with numerical sediment quality guidelines. *Environ Toxicol Chem* 17(4):714–727.

Long ER, Hameedi MJ, Sloane GM, Read LB. 2002. Chemical contamination, toxicity, and benthic community indices in sediments of the lower Miami River and adjoining portions of Biscayne Bay. *Estuaries* 25:622–637.

Long ER, MacDonald DD, Cubbage JC, Ingersoll CG. 1998. Predicting toxicity of sediment-associated trace metals with SEM:AVS concentrations and dry weight-normalized concentrations: A critical comparison. *Environ Toxicol Chem* 17(5): 972–974.

Long ER, MacDonald DD, Severn CG, Hong CB. 2000. Classifying the probabilities of acute toxicity in marine sediments with empirically-derived sediment quality guidelines. *Environ Toxicol Chem* 19:2598–2601.

Long ER, MacDonald DD, Smith SL, Calder FD. 1995. Incidence of adverse biological effects within ranges of chemical concentrations in marine and estuarine sediments. *Environ Manag* 19:81–97.

Long ER, Morgan LG. 1990. The potential for biological effects of sediment-sorbed contaminants tested in the National Status and Trends Program. Seattle (WA): National Oceanic and Atmospheric Administration. NOAA Tech Memo NOS OMA 52. 175 p + appendices.

Long ER, Robertson A, Wolfe DA, Hameedi J, Sloane GM. 1996. Estimates of the spatial extent of sediment toxicity in major U. S. estuaries. *Environ Sci Technol* 30:3585–3592.

Lorenzato SG, Gunther AJ, O'Connor JM. 1991. Summary of a workshop concerning sediment quality assessment and development of sediment quality objectives. Sacramento (CA): California State Water Resources Control Board. 31 p.

Loring DH. 1991. Normalization of heavy-metal data from estuarine and coastal sediments. *ICES J Mar Sci* 48:101–115.

MacDonald DD. 1994. Approach to the assessment of sediment quality in Florida coastal waters. (MacDonald Environmental Sciences, Ltd.) Tallahassee (FL): Florida Dept. of Environmental Protection.

MacDonald DD, Carr RS, Calder FD, Long ER, Ingersoll CG. 1996. Development and evaluation of sediment quality guidelines for Florida coastal waters. *Ecotoxicology* 5:253–278.

MacDonald DD, DiPinto LM, Field J, Ingersoll CG, Long ER, Swartz RC. 2000. Development and evaluation of consensus-based sediment effect concentrations for polychlorinated biphenyls. *Environ Toxicol Chem* 19:1403–1413.

MacDonald DD, Ingersoll CG. 2002a. A guidance manual to support the assessment of contaminated sediments in freshwater ecosystems. Volume I, An ecosystem-based framework for assessing and managing contaminated sediments. Chicago: USEPA Great Lakes National Program Office. EPA-905-B02-001-A.

MacDonald DD, Ingersoll CG. 2002b. Guidance manual to support the assessment of contaminated sediments in freshwater ecosystems. Volume II, Design and implementation of sediment quality investigations. Chicago: USEPA Great Lakes National Program Office. EPA-905-B02-001-B.

MacDonald DD, Ingersoll CG, Berger TA. 2000. Development and evaluation of consensus-based sediment quality guidelines for freshwater ecosystems. *Arch Environ Contam Toxicol* 39:20–31.

MacDonald DD, Ingersoll CG, Smorong DE, Lindskoog RA, Sloane G. 2002. Development and evaluation of numerical sediment quality assessment guidelines for Florida inland waters. Tallahassee (FL): Prepared for the Florida Department of Environmental Protection.

MacDonald DD, Ingersoll CG, Smorong DE, Lindskoog RA, Sparks DW, Smith JR, Simon TP, Hanacek MA. 2002. Assessment of injury to sediments and sediment-dwelling organisms in the Grand Calumet River and Indiana Harbor Area of Concern, USA. *Arch Environ Contam Toxicol* 43:141–155.

Malueg KW, Schuytema GS, Krawczyk DF, Gakstatter JH. 1984. Laboratory sediment toxicity tests, sediment chemistry and distribution of benthic macroinvertebrates in sediments from the Keenaw Waterway, Michigan. *Environ Toxicol Chem* 3:233–242.

McGee BL, Schlekat CE, Reinharz E. 1993. Assessing sublethal levels of sediment contamination using the estuarine amphipod *Leptocheirus plumulosus*. *Environ Toxicol Chem* 12:577–587.

Mearns A, Swartz R, Cummins J, Dinnell P, Plesha P, Chapman PM. 1986. Inter-laboratory comparison of a sediment toxicity test using the marine amphipod *Rhepoxynius abronius*. *Mar Environ Res* 18:13–37.

Milbrink G. 1987. Biological characterization of sediments by standardized tubificid bioassays. *Hydrobiologia* 155:267–275.

Nebeker AV, Cairns MA, Gakstatter JH, Malueg KW, Schuytema GS, Krawczyk DF. 1984. Biological methods for determining toxicity of contaminated freshwater sediments to invertebrates. *Environ Toxicol Chem* 3:617–630.

Neff JM, Bean DJ, Cornaby BW, Vaga RM, Gulbransen TC, Scanlon JA. 1986. Sediment quality criteria methodology validation: Calculation of screening level concentrations from field data. Washington DC: United States Environmental Protection Agency. Work Assignment 56, Task IV. 225 p.

Neff JM, Word JQ, Gulbransen TC. 1987. Recalculation of screening level concentrations for nonpolar organic contaminants in marine sediments. Washington DC: United States Environmental Protection Agency Region V. Final report. 18 p.

[NRC] National Research Council. 1989. Contaminated marine sediments: Assessment and remediation. Washington DC: National Academy Pr.

O'Connor TP, Daskalakis KD, Hyland JL, Paul JF, Summers JK. 1998. Comparisons of measured sediment toxicity with predictions based on chemical guidelines. *Environ Toxicol Chem* 17:468–471.

Oshida PS, Word LS, Mearns AJ. 1981. Effects of hexavalent and trivalent chromium on *Neanthes arenaceodentata* (Polychaeta). *Mar Environ Res* 5:41–49.

Pavlou SP, Weston DP. 1983. Initial evaluation of alternatives for development of sediment related criteria for toxic contaminants in marine waters (Puget Sound). Phase I: Development of conceptual framework. Bellevue (WA): JRB Assoc. Final report. 56 p.

Penny C, Adams C. 1863. Royal Commission on Pollution of Rivers in Scotland. London. Fourth report. 23:377–391.

Persaud D, Jaagumagi R, Hayton A. 1992. Guidelines for the protection and management of aquatic sediment quality in Ontario. Toronto: Ontario Ministry of the Environment. 23 p. ISBN 0-7729-9248-7.

Phillips FE. 1981. Deputy Regional Administrator; Region VI Sediment Quality Indicators Memorandum of 19 Aug 1981 to J. Hernandez, Deputy Administrator, U.S. EPA; Dallas, TX.

Prater BL, Anderson MA. 1977. A 96-hour sediment bioassay of Duluth and Superior Harbor basins (Minnesota) Using *Hexagenia limbota*, *Asellus communis*, *Daphnia magna*, and *Pimephales promelas* as test organisms. *Bull Environ Contam Toxicol* 18:159–169.

[PSDDA] Puget Sound Dredged Disposal Analysis Program. 1989. Management plan report: Unconfined open-water disposal of dredged material, Phase 2 (north and south Puget Sound). Seattle (WA): US Army Corps of Engineers.

Reish DJ, Barnard JL. 1960. Field toxicity test in marine waters suing the polychaetous annelid *Capitella capitata* (Fabricius). *Pacific Naturalist* 1:1–12.

Reish DJ, Lemay JA. 1988. Bioassay manual for dredged sediments. Los Angeles (CA): US Army Corps of Engineers, Los Angeles District.

Reynoldson TB, Thompson SP, Bamsey, JL. 1991. A sediment bioassay using the tubificid oligochaete worm *Tubifex tubifex*. *Environ Toxicol Chem* 10:1061–1072.

Schiewe MH, Hawk EG, Actor DI, Krahn MM. 1985. Use of a bacterial bioluminescence assay to assess toxicity of contaminated marine sediments. *Can J Fish Aquat Sci* 42:1244–1248.

Schlekat CE, McGee BL, Reinharz E. 1992. Testing sediment toxicity in Chesapeake Bay with the amphipod *Leptocheirus plumulosus*: An evaluation. *Environ Toxicol Chem* 11:225–236.

Schropp SJ, Lewis FG, Windom HL, Ryan JD, Calder FD, Burney LC. 1990. Interpretation of metal concentrations in estuarine sediments of Florida using aluminum as a reference element. *Estuaries* 13:227–235.

Scott KJ, Redmond MS. 1989. The effects of a contaminated dredged material on laboratory populations of the tubicolous amphipod, *Ampelisca abdita*. In: Cowgill UM, Williams LR, editors. Aquatic toxicology and hazard assessment: Twelfth volume. Philadelphia: American Soc Testing and Materials. ASTM STP 854. p 284–307.

Smith SL, MacDonald DD, Keenleyside KA, Ingersoll CG, Field LF. 1996. A preliminary evaluation of sediment quality assessment values for freshwater ecosystems. *J Gt Lakes Res* 22(3):624–638.

Swartz RC. 1999. Consensus sediment quality guidelines for polycyclic aromatic hydrocarbon mixtures. *Environ Toxicol Chem* 18:780–787.

Swartz RC, DeBen WA, Cole FA. 1979. A bioassay for the toxicity of sediment to the marine macrobenthos. *J Water Pollut Control Fed* 51:944–950.

Swartz RC, DeBen WA, Jones JKP, Lamberson JO, Cole FA. 1985. Phoxocephalid amphipod bioassay for marine sediment toxicity. In: Cardwell RD, Purdy R, Bahner RC, editors. Aquatic Toxicology and Hazard Assessment: 7th Symposium. Philadelphia: American Soc for Testing and Materials. ASTM STP 854. p 284–307.

Swartz RC, DeBen WA, Sercu KA, Lamberson JO. 1982. Sediment toxicity and the distribution of amphipods in Commencement Bay, Washington, USA. *Mar Pollut Bull* 13:359–364.

Swartz RC, Schults DW, Ozretich RJ, Lamberson JO, Cole FA, DeWitt TH, Redmond MS, Ferraro SP. 1995. Sigma PAH: A model to predict the toxicity of polynuclear aromatic hydrocarbon mixtures in field-collected sediments. *Environ Toxicol Chem* 14:1977–1987.

Turgeon DD, Hameedi J, Harmon MR, Long ER, McMahon KD, White HH. 1998. Sediment toxicity in U. S. coastal waters. Special Report. Silver Spring (MD): National Ocean Service, National Oceanic and Atmospheric Administration.

[USACE] US Army Corps of Engineers. 1976. Ecological evaluation of proposed discharge of dredged of fill material into navigable waters. Vicksburg (MS): US Army Engineer Waterways Experiment Station, CE. Miscellaneous paper D-76-17.

US Army Office, Chief of Engineers. 1971. Disposal of dredged materials. Washington DC: US Government Printing Office. Engineering Circular 1165-2-97.

[USEPA] US Environmental Protection Agency. 1977a. Ocean dumping: Final revisions of regulations and criteria. *Federal Register* Part IV, Vol. 42, No. 7, Tuesday, 11 January 1977.

[USEPA] US Environmental Protection Agency. 1977b. Proposed revisions to ocean dumping criteria. Final environmental impact statement. Volume 1. Washington DC: USEPA.

[USEPA] US Environmental Protection Agency. 1989a: Guidance manual: Bedded sediment bioaccumulation tests. Narragansett (RI): USEPA. EPA 600/x-89/302.

[USEPA] US Environmental Protection Agency. 1989b. A science advisory report: Evaluation of the apparent effects threshold (AET) approach for assessing sediment quality. Washington DC: USEPA. EPA-SAB-EPEC-89-027.

[USEPA] US Environmental Protection Agency. 1990a. Report of the Sediment Criteria Subcommittee of the Ecological Processes and Effects Committee. Evaluation of the equilibrium partitioning (EqP) approach for assessing sediment quality. Washington DC: USEPA Science Advisory Board. EPA-SAB-EPEC-90-006.

[USEPA] US Environmental Protection Agency. 1990b. A science advisory report: Evaluation of the sediment classification methods compendium. Washington DC: USEPA. EPA-SAB-EPEC-90-018.

[USEPA] US Environmental Protection Agency. 1991a. Proposed sediment quality criteria for the protection of benthic organisms: Fluoranthene. Washington DC: USEPA Office of Water. 54 p.

[USEPA] US Environmental Protection Agency. 1991b. Proposed sediment quality criteria for the protection of benthic organisms: Acenaphthene. Washington DC: USEPA Office of Water. 57 p.

[USEPA] US Environmental Protection Agency. 1991c. Proposed sediment quality criteria for the protection of benthic organisms: Phenanthrene. Washington DC: USEPA Office of Water. 49 p.

[USEPA] US Environmental Protection Agency. 1991d. Sediment toxicity identification evaluations: Phase I. Toxicity characterization procedures. Duluth (MN): USEPA. EPA/600/3-88/034.

[USEPA] US Environmental Protection Agency. 1992. An SAB report: review of sediment criteria development methodology for non-ionic organic contaminants. Prepared by the Sediment Quality Subcommittee of the Ecological Processes and Effects Committee. Washington DC: USEPA. EPA-SAB-EPEC-93-002.

[USEPA] US Environmental Protection Agency. 1993a. Sediment quality criteria for the protection of benthic organisms: Fluoranthene. Washington DC: USEPA. EPA 822R/93/012.

[USEPA] US Environmental Protection Agency. 1993b. Sediment quality criteria for the protection of benthic organisms: Acenaphthalene. Washington DC: USEPA. EPA 822R/93/013.

[USEPA] US Environmental Protection Agency. 1993c. Sediment quality criteria for the protection of benthic organisms: Phenanthrene. Washington DC: USEPA. EPA 822R/93/014.

[USEPA] US Environmental Protection Agency. 1994a. Methods for measuring the toxicity and bioaccumulation of sediment-associated contaminants with freshwater invertebrates. Duluth (MN): USEPA. EPA 600/R-94/024.

[USEPA] US Environmental Protection Agency. 1994b. Methods for measuring the toxicity of sediment-associated contaminants with estuarine and marine invertebrates. Washington DC: USEPA. EPA 600/R-94/025.

[USEPA] US Environmental Protection Agency. 1994c. Assessment and remediation of contaminated sediments (ARCS) program. Chicago: Great Lakes National Program Office. EPA 905/B-94/002.

[USEPA] US Environmental Protection Agency. 1995. An SAB report: Review of the agency's approach for developing sediment criteria for five metals. Prepared by the Sediment Quality Subcommittee of the Ecological Processes and Effects Committee. EPA-SAB-EPEC-95-020.

[USEPA] US Environmental Protection Agency. 1997. The incidence and severity of sediment contamination in surface waters of the United States. Vol. 1. National sediment quality survey. Washington DC: USEPA. EPA 823-R-97-006.

[USEPA] US Environmental Protection Agency. 2002a. Procedures for deriving equilibrium partitioning sediment benchmarks (ESBs) for the protection of benthic organisms: Endrin. Washington DC: USEPA Office of Research and Development. (AED-02-046) EPA-600-R-02-009.

[USEPA] US Environmental Protection Agency. 2002b. Procedures for deriving equilibrium partitioning sediment benchmarks (ESBs) for the protection of benthic organisms: Dieldrin. Washington DC: USEPA Office of Research and Development. (AED-02-047) EPA-600-R-02-010.

[USEPA] US Environmental Protection Agency. 2002c. Procedures for deriving equilibrium partitioning sediment benchmarks (ESBs) for the protection of benthic organisms: Metal mixtures (cadmium, copper, lead, nickel, silver, and zinc). Washington DC: USEPA Office of Research and Development. (AED-02-048) EPA-600-R-02-011.

[USEPA] US Environmental Protection Agency. 2002d. Procedures for deriving site-specific equilibrium partitioning sediment benchmarks (ESBs) for the protection of benthic organisms: Nonionic organics. Washington DC: USEPA Office of Research and Development. (AED-02-049) EPA-600-R-02-012.

[USEPA] US Environmental Protection Agency. 2002e. Procedures for deriving equilibrium partitioning sediment benchmarks (ESBs) for the protection of benthic organisms: PAH mixtures. Washington DC: USEPA Office of Research and Development. (AED-02-050) EPA-600-R-02-013.

[USEPA] US Environmental Protection Agency. 2002f. Technical basis for the derivation of equilibrium partitioning sediment benchmarks (ESBs) for the protection of benthic organisms: nonionic organics. Washington DC: USEPA Office of Research and Development. (AED-02-051) EPA-600-R-02-014.

[USEPA] US Environmental Protection Agency. 2002g. Procedures for deriving equilibrium partitioning sediment benchmarks (ESBs) for the protection of benthic organisms: Nonionics compendium. Washington DC: USEPA Office of Research and Development. (AED-02-052) EPA-600-R-02-016.

[USEPA/USACE] US Environmental Protection Agency/US Army Corps of Engineers. 1973. Technical Committee on Criteria for Dredged and Fill Material. Ecological evaluation of proposed discharge of dredged material into ocean waters: Implementation manual for Section 103 of Public Law 92-532. (Marine Protection, Research and Sanctuaries Act of 1972). Vicksburg (MS): US Army Engineer Waterways Experiment Station, CE.

[USEPA/USACE] US Environmental Protection Agency/US Army Corps of Engineers. 1977. Ecological evaluation of proposed discharge of dredged material into ocean waters. Vicksburg (MS): USEPA/USACE Technical Committee on Criteria for Dredged and Fill Material, Environmental Effects Laboratory, US Army Engineer Waterways Experiment Station.

[USEPA/USACE] US Environmental Protection Agency/US Army Corps of Engineers. 1991. Evaluation of dredged material proposed for ocean disposal (testing manual). Washington DC: USEPA/USACE. EPA-503/8-91/001.

[USEPA/USACE] US Environmental Protection Agency/US Army Corps of Engineers. 1998. Evaluation of dredged material proposed for discharge in waters of the US (testing manual). Washington DC: USEPA/USACE. EPA-823-B-98-004.

[USEPA/USACE] US Environmental Protection Agency/US Army Corps of Engineers. 2001. Method for assessing the chronic toxicity of marine and estuarine sediment-associated contaminants with the amphipod *Leptocheirus plumulosus.* Washington DC: USEPA/USACE. EPA/600/R-01/020.

US House of Representatives, Committee on Public Works. 1973. Laws of the United States relating to water pollution control and environmental quality, 93-1. Washington DC: US Government Printing Office.

[WAC] Washington Administrative Code. 1991. Chapter 173-204, Sediment Management Standards. Olympia, (WA): State of Washington.

[WEF] Water Environment Federation. 2002. Handbook on sediment quality. Whittemore RC, editor. Alexandria (VA): WEF. 372 p.

[WDOE] Washington Department of Ecology. 1995. Sediment management standards. Olympia (WA): WDOE. Publication Nr. 96-252.

Wenning RW, Ingersoll CG, editors. 2002. Use of sediment quality guidelines (SQGs) and related tools for the assessment of contaminated sediments: Summary from a SETAC Pellston Workshop. Pensacola (FL): Society of Environmental Toxicology and Chemistry (SETAC).

Wentsel R, McIntosh A, Anderson V. 1977. Sediment contamination and benthic invertebrate distribution in a metal-impacted lake. *Environ Pollut* 14:187–193.

Wentsel R, McIntosh A, Atchison G. 1977. Sublethal effects of heavy metal contaminated sediment on midge larvae (*Chironomus tentans*). *Hydrobiologia* 56: 53–156.

Wentsel RA, McIntosh A, McCafferty PC. 1978. Emergence of the midge *Chironomus tentans* when exposed to heavy metal contaminated sediments. *Hydrobiologia* 57:195–196.

International overview of sediment quality guidelines and their uses

9

MARC P BABUT, WOLFGANG AHLF, GRAEME E BATLEY, MARINA CAMUSSO, ERIC DE DECKERE, PIETER J DEN BESTEN

Most industrialized countries have to deal with the regulation and management of contaminants in sediments and dredged materials in their waterways. As a consequence, a variety of approaches have been or are being developed. In many cases, sediment quality guidelines (SQGs) form part of these approaches. This chapter provides an overview of the methods used for deriving SQGs, in countries other than those in North America, whose developments have been discussed in considerable detail elsewhere in this volume. How these guidelines are used within assessment frameworks is also discussed. In so doing, the following questions are addressed:

- How do countries outside North America evaluate sediments (rationale and tools)?
- What common technical ground is shared on this subject? Are there some differences, and where do they originate?
- How valid are existing SQGs? Can SQGs developed in a given region or country be used in another?
- More generally, are the assessment approaches using SQGs relevant?

Sediment Quality Guidelines in Europe, Australia, and Hong Kong: Methods Applied for Derivation

Overview of approaches

Two distinct approaches to SQGs have been adopted in countries outside of North America. Particularly in Europe, the application of guidelines based

Use of Sediment Quality Guidelines and Related Tools for the Assessment of Contaminated Sediments
Wenning RJ, Batley GE, Ingersoll CG, Moore DW, editors.
 ISBN 1-880611-71-6

on reference conditions has been widely practiced. In this context, the reference condition can be defined either as "background concentrations" or as an array of chemical and biological parameters measured at reference sites. This approach has, at least for some time, hampered the application of empirical or mechanistic SQGs (Chapter 3) that have achieved wide use in North America in recent years.

The reference condition approach is, or was, used by Flanders in Belgium, France, Germany, and Italy. However, several of these countries have begun to develop effects-based SQGs, which are or will be used instead of, or sometimes in combination with, the reference condition approach.

Australia and New Zealand

Australia and New Zealand have recently revised their guidelines for fresh and marine water quality (ANZECC/ARMCANZ 2000) and have included, for the first time, a consideration of sediment quality. Australian guidelines for ocean disposal of dredged materials have also been developed recently (Environment Australia 2002). Both use a common approach to SQG development: an ecotoxicological effects database. Faced with a paucity of local sediment effects data, the effects range low (ERL) values derived from the North American database were used as the basis for interim Australian and New Zealand guidelines. In the case of Cu and As, the Hong Kong guideline values (Chapman et al. 1999) were used, rather than the more conservative published ERL values (Long et al. 1995). The deficiencies of the ERL and effects range median (ERM) approach were, however, implicitly acknowledged. As discussed in Chapter 3, these empirical SQGs are derived from a large effects database in which toxicity frequently is due to multiple co-occurring contaminants. Toxicity testing for these used mainly amphipod species, supplemented by a number of other tests on both pore waters and whole sediments. The lower SQG values were termed "trigger values" to imply that further action is triggered if the values are exceeded. These values imply a low level of biological effects and therefore are protective of most benthic biota, representing as they do, the lower 10th percentile of effects data. Even so, the derived SQGs may not be reflecting toxicity to the contaminant of concern. Limited value was seen in using trigger values such as ERMs that are said to be "predictive of toxicity" (Chapter 3); rather, values were required that were protective. The upper ERM values (termed "interim SQG-high") were, however, included in the guideline documentation to provide an indication of a value at which toxicity is more likely.

Belgium (Flanders)

In Belgium, sediment quality was assessed on the basis of comparisons with reference sites (Bervoets and Verheyen 1994). This has been undertaken for Belgium over the period 1994 to 2000, but no such developments have yet been undertaken for Wallonia and Brussels. (Belgium is a federal country that includes 2 provinces, Flanders and Walloon; the federal capital, Brussels, has a special status.)

Initially, 5 locations were chosen as reference sites. Previous ecological studies showed that these sites were in relatively undisturbed areas with a high ecological quality. The chemistry at these sites was considered close to reference values used in other European countries. The metals concentrations were converted to those in a standard sediment with 11% clay and 2% organic matter, using the following equation:

$$C_{std} = C_{meas} \times (A + (B \times \text{clay}) + (C \times \text{OM})) / (A + 11B + 2C),$$

where C_{std} is the standardized concentration, C_{meas} is the measured concentration, *clay* is the clay content in %, OM is the organic matter content in %, and *A, B,* and *C* are constants, which were derived by multiple regression applied to a dataset that included 328 points. For organic contaminants, the standardization was based only on the organic matter content.

The geometric average of the standardized concentrations at reference sites was calculated and used. The reference values were recalculated in 1999 (de Deckere et al. 2000) on the basis of a dataset of 360 sampling locations from which 12 new reference sites were selected. These locations were situated in biologically valuable areas that had good biological sediment quality, according to benthic community assessments (biotic sediment index, mouth deformities of chironomids larvae) and to toxicity tests (*Pseudokirchneriella subcapitata* [previously known as *Raphidocelus capitata,* which is commonly used in Flemish documents] growth inhibition test; *Thamnocephalus platyurus* survival and *Hyalella azteca* survival). This time, standardization was based on a standard sediment containing 11% clay and 5% silt, which was the average value for the sampled sediments. The geometric means are currently used as the reference contents for sediments in Flanders. The reference values (Table 9-1) are then used as SQG values, but an extensive evaluation of all the data collected (1220 sampling points) should result in new SQGs for the future.

Guidelines for dredged material are based on guidelines for the reuse of contaminated soil and differ from the reference values. Walloon and Brussels do not have SQGs yet.

Table 9-1 Reference values for Belgian (Flanders) sediments based on 5 (1994 to 1999) and 12 (1999 to present) reference sites respectively

Chemical	Concentration unit	1994 to 1999	1999 to present
As	mg/kg dw	21	11
Cd	mg/kg dw	0.40	0.38
Cr	mg/kg dw	13	17
Cu	mg/kg dw	9	8
Hg	mg/kg dw	0.12	0.05
Ni	mg/kg dw	9	11
Pb	mg/kg dw	21	14
Zn	mg/kg dw	115	67
NPHCs	mg/kg dw	15	37
EOX	mg Cl/kg dw	14	31
Sum OCP	µg/kg dw	1	3.9
Sum 7 PCBs	µg/kg dw	5	5.1
6 Borneff PAHs	mg/kg dw	0.255	0.220

NPHCs = nonpolar hydrocarbons or mineral oils
EOX = extractable organohalogens
OCP = organochlorine pesticides (sum of the concentration of α–hexachlorohexane [HCH], ß-HCH, γ-HCH [lindane], hexachlorobenzene, heptachloor, heptachloroepoxide, op'DDD, pp'DDD, op'DDE, pp'DDE, op'DDT, pp'DDT, aldrin, dieldrin, isodirn, endrin, and alpha endosulfan)
PCB = polychlorinated biphenyl (sum of PCB28, PCB52, PCB101, PCB118, PCB138, PCB153, and PCB 180)
PAH = polycyclic aromatic hydrocarbons (sum of the PAHs of Borneff: fluoranthene, benzo(*b*)fluoranthene, benzo(*k*)fluoranthene, benzo(*a*)pyrene, benzo(*g,h,i*)perylene, and indeno(1,2,3,*c,d*)pyrene)

France

Freshwater sediments

Before 1996, freshwater sediment quality in France was assessed for only 8 trace elements (As, Cd, Cr, Cu, Hg, Ni, Pb, and Zn). Measured concentrations allowed classification of sediments into 5 classes of quality, based on geochemistry and statistics. A global overhaul of assessment systems for different components of waterway quality led to the abandonment of this geochemistry approach, which became inappropriate when the surveillance programs began to include organic chemicals.

The French water quality assessment system (Système d'Evaluation de la Qualité de l'Eau, or SEQ-Eau), which includes sediments, defines 4 threshold levels based on the population and the abundance of aquatic and benthic organisms. In the case of organic chemicals in water, definitions have been adapted to take available data into account (Table 9-2).

A preliminary set of criteria for some 150 organic chemicals and metals was first published in 1997 (Babut 1997) and completed in 1999 (Oudin and Maupas 1999). Two thresholds out of the required 4 have been determined for sediments: 1) a co-occurrence approach (determination of threshold effects levels [TELs] and probable effects levels [PELs]) applied to trace elements and a few organic compounds (DDE, dieldrin, fluoranthene, lindane, polychlorinated biphenyls [PCBs]), and 2) the equilibrium partitioning (EqP) approach for the majority of organic substances. TEL and PEL values were calculated from an extract of the biological effects database for sediments (BEDS) (Smith et al. 1996).

Table 9-2 Quality threshold definitions for chemicals in the SEQ-Eau framework

Threshold	Risk	Definition
	Reduction of diversity and abundance	
4		Geometric mean of the lowest EC50 or LC50 by trophic level
	Lethal effects on the most sensitive species, abundance decrease	
3		Lowest EC50 or LC50
	Chronic sub-lethal effects, reduction in abundance, predominance of tolerant species	
2		Lowest chronic NOEC (at least 3 trophic levels) or lowest EC50 or LC50/10
	Chronic sublethal effects, particularly for the most sensitive species and young individuals	
1		Lowest chronic NOEC10 (at least 3 trophic levels) or lowest EC50 or LC50/100

A further attempt at defining the 4 required thresholds was made in 1998 (Garric, Baligand, et al. 1998). A summary of the proposal is given in Table 9-3. However, there was no direct follow-up of this proposal.

Table 9-3 Proposed approach to complete the sediment thresholds in France

Threshold	Definition
4	Geometric mean of ECx
3	PEL or minimum for ECx (x ≥ 50)
2	TEL, or minimum value of NOEC out of all available species, divided by 10 if only lethal effects are considered
1	TEL/10 or, if a TEL value does not exist, the lowest value of NOEC divided by 10 if chronic effects are considered, divided by 100 if only lethal effects are considered

In parallel, the SEQ-Eau was updated for polycyclic aromatic hydrocarbons (PAHs), when the list of monitored compounds of that category was extended in 1997. For PAHs, sediment thresholds (Bisson et al. 2000) were based on a comparison of several approaches, that is, TELs and PELs (Smith et al. 1996), EqP (Di Toro et al. 1991), and the distribution of no-observed-effects concentrations (NOECs) or effects concentrations (Aldenberg and Slob 1993) combined with the assumption that PAHs act through a narcotic mode. If the narcotic assumption is correct, then 1) the toxicity of all PAHs, expressed as an internal concentration, is constant, and 2) this internal concentration is proportional to the sediment concentration because the toxicity depends only on the accumulation of the compounds. This permits the determination of "toxicity thresholds" either for single compounds or for total PAHs. A literature review provided sediment toxicity data, and the concentration hazardous to 5% of the species was calculated, yielding a value of 1105 µg/kg dry weight.

The EqP approach was applied to the original aquatic toxicity data provided in the same study (Bisson et al. 2000). Again, the narcotic assumption was used, permitting the determination of a single threshold value of 1258 µg/kg.

TEL values (Smith et al. 1996) were available for only 6 compounds. A comparison with the values obtained from the 2 other approaches showed large differences when focused on single compounds. However, the narcotic assumption leads also to the adoption of an assumption of additivity. Swartz (1999) also adopts this latter assumption and stresses that ERLs for PAHs are close to guidelines determined by other methods, including EqP, provided PAHs are considered as a group. In conclusion, it was proposed that the guidelines for PAHs as a group be adopted, being calculated by the method that provided the complete set of compounds and thresholds, that is, EqP.

Therefore, a complete set of thresholds was derived on the basis of the EqP approach, combined with assessment factors. Two groups of PAHs were distinguished on the basis of their genotoxicity. Later SEQ-Eau updates all relied upon the EqP approach (Babut, Bedell, et al. 2001; Babut, Bonnet, et al. 2001).

The development of protocols for the management of dredged materials, commenced in 1999. These protocols will probably recommend the use of SQGs within a tiered approach; these SQGs might be different from those used for monitoring purposes. A proposal for this was submitted to the Ministry of Transport (Babut and Perrodin 2001). The SQGs will be obtained from a literature survey because it would not be possible to determine original values within an acceptable time frame. Moreover, it appeared more relevant to assess the predictive ability of existing SQGs with regional datasets.

Coastal and marine sediments

There is currently no classification scheme available for coastal sediments, although the implementation of the Water Framework Directive will probably result in such a scheme. Some guidelines have been promulgated, however, for harbor sediments to comply with the Convention for the Protection of the Marine Environment of the North-East Atlantic (OSPAR 1998) requirements. Two sets of guidelines were derived by a geochemical approach. The Level 1 values correspond to the medians of natural concentrations, while Level 2 values correspond to the 95th percentile. The latter are actually extrapolated from the cumulative frequency curve, assuming a normal distribution. As a consequence, these guidelines are not effects based, which is rather uncommon.

Germany

The Federal Republic of Germany does not yet have common national regulations for the management of contaminated sediments. Federal waterways are under the jurisdiction of the Federal Ministry of Transport (BMV), while the Federal Institute for Hydrology (BfG) provides conceptual guidance and project monitoring. All other inland waterways are under the responsibility of Länder (the German word for “state”).

Regulations for water, soil, and waste, including dredged materials, are, however, treated at a common and coordinated level through technical working groups. Apart from official regulations, some unofficial guidance was also provided, for example, by the German Association for Water Pollution Control (ATV), which gave recommendations for the handling of dredged materials.

Conversely, the conference of Länder Ministers for the Environment gave a mandate to a working group (Arbeitsgemeinschaft für die Reinhaltung der Elbe [ARGE Elbe]), which issued a recommendation for the handling of contaminated dredged materials on the River Elbe. This recommendation also included contaminant thresholds for the relocation (open water disposal) of dredged materials in the river. The thresholds were derived from data in the literature, combining the equilibrium approach and information from toxicity tests. Some Länder, for example, Schleswig-Holstein, have also issued their own recommendations, including contaminant thresholds. In most cases, however, because no regulation specific to dredged materials or sediments was available, regulations applying to soils or wastes were used (Peters and Hagner 2001). A summary of these is given in Table 9-4.

Table 9-4 Summary of dredged materials management in Germany

Responsible party	Target	Organization	Regulation
Federal government	Federal waterways	• Ministry of Transport • Waterways and shipping administration • BfG	• Directive for the handling of dredged materials in federal coastal waterways • Directive for the handling of dredged materials in federal inland waterways
Joint Länder working group	River catchments	Specific (e.g., ARGE Elbe)	Recommendations for the handling of contaminated dredged materials in the River Elbe
Länder and municipalities	Länder water bodies	Land government	Dredged material management concept

Because Germany is involved in several international organizations dealing with the management of transboundary water bodies (including the OSPAR Commission, the Helsinki Commission, and the London Convention for the marine environment), the directives issued by these institutions are accounted for in the national regulations. (The Helsinki Commission [HELCOM] aims to protect the marine environment of the Baltic Sea from all sources of pollution through intergovernmental cooperation between Denmark, Estonia, the European Community, Finland, Germany, Latvia, Lithuania, Poland, Russia, and Sweden.) For federal coastal waterways, the resulting directive is based on a reference-level approach. Two action levels have been defined for coastal sediments. The first reference value (RW1) is based on prevailing contaminant concentrations in the Wadden Sea (part of the North Sea, mainly along the

German coast) between 1982 and 1992. The second level (RW2) is defined as 5× the reference value for metals and 3× this value for organic compounds. Thus, they are derived by convention and are not effects based. The contaminant list includes 8 trace elements (As, Cd, Cr, Cu, Pb, Hg, Ni, and Zn) and 17 organic compounds (PCBs, 6 PAHs, hexachlorocyclohexane derivatives, penta- and hexachlorobenzene, DDT and related metabolites, and tributyltin [TBT] and dibutyltin). In addition, standard ecotoxicological tests on pore water and elutriates (bioluminescence inhibition following the standard ISO EN DIN 11348-1-3, and marine algae growth inhibition following ISO EN DIN 10253) are applied. Whole sediment tests (amphipod survival ISO DIN 16712) will be implemented in the future (Peters et al. 2002).

For the regulation of dredged sediments in German inland waterways, the chemical substances monitored are basically the same as those required for coastal sediments, except that iron and mineral oil were added, while pentachlorobenzene, TBT, and dibutyltin were not included. Instead of fixing reference values and action levels, this directive determined site-specific values based on the median of contaminant concentrations in the suspended matter over a 3-y period. The reference value was set at 1.5 times the median and the action level at 3 times the median.

Hong Kong

Sediment quality issues for Hong Kong were largely focused on marine dredged sediments. These were first managed according to 3 classes of contamination. The criteria delineating these classes were considered conservative and were too close to one another, given measurement uncertainty. Chapman et al. (1999) recommended the development of interim sediment quality values (ISQVs), on the basis of published papers combined with the collection of site-specific information (e.g., through biological testing). Thus 2 sets of ISQVs were determined, that is, an ISQV-low, below which adverse effects were unlikely, and an ISQV-high, above which severe adverse effects were very likely. In practice, ISQV-low were set equivalent to ERL and ISQV-high to ERM (as published by Long et al. 1995). There were, however, exceptions for Cu and Ni, for which higher ISQV-low guideline values were chosen on the basis of high local reference analyses. Only the ISQV-low was proposed for decision-making. Implementation of the ISQV-high values would follow the incorporation of regional data (toxicity test results and chemical concentrations).

Italy

In Italy, there are no national SQGs, although some work has been undertaken on the development of guidelines for freshwater sediments, dredged material, and sediments for marine disposal. The general structure of the "Protection of Waters from Pollution" decrees (Italian Legislative Decrees 152/1999 and 258/2000) has a common framework for surface and ground waters in fresh, transition, and coastal marine areas, which requires monitoring and classification of water bodies according to environmental quality objectives (Water Framework Directive, 2000/60/EC, 23/X/2000). Environmental quality is evaluated on the basis of ecological and chemical status of a given body compared to a reference one, and water bodies are classified in 5 classes of quality. The ecological status is assessed on the basis of the diversity and abundance of certain types of living organisms (e.g., by applying the extended biotic index [EBI], or Indice Biotico Esteso [IBE]; Ghetti 1997) and morphological conditions, while the chemical status is based on the occurrence and concentrations of dangerous substances (priority list) in compliance with environmental quality standards (EQSs). The list of priority substances is given in Table 9-5; these substances were selected from other lists issued by organizations such as the US Environmental Protection Agency (USEPA), the United Nations Environment Programme (UNEP), and the European Commission. Other hydrological and morphological variables are also included in the monitoring requirements because of the potential impacts of human activity and water uses and the affinities of contaminants for sediments. It is recommended, in case of further investigations, that toxic effects be assessed using a multispecies approach, short- and long-term toxicity tests on sediment extracts, whole sediments, or pore water with test organisms such as *Oncorhynchus mykiss, Daphnia magna, Ceriodaphnia dubia, Chironomus tentans, Chironomus riparius, Pseudokirchneriella subcapitata* and luminescent bacteria (APAT 2002; case study in Viganò et al. 2003).

Methods and criteria for the assessment of sediment quality will be developed by the Ministry of Environment, in collaboration with research institutions and national and regional agencies. Research is underway to develop SQGs (EQSs) for sediments in fresh and marine waters.

Dredged material regulation commenced in 1980, with additions in 1996 and 1999 (Italian Legislative Decree 471). For these sites, quality criteria for surface waters have been developed, including a recommendation for studies to assess risk posed by contaminants in sediments and contaminant bioaccumulation, but numerical sediment criteria have not yet been developed.

Table 9-5 List of priority pollutants in Italy

Inorganic compounds	Organochlorine compounds	PAHs
As	DDT homologues	Naphthalene
Cd	Σ-HCH	Acenaphthene
Cr	Σ-drins	Phenanthrene
Cu	Hexachlorobenzene	Fluoranthene
Hg	PCB52	Benzo(*a*)anthracene
Ni	PCB77	Chrysene
Pb	PCB81	Benzo(*b*)fluoranthene
Zn	PCB128	Benzo(*k*)fluoranthene
	PCB138	Benzo(*a*)pyrene
	PCB153	Dibenzo(*a,h*)anthracene
	PCB169	Benzo(*ghi*)perylene
		Anthracene
		Pyrene
		Indeno(1,2,3,*cd*)pyrene
		Acenaphthylene
		Fluorene

The assessment of dredged materials from harbors and marine waters relies on the same jurisdictional decree as freshwater sediments (i.e., decrees 152/1999 and 258/2000), but the monitored parameters and sampling procedures are slightly different. The priority list, as for freshwater sediments, includes metals, PAHs, organochlorine compounds (PCBs, DDT and related compounds, aldrin, dieldrin, hexachlorobenzene and hexachlorocyclohexane derivatives), together with TBT. In the technical proposal prepared by Istituto Centrale per la Ricerca Scientifica e Tecnologica Applicata al Mare (ICRAM) (Pellegrini et al. 2002), toxicity tests would also be required, without specification of species; however, the use of indigenous species (e.g., echinoderms) or standard procedures (ASTM, USEPA, USACE, PARCOM, ISO) is recommended. Two SQG levels are proposed, a base chemical level (BCL) and a threshold chemical level (TCL). The BCL applies to sediments for marine disposal, including offshore sites, while the TCL is the quality for a port environment. The respective proposed definitions for BCL and TCL are a background level and a threshold beyond which toxic effects are likely to occur. In the derivation of BCLs, the absence of toxicity in a given test species would be checked. For organic compounds, the TELs are used as the source for BCLs (CCME 1999, 2001; MacDonald et al. 1996), while PELs are used as the source for all TCLs. Pellegrini et al. (2002) stress that these SQGs, if adopted, should be considered temporary, that is, subject to periodic revision. These revisions could be related to the harmonization expected with the implementation of recent European regulations (e.g., European Directive EC 2000/60) or to an improved knowledge of background concentrations (BCs) at coastal reference sites.

A specific regulation applies to Venice Lagoon sediments because of the peculiarities of this ecosystem. According to Italian law 360/91 and more specifically its Article 4 completed and signed April 8, 1993, three SQGs had to be implemented for the management of dredged materials, depending of the intended use of the materials. This implementation was done on the basis of marine TELs (MacDonald et al. 1996). The adopted guidelines cover a range of metals, PAHs (as a sum), PCBs (as a sum), and organochlorine pesticides (as a sum).

The Netherlands

The development of water quality criteria (WQC) in the Netherlands was one of the elements of a plan that began in 1987 (Van Der Gaag et al. 1991). The approach adopted for sediments was first presented by Van Der Kooij et al. (1991), summarized in Figure 9-1. It consisted of calculating quality criteria for water and sediment from existing ecotoxicity test data (usually concentrations of toxicants in water) or product standards (in food) based on the EqP theory. The sediment–water partition coefficients (Ksw) for organic chemicals were derived theoretically, while for metals, they were obtained from field measurements of concentrations in water before and after filtration.

Quality criteria for suspended solids were derived from the SQGs for sediment on the assumption that concentrations in suspended solids are related to sediment concentrations through empirical concentration ratios (concentration in suspended matter–concentration in sediment) of 1.5 for metals and 2.0 for organic compounds.

The derivation procedure relies upon the EqP concept, including biota. This allows harmonizing of water and sediment quality criteria with other kinds of standards, such as those that apply to the consumption of fish or mussels.

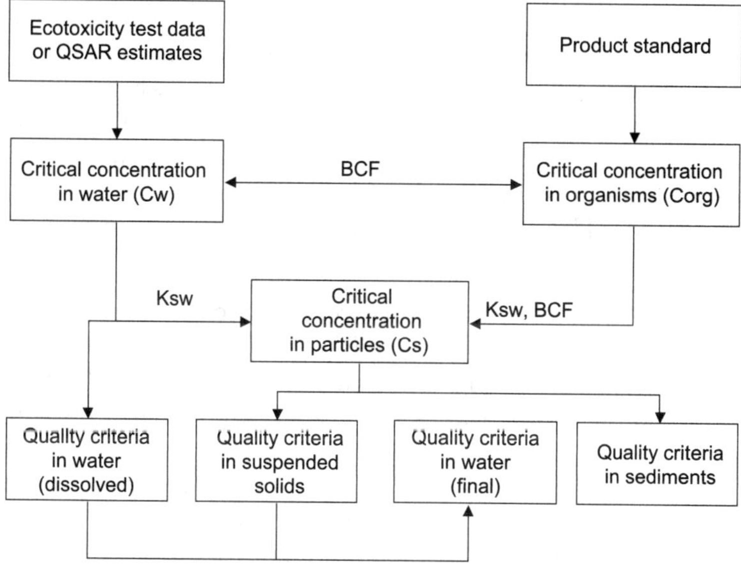

Figure 9-1 Derivation of sediment quality criteria in the Netherlands (Reprinted from *Water Research* 25(6), Van Der Kooij et al., Deriving quality criteria for water..., p 697–705, copyright 1991, with permission from Elsevier). BCF = bioconcentration factor; C_i = concentration in medium *i*; w = water; org = organism; s = sediment particle; Ksw = partition coefficient between sediment and water)

The determination of WQC is a critical step in the process. It can be done either through a conventional procedure using assessment factors or through more sophisticated approach using a statistical distribution approach (Aldenberg and Slob 1993; Aldenberg and Jaworska 2000). Using the latter model, 3 SQGs were calculated for each contaminant: 1) a high risk level (HRL) based on 50% protection of all species theoretically present in the aquatic ecosystem, 2) a maximum permissible concentration (MPC) based on 95% protection, and 3) a long-term quality target that is set at $1/100^{th}$ the MPC (except for metals, for which it is equal to the background level). The HRLs for soil and sediment were used as "intervention values," but they were replaced by MPCs for human exposure if these were lower. Sediments or soils with contaminant concentrations higher than their respective intervention values are listed for future remediation.

In theory, alternative approaches might be applied for the derivation of SQGs, either by applying assessment factors to spiked tests results or by statistical extrapolation. Nevertheless, the criteria should then be harmonized among water and sediment compartments by applying EqP theory. In the Netherlands, most of the published SQGs are obtained by EqP (Sijm et al. 2001).

The SQG values obtained represent sediment quality objectives that were initially established for the year 2000. Long-term quality objectives have been derived from these early values by applying a safety factor. The implementation of this method has sometimes created particular problems, for example, in the case of certain PAHs. The aquatic toxicity data used were in fact obtained by modeling, based on the K_{ow}, which was used again in the calculation of the K_{oc} (Van De Plassche, Polder, Canton 1992; Kalf, Crommentuijn, Van De Plassche 1997).

When the SQGs derived for metals were compared to presumed BCs, it was found that some values were lower than BCs. Because this was regarded as unrealistic, an approach was proposed that included the BCs in the derivation process (the "added risk approach"). In this approach, the hazardous concentration for 5% of the species, calculated according to Aldenberg and Slob (1993), sums the available part of the BCs with a maximum permissible addition (MPA). The criterion MPC is set equal to BC + MPA (Crommentuijn et al. 2000). The BCs used in the Crommentuijn et al. paper were based on model estimations or on literature studies and differ from the currently accepted BCs (i.e., the 90th percentile value).

Recently, a new assessment system, the chemistry toxicity test (CTT), was proposed for marine dredged materials (Stronkhorst et al. 2001). It has not yet been adopted formally. The system is based on a list of parameters, numerical values, and 3 toxicity tests. Most chemical parameters on the list have been monitored since the early 1980s (metals, aldrin, dieldrin, endrin, lindane, hexachlorobenzene, PAHs [as a group, including 10 individual compounds], DDT and related compounds [as a group], and PCBs [again, as a group based on 7 congeners]). In addition, TBT was included because of its frequent occurrence in harbor and coastal sediments. For each parameter, a numerical value is proposed as a pass/fail criterion (disposal at sea or not). Most of these values are derived from already existing standards. The proposed value for TBT (100 µg Sn/kg dry weight) was derived from toxicity tests on marine species.

United Kingdom

The UK began to consider regulations for sediment quality in 1990 and later in 1995, particularly in relation to dredged sediment disposal. These materials were considered as waste, subject to regulation based on contaminant concentrations. Because there was a unified classification system, the authorization for disposal was examined on a case-by-case basis, relying on systems of classification put forward under a different context (particularly the one suggested for the redeployment of contaminated sites, Interdepartmental Committee on the Redevelopment of Land [ICRCL]) (Challinor and John 1997). At the same time, research was being undertaken to develop EQSs for sediments.

More recently, because there were growing concerns within the Environment Agency for developing national SQGs in England and Wales, it was proposed that the effects range approach of Long et al. (1995) be adopted after validation with data on UK sediments (Rowlatt et al. 2002).

International organizations

International Commission for the Protection of the Rhine

In 1991, as part of the Rhine restoration program (so-called "PAR"), the member states (France, Germany, Luxembourg, the Netherlands, and Switzerland) adopted the concept of quality objectives for hazardous substances (Document PLEN 3/91, adopted by the plenary commission in Lenzburg on 2 June 1991 and further revised on 9 July 1992 and 1 April 1993). This concept defined 4 values to be protected, including sediments and suspended particles.

For sediments, the quality objectives must warrant the protection of

1) receiving soils (e.g., in case of deposits occurring during flood events),
2) seas beyond River Rhine's mouth, and
3) sediment-dwelling organisms (ICPR 1995).

The national experts involved in the determination of these objectives could not reach agreement on the methodology to be applied to sediment and suspended matter. This group finally considered that the current state of knowledge was insufficient to provide scientifically sound methods for protecting marine water bodies and sediment-dwelling organisms. Therefore, they restricted the quality objective setting to the first category of protected values (i.e., soils).

Consequently, the Rhine Commission adopted quality objectives for metals and arsenic in suspended particles. These values derived from the regulations applying to the disposal of sewage sludge on agricultural soils.

OSPAR Convention

In 1997, the Prevention of Marine Pollution Convention (this convention was adopted in 1992 by countries and organizations involved in Oslo and Paris Conventions, namely: Belgium, European Commission, Denmark, Finland, Germany, Iceland, Ireland, the Netherlands, Norway, Portugal, Spain,

Sweden, and the United Kingdom. Two more countries are also involved in the 1997 Convention, although they have no seashore: Luxemburg and Switzerland), currently named "OSPAR," adopted ecotoxicological assessment criteria (EAC) for trace metals, PCBs, PAHs, TBT, and several organochlorine pesticides (OSPAR 1998). A group of experts established the criteria as a range of values based on available ecotoxicological data. They were adopted for water, sediments, fish flesh, and mussels. They are used to identify potentially polluted areas (potential areas of concern) in the continuous surveillance program of the area monitored by the Convention (from the Azores Islands to Greenland). The EAC values are temporarily or definitively classified and are reviewed every 5 to 10 years. Because of chemicals, long-term effects, such as mutagenesis, carcinogenesis, and disturbed endocrine mechanisms of reproduction, it is admitted that the EAC values must not be used as standards without a scientific assessment.

Despite their limited application, the EAC values are seen as guideline values for specific contaminants in the following fields: water, sediment, and living organisms. They have no regulatory implication and simply help to define whether a monitored area should be considered of concern or not.

In addition to these monitoring guidelines, OSPAR also required member countries to develop criteria for the management of dredged sediments in their respective jurisdictions.

Use of Sediment Quality Guidelines in Sediment Quality Assessment Approaches

Introduction

There are 2 main goals for sediment quality assessments: the first is related to environment quality monitoring and ecological status maintenance and the second to dredging. Approaches related to the first goal focus on in situ sediments. In the case of dredging, the decision for removal has been already taken, and the assessment has to focus on dredged material management options.

The 2 assessment objectives are different in nature and therefore are structured differently; however, both types of approaches support decisions, often rely on tiered processes, and share the same purposes (USEPA 1998):

- assessment of the likelihood of adverse effects related to management scenarios,
- proof of causality between effects and sediment pollution, and
- increasing confidence (or uncertainty reduction).

The main difference between these approaches lies in the kind of biological assessment tools involved: more field-oriented (benthic community assessment, field toxicity tests) in the case of in situ sediments and based on laboratory toxicity tests for dredged materials, although this is not always the case.

Australia and New Zealand

The application of the Australian and New Zealand SQGs for both in situ and dredged sediments follows a tiered system, in the form of a risk-based decision tree that progresses through a hierarchy of measurements that get closer to a bioavailable fraction (Figure 9-2), in the same manner as the tiered system has adopted for water quality. For metals, the use of a dilute acid-soluble metals fraction was seen as more useful than a total measurement, although total concentrations are used in the North American guidelines from which the tiered system was adapted (Chapter 3). If the acid-soluble metals concentration exceeds the guideline trigger value, a comparison is made with BCs because in many cases the sediments are naturally mineralized. The next level of investigation for metals might be a consideration of acid volatile sulfide

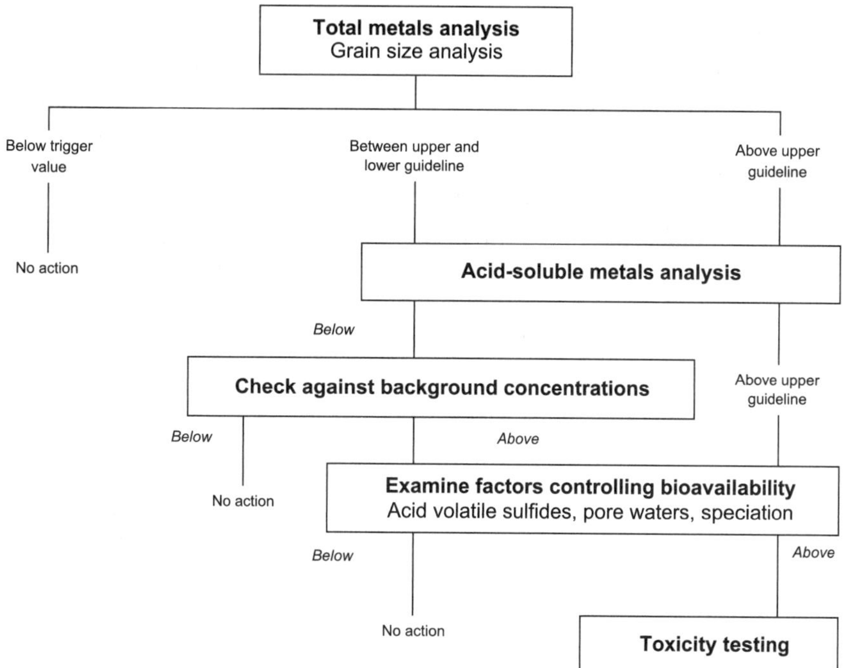

Figure 9-2 Preliminary hazard assessment decision tree for metals in sediments (ANZECC/ARMCANZ 2000)

(AVS) to simultaneously extracted metals (SEM) concentrations or porewater measurements. All of these measurements require interpretation in terms of chemical processes. If the trigger value is still exceeded, then toxicity testing is recommended.

For organic contaminants, normalization of the measured concentrations to organic carbon content is recommended, but only in the range 0.2% to 10% organic carbon because other factors might dominate bioavailability outside this range. Organic carbon normalization is not typically applied in empirical guidelines but is applied in mechanistic guidelines (Chapters 3 and 4).

The development of appropriate sediment toxicity tests is universally a high priority. While water-only exposures may be considered the most important, it will be essential that tests cover all uptake routes (pore water, overlying water, sediment ingestion, and food uptake). A suite of whole-sediment tests that consider different routes of exposure is preferred. In Australia and New Zealand, amphipods, benthic algae, bivalves, and polychaetes currently form the basis of sediment tests in use or being developed, and these can be supplemented by porewater tests using appropriate water column organisms.

In addition to their toxic potential, bioavailable contaminants have the potential to bioaccumulate (and in some cases biomagnify) as well as affect the benthic ecology. Just as an integrated chemistry–biology approach is recommended in the revised water quality guidelines, so too is it appropriate for sediments. If required, a more complete investigation using a weight-of-evidence (WOE) approach may be undertaken after the preliminary hazard assessment, although the application of this approach is not explicitly discussed in the guideline documents (ANZECC/ARMCANZ 2000). The WOE approach would combine chemical, ecotoxicological (toxicity and bioaccumulation or biomagnification), and ecological approaches but the need for a more complete WOE investigation would depend on the site being investigated, the extent of the contamination, and the management question.

The application of the interim Australian and New Zealand SQGs requires a better understanding of the processes associated with contaminant partitioning and uptake. It has been difficult to communicate an understanding of the uncertainties associated with the derived trigger values, and these uncertainties still need reinforcing (Chapter 3).

The use of acid-soluble metals is recommended, but there have been concerns that, despite this measurement being more environmentally relevant, it is not compatible with existing datasets. When the acid-soluble metals measurement is not compatible with existing databases, collecting both total and acid-soluble metals data was recommended.

It is difficult in most cases to ascribe blame for sediment contamination, except where the contamination has a point rather than a diffuse source. Equally difficult is discriminating between recent and historical contamination, the latter having occurred when environmental regulations were absent. Further, it is unlikely, given the costs involved, that remediation might be pursued. The application of the guidelines is therefore targeted more at identifying uncontaminated areas for protection, as well as noting areas of contamination. The approach has also been adopted in the Australian National Ocean Disposal Guidelines for Dredged Material (Environment Australia 2002).

The weak link in the decision tree at this stage is the assessment of impacts by toxicity testing. Often we are dealing with multiple toxicants, where the test species may be especially sensitive to only some classes of toxicants. It is important that tests encompass the range of uptake routes and include sensitive species. The expansion of the risk-based approach to include ecological impacts as well as bioaccumulation is a logical extension of the Figure 9-2 framework, along the lines of a WOE approach. This inclusion of ecological impacts is already addressed explicitly in the Australian and New Zealand guidelines that recommend a combination of biological (biodiversity) and chemical assessments. The adoption of this integration of chemistry and biology for sediments is equally desirable.

Belgium (Flanders)

Sediment quality assessment in Belgium was incorporated in a monitoring network by the Flemish Environment Agency in 2000 (de Deckere et al. 2000). Every year, 150 locations are sampled, and every 4 years, the same locations are sampled again, so in total 600 locations are included in the monitoring program. The assessment is based on a triad approach. An equal weight is assigned to each of the 3 measurements. The underlying principle of the classification of sediment quality rests on the comparison with a reference condition. This means that, for each type of measurement (physicochemical, toxicity tests, invertebrate communities), a desirable reference condition has first to be defined. This definition permits classification of sediments in the absence of fixed guidelines.

1) Physicochemical assessment

   The chemical parameters that are included in the assessment are nonpolar hydrocarbons (NPHCs), extractable organohalogens (EOX), sum of the pesticides (SOCP), sum of 7 PCBs (PCB7), sum of 6 Borneff's

PAHs (PAH6), and heavy metals (Cd, Cr, Cu, Ni, Pb, Hg, Zn, and As). The concentrations are converted to standard conditions. Based on the ratios of the measured concentrations to reference values, the site is classified. The highest class of all the contaminants becomes the overall sample class.

2) Ecotoxicological assessment

   A battery of 3 tests is used for the ecotoxicological assessment, comprising 2 porewater tests, namely a 72-h growth inhibition test with *Pseudokirchneriella subcapitata* (OECD 1984) and a 24-h acute test with the rotifer *Thamnocephalus platyurus* (de Deckere et al. 2000), and a whole sediment test, namely a 10-d acute test with *Hyalella azteca* (ASTM 2004). The porewater tests are carried out with a range of porewater concentrations. The pore water is diluted with deionized water, and nutrient medium is added to all. A porewater concentration of 0 % is used as the reference in all test series. A reference sediment is used for the whole sediment test. The ultimate ecotoxicological class is determined by the highest class of the 2 assessments (pore water and bulk sediment) and the acute impact measured on aquatic life forms.

3) Biological assessment

   Two indexes are used for the biological quality of watercourse sediments, namely the Biotic Sediment Index (De Pauw and Heylen 2001) and the percentage of mouth deformities of *Chironomus* sp. (Heylen and De Pauw 2001). The sediment is assumed to be in the good quality class if it contains species of Trichoptera, Bivalvia, and/or Gammaridae and a high diversity. The percentage of mouth deformities in reference sediments should be lower than 8%.

Finally, the results are integrated on the basis of the 3 approaches: Results obtained from each approach are converted into quality classes (from 1 = good to 4 = bad), and an overall classification is determined. This assessment method results in a rough indication of the sediment quality. Based on this indication, a priority list can be made for remediation; however, other factors such as technical and economical feasibility must be included for a final decision.

France

In France, sediment quality assessments are undertaken for 2 purposes: 1) the monitoring of water bodies, sometimes followed by more detailed diagnostics at impaired locations, and 2) the management of dredged materials.

Monitoring

Sediment quality monitoring was until recently based on analyses of priority pollutants and application of numerical SQGs. The use of toxicity tests is now seriously envisaged, following several demonstrative studies (Garric, Bonnet, et al. 1998). Two toxicity tests have been selected, the midge *Chironomus riparius* (10-d survival and growth) and the amphipod *H. azteca* (14-d survival and growth) and will be applied after the formal adoption of a standard. Observations of invertebrate communities are also done at almost all the stations of the monitoring network, but their results were usually not matched with sediment contaminants measurements. In 2001, an attempt was made to identify sensitive benthic organisms and reliable descriptive variables and possibly underlying contamination or biological response patterns (Garric et al. 2002). This first study appears interesting but should be extended with more powerful multivariate methods and a more selective approach because it appears that the impacts of contaminants are stronger in low-current sections. (The standard method for invertebrate communities' assessment is based on sampling of various habitats; the abovementioned approach looking for relationship between richness or abundance and sediment contamination did not discriminate between the habitat types.) Ultimately, either the use of toxicity tests or matching benthic observations with sediment chemistry should help to consolidate or refine the existing SQGs.

Dredged materials management

The regulatory framework for dredged materials is still under discussion for fresh waters. From a legal point of view, these materials are classified as wastes but not necessarily as hazardous wastes. Another confusing issue is the destination of the materials: A land-based deposition is covered by a different set of regulations than is disposal in waters. Guidance is needed at various levels of the management process. Some has already been introduced, for example, for the overall management process (Imbert et al. 1998), and this needs completion with the addition of specific frameworks for the evaluation of the dredged materials. Such a framework was recently proposed for the ecological part of the assessment (Figure 9-3; Babut and Perrodin 2001) by the Ministry of Equipment and Transport.

In this proposed framework, SQGs would be used at Tier 1 to determine a risk score, which may trigger 3 decisions:

1) Below a given value, the sediments can be disposed of without specific requirements.

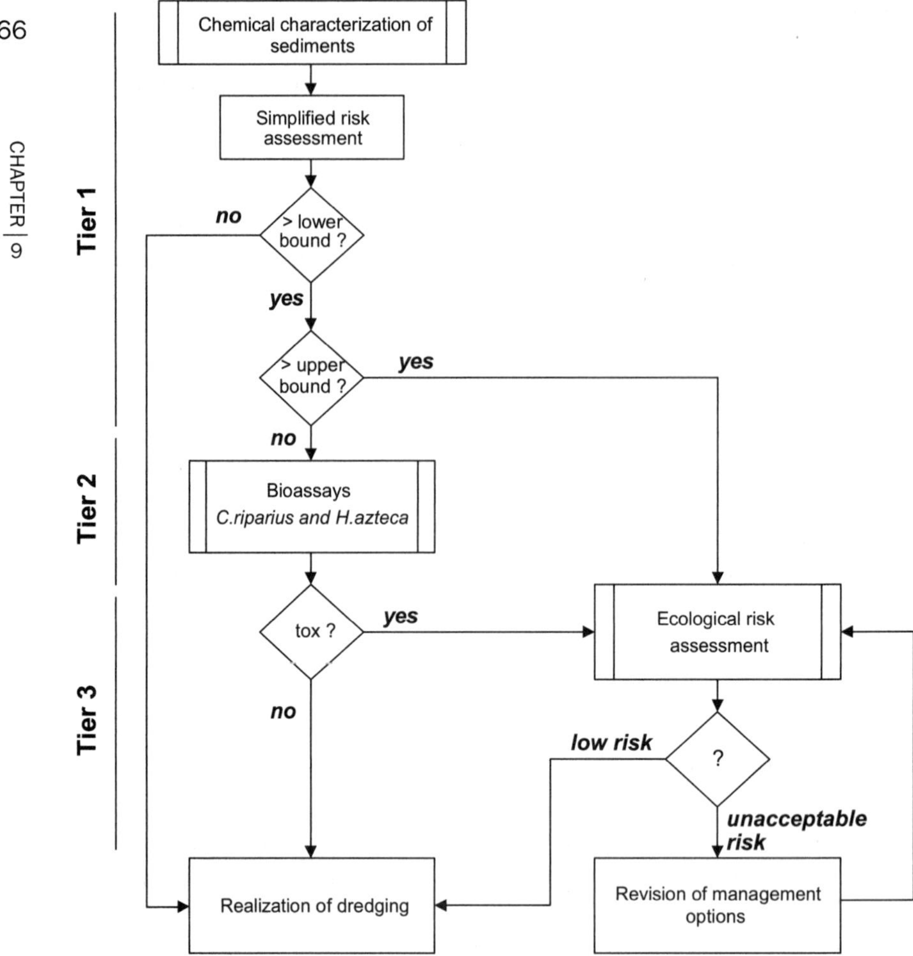

Figure 9-3 Tiered framework proposed for dredged materials ERA (France)

2) Above another given value, the intended disposal should be further assessed through a detailed ecological risk assessment (ERA), using several toxicity and elutriation tests.
3) Between these values, 2 toxicity tests should be performed. Depending on the test results, the sediments would be disposed of as in item 1) above or would be subject to an ERA.

Neither the SQGs nor the Tier 1 trigger values have yet been formally adopted; moreover, the process at Tier 2 has not been designed in detail or tested. The Tier 1 trigger values could be derived from the PECs proposed by MacDonald et al. (2000), but their appropriateness for French (or at least European) freshwater sediments would first have to be studied. In addition, these trigger values could be either the mean quotient values of 0.1 and 0.5 or different ones relevant to the distribution of toxicity incidence and chemicals concentrations in region-specific datasets, together with other considerations (Chapter 12).

The management of harbor or coastal sediments is a less complicated issue because the only disposal is by dumping at sea. The 2 guidelines derived to comply with OSPAR Convention requirements are used as follows:

- Level 1: Below this level, dredged materials could be dumped without further assessment; above it, and up to Level 2, a risk assessment should be done.
- Level 2: Above this concentration, sediments could not be disposed of by dumping at sea.

The mentioned risk assessment approach relies upon several toxicity tests: bivalve embryo toxicity (oyster *Crassostrea gigas* or mussel *Mytilus edulis*), solid-phase (Microtox), *Corophium* sp. (amphipod) mortality, and the copepod *Tigriopus brevicornis* mortality and acetyl cholinesterase inhibition. An overall risk score is determined from individual scores in each toxicity test (Alzieu et al. 2001).

Germany

The management of dredged materials in coastal areas in Germany is driven by the reference values described in "Germany" (under "Sediment Quality Guidelines in Europe, Australia, and Hong Kong," p 345) combined with ecotoxicological criteria based on the dilution factor corresponding to a no-effect level in tests on elutriates (Peters and Ahlf 2000). Three situations may occur:

1) contaminant concentrations below the RW1 reference value and low toxicity;
2) contaminant concentrations between RW1 and RW2, along with moderate toxicity; and
3) contaminant concentrations above RW2 and high toxicity.

Sediments belonging to Situation 1 may be disposed of at sea without restriction; those belonging to Situation 2 can be disposed of at sea or on land with restrictions, while source reduction should be sought. Sediments in the third category can be disposed of at sea or on land after treatment or can be prohibited from disposal (Peters and Hagner 2001).

A tiered approach combining ecotoxicological, chemical, and ecological measurements was also recently proposed by Ahlf et al. (2002). Three successive tiers are possible, depending of the goals of the assessment and the results obtained at a given tier (Figure 9-4). In contrast to former recommendations, this approach emphasizes ecotoxicological assays at the first tier, in conjunction with a standard set of chemical analyses. Analysis of specific toxicants is envisaged at the second tier, if necessary. The recommended toxicity tests at the first tier include an algal growth test, a bioluminescence inhibition assay (both applied to elutriates), and a bacterial contact assay on the whole sediment. The extended battery at Tier 3 is based on 2 series of assays applied either to the whole sediment (chironomid survival and growth, nematode growth and reproduction, and teratogenicity observed on fish eggs) or to elutriates (*Lemna minor* growth, mutagenicity with *Salmonella typhimurium*, or the Ames test).

In the tiered approach, the ecotoxicological classification was derived from a fuzzy logic model developed by Heise et al. (2000). Chemical classes are basically determined on the basis of a regional classification scheme (ARGE Elbe) or from the standards used for the management of sewage sludge (ATV standards).

The Netherlands

Freshwater sediments

Sediment quality assessments in the Netherlands also follow a tiered approach.

1) 1st tier assessment: comparison of levels of priority pollutants with national standards or guidelines. Contaminant levels are normalized according to the approach described by CUWVO (1990), to compensate for differences in sorption characteristics between sediments (standard sediment is defined as having a 25% particle fraction < 2µm and 10% organic matter on a dry weight basis). Normalized contaminant levels are then compared with the Dutch sediment quality criteria. According to the resulting classification, most polluted sediments (class 4 on a scale from 0 to 4) require a risk assessment (2nd tier).
2) 2nd tier assessment: The primary statement for this second level assessment is that, if a priority pollutant exceeds the intervention value (indicating potential risks), then the site needs to be remediated urgently, unless it is shown that there are no major risks indicated at that particular site. When the supplied data from the 2nd tier show that there is actually no high risk at a site where a priority pollutant exceeds the

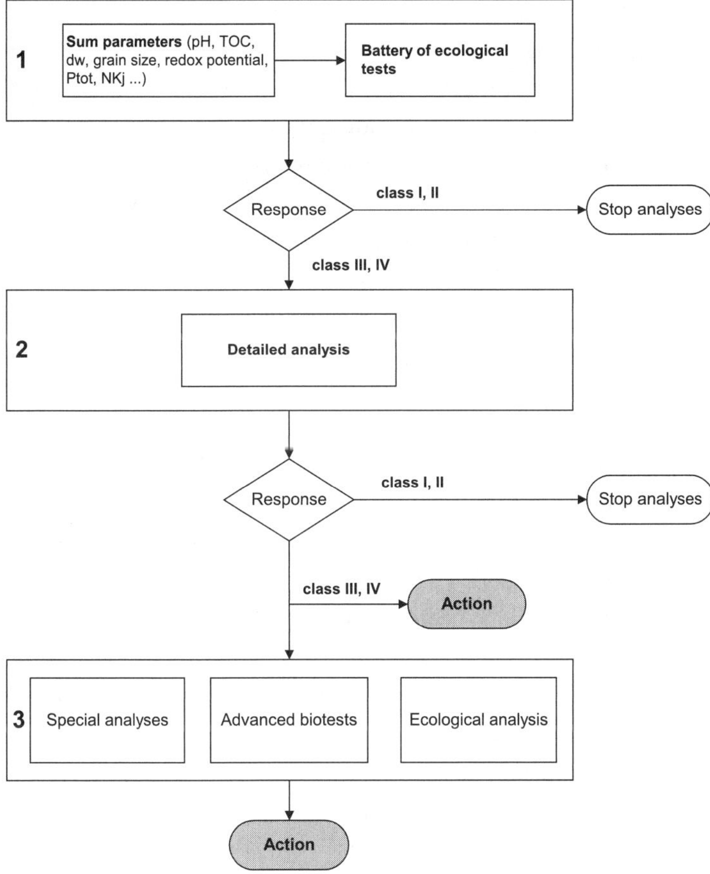

Figure 9-4 Tiered approach proposed by Ahlf et al. (2002) in Germany (Reprinted with permission from *Journal of Soil and Sediments*, copyright ecomed verlag)

intervention value, the need for remediation is no longer considered urgent. Conversely, if actual high risk were confirmed, the next step (3rd tier) would be to review different remedial options in order to achieve a risk reduction. Three main pathways are considered within the 2nd tier for achieving a complete risk assessment:

1) Human exposure: model calculations are carried out to quantify the extent to which humans (adults or children) can be exposed to contaminant via food consumption or via recreation activities in water. When the exposure exceeds maximum permissible risk criteria, actual risk is concluded.

2) Triad approach: This evaluation of risks for the ecosystem can be done using the triad assessment (Chapman 1990; Chapters 5 and 6). In the Dutch version of the triad, bioaccumulation measurements are also considered, using the results of laboratory tests or preferably using measurements on indigenous organisms (Den Besten et al. 1995). On the basis of the most sensitive parameter, sediments are classified for the categories of field observations and toxicity tests as either no effect/low risk (–), moderate effect/risk (±), or strong effect/high risk (+). The goal is to elucidate the relationship between effects on macrozoobenthos and responses of toxicity tests that, in turn, can be related to levels of chemical pollution. For that purpose, chemical concentrations are converted into toxic units (TUs), the ratio between the chemical's normalized concentration and the lowest NOEC reported in the literature among the toxicity tests included in the battery (Den Besten et al. 1995). High risk is inferred when strong effects are observed in field surveys and/or toxicity tests that can be related to chemicals present in the sediment. The information from the triad assessment can also be used in setting remediation priorities among sites (Den Besten et al. 1995).
3) Risk for transport of contaminants from the sediment to groundwater or to surface water. Model calculations are carried out in order to quantify the extent to which these processes occur. When contaminant fluxes exceed high risk criteria, actual risk is inferred.

Marine sediments

The triad approach has also been used for the characterization of in situ risks of polluted sediments. However, in the Netherlands, the reason for dredging marine sediments is usually the maintenance of shipping routes and harbors, which means that the decision on dredging is independent of the current risk observed in situ. The disposal of marine dredged material in the coastal waters of the Netherlands is assessed through a hazard assessment approach; recently, a new procedure was developed, called the "chemistry toxicity test" (CTT; Stronkhorst et al. 2001). Three toxicity tests have been selected for routine application in the CTT approach:

1) a mud shrimp toxicity test,
2) a bacterial test (solid-phase Microtox), and

3) the DR-CALUX assay, which is specific for dioxin-type compounds.

Criteria proposed for the tests are 35% mortality in the mud shrimp test, 100 TU (1/EC50) in the Microtox test, and 50 ng TEQ/kg dry weight in the DR-CALUX assay (Stronkhorst et al. 2001).

In this approach, in order to allow free disposal of sediments, SQGs need to be met both for the concentrations of listed chemicals and for the degree of effect observed in the toxicity tests.

United Kingdom

In the UK, Rowlatt et al. (2002) stressed that SQGs are insufficiently reliable to support automatic regulatory action, should guideline concentrations in sediments be exceeded. Exceedance of SQGs should always trigger investigative actions that seek to confirm or deny the predicted risk. Furthermore, they mentioned that SQGs could be useful in the UK, provided 1) they minimize false negatives (samples that are considered nontoxic when actually they are toxic), and 2) their exceedance is not the sole reason for regulatory action. SQGs should rather be used as a first screening, along with considerations of BCs.

Although tiered approaches similar to that recommended by Chapman and Mann (1999) are supported, a definite framework has not yet been developed. It is proposed that existing international approaches be studied and the chosen approach validated by establishing that the toxicity measured in toxicity tests is a good predictor of impacts on benthic communities in the UK.

Discussion

Topics related to guidelines

Comparison of published guidelines

A range of international guidelines for metals have been compared in Figure 9-5, to illustrate the differences that exist. Considering only the SQGs indicating a low effects range, there is a good agreement for Cd, Zn, and Hg (values within a range maximum to minimum of 10 or less). Conversely, the agreement appears rather poor for Ni (22), and Cu (34). Moreover, agreement is very poor (bad) for Cr (73), As (79), and Pb (129) (Figure 9-5).

For Pb, and Cr to a lesser extent, there are both high (MPC) and low values (S1). For As, there is only 1 divergent value (S1); others are within a factor of 10. Again, S1 values appear to be much lower than the others for Ni and Cu.

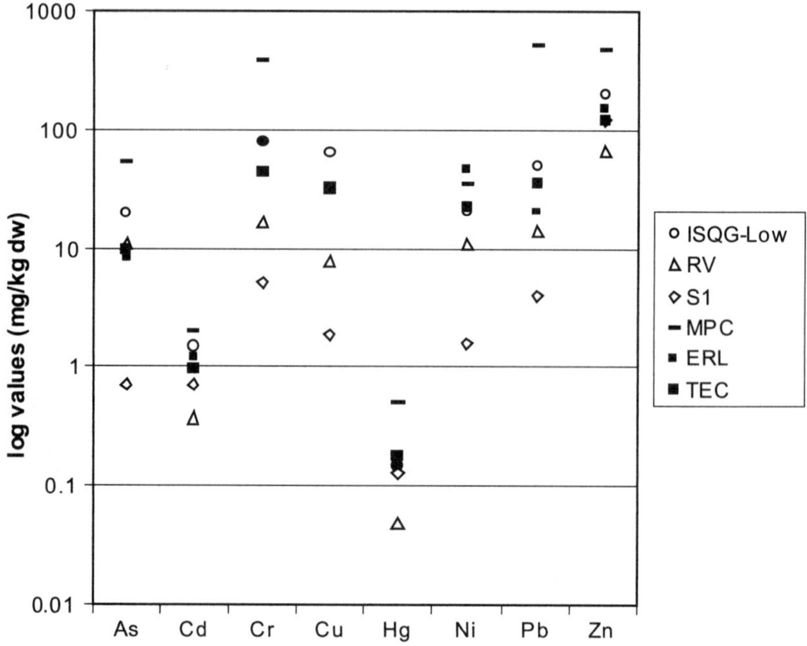

Figure 9-5 Comparison of some low-range SQGs for selected trace elements (ISQG-low: Australia; RV: reference values, Belgium; S1: France until 2002; MPC: the Netherlands; ERL: Long et al. 1995; TEC: MacDonald et al. 2000)

The causes of these discrepancies might be related to methodological aspects, to the sets of data involved in the derivation, or to both. For example, S1 values (French guideline, first-quality class boundary) are based on recalculated TEL values to which an assessment factor of 10 was applied. These TELs were derived from a subset of the BEDS database, including estuarine organisms. The decision to apply an assessment factor was justified by the uncertainties in using TEL values out of their original context (Oudin and Maupas 1999). The assessment factor explains most of the gap between S1 values and those of other guidelines.

Furthermore, the fact that the MPC values are generally higher than the other SQGs in this exercise is probably due to the derivation process: MPCs are derived by applying EqP theory, while other SQGs are empirical guidelines. Consequently, while the former solely considers the toxicity of single compounds, the latter include some implicit considerations to other chemicals eventually associated to sediments (for further discussion on that point, see Chapter 3).

It is more difficult to compare guidelines from various countries or institutions for organic compounds, although most focus on (more or less) the same list of chemicals. There are several drawbacks hampering such comparisons: For example, for PAHs, the comparison should deal with PAH mixtures because it does not make sense to consider individual compounds (Swartz 1999). However, not all countries have adopted SQGs for PAH mixtures, and those that have do not always use the same list of compounds for the mixture (some deal with 6, others with 10; still others distinguish between low and high molecular weight). Likewise, DDT compounds and PCBs should be considered mixtures, but the definitions of the mixtures may vary between countries. A tentative comparison of Sum-DDT and total-PCB SQGs is presented in Figure 9-6. While available DDT SQGs display a relatively narrow range of values, PCB SQGs are distributed over several orders of magnitude. In both examples, the highest value (MPC, Netherlands) was issued from the EqP approach.

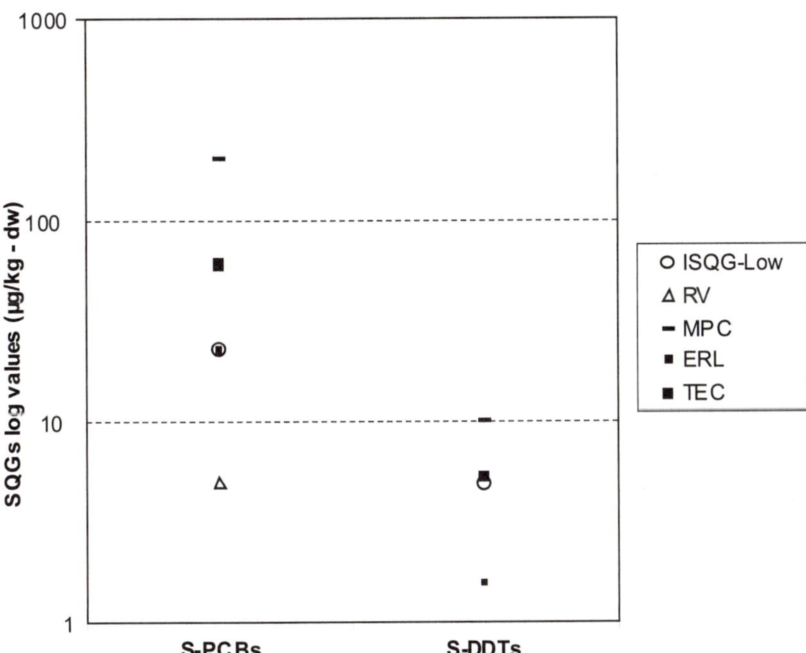

Figure 9-6 Comparison of some low-range SQGs for PCBs and DDTs (ISQG-low: Australia; RV: reference values, Belgium; MPC: the Netherlands; ERL: Long et al. 1995; TEC: MacDonald et al. 2000), added for the sake of comparison)

Methodological aspects

Three main approaches were used for determining all of the above SQGs: the EqP approach, empirical approaches, and adaptation of already published guidelines (Chapman et al. 1999; ANZECC/ARMCANZ 2000). In the latter case, the rationale behind the adaptations is not always transparent.

Most of the SQGs are not normalized to organic carbon (see discussions in Chapters 3, 13, and 17). The Dutch SQGs are normalized to organic matter (loss on ignition) and grain size (CUWVO 1990). Considerations of BCs are less commonly incorporated in the derivation process of effects-based SQGs. Except in the case of the Netherlands, few countries explicitly address the issue. Basically, it should be covered in empirical guidelines, if the database used for deriving these SQGs includes samples from a wide range of sites, because SQGs will also encompass a varied geochemical context. As a consequence, it seems likely that the more extended the database, the less BCs will be actually accounted for. Thus, it seems possible that some unexpected effects are in fact related to areas with lower BCs as compared to mean BCs in the whole database. There may be some overlaps between anthropogenic and natural concentrations, depending of the selected (statistical) descriptor of BC.

Predictive ability of SQGs

Although some of the abovementioned SQGs were published several years ago, testing of their predictive ability has not been considered until recently (Chapters 4 and 12). The most recent approaches (e.g., Chapman et al. 1999; ANZECC/ARMCANZ 2000; Rowlatt et al. 2002) all mention the need to develop databases matching chemistry and toxicity data, allowing either testing of their predictive ability or development of new SQGs. Some projects are underway in Australia and some parts of Europe, but they have not yet been published.

Nevertheless, the extent to which such databases will address new (and relevant) chemicals appears somewhat questionable. Most existing guidelines, and monitoring programs accordingly, focus on a limited number of substances (metals, DDT and its metabolites, PCBs, and PAHs). Obviously, some of these chemicals will remain a threat to ecosystems for several decades at least. However, in the meantime, many other chemicals are currently being released to the aquatic environment (pharmaceutical drugs, new domestic products, etc.). This release could explain a number of observations of unexpected toxicity.

Other pending questions

The initial purpose of a number of SQG approaches was to provide screening levels in monitoring programs or extensive field studies. Nevertheless, there have always been demands for using them as triggers for action or for regulatory applications. Such demands may lead to the adoption of criteria (pass/fail decisions) with a weaker scientific basis or at least with an insufficient consideration of uncertainties, rather than leading to the adoption of guidelines.

A good example of this kind of drift is shown by the European Water Framework Directive (Directive 2000/60/EC of the European parliament and the Council of October 23, 2000 establishing a framework for community action in the field of water policy, OJ L327, p 1–51) implementation process. In its Article 2 and Annex 5, this directive provides a series of definitions, while Article 4 sets up the goal of achieving at least a "good ecological status" within some years. This "ecological status" includes consideration of the chemical status, along with other indicators. Consequently, EQS will be derived for each priority chemical in surface water, sediments, and biota. Similar dispositions are included for estuarine and coastal waters. In brief, the boundary for the "excellent" class corresponds to background levels, while the class limit for the "good" status will be set at the EQS. Some remediation measures should be taken in case of exceedance of any standard. The main approach proposed for deriving sediment EQSs relies upon a modified EqP theory and classical no-effect concentrations for aquatic organisms (Lepper 2002). The modification of the original EqP approach seems due to a misunderstanding of its scientific basis because the uptake is supposed to occur through the water phase. An assessment factor is thus introduced for chemicals with log Kow values above 5.

Besides this typical flaw, which merely reflects the current European procedure used for assessing the ecological risk of chemicals prior to their release to the market, the underlying assessment framework appears questionable. A less stringent approach, using SQGs, whatever the derivation method, for screening purposes and eventually completing the assessment by biological means would be highly likely to be more relevant and would not afford less confidence in the decisions taken at the end of the process.

Topics related to assessment frameworks

Most, if not all, assessment frameworks described in "Use of SQGs in Sediment Quality Assessment Approaches" (p 360) are tiered, and combine chemical and biological approaches. They differ in the media component (pore water or whole sediment) used either for chemical analyses or biological testing, which means implicitly different assumptions on the hazards generated by contaminated sediments. In addition, biological methods include either toxicity tests or field invertebrate communities observations, due in part to differences in purposes.

Few frameworks, if any, take bioavailability into consideration, but parameters such as organic carbon, grain size, and AVS are often included. Whether these additional measurements actually help to refine the assessments is not clear (Chapter 3).

Finally, it appears that there is a preference for decisions relying on biological assessments rather than on numerical SQGs. When the documentation is examined in detail, it appears that SQGs should not be used alone for decisions such as prohibiting open water disposal (when the guideline is exceeded), but they could be, and in fact are, used for the opposite decision (when the guideline is not exceeded). This asymmetry means that, in practice, the risk of ecological damage is given less consideration than the risk of economic losses if the SQG-based decision to prohibit the disposal is wrong.

Conclusions

This chapter stresses that many countries throughout the world share similar concerns about sediment management. Monitoring approaches along with assessment frameworks aim at properly managing contaminated sediments and dredged materials. SQGs appear generally as a cornerstone in both approaches; however, there are also many concerns about the uncertainty associated with SQGs (Chapters 3, 16, and 17). This uncertainty is further addressed by several means: Some countries may account for it by using assessment factors (for SQGs used in monitoring approaches); others combine SQGs with toxicity tests or other biological approaches (in particular, but not only, in assessment frameworks).

Three main derivation processes were used for determining these SQGs: the EqP approach, empirical approaches, and adaptation of existing (published) guidelines. In countries other than those in North America, few attempts have been made to assess the predictive ability of SQGs, although some intend to do so. Any such assessments need to take into account the potentially confounding factor that empirical guidelines are generally based on co-occurrence of chemicals (Chapter 3). The use by many countries of SQGs as a screening tool in a tiered assessment (as discussed in Chapters 6 and 10) currently remains the best option.

References

Ahlf W, Hollert H, Neumann-Hensel H, Ricking M. 2002. A guidance for the assessment and evaluation of sediment quality: A German approach based on ecotoxicological and chemical measurements. *J Soils Sediments* 2:37–42.

Aldenberg T, Slob W. 1993. Confidence limits for hazardous concentrations based on logistically distributed NOEC toxicity data. *Ecotoxicol Environ Saf* 25:48–63.

Aldenberg T, Jaworska JS. 2000. Uncertainty of the hazardous concentration and fraction affected for normal species sensitivity distributions. *Ecotoxicol Environ Saf* 46:1–18.

Alzieu C, Casanova C, Quiniou F, Erard-Le Den E, Le Grand J, Gourmelon M, Pommepuy M. 2001. Geodrisk - Evaluation des risques liés à l'immersion de déblais de dragage. IFREMER

[APAT] Agenzia per la Protezione dell'Ambiente e per i servizi Tecnici. 2002. Messa a punto di protocolli per valutare tossicità e bioaccumulabilità. Roma: APAT. 25/2002. 53 p.

[ASTM] American Society for Testing and Materials. 2004. Standard methods for measuring the toxicity of sediment-associated contaminants with freshwater invertebrates. Method E1706-00. Annual book of ASTM standards, Volume 11.05. West Conshohocken (PA): ASTM.

[ANZECC/ARMCANZ] Australian and New Zealand Environment and Conservation Council/ Agriculture and Resource Management Council of Australia and New Zealand. 2000. Australian and New Zealand guidelines for fresh and marine water quality. Canberra, ACT (AU): ANZECC/ARMCANZ.

Babut M. 1997. Seuils de qualité pour les micropolluants organiques et minéraux dans les eaux superficielles. Agences de l'Eau, Etude nr 53. 19 p.

Babut M, Bedell JP, Bray M, Clement B, Devaux A, Delolme C, Durrieu C, Garric J, Montuelle B, Perrodin Y, Vollat B. 2001a. Evaluation écotoxicologique de sédiments contaminés ou de matériaux de dragage. Synthèse de rapport d'étude. Centre d'Etudes Techniques Maritimes et Fluviales (CETMEF), Voies Navigables de France (VNF). 12 p.

Babut M, Bonnet C, Bray M, Flammarion P, Garric J. 2001b. 2. Complément au SEQ-Eau. Seuils d'aptitude à la vie aquatique pour 28 substances phytosanitaires. Fiches substances. Ministère de l'Aménagement du Territoire et de l'Environnement. 76 p.

Babut M, Perrodin Y. 2001. Evaluation écotoxicologique de sédiments contaminés ou de matériaux de dragage. (I) Présentation et justification de la démarche. Voies Navigables de France (VNF) Centre d'Etudes Techniques Maritimes et Fluviales (CETMEF). 43 p.

Bisson M, Dujardin R, Flammarion P, Garric J, Babut M, Lamy MH, Porcher JM, Thybaud E, Vindimian E. 2000. Complément au SEQ-Eau: Méthode de détermination des seuils de qualité pour les substances génotoxiques. INERIS / Agence de l'Eau Rhin-Meuse, DRCG-EVi-00-20567. 152 p.

Bervoets L, Verheyen RF. 1994. Methodologische studie naar de inventaritie, de ecologische effecten en de sanering van de waterbodems in Vlaanderen. Rapportering fase 1: methodologische fase, Syntheseverslag. Antwerpen: Ministerie van de Vlaamse Gemeenschap - AMINAL, University of Antwerpen.

[CCME] Canadian Council of Ministers of the Environment. 1999 [rev. 2001]. Canadian environmental quality guidelines, Chap. 6: Canadian sediment quality guidelines for the protection of aquatic life. Available from: http://www.ccme.ca/publications/pubs_updates.html. Accessed 27 Jul 2004.

Challinor SL, John SA. 1997. National policies and strategies: British licensing for the disposal of dredged material to land. *Int Conf Contam Sediment* 1:171–187.

Chapman PM. 1990. The Sediment Quality Triad approach to determining pollutant induced degradation. *Sci Total Environ* 97:815–825.

Chapman PM, Allard PJ, Vigers GA. 1999. Development of sediment quality values for Hong Kong special administrative region: A possible model for other jurisdictions. *Mar Pollut Bull* 38:161–169.

Chapman PM, Mann GS. 1999. Sediment quality values (SQVs) and ecological risk assessment (ERA). *Mar Pollut Bull* 38:339–344.

Crommentuijn T, Polder M, Sijm D, De Bruijn J, Van De Plassche FJ. 2000. Evaluation of the Dutch environmental risk limits for metals by application of the added risk approach. *Environ Toxicol Chem* 19:1692–1701.

[CUWVO] coödinatie uitvoering wet verontreiniging oppervlaktewater (Dutch Commission for the Implementation of the Act on Pollution of Surface Waters). 1990. Recommendations for the monitoring of compounds of the M-list of the national policy document on water management: Water in the Netherlands: A time for action (in Dutch). The Hague: Commission for the Implementation of the Act on Pollution of Surface Waters. 90 p.

de Deckere E, de Cooman W, Florus M, Devroede-Vanderlinden M-P. 2000. A manual for the characterisation of sediments in Flemish watercourses: a TRIAD approach. Brussels: Ministry of the Flemish Community - AMINAL.

Den Besten PJ, Schmidt CA, Ohm M, Ruys M, Van Berghem JW, Van De Guchte C. 1995. Sediment quality assessment of the delta of rivers Rhine and Meuse based on field observations, bioassays and food chain implications. *Aquat Ecosyst Health Manag* 4:257–270.

De Pauw N, Heylen S. 2001. Biotic index for sediment quality assessment of watercourses in Flanders, Belgium. *Aquat Ecol* 35:121–133.

Di Toro DM, Zarba CS, Hansen DJ, Berry WJ, Swartz RC, Cowan CE, Pavlou SP, Allen HE, Thomas NA, Paquin PR. 1991. Technical basis for establishing sediment quality criteria for nonionic organic chemicals using equilibrium partitioning. *Environ Toxicol Chem* 10:1541–1583.

Environment Australia 2002. National ocean disposal guidelines for dredged material. Canberra: Commonwealth of Australia. 153 p. Available from: http://www.ea.gov.au/coasts/pollution/dumping. Accessed 27 Jul 2004.

Garric J, Baligand MP, Flammarion P, Gouy V, Montuelle B, Roulier JL. 1998. Etude d'établissement de seuils de qualité pour les sédiments. Agence de l'Eau Rhin-Meuse. 83 p.

Garric J, Bonnet C, Bray M, Migeon B, Mons R, Vollat B. 1998. Bioessais sur sédiments: Méthodologie et application à la mesure de la toxicité de sédiments naturels. Agence de l'Eau Rhône Mediterranée Corse 97.9004. 75 p.

Garric J, Flammarion P, Bonnard R, Roger MC. 2002. Etude de l'impact de la contamination toxique du sédiment sur les biocénoses benthiques: Mise en relation micropolluants - liste faunistique (IBGN). Agence de l'Eau Rhône-Méditérannée-Corse. 24 p.

Ghetti PF. 1997. Manuale di applicazione Indice Biotico Esteso (I.B.E.): I macroinvertebrati nel controllo della qualità degli ambienti di acque correnti. Trento: A.P.P.A., Provincia Autonoma di Trento. 222 p.

Heise S, Maas V, Gratzer H, Ahlf W. 2000. Ecotoxicological sediment classification: Capabilities and potential. Gewässerkunde MdBf Elbe River sediments. Bundesanstalt für Gewässerkunde. p 96–104.

Heylen S, De Pauw N. 2001. Mentum deformities in Chironomus larvae for assessment of freshwater sediments in Flanders, Belgium. Paper presented at 28th SIL Congress; 2001 Feb 4–10; Melbourne, Australia.

[ICPR] International Commission for the Protection of the Rhine. 1995. Rhine Action Program: Data sheets for the quality objectives. Koblenz: International Commission for the Protection of the Rhine. 224 p.

Imbert T, Py C, Duchene M. 1998. Enlèvement des sédiments - Guide méthodologique - Faut-il curer ? Pour une aide à la prise de décision. Douai (FRA): Pôle de compétence sur les sites and sols pollués Nord/Pas de Calais - Agence de l'Eau Artois-Picardie. 161 p.

Kalf DF, Crommentuijn T, Van De Plassche EJ. 1997. Environmental quality objectives for 10 polycyclic aromatic hydrocarbons (PAHs). *Ecotoxicol Environ Saf* 36:89–97.

Koethe HF, Bertsch W. 1997. Directive for the handling of dredged material on federal inland waterways in Germany. *Int Conf Contam Sediment* 1:161–170.

Lepper P. 2002. Towards the Derivation of Quality Standards for Priority Substances in the Context of the Water Framework Directive: Identification of quality standards for priority substances in the field of water policy. Fraunhofer-Institute Molecular Biology and Applied Ecology. B4-3040/2000/30637/MAR/E1. 131 p.

Long ER, Macdonald DD, Smith SL, Calder FD. 1995. Incidence of adverse biological effects within ranges of chemical concentrations in marine and estuarine sediments. *Environ Manag* 19:81–97.

MacDonald DD, Carr RS, Calder FD, Long ER, Ingersoll CG. 1996. Development and evaluation of sediment quality guidelines for Florida coastal waters. *Ecotoxicology* 5:253–278.

MacDonald DD, Ingersoll CG, Berger TA. 2000. Development and evaluation of consensus-based sediment quality guidelines for freshwater ecosystems. *Arch Environ Contam Toxicol* 39:20–31.

[OECD] Organisation for Economic Co-operation and Development. 1984. Algal growth inhibition test. OECD guidelines for the testing of chemicals, Nr 201. Paris: OECD.

[OSPAR] (Ad Hoc Working Group on Monitoring). 1998. Ecotoxicological assessment criteria endorsed by OSPAR 1997 for trace metals, PCB, PAHs, TBT and some organochlorine pesticides. OSPAR, MON 98/2/Info.2-E.

Oudin LC, Maupas D. 1999. Système d'évaluation de la qualité de l'eau des cours d'eau. Rapport de présentation SEQ-Eau (version1). Agences de l'eau Orléans, Etude Inter Agences, Etude Inter Agences, n° 52 / 2 volumes d'annexes. 59 p.

Pellegrini D, Onorati F, Virno Lamberti C, Merico G, Gabellini M, Ausili A. 2002. Aspetti tecnico-scientifici per la salvaguardia ambientale nelle attività di movimentazione dei fondali marini: dragaggi portuali. *Quaderno ICRAM* 1: 1–201.

Peters C, Ahlf W. 2000. The necessity of a marine bioassay test-set to assess marine sediment quality and its approach. River sediments and related dredged material in Europe: Scientific background from the viewpoints of chemistry, ecotoxicology and regulations. GKSS Research Centre, Geesthacht, Germany.

Peters C, Hagner HH. 2001. Current and future policies and regulatory framework: The National Policy Framework in Germany. In: Gandrass J, Salomons W, editors. Dredged materials in the Port of Rotterdam: Interface between Rhine Catchment Area and the North Sea. Geeshacht (Germany). p 160–183.

Peters C, Becker S, Noack U, Pfitzner S, Bülow W, Barz K, Ahlf W, Berghahn R. 2002. A marine bioassay test set to assess marine water and sediment quality: Its need, the approach and first results. *Ecotoxicology* 11:379–384.

Rowlatt S, Matthiessen P, Reed J, Law R, Mason C. 2002. Recommendations on the Development of sediment quality guidelines for the Environment Agency of England and Wales. Bristol (UK): Environment Agency.

Sijm D, De Bruijn J, Crommentuijn T, van Leeuwen K. 2001. Environmental quality standards: Endpoints or triggers for a tiered ecological effect assessment approach? *Environ Toxicol Chem* 20:2644–2648.

Smith SL, MacDonald DD, Keenleyside KA, Ingersoll CG, Field J. 1996. A preliminary investigation of sediment quality assessment values for freshwater ecosystems. *J Gt Lakes Res* 22:624–638.

Stronkhorst J, Schipper CA, Honkoop J, van Essen K. 2001. Disposal of dredged material in Dutch coastal waters: A new, effect-oriented assessment framework. The Hague: National Institute for Coastal and Marine Management (RIKZ),

Ministry of Transport Public Works and Water Management. RIKZ/2001.030. 77 p.

Swartz RC. 1999. Consensus sediment quality guidelines for polycyclic aromatic hydrocarbon mixtures. *Environ Toxicol Chem* 18:780–787.

[USEPA] US Environmental Protection Agency. 1998. Guidelines for ecological risk assessment. Washington DC: USEPA. EPA-630/R-95/002F. 159 p.

Van De Plassche EJ, Polder MD, Canton JH. 1992. Maximum permissible concentrations for water, sediment and soil derived from toxicity data for nine trace metals. Bilthoven (NL): RIVM. Report nr 679101002. 93 p.

Van Der Gaag MA, Stortelder PBM, an Der Kooij LA, Bruggeman WA. 1991. Setting environmental quality criteria for water and sediment in The Netherlands: A pragmatic ecotoxicological approach. *Eur Water Pollut Control* 1:13–20.

Van Der Kooij LA, Van De Meent D, Van Leeuwen CJ, Bruggeman WA. 1991. Deriving quality criteria for water and sediment from the results of aquatic toxicity tests and product standards: application of the equilibrium partitioning method. *Water Res* 25:697–705.

Viganò L, Arillo A, Buffagni A, Camusso M, Cianarella R, et al. 2003. Quality assessment of bed sediments of the Po River (Italy). *Water Res* 37:501–518.

Use of sediment quality guidelines in existing assessment frameworks

10

WALTER J BERRY, TODD S BRIDGES, STEPHEN J ELLS, THOMAS H GRIES, D SCOTT IRELAND, EILEEN M MAHER, CHARLES A MENZIE, LINDA M POREBSKI, JOOST STRONKHORST

This chapter provides background on how assessment frameworks have been used to guide management decisions with respect to contaminated sediments. It also indicates how sediment quality guidelines (SQGs) have been used within these frameworks. The authors reviewed these frameworks to gain insights into the relationships between management objectives and frameworks structures and to provide the background for a suggested framework approach that would have broad application (Chapter 6). Sediment assessment frameworks aid management decisions with an orderly and logical match between the objectives of a sponsoring program and available assessment tools. Because management objectives related to sediments vary among programs (e.g., guiding dredging decisions versus site remediation), differences in objectives influence framework structure. It follows that differing objectives also influence the manner in which the assessment tools, such as SQGs, are used within that framework. This chapter is a discussion of the elements and structure of a number of sediment assessment frameworks used around the world. Chapter 9 provides more information on European, Australian, and Hong Kong frameworks.

The major applications of the reviewed frameworks are these:

- dredged material assessment in which the management decision often focuses on the risk of various disposal options rather than on whether to dredge,
- sediment cleanup in which the decision commonly involves whether dredging is the most feasible remedial alternative and how much volume to remove,

Wenning RJ, Batley GE, Ingersoll CG, Moore DW, editors.
 ISBN 1-880611-71-6

- sediment monitoring related to previous cleanup or dredging activities,
- support for source reduction or total maximum daily load (TMDL) programs in which contaminated sediments have been determined to be a "cause of impairment," and
- state of the environment monitoring programs (impact and trend detection).

Among the reviewed sediment assessment frameworks, only one did not employ a tiered structure. A tiered structure allows an a priori set of inclusionary and/or exclusionary criteria to evaluate whether a specific environmental condition meets one of several conclusions, usually no risk, high risk, or potential but unquantified risk. In the last case, the evaluation proceeds to the next tier for evaluation against either modified criteria or application of a more rigorous evaluative scheme. There can be any number of tiers, each with their own evaluation criteria, but the limit is 3 or 4 in most frameworks.

The reviewed frameworks incorporated 6 basic types of data:

1) site descriptions,
2) sediment physical characteristics,
3) chemical,
4) toxicity tests,
5) bioaccumulation, and
6) benthic community.

The first 4 types of data are common to all reviewed frameworks. Bioaccumulation and benthic community data are included much less frequently. Chemistry is typically performed in an earlier tier relative to biological testing.

Dredged Material Assessment Frameworks

1996 Protocol to the London Convention on the Prevention of Pollution from the Disposal of Waste or Other Matter at Sea

Approximately 80 nations are parties to The London Convention 1972, which is the international treaty on the prevention of marine pollution caused by dumping of wastes and other matter (LC 1972). In 1996, members developed guidelines for the assessment of wastes intended for disposal at sea. These "guidelines" have been placed into a new international agreement or treaty called the "1996 Protocol," which, when it comes into force, will eventually supersede the London Convention. The protocol calls upon parties to control disposal in their marine waters by establishing a permit system, assessing wastes according to the guidelines, using a precautionary approach, reducing marine pollution at the source, and reporting on these activities. The guidelines have also been adapted specifically for dredged material (Figure 10-1). Countries (including Canada, Hong Kong China, and the United States [US]) that are parties to one or both treaties are invited to implement the general and the specific guidelines.

The guidelines form an assessment framework; Figure 10-1 illustrates the tiered nature of the assessment. After some initial exclusionary criteria (exempting sand and material away from known sources of pollution from extensive testing), the framework proceeds with the following:

- Evaluation of alternatives to ocean disposal, including beneficial uses: For dredged material, the goal of waste management is to identify and control the sources of contamination.
- Waste characterization: Data collection and evaluation of sources and spills and historical use are required. Information concerning the physical and chemical and available biological characteristics of the sediment is needed. When the above information is insufficient to assess potential impacts, biological assessment is required to determine the potential for acute and chronic toxicity and the potential for bioaccumulation and tainting.
- Decisions on acceptability of the material: The guidelines instruct the national authorities to construct a national action list that provides action levels for contaminants of concern. The guidelines provide examples of contaminants of concern from anthropogenic sources such as Cd, Hg, organohalogens, petroleum hydrocarbons, As, Pb, Cu, Zn, Be, Cr, Ni, V, organosilicon compounds, cyanides, fluorides, pesticides or their by-products, but each country may make their own list of contaminants that are toxic, persistent, and bioaccumulative.
- Comparison to action lists: The action list must include an upper level that avoids acute or chronic effects on human health or on sensitive organisms representative of the marine ecosystem and may also include a lower level, below which effects are unlikely to occur. Action levels may be concentration limits, biological responses (e.g., toxicity limits), environmental quality standards, flux considerations, bioaccumulation potential, potential for tainting food resources, and other reference values.

Under this framework, SQGs could be used at upper or lower levels, depending on the derivation and narrative intent of the guidelines chosen. Lower

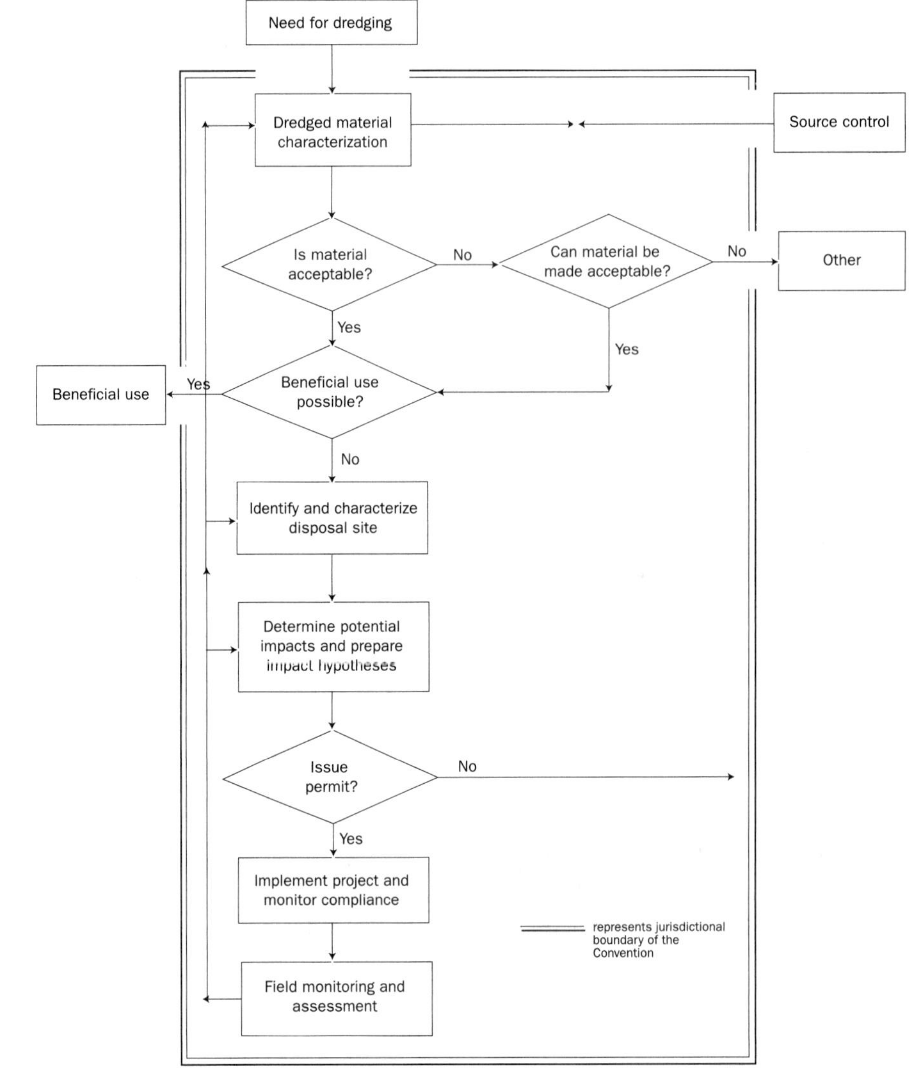

Figure 10-1 Waste assessment framework for the London Convention (after LCSG 2001)

action levels may be SQGs or biological responses, predictive of no effect and used as screening criteria, whereas lower levels would either trigger more detailed assessment or indicate little concern. Upper action levels may be SQGs, biological responses predictive of effects, or a combination of these tools. When upper levels are exceeded, material shall not be disposed at sea unless it is made acceptable for dumping through the use of management techniques or processes.

- Decisions on disposition of the material: Following assessment of the sediment is the selection and assessment of a suitable disposal site, an overall statement on the expected consequences of disposal, and the issuing of a permit where appropriate.
- Monitoring and follow-up: Monitoring is used to verify that permit conditions were met and that the assumptions made were correct and sufficient to protect human health and the environment. SQGs may also be used during this follow-up or management phase.

The following sections describe the implementation of the guidelines for dredged material in the US navigation dredging program, the Canadian dredging program, and the Hong Kong, People's Republic of China, dredging program.

US Navigation Dredging Program

In the US, Section 103 of the Marine Protection, Research and Sanctuaries Act (MPRSA) regulates dredged material disposal at ocean sites. The Ocean Testing Manual (OTM) provides guidance for conducting evaluations of material proposed for ocean disposal (USACE/USEPA 1991). Section 404 of the Clean Water Act (CWA) regulates dredged material management in the aquatic environment landward of the baseline of the territorial sea. The Inland Testing Manual (ITM) provides guidance for conducting evaluations of material proposed for management in these territorial waters (USACE/USEPA 1998).

US navigation dredging program assessment framework

The OTM and ITM use the same 4-tiered evaluative framework to characterize exposure and effects. Each tier has explicit data requirements that inform the decision to either accept the dredged material for a specific management option or to move to the next tier with its broader data requirements. The framework specifies the data requirements and decision criteria at each tier (Figure 10-2).

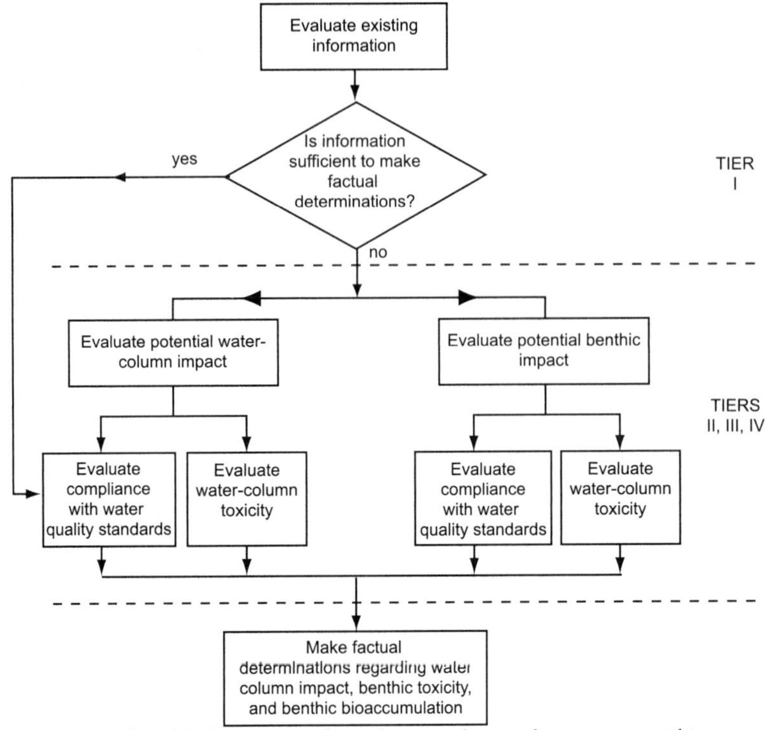

Figure 10-2 Simplified overview of tiered approach to evaluating potential impact of aquatic disposal of dredged material in the US dredging program (after USEPA/USACE 1998)

Tier I involves the collection and analysis of existing information on the physical, chemical, and biological properties of the dredged material. This includes contaminant sources, examination of recent chemical and biological testing results, and limited optional testing to substantiate that the material can be excused from further testing on the basis of relevant exclusion criteria (e.g., course-grained material, geographically removed from sources of contamination).

Tier II involves the collection and use of whole sediment or elutriate chemistry data in combination with numerical mixing models to determine compliance with existing and relevant water quality standards during disposal operations (Figure 10-3). Whole-sediment chemistry data for nonpolar organic contaminants can be used within Tier II to develop estimates of bioaccumulation potential in benthic receptors using sediment total organic carbon (TOC), organism lipid content, and chemical-specific biota–sediment accumulation

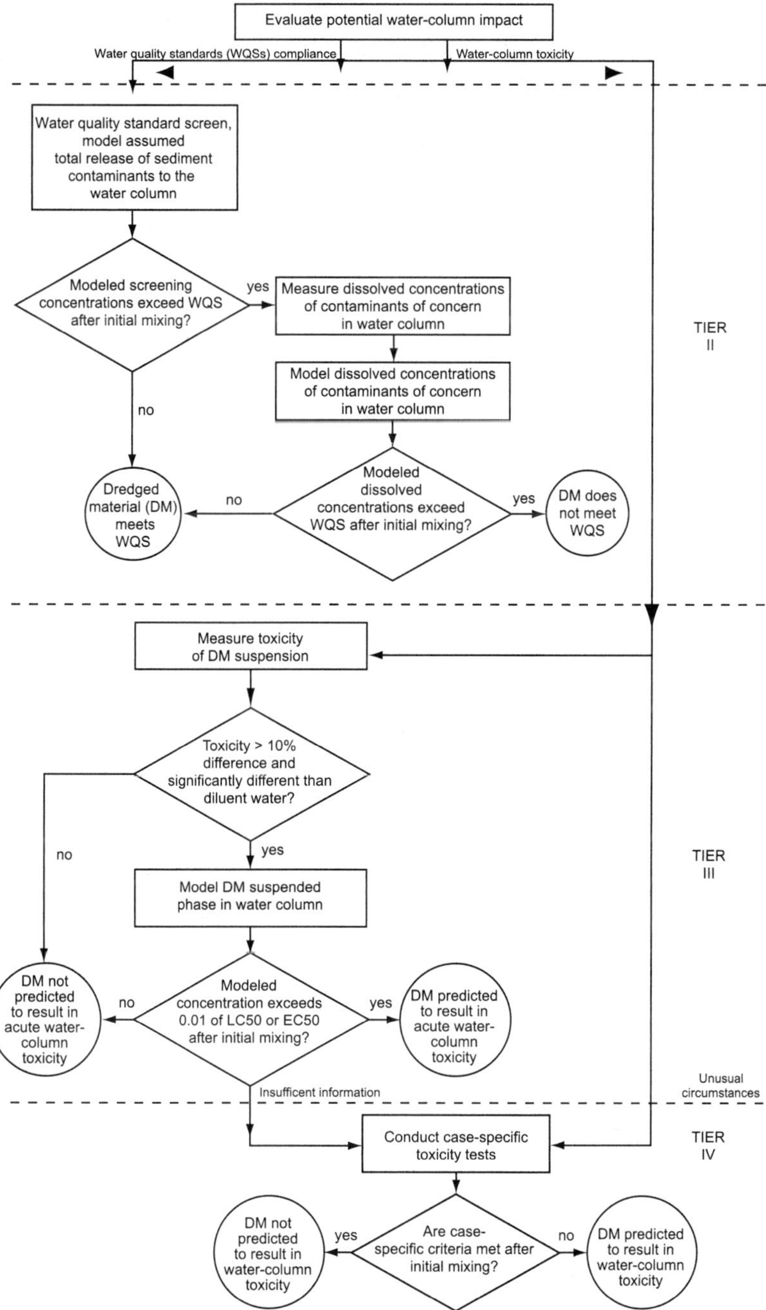

Figure 10-3 Tiered approach to evaluating potential water column impacts of deposited dredged material (after USEPA/USACE 1998)

factors (BSAFs) (Figure 10-4). Acceptance of material for disposal options under Tier II is likely only in cases where concerns are limited to potential water column impacts during disposal operations.

Tier III data include toxicity and multiple-species bioaccumulation testing. Figure 10-3 shows the use of toxicity tests of dredged material suspensions in

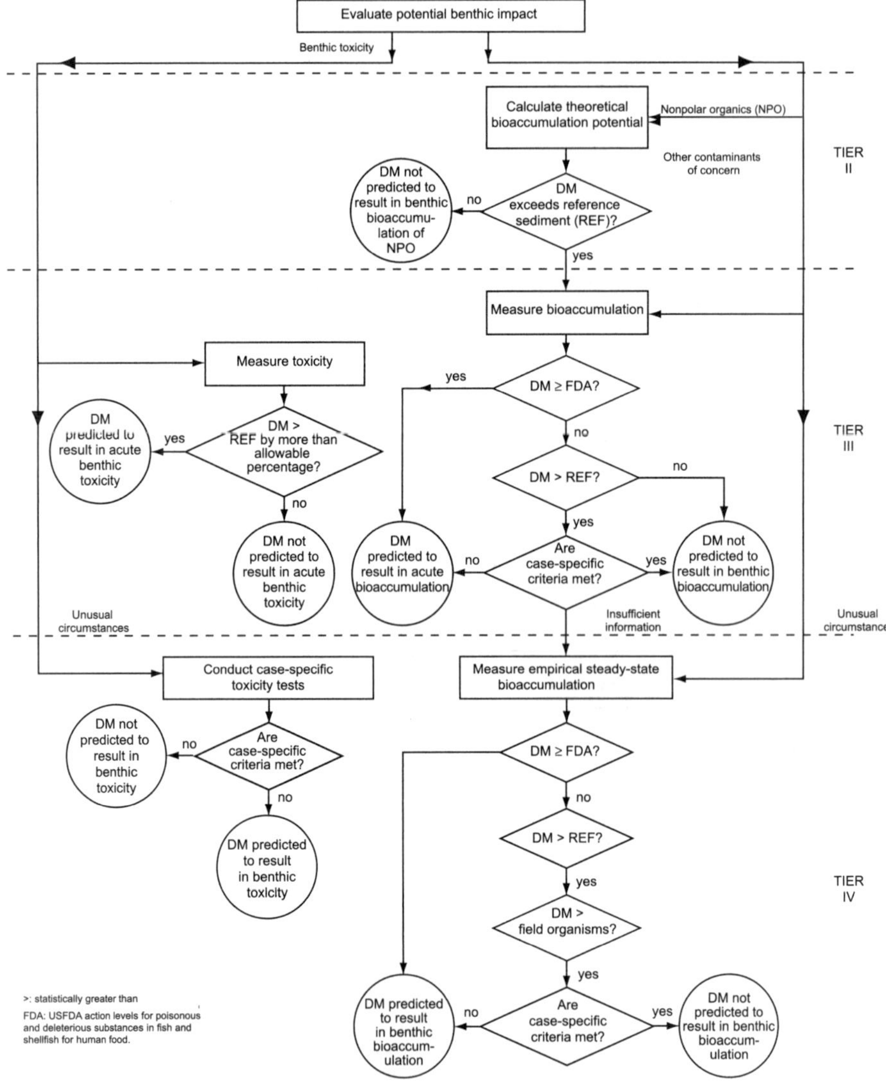

Figure 10-4 Tiered approach to evaluating potential benthic impacts of deposited dredged material (after USEPA/USACE 1998). REF = Reference. FDA = USDA action levels for poisonous and deleterious substances in fish and shellfish for human food. DM = dredged material.

combination with mixing models to compare to acceptance criteria, which includes comparison to toxicity of diluent water. Figure 10-4 shows the evaluation of benthic exposures and effects using solid-phase toxicity and bioaccumulation tests. The toxicity test acceptance criteria employ comparisons between survival in test sediments and reference sediment (reference sediments are relatively free of contaminants and represent conditions near the disposal site absent any disposal activity). The acceptance criteria for bioaccumulation test results include comparing tissue concentrations in test animals exposed to dredged material for 28 d to relevant action levels (e.g., US Environmental Protection Agency [USEPA] or US Food and Drug Administration [USFDA] advisory levels) and then to tissue concentrations in test animals exposed to reference sediments. Most dredged material evaluations are concluded with a compliance decision by the end of Tier III testing.

Tier IV applies to exceptional cases in which compliance decisions could not be made within the preceding tiers and includes any case- or project-specific evaluations that may be required to address unanswered questions. Past examples of Tier IV evaluations have included the use of chronic, sublethal toxicity tests, steady-state bioaccumulation testing, and the application of risk assessment to guide interpretation of bioaccumulation results (Figures 10-3 and 10-4).

Use of sediment quality guidelines in the US dredging program

Neither the ocean disposal nor inland waters disposal regulations specifically require or endorse the use of sediment quality standards to make compliance determinations for unrestricted open water disposal of dredged material. The OTM explicitly states, "While sediment chemistry cannot be used to predict biological effects, it can be used to identify contaminants of concern" (USACE/USEPA 1991, p 2–7).

A greater role for SQGs is possible in the inland waters disposal program, as indicated in a US Army Corps of Engineers (USACE) 1998 policy memorandum (Fuhrman 1998):

> It is the policy of the Corps that SQGs may be used only as an initial screen for determining if higher "effects based" tiers are needed. If available SQGs and other information indicate that there is no "reason to believe" contaminants are present, no further chemical or toxicological evaluations at higher tiers are necessary…

This policy further explains that regulatory decisions made using effects-based evaluations will rely "…on a preponderance of evidence derived from biological, physical and chemical assessments" of the material in question. USACE (1998) published a Technical Note on the use of SQGs as a guidance vehicle for field personnel, which states that

> SQGs may be useful as initial screening values in Tier I or Tier II of dredged material evaluations ...as a part of the reason to believe assessment for the presence of sediment contaminants, if appropriate consideration is given to the uncertainties previously described.

This Technical Note describes a number of existing SQGs but does not endorse or encourage the use of a specific set of SQGs in the dredging program. Since 1998, little if any direction or guidance regarding the use of SQGs has been issued at the national level in the US dredging program.

Canadian Dredging Program

Environment Canada controls ocean or estuarine disposal of waste under the Canadian Environmental Protection Act (CEPA 1999). A permit system is in place for dredged material destined for open water disposal. Permit application criteria and information requirements are regulated under the Regulations Respecting Applications for Disposal at Sea (CEPA 2001). Application guidance is provided through the national and regional disposal at sea websites (http://www.ec.gc.ca/seadisposal/main/index_e.htm). All permits are published, and the public has a 30-d review period in which to comment. Material not suitable for disposal at sea may be left in place or disposed of or treated on land under various other jurisdictions. Provincial bodies and other federal departments (Environment Canada, Transport Canada, Fisheries and Oceans Canada) control dredging and disposal in internal fresh waters (see, e.g., Persaud et al. 1992).

With respect to ocean disposal, maintenance dredging is the predominant activity, with smaller amounts of capital (new work) dredging. In general, levels of contamination are low, and approximately 2 to 5 million cubic meters are disposed of annually under permit. The CEPA legislation does not exempt sand from consideration, nor is there a de minimus quantity below which a permit is not needed. Canada is a party to the 1996 Protocol and has integrated its waste assessment guidance into CEPA 1999. The program specifies the data requirements and decision criteria at each tier.

Canadian assessment framework

The framework for dredged material assessment mirrors the specific 1996 Protocol guidelines in Figure 10-1. Because each 1996 Protocol party must develop a national action list to make decisions on the acceptability of their wastes for ocean disposal, Canada has placed its action list requirements in the *Disposal at Sea Regulations* (CEPA 2001, Figure 10-5). The data requirements in Tier 1 include an assessment of the presence of contaminants and selection of contaminants of concern through an evaluation and mapping of sources and spills, historical use, and chemical and biological data. Minimum information requirements for chemistry are Cd, Hg, total polycyclic aromatic hydrocarbon (PAH), and polychlorinated biphenyls (PCBs). Assessment proceeds to Tier 2 when contaminants of concern exceed regulated lower action levels (CEPA 2001), interim SQGs (CCME 1999), or more site-specific no-effects levels.

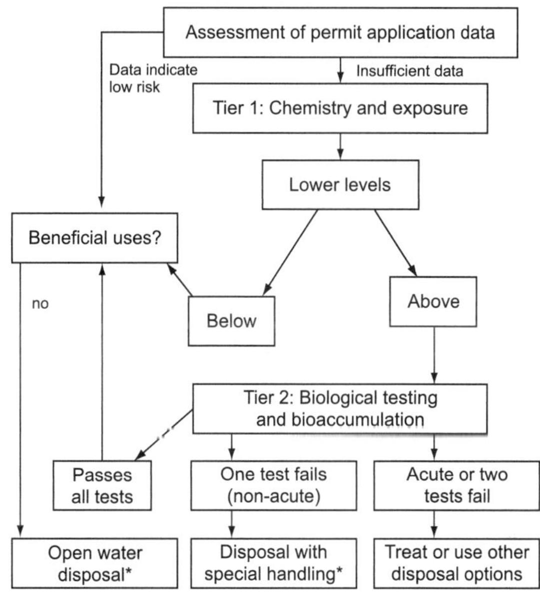

Figure 10-5 Evaluation flowchart for the Characterization of Dredged Material section of the waste assessment guidance of the 1996 Protocol as used in the Canadian dredged material disposal program (adapted from the CEPA 1999, 2001). DMAF = Dredged material assessment framework.

Tier 2 includes evaluating acute and sublethal toxicity and bioaccumulation to address exposure and effects (in the sediment and pore water) and uptake from ingestion, dermal, or respiratory routes. Program policy that favors the use of sessile, representative species at the low end of the food web, under controlled conditions, is thought to be conservative and appropriate (L. Porebski, personal communication). Toxicity tests in the analysis framework include

- 10-d survival tests with amphipods in whole sediment,
- Microtox solid-phase test measuring metabolic light reduction,
- 20-min fertilization test with echinoids in pore water,
- 14-d polychaete growth and survival test in whole sediment (pending), and
- 28-d bivalve bioaccumulation in whole sediment.

The above testing must be done using standardized test methods with acceptability criteria and control or reference sediments.

Benthic assessment is used in disposal site monitoring to assess recovery at the site or to gauge the potential for effects on the area surrounding the disposal site. If material with a chemical exceedance in Tier 1 "passes" all the biological tests in Tier 2, the program regards the chemical contamination as being non-bioavailable, and the sediment is acceptable for disposal. If the sediment fails only one of the biological tests, the sediment can be disposed of with special handling, such as capping. Interpretation criteria appear either within the test method or in separate guidance (Porebski and Osborne 1998).

Two biological test failures plus chemical exceedance of a guideline are deemed sufficient to conclude that environmental effects are likely, and a permit would be refused unless the dredged material is made acceptable for disposal through the use of management techniques or processes.

Use of sediment quality guidelines in Canadian dredging program

In Tier 1, the Canadian program uses interim SQGs (ISQGs; CCME 1999) to screen contaminants of concern other than Cd, Hg, total PAH, and PCBs, which are regulated (CEPA 2001; Table 10-1). All the national levels used are in dry weight and are not normalized in any way.

The procedures used in deriving Canadian SQGs for the protection of aquatic life are described in the Protocol for the Derivation of Canadian Sediment Quality Guidelines for the Protection of Aquatic Life (CCME 1995).

The SQG protocol balances information from 2 approaches. The modified National Status and Trends Program approach relies on field data that demonstrate associations between chemicals and biological effects. The Spiked-Sediment Toxicity Test approach establishes cause–effect relationships. Spiked-sediment toxicity data are currently available for only a few substances such as Cd, Cu, fluoranthene, and pyrene. Therefore, the threshold effect levels (TELs) calculated using the modified National Status and Trends Program approach are most likely to be adopted ISQGs. The current ISQGs are similar to the TELs developed for Florida (MacDonald 1994). Environment Canada and the Canadian Council of Ministers of the Environment (CCME) are currently conducting research on the feasibility of upgrading some of these guidelines (Porebski personal communication).

The use of site-specific guidelines may be considered on a case-by-case basis for disposal at sea. The SQG protocol allows for development of site-specific levels that can take into account local conditions and background concentra-

Table 10-1 Canadian interim sediment quality guidelines and regulated lower and upper levels for ocean disposal of sediments

Contaminant [a]	Lower level	Upper level
Metals (mg kg^{-1} dry weight)		
Cadmium [b]	0.6	Bioassays [c]
Chromium	52.3	
Copper	18.7	
Mercury [b]	0.75	
Nickel [b]	NA	
Lead	30.2	
Silver	NA	
Zinc	124	
Metalloid (mg kg^{-1} dry weight)		
Arsenic	7.2	
Organics (µg kg^{-1} dry weight)		
Chlordane	2.26	
DDD	1.22	
DDE	2.07	
DDT	1.19	
Dieldrin	0.71	
Endrin	2.67	
Heptachlor epoxide	0.6	
Lindane	0.32	
PCDD/Fs	0.85 ngTEQ·kg^{-1} dw	
Toxaphene	0.1	
Total PAHs	2500	
Total PCBs	100	
Organometallics (mg TBT L^{-1} in interstitial water)		
Tributyltin [b]	NA	

[a] To ensure that the chemistry-based assessment is conducted in a protective fashion the guidelines are compared to the upper 95% confidence limit of the mean of the sediment area being considered. This allows for some variability and heterogeneity and acknowledges that the material will be mixed upon dredging and disposal. The guidelines themselves are taken at face value to enable consistent decision making. Biological testing could be used to integrate the responses caused by chemicals for which no guidelines yet exist, or which have not been measured, but to date this has not been done. Sources: CCME 1999; CEPA 2001.

[b] Regulated levels. Source: CEPA 2001. Note: ISQG values differ.

[c] Results from bioassay tests are used in lieu of chemical measurements for upper action levels.

tions. It is the policy of the CCME, however, that modifications of guidelines to site-specific objectives should not be made on the basis of aquatic ecosystem characteristics that have arisen as a direct result of previous human activities.

For contaminants for which no final guidelines have been derived or in areas where intense efforts to characterize the sediments have taken place, other marine sediment guideline values may be used if they are deemed appropriate for the area under assessment. These guidelines would use various approaches but would retain the same narrative definition of levels below which effects on sensitive organisms are not likely.

The regulated lower action levels and SQGs are used as screening levels, which contribute to a weight-of-evidence (WOE) decision. Where exposure assessment reveals no proximity to sources of pollution, spills, or historical contamination and sediment analysis indicates that lower action levels are not exceeded, the sediment is assessed as being a low risk to the marine environment. A single exceedance of a lower action level is sufficient, under the current approach, to trigger the need for biological testing. There is no chemical level above which a permit would be refused on the basis of chemistry alone; however, applicants often seek other disposal options when their sediment exceeds lower levels in order to avoid performing costly biological tests with uncertain outcomes (L. Porebski, personal communication).

Hong Kong Dredging Program

The Director of Environmental Protection (DEP) and the Director of Civil Engineering (DCE) regulate ocean disposal of sediment in Hong Kong, one of the world's busiest ports. The DEP is in charge of dumping at sea, and DCE is in charge of sediment disposal for land reclamation. DEP is the authority for the licensing and statutory control of marine disposal of dredged sediment, but the Fill Management Committee (FMC), of which DCE is the Chair, is responsible for the management of sediment disposal capacity at different disposal sites. Once sediments have been dredged, they are disposed of either in marine borrow pits, confined disposal facilities, or open seafloor disposal sites, depending on the degree of sediment contamination.

The SQG framework is designed to ensure that only those sediments with low contaminant levels or those that do not cause significant biological effects are approved for open water disposal. It also ensures that sediments with high contaminant levels or that cause significant biological effects would be isolated from the marine environment by confined disposal or subjected to other special disposal arrangements under special circumstances.

Hong Kong assessment framework

The Hong Kong framework uses chemical analysis to detect contaminants and bioassays to assess sediment toxicity in a 3-tiered system (Figure 10-6) (ETWB 2000). Tier I is a review of existing information to see if there is sufficient information and data to determine whether the sediment is uncontaminated and suitable for open sea disposal. Otherwise, the framework proceeds to Tier II. Tier II is a chemical screening using a prescribed list of chemical contaminants to classify the sediments into 3 categories: L (low contaminant level), M (moderate contaminant level), or H (high contaminant level). Category L sediments are suitable for open water disposal, while categories M and H sedi-

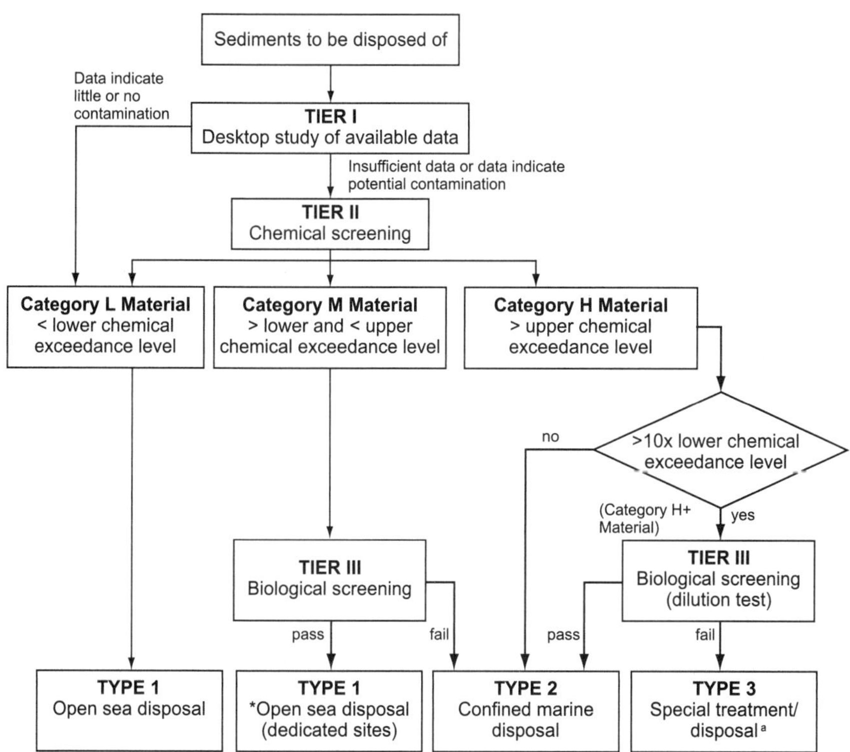

Figure 10-6 Hong Kong dredged material evaluation flow chart (ETWB 2000).
[a] Special treatment or disposal includes stabilization or containment of sediments prior to landfilling or confined marine disposal.

ments must go through Tier III assessment. Tier III uses bioassays to determine the most appropriate disposal options for categories M and H sediments.

Use of sediment quality guidelines in Hong Kong dredging program

The Hong Kong dredging program uses SQGs and bioassays to classify the sediments evaluated in Tiers II and III. The SQG values were derived from an extensive international database including effects range low (ERL), effects range median (ERM), and the Puget Sound Estuary Program. The chemical classes currently measured include metals, metalloids, PAHs, PCBs, and tributylin (TBT) (Table 10-2). Two SQGs were developed for each chemical or class of chemicals. The lower chemical exceedance level (LCEL) represents a value below which contaminants in the sediment are not expected to have adverse biological effects, whereas the upper chemical exceedance level (UCEL) represents a value above which toxicity is likely (Table 10-2).

The bioassays used in the Hong Kong dredging program are based on methods developed in the US for the Puget Sound Estuary Program. The organisms used include a burrowing amphipod (*Eohaustorius estuarius*), an epibenthic tube-dwelling amphipod (*Ampelisca abdita*), a burrowing polychaete (*Neanthes arenaceodentata*), and either free-swimming bivalve larvae (*Mytilus galloprovincialis*) or echinoderm larvae (*Dendraster excentricus* or *Strongylocentrotus* spp).

Definitions for the categories of material are as follows:

- Category L: Sediment with all contaminant levels not exceeding the LCEL. The material must be dredged, transported, and disposed of in a manner that minimizes the loss of contaminants, either into solution or by resuspension.
- Category M: Sediment with any 1 or more contaminant levels exceeding the LCEL and none exceeding the UCEL. The material must be dredged and transported with care and must be effectively isolated from the environment upon final disposal, unless appropriate biological tests demonstrate that the material will not adversely affect the marine environment.
- Category H: Sediment with any 1 or more contaminant levels exceeding the UCEL. Category H material requires Tier III biological screening before the appropriate disposal-at-sea option is determined. Highly contaminated material (>10× LCEL) that fails bioassays requires treatment (e.g., incineration) to stabilize the material before final disposal (Figure 10-6). The material must be dredged and transported with great care and must be effectively isolated from the environment upon final disposal.

Table 10-2 Sediment quality criteria: lower and upper chemical exceedance levels (LCEL, UCEL) used in the Hong Kong dredging program (ETWB 2000)

Contaminant	LCEL	UCEL
Metals (mg kg^{-1} dry weight)		
Cd	1.5	4
Cr	80	160
Cu	65	110
Hg	0.5	1
Ni [a]	40	40
Pb	75	110
Ag	1	2
Zn	200	270
Metalloid (mg kg^{-1} dry weight)		
As	12	42
Organics: PAHs (mg kg^{-1} dry weight)		
Low molecular weight PAHs	550	3160
High molecular weight PAHs	1700	9600
Organics: non-PAHs (mg kg^{-1} dry weight)		
Total PCBs	23	180
Organometallics (mg TBT L^{-1} in interstitial water)		
Tributyltin [a]	0.15	0.15

[a] The contaminant is considered to have exceeded the UCEL if it is greater than the value shown.

Sediment classified as Category M shall be subjected to the following 3 toxicity tests on each composite sample:

1) 10-d burrowing amphipod toxicity test,
2) 20-d burrowing polychaete toxicity test, and
3) 48- to 96-h larvae (bivalve or echinoderm) toxicity test.

Sediment classified as Category H and with 1 or more contaminant levels exceeding 10× LCEL shall also be subjected to the above 3 toxicity tests but in a diluted manner (dilution test). A sediment is deemed to have failed the biological test if it fails in any 1 of the 3 toxicity tests. In cases of emergency dredging for safety reasons and for small-scale maintenance dredging less than 5000 cubic meters in situ, DEP may waive the sediment toxicity testing procedures.

Comparison of Dredging Frameworks

The 3 frameworks proceed to a decision through a sequence of tiers. The process starts with an initial tier that relies on existing information and proceeds, if necessary, to more data-rich (and expensive) testing in subsequent tiers. However, there are important differences among the frameworks, despite their common relationship to the general framework of the London Convention. For example, the US dredging program uses a comparison with a reference sediment. The Canadian program uses a comparison with reference sediment for amphipod survival, polychaete growth, or bioacuumulation testing. To date, it allows the use of control sediments or waters with Microtox solid-phase and echinoid survival tests for practical reasons. Reference sediments are not always required for chemical assessment because the lower action levels are fixed values. However, background concentrations of metals in sediment can be considered in the interpretation of the chemistry results, and reference sediments may be useful to that end. The use of reference sediments in the Hong Kong program is unclear at this time.

One of the crucial differences between the frameworks is how they use SQGs. In the Canadian and Hong Kong frameworks, chemistry guidelines can trigger a "pass" (sediment can be deemed acceptable for unrestricted open water disposal), while in the US dredging program it is not possible to get a "pass" using chemistry guidelines alone. It is difficult to know if the biological testing requirement makes the US dredging program more or less conservative than the other programs, without specific comparisons between the sensitivities of the biological tests and the chemical guidelines (Chapter 3). One consequence of the biological testing requirement, however, is that the dredged material assessment process can take much longer and be much more expensive than it is under the other 2 dredging programs. The delay and cost of biological testing can be great enough that it influences business decisions on the part of dredging applicants in the US and other countries. For example, some Canadian applicants choose to forgo open water disposal rather than go through biological testing, if their sediments fail the chemical criteria (L. Porebski personal communication).

Contaminated Sediment Remediation and Monitoring Assessment Frameworks

US Superfund Program

The Comprehensive Environmental Response, Compensation, and Liability Act of 1980 (CERCLA, often referred to as "Superfund") as amended by the

Superfund Amendments and Reauthorization Act of 1986 (SARA) provides one of the most comprehensive authorities available to the USEPA to require sediment cleanup, reimbursement of USEPA cleanup costs, and compensation to natural resource trustees for damages to natural resources affected by contaminated sediments. To assist in identifying sites in the US where the risks to the environment are unacceptable, the USEPA developed guidance for conducting remedial investigations and feasibility studies under CERCLA (USEPA 1988) and guidance for conducting ecological risk assessments (ERAs) within the Superfund program (USEPA 1997). These ERAs are used to 1) identify and characterize the current and potential threats to the environment from a hazardous substance release and 2) identify cleanup levels that would protect those natural resources from unacceptable risk. ERAs incorporate a wide range of tests and studies to determine whether or not there are unacceptable ecological risks at a site. These include plant and animal tissue residue data, toxicity test data, bioavailability factors, and population- or community-level effects studies (USEPA 1999). Steps 1 and 2 of the ERA involve the compilation of existing information; Steps 3 through 6 involve data collection. Step 7 is a risk characterization in which exposure and effects data are integrated into a statement about risk to the assessment endpoints established during the previous phases. Finally, Step 8 is risk management in which risk reductions associated with the cleanup are balanced against the potential impacts of the remedial actions themselves. See Figure 10-7 for more details of the Superfund process.

Step 1 of a Superfund ERA typically includes a screening-level step to determine which, if any, of the contaminants found at a site are present in concentrations that may be harmful to ecological receptors. This typically involves comparing measured contaminant concentrations (often the maximum) at a site to an ecotoxicologically based benchmark. These are commonly referred to as "ecotox thresholds" and include values developed for sediments (USEPA 1996). Ecotox thresholds are derived using EqP or empirically based SQGs. If the concentration in the sediment exceeds an ecotox threshold, further assessment is warranted. This screening-level step is useful for focusing the assessment on those chemicals at Superfund sites that may pose a risk; this focus is particularly helpful when a long list of chemicals has been identified at a site. The ecotox thresholds provide Superfund site managers with a tool to efficiently identify contaminants and are meant to be used for screening purposes only; they are not regulatory criteria, site-specific cleanup standards, or remediation goals (USEPA 1996).

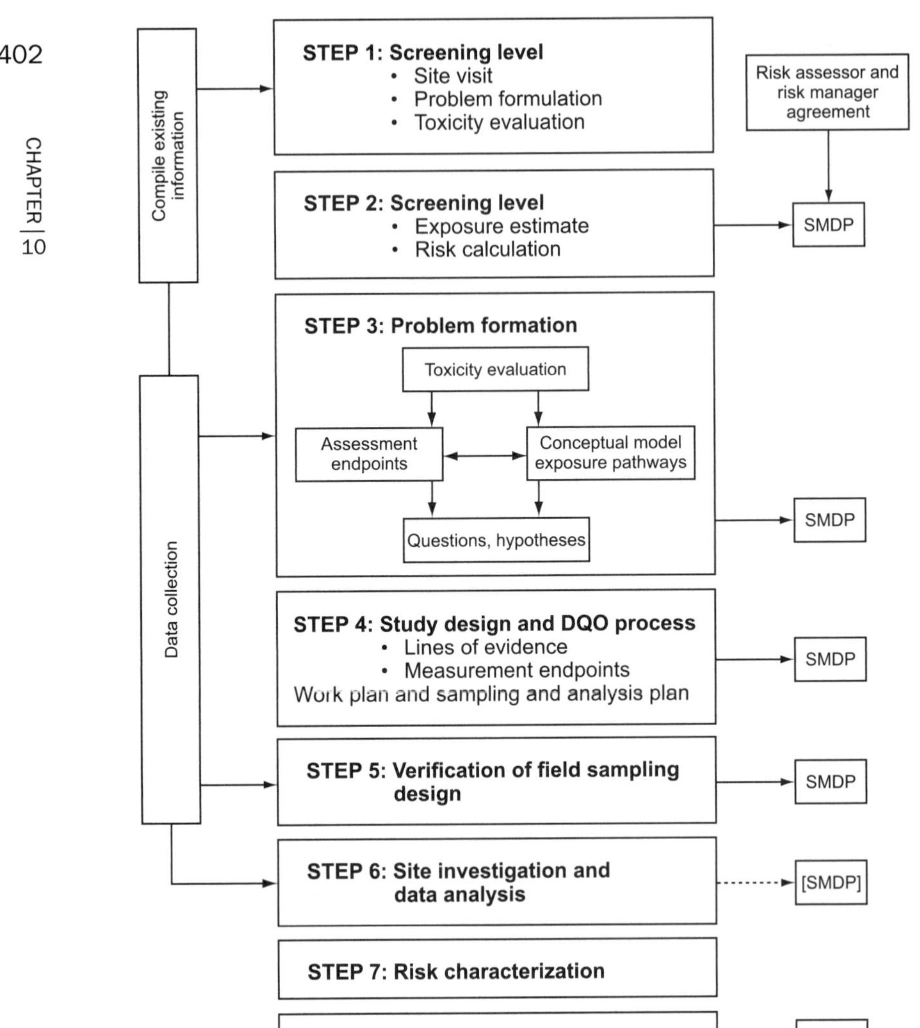

Figure 10-7 Ecological risk assessment guidance for USEPA's Superfund Program: Process for designing and conducting ecological risk assessments (from USEPA 1997) SMDP = Scientific/management decision point.

The assessment framework specified by the Superfund guidance is less prescriptive than the dredging frameworks described earlier. This is true in part because the Superfund framework is designed to be flexible for use at a number of different kinds of sites where there may be a range of management goals. Sediment management during dredging operations is typically a more focused activity that needs to be implemented on a regular basis in many water bodies. Another reason that may be even more important in the design of the Superfund ERA framework is that site specificity is critical to the success of a Superfund assessment. Therefore, the site-specific conceptual model developed during the problem formulation step determines what testing is appropriate (Figure 10-7). The dredging programs, in contrast, are working with a more general conceptual model (Chapter 6) and usually with fewer management options.

US National Sediment Quality Survey

Sediment assessment frameworks are also used for routine sediment monitoring to assess the relative state of the aquatic environment apart from any specific management action (e.g., navigational dredging, sediment remediation). An example of this type of monitoring activity is the USEPA's National Sediment Quality Survey (NSQS; USEPA 2002). A Congressional mandate (Section 503 of the Water Resources Development Act [WRDA] 1992) requires USEPA to compile all existing information regarding the quality of sediments in the US and report to Congress the findings (NSQS). The objective of the NSQS report is to use screening-level assessment protocols to identify potentially contaminated sediment; it is produced biennially for Congress as well as for the USEPA regions, states, and tribes on the incidence and severity of sediment contamination nationwide. The purpose of this report is to depict and characterize the incidence and severity of sediment contamination based on the probability of adverse effects to human health and the environment. To accomplish this objective, USEPA applies assessment protocols to existing available data in a uniform fashion (i.e., sediment assessment framework). One of the protocols used to evaluate the sediment chemistry data compiled in the National Sediment Inventory (NSI, the database developed to support this report) is SQGs.

The sediment chemistry screening values applicable to aquatic life include theoretically and empirically derived values. USEPA notes that these are not regulatory criteria, site-specific cleanup standards, or remediation goals. Instead, these values are reference values above which a sediment ecotoxicological assessment might indicate a potential threat to aquatic life. The theoretically based values rely on the physical or chemical properties of sediment and contaminants to predict the level of contamination that would not cause an adverse effect on aquatic life. The theoretically based screening values used in the evaluation of NSI data include draft equilibrium partitioning sediment guidelines (ESGs) developed by USEPA. These include dieldrin, endrin, 32 nonionic organics, mixtures of PAHs, and metal mixtures. The empirically and statistically derived screening values developed by Field et al. (2002) are based on estimating the probability that a sediment toxicity test would indicate significant toxicity using multiple measures of 37 target chemicals.

Using the sediment benchmarks (including SQGs), USEPA sorts the sediments into 1 of 3 categories (tiers):

1) associated adverse effects on aquatic life or human health probable (Tier 1),
2) associated adverse effects on aquatic life or human health possible (Tier 2), or
3) no indication of associated adverse effects (Tier 3).

Recognizing the imprecise nature of some assessment benchmarks used in the framework for this report, Tier 1 sampling stations are distinguished from Tier 2 sampling stations based on the magnitude of a contaminant concentration in sediment using the degree of corroboration among the different types of sediment quality measures (Figure 10-8).

The methodology used in the NSQS was designed for screening sediment quality across a large number of locations. Because there is considerable uncertainty associated with the site-specific measures, assessment techniques, exposure scenarios, and default parameter selections, the survey report recommends further site-specific evaluations to achieve more accurate information concerning sediment quality.

The NSQS does not rely on a tiered approach but instead involves a set of measurements at each location. The data are used independently to make inferences of sediment quality at each individual sampling station. Tiered monitoring strategies could be designed to involve more elaborate monitoring at sites that do not pass screening levels or to involve strategies that dictate more intensive sampling at such sites. In fact, many monitoring programs have an implied tiering; as stated in the NSQS "when appropriate, the results of evaluating particular sampling stations based on this methodology should be followed up with more intensive assessment efforts" (USEPA 2002).

Other national monitoring programs in the US use similar sorts of frameworks. USEPA's Environmental Monitoring and Assessment Program (EMAP) uses sediment chemistry, sediment toxicity, and benthic community analyses (USEPA 2002). The US National Oceanic and Atmospheric Administration's National Status and Trends program also uses sediment chemistry, sediment toxicity tests, and benthic community analysis (Long and Morgan 1991).

Sampling station classification	Data used to determine classifications				
	Sediment chemistry		Tissue residue		Toxicity
Tier 1: Associated adverse effects on aquatic life or human health are probable	Sediment chemistry value exceeds a draft equilibrium partitioning sediment guideline (ESG) derived from a final or secondary acute value (FAV or SAV)[a]	OR	Tissue levels of chemicals with a log $K_{ow} \geq 5.5$ in samples[d] that exceed USEPA's human health cancer risk of 10^{-5}, a noncancer HQ of 1, or USFDA's guidance, action, or tolerance levels	OR	Toxicity demonstrated by 1 solid-phase sediment text resulting in 1) <75% control-adjusted survival, 2) freshwater invertebrate (*Hyalella azteca*) sublethal toxicity <90% control-adjusted length, or 3) freshwater invertebrate (*H. azteca*, *Chironomus tentans*, and *C. riparius*) sublethal toxicity <70% control-adjusted weight
	OR [SEM]–[AVS] > 5 for the sum of molar concentrations of Cd, Cu, Ni, Pb, Zn, and ½ x Ag[b]				OR Any sample meeting the benchmark described under Tier 2 for toxicity using at least 2 different species
	OR Any sample with a predicted proportion toxic ≥ 0.5 using a logistic regression model				
	OR Sum PAH ESG toxicity unit (draft) derived from FAV > 1[a,c]	AND	Tissue levels of chemicals with a log Kow < 5.5 in samples[d] that exceed USEPA's human health cancer risk of 10^{-5}, a noncancer HQ of 1, or USFDA's guidance, action, or tolerance levels		
Tier 2: Associated adverse effects on aquatic life or human health are possible	Sediment chemistry value exceeds a draft ESG derived from a final or secondary chronic value (FCV or SCV)[a]	OR	Tissue levels of chemicals with a log Kow < 5.5 in samples[d] that exceed USEPA's human health cancer risk of 10^{-5}, a noncancer HQ of 1, or USFDA's guidance, action, or tolerance levels	OR	Toxicity demonstrated by one solid-phase sediment test resulting in 1) <90% control-adjusted survival (but ≥75% control-adjusted survival), 2) freshwater invertebrate (*H. azteca*) sublethal toxicity <95% control-adjusted length (but ≥90% control-adjusted length), or 3) freshwater invertebrate (*H. azteca, C. tentans,* and *C. riparius*) sublethal toxicity <90% control-adjusted weight (but ≥70% control-adjusted weight)
	OR [SEM]-[AVS] = 0-5 for the sum of molar concentrations of Cd, Cu, Ni, Pb, Zn, and ½ x Ag[b]				
	OR Any sample with a predicted proportion toxic ≥0.25 but < 0.5 using a logistic regression model				
	OR Sum PAH ESG toxicity unit (draft) derived from FCV > 1[a,c]				
	OR Sediment chemistry TBP[e] exceeds USEPA's human health cancer risk of 10-5, a noncancer HQ of 1, or USFDA's guidance, action, or tolerance levels[a]				
Tier 3: No indication of associated adverse effects	Any sampling station not categorized as Tier 1 or Tier 2. Available data (which may be very limited or quite extensive) do not indicate a likelihood of adverse effects on aquatic life or human health.				

[a] If total organic carbon (TOC) is not reported, a default value of 1% was assumed. For ESG-based methods if the reported TOC is less than 0.2%, a default TOC value of 0.2% was used.

[b] Metals: Cd = cadmium, Cu = copper, Ni = nickel, Pb = lead, Zn = zinc, Ag = silver.

[c] Acenaphthene, acenaphthylene, anthracene, benzo(*a*)anthracene, benzo(*a*)pyrene, benzo(*b*)fluoranthene, benzo(*k*)fluoranthene, chrysene, fluoranthene, fluorene, naphthalene, phenanthrene, and pyrene used to compute ESG toxicity unit.

[d] Only those species considered benthic (demersal), nonmigratory (resident), and edible by human populations are included in human health assessments.

[e] TBP = Theoretical bioaccumulation potential

Figure 10-8 Sample tiering matrix for USEPA's National Sediment Quality Survey (2002)

Other Sediment Assessment Frameworks

Appendix 1 is a representative but not exhaustive compilation of the sediment assessment frameworks currently in use around the world. Appendix 2 is a compilation of the citations for the documents available for the framework information summarized in Appendix 1. Most of the frameworks discussed in this chapter come from North America because of the experience of the authors. More information on European programs, as well as those used in Australia and Hong Kong, can be found in Chapter 9. Because many of the frameworks in Appendix 1 come from individual US states, they will be discussed separately below.

Overview of US state assessment frameworks

Most state frameworks are based either on the USEPA's Risk Assessment Framework and Guidelines (USEPA 1992) or on the USEPA's Superfund guidelines. Some states have adopted features developed in the American Society for Testing and Materials (ASTM) risk-based corrective action (RBCA) framework (ASTM 2004). The USEPA's risk assessment framework guidelines also provide a widely accepted body of general procedures and terminology.

Screening procedures are a common element in existing guidelines. Interestingly, it appears that state guidelines have developed more detailed, early screening-level procedures than are found in federal guidelines. This may be because states have a wider variety (in terms of scale) of sites and situations to address, and more detailed screening procedures serve to help sort these into manageable categories.

Screening procedures vary somewhat among states, but the following seem common: use considerations, determining if receptors and pathways are present, and screening against guidelines. Other factors that occur during screening steps include size of exposure area (e.g., specifying areas of contamination for sediments or sediment impact zones), distance of receptors from release location, size of potentially affected habitats, and determination of local conditions (i.e., presence of regional pollution).

An important aspect of screening sites and applying any sort of sediment guideline is the sediment depth. A few states have defined surface and subsurface sediments for this purpose.

Most state frameworks use tiered assessment approaches. The number of tiers may vary, but the basic concept is the same. Typically, sediment guidelines, criteria, or benchmarks are used in an early tier as part of a screening evaluation. However, they may also be used in later tiers as part of an overall WOE approach.

Most of the state approaches reviewed do not apply to management of dredged material. Instead they are intended for use in establishing inventories of sediment contamination, evaluating sediments as part of a cleanup program, and TMDL or other permit strategies. The types of approaches include the following:

1) No uniform state guidance: Regions in the state may apply guidelines that they consider most appropriate.
2) State guidance specifies guideline values for fresh and/or saline waters. These guideline values are taken from published and widely available documents (e.g., ERLs, ERMs, or USEPA's EqP guidance) Some states may have listed these guideline values in the order of priority for use. Others simply list these guideline values.
3) Guideline values are based on a review of available published databases and are developed specifically for the state (e.g., Florida).
4) Guideline values or criteria are based on historical studies conducted on sediments in the state (e.g., State of Washington).

Conclusions

The type of guidelines and testing required in a framework is determined by the framework's legal mandate and management goals. At least 3 broad categories of frameworks exist:

1) management of dredged materials for navigation and construction projects,
2) investigation and remediation of contaminated sediments at specific sites, and
3) monitoring of sediment conditions to assess the quality of the environment and possibly identify impairments.

These categories vary in specificity, as might be expected given their different purposes.

For example, the US dredging program must be protective of both water column and sediment effects, in regards to both human health and ecological effects. As a result, water-only, sediment phase, and bioaccumulation testing are all required by the US navigational dredging program. In contrast, the

Congressional mandate (Section 503 of WRDA 1992) required the USEPA to conduct a national survey of data regarding the quality of sediments in the US and report the findings (NSQS). A major difference between the 2 frameworks is that the US navigational dredging program is used to support the regulatory action (under either Section 404 of the CWA or the MPRSA) of the disposal of material resulting from navigational dredging in the nation's waters. In contrast, the framework used for the NSQS is for assessment purposes only, providing environmental managers with valuable information by identifying locations where available data indicate that direct or indirect exposure of sediment could be associated with adverse effects to aquatic life and/or human health. The NSQS does not recommend actions to be taken (e.g., dredging) based on the report's findings; rather, it identifies areas where further evaluations may be warranted. Additionally, this is the type of approach used for routine sediment monitoring programs employed by local, state, and federal programs in which a framework is used to determine the state of the environment as it relates to sediment quality rather than for regulatory actions (e.g., USEPA's EMAP and the NOAA Status and Trends Program).

Despite the differences that exist among the frameworks discussed here, some general patterns emerge. Tiering is the norm, and the tiers are generally set up as

1) site history or sources of contamination,
2) chemistry,
3) toxicity, and
4) bioaccumulation and risk assessment (Table 10-3).

SQGs are often used at an early tier to sort sediments into different categories. The weight or reliance placed on use of SQGs for decision making varies from framework to framework. In some cases, SQGs are used as indicators of general conditions, while in other cases, they support specific management actions. To a large degree, this variation reflects differences in policies and management goals rather than technical judgments. In general, most frameworks use SQGs as a screening tool. If SQGs are exceeded, a potential sediment problem is indicated; however, more site-specific investigation may be warranted in order to reduce the uncertainty associated with applying SQGs. A few frameworks rely on SQGs to guide specific decisions; most of these frameworks relate to the management of dredged sediment.

Conceptual models are a common feature of frameworks used to assess sediments. While these models may not always consider all pathways and receptors, they have provided a useful tool for representing the linkages between

Table 10-3 Similarities and differences in tiered sediment evaluation/decision frameworks

	Regulatory program and purpose			
	Navigation dredging	Contaminated sediment cleanup	Source control	Monitoring
Tier I	Use existing information to determine the need for sediment and/or porewater chemistry analysis, and/or biological testing ("reason to believe").	Use existing information to determine the need for sediment and/or porewater chemistry analysis, and/or biological testing ("reason to believe").	Use existing information to determine the need for sediment and/or porewater chemistry analysis, and/or biological testing ("reason to believe").	Develop a sediment sampling and analysis strategy based on one of various program goals, e.g., to assess ambient conditions, sediment quality near discharge outfalls, effectiveness of remedial actions and/or health of open-water disposal sites.
Tools	Site use history, known or suspected contaminants, ongoing sources of contamination, physical characteristics of sediment, existing sediment quality data, etc.	Site use history, known or suspected contaminants, ongoing sources of contamination, physical characteristics of sediment, existing sediment quality data, etc.	Site use history, known or suspected contaminants, ongoing sources of contamination, physical characteristics of sediment, existing sediment quality data, etc.	"Tools" that reflect specified goals/strategy and can be used in a routine manner, e.g., relatively inexpensive chemical and biological screening analyses.
Tier II	Measure concentrations of contaminants in whole sediment and/or pore water in order to predict incidence of toxicity or benthic community effects.	Measure concentrations of contaminants in whole sediment and/or pore water in order to predict incidence of toxicity or benthic community effects.	Measure concentrations of contaminants in whole sediment and/or pore water in order to predict incidence of toxicity or benthic community effects.	Measure concentrations of contaminants in whole sediment and/or pore water in order to predict incidence of toxicity or benthic community effects.
Tools	Appropriate and consistent analytical protocols (with appropriate detection limits and QA/QC) that measure "reason to believe" contaminants of concern in composited samples representing future dredge prisms and/or surface grab samples from open-water disposal sites.	Appropriate and consistent analytical protocols, etc., that measure "reason to believe" contaminants of concern in the "biologically active zone", e.g., 0-15 cm, and/or at depth.	Appropriate and consistent analytical protocols, etc., that measure contaminants of concern in the "biologically active zone", e.g., 0–15 cm, near a discharge and/or impact zone. Protocols may need to measure less common chemicals related to the known source(s).	Appropriate and consistent analytical protocols, etc., that measure contaminants of concern in recently deposited sediment, e.g., 0–3 cm. Protocols may need to measure a more exhaustive list of chemicals.

Table 10-3 *contd*

	Regulatory program and purpose			
	Navigation dredging	Contaminated sediment cleanup	Source control	Monitoring
Tier III	Evaluate the potential for contamination to cause significant effects in the benthic community.	Evaluate the potential for contamination to cause significant effects in the benthic community.	Evaluate the potential for contamination to cause significant effects in the benthic community.	Evaluate the potential for contamination to cause significant effects in the benthic community.
Tools	• Protocols that measure laboratory toxicity as a surrogate for benthic community effects. Toxicity protocols should be ecologically relevant, sensitive, reliable, etc., and simulate conditions associated with disposal activity and/or disposal sites. • Protocols that directly measure relative health of the benthic community found at disposal sites. Multiple field/lab replicates and multivariate/trend analyses may be needed to assess effects caused by contaminants in dredged material.	• Protocols that measure laboratory toxicity as a surrogate for benthic community effects. Toxicity protocols should be ecologically relevant, sensitive, reliable, etc., and simulate in situ conditions. • Protocols that measure relative health of benthic communities found at cleanup sites. Multiple field/lab replicates and multivariate/trend analyses may be needed to assess in situ effects caused by sediment contaminants.	• Protocols that measure laboratory toxicity as a surrogate for benthic community effects. Toxicity protocols should be ecologically relevant, sensitive, reliable, etc., and simulate in situ conditions. • Protocols that measure relative health of benthic communities found at cleanup sites. Multiple field/lab replicates and multivariate/trend analyses may be needed to assess in situ effects caused by sediment contaminants.	• Protocols that measure laboratory toxicity as a surrogate for benthic community effects. Toxicity protocols should be ecologically relevant, sensitive, reliable, consistent, etc., and simulate in situ conditions. Rapid assessment and/or screening-level toxicity tests may be appropriate. • Protocols that measure relative health of benthic communities found at routine sampling locations. Multiple field/lab replicates may or may not be needed.

Table 10-3 *contd*

	Regulatory program and purpose			
	Navigation dredging	Contaminated sediment cleanup	Source control	Monitoring
Tier IV	Evaluate potential risk to the environment and/or human health caused by exposure to bioaccumulative chemicals found in sediment and/or animal tissues.	Evaluate potential risk to the environment and/or human health caused by exposure to bioaccumulative chemicals found in sediment and/or animal tissues.	Evaluate potential risk to the environment and/or human health caused by exposure to bioaccumulative chemicals found in sediment and/or animal tissues. Evaluate potential risk to the environment and/or human health caused by exposure to bioaccumulative chemicals found in sediment and/or animal tissues.	Evaluate potential risk to the environment and/or human health caused by exposure to bioaccumulative chemicals found in sediment and/or animal tissues.
Tools	• Protocols that measure potential bioaccumulation, e.g., concentrations of contaminants in tissues of standard test organisms after short-term exposure to in situ sediment. • Results can be compared to effects-based target tissue levels as means of estimating relative risk to benthic communities found at disposal sites (less commonly risk to human health).	• Protocols that measure potential bioaccumulation, e.g., concentrations of contaminants in tissues of standard test organisms after short-term exposure to in situ sediment. • Protocols that measure actual bioaccumulation, e.g., concentrations of bioaccumulative contaminants in the in situ benthic community, fish and/or shellfish tissue may also be used. • Results may be reported in terms of the risk to benthic communities, wildlife and humans posed by continued exposure to site sediments.	• Protocols that measure potential bioaccumulation, e.g., concentrations of contaminants in tissues of standard test organisms after short-term exposure to in situ sediment. • Protocols that measure actual bioaccumulation, e.g., concentrations of bioaccumulative contaminants in shellfish and/or fish tissue may also be used. • Results may be reported in terms of the risk to wildlife and humans posed by continued exposure to site sediments.	Most sediment monitoring efforts do not measure either potential or actual bioaccumulation. However, fish and shellfish monitoring efforts may estimate risks to wildlife and human health posed by exposure to contaminated sediment.

contaminant sources, pathways or exposure, and ecological or human receptors (Chapter 6). The development of site-specific conceptual models is often a key component when site-specific risk assessments are used to evaluate hazardous waste sites with contaminated sediments. Generic conceptual models are often used to consider exposure pathways for navigational dredging projects or to develop guidance on how to evaluate disposal alternatives for dredged material.

The degree to which frameworks identify stakeholder involvement as a component depends on when the framework was developed as well as the purpose of the framework. Older frameworks tended to focus almost exclusively on the technical aspects of the process. While there may have been involvement by a wide array of interested parties, this involvement was not explicitly identified as part of the process. More recent frameworks have explicitly included stakeholders in recognition of the importance of stakeholder involvement for certain processes. How and when to involve stakeholders is still a subject that is evolving. While the emphasis on stakeholder involvement may vary depending on management goals and the purpose of the assessment, the subject has come to the fore and will be given increased attention in the future.

Risk-based frameworks are emerging as a useful way to structure and assess contaminated sediments as well as to distinguish among options (for either remediation or disposal of dredged material). The link between assessment and management decisions is treated differently among the frameworks. In some cases, there may be a risk assessment framework followed by a management framework. However, because these processes are intricately related, it may be most useful to consider a more holistic framework that encompasses all the key aspects of assessing and managing the specific sediment contamination problem (Chapter 6).

References

[ASTM] American Society for Testing and Materials. 2004. Standard guide for risk-based corrective action. West Conshohocken (PA): ASTM.

[CCME] Canadian Council of Ministers of the Environment. 1995. Protocol for the derivation of Canadian sediment quality guidelines. Ottawa (ON): CCME. CCME EPC-98E. ISBN 1-895925-58-4. 38 p. Available from: http://www.ec.gc.ca/CEQG-RCQE/English/Ceqg/Sediment/default.cfm. Accessed 16 Aug 2004.

[CCME] Canadian Council of Ministers of the Environment. 1999. Canadian environmental quality guidelines. Winnipeg (MB): CCME. ISBN 1-896997-34-1. Publication Nr. 1299.

[CEPA] Canadian Environmental Protection Act. 1999. Statutes of Canada, Chapter 33.

[CEPA] Canadian Environmental Protection Act. 2001. Disposal at sea regulations and regulations respecting applications for permits for disposal at sea. SOR/2001-275 and 276.

[ETWB] Hong Kong Environment Transport and Works Bureau. 2000. Management of dredged/excavated sediment. Works Bureau Technical Circular No. 3/2000. 27 April 2000.

Field LJ, MacDonald DM, Norton SB, Severn CG, Ingersoll CG, Smorong D, Lindskoog R. 2002. Predicting amphipod toxicity from sediment chemistry using logistic regression models. *Environ Toxicol Chem* 21(9):1993–2005.

Fuhrman RL. 1998. Memorandum. Policy Guidance Letter (PGL) No. 49, Section 312 of the Water Resources Development Act of 1990, Environmental Dredging, as Amended by Section 205 of the Water Resources Development Act of 1996. 7 p.

[IMO] International Maritime Organization. 2003. London Convention 1972, Convention on the Prevention of Marine Pollution by Dumping of Wastes and Other Matter, 1972. 39 p.

[LCSG] London Convention Scientific Group. 2000. 23rd meeting, Waste Assessment Guidance: Experience Gained in Implementing the Waste Assessment Guidance: The New Sediment Assessment Framework for Hong Kong; 2000 May.

[LCSG] London Convention Scientific Group. 2001. Dredged Material Assessment Framework, in Report of the 24th Meeting of the Scientific Group, International Maritime Organization; London. Document LC/SG/24/11, Annex 3.

Long ER, Morgan LG. 1991. The potential for biological effects of sediment-sorbed contaminants tested in the National Status and Trends Program. Seattle (WA): National Oceanic and Atmospheric Administration. NOAA Technical Memorandum NOS OMA 52. 175 p + appendices.

Meador JP, Robisch PA, Clark Jr RC, Ernest DW. 1999. Elements in fish and sediment from the Pacific Coast of the United States: Results from the National Benthic Surveillance Project. *Mar Pollut Bull* 37:56–66.

Peddicord R, Chase T, Dillon T, McGrath J, Munns WR, van de Gutche K, van de Schalie W. 1995. Workgroup summary report on navigational dredging. In: Ingersoll CG, Dillon T, Biddinger GR, editors. Ecological risk assessment of contaminated sediments. Pensacola (FL): Society of Environmental Toxicology and Chemistry (SETAC).

Persaud D, Jaagumagi R, Hayton A. 1992. Guidelines for the protection and management of aquatic sediment quality in Ontario. Toronto: Ontario Ministry of the Environment. 25 p.

Porebski LM, Osborne JM. 1998. The application of a tiered testing approach to the management of dredged sediments for disposal at sea in Canada. *Chem Ecol* 96/14.

Stronkhorst J, Schipper CA, Honkoop J, van Essen K. 2001. Disposal of dredged material in Dutch coastal waters. A new effect-oriented assessment framework. The Hague: National Institute for Coastal and Marine Management/RIKZ. Report RIKZ 2001.030.

[USEPA] US Environmental Protection Agency. 1988. Guidance for conducting remedial investigations and feasibility studies under CERCLA. Washington DC: USEPA Office of Emergency and Remedial Response. EPA/540/G-89/.

[USEPA] US Environmental Protection Agency. 1992. Framework for ecological risk assessment. Washington DC: USEPA. EPA/630/R-92/001.

[USEPA] US Environmental Protection Agency. 1996. Ecotox thresholds. ECO update, Interim bulletin, Volume 3, Number 2. Washington DC: USEPA Office of Emergency and Remedial Response, Hazardous Site Evaluation Division. EPA/540/F-95/038.

[USEPA] US Environmental Protection Agency. 1997. Ecological risk assessment guidance for Superfund: Process for designing and conducting ecological risk assessments. Washington DC: USEPA Office of Solid Waste and Emergency Response. EPA540-R-97-006.

[USEPA] US Environmental Protection Agency. 1999. Ecological risk assessment and risk management principles for Superfund sites. Washington DC: USEPA Office of Emergency and Remedial Response. OSWER Directive 9285.7. 28 p.

[USEPA] US Environmental Protection Agency. 2002. The incidence and severity of sediment contamination in surface waters of the United States. National sediment quality survey. 2nd ed. Washington DC: USEPA Office of Science and Technology. EPA 823-R-01–01.

[USEPA] US Environmental Protection Agency. 2002. Research Strategy, Environmental Monitoring and Assessment Program. EPA 620/R-02/002. Research Triangle Park (NC): USEPA Office of Research and Development.

[USEPA/USACE] US Environmental Protection Agency/US Army Corps of Engineers. 1991. Evaluation of dredged material proposed for ocean disposal: Testing manual. Washington (DC): USEPA. EPA 503/8-91/001.

[USEPA/USACE] US Environmental Protection Agency/US Army Corps of Engineers. 1998. Evaluation of dredged material proposed for discharge in waters of the U.S.: Testing manual. Washington (DC): USEPA. EPA-823-B-98-004.

Acknowledgements

This manuscript profited greatly from comments by R. Burgess, J. Cura, J. Keating, and W. Nelson. This is contribution AED-03-090. This document has been reviewed in accordance with US Environmental Protection Agency, Office of Research and Development, National Health and Environmental Effects Research Laboratory, Atlantic Ecology Division policy and approved for publication. Although the research described in this publication has been funded in part by the US Environmental Protection Agency, it has not been subjected to Agency level review. Therefore, it does not necessarily reflect the views of the Agency. Mention of trade names or commercial products does not constitute endorsement or recommendation for use.

Appendix 1: Sediment Management Frameworks in the US

Name/ jurisdiction	Program/ purpose	Nr tiers	Used in what Tier? Chemistry (SQGs)	Toxicity	Bioaccumulation	Benthic analysis
US dredging program	Navigational dredging (CWA 404/ 401; Section 103 of MPRSA)	4	Tier 1 (existing data) and Tier 2	Tier 3: Mostly acute testing Tier 4: Chronic, sublethal Possibly in existing data in Tier 1	Tier 2: Estimates made from sediment chemistry, TBP Tier 3: Bioaccumulation testing	May be specified in Tier 4 (possibly in existing data in Tier 1)
Canada dredging program	Navigational dredging (CEPA 1999)	4	Tier 1 (existing data) and Tier 2	Tier 3 (possibly in existing data in Tier 1)	Tier 3	May be specified in Tier 4 (monitoring)
Hong Kong	Navigational dredging (DEP; DCE)	3	Tier 1 (existing data) and Tier 2	Tier 3	No	No
Netherlands	Regulation of dredged material (CUWVO 1990)	3	Tier 1 and Tier 2	Tier 2	Tier 2	No
London Protocol	Navigational dredging: Pending international agreement	4	4 tiers using chemical and/or biological testing	Use at any tier	Use at any tier	Use at any tier
Superfund	Remediation dredging: Regulatory		Preliminary site assessment/investigation: Existing data/data collected (chemistry and/or biology if available) Site/remedial investigation (chemistry and/or biology)	Both preliminary and site investigation	—	Both preliminary and site investigation
GIPME		3	3 Tiers: Tier 1: Initial screen, existing Tier 2: Chemical screening Tier 3: Toxicity testing	Tier 2 (not specified)	—	Tier 3 (biological testing)

[a] Note: This table spans 2 facing pages

and Other Countries[a]

Additional information[1]	Decision made from		Reference	Comments
	Chemistry	Biology		
—	Could be used for reason to believe in Tier 1 but not responsible for final decision on disposal of dredged material	—	Ocean testing manual (OTM); Inland testing manual (ITM)	—
—	Can be a pass if no guideline exceeded. Fail: move to next Tier, additional testing trigger. Doesn't make the final "fail" decision	If no hit on toxicity, pass; if fail only one, dispose with special handling	CEPA 1999	SQGs are no-effects levels
—	Move to next Tier, additional testing trigger. Doesn't make the final decision	—	See Appendix 2	SQGs based on UCEL and LCEL, empirically derived
—	Pass/fail	—	See Appendix 2	Changes proposed
Yes, guidance	Can be pass/fail	Can be pass/fail	—	Generic framework adapted as needed
Yes, for both preliminary and site investigation	—	—	USEPA 1997	—
No	—	—	See Appendix 2	—

Name/ jurisdiction	Program/ purpose	Nr tiers	Used in what Tier?			
			Chemistry (SQGs)	Toxicity	Bioaccumulation	Benthic analysis
National Sediment Quality Survey (NSQS)	Monitoring (uses existing data) Required under Section 503 of WRDA 1992	1	Tiers 1 and 2: EqP and LRM	Tiers 1 and 2: Sediment toxicity, acute and chronic	Tier 1 and Tier 2: TBP: cancer and noncancer guidelines Tissue residue: cancer and noncancer guidelines	No
Australia/ New Zealand		4	3 Tiers: Tier 1: Reason to believe Tier 2: Chemical (SQGs and background; bioavailability, AVS, etc. Tier 3: Biological acute Tier 4: Biological chronic	Tier 2 (ERLs and ERMs)	—	Tier 3: Acute sediment tox Tier 4: Chronic sediment tox
Washington	Navigational dredging, source control, cleanup, monitoring	4	4 Tiers: Tier 1: Existing data Tier 2:	Tier 3	Tier 3, as appropriate	Tier 3: Suite of toxicity, acute and chronic and bioaccumulation
Alaska	Not listed (not sure if for 401 or in WQS program)	2	2 Tiers: Tier 1: Use existing data and literature values; not specified as to chemistry and/or biology based Tier 2: Detailed assessment as dictated from Tier 1 results OR chemicals of concern are bioaccumulative	Not specified	—	Tier 3 or 4
California	Not listed (not sure if for 401 or in WQS program)	4	4 Tiers: Tier 1: Scoping assessment Tier 2: Predictive assessment Tier 3: Validation study Tier 4: Impact assessment In none of the tiers is it clear if these are chemistry and/or biology based.	Not specified	—	Not specified

Additional information[1]	Decision made from: Chemistry	Biology	Reference	Comments
Tier 1: associated adverse effects to aquatic life or human health are probable Tier 2: associated adverse effects on aquatic life or human health are possible.	Chemistry can be used to place sampling stations into Tier 1 or Tier 2. Chemistry data are most commonly used to trigger biological evaluations but can be used alone for certain decisions.	Biology can be used to place sampling stations into Tier 1 or Tier 2. Biological data are most commonly used to confirm or refute chemical predictions of adverse effects but can be used alone for certain decisions	USEPA 2001 (draft)	—
No	—	—	See Appendix 2	—
Yes, If Tier 4	—	—	See Appendix 2	—
Unknown	—	—	—	—
Unknown	—	—	—	—

Name/ jurisdiction	Program/ purpose	Nr tiers	Used in what Tier?			
			Chemistry (SQGs)	Toxicity	Bioaccumu-lation	Benthic analysis
Illinois	Not listed (not sure if for 401 or in WQS program)	3	3 Tiers: Tier 1: A lookup table; assume this is chemistry based Tier 2: Allows user to define site-specific inputs and to recalculate objectives by incorporating them into Agency-provided equations Tier 3: Allows user of site-specific information and alternative models for estimating exposure	Not specified	—	Not specified
Indiana	Monitoring		Use SQGs for 305b reporting (empirical guidelines)	Yes, empirical	—	No
Louisiana	Not listed (not sure if for 401 or in WQS program)	3	3 Tiers: Tier 1: ERA checklist Tier 2: Screening-level ERA Tier 3: Baseline ERA	Not specified	—	Not specified
Massa-chusetts	Not listed (not sure if for 401 or in WQS program)	2	2 Tiers: Tier 1: Comparison values, chemistry based Tier 2: Not discussed	Yes, ERLs for saltwater	—	Not specified
New Jersey	Not listed (not sure if for 401 or in WQS program)	2	2 Tiers: Tier 1: Baseline ecological evaluation, screen, chemistry based Tier 2: If fail Tier 1	Site both empirical and theoretical - based documents)	—	Not specified
New Hampshire	WQS	2	2 Tiers: Tier 1: Chemical guideline Tier 2: Tox and benthos can be used Tier 2: If bioaccumulation, tissue residue, or TBP	Yes, TEL	—	Not specified

	Decision made from			
Additional information[1]	Chemistry	Biology	Reference	Comments
Unknown	—	—	—	—
Unknown	—	—	—	—
Unknown, but probably so since ERA	—	—	—	—
Unknown	—	—	See Appendix 2	—
Unknown	—	—	See Appendix 2	—
Unknown	—	—	See Appendix 2	—

Name/ jurisdiction	Program/ purpose	Nr tiers	Used in what Tier? Chemistry (SQGs)	Toxicity	Bioaccumulation	Benthic analysis
New York	Not listed (not sure if for 401 or in WQS program)	5	5 Tiers: No more detail provided	Not specified		Not specified
Ohio	Not listed (not sure if for 401 or in WQS program)	2	2 Tiers: Tier 1: Chemical guidelines Tier 2: Additional analysis; risk assessment may be performed. No more detail on additional analysis or risk assessment	Not specified		Not specified
Oregon	Not listed (not sure if for 401 or in WQS program)	4	4 Tiers: No more detail provided	Not specified	—	Not specified
Pennsylvania	Not listed (not sure if for 401 or in WQS program)	3	3 Tiers: Tier 1: Screen Tier 2: More detailed step requiring a site visit Tier 3: Site-specific evaluation to determine impact to receptors. No information on whether these are chemistry and/or biology based.	Not specified	—	Not specified
Texas	Not listed (not sure if for 401 or in WQS program)	3	3 Tiers: Conceptual model identifying exposure pathways; no detailed information Tier 2: Screening-level assessment to identify contaminants that pose an ecorisk and eliminate those that do not Tier 3: Quantitative ecorisk through ERA process	Not specified	—	Not specified

Additional information[1]	Decision made from			
	Chemistry	Biology	Reference	Comments
Unknown	—	—	—	—
Unknown, but if RA then probably	—	—	—	—
Unknown	—	—	—	—
Unknown	—	—	—	—
Unknown, but if RA then probably	—	—	—	—

Name/ jurisdiction	Program/ purpose	Nr tiers	Used in what Tier?			
			Chemistry (SQGs)	Toxicity	Bioaccumu-lation	Benthic analysis
West Virginia	Not listed (not sure if for 401 or in WQS program)	3	3 Tiers: Tier 1: De minimus screening Tier 2: Uniform, risk-based screening Tier 3: Site-specific, eco-risk assessment No information on whether these are chemistry and/or biology based.	Not specified	—	Not specified
Wisconsin	Not listed (not sure if for 401 or in WQS program)	—	Tiered approach: Number of Tiers uncertain; use sediment chemistry to develop 4 zones (upper, lower, midway between upper and lower). Later tiers would be chemistry as well as biology.	Consensus based	—	Not specified

Additional information[1]	Decision made from Chemistry	Biology	Reference	Comments
Unknown, but if RA, then probably	—	—	—	—
Unknown	—	—	See Appendix 2	—

Appendix 2: Available sources for Appendix 1

Australia and New Zealand

- Australia and New Zealand Guidelines for Fresh and Marine Water Quality, 2000.

Canada

- CCME (Canadian Council of Ministers of the Environment) 1999. Canadian Environmental Quality Guidelines. Canadian Council of Ministers of the Environment, Winnipeg. http://www.ec.gc.ca/ceqg-rcqe/English/Ceqg/Sediment/default.cfm
- Persaud D, Jaagumagi R, Hayton A. 1992. Guidelines for the protection and management of aquatic sediment quality in Ontario. Ontario Ministry of the Environment. 25 p.
- Canadian Disposal at Sea Program website. 2003. The disposal of dredged material and other wastes under the Canadian Environmental Protection Act 1999. http://www.ec.gc.ca/seadisposal/main/index_e.htm

Florida

- MacDonald DD. 1994. Approach to the Assessment of Sediment Quality in Florida Coastal Waters. Volume 1, Development and Evaluation of Sediment Quality Assessment Guidelines. Prepared for: Florida Department of Environmental Protection Office of Water Policy, 3900 Commonwealth Boulevard, MS46 Tallahassee, Florida, 32399-3000. http://www.dep.state.fl.us/waste/quick_topics/publications/documents/sediment/volume1.pdf

GIPME

- Programme of Global Investigation of Pollution in the Marine Environment (GIPME). 1999. Guidance on assessment of sediment quality. International Maritime Organization/IMO, London, UK.

Hong Kong

- Hong Kong Environment Transport and Works Bureau. (ETWB) 2000. Management of dredged/excavated sediment. Works Bureau Technical Circular No. 3/2000. 27 April 2000. http://www.etwb.gov.hk/UtilManager/tc/2000/wb0300.doc

London Convention and 1996 Protocol

- London Convention Scientific Group. 2001. Dredged Material Assessment Framework, in Report of the 24th Meeting of the Scientific Group, International Maritime Organization, London, Document LC/SG/24/11, Annex 3. http://www.londonconvention.org/

Massachusetts

- Massachusetts Department of Environmental Protection. 1999. 310 CMR 40.0000: Massachusetts Contingency Plan. http://www.state.ma.us/dep/bwsc/files/310cmr40.pdf
- Massachusetts Department of Environmental Protection. 2002. Update to: Section 9.4 of Guidance for Disposal Site Risk Characterization – In Support of the Massachusetts Contingency Plan (1996). http://www.state.ma.us/dep/ors/files/sedscrn.doc

Netherlands

- Stronkhorst J, Schipper CA, Honkoop J, van Essen K. 2001. Disposal of dredged material in Dutch coastal waters. A new effect-oriented assessment framework. National Institute for Coastal and Marine management/RIKZ, The Hague, the Netherlands. Report RIKZ 2001.030.
- Commissie Integraal Waterbeheer (CIW). 2000. Standards for water management. Secretariat of CIW, The Hague, the Netherlands. (in Dutch).

New Hampshire

- New Hampshire Department of Environmental Services. 2002. Evaluation of Sediment Quality. NH-DES-WMB Policy Number 002. Last Revised 4/9/02. 7 p.

New Jersey

- New Jersey Department of Environmental Protection. 2002. Draft Technical Requirements for Site Remediation. N.J.A.C. 7:26E. http://www.state.nj.us/dep/srp/regs/techrule/proposed/tech_rule_text.doc

USEPA Superfund Program

- USEPA. 1997. Ecological Risk Assessment Guidance for Superfund: Process for Designing and Conducting Ecological Risk Assessments. EPA540-R-97-006. U.S. Environmental Protection Agency, Office of Solid Waste and Emergency Response, Washington, DC. http://www.epa.gov/superfund/programs/risk/ecorisk/intro.pdf

US Navigational Dredging Program

- USEPA/USACE. 1991. Evaluation of dredged material proposed for ocean disposal: testing manual. EPA 503/8-91/001. US Environmental Protection Agency, Washington, DC, USA.
- USEPA/USACE. 1998. Evaluation of dredged material proposed for discharge in waters of the U.S. Testing manual. EPA-823-B-98-004. US Environmental Protections Agency, Washington, DC, USA.

Wisconsin

- State of Wisconsin Department of Natural Resources. 2002. Consensus-Based Sediment Quality Guidelines; Recommendations for Use and Application. WT-732-2002. Developed by the Contaminated Sediment Standing Team. February, 2002. 32 p.

Washington

- Washington State Department of Ecology Toxics Cleanup Program. 1995. Sediment Management Standards. Chapter 173-204 WAC. http://www.ecy.wa.gov/pubs/wac173204.pdf.
- Washington State Department of Ecology Toxics Cleanup Program: Sediment Quality Chemical Criteria. http://www.ecy.wa.gov/programs/tcp/smu/sed_chem.htm.
- US Army Corps of Engineers, Dredged Material Management Office. http://www.nws.usace.army.mil/PublicMenu/Menu.cfm?sitename=dmmo&pagename=home

Bioaccumulation 11
in the assessment of sediment quality: uncertainty and potential application

DAVID W MOORE, RENATO BAUDO, JASON M CONDER, PETER F LANDRUM, VICTOR A MCFARLAND, JAMES P MEADOR, RICHARD N MILLWARD, JAMES P SHINE, JACK Q WORD

The desire for cost-effective screening tools for contaminated sediments has resulted in the development of a variety of numerical sediment quality guidelines (SQGs). Currently developed guideline values can be categorized as either mechanistically based (e.g., equilibrium partitioning [EqP]) or empirically based (threshold effects levels [TELs], probably effects levels [PELs], effects range low [ERLs], effects range median [ERMs], apparent effects thresholds [AETs]) (Chapters 3 and 4). None of the existing approaches in either of these categories were designed or intended to be protective of indirect effects through bioaccumulation Thus, there is a need (and under certain United States [US] regulatory programs, a requirement, e.g., the Marine Protection, Research and Sanctuaries Act [MPRSA] and the Clean Water Act [CWA]) to assess the potential for sediment associated contaminants to bioaccumulate and to evaluate any potential effects (direct and indirect [i.e., foodchain]) associated with that bioaccumulation. While the technical basis, utility, and accuracy of existing SQG approaches in predicting direct effects to benthic infaunal organisms is discussed in other chapters (e.g., 4, 12, and 13), the intent of this chapter is to provide an overview of the potential for guideline values to predict effects through bioaccumulation.

Bioavailability and organism physiology are the two most important variables affecting chemical contaminant body burdens (Landrum et al. 1996). Organic carbon (both content and composition), contact time (aging), source (e.g., polycyclic aromatic hydrocarbons [PAHs] that are petrogenic in origin versus byproducts of combustion), and sediment surface area, howev-

Use of Sediment Quality Guidelines and Related Tools for the Assessment of Contaminated Sediments
Wenning RJ, Batley GE, Ingersoll CG, Moore DW, editors.
 ISBN 1-880611-71-6

er, can also affect the proportion of nonionic hydrophobic organic compounds available for uptake by organisms. For organic compounds, the octanol–water partition coefficient (K_{ow}) may be the single best predictor of partitioning, but it is insufficient for completely determining contaminant behavior and its bioavailability in the environment. For compounds with ionizable groups, the pH of the environment also affects contaminant bioavailability. Thus, 2 factors, organism lipid content and sediment organic carbon, control to a large extent the partitioning behavior of organic compounds between sediment, water, and tissue. The more hydrophobic a compound, the more likely it is to associate with nonpolar (nonionic) matrices (lipid and organic carbon).

Metals and ionizable organics, on the other hand, are more complicated because their bioavailability can be controlled by a multitude of inorganic and organic ligands, such as iron and manganese oxyhydroxides, acid volatile sulfides (AVS), organic carbon, and others (Chapters 4 and 13). The degree to which some of these factors control metal bioavailability can vary greatly as a result of changes in redox state and hydrogen ion activity (Tessier and Turner 1995). One active area of research is focused on the relationship between AVS in sediment and concentrations of metals in water that are presumably available for uptake (e.g., Di Toro et al. 1990; Ankley et al. 1993; Hare et al. 2001; Sundlin and Eriksson 2001; Chapter 13).

While all organisms bioaccumulate (primarily organics), there are often large differences in tissue residue concentrations between species exposed to the same sediment. Similarly, there can be large differences for the same species exposed to different sediments or the same sediment at different times. The tissue residue concentration of a contaminant is a function of the environmental concentration (including route and source of exposure), duration of exposure, and a given species' ability to metabolize accumulating compounds. Organismal factors, such as cellular lipid level and rate of uptake and elimination (metabolism, diffusion, and excretion), are the primary determinants of tissue residue concentration and time to reach steady state. Often, the highest concentrations can be found in internal organs, such as the hepatopancreas, because of the high lipid content. In addition, tissue concentrations tend to follow seasonal cycles, which may be related to variations in lipid content, spawning cycles, or environmental flux (Jovanovich and Marion 1987; Maruya et al. 1997; Miles and Roster 1999).

Bioaccumulation values can be useful for gauging the degree to which contaminants have been bioaccumulated across species and locations. Many studies (e.g., Lake et al. 1990; Bartell et al. 1998) have shown that biota-to-sediment accumulation factor (BSAF) values at steady state reduce the variability inherent with simple bioaccumulation factors (BFs) based on the ratio tissue concentration to sediment concentration. In the BSAF approach, the ratio is derived using a sediment concentration normalized to organic carbon and a tissue concentration normalized to its lipid content. However, there is a substantial amount of data, particularly in the freshwater literature, demonstrating that the BSAF approach is not adequate to describe the extent a compound will accumulate (e.g., Van Hoof et al. 2001). The recent appendix to the USEPA (2000) bioaccumulation guidance document shows BSAF values exceeding 50 for some chlorinated pesticides. Such large numbers suggest that factors other than the amount of organic carbon in the sediment are responsible for controlling the bioaccumulation, and simple predictions of BSAF should be used with caution. Additionally, there has been a tendency to misuse the BSAF. The BSAF approach was never intended for certain applications such as the normalization of fish tissue concentrations or the assessment of bioaccumulation of metals. For fish, this approach may be applicable, depending on the compound, species, and its site fidelity, and on the degree to which different matrices are characterized. In general, this application ignores the impact of trophic transfer for some compounds even when the primary source is sediment. Finally, it has been shown that normalizing to the amount of organic carbon, which presumably governs bioavailability, can be ineffective in improving the predictive ability of the empirically based SQGs (e.g., Ingersoll et al. 1996; Chapters 4 and 12).

In addition to the difficulties in estimating tissue residue concentrations from sediment chemistry data, interpreting the significance of these predicted residue concentrations has also been problematic. Interpretation of tissue residue concentrations has historically been limited to comparison to a fixed standard (e.g., US Food and Drug Administration [USFDA] action level or comparison to reference). Recent efforts on the part of the US Environmental Protection Agency [USEPA] and US Army Corps of Engineers [USACE] have helped to establish more formalized guidance for interpreting tissue residue data. Using such tools as the critical body residue (CBR) model for nonionic organics (McCarty and Mackay 1993), the Environmental Residue Effects Database (ERED) (USACE/USEPA 2002a) and the Jarvinen and Ankley (1998) database have enabled a more thorough assessment of the potential effects associated with measured tissue concentrations. However, because of the paucity of residue-effects data and the need to assess potential effects to higher trophic levels, extrapolations between compounds and species and applications of models are sometimes employed to assess the potential effects of bioaccumulation. Tissue residue data have been used for establishing trophic transfer models to estimate the potential for contaminant exposure from sediments on higher-level organisms (e.g., Thomann et al. 1992; Thomann and Komlos 1999). It may be possible to link these tools for purposes of estimating sediment concentrations expected to be deleterious to organisms through bioaccumulation. For example, when tissue concentrations associated with adverse effects have been identified, it may be possible to use site-specific BSAF values to calculate sediment concentrations that would be expected to produce a toxic tissue concentration (Meador et al. 2002a). Currently, there appear to be 2 potential types of SQGs that can be developed for assessing potential effects of contaminant bioaccumulation from sediments, namely direct guidelines based on tissue residue effects data and guidelines that incorporate the indirect effects through the action of trophic transfer. The primary concern in developing meaningful guideline values is addressing the uncertainty associated with either approach. Any bioaccumulation-based SQG value must possess a clear ability to predict the bioaccumulation potential for contaminants from sediments as well as the subsequent effects of that exposure in the receptors of concern. The following sections focus on questions fundamental to understanding and developing SQG values predictive of bioaccumulative effects.

Are Results from Laboratory Bioaccumulation Studies Predictive of Tissue Residues in Field-Collected Organisms?

One important question concerns the fidelity between laboratory and field assessment of bioaccumulation. Many bioaccumulation tests are performed in the laboratory with the assumption that the results will reflect the bioaccumulation potential of organisms in the field. It is clear that the conditions under which toxicity tests are normally performed in the laboratory reflect only a snapshot of the possible exposure conditions for organisms in the field. However, if the laboratory conditions are varied sufficiently, it may be possible to predict the range of results expected in the field (e.g., Landrum et al. 2001). Field-collected BSAF values have generally followed predicted values (Wong et al. 2001). It is also important to note that bioaccumulation studies are generally focused on benthic infauna (i.e., animals with limited mobility and in direct contact with sediment). Consequently, the application of BSAFs to fish is probably not meaningful because of variable degrees of site fidelity and the potential impact of trophic transfer.

In Table 11-1, several studies that examined bioaccumulation in invertebrates in the laboratory and field are listed. In general, a comparison of values for a given compound or class of compounds shows substantial variability; however, when means are considered, relatively consistent patterns emerge. For the PAHs, BSAFs are consistently in the 0.01 to 0.5 range across species and are not appreciably different between those determined in the laboratory and those measured in field-collected organisms. The BSAFs for chlorinated hydrocarbons (CHs) are generally much higher than those for the PAHs, ranging from 0.2 to 5. For both groups, values much lower, and sometimes much higher than these, have been reported. The lower values generally include the more highly hydrophobic compounds or those that can be metabolized. If total polychlorinated biphenyls (PCBs) are considered, it appears that there is very little difference between the laboratory and field-based BSAF values (Table 11-1), leading to the conclusion that lab-generated BSAFs can be predictive of those in the field.

The differences between PAHs and CHs are highlighted in studies that examined these compounds in the same species and sediments (Landrum and Faust 1991; Hickey et al. 1995; Meador et al. 1995; Meador, Adams et al. 1997). The differences between PAHs and CHs may be due to the metabolic transformation of PAHs by many of the species tested and the lack of metabolism for PCBs by most species. Other factors for the observed differences include, but are not limited to, differences in the bioavailable fraction for each compound, differential toxicokinetics for chemicals, and insufficient time for accumulation.

In general, many studies have demonstrated that BSAFs for PAHs in benthic invertebrates appear to occur at values approximating 1 order of magnitude below those BSAF values (generally ranging from 1 to 4) expected at equilibrium (Bierman 1990; Di Toro et al. 1991; Boese et al. 1995). Without detailed analysis, it is generally not possible to determine which factors (e.g., reduced exposure time, metabolism, reduced bioavailability) are most important for this observation. When nonmetabolized, nonionic, organic compounds are examined, measured BSAF values are often close to expected values (Bierman 1990; Tracey and Hansen 1996; Wong et al. 2001), but deviations can occur (USEPA 2000).

One analysis of BSAFs from the literature found no differences between laboratory-and field-exposed benthic invertebrates for PAHs (USACE/USEPA 2002). The mean and standard deviation (sd) PAH BSAF values for laboratory-exposed and field-exposed benthic invertebrates were 0.34 (0.95), n = 167, and 0.36 (0.98), n = 183. The median values for each group were identical (= 0.077). Based on Dunn's test on medians, the median BSAF values for CHs were significantly different (USACE/USEPA 2002). The lab-exposed benthic

Table 11-1 Bioaccumulation factors for polycyclic aromatic and chlorinated hydrocarbons (PAHs and CHs) by invertebrates

Species	Feeding type[a]	Area, type[b]	Compounds[c]	Total conc (ng/g) dry	BAF (dry wt)	BSAF (range)	BSAF (mean)	Time (d)	Ref
Abarenicola pacifica	DF	Northwest USA, LS	3 PAHs			1.5–3.7	2.4	60	Augenfield et al. 1982
Arenicola marina	DF	Netherlands, LF	7 PAHs	370–3;100	0.76			60–90	Kaag et al. 1997
Armandia brevis	DF	New York USA, LF	Sev CHs[d]			0.2–6.3	0.44[e]	10	Meador et al. 1997a
Armandia brevis	DF	New York USA, LF	24 PAHs			0.002–0.9	0.18	10	Meador et al. 1995
Austrovenus stutchburyi	FF	No. New Zealand, F	Sev CHs[d]	1.2–16	1.2–51.5	0.2–3.8	1.45[e]		Hickey et al. 1995
Austrovenus stutchburyi	FF	Northern New Zealand, F	9 PAHs	9–47	0.01–0.58	0.002–0.05	0.04		Hickey et al. 1995
Chlamys septemradiata	FF	Norway, F	12 PAHs	75–304		0.007–0.43	0.11		Naes et al. 1999
Corophium volutator	Detr	Netherlands, LF	8 PAHs			0.5–1.7		25	Kraaij et al. 2001
Coullana sp.	Omn	Louisiana USA, LS	fluoranth			0.22–0.67	0.43	1	Lotufo 1998
Diporeia spp.	DF	Lake Michigan USA, LS	DDT			0.03–0.31	0.15	28	Lotufo et al. 2001
Eohaustorius washingtonianus	Detr	Northwest USA, LF	16 PAHs		0.1–0.45			7	Varanasi et al. 1985
Hyalella azteca	Omn	Lake Michigan USA, LS	DDT			0.44–2.1	0.97	28	Lotufo et al. 2001
Leptocheirus plumulosus	Detr, FF	Chesapeake Bay USA, LS	fluoranth				0.32	26	Kane Driscoll et al. 1998
Lumbriculus variegatus	DF	Green Bay, Wisconsin USA, LF	Total PCBs			0.21 - 1.35	0.84[e]	30	Ankley et al. 1992
Macoma balthica	DF, FF	Chesapeake Bay USA, LS	chyr, naph			0.17–0.78	0.17	12	Foster et al. 1987

Table 11-1 *contd*

Species	Feeding type[a]	Area, type[b]	Compounds[c]	Total conc (ng/g) dry	BAF (dry wt)	BSAF (range)	BSAF (mean)	Time (d)	Ref
Macoma inquinata	DF, FF	Northwest USA, LS	3 PAHs			0.6–2.4	1.3	60	Augenfield et al. 1982
Macoma nasuta	DF	Southern California USA, LF	11 PCBs			0.21–2.1[f]	0.91	28	Ferraro et al. 1991
Macoma nasuta	DF, FF	Oregon USA, LS	PCBs, HCB		0.9–30.3	0.06–2.0	1.2[e]	119	Boese et al. 1995
Macoma nasuta	DF, FF	Northwest USA, LF	16 PAHs		0.06–0.19			28	Varanasi et al. 1985
Macoma nasuta	DF, FF	Los Angeles USA, LF	5 PAHs			0.2–1.0	0.4	28	Ferraro et al. 1990
Macomona liliana	DF	Northern New Zealand, F	Sev CHs[d]	1.3–47	3.1–61.3	1.4–23.2	3.5[e]		Hickey et al. 1995
Macomona liliana	DF	Northern New Zealand, F	9 PAHs	18–203	0.09–1.7	0.04–0.13	0.10		Hickey et al. 1995
Mya arenaria	FF	Chesapeake Bay USA, LS	chyr, naph			nc		12	Foster et al. 1987
Nephtys incisa	Omn	Northeast USA, F	Total PCBs			3.2–4.3	3.8		Lake et al. 1990
Nereis virens	Omn	California and New Jersey USA, LF	fluoranth			0.8–3.3		15	Brannon et al. 1993
Oligochaetes	DF	Green Bay, Wisconsin USA, F	Total PCBs	≈ 1800		0.87–2.59	0.87[e]		Ankley et al. 1992
Palaemonetes pugio	Omn	Louisiana USA, LS	BaP, phen			nd–1.6	0.23	14	Mitra et al. 2000
Polychaetes (several spp)	DF, Omn	San Francisco USA, F	18 PAHs	310–1790		0.04–2.0	0.2		Mayrua et al. 1997
Potamocorbula amurensis	FF	San Francisco USA, F	18 PAHs	130–860		0.6–5.4	0.3		Mayrua et al. 1997
Rangia cuneata	FF	Louisiana USA, LS	BaP, phen			nd–1.5	0.42	14	Mitra et al. 2000

Table 11-1 *contd*

Species	Feeding type[a]	Area, type[b]	Compounds[c]	Total conc (ng/g) dry	BAF (dry wt)	BSAF (range)	BSAF (mean)	Time (d)	Ref
Rhepoxynius abronius	Omn	New York USA, LF	Sev CHs[d]			0.02–0.95	1.6[e]	10	Meador et al. 1997a
Rhepoxynius abronius	Omn	New York USA, LF	24 PAHs			0.001–0.5	0.052	10	Meador et al. 1995
Rhepoxynius abronius	Omn	Northwest USA, LF	16 PAHs		0.09–0.5			7	Varanasi et al. 1985
Schizopera knabeni	DF	Louisiana USA, LS	fluoranth			0.51–0.80	0.62	1	Lotufo 1998
Stichopus tremulus	DF	Norway, F	12 PAHs	237–797		0.004–0.67	0.18		Naes et al. 1999
Streblospio benedicti	DF	South Carolina USA, F	3 PAHs	860–2000	0.2–1.4	0.08–0.4	0.22		Ferguson and Chandler 1998
Tapes japonica	FF	San Francisco USA, F	18 PAHS	95–450		0.007–2.7	0.15		Mayrua et al. 1997
Yoldia limatula	DF	Northeast USA, F	Total PCBs			4.1–4.8	4.4		Lake et al. 1990

[a] DF = deposit feeder; FF = filter feeder; Omn = omnivore; Detr = detrivore
[b] Where found or where research was conducted, type of exposure (F = samples from field; L = exposures in lab to field-contaminated sediment; LS = exposures in lab to spiked sediments)
[c] Fluoranth = fluoranthene; phen = phenanthrene; BaP = benzo[*a*]pyrene; BA = benz[*a*]anthracene; HCB = hexachlorobenzene; CH = chlorinated hydrocarbons). All analyses conducted with whole organisms or soft tissue of clams. Mean BSAFs includes all compounds and locations studied.
[d] PCBs, DDTs, and chlorinated pesticides (e.g., lindane, dieldrin, chlordane). A conversion factor of 5 was used for wet weight to dry weight concentrations.
[e] Total PCBs
[f] 0 to 2 cm depth

invertebrates exhibited a higher median BSAF of 0.85, *n* = 280 compared to field-exposed benthic invertebrates of 0.32, *n* = 414. The mean (sd) values were closer at 1.4 (1.7) for lab-exposed invertebrates and 1.2 (2.4) for field-exposed invertebrates. The field-exposed invertebrates were expected to have the higher BSAFs resulting from increased bioaccumulation time. However, these results may indicate that field conditions are more heterogeneous than those found in the laboratory and may be influenced more by external factors such as fresh food or alternate sources of contaminants. The PAH results may also indicate the same mechanism. However, interspecies differences in metabolism may obscure any such pattern.

A few studies have addressed the laboratory versus field question directly. Ankley et al. (1992) compared BSAF values for oligochaete worms exposed to PCBs in sediment. They measured concentrations in worms from the field and then exposed *Lumbriculus variegatus* in the laboratory to these same sediments for 30 days. For 3 of the PCB congener classes (dichloro, heptachoro, and octachloro), oligochaetes from the field exhibited significantly higher BSAF values than those exposed in the laboratory. Additionally, Brunson et al. (1998) found good relationships between laboratory-exposed oligochaetes and field-collected organisms for PAHs. Ninety percent of the paired PAH concentrations between laboratory and field samples fell within a factor of 3. Oliver (1987) reached a similar conclusion with regard to persistent organic compounds while examining BAFs in laboratory exposures and field collections of oligochaetes. Ingersoll et al. (2003) reported that concentrations of DDT, DDD, and DDE, and PAHs measured in native oligochaetes which were collected at the same time that sediment was collected from the field were similar to the steady-state concentrations estimated from the laboratory exposures with the oligochaete *L. variegatus*. However, concentrations of low-K_{ow} PAHs in native oligochaetes were biased higher than the steady-state concentrations estimated from the laboratory exposures with *L. variegatus,* while concentrations of high-K_{ow} PAHs in native oligochaetes were biased lower. In another laboratory study, Landrum et al. (2001) examined the BSAF of PCB congeners in *Diporeia* spp. and found that the BSAF was not constant with log K_{ow} as suggested from EqP theory. They concluded, however, that field BSAF values could be predicted from laboratory data, provided the organism size and exposure temperature were taken into consideration.

Studies in which organisms are placed in cages or on racks and left in the field for a predetermined period of time can be a useful way to assess bioaccumulation. These techniques are useful because they allow researchers to get information from species not found at the site, compare results across sites for the same species, and control several temporal and spatial factors that potentially affect bioaccumulation. The technique of placing bivalves at a site has been used for years to assess bioaccumulation. Mussels are commonly used for this type of experiment because of their high rate of water filtration and low biotransformation capacity. Bivalves can also be sampled from the field and assessed for bioaccumulation (e.g., Mussel Watch program); however, in situ placement allows greater control of important variables. One drawback to using mussels to monitor bioaccumulation is that they do not interact directly with the sediment and may exhibit much less bioaccumulation than other benthic species.

Other species, including benthic invertebrates and fish, have been placed in cages left on site for various periods of time to monitor contaminant accumulation under more realistic field exposure conditions. These in situ toxicity tests also can be used to measure biological effects such as mortality or reduced growth. Some of the biggest challenges facing these studies include the selection of comparable reference sites, effects due to caging, loss of experimental units, and variability in environmental parameters that can affect toxicokinetics and organism health (e.g., temperature, food, dissolved oxygen, pH). Details for caging and transplant studies can be found in the literature (e.g., Chappie and Burton 1997; Salazar and Salazar 1998; Forrester et al. 2003).

What Factors Affect Assessments of Bioaccumulation?

Several factors can affect the results in bioaccumulation assessments, regardless of the origin of the exposure source (field contaminated or spiked) or location (field versus laboratory). A number of these factors and their potential effects on bioaccumulation are discussed below.

Route of exposure: Field versus laboratory sources

Bioaccumulation of contaminants in the field by aquatic organisms results from exposure to multiple potential sources. For benthic organisms, these include sediment, sediment pore water, ingested particles including fresh detritus and bacteria, and overlying water. Many of these sources are out of chemical equilibrium with each other. Tissue residue concentrations provide a measure of contaminant exposure from all potential sources and metabolism by the organism. Further, the concentration in the environment of the organism may be different from that taken for analytical measurements, which would impact attempts to establish relationships between single sources and measured bioaccumulation. When exposure occurs in the laboratory setting, the resultant accumulation may not reflect the environmental measurements because the sources are not reflective of those in the environment. The experimental conditions, such as temperature and population structure, that affect the physiology and thus the toxicokinetics are not reflected in the laboratory environment. The exposures in the laboratory tend to be snapshots of specific conditions that may be found in the field but cannot reflect the overall exposure of organisms over temporal and spatial scales with multiple sources of exposure.

Accumulation of nonionic hydrophobic compounds is generally passive and is driven by the chemical activity of the source and sink compartments. However, the rate may well be dictated by the route and volume of source compartment that the organism experiences. Uptake from water is generally accomplished by ventilation over the gill structure; however, diffusion through the integument may also contribute to tissue concentrations (Landrum and Stubblefield 1991). Ingestion of prey organisms, detritus, and sediment is also important for accumulation. In the short term (before steady state), the degree to which each route contributes to the total body residue is difficult to determine without well-designed experiments. According to EqP theory, when all phases are in equilibrium, the route of uptake is immaterial because no matter which route dominates, the resulting tissue concentration is always the same (Chapter 13). However, systems are rarely in chemical equilibrium such that EqP would apply (Lee 1990). For instance, the overlying water does not attain expected equilibrium with surficial sediments for PCB partitioning in Lake Michigan (P.F. Landrum, personal communication NOAA GLERL, Ann Arbor, MI), and while individuals of the amphipod *Diporeia* spp. are at steady state with the sediment concentrations based on the toxicokinetics and field measurements for PCBs, the BSAF values deviate substantially from expected EqP theory, ranging from approximately 0.2 to greater than 10 (Landrum et al. 2001).

Bioaccumulation of metals is more complicated than that of organic compounds. The control of water concentrations is complex and often not correlated in a linear fashion to the ligands that control their solubility. Many studies indicate that porewater concentrations are controlled by AVSs, which in turn are correlated to the degree of toxicity (Ankley et al. 1996) and presumably the amount bioaccumulated. Other studies, however, have demonstrated that dietary uptake may also be an important route for bioaccumulation of metals (Weeks and Rainbow 1993; Roy and Hare 1999).

The exposure matrix can be an important determinant for bioaccumulation, especially under nonequilibrium conditions. Benthic species are likely to exhibit higher tissue residues because they often ingest sediment and are exposed to pore water. There are a variety of feeding modes for benthic species, including but not limited to sediment ingestion (selective and nonselective), detritus feeding, predation, and filter feeding. Each of these modes may have an impact on the degree that the organism is exposed to contaminants and final bioaccumulation values, especially if disequilibrium prevails among water, tissue, and sediment. Some infauna are exposed to pore water, and some build tubes and pump overlying water through their burrows. Studies that compare different species under identical conditions can be very informative when the bioaccumulation potential is being determined.

A number of studies have examined the mode of feeding by invertebrates in relation to bioaccumulation (Foster et al. 1987; Hickey et al. 1995; Meador et al. 1995; Kaag et al. 1997). For example, Foster et al. (1987) demonstrated large differences in tissue concentrations between 2 clams, one a deposit feeder and the other a filter feeder (Table 11-1). Another study comparing *Rhepoxynius abronius* (an infaunal amphipod that does not ingest sediment) and *Armandia brevis* (an infaunal nonselective deposit-feeding polychaete) found similar accumulations of LPAHs by the 2 species but substantially more accumulation of HPAHs by the polychaete (Meador et al. 1995). Because these were lab toxicity tests, the amphipod likely had little prey available, which led to the conclusion that deposit and nondeposit-feeding infaunal invertebrates will acquire most of their body burden of LPAHs through pore water, regardless of feeding strategy; however, ingestion (of sediment or food) may be the dominant route of uptake for hydrophobic compounds exceeding a log K_{ow} of approximately 5.5. If sufficient prey were available to *R. abronius* in this experiment, it is likely that BSAF values for the HPAHs would have been higher.

Temporal issues

The amount of time allowed for bioaccumulation can have a large effect on the tissue residues and the degree to which the organism reaches steady state. Several studies have examined the temporal aspects of bioaccumulation. Oliver (1987) found in a study of 37 CHs that oligochaete worms reached steady-state tissue residues within 2 weeks for most compounds. Accumulation of PAHs and DDT from sediment by the oligochaete *L. variegatus* typically reached steady state in 14 to 28 d during 56-d laboratory exposures (Ingersoll et al. 2003). Several of the more hydrophobic compounds ($\log_{10} K_{ow} > 6.0$) exhibited relatively long half-lives, indicating that steady state would take longer. Most bioaccumulation studies are conducted for 28 d because of the recommendations and research by Lee et al. (1993) showing this period of time to be sufficient for many infaunal benthic species and chemicals to reach at least 80% of steady-state tissue concentrations. At least 1 study has found that very small benthic invertebrates achieve steady state within hours to a few days when exposed to sediment-associated fluoranthene (Lotufo 1998), which may be related to their large surface area to volume ratio.

The time to achieve steady state is dictated by the magnitude of the elimination rate constant. The elimination rate constant can be substantially influenced by the lipid content of the organisms. For instance, the elimination rate of *Hyalella azteca* for fluoranthene is relatively rapid, such that steady state would be achieved in approximately 30 h, while the more lipid-rich *Diporeia* has substantially slower rates of elimination that would not result in achievement of steady state until upward of 35 to 130 d (Kane-Driscoll et al. 1997). While some of the difference was attributed to the biotransformation capability of *H. azteca,* the majority of the effect was thought to result from the very high lipid content of *Diporeia.*

Seasonality

Organisms show changes in their lipid content with the season, and these changes can result in either increased or reduced storage capacity, depending on whether the organisms are accumulating or consuming lipid for energy or for transfer to offspring. The influence of lipids on bioaccumulation and trophic transfer of organic contaminants has been reviewed by Landrum and Fisher (1998). Another study was designed to specifically examine the seasonality of PCB bioaccumulation by *Diporeia* spp. (Robinson et al. 2000). While seasonality was not strong, that is, BSAF values varied only within a factor of 2 for a specific congener, 2 factors appeared to dominate the limited seasonal trends in the data. One important factor was the accumulation of lipids with ingestion of the spring diatom bloom resulting in apparent concentration dilution in the tissues, which reduced the BSAF. A second factor was alterations in the organic carbon–normalized PCB concentration in source. Alterations also occurred in the organic carbon–normalized PCB concentrations in the source. This dilution of the source with fresh carbon had the net effect of reducing the overall source concentration, thus elevating the BSAF. Similarly, the influence of lipids and the spawning cycle on the bioaccumulation of PCB in *Mytilus edulis* was found for exposure in Buzzard's Bay, resulting in a 2- to 4-fold difference in the bioaccumulation (Capuzzo et al. 1989). Other factors such as organism feeding behavior, food composition, and the population size distribution of food sources can also influence seasonal exposure to contaminants, both through changes in the rate of feeding and in the choice of food.

Uptake efficiency

Some studies indicate that uptake efficiency generally declines with increasing chemical hydrophobicity, which may be due to a combination of slow desorption kinetics and short residency time in the gut. Additionally, because the role of metabolism often is not addressed, the apparent uptake efficiency may be underestimated because of the effective loss of the parent compound. For some compounds, it has been demonstrated that the rates and efficiencies of uptake vary slightly over K_{ow} and are therefore not strongly linked to chemical hydrophobicity (McKim et al. 1985; Bender et al. 1988; Landrum et al. 2001). Other studies demonstrate that aquatic organisms exhibit strong relationships between the rate of uptake and log K_{ow} (e.g., Landrum 1988).

In order to avoid confounding the estimate of uptake efficiency for compounds that are metabolized, the parent compound plus metabolites should be determined. For example, research that examines uptake and elimination kinetics would be needed to better assess uptake efficiency of PAHs for the different routes of uptake, especially the dietary route. These data would help greatly in predicting bioaccumulation from different environmental matrices.

Metabolism

Metabolism of xenobiotic compounds is a crucial factor in determining and predicting bioaccumulation. Some hydrophobic organic compounds are poorly metabolized by invertebrates and fish (e.g., PCBs), while others are biotransformed. These processes, however, vary widely among different taxa. Elements are not metabolized but are often rendered less toxic by complexation with metallothionein or are incorporated into granules, shell, or bone. Among benthic invertebrates, metabolism of PAHs can be highly variable, even within taxonomic groups. For example, large differences in metabolic transformation of PAHs were found for different species of polychaetes by Kane-Driscoll and McElroy (1996). Among crustaceans, some species such as *R. abronius* have relatively active P450 systems (Reichert et al. 1985), and others such as *Chironomus tentans* show high biotransformation while the P450 system has not been directly evaluated (Lydy et al. 2000). Conversely, other crustaceans, such as the American lobster (*Homarus americanus*) (Foureman et al. 1978; Bend et al. 1981) and *Diporeia* spp. (Landrum 1988), have very limited biotransformation capabilities for some compounds such as PAHs. In general, mollusks accumulate high levels of contaminants, including PAHs because of their high rate of filtration and low metabolic capacity at steady state despite having low lipid contents (Livingstone 1994).

Sometimes the assessment of metabolism in determining bioaccumulation is not straightforward. For example, large differences were observed in BAF values for 2 infaunal amphipods (*Eohaustorius washingtonianus* and *R. abronius*) and a deposit-feeding clam (*Macoma nasuta*) exposed to benzo[*a*]pyrene (Varanasi et al. 1985). The amphipods exhibited higher BAFs than the clam, although their metabolic capacity for this compound was much greater. Several factors may explain these unexpected results, including a higher rate of uptake leading to higher body burdens or exposure to additional sources of contaminant such as ingestion. Additionally, the clam may have simply interacted more with overlying water or it may have closed its valves, thus reducing contaminant uptake. The BAFs for the amphipods at 7 d of exposure were several-fold higher than the BAFs observed for the clam at 28 d, a time sufficient for near steady-state accumulation to occur (Lee et al. 1993).

Aging of Sediment-Associated Contaminants

A limited number of studies have identified the effects of aging and the change in bioavailability of a contaminant with increasing contact time between the contaminant and sediment particle (e.g., Landrum 1989). The impact seems to be larger and more rapid with compounds that are less hydrophobic, likely because they fit better into sediment particle pores. The difficulty has been to assess the impact of this mechanism on field samples that presumably have long contact times. In the laboratory, long contact times have been used as a way to ensure equilibrium prior to any bioaccumulation testing. This allows for better comparisons with the field condition (USEPA 2000). Studies associating the rate and extent of desorption from the rapidly desorbing fraction to the extent of biodegradation and bioaccumulation of contaminants have shown that the rapidly desorbing fraction, whether in laboratory-dosed sediment or in native sediments, correlates with bioavailability (Cornelissen et al. 1998; Kraaij et al. 2001). This approach is new and the data are sparse; however, this does suggest an approach for moving from total contaminant concentration to a practical measure of the bioavailable contaminant without necessarily performing toxicity tests. Recent work with organic contaminants suggests that it is not the fraction of rapidly desorbing contaminant that is responsible for bioavailability; rather it is the flux (i.e., the concentration in the rapidly desorbed fraction times the desorption rate constant) off the particles that dictates the bioaccumulation (Kukkonen et al. 2001). Research focused on the role of desorption kinetics shows great promise in providing a method for evaluating the bioavailability through chemical means, but more research is required.

Physical and chemical factors

The amount of organic carbon for sorbing organic contaminants and the amount of ligands for binding metals are widely recognized as modifying the bioavailability of sediment-associated contaminants. The quality of organic matter is also recognized as critical to the bioavailability of organic contaminants. The polarity of organic matter has been demonstrated to alter the rate of accumulation of PAHs (Landrum et al. 1997). However, the same work showed that differences in bioaccumulation for PCBs were most readily attributed to total organic carbon content and polarity was not a major factor. Thus, it appears that the chemical characteristics of the contaminant and the manner in which it interacts with the sediment organic matter are critical for dictating bioavailability. Because of selective feeding and the potential role of feeding on the bioaccumulation of contaminants, the uneven distribution of the contaminants on particles of varying sizes could have a large influence on the bioaccumulated fraction. The bioaccumulation and assimilation efficiency for PAH versus PCB congeners were tied to this unequal distribution and the selective feeding of *Diporeia* spp. (Harkey, Lydy, et al. 1994). In the field, there is some indication that such differential distribution of compounds among particles of differing sizes does exist (Umlauf and Bierl 1987; Evans et al. 1990; Pierard et al. 1996; Van Hoof and Eadie 1999).

For metals, the extent of binding to ligands, particularly AVS, has been demonstrated to reduce bioavailability (Di Toro et al. 1990; Chapman et al. 1998; Chapter 13). However, many other ligands can participate in the reduction in bioavailability, depending on the composition of the system (Newman and Jagoe 1994). From a practical perspective, it appears that the porewater concentration more accurately predicts toxicity (e.g., Ankley et al. 1993). It is clear, however, that environmental exposure to metals remains complex, and additional work to establish a useful practical approach to evaluating metal bioavailability remains to be developed.

Contaminant source

The bioavailability of some contaminants can vary considerably within a limited region, likely resulting from differences in sources. For example, the source for PAH exposure varies widely, with some sites containing mainly petroleum-based PAHs while others are contaminated with combustion PAHs. Several studies have shown reductions in water or tissue concentrations of PAHs as they may relate to PAH type and source (Farrington et al. 1983; McGroddy et al. 1995; Meador et al. 1995; Maruya et al. 1997; Naes et al. 1999). For example, PAHs from different combustion sources (e.g., soot, coal, or an aluminum smelter) may produce very different bioaccumulation patterns. Likewise, fresh petroleum would likely result in different bioaccumulation patterns compared to weathered petroleum because of proportional change among congeners. One interesting study (Naes et al. 1999) examined BSAF values in 3 different benthic invertebrates and found a gradient of decreasing values towards an aluminum smelter. Depending on the species, BSAFs decreased from 5- to 10-fold from one end of a fjord to the other end where the smelter was located and may be related to the types and sources of PAHs found along the gradient. Another study proposed that the lower PAH BSAF values observed by Landrum (1988) for the amphipod *Diporeia* spp., a species that is not able to biotransform PAHs, was due to the importance of soot on the partitioning of PAHs in sediment systems (Van Hoof et al. 2001). The impact of soot carbon has also been implicated in limiting the bioaccumulation of PCBs in Ashtabula Harbor, Ohio, USA (Pickard et al. 1998).

Organism behavior

Organism activity can also affect the sorption dynamics of contaminants associated with sediment. For example, the suite of organisms nearby the target species in the laboratory or the field may have a significant effect (approximately a factor of 2) on the amount of a contaminant that is bioaccumulated (Schuler et al. 2002). Bioturbation by organisms can increase contaminant concentrations in overlying water, which may be an important factor in assessing bioaccumulation for some species. For example, McElroy et al. (1990) demonstrated that the presence of a tubiculous polychaete (*Nereis virens*) can enhance the flux of sediment-sorbed benz[*a*]anthracene to the water column. This increased flux to the water column could elevate tissue concentrations in those animals that take in contaminants through gill membranes by ventilating overlying water. This was also confirmed by Ciarelli et al. (1999), who found that bioturbation by 1 species enhanced bioaccumulation in another. In this study, the authors showed a linear relationship between fluoranthene in *Mytilus edulis* and amphipod *Corophium volutator* density. Additionally, benthic organisms can inhibit the desorption of chemicals into water. For example, it has been observed that oligochaete fecal pellets can dramatically reduce the amount of compound that will desorb into water (Karickhoff and Morris 1985). These mechanisms are important because severe alterations to sediment–water partitioning and equilibrium can occur, thus confounding comparisons of bulk sediment and tissue concentrations for bioaccumulation assessment.

Some species may change their behavior as tissue residues approach toxic levels. In some cases, animals may exhibit adverse effects and alter their rates of ingestion or ventilation, causing tissue concentrations to increase or decrease dramatically. This phenomenon has been observed by Landrum et al. (1994) and Meador and Rice (2001). Exposure avoidance is another approach for reduction in observed toxicity (Kukkonen and Landrum 1994). Basing the toxicity on the concentration in the organism, therefore, rather than on the exposure environment, allows improved interpretation of the toxicity data.

Species differences

In general, it is believed that species within a given taxonomic family will exhibit similar responses to toxicants (Suter and Rosen 1986). Several studies have shown that relatively closely related species can exhibit very different toxicity responses as a result of differential bioaccumulation. For example, Meador, Krone, et al. (1997) found that 2 species of amphipods from closely related families (Superfamily Haustorioidea) exhibited LC50s to tributyltin (TBT) that differed by 14-fold in 10-d water exposures. When compared to *E. estuarius, R. abronius* was the more tolerant species. Even larger differences were observed in long-term sediment exposures with *R. abronius* and *E. washingtonianus*, a species that responds identically to TBT as *E. estuarius* (Meador, Krone, et al. 1997). When these same species were exposed to Cd in water, *R. abronius* was the more sensitive of the two, exhibiting an LC50 that was 10× lower (ASTM 2004). Interestingly, another study found that these 2 species responded similarly to sediments contaminated with PAHs (Pastorok and Becker 1990).

Certainly, the use of body residue as a dose metric suggests that for nonpolar narcosis (anesthesia), the toxic response for acute mortality has a narrow range for fish (McCarty and Mackay 1993). The difference for this mechanism of action between species depends strongly on the lipid content of the organisms (van Wezel and Opperhuizen 1995). However, the relative species response may well depend on the mechanism of toxic action. For specific acting chemicals, larger variability can be expected because of the need for toxicant to fit to a specific receptor that is likely different for different species, even closely related ones. In a review by Barron et al. (2002), the issue of species differences and particularly differences in mechanisms of action contributes significantly to the observed body residue to produce a toxic response. Thus, toxicity tests or field assessments with only a few species may underestimate the toxicity potential of a given sediment.

Lipid content

Lipid content in organisms is an important factor for assessing bioaccumulation of nonionic hydrophobic compounds. Even though the lipid content in many invertebrates approximates 5% (dry weight) (Boese and Lee 1992), many exceptions occur (e.g., *Diporeia*; Landrum and Nalepa 1998) and can range up to 50%. The ability of the lipid to sequester contaminants away from the site of toxic action is critical to the observed response and has led to the hypothesis of the survival of the fattest (Lasiter and Hallam 1990). To adequately test the hypothesis that lipids control the bioaccumulation of nonionic hydrophobic compounds, lipid content would need to be varied for a given species and toxicokinetic rates measured in order to ensure that they are comparable between groups. One study (Bruner et al. 1994) found that high-lipid zebra mussels had greater BCFs and faster uptake kinetics for highly hydrophobic compounds.

For some hydrophobic compounds, lipid may not be important for bioaccumulation. For example, the bioaccumulation of TBT was found to be far more extensive than that predicted by its K_{ow}, leading to the conclusion that lipid does not play a role in determining tissue residues for TBT (Meador 2000). It is expected, however, that for a given whole-body tissue concentration, the magnitude of the toxic response from TBT may be a function of the organismal lipid content (Meador 1993, 2000), although the data are not sufficient for a rigorous test of this hypothesis.

Spatial issues

The relationship between the measured contaminant concentrations found in sediment and those found in the organisms may be complicated by sampling errors. If the matrix that produces the exposure is not the matrix analyzed, then there will be a disconnect in the relationship between concentrations in sediment and the organism. For instance, Lee (1991) points out that it is typical for analytical chemists to select the top 2 cm of sediment, while organisms may feed on the very surficial material (fresh-falling detritus, e.g., *Macoma*) or may feed at depth (e.g., oligochaete or polychaete worms). Additional issues include spatial variability and could include temporal variability for labile contaminants (Holland et al. 1993; Sarda and Burton 1995). The extent of the variability can result in variation in exposure conditions and in a poor match between the measured exposure concentration and that accumulated by the organism, resulting in additional variability in observed relationships.

Organism size

Clearly, even within a species, there are characteristics that relate organism size to the extent of accumulation. In some organisms, the mode of feeding changes as the organism metamorphoses from larval to juvenile to adult forms. Further, feeding rates, filtering rates, etc., are allometric, and they change the resultant exposure and toxicokinetics of organisms. For instance, in *Diporeia* spp., the accumulation of PCB congeners is greater in small organisms compared to larger ones, primarily because of changes in the respiration relationship to contaminant accumulation (Landrum and Stubblefield 1991) and the higher feeding rate for smaller organisms (Lozano et al. 2003). The resultant bioavailability in small organisms is not due to preferential ingestion of small particles because distribution to smaller particles with higher organic carbon apparently reduces bioavailability (Harkey, Lydy, et al. 1994). The result is that small *Diporeia* show greater accumulation in the field compared to even older, larger organisms (P.L. Van Hoof, personal communication, Monrovia, MD). These higher rates are manifest in larger BSAF values for smaller amphipods.

Habitat: Laboratory versus field

Biomass loading of the experimental systems in the laboratory may not reflect that in the field. This may create variation in the observed bioaccumulation of contaminants. For the oligochaete *L. variegatus*, increasing the biomass loading relative to the amount of organic carbon increased the exposure of the organisms (Kukkonen and Landrum 1994). Additionally, differences in organism abundance can have an effect on sediment geochemistry and the amount of contaminant available for uptake.

Feeding

The feeding behavior of the species can be important in assessing its bioaccumulation of sediment-associated contaminants. Benthic species exhibit several feeding strategies, some of which maximize their exposure to contaminants and others that allow very little exposure. When an organism is in contact with sediment but does not ingest sediment, equilibrium or steady-state tissue concentrations may be very different, reflecting only its interaction with overlying water. A clam living in the sediment might experience little exposure because it filters overlying water that will never be in equilibrium with the sediment. Support for this comes from Foster et al. (1987), who found that a deposit-feeding clam (*Macoma balthica*) accumulated more PAH than a filtering feeding clam (*Mya arenaria*). To be complete in such analyses, potential metabolism of these compounds also needs to be considered to avoid confounding the conclusions.

Because most bioaccumulation toxicity tests and sublethal toxicity tests can be several weeks in duration, adequate nutrition is essential for the health of the test species. Consequently, if the species is not a deposit feeder, some form of nutrition has to be added. For some toxicity test species, a food supplement is added (e.g., TetraMarin flakes), and in some long-term bioaccumulation toxicity tests with deposit feeders, new sediment is added periodically to ensure an adequate food supply. Additionally, some common toxicity test species (e.g., *R. abronius*) are predators, and standard testing protocols do not provide prey other than what may be resident in the sediment being tested.

The presence of fresh food, even for sediment-ingesting organisms, can affect the observed exposure. For the most part, bioaccumulation tests are performed in the absence of feeding, but some require fresh food. For example, bioaccumulation of PAH congeners increased significantly (e.g., approximately by a factor of 2) with external feeding for both *H. azteca* and *C. riparius,* while the bioaccumulation of chlorinated compounds was not affected (Harkey, Landrum, Klaine 1994; Harkey et al. 1997). Thus, the addition of fresh food in laboratory toxicity tests can impact the observed bioaccumulation.

The results of bioaccumulation or toxicity studies in which additional food is required can be compromised if the food does not come into equilibrium with sediment and water concentrations. Individuals will ingest the added food, which may contain lower concentrations of the contaminant, and not the sediment or detritus that likely contain higher concentrations. For example, Bridges et al. (1997) examined the effect of food ration on a common toxicity test species, *Neanthes arenaceodentata*. The results of this study demonstrated strong effects on growth and survival as a function of the amount of food added to toxicity test chambers. Improved food quality and ration can lead to increased lipid, which in turn can increase bioaccumulation but also can provide storage tissue that can sequester lipophilic contaminants away from the site of toxic action. In addition, improved food quality can lead to increased growth, which will also reduce concentrations at the site of toxic action, having a similar impact as added lipid. These results are particularly important for the growth endpoint because variable food ration can be an important confounding factor when growth effects from toxicity are being determined.

Effect of contaminant interaction

When contaminants are present in low concentration, there does not appear to be any impact of 1 contaminant on the accumulation of others (Landrum 1989). However, there are cases in which the presence of certain compounds has acted to reduce the bioaccumulation of other contaminants. Specifically, the presence of polydimethylsiloxane (PDMS) was found to reduce the bioavailability of benzo[*a*]pyrene to *L. variegatus* (Kukkonen and Landrum 1995). This was the result of creating an additional sorption phase for the contaminant while the PDMS was not bioavailable. Even when a few compounds in a mixture are present in toxic concentrations, there is no evidence that these toxic compounds affect the relative accumulation of other compounds in the mixture (Landrum et al. 1989). Thus, when conditions are such that a second sorptive phase is created in a sample, then bioavailability can be reduced. However, when all of the compounds are sorbed to the same matrix, although some are producing toxicity, the relative accumulation does not appear to be affected. It should also be noted that the response to toxic compounds can affect the overall toxicokinetics.

Measurement artifacts

Artifacts in assessing bioaccumulation can arise in field or laboratory studies. Examination of Table 11-1 shows that some of the BSAF values for a given species (e.g., *Macoma* spp., for PAHs) can be highly variable (ranging from 0.1 to 1.3) across studies. This may be due to artifacts such as lack of gut purging, field versus laboratory sources of contamination, static versus flow-through testing conditions, differing organism sizes or ages, differential contaminant bioavailability, and/or variable exposure time.

Performing bioaccumulation tests in the laboratory can be difficult and inconclusive if conditions found in the laboratory are not reflective of those in the field. Of course, many factors that vary in the field, such as temperature, pH, salinity, and oxygen content, are kept constant in the laboratory. Thus, the laboratory bioaccumulation test is a snapshot of one set of conditions found in the field. Other factors that are not often considered can be very important to the outcome. For example, a bioaccumulation test can be static (without water changes) or flow-through (with water exchange at various turnover rates). The importance of this variable depends in part on the species and chemical being tested. Some benthic invertebrates interact extensively with the water column and may derive a large proportion of their acquired tissue burden from filtering overlying water. However, for a deposit-feeding species such as *Diporeia*, the accumulation of PAH was the same whether the organisms were exposed under flow-through conditions or under static conditions (Landrum 1989). This was not the case for the amphipod *H. azteca*, where the observed toxicity was greater under static conditions than under flow-through conditions, reflecting differences in the accumulated toxicant, likely because of the epibenthic nature of the amphipod (Spehar et al. 1998). Additionally, the degree to which a chemical partitions between water and sediment may also be important. For very hydrophobic compounds, most of the accumulation will come from ingestion of sediment or prey (Landrum and Robbins 1990; Leppänen and Kukkonen 1998; Selck et al. 2003), so flow-though conditions may not affect the results.

Several important areas of research currently are aimed at enhancing our understanding of the nature of bioaccumulation. Studies on uptake efficiency from different sources, variability in digestive physiology among species, and qualitative and quantitative differences in organic carbon in determining bioavailability are but a few of the ongoing research projects leading the way to predicting both direct (tissue concentration–adverse effects relationships) and indirect (trophic transfer) effects more accurately and effectively in the future. Steps are also being taken to uncover the role of sediment particle size and relative contaminant distribution on bioavailability, the degree of equilibrium in field environments, the efficiency in trophic transfer of parent compounds and metabolites, and the intra- and interspecific toxicokinetic differences among compounds. Research in these and related areas will lead to enhancements in the current understanding of the processes governing contaminant accumulation and resulting effects in aquatic organisms.

Can Tissue Residue Concentrations Be Accurately Estimated using Theoretical Bioaccumulation Potential?

An EqP model, Theoretical Bioaccumulation Potential (TBP), has been used for nearly 2 decades as a screening tool to estimate the potential levels of bioaccumulation of persistent nonionic organic chemicals that could result in benthic organisms exposed to contaminated sediments. TBP has become a standard test in Tier II evaluations of dredged sediments proposed for open-water disposal (USEPA/USACE 1991, 1998; Chapter 6). A point estimate is obtained by normalizing chemical concentration in sediments (C_s) on organic carbon content (f_{OC}) and by normalizing the expected concentration in an exposed organism's tissues (TBP) on lipid content (f_L) at steady state; the difference between the two is described by a coefficient, the BSAF:

$$TBP = BSAF \times (C_s / f_{OC}) \times f_L.$$

The empirically derived BSAF itself is calculated from normalized tissue and sediment data for the nonionic organic chemical of interest and is

$$BSAF = (C_t / f_L) / (Cs / f_{OC}),$$

where C_t is the measured tissue residue concentration. As risk assessment has gradually been introduced into the ecological sphere, it has become increasingly obvious that point estimates made with no associated measure of variability or uncertainty have limited utility. TBP is the simplest and most easily understood model for estimating bioaccumulation, but as such, it is also subject to a large degree of uncertainty. The model takes no account of the influences of kinetic processes that determine chemical bioavailability from sediments or retention, metabolic degradation, or elimination from organisms. These processes are integrated in the BSAF, which presumably will permit a good estimation of bioaccumulation, provided the conditions to which it is applied are similar to those from that it was derived. Although BSAFs are reported in the literature for organisms which have no direct association with sediments, the concept applies best to those that are benthically coupled, particularly sediment-processing infauna with relatively low xenobiotic metabolizing capability. This is because EqP derives from thermodynamics based on a closed system in equilibrium. The concentrations of a chemical in contiguous environmental compartments of such a system are a function of chemical potential (or its ancillary, fugacity) and ideally are measured when kinetics are at steady state (i.e., when there is no further net exchange of chemical among the compartments). Therefore, whatever contributes to disequilibrium, such as metabolic degradation or distance from source to sink, necessarily contributes to uncertainty.

High-quality TBP estimations are necessary, first, because they are the simplest and most easily understood means of estimating biotic exposure to sediment contamination and, second, because they represent the first step in any risk assessment involving sediment as a source. Thus, the quality of TBP estimations is highly dependent on the quality of selected BSAFs, whose variability may reach several orders of magnitude within a given classification.

Table 11-2 contains descriptive statistics calculated for BSAFs in a BSAF/lipid database (USACE/USEPA 2002b). The database (http://www.wes.army.mil/el/bsaf/bsaf.html) has been compiled from the open literature and from government reports and presently contains more than 1300 BSAF entries for organic chemicals. Lipid data for numerous species can also be found in the database. The BSAF data have been analyzed statistically by chemical class, specific chemical, laboratory or field exposure, organism habitat, and class of organism. Of course, the data may be grouped in any way desired, and there may be advantages to be gained, for example, by grouping by specific organism or specific chemical (or both). However, in the examples that follow, the statistics in line 6 of Table 11-2 were used to calculate uncertainty in TBP estimation for a benthically coupled fish exposed to PCB-contaminated sediments. In order to calculate uncertainty, replication in all input parameters is necessary.

Two methods for calculating the uncertainty associated with PAH TBP estimates have been suggested (Clarke and McFarland 2000) a root sum of squares (RSS) method and a statistical bootstrap method. The resulting uncertainty measures can be used in 3 ways:

1) to gauge the precision of TBP estimates,
2) to enable statistical tests of significance comparing TBP predictions with observed contaminant bioaccumulation, and
3) to provide inputs for risk assessments.

The RSS method computes total error from component method error and TBP data input propagated error. The bootstrap method uses computer-intensive resampling of the TBP data inputs to generate statistical uncertainty measures (e.g., standard error and confidence limits) and to perform tests of significance. Both methods are described in Clarke and McFarland (2000), and computational formulas are given for the RSS method. The methods are described in detail on the USACE/USEPA website in Technical Note EEDP-04-32 (http://www.wes.army.mil/el/dots/eedptn.html).

Recently, uncertainty analysis using the 2 methods was applied to estimating PCB bioaccumulation potential in a bottom-feeding freshwater catfish, *Ameiurus melas*, inhabiting a confined disposal facility (CDF) in Calumet Harbor, MI. The CDF covers 43 acres and is lined with a synthetic membrane of low permeability. Dredged material that has been disposed in the CDF is primarily contaminated with low levels of PCBs. Because there is virtually no advection of water through the system, it was considered that fish inhabiting the CDF pond would essentially be at steady state with the sediments,

Table 11-2 Descriptive statistics for BSAFs from the USACE/USEPA BSAF database [a]

Description	BSAF mean	*n*	SD	SE	CI of mean	Range	Max	Min	Median	25th percentile	75th percentile
1) All BSAFs in database	1.39	1179	2.78	0.08	0.16	41.47	41.47	0.001	0.48	0.10	1.57
2) All field studies	1.67	737	3.35	0.12	0.24	41.47	41.47	0.001	0.44	0.09	2.00
3) All lab studies	0.94	442	1.28	0.06	0.12	13.34	13.34	0.002	0.52	0.15	1.17
4) All benth[b] field	1.27	642	2.59	0.10	0.20	41.47	41.47	0.001	0.31	0.08	1.48
5) All benth[b] lab	0.97	428	1.29	0.06	0.12	13.34	13.34	0.003	0.53	0.18	1.29
6) All PCB benth[b] field	1.70	293	2.10	0.12	0.24	10.99	11.00	0.007	0.79	0.13	3.03
7) All PCB benth[b] lab	1.18	206	0.97	0.07	0.13	4.70	4.74	0.040	0.80	0.50	1.63
8) All PCDD/F invert[c] field	0.37	103	0.31	0.03	0.06	1.44	1.46	0.020	0.29	0.15	0.50
9) All PCDD/F benth[b] lab	0.24	30	0.23	0.04	0.09	0.90	0.90	0.003	0.17	0.13	0.27
10) All PAH invert[c] field	0.36	161	1.00	0.08	0.16	8.80	8.80	0.001	0.07	0.01	0.30
11) All PAH benth[b] lab	0.61	154	1.51	0.12	0.24	13.34	13.34	0.003	0.13	0.03	0.48
12) All pest benth[b] field	2.59	83	5.41	0.59	1.18	41.47	41.47	0.004	1.00	0.22	2.56
13) All pest invert[c] lab	1.91	31	1.74	0.31	0.64	5.49	5.88	0.390	0.80	0.58	3.88

[a] http://www.wes.army.mil/el/bsaf/bsaf.html
[b] benth = benthically coupled and refers to invertebrates and bottom-feeding fish, whole organism, muscle, or filet
[c] invert = invertebrates and refers to mollusks and polychaetes

in terms of contaminants. Approximately 30 sediment core composites were taken for PCBs as total PCB, Aroclors, and specific congeners and TOC. A similar number of fish samples were taken by electroshock for the same PCB analytes and for lipids. RSS uncertainty (Table 11-3) was calculated using the mean, n, and SE of line 6, Table 11-1 as the BSAF input data. Average method error (ME), the error inherent in the model equation itself if all measurements could be made with perfect accuracy, was found to be 49%. This value is similar to the average ME previously determined for PAHs (Clarke and McFarland 2000). Propagated error was evaluated separately and total error (TE) was calculated as the square root of the sum of the squared ME and the propagated error. The measured tissue concentration was found to lie within the limits of TBP ± TE 69% of the time and to overestimate TBP ± TE the remainder of the time.

The RSS method is relatively simple to perform and gives an estimation of the uncertainty surrounding estimations of bioaccumulation potential, but the bootstrap method has the advantage of calculating statistical measures, including confidence limits and probability. Using the bootstrap method on the same data set (Table 11-4), TBP overestimated the measured tissue concentration 6 times, the confidence interval contained the measured concentration 8 times, and TBP overestimated the mean measured concentration twice. However, when a test of significance was done on the analysis, there was only one significant overestimation ($P = 0.03800$) and 1 significant underestimation ($P = 0.03588$). All other comparisons found no significant difference between TBP and Ct.

What Other Methods are Available for Assessing Contaminant Bioavailability?

Semipermeable membrane devices and solid-phase microextractions fibers

Semipermeable membrane devices (SPMDs) and solid-phase microextraction fibers (SPMEs) have received considerable attention as useful tools for measuring organic compounds in environmental matrices. Both types of samplers are commonly referred to as "biomimetic devices" because they can be used as surrogates for organisms to estimate the bioavailability of organic contaminants. With regard to contaminant uptake, biomimetic devices share 2 features in common with organisms: 1) They selectively absorb "available" compounds, and 2) they can concentrate them to very high levels in comparison

Table 11-3 Computational (RSS) uncertainty analysis of PCB bioaccumulation potential in freshwater catfish *A. melas*

PCB	Mean BSAF[a]	Cs mean[b]	Cs CI[c]	TBP[d]	Propagated error	Total error	Range of total error: TBP – TE[e]	Range of total error: TBP + TE	Ct mean[f]	TBP/Ct	Ct within range of error?
Total	1.7	0.175	0.070	0.247	0.112	0.165	0.082	0.412	0.413	0.6	above
A1254	1.7	0.150	0.056	0.211	0.092	0.138	0.073	0.350	0.238	0.9	within
A1260	1.7	0.036	0.020	0.048	0.027	0.036	0.012	0.084	0.159	0.3	above
44	1.7	0.004	0.004	0.005	0.004	0.005	0.000	0.010	0.010	0.6	within
52	1.7	0.005	0.005	0.007	0.006	0.007	-0.000	0.014	0.011	0.6	within
77	1.7	0.004	0.004	0.006	0.004	0.005	0.001	0.011	0.009	0.7	within
86	1.7	0.003	0.001	0.004	0.002	0.002	0.001	0.006	0.002	2.4	within
101	1.7	0.005	0.003	0.007	0.004	0.005	0.001	0.012	0.013	0.5	above
128	1.7	0.002	0.001	0.003	0.001	0.002	0.001	0.005	0.003	1.2	within
138	1.7	0.010	0.004	0.013	0.006	0.009	0.004	0.022	0.006	2.3	within
141	1.7	0.002	0.001	0.003	0.001	0.002	0.001	0.005	0.003	1.0	within
156	1.7	0.002	0.000	0.002	0.001	0.001	0.001	0.004	0.004	0.6	within
170	1.7	0.002	0.001	0.003	0.001	0.002	0.001	0.006	0.002	1.4	within
171	1.7	0.002	0.000	0.002	0.001	0.001	0.001	0.004	0.002	1.2	within
180	1.7	0.003	0.002	0.004	0.002	0.003	0.001	0.007	0.010	0.4	within

[a] BSAF from Table 11-2
[b] Concentration in sediment, ppm.
[c] 95% confidence interval.
[d] Theoretical bioaccumulation potential, ppm
[e] Total error
[f] Measured concentration in tissue, ppm ("fresh weight," same basis as measurement of lipids)

Table 11-4 Bootstrap uncertainty analysis of PCB bioaccumulation in the freshwater catfish *A. melas*

PCB	Bootstrap mean TBP	Bootstrap std. error	Bootstrap CV	DF for CL[a]	Lower 95% CL	Upper 95% CL	Ct mean[b]	Ct within confidence limits?
Total	0.248	0.063	0.765	8	0.131	0.366	0.413	above
A1254	0.213	0.052	0.737	8	0.116	0.311	0.238	within
A1260	0.049	0.015	0.905	8	0.021	0.076	0.159	above
44	0.005	0.002	1.354	8	0.001	0.010	0.010	within
52	0.007	0.004	1.538	8	0.000	0.014	0.011	within
77	0.006	0.003	1.234	8	0.001	0.011	0.009	within
86	0.004	0.001	0.701	8	0.002	0.006	0.002	within
101	0.007	0.002	1.080	8	0.002	0.011	0.013	above
128	0.003	0.001	0.724	8	0.002	0.005	0.003	within
138	0.013	0.004	0.791	8	0.007	0.020	0.006	below
141	0.003	0.001	0.725	8	0.002	0.004	0.003	within
156	0.002	0.000	0.580	8	0.002	0.003	0.004	above
170	0.003	0.001	0.644	8	0.002	0.005	0.002	within
171	0.002	0.000	0.528	8	0.002	0.003	0.002	within
180	0.004	0.001	0.944	8	0.002	0.006	0.010	above

[a] DF for input parameter having smallest sample size (lipid, $n = 9$)
[b] Measured concentration in tissue, ppm (fresh weight)

to the surrounding matrix. In contrast to vigorous strong-solvent extractions, which attempt to sample 100% of the compound present, biomimetic devices sample only what is dissolved in solution or, in some cases, easily disassociated from other matter. Much like the organisms they attempt to mimic, SPMDs and SPMEs have large affinities for organic compounds, concentrating them to several orders of magnitude higher than the surrounding matrix. Similar to restrictions placed upon experiments using organisms to measure bioavailable contaminants, device–matrix volume ratios must be large to ensure the device does not deplete the "available" pool of compounds present in the matrix, and uptake should be measured preferably after reaching steady-state equilibrium.

The primary driving force behind the development of biomimetic devices is to provide a better estimate of exposure than is gleaned from vigorous strong-solvent extractions, while complimenting the use of live organisms for evaluating contamination in soils, sediments, and waters. Biomimetic devices are not substitutes for organisms but can provide complementary or preliminary information that allows a more efficient use of biota and a more refined understanding of toxicant exposure. Several key characteristics of biomimetic devices enable them to be easily integrated with an exposure assessment scheme. They are generally less expensive than using laboratory-reared organisms with a known exposure history. In some cases, exposure times are less than required for laboratory-reared organisms (deployed caged in situ or in laboratory toxicity tests). SPMEs and SPMDs can be deployed in most situations without regard to the myriad noncontaminant issues associated with organism needs for a matrix, such as nutrient or food quality, physical structure, and physicochemical conditions (pH, oxygen, light, etc.). Organisms caged in the field may die and/or be subject to predation. Contaminant uptake by an organism is influenced by a suite of variables, including genetics, age, behavior, handling, and overall health, each of which may contribute inherent variability compared to that of biomimetic devices.

Biomimetic devices are subject to a number of general disadvantages for measuring bioavailable organic compounds, in addition to device-specific disadvantages. The primary criticism of biomimetic devices is that they are overly simplistic (Salazar and Salazar 2001). SPMDs and SPMEs only simulate the dermal or respiratory (gills) route of exposure. For organics with log $K_{ow} > 5$, dietary routes of exposure become increasingly important (Leppännen and Kukkonen 1998); biomimetic devices may underestimate the bioavailability of such compounds. Biomimetic devices cannot account for biomagnification of persistent (usually very high log K_{ow}) compounds across trophic levels. Furthermore, biomimetic devices are simple, 1-compartment sinks for organic compounds. They cannot simulate the complex compartmentalization of organics among different tissues, organs, and/or biomolecules, nor can they mimic biotransformation that results from metabolic reactions with biomolecules or detoxification systems (e.g., P450). Although these devices can lose accumulated molecules if external activity decreases, they cannot actively excrete compounds.

Though both SPMDs and SPMEs passively sample organic compounds, their designs are quite different. SPMDs are designed as high-capacity samplers and are usually used to detect trace or ultra-trace levels of compounds in air or water. SPMDs are layflat, thin-walled bags or tubes (standard size 106 × 2.5 cm) comprised of low-density polyethylene (LDPE) containing a thin film of the nonionic lipid triolein (standard volume 1 mL). Organic compounds that cross the LDPE membrane are sequestered in the triolein, mimicking storage of organics in fat. After exposure, the SPMD is subjected to a somewhat lengthy (1 to 2 d) procedure involving dialysis, size-exclusion chromatography, and fractionation or preconcentration before analysis via traditional gas chromatography (GC), gas chromatography–mass spectrometry (GC–MS), or high-performance liquid chromatography (HPLC) methods (Petty et al. 2000). Although smaller-sized SPMDs can be used in laboratory situations (Wells and Lanno 2000), standard-sized SPMDs sample large volumes (e.g., 3 L/d for phenanthrene in water) and generally need a longer time period to achieve steady state, making in situ deployment the norm (Petty et al. 2000). Because of their long exposure times, in many environmental matrices, the LDPE membrane can become fouled with silt, periphyton, or bacteria, thus decreasing sampling rate (Petty et al. 2000). SPMDs usually are used to measure contaminant availability at the sediment–water interface (Echols et al. 2000; Petty et al. 2000; Voie et al. 2002). However, Leppänen and Kukkonen (2000) were able to bury small (2.7 × 2.5 cm) SPMDs in sediments spiked with pyrene and benzo[*a*]pyrene. Although neither worms nor SPMDs achieved steady state during the short (12 h) exposure periods, SPMD uptake of these compounds was well related to nonfeeding organism uptake and depicted a general decrease in contaminant availability in the sediments over time (Figure 11-1). However, SPMDs were not as effective in predicting uptake of organisms that were actively feeding (Leppänen and Kukkonen 1995). The intestinal route of uptake was found to be a significant exposure route, and therefore SPMDs underestimated bioavailability of the compounds to these organisms. Though SPMDs were able to mimic dermal or body wall exposure routes, caution should be exercised when using SPMDs if dietary exposure routes could be significant.

SPMEs are thin silica fibers (about 110 μm diameter, 1 cm long) coated with a microlayer (5 to 100 μm thick) of organic polymer. Unlike the triolein lipid found in SPMDs, the absorptive phase of the SPME is the polymer coating. Seven polymer coatings are available, ranging in analyte selectivity from polydimethylsiloxane (PDMS), useful for measuring very hydrophobic organics, to polyacrylate (PA), useful for measuring more hydrophilic organics. In contrast to SPMDs, SPMEs were developed for laboratory use in a wide variety of analytical applications, with their primary advantage being a reduction in solvent usage and freedom from chromatographic interferences (Pawliszyn 1997). SPME fibers are traditionally used mounted on a syringe applicator, enabling one to immerse the fiber in solution or in the headspace above a solution, retract the fiber, and directly inject the fiber into a custom GC or HPLC interface. In most cases, sample extraction, preconcentration, and purification are achieved in 1 step: No solvent is used, no cleanup is necessary, and the fiber may be reused. Although the syringe applicator method is very fast (steady state reached in minutes to hours in an agitated solution or slurry) and convenient, its use is primarily restricted to the laboratory bench top. With regard to sediments and soils, this application has been promising for predicting bioavailability in stirred pore water, overlying

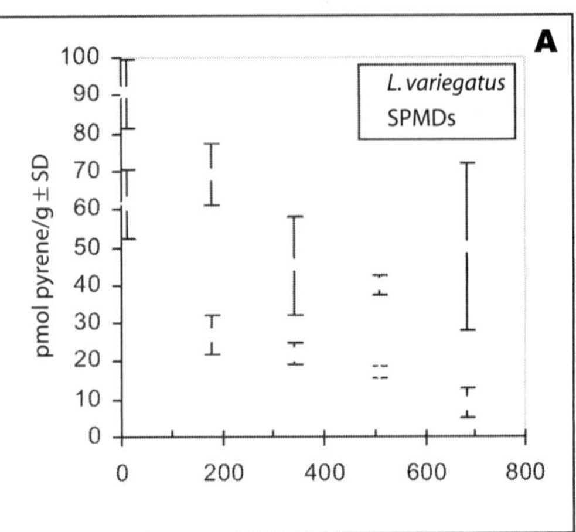

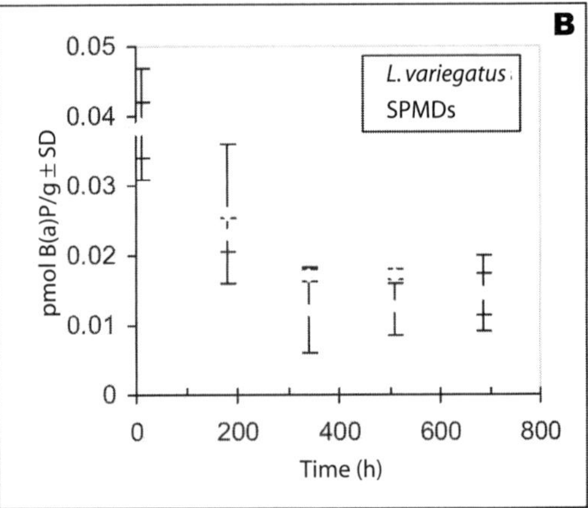

Figure 11-1 Accumulation of 14C-pyrene (A) and 3H-benzo[*a*]pyrene (B) in spiked Lake Höytiäinen (Finland) sediment to semipermeable membrane devices (SPMDs) and to nonfeeding *Lumbriculus variegatus* after five 12-h exposures in relation to sediment–chemical contact time. (Reprinted from *Aquatic Toxicology* 49:227–241, Leppänen and Kukkonen, Effect of sediment-chemical contact time…, copyright 2000, with permission from Elsevier.)

water, and slurries (Urrestarazu Ramos et al. 1998; Leslie et al. 2000, 2002; Wells and Lanno 2001). In a water-only exposure to *Chironomus riparius*, Leslie et al. (2002) derived 96-h LC50s based on water, organism, and SPME concentrations for a polar (2,4,5-trichloroaniline, TCA) and a nonpolar (1,2,3,4-tetrachlorobenzene, TeCB) narcotic compound (Table 11-5). A 3- to 4-fold difference was found for LC50s based on organism concentrations (CBRs). Assuming that the dose at the site of toxic action (cell membrane for narcosis) is similar for both compounds, the authors hypothesized that this large discrepancy in CBRs may be due to a higher toxicological bioavailability for the more polar TCA. That is, because polar organics tend to accumulate in the membrane, a lower CBR for TCA is to be expected. By comparison, LC50s based on SPME uptake for both chemicals were not only less variable than CBRs, they were nearly identical. Binding to this SPME coating (PA) may represent partitioning to cell membranes, rather than to the whole organism (Vaes et al. 1997, 1998; Verbruggen et al. 2000). In this case, SPMEs provided a more precise, possibly more accurate estimate of the dose at the site of toxic action than did whole-body residues, which misrepresented the total exposure.

Table 11-5 Median-lethal doses (LC50), with 95% confidence limits, calculated using concentrations in water, organisms, or polyacrylate (PA)-coated solid-phase microextraction fibers (SPMEs) for *Chironomus riparius* exposed for 96 h in water spiked with 2,4,5-trichloroaniline (TCA) or 1,2,3,4-tetrachlorobenzene (TeCB)[a]

	LC50	
Dose metric	TCA	TeCB
[Water] (μM)	8.87 (7.31–10.8)	2.30 (2.13–2.49)
[Organism] (mmol/kg lipid)	38.7 (29.1–51.6)	144.2 (109.7–189.4)
[SPME] (mmol/L PA)	32.3 (29.4–35.6)	30.5 (29.7–31.3)

[a] Reprinted with permission from Leslie et al. 2002; copyright Society of Environmental Toxicology and Chemistry (SETAC)

Although there are published toxicological studies with SPMEs, all have been limited to the overlying water or water-only exposures of sediment. Slower static exposures via direct insertion of the fiber into whole-sediment samples have been attempted (Mayer et al. 2000). Because of the fiber's fragility, this technique was found to be problematic. Researchers at the University of North Texas (Conder et al. 2003) have developed a method to bury SPME fiber pieces (not mounted on syringe, cut to custom length) within steel mesh envelopes to measure compounds directly within the sediment of a toxicity test without damaging the fiber. Currently underway is the development of a method to bury these envelopes within sediment in situ. Exposure times for static SPME are much longer than for traditional stirred SPME, taking 2 d for nitroaromatics sampled with PA-coated SPME to reach steady state and longer for more hydrophobic compounds.

SPME fibers have a few advantages over SPMDs for measuring available sediment-associated organic compounds, including costs and ease of use. Syringe-mounted SPME fibers are reusable and avoid costly solvent usage and disposal. For nitroaromatics, the mesh envelope burial method was found to be even less expensive than a traditional liquid–liquid solvent extraction to determine "total" organics. SPMEs need much smaller amounts of sample and achieve steady-state equilibrium much more rapidly than SPMDs. In the only published study comparing biomimetic SPMD and SPME, SPMD uptake was found to be roughly an order of magnitude higher than SPME uptake when phenanthrene was sampled from artificial soil (Wells and Lanno 2001). This increase in detection limit requires a larger sample size (i.e., 200 g for SPMD v. 1 to 2 g for SPME). However, because both the triolein in SPMDs and most SPME fiber coatings have extremely large affinities for organic compounds, both method detection limits sink well below levels producing toxicological effects for most toxicants. In some cases, sampling a small amount of sediment with SPMEs may be a disadvantage, particularly for heterogeneously distributed contaminants. In such cases, SPMDs may be able to provide a more robust integration of compound within the sample.

Dependent on study objectives, the use of biomimetic SPMD and SPME devices to estimate the bioavailability of sediment-associated organic compounds can provide a useful surrogate measure of contaminant availability. In general, biomimetic devices are easier to use, less expensive, less variable, and involve simpler chemical analysis, compared to measuring organic concentrations in organisms (Table 11-6). While SPMD and SPMEs do have limitations in predicting toxicity, the preliminary studies conducted to date have demonstrated their merit in investigating contaminant availability. SPMDs and SPMEs deserve more attention as complementary investigative tools for assessing the bioavailability of organics in sediment.

Deposit-feeder gut fluid extraction of sediment-associated contaminants

Deposit-feeding and some suspension-feeding organisms accumulate many heavy metals and hydrophobic organic compounds via the ingestion of sediment (Landrum and Robbins 1989; Wang and Fisher 1999; Lee et al. 2000;

Table 11-6 Comparison of organisms, semipermeable membrane devices (SPMDs), and solid phase micro-extraction fibers (SPMEs) for measuring the bioavailability of sediment-associated organic compounds

Sampler	Expense	Correlation with toxicity	Sampling time	Ease of use and analysis	Variability	Multiple exposure routes
Organisms	Higher	Better	Days to weeks[a] Hours[b]	More difficult or complex	More variable	Yes (ingestion, absorption)
SPMDs	Lower	Good	Hours to weeks	Less difficult or complex	Less variable	No (only absorption)
SPMEs	Very low	Good	Minutes to days	Less difficult or complex	Less variable	No (only absorption)

[a] Laboratory-reared organisms
[b] Site-collected organisms

Weston et al. 2000). However, a substantial, often major, proportion of any given contaminant is not desorbed from the particles while in the gut and passes out of the organism via the feces. Environmental management decisions pertaining to contaminated sediments must include consideration of the bioavailable fraction rather than the total contaminant concentration. Chemical approaches to toxicant exposure have some advantages over biological ones, but existing chemical methods of analysis generally extract all of the targeted contaminant from sediments by using a strong acid or strong organic solvent. As a result, these approaches can overestimate the bioavailable fraction, by varying degrees, and can lead to misleading interpretations or noisy data.

Digestive fluids of benthic deposit-feeding organisms have been used as an extraction medium to provide a better assessment of the potential bioavailability of particle-associated contaminants (Mayer et al. 1996). Several investigators have attempted to improve human health risk assessment by developing fluids that mimic human stomach fluid and to use these fluids as in vitro extractants to estimate how much contaminant would be bioavailable from soil if incidentally ingested by humans (Ruby et al. 1993; Hack and Selenka 1996; Jin et al. 1999; Oomen et al. 2000). In the new method, digestive fluid of a deposit-feeding organism is removed from the gut lumen, and the sediments of concern are incubated with that fluid in vitro. The amount of the particle-associated contaminant that is desorbed in the fluid is then quantified on the presumption that sediment-associated contaminants must first be solubilized in order to be bioavailable (excluding the potential for intracellular digestion in some taxa). While the approach does not address the subsequent absorption of the solubilized contaminant across the gut wall, the method at least places an upper limit on the contaminant that is likely to be made bioavailable from a given sediment during gut passage. The approach has the simplicity of a chemical extraction, but by using digestive fluid rather than an exotic solvent, the approach may provide more biological realism than is achieved by conventional chemical methods. The digestive fluid extraction approach is probably not useful for compounds for which ingestion is likely to be a minor route of uptake (e.g., hydrophilic organic compounds; Weston and Mayer 1998) or for contaminants in which intestinal absorption rather than solubilization constrains uptake (e.g., Cr).

Recent attempts to assess sediment risk using in vitro digestive fluid extraction have illustrated some advantages of the approach over conventional measures of bioavailability involving exposure of live organisms (Weston and Maruya 2002). First, it can be done much faster than conventional bioaccumulation testing (a few hours versus nearly a month), with associated cost savings and faster data availability. Second, the digestive fluid approach to predict bioavailability eliminates the potential effects of biotransformation. A comparison of results from digestive fluid extraction to measured tissue concentrations can help to elucidate the role and importance of biotransformation in a particular organism. Third, the technique allows evaluation of sediments by a consistent method over a wider range of abiotic parameters (e.g., grain size, salinity) than would be tolerated by any single bioaccumulation test species. While reliance upon natural populations of deposit feeders as a source of digestive fluid limits widespread use of the approach, the use of commercially available substances having extraction properties similar to the natural constituents shows promise (Chen and Mayer 1999; Ahrens et al. 2001).

It is theoretically defensible to suggest that in vitro gut-fluid contaminant extraction might provide a direct measurement of contaminant bioavailability to deposit feeders. In addition, the method may be useful as a predictive tool for contaminant bioaccumulation in sediment-dwelling invertebrates, dependent upon certain conditions. To illustrate this extrapolation, one must first consider the steps whereby a contaminant is released from sediment and accumulated ultimately within the tissues of a benthic organism. The concentrations or fluxes of sediment-associated contaminants bioaccumulated via ingestion by deposit feeders can be considered using a multi-step model (Figure 11-2). The first 2 functions (a and b, Figure 11-2) consider the amount of contaminants transferred from the sediment into the gut fluids, calculated as the amount of contaminant solubilized by the gut fluid (a) minus the amount subsequently reabsorbed into the sediment matrix (b). The difference (a – b) equates to the net in vitro gut-fluid extracted contaminant. The amount of contaminant absorbed from gut fluids into tissues (c, Figure 11-2) is proportional to the absorption efficiency. Final tissue burdens are a product of the absorbed fraction minus losses due to metabolism and elimination (d, Figure 11-2). Using this construct, it is evident that bioaccumulation is proportional to, and therefore theoretically predictable by, bioavailability. However, if a contaminant is rapidly biotransformed, digestive fluid extraction may predict high bioavailability, and the compound may indeed be taken up quite readily, but biotransformation of the compound could result in little or none of the substance being measured in the tissues. Bioaccumulation at steady state should be a correlate of in vitro solubilization for substances that are not biotransformed (e.g., DDE), for taxa having poor biotransformation capabilities (e.g., bivalves), or when values for biotransformation can be empirically estimated. Regression analyses of gut-fluid extracted contaminant concentrations versus bioaccumulated body burdens reveal strong positive correlations for a number of contaminants, suggesting that gut-fluid extractions might be considered as predictors of bioaccumulation (Lawrence et al. 1999; Weston and Maruya 2002).

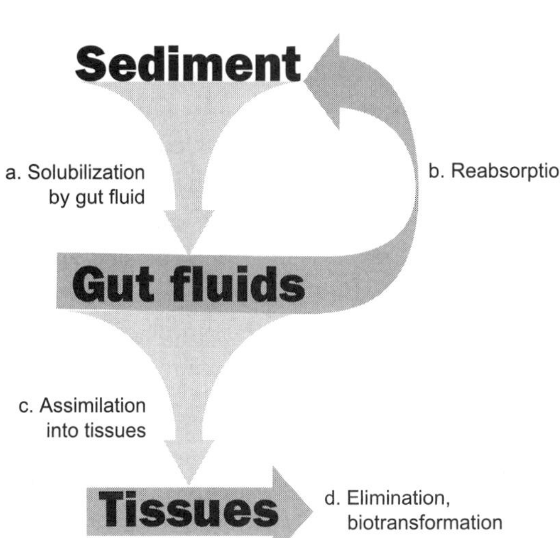

Figure 11-2 Conceptual model showing bioaccumulation of contaminants from sediments via deposit feeding

Can Bioaccumulation at Higher Trophic Levels Be Predicted Using Current Modeling Techniques?

The preceding sections have generally dealt with factors affecting bioaccumulation at lower trophic levels, such as polychaete worms or mollusks. This is important because in most ecosystems, the overall biomass of organisms at lower trophic levels is greater than the biomass at higher trophic levels (Valiela

1995). Thus, initial bioaccumulation of contaminants at lower trophic levels generally represents the major route of entry for contaminants into biotic systems. To understand the effects of contaminants on overall ecosystem form and function, however, one must also understand the efficiency with which contaminants that initially accumulated at lower trophic levels are transferred to higher trophic levels. Whereas contaminant uptake modeling at lower trophic levels focuses primarily on contaminant bioconcentration from the surrounding media (sediment particles, pore water, overlying water, etc.), biomagnification modeling at higher trophic levels attempts to more explicitly consider food as the primary route of exposure. Trophic transfer models range from kinetic and equilibrium modeling at the individual level to whole ecosystem approaches such as radiolabeling in order to assess trophic structure and subsequent potential for contaminant biomagnification.

The models that currently are used to estimate contaminant biomagnification at higher trophic levels will be briefly discussed in the following text. Uncertainties associated with each approach will also be addressed. It is important to note that although parameter or model uncertainty can potentially be reduced with additional research, uncertainty due to the underlying inherent variability in uptake or contaminant response between individuals or between species cannot be reduced. This type of uncertainty can only be quantified, understood, and incorporated into risk management decisions in determining the acceptability of risks due to the presence of contaminants in sediment.

Individual organism: Equilibrium-based approaches

Although field studies show increases in tissue concentrations in organic contaminants at higher trophic levels (e.g., Oliver and Niimi 1988), animals higher up on the food chain also tend to have greater lipid content than organisms in lower trophic levels (LeBlanc 1995). In fact, for some organic contaminants, lipid-normalized concentrations do not vary widely at different trophic levels (LeBlanc 1995), indicating that although food can be a major route of exposure, aquatic organisms may remain in simple equilibrium with surrounding media. Alternatively, while contaminant concentration in ingested food may be high, exposure from other routes may be more important. For example, although contaminant concentration in water or suspended particles may be low, the large volume of water processed by filter feeders or passed through the gills of fish may lead to these media being the dominant pathways of exposure.

If the contaminant assimilation efficiency is inversely related to the organic carbon content of the water or particles, and the elimination rate is inversely related to the lipid content of the organism (LeBlanc 1995), then at steady state the lipid-normalized tissue concentration of organic contaminants will be related to the organic carbon–normalized contaminant concentration in the surrounding media. In both cases, the lipid-normalized concentration of organic contaminants may not vary significantly between different trophic levels and can be considered to be in equilibrium with the organic carbon–normalized concentration in the surrounding sediments or water. If this is the case, evidence for true biomagnification (food ingestion leading to an increase in tissue concentration over simple equilibrium bioconcentration) is not always observed. In a model of field data by LeBlanc (1995), evidence that biomagnification is due to food ingestion was observed only for compounds with a log $K_{ow} > 6.3$. If this is true, modeling of many contaminants at higher trophic levels can use the same equilibrium bioconcentration models used at lower trophic levels.

Individual organism: Kinetic approaches

The simplicity of equilibrium models makes them an attractive choice as predictive tools for determining bioaccumulation of organic contaminants at higher trophic levels. However, trace metals are rarely in equilibrium with aquatic sediments, either spatially or temporally. Interactions between inorganic sediment particles (comprised in part by both macrofauna and microbial populations) result in distinct concentration gradients and relatively rapid geochemical cycling of associated metals. Thus, within complex sedimentary systems, exposure of an organism to contaminant metals is determined by the concentration and bioavailability of the trace metals at a scale proximate to the organism. Therefore, depending on a specific animal's feeding and living habits, metal exposure to a benthic organism can be substantially different than that predicted by metal concentrations measured in bulk sediment or food alone. In this regard, equilibrium models applied to benthic ecosystems may be inappropriate.

Metals may be accumulated from both dissolved and dietary sources (Wang et al. 1996). Recent work has focused on developing a mechanistic understanding of metal uptake by separately evaluating the contribution of each potential source to the total metal accumulation in aquatic organisms. For site-specific exposure assessment, Luoma and Fisher (1997) defined a simple kinetic model approach that includes empirically derived species-specific rate constants of uptake and loss. Key parameters determined experimentally include metal influx rates from solution, influx rates from ingestion, and efflux rates. These adjustable constants are combined with environmental data (metal concentrations in pore water, overlying water, surface sediment and/or suspended particulates) representative of a range of conditions in the system of interest. Kinetic modeling results so far appear to predict metal concentrations in the few species studied (Luoma et al. 1992; Wang et al. 1996, 1999; Roditi et al. 2000; Griscom et al. 2002) and further studies hopefully will find ways to improve the model.

For many aquatic invertebrates, ingested food accounts for a major proportion of total trace metal accumulation. For metals accumulated primarily from food, trophic transfer models are useful in quantifying the potential for the specific contaminant to be magnified or minimized from 1 trophic level to the next. The trophic transfer potential (TTP) can be defined by the equation:

$$\mathrm{IR} \times \mathrm{AE} \,/\, [k_e + g]$$

where IR is the weight-specific ingestion rate, AE is the assimilation efficiency of the metal from the food item, k_e is the physiological loss rate constant, and g is the weight-specific growth rate. Recent work devoted to the quantification of AE and k_e for a variety of organisms and metals has provided the data needed for comparative studies among different species (Reinfelder et al. 1998). Results suggest that Ag and Cr are not biomagnified; some elements, Se and Cd in particular, may be somewhat biomagnified in bivalves consuming phytoplankton; and methyl-mercury is highly biomagnified in all aquatic organisms studied.

Kinetic models have been used to predict trace element concentrations in aquatic organisms for more than 10 years. However, these methods are not yet incorporated into SQGs. Problems that are of concern regarding the viability of kinetic models as a predictive tool may include, for example, the large number of required parameters. Kinetic models are not one-size-fits-all, but instead are best at describing what is, in fact, a complex and variable system and are site-specific. In addition, variations in food sources can have large effects on the AE (Lee and Luoma 1998; Griscom et al. 2000), and altered ingestion rates can effect the loss rate k_e (Reinfelder et al. 1998). Both of these variables can change the relative importance of the route of uptake, and the loss rate term k_e may not be best described by a single value. For example, exposure of organisms to high metal concentrations can trigger the production of detoxification products (e.g., metallothioneins or phosphate granules), which depending on the amount of metal concentrated in each fraction, may require additional loss-rate terms to be added to the model. Nevertheless, aside from the aforementioned issues, the model is designed to be adjustable, and thus it can be used not only to predict bioaccumulation but also to quantify the route of uptake.

The kinetic models described above treat the animal as a single compartment. However, 1-box kinetic models may not be appropriate for larger organisms such as fish or other organisms in which metals are subject to feedback among various internal compartments. In general, models that explicitly incorporate mechanistic observations will prove most useful in the long term, and advanced compartmental models (Thomann et al. 1997, for Cd in trout) represent an improvement in assessing trace element accumulation. Concerns regarding the extensive data requirements for kinetic models may be less of an issue as the importance of specific variables becomes better understood. Eventually, it may be possible to estimate some of the species-specific parameters for ecologically similar organisms as more modeling studies are completed and general patterns emerge.

Whole ecosystem approaches

Even when all the kinetic or equilibrium parameters are known for a given species, assessing biomagnification in higher trophic levels can prove difficult. In order to estimate biomagnification at higher trophic levels, one must know the steady-state (or equilibrium) concentration of contaminants at lower trophic-level organisms serving as the food source in addition to knowing the feeding strategy used by the higher trophic-level organisms. Historically, this has been done by attempting to assign organisms to discrete secondary consumer, or higher, trophic levels and applying trophic transfer factors as described in the preceding section. However, organisms can feed at different trophic levels, for example, feeding on both zooplankton and small planktivorous fish, and thus do not fall neatly into specific classes. In a study of the biomagnification of Hg in lake trout, the concentration of Hg in this top predator was highly variable between a number of lakes (Cabana and Rasmussen 1994). Food chain length, and thus trophic class and Hg concentration in food sources, varied among the lakes. Not surprisingly, Hg concentration in lake trout, the top predator, was higher in lakes with longer food chains. This result is similar to previous work (Rasmussen et al. 1990) for PCBs in Ontario lakes.

Stable nitrogen and carbon isotopes have been used as markers of trophic status in aquatic ecosystems (Peterson and Fry 1987). This is because the $\delta^{13}C$ or $\delta^{15}N$ of an organism is slightly enriched over the isotopic ratio in the food source (Peterson and Fry 1987). Therefore, organisms at higher trophic levels are enriched in the heavier isotope. Rather than constraining organisms to an arbitrary trophic position based on assumptions of feeding behavior, the isotopic ratio serves as a continuous variable, integrating feeding behavior and thus trophic status. In the study by Cabana and Rasmussen (1994), although the lake trout Hg concentration was variable between lakes because of differences in food chain length, the Hg concentration was correlated with $\delta^{15}N$. That is, trophic status, and thus biomagnification potential, is reflected in the isotopic content of the trout. Nitrogen and carbon isotopic content have been used to explain the biomagnification of contaminants at higher trophic levels in both freshwater and marine systems (Jarman et al. 1996; Vander Zanden and Rasmussen 1996).

Although the concept of stable isotopic modeling of contaminants is sound, there are problems in its application. In a study on the concentration of Hg in yellow perch, the authors concluded that between-lake variability was better described by within-lake differences in chemistry, and not by differences in trophic position. In addition, the decidedly nonconservative behavior of nitrogen in marine ecosystems often complicates interpretation of isotopic data in higher organisms (Cloern et al. 2002). Finally, while the isotopic composition of higher organisms may "explain" the biomagnified concentration of contaminants, it is not possible to set up a priori models predicting trophic structure, biomagnification potential, and adverse outcomes at the ecosystem level. Food chain length and breadth are likely site specific, thus no overall model will apply universally to all ecosystems.

How Does One Determine Acceptability of Predicted or Measured Chemical Residues Using a Risk-Based Approach?

Knowledge of the biogeochemical factors that affect the partitioning and fate of contaminants in sediments is improving. As a result, a variety of models and measurement tools have been developed to evaluate the potential for toxic chemicals to bioaccumulate from sediment into aquatic organisms, including biomagnification along a food chain. However, bioaccumulation itself is not an adverse effect. Bioaccumulation becomes an adverse effect only when contaminants accumulate within an organism to levels that elicit an adverse response. Therefore, in order to incorporate bioaccumulation potential into the decision-making process, decision makers need to understand the extent to which any observed bioaccumulation in aquatic organisms will adversely affect ecological or human health.

In recent years, the field of risk assessment has been used to assess the likelihood that exposure to contaminants will result in unacceptable outcomes. This requires both knowledge of how organisms are exposed to contaminants and an understanding of the toxicological effects associated with that level of exposure. Risk assessments can be used retrospectively to determine whether unacceptable risks have resulted from previous exposures. This type of assessment is used, for example, to determine if current levels of contaminants in a sediment are of concern and thus merit remediation. In addition, risk assessments can be used to predict future risks on the basis of future exposure scenarios. This is the type of assessment that is used to set site-specific cleanup goals for a contaminated site or to develop more widely applicable SQGs predictive of risk (or lack thereof) to either human or ecological receptors.

The 3 basic phases of a risk assessment are

1) determining important receptors, endpoints, and exposure pathways of concern,
2) collecting and analyzing exposure and effects data, and
3) characterizing risk (USEPA 1998; Chapter 6).

The first phase, identifying endpoints of concern, entails defining valuable aspects of the environment considered to be at risk. Important questions in Phase 1 include, What species do we care about? How much do we care about them? How are they exposed to the contaminants? What contaminants do we care about? What models or measurements will we need? How will the data relate to these endpoints? Issues such as societal or biological relevance and accessibility to prediction or measurement are important. Failure to unambiguously address these issues up front often results in a program that fails to adequately characterize risk to either human or ecological receptors (Chapter 6). How bioaccumulation potential of sediment-bound contaminants is used to assess ecological risks, therefore, is primarily a function of how risk managers choose to define ecological health. In other words, it is impossible to determine how bioaccumulation potential provides answers without first figuring out what the important questions are. It is beyond the scope of this chapter to discuss in detail the types of questions and endpoints required of an ecological or human health risk assessment. A more detailed discussion of the design of risk assessment programs, especially ecological risk assessments, is given in Schmitt and Osenberg (1996) and Suter (1993).

The second part of a risk assessment, characterizing exposure, is what this chapter has discussed. With respect to exposure in ecological receptors, this chapter has reviewed the various models and approaches used to estimate the uptake of toxic contaminants into organisms. As will be discussed, exposure estimates must be used to model and act upon estimated risks of unacceptable toxic outcomes. For human health risk assessments, the primary route of exposure to sediment-bound contaminants is through oral consumption of fish or shellfish that has bioaccumulated those contaminants, either through direct contact with the sediment or via biomagnification along a food chain. For these assessments, models need to address not only contaminant concentration in the food source but also consumption rates.

Finally, the last part of a risk assessment entails using the exposure data to characterize actual risk. This implies knowledge of dose–response relationships for each contaminant and each species of interest. With respect to bioaccumulation of contaminants into aquatic organisms, how tissue levels of contaminants correspond to doses associated with adverse effects needs to be understood, including the temporal aspects of the dose–response relationship (Lee et al. 2002). For human health, a relation to the ingestion rate of contaminants with the likelihood of adverse cancer and noncancer outcomes is needed. The following briefly describe current risk assessment methodologies for both ecological and human health outcomes. The goal is to demonstrate major uncertainties in the application of these risk assessment models.

Risk assessment for ecological health endpoints

An obvious concern with contaminated sediments is that the presence of contaminants does not cause unacceptable environmental degradation. Determining unambiguous measurement endpoints that assess "environmental degradation" is difficult. An advantage of ecological risk assessment over human health risk assessment is the fact that it is possible to test the effects of contaminants directly on a species of interest. However, unlike human health assessments, there are usually more than 1 species to protect. The degree of extrapolation from tests on a few indicator or standard test species to other components of the ecosystem is the subject of considerable debate (Suter 1993). Similarly, extrapolation from the results of a few acute or chronic toxicity tests to overall ecosystem health is of concern. When levels of biological responses are considered, ranging from subcellular and cellular responses, through individual and population responses, to responses altering overall ecosystem form and function, using bioaccumulation to assess ecological risk is extrapolating from the level of individual organisms to assess ecosystem health. Some monitoring programs extrapolate all the way from biomarkers at the subcellular level (such as enzyme induction, gene expression, or lysosome integrity) to indicate overall ecosystem health (IOC 1996). While this may be excessive, it may be possible to develop quantitative or qualitative subcellular biomarkers of individual health, such as biomarkers indicative of exposure and thus bioaccumulation, to aid in assessing overall ecosystem health. However, much more research is needed in this area before these approaches can readily be applied routinely in ecological risk assessments.

Assuming an ecological endpoint of interest is acute or chronic toxicity in native organisms, the CBR approach has been used to model dose–response relationships in aquatic organisms (McCarty and Mackay 1993). In classic mammalian toxicology, "dose" is often defined as an amount (or rate) of a chemical delivered to an organism, with units such as mg contaminant/kg of body weight/day. Adverse toxicological responses are estimated as a function of the magnitude of the dose. However, in ecotoxicology it is often difficult to measure dose in this way, especially in field studies. Commonly, the concentration of contaminants in the surrounding medium is used as a surrogate for dose in ecotoxicological studies (Rand et al. 1995). Toxicity, therefore, is measured as a function of concentration, not as a function of the classic definition of dose. The CBR approach requires an understanding of the critical dose to an organism that elicits an unacceptable adverse effect (measured as a concentration in the surrounding medium). Models discussed in this chapter can then be used to additionally estimate the body burden of contaminants at that critical dose. Essentially, the body burden of contaminants can be used as an estimate of adsorbed dose. If the biogeochemical factors controlling bioaccumulation are understood, the concentrations of contaminants leading to unacceptable responses can be determined.

A CBR type of approach has a number of strengths. Primary strengths are that bioavailability, exposure via food, and accumulation and depuration rate kinetics can be explicitly addressed. More sophisticated modeling of contaminant uptake from the environment to estimate subsequent tissue levels indicative of adverse doses. In general, CBR models for nonionic organics tend to be more developed than CBR approaches for heavy metals. This is because the mode of action for these contaminants is a narcosis, or nonspecific effect on lipid membrane integrity (Escher and Schwarzenbach 2002). The bioaccumulation of nonionic organics into lipid tissue is somewhat easier to model than the bioaccumulation of metals, which can be more effectively regulated by the organism. However, although CBR approaches are generally used for organic contaminants, the approach can be used for heavy metals as well (McCarty and Mackay 1993; Shephard 1998; Hamilton 2002).

A major uncertainty in the use of the CBR approach is determining a dose or response protective of ecological health. A large amount of acute toxicity data is on acute mortality. However, using an acutely toxic dose to calculate a corresponding threshold CBR may not be protective of ecological health. One approach circumventing this problem is to use USEPA water quality criteria (WQC) to estimate target CBRs. This is based on the assumption that WQC are protective of a large fraction of species present in the environment. By determining the bioconcentration of dissolved contaminants into soft tissue, one can estimate a body burden, or CBR, of contaminants corresponding to a biological dose presumed to be protective of ecosystem health (Shephard 1998). This CBR value can be subsequently applied to sediments, where multiple routes of exposure (water, particles, food) can be modeled to determine sediment concentrations that yield an anticipated body burden less than the CBR indicative of an unacceptable dose. In short, tissue levels of contaminants can be a very useful tool to measure dose from multiple exposure pathways and can be compared with CBR-based estimates of unacceptable doses. This is especially useful for sediments, for which there are often multiple routes of exposure. However, as plagues all ecological risk assessments, what one chooses to define as an unacceptable outcome protective of ecosystem health is very difficult to determine.

Establishing the tissue residue that is protective as suggested by Shephard (1998) or establishing a CBR that reflects a toxic response allows the potential for establishing an SQG through the use of the BSAF. The approach to establishing the body residue can use ambient WQC and an appropriate BCF to establish a protective body residue (Shephard 1998). Mixtures are then addressed in this approach by the use of a toxic units model that assumes additivity. Similarly, it is possible to establish the body residue at some level of toxic response by using the LC50 or EC50 or other appropriate level of response and the corresponding toxicokinetics. The simplest approach would be to assume toxicokinetics at steady state and then use the product of the LC50 and the BCF to yield a body residue corresponding to the LR50 (50% mortality based on tissue residue). If the species-specific toxicokinetics are known, then temporal variations in the body residue–response relationship could be addressed at other than steady state. However, steady state in the environment would likely dominate the expected exposure for most cases. Finally, the body residue could also be established through experimental measurements (note: this is becoming more common and data are available in published databases [e.g., Jarvinen and Ankley 1998; USACE/USEPA ERED 2002a]). All of these approaches to establishing the body residue–response relationship can be applied through the BSAF to relate the body residue to the concentration in the sediment, assuming steady state. The BSAF value needs to be applied based on the value for the specific compound class (Table 11-2). Alternatively, site-specific BSAF values can be generated and applied to account for site-specific factors. To develop site-specific BSAF values, data should be obtained from several species from diverse taxa. A BSAF is then selected from the upper end of the distribution (e.g., 95th percentile) to protect the most sensitive species. It may be that a set of BSAF values will be developed for a site-specific target organism to insure the protection of that species. The BSAF values are then used in the following equation to establish a compound-specific sediment concentration:

$$C_s/f_{OC} = \frac{[\text{tissue}]}{\text{BSAF} \times f_L}$$

where [tissue] is the concentration (e.g., lethal residue [LR50], effective residue [ER50], lowest effect residue [LOER], or other tissue residue response value as suggested above), BSAF is the biota–sediment accumulation factor, f_L is the fraction of lipid in the organism, and C_s is the chemical concentration in sediment and f_{OC} is the fraction of organic carbon in the sediment. Because this sediment value is for a specific compound, mixtures would need to be addressed, again using a toxic unit approach. This then leads to the possibility of establishing bioaccumulation-based sediment concentrations protective of toxic responses for a wide range of endpoints, both acute and chronic.

Two recent examples of this approach can be found in Meador et al. (2002a, 2002b) . Several studies with sufficient data to determine the tissue concentration associated with effects of tributyltin exposure were used to calculate the LR50, a growth LOER, and a threshold residue for sterility in stenoglossan gastropods (Meador et al. 2002b). This information, coupled with BSAFs for several species, was then used to determine the sediment concentrations expected to produce the adverse effects. These sediment concentrations were generated as guideline values for use in assessing sites contaminated with this compound. A similar approach was used to determine sediment concentrations of total PCBs that would likely cause adverse effects in juvenile salmonids (Meador et al. 2002a).

Human health–based risk assessments

In addition to protecting ecological health, environmental decision makers must also protect human health. In contrast to ecological risk assessments, in a human health risk assessment there is only 1 species of interest. Furthermore, whereas an ecological risk assessment is often aimed at maintaining a stable population of a species, a human health assessment is generally trying to protect the health status of each individual. Although routes of exposure can include dermal contact or direct sediment ingestion, human exposure to sediment-borne contaminants is generally via consumption of contaminated food from the aquatic environment. Assuming the biogeochemical factors controlling the bioaccumulation of contaminants from sediments into aquatic organisms are understood, a human health risk assessment focuses on using these data to relate contaminant intake via food consumption (dose) with the likelihood of adverse health outcomes (response). The 2 major aspects of a human health risk assessment, therefore, are estimating aquatic food consumption rates and understanding the potential for adverse effects associated with a given dose of contaminants.

A variety of approaches are used to estimate risks from consumption of contaminated seafood. A common approach relies on toxicity data summarized in the USEPA's Integrated Risk Information System (IRIS) Database, which provides data on doses of contaminants associated with risks of both cancer and noncancer outcomes (USEPA 2002). For noncancer outcomes, the IRIS database provides a threshold contaminant dose (mg contaminant/kg body wt/time), below which the likelihood of a specific noncancer outcome is acceptably low. For cancer outcomes, the IRIS database provides an estimate of the incremental increase in the likelihood of cancer associated with a given rate of exposure (e.g., the cancer slope factor). As valuable as these consensus dose–response factors are in estimating risk, they are the subject of considerable debate. Because direct testing of humans is considered unacceptable, much of the data on human susceptibility to contaminants are based on extrapolation from mouse or rat studies. Some toxicologists feel that rat or mouse extrapolations to humans are unreasonable, while others feel that there is such underlying interindividual variability in susceptibility to these toxicants as to render the fixed thresholds meaningless (Crouch 1983). Thus, our ability to reliably estimate human health risks from the bioaccumulation of sediment contaminants into seafood is already compromised by our inability to precisely predict the effects of these contaminants on humans, either because of imperfect extrapolations from mouse or rat studies or by large variability in interindividual responses to contaminant exposure.

Putting aside uncertainties with the USEPA's cancer slope factors or non-cancer reference doses, risks from seafood consumption are modeled as a function of contaminant ingestion rate. Contaminant ingestion rates, sometimes called a "lifetime average daily dose" (LADD) are derived as follows:

$$\text{LADD (mg/kg/d)} = (\text{CF} \times \text{IR} \times \text{ED}) / (\text{BW} \times \text{AT}),$$

where CF = concentration of the contaminant in fish (mg/kg wet weight), IR= fish ingestion rate (kg fish/d), ED = exposure duration (y), BW = body weight (kg), and AT = averaging time (y). Guidance on default values to use for ED, AT, and BW can be found in documents such as the USEPA Guidance Manual (1989). The CF term can be estimated using the tools described in this chapter, although there are certain caveats with respect to the portion of the fish consumed.

Deserving further discussion is the IR, which can be highly variable. One must understand the distribution of fish consumption rates in the population under consideration in order to adequately protect that population from adverse risks. Risk managers must decide on a consumption rate that is representative of an acceptably high proportion of that population. It is clear that risks will vary with specific subpopulations that have markedly different dietary intakes, for example, subsistence fishers. Variability in the IR term that results in interindividual variability in fish consumption rates can lead to imprecision of final risk estimates. However, this source of uncertainty is due to stochasticity, or natural variability, and cannot be reduced through further research on fish consumption patterns. Uncertainty in the IR term must be understood and used in determining a target IR rate for the exposure models. Alternatively, probabilistic Monte Carlo models can be used to propagate this uncertainty and determine a distribution (rather than a point estimate) of exposure doses to the population of interest.

Once an LADD is calculated, one determines an excess cancer probability from that exposure by multiplying the LADD by the cancer slope factor. For noncancer outcomes, the LADD is divided by the "safe" reference dose to get a hazard quotient. As with ecological risk assessment, before one can back-calculate a sediment concentration based on a target risk level or calculate risks due to current conditions, an acceptable level of risk is established. A cancer probability less than 10^{-6} is generally considered to be acceptable, while a probability greater than 10^{-4} is generally thought to be unacceptable. Selection of acceptable criteria between these 2 extremes is a subjective risk management issue. Given that what can be considered an acceptable cancer risk can vary by a factor of 10 to 100, allowable amounts of contaminants in sediments can vary widely based on the risk threshold selected. For non-cancer outcomes, a hazard quotient less than 1.0 is considered acceptable. When values are greater than 1.0, establishing at what level risks become unacceptable is a risk management decision. As an example, the state of Washington has a program incorporating human health risk based criteria into setting cleanup goals for contaminated sediments (Washington Department of Ecology

1997). Although the target cancer risk threshold is 10^{-5}, consideration of other criteria, such as background concentrations of contaminants, has led to cases where the estimated cancer risks associated with final cleanup goals are higher than the desired 10^{-5} risk threshold (Washington DOE 1997).

A second type of threshold used to estimate human health risks that result from food consumption in aquatic environments: comparison to USFDA action levels of allowable amounts of contaminants in food. The USFDA has published a list of allowable amounts of contaminants in food and seafood (USFDA 2002). One can therefore consider sediment concentrations unacceptable for humans if they result in fish tissue levels above the USFDA action levels. However, it should be pointed out that USFDA action levels are not solely based on estimates of human health risk. Economic factors (e.g., availability of alternative food sources) are also incorporated into the establishment of the action levels (21 CFR 109; 21 CFR 509). In certain instances, the health risks associated with a USFDA action level may be higher than desired. For example, a recent report by the National Academy of Sciences investigated human health risks associated with the consumption of Hg-contaminated seafood (NRC 2000). The report indicated that the USFDA action level of 1.0 µg/g may not be protective of human health and recommended lower tolerance levels for Hg in fish tissue (NRC 2000).

With respect to the exposure equation given above, another source of uncertainty is the fish concentration of contaminants that should be input into the exposure model. Much of the discussion in this chapter has described bioaccumulation of contaminants into the whole body of organisms. In many instances, however, contaminants are not distributed uniformly throughout the body. For example, in the American lobster (*Homarus americanus*), the body burden of bioaccumulated organic and inorganic contaminants is often concentrated in the hepatopancreas and these tissues may or may not be consumed by humans, depending on individual or cultural preference (Canli and Furness 1993; James et al. 1995). In fish such as salmon, bluefish, and carp, it has been shown that the concentration of nonionic organics in skin is higher than in muscle tissue (Hora 1981; Armbruster et al. 1989; Zabik et al. 1995). Thus an exposure model that assumes consumption of a fish filet with the skin on will lead to higher exposure estimates than will a consumption model assumes skin-off filets. Although we may be able to accurately model the bioaccumulation of contaminants into aquatic organisms, we must further be able to understand the partitioning within the organism to adequately understand the potential for human exposure and adverse human health outcomes.

A further difficulty in estimating the concentration of contaminants in fish is the fact that we often conservatively assume that an organism of interest spends all of its life in proximity to the area containing contaminated sediments. Thus the modeled concentration of contaminants in migratory species at the top of the food chain may be higher than actually observed because of the fact that these species spend only a fraction of time at the contaminated site. There are current models that attempt to account for this phenomenon and adjust risk estimates accordingly (Linkov et al. 2002; Von Stackelberg et al. 2002). These models result in lower risk estimates associated with a given level of contaminants, leading to higher sediment cleanup levels. A practical difficulty in applying these models is that the cost of collecting sufficient data to adjust for migration patterns (such as fish tagging studies) may be so high as to offset any cost savings from a reduced amount of sediments requiring remediation. Furthermore, assuming that conditions at the "other locations" are pristine, while allowing higher site-specific cleanup goals, may slowly raise the regional baseline concentration of contaminants to potentially undesirable levels.

In summary, human health can be used as an endpoint of concern in assessing risks that result from contaminated sediments. Using our knowledge of the bioaccumulation of sediment contaminants and subsequent biomagnification along a food chain, we can use the exposure–response models described in this section to estimate human health risks. Risk assessment models can be run in a forward direction to determine risks associated with current sediment contaminant levels. Alternatively, the models can be run in reverse, using an acceptable risk threshold to back calculate a sediment concentration protective of human health. However, our ability to precisely estimate human health risks is limited. Much of the variance in our ability to estimate these risks is not due to model or parameter uncertainty, but rather to natural variability in human exposure and susceptibility to these contaminants. We could have perfect models of bioaccumulation and still have highly variable risk estimates. A danger in running a risk assessment model in reverse is that if these uncertainties are propagated, either through selection of conservative point estimates or by using Monte Carlo simulations, there are likely to be allowable concentrations of contaminants that are either at or below naturally occurring or regional diffuse background levels for inorganic and organic contaminants respectively. Because much of this variance is due to natural variability, only an extraordinary amount of research can improve our ability to precisely estimate human health risks. The goal, therefore, is to adequately characterize this variability and to use this knowledge to make management decisions protective of human health.

Summary

Existing effects-based SQG approaches were not designed or intended to be protective of effects through bioaccumulation, either by interpreting risk of bioaccumulated contaminants or through food web transfer. While it is possible to develop SQGs protective of bioaccumulative effects, a number of issues must be considered to reduce uncertainty and ensure that the resulting guidelines are meaningful. One issue relates to potential discrepancies between laboratory and field-collected data. In general, laboratory-based bioaccumulation studies provide information that is representative of tissue residues obtained from field-collected organisms. However, differences in route and duration of exposure, seasonality, lipid content, etc. can lead to differences and reduce the reliability of laboratory to field comparisons. Appropriate consideration of such factors can be used to reduce uncertainty and improve the predictive ability of laboratory estimates. While other tools exist for estimating bioaccumulative potential of sediment-associated contaminants such as TBP modeling for nonionic organics, the use of biomimetic devices (e.g., SPMDs and SPMEs), or specialized extraction techniques (e.g., gut juice), these currently result in estimates with higher degrees of uncertainty than either laboratory- or field-based exposures. Another important issue is interpreting the significance of measured tissue residues. Understanding the potential consequences of bioaccumulation requires linking measured tissue residue levels with associated effects. While a number of databases summarize available residue effect information, there is a general paucity of available data, which limit the usefulness of this approach. One final issue relates to the reliability of risk-based approaches for establishing the potential for effects in higher trophic levels. Currently, the best estimates of bioaccumulation potential and resulting effects are for organisms in direct contact with the sediment. Generally, the further removed from direct exposure (i.e., higher in the food chain), the greater the uncertainty in the estimate of bioaccumulation and hence the greater the uncertainty in the estimate of risk. To ensure environmental protection of higher trophic levels (including humans), often overly conservative assumptions are used. Accuracy of these risk based estimates can be improved with development and application of region or site-specific data relating to food web structure (including area use factors) and trophic transfer coefficients.

Currently, there appear to be 2 potential types of SQGs that could be developed for assessing potential effects of contaminant bioaccumulation from sediments: 1) direct guidelines based on tissue residue effects data and 2) guidelines that incorporate the indirect effects through trophic transfer of contaminants. A primary concern in developing any meaningful guideline is addressing uncertainty. While it is possible to develop site-specific SQG values protective of effects through bioaccumulation, any approach should be adequately validated with field-collected data prior to implementation.

References

Ahrens MJ, Hertz J, Lamoureux EM, Lopez GR, McElroy AE, Brownawell BJ. 2001. The role of digestive surfactants in determining bioavailability of sediment-bound hydrophobic organic contaminants to two deposit-feeding polychaetes. *Mar Ecol Progr Ser* 212:145–157.

[ASTM] American Society for Testing and Materials. 2004. Standard guide for conducting sediment toxicity tests with marine and estuarine invertebrates. E 1367-03. In: Annual book of ASTM standards, Volume 11.5. West Conshohocken (PA): ASTM. p 1138–1163.

Ankley GT, Di Toro DM, Hansen DJ, Berry WJ. 1996. Technical basis and proposal for deriving sediment quality criteria for metals. *Environ Toxicol Chem* 15:2056–2066.

Ankley GT, Mattson VR, Leonard EN, West CW, Bennett JL. 1993. Predicting the acute toxicity of copper in freshwater sediments: Evaluation of the role of acid-volatile sulfide. *Environ Toxicol Chem* 12:315–320.

Ankley GT, Cook PM, Carlson AR, Call DJ, Swenson JA, Corcoran HF. 1992. Bioaccumulation of PCBs from sediments by oligochaetes and fishes: comparison of laboratory and field studies. *Can J Fish Aquat Sci* 49:2080–2085.

Armbruster G, Gall KL, Gutenmann WH, Lisk DJ. 1989. Effects of trimming and cooking by several methods on polychlorinated-biphenyls (PCB) residues in bluefish. *J Food Safety* 9:235–244.

Barron MG, Hansen JA, Lipton J. 2002. Association between contaminant tissue residues and effects in aquatic organisms. *Rev Environ Contam Toxicol* 173:1–37.

Bartell SM, LaKind JS, Moore JA, Anderson P. 1998. Bioaccumulation of hydrophobic organic chemicals by aquatic organisms: A workshop summary. *Int J Environ Pollut* 9:3–25.

Bend JR, James MO, Little PJ, Foureman GL. 1981. In vitro and in vivo metabolism of benzo[*a*]pyrene by selected marine crustacean species. In: Dawe CJ, editor. Phyletic approaches to cancer. Tokyo: Japan Scientific Soc Pr. p 179–194.

Bender ME, Hargis WJ, Huggett RJ, Roberts MH. 1988. Effects of polynuclear aromatic hydrocarbons on fishes and shellfish: an overview of research in Virginia. *Mar Environ Res* 24:237–241.

Bierman Jr VJ. 1990. Equilibrium partitioning and biomagnification of organic chemicals in benthic animals. *Environ Sci Technol* 23:1407–1412.

Boese BL, Lee II H. 1992. Synthesis of methods to predict bioaccumulation of sediment pollutants. Washington DC: US Environmental Protection Agency. ERL-N Contribution nr N232.

Boese BL, Winsor M, Lee II H, Echols S, Pelletier J, Randall R. 1995. PCB congeners and hexachlorobenzene biota sediment accumulation factors for *Macoma nasuta* exposed to sediments with different total organic carbon contents. *Environ Toxicol Chem* 14:303–310.

Bridges TS, Farrar JD, Duke BM. 1997. The influence of food ration on sediment toxicity in *Neanthes arenaceodentata* (Annelida:Polychaeta). *Environ Toxicol Chem* 16:1659–1665.

Bruner KA, Fisher SW, Landrum PF. 1994. The role of the zebra mussel, *Dreissena polymorpha,* in contaminant cycling: I. The effect of body size and lipid content on the bioconcentration of PCBs and PAHs. *J Gt Lakes Res* 20:725–734.

Brunson EL, Canfield TJ, Dwyer FJ, Ingersoll CG, Kemble NE. 1998. Assessing the bioaccumulation of contaminants from sediments of the upper Mississippi River using field-collected oligochaetes and laboratory-exposed *Lumbriculus variegatus. Arch Environ Contam Toxicol* 35:191–201.

Cabana G, Rasmussen JB. 1994. Modeling food chain structure and contaminants bioaccumulation using stable nitrogen isotopes. *Nature* 372:255–257.

Canli M, Furness RW. 1993. Toxicity of heavy metals dissolved in water and influences of sex and size on metal accumulation and tissue distribution in the Norway lobster *Nephrops Norvegicus. Mar Environ Res* 36:217–236.

Capuzzo JM, Farrington JW, Rantamaki P, Clifford CH, Lancaster BA, Leavitt DF, Jia X. 1989. The relationship between lipid composition and seasonal differences in the distribution of PCBs in *Mytilus edulis* L. *Mar Environ Res* 28:259–264.

21 CFR 109. 2002. Unavoidable contaminants in food for human consumption and packaging material. Title 21, Part 109, *Code of Federal Regulations.* Washington DC: US Government Printing Office.

21 CFR 509. 2002. Unavoidable contaminants in food for human consumption and packaging material. Title 21, Part 509, *Code of Federal Regulations.* Washington DC: US Government Printing Office.

Chapman PM, Wang F, Janssen C, Persoone G, Allen HE. 1998. Ecotoxicology of metals in aquatic sediments: binding and release, bioavailability, hazard, risk and remediation. *Can J Fish Aquat Sci* 55:221–2243.

Chappie DJ, Burton Jr GA. 1997. Optimization of in situ bioassays with *Hyalella azteca* and *Chironomus tentans. Environ Toxicol Chem* 16:559–564.

Chen Z, Mayer L. 1999. Sedimentary metal bioavailability determined by the digestive constraints of marine deposit feeders: gut retention time and dissolved amino acids. *Mar Ecol Progr Ser* 176:139–151.

Ciarelli S, van Straalen NM, Klap VA, van Wezel AP. 1999. Effects of sediment bioturbation by the estuarine amphipod *Corophium volutator* on fluoranthene resuspension and transfer into the mussel (*Mytilus edulis*). *Environ Toxicol Chem* 18:318–328.

Clarke JU, McFarland VA. 2000. Uncertainty analysis for an equilibrium partitioning-based estimator of polynuclear aromatic hydrocarbon bioaccumulation potential in sediments. *Environ Toxicol Chem* 19:360–367.

Cloern JE, Canuel EA, and Harris D. 2002. Stable carbon and nitrogen isotope composition of aquatic and terrestrial plants of the San Francisco Bay Estuary. *Limnol Oceanogr* 47:713–729.

Conder JM, La Point TW, Lotufo GR, Steevens JA. 2003. Nondestructive, minimal-disturbance, direct-burial solid phase microextraction fiber technique for measuring TNT in sediment. *Environ Sci Technol* 37:1625–1632.

Cornelissen G, Rigterink H, Ferdinandy M, van Noort P. 1998. Rapidly desorbing fractions of PAHs in contaminated sediments as predictor of the extent of bioremediation. *Environ Sci Technol* 32:966–970.

Crouch EAC. 1983. Uncertainties in interspecies extrapolation of carcinogenicity. *Environ Health Persp* 50:321–327.

Di Toro DM, Mahoney JD, Hansen DJ, Scott KJ, Hicks MB, Mayr SM, Redmond MS. 1990. Toxicity of cadmium in sediments: The role of acid volatile sulfide. *Environ Toxicol Chem* 9:1487–1502.

Di Toro DM, Zarba CS, Hansen DJ, Berry WJ, Swartz RC, Cowan CE, Pavlou SP, Allen HE, Thomas NA, Paquin PR. 1991. Technical basis for establishing sediment quality criteria for nonionic organic chemicals using equilibrium partitioning. *Environ Toxicol Chem* 10:1541–1583.

Echols KR, Gale RW, Schwartz TR, Huckins JN, Williams LL, Meadows JC, Morse D, Petty JD, Orazio CE, Tillitt DE. 2000. Comparing polychlorinated biphenyl concentrations and patterns in the Saginaw River using sediment, caged fish, and semi-permeable membrane devices. *Environ Sci Technol* 34:4095–4102.

Escher BI, Schwarzenbach RP. 2002. Mechanistic studies on baseline toxicity and uncoupling of organic compounds as a basis for modeling effective membrane concentrations in aquatic organisms. *Aquat Sci* 64:20–35.

Evans KM, Gill RA, Robothan PWJ. 1990. The PAH and organic carbon content of sediment particle size fractions. *Water Air Soil Pollut* 51:13–31.

Farrington JW, Goldberg ED, Risebrough RW, Martin JH, Bowen VT. 1983. U.S. "mussel watch" 1976-1978: An overview of the trace-metal, DDE, PCB, hydrocarbon, and artificial radionuclide data. *Environ Sci Technol* 17:490–496.

Forrester GE, Fredericks BI, Gerdeman D, Evans B, Steele MA, Zayed K, Schweitzer LE, Suffet IH, Vance RR, Ambrose RF. 2003. Growth of estuarine fish is associated with the combined concentration of sediment contaminants and shows no adaptation or acclimation to past conditions. *Mar Environ Res* 56:423–442.

Foster GD, Baksi SM, Means JC. 1987. Bioaccumulation of trace organic contaminants from sediment by Baltic clams (*Macoma balthica*) and soft shelled clams (*Mya arenaria*). *Environ Toxicol Chem* 6:969–976.

Foureman GL, Ben-Zvi S, Dostal L, Fouts JR, Bend JR. 1978. Distribution of ^{14}C-benzo[a]pyrene in the lobster, Homarus americanus, at various times after a single injection into the pericardial sinus. *Bull Mt Desert Island Biol Lab* 18:93–96.

Griscom SB, Fisher NS, Luoma SN. 2000. Geochemical influences of assimilation of sediment-bound metals in clams and mussels. *Environ Sci Technol* 34:91–99.

Griscom SB, Fisher NS, Luoma SN. 2002. Kinetic modeling of Ag, Cd, and CO, bioaccumulation in the clam *Macoma balthica*: Quantifying dietary and dissolved sources. *Mar Ecol Prog Ser* 240:127–241.

Hack A, Selenka F. 1996. Mobilization of PAH and PCB from contaminated soil using a digestive tract model. *Toxicol Lttrs* 88:199–210.

Hamilton SJ. 2002. Rationale for a tissue-based selenium criterion for aquatic life. *Aquat Toxicol* 57:85–100.

Hare L, Tessier A, Warren L. 2001. Cadmium accumulation by invertebrates living at the sediment-water interface. *Environ Toxicol Chem* 20:880-889.

Harkey GA, Driscoll SK, Landrum PF. 1997. Effect of feeding in 30-day bioaccumulation assays using *Hyalella azteca* in fluoranthene-dosed sediment. *Environ Toxicol Chem* 16:762–769.

Harkey GA, Lydy MJ, Kukkonen J, Landrum PF. 1994. Feeding selectivity and assimilation of PAH and PCB in *Diporeia* spp. *Environ Toxicol Chem* 13:1445–1455.

Harkey GA, Landrum PF, Klaine SJ. 1994. Preliminary studies on the effect of feeding during whole sediment bioassays using *Chironomus riparius* larvae. *Chemosphere* 28:597–606.

Hickey CW, Roper DS, Holland PT, Trower TM. 1995. Accumulation of organic contaminants in two sediment-dwelling shellfish with contrasting feeding modes: deposit- (*Macomona liliana*) and filter-feeding (*Austrovenus stutchburyi*). *Arch Environ Contam Toxicol* 29:221–231.

Holland PT, Hickey CW, Roper DS, Trower TM. 1993. Variability of organic contaminants in inter-tidal sandflat sediments from Manukau Harbour, New Zealand. *Arch Environ Contam Toxicol* 25:456–463.

Hora ME. 1981. Reduction of polychlorinated biphenyl (PCB) concentration in Carp (*Cyprinus-Carpio*) fillets through skin removal. *Bull Environ Contam Toxicol* 26: 364–366.

Ingersoll CG, Brunson EL, Wang N, Dwyer FJ, Huckins J, Petty J, Ankley GT, Mount DR, Landrum PF. 2003. Uptake and depuration of sediment-associated contaminants by the oligochaete, *Lumbriculus variegates. Environ Toxicol Chem* 22:872–885.

Ingersoll CG, Haverland PS, Brunson EL, Canfield TJ, Dwyer FJ, Henke CE, Kemble NE, Mount DR, Fox RG. 1996. Calculation and evaluation of sediment effect concentrations for the amphipod, *Hyalella azteca* and the midge *Chironomus riparius. J Gt Lakes Res* 22:602–606.

Intergovernmental Oceanographic Commission (UNESCO). 1996. A strategic plan for the assessment and prediction of the health of the ocean: A module of the Global Ocean Observing System. Paris: UNESCO. Document IOC/INF-1044. 39 p.

James MO, Altman AH, Li CLJ, Schell JD. 1995. Biotransformation, hepatopancreas DNA-binding and pharmakokinetics of benzo(*a*)pyrene after oral and parenteral administration to the American lobster, *Homarus Americanus. Chemico-Biol Interact* 95:141–160.

Jarman WM, Hobson KA, Sydeman WJ, Bacon CE, McClaren EB. 1996. Influence of trophic position and feeding location on contaminant levels in the Gulf of Farallones food web revealed by stable isotope analysis. *Environ Sci Technol* 30: 654–660.

Jarvinen AW, Ankley GJ. 1998. Linkage of effects to tissue residues: development of a comprehensive database for aquatic organizer exposed to inorganic and organic chemicals. Pensacola (FL): Society of Environmental Toxicology and Chemistry (SETAC).

Jin Z, Simkins S, Xing B. 1999. Bioavailability if freshly added and aged naphthalene in soils under gastric pH conditions. *Environ Toxicol Chem* 18:2751–2758.

Jovanovich MC, Marion KR. 1987. Seasonal variation in uptake and depuration of anthracene by the brackish water clam *Rangia cuneata. Mar Biol* 95:395–403.23.

Kaag NHBM, Foekema EM, Scholten MC Th, van Straalen NM. 1997. Comparison of contaminant accumulation in three species of marine invertebrates with different feeding habits. *Environ Toxicol Chem* 16:837–842.

Kane-Driscoll S, Harkey GA, Landrum PF. 1997. Accumulation and toxicokinetics of fluoranthene in sediment bioassays with freshwater amphipods. *Environ Toxicol Chem* 16:742–753.

Kane-Driscoll S, McElroy AE. 1996. Bioaccumulation and metabolism of benzo[*a*]pyrene in three species of polychaete worms. *Environ Toxicol Chem* 15: 1401–1410.

Karickhoff SW, Morris KR. 1985. Impact of tubificid oligochaetes on pollutant transport in bottom sediments. *Environ Sci Technol* 19:51–56.

Kraaij R, Ciarelli S, Tolls J, Kater BJ, Belfroid A. 2001. Bioavailability of lab-contaminated and native polycyclic aromatic hydrocarbons to the amphipod

Corophium volutator relates to chemical desorption. *Environ Toxicol Chem* 20: 1716–1724.

Kukkonen J, Landrum PF. 1995. Measuring assimilation efficiency for sediment-bound PAH and PCB congeners by benthic organisms. *Aquat Toxicol* 32:75–92.

Kukkonen J, Landrum PF. 1994. Toxicokinetics and toxicity of sediment-associated pyrene to *Lumbriculus variegatus* (Oligochaeta). *Environ Toxicol Chem* 13:1457–1468.

Kukkonen JV, Gunnarsson J, Gossiaux DC, Landrum PF, Weston DP. 2001. Bioavailability of pollutants in relation to sediment characteristics and chemical extractions. Abstract 133. 22nd Annual Meeting of the Society of Environmental Toxicology and Chemistry, 2001 Nov 11–15; Baltimore, MD, USA.

Lake JL, Rubinstein NI, Lee H II, Lake CA, Heltshe J, Pavingano H. 1990. Equilibrium partitioning and bioaccumulation of sediment-associated contaminants by infaunal organisms. *Environ Toxicol Chem* 9:1095–1106.

Landrum PF. 1988. Toxicokinetics of polycyclic aromatic hydrocarbons in the amphipod, *Pontoporeia hoyi*: Role of season and temperature. *Aquat Toxicol* 12: 245–271.

Landrum PF. 1989. Bioavailability and toxicokinetics of polycyclic aromatic hydrocarbons sorbed to sediments for the amphipod, *Pontoporeia hoyi*. *Environ Sci Technol* 23:588–595.

Landrum PF, Dupuis WS, Kukkonen J. 1994. Toxicokinetics and toxicity of sediment-associated pyrene and phenanthrene in *Diporeia* spp.: Examination of equilibrium-partitioning theory and residue-based effects for assessing hazard. *Environ Toxicol Chem* 13:1769–1780.

Landrum PF, Faust WR. 1991. Effect of variation in sediment composition on the uptake clearance of selected PCB and PAH congeners by the amphipod, *Diporeia* sp. Aquatic toxicology and risk assessment: Fourteenth. In: Mayes MA, Barron MG, editors. Philadelphia: American Soc for Testing and Materials. ASTM STP 1124. p 263–279.

Landrum PF, Faust WR, Eadie BJ. 1989. Bioavailability and toxicity of a mixture of sediment-associated chlorinated hydrocarbons to the amphipod, *Pontoporeia hoyi*. In: Cowgill UM, Williams LR, editors. Aquatic toxicology and hazard assessment: 12th Volume. Philadelphia: American Soc for Testing and Materials. ASTM STP 1027. p 315–329.

Landrum PF, Fisher SW. 1998. Influence of lipids on the bioaccumulation and trophic transfer of organic contaminants in aquatic organisms. In: Arts M, Wainman B, editors. Lipids in freshwater ecosystems. New York: Springer-Verlag. p 203–234.

Landrum PF, Gossiaux DC, Kukkonen J. 1997. Sediment characteristics influencing the bioavailability of nonpolar organic contaminants to *Diporeia* spp. *Chem Spec Bioavail* 9:43–55.

Landrum PF, Harkey GA, Kukkonen J. 1996. Evaluation of organic contaminants exposure in aquatic organisms: The significance of bioconcentration and bioaccumulation. In: Newman MC, Jagoe CH, editors. Ecotoxicology: A hierarchial treatment. Boca Raton (FL): Lewis. p 85–131.

Landrum PF, Nalepa TF. 1998. A review of the factors affecting the ecotoxicology of *Diporeia* spp. *J Gt Lakes Res* 24:889–904.

Landrum PF, Robbins JA. 1989. Bioavailability of sediment-associated contaminants to benthic invertebrates. In: Baudo R, Giesy JP, Muntau H, editors. Sediments: Chemistry and toxicity of in-place pollutants. Chelsea (MI): Lewis. p 237–263.

Landrum PF, Robbins JA. 1990. Bioavailability of sediment associated contaminants: A review and simulation model. In: Baudo R, Giesy JP, Muntau H, editors. Sediments: Chemistry and toxicity of in-place pollutants. Chelsea (MI): Lewis. Chapter 8, p 237–263.

Landrum PF, Stubblefield CR. 1991. Role of respiration in the accumulation of organic xenobiotics by the amphipod, *Pontoporeia hoyi*. *Environ Toxicol Chem* 10: 1019–1028.

Landrum PF, Tigue EA, Driscoll SK, Gossiaux DC, Van Hoof PL, Gedeon ML, Adler M. 2001. Bioaccumulation of PCB congeners by *Diporeia* spp.: Kinetics and factors affecting bioavailability. *J Gt Lakes Res* 27:117–133.

Lasiter RR, Hallam TG. 1990. Survival of the fattest: Implications for acute effects of lipophilic chemical on aquatic populations. *Environ Toxicol Chem* 9:585–595.

Lawrence AL, McAloon KM, Mason RP, and Mayer LM. 1999. Intestinal solubilization of particle-associated organic and inorganic mercury as a measure of bioavailability to benthic invertebrates. *Environ Sci Tech* 33:1871–1876.

LeBlanc GL. 1995. Trophic level differences in the bioconcentration of chemicals: implications in assessing environmental biomagnification. *Environ Sci Technol* 29: 154–160.

Lee BG, Luoma SN. 1998. Influence of microalgal biomass on adsorption efficiency of Cd, Cr, and Zn by two bivalves from San Francisco Bay. *Limnol Oceanogr* 43: 1455–1466.

Lee BG, Griscom SB, Lee J-S, Choi HJ, Koh C-H, Luoma SN, Fisher NS. 2000. Influences of dietary uptake and reactive sulfides on metal bioavailability from aquatic sediments. *Science* 287:282–284.

Lee II H. 1990. A clam's eye view of the bioavailability of sediment-associated pollutants. In: Baker RA, editor. Organic substances and sediments in water. Volume 3, Biological. Boca Raton (FL): Lewis. p 73–93.

Lee II H. 1991. A clam's eye view of the bioavailability of sediment-associated pollutants. In: Baker RA, editor. Organic substances and sediments in water. Volume 3, Biological. Boca Raton (FL): Lewis. p 73–94.

Lee II H, Boese BL, Pelletier J, Winsor M, Specht DT, Randall RC. 1993. Guidance manual: Bedded sediment bioaccumulation tests. ERL-Narragansett (RI): US Environmental Protection Agency. USEPA Report No 600/R-93/183. 231 p.

Lee J-H, Landrum PF, Koh C-H. 2002. Prediction of time-dependent PAH toxicity in *Hyalella azteca* using a damage assessment model. *Environ Sci Technol* 36: 3131–3138.

Leppänen MT, Kukkonen JVK. 1998. Relative importance of ingested sediment and pore water as bioaccumulation routes for pyrene to Oligochaete (*Lumbriculus variegatus*, Muller). *Environ Sci Technol* 32:1503–1508.

Leppänen MT, Kukkonen JVK. 2000. Effect of sediment-chemical contact time on availability of sediment-associated pyrene and benzo[*a*]pyrene to oligochaete worms and semipermeable membrane devices. *Aquat Toxicol* 49:227–241.

Leslie HA, van de Wal LR, ter Laak T, Oosthoek A, Busser F, Kraak MH, Hermens JL. 2000. Validation of the biomimetic SPME approach to the estimation of body residues and narcotic toxicity. Society of Environmental Toxicology and Chemistry 21st Meeting (North America), 2000 Nov 12–16; Nashville, TN, USA.

Leslie HA, Oosthoek AJP, Busser FJM, Kraak MHS, Hermens JLM. 2002. Biomimetic solid-phase microextraction to predict body residues and toxicity of chemicals that act by narcosis. *Environ Toxicol Chem* 21:229–234.

Linkov I, Burmistrov D, Cura J, Bridges TS. 2002. Risk-based management of contaminated sediments: Consideration of spatial and temporal patterns in exposure modeling. *Environ Sci Technol* 36:238–246.

Livingstone DR. 1994. Recent developments in marine invertebrate organic xenobiotic metabolism. *Toxicol Ecotoxicol News* 1:88–95.

Lotufo GR. 1998. Bioaccumulation of sediment-associated fluoranthene in benthic copepods: uptake, elimination and biotransformation. *Aquat Toxicol* 44:1–15.

Lozano SJ, Gedeon ML, Landrum PF. 2003. The effects of temperature and organism size on the feeding rate and modeled chemical accumulation in *Diporeia* spp. For Lake Michigan sediments. *J Gt Lakes Res* 29:79–88.

Luoma SN, Fisher NS. 1997. Uncertainties in assessing contaminant exposure from sediments. In: Ingersoll CG, Dillon T, Biddinger GR, editors. Ecological risk assessment of contaminated sediments. Pensacola (FL): Society of Environmental Toxicology and Chemistry (SETAC). p 211–237.

Luoma SN, Johns C, Fisher NS, Steinberg NA, Oremland RS, Reinfelder JR. 1992. Determination of selenium bioavailability to a benthic bivalve from particulate and solute pathways. *Environ Sci Technol* 26:485–491.

Lydy MJ, Lasater JL, Landrum PF. 2000. Toxicokinetics of DDE and 2-chlorobiphenyl in *Chironomus tentans*. *Arch Environ Contam Toxicol* 38:163–168.

Maruya KA, Risebrough RW, Horne AJ. 1997. The bioaccumulation of polynuclear aromatic hydrocarbons by benthic invertebrates in an intertidal marsh. *Environ Toxicol Chem* 16:1087–1097.

Mayer L, Chen Z, Findlay R, Fang J, Sampson S, Self L, Jumars P, Quetél C, Donard O. 1996. Bioavailability of sedimentary contaminants subject to deposit-feeder digestion. *Environ Sci Technol* 30:2641–2645.

Mayer P, Vaes WHJ, Wijnker F, Legierse KCHM, Kraaij R, Tolls J, Hermens JLM. 2000. Sensing dissolved sediment porewater concentrations of persistent and bioaccumulative pollutants using disposable solid-phase microextraction fibers. *Environ Sci Technol* 34:5177–5183.

McCarty LS, Mackay D. 1993. Enhancing ecotoxicological modeling and assessment: body residues and modes of toxic action. *Environ Sci Technol* 27:1719–1728.

McElroy AE, Farrington JW, Teal JM. 1990. Influence of mode of exposure and the presence of a tubiculous polychaete on the fate of benz[*a*]anthracene in the benthos. *Environ Sci Technol* 24:1648–1655.

McGroddy SE, Farrington JW. 1995. Sediment pore water partitioning of polycyclic aromatic hydrocarbons in three cores from Boston Harbor, Massachusetts. *Environ Sci Technol* 29:1542–1550.

McKim J, Schmieder P, Veith G. 1985. Absorption dynamics of organic chemical transport across trout gills as related to octanol-water partition coefficient. *Toxicol Appl Pharmacol* 77:1–10.

Meador JP. 1993. The effect of laboratory holding on the toxicity response of marine infaunal amphipods to cadmium and tributyltin. *J Exper Mar Biol Ecol* 174: 227–242.

Meador JP. 2000. Predicting the fate and effects of tributyltin in marine systems. *Rev Environ Contam Toxicol* 166:1–48.

Meador JP, Adams NG, Casillas E, Bolton JL. 1997. Comparative bioaccumulation of chlorinated hydrocarbons from sediment by two infaunal invertebrates. *Arch Environ Contam Toxicol* 33:388–400.

Meador JP, Casillas E, Sloan CA, Varanasi U. 1995. Comparative bioaccumulation of polycyclic aromatic hydrocarbons from sediment by two infaunal invertebrates. *Mar Ecol Progr Ser* 123:107–124.

Meador JP, Collier TK, Stein JE. 2002a. Use of tissue and sediment based threshold concentrations of polychlorinated biphenyls (PCBs) to protect juvenile salmonids listed under the U.S. Endangered Species Act. *Aquat Conserv: Mar Freshw Ecosyst* 12:493–516.

Meador JP, Collier TK, Stein JE. 2002b. Determination of a tissue and sediment threshold for tributyltin to protect prey species for juvenile salmonids listed by the U.S. Endangered Species Act. *Aquat Conserv: Mar Freshw Ecosyst* 12: 539–551.

Meador JP, Krone CA, Dyer DW, Varanasi U. 1997. Toxicity of sediment-associated tributyltin to infaunal invertebrates: species comparison and the role of organic carbon. *Mar Environ Res* 43:219–241.

Meador JP, Rice CA. 2001. Impaired growth in the polychaete *Armandia brevis* exposed to tributyltin in sediment. *Mar Environ Res* 51:113–129.

Miles AK, Roster N. 1999. Enhancement of polycyclic aromatic hydrocarbons in estuarine invertebrates by surface runoff at a decommissioned military fuel depot. *Mar Environ Res* 47:49–60.

Naes K, Hylland K, Oug E, Forlin L, Ericson G. 1999. Accumulation and effects of aluminum smelter-generated polycyclic aromatic hydrocarbons on soft-bottom invertebrates and fish. *Environ Toxicol Chem* 18:2205–2216.

[NRC] National Research Council. 2000. Toxicological effects of methylmercury. Washington DC: National Academy Pr.

Newman MC, Jagoe CH. 1994. Ligands and the bioavailability of metals in aquatic environments. In: Hamelink JL, Landrum PF, Bergman HL, Benson WH, editors. Bioavailability: physical, chemical, and biological interactions. Boca Raton (FL): Lewis. p 39–61.

Oliver BG. 1987. Biouptake of chlorinated hydrocarbons from laboratory-spiked field sediments by oligochaete worms. *Environ Sci Technol* 21:785–790.

Oliver BG, Niimi AJ. 1988. Trophodynamic analysis of polychlorinated biphenyl congeners and other chlorinated hydrocarbons in the Lake Ontario ecosystem. *Environ Sci Technol* 22:388–397.

Oomen AG, Sips AJAM, Groten JP, Sijm DTHM, Tolls J. 2000. Mobilization of PCBs and lindane from soil during in vitro digestion and their distribution among bile salt micelles and proteins of human digestive fluid and the soil. *Environ Sci Technol* 34:297–303.

Pastorok RA, Becker DS. 1990. Comparative sensitivity of sediment toxicity bioassays at three Superfund sites in Puget Sound. In: Landis WG, van der Schalie WH, editors. Aquatic toxicology and risk assessment: 13th Volume. Philadelphia: American Soc for Testing and Materials. STP 1096. p 123–139.

Pawliszyn J. 1997. Solid phase microextraction: Theory and practice. New York: Wiley-VCH.

Peterson BJ, Fry B. 1987. Stable isotopes in ecosystem studies. *Ann Rev Ecol Syst* 18: 293–320.

Petty JD, Orazio CE, Huckins JN, Gale RW, Lebo JA, Meadows JC, Echols KR, Cranor WL. 2000. Considerations involved with the use of semipermeable membrane devices for monitoring environmental contaminants. *J Chromat* A 879:83–95.

Pickard SW, Yaksich SM, Ervine KN, McFarland VA. 1998. Bioaccumulation potential of sediment-associated polychlorinated biphenyls (PCBs) in Ashtabula Harbor, Ohio. *J Gt Lakes Res* 27:44–59.

Piérard C, Budzinski H, and P Garrigues. 1996. Grain-size distribution of polychlorobiphenyls in coastal sediments. *Environ Sci Technol* 30:2776–2783.

Rand GM, Wells PG, McCarty LS. 1995. Introduction to Aquatic Toxicology. In: Rand GM, editor. Fundamentals of aquatic toxicology: Effects, environmental fate, and risk assessment. Briston (PA): Taylor & Francis. p 3–67.

Rasmussen JB, Rowan DJ, Lean DRS, Carey JH. 1990. Food chain structure in Ontario lakes determine PCB levels in lake trout *Salvelinus namaycush* and other pelagic fish. *Can J Fish Aquat Sci* 47:2030–2038.

Reichert WL, Eberhart B-T, Varanasi U. 1985. Exposure of two species of deposit-feeding amphipods to sediment-associated [^{3}H]benzo[*a*]pyrene: uptake, metabolism and covalent binding to tissue macromolecules. *Aquat Toxicol* 6: 45–56.

Reinfelder JR, Fisher NS, Luoma SN, Nichols JW, Wang WX. 1998. Trace element trophic transfer in aquatic organisms: A critique of the kinetic model approach. *Sci Tot Environ* 219:117–135.

Robinson SD, Landrum PF, Van Hoof PL. 2000. Seasonal variation of polychlorobiphenyl accumulation in the amphipod, *Diporeia* spp. Abstract 127. 21st Annual Meeting of the Society of Environmental Toxicology and Chemistry; 2000 Nov 12–16; Nashville, TN; USA.

Roditi HA, Fisher NS, Sanudo-Wilhelmy SA. 2000. Field testing a metal bioaccumulation model for zebra mussels. *Environ Sci Technol* 34:2817–2825.

Roy I, Hare L. 1999. Relative importance of water and food as cadmium sources to the predatory insect *Sialis velata* (Megaloptera). *Can J Fish Aquat Sci* 56:1143–1149.

Ruby MV, Davis A, Link TE, Schoof R, Chaney RL, Freeman GB, Bergstrom P. 1993. Development of an in vitro screening test to evaluate the in vivo bioaccessibility of ingested mine-waste lead. *Environ Sci Technol* 27:2870–2877.

Salazar MH, Salazar SM. 2001. A pickle jar is a pickle jar and a fish is a moving target but a fish is not a fat bag and bivalves don't eat pickles. 27th Annual Aquatic Toxicity Workshop; 2001 Oct 1–4; St. Johns, Newfoundland.

Salazar MH, Salazar SM. 1998. Using caged bivalves as part of an exposure-dose-response triad to support an integrated risk assessment strategy. In: de Peyster A, Day K, editors. Ecological risk assessment: A meeting of policy and science. Pensacola (FL): Society of Environmental Toxicology and Chemistry. p 167–192.

Sarda N, Burton Jr GA. 1995. Ammonia variation in sediments: spatial, temporal, and method-related effects. *Environ Toxicol Chem* 14:1499–1506.

Schmitt RJ, Osenberg CW, editors. 1996. Detecting ecological change: Concepts and applications in coastal habitats. San Diego: Academic Pr. 401 p.

Schuler LJ, Heagler MG, Lydy MJ. 2002. Bioavailability within single versus multiple species systems. *Arch Environ Contam Toxicol* 42:199–204.

Selck H, Palmqvist A, Forbes VE. 2003. Uptake, depuration, and toxicity of dissolved and sediment-bound fluoranthene in the polychaete, *Capitella* sp. I. *Environ Toxicol Chem* 22:2354–2363.

Shephard B. 1998. Quantification of ecological risks to aquatic biota from bioaccumulated chemicals. In: National Sediment Bioaccumulation Conference Proceedings. Washington DC: US Environmental Protection Agency Office of Water. EPA 823-R-98-002.

Spehar RL, Mount DR, Lukasewycz MT, Mattson VR, Leonard EN, Burkhard LP, Burgess RM, Serbst JR, Berry WJ. 1998. Chronic exposure of freshwater and saltwater invertebrates to sediment containing a mixture of high K_{ow} PAHs. Abstract 119. 19th Annual Meeting of the Society of Environmental Toxicology and Chemistry; 1998 Nov 15–19; Charlotte, NC, USA.

Suter GW. 1993. Ecological risk assessment. Boca Raton (FL): Lewis. 538 p.

Suter GW, Rosen AE. 1986. Comparative toxicology of marine fishes and crustaceans. Rockville (MD): National Oceanic and Atmospheric Administration. NTIS No. PB87-151916. 25 p.

Sundelin B, Eriksson A-K. 2001. Mobility and bioavailability of trace metals in sulfidic coastal sediments. *Environ Toxicol Chem* 20:748–756.

Tessier A, Turner DR. 1995. Metal speciation and bioavailability in aquatic systems. IUPAC Series on analytical and physical chemistry of environmental systems, Volume 3. New York: J Wiley. 679 p.

Thomann RV, Komlos J. 1999. Model of biota-sediment accumulation factor for polycyclic aromatic hydrocarbons. *Environ Toxicol Chem* 18:1060–1068.

Thomann RV, Connolly JP, Parkerton TF. 1992. An equilibrium model of organic chemical accumulation in aquatic food webs with sediment interaction. *Environ Toxicol Chem* 11:615–629.

Thomann RV, Shkreli F, Harrison S. 1997. A pharmacokinetic model of cadmium in rainbow trout. *Environ Toxicol Chem* 16:2268–2274.

Tracey GA, Hansen DJ. 1996. Use of biota-sediment accumulation factors to assess similarity of nonionic organic chemical exposure to benthically-coupled organisms of differing trophic mode. *Arch Environ Contam Toxicol* 30:467–475.

Umlauf G, Bierl R. 1987. Distribution of organic micropollutants in different size fractions of sediment and suspended solid particles of the River Rotmain. *Z Wasser Abwasser Forsch* 20:203-209.

Urrestarazu RE, Meijer SN, Vaes WHJ, Verhaar HJM, Hermens JLM. 1998. Using solid-phase microextraction to determine partition coefficients to humic acids and bioavailable concentrations of hydrophobic chemicals. *Environ Sci Technol* 32:3430–3435.

[USACE/USEPA] US Army Corps of Engineers/US Environmental Protection Agency. 1991. Evaluation of dredged material proposed for ocean disposal: Testing manual. Washington DC: USEPA Office of Water. EPA 503/8-91/001.

[USACE/USEPA] US Army Corps of Engineers/US Environmental Protection Agency. 1998. Evaluation of dredged material proposed for discharge in waters of the U.S.: Testing manual. Washington DC: USEPA Office of Water. EPA-823-B-98-004.

[USACE/USEPA] US Army Corps of Engineers/US Environmental Protection Agency. 2002a. Environmental Residue Effects Database (ERED). Available from: http://www.wes.army.mil/el/ered/. Accessed 25 Oct 2004.

[USACE/USEPA] US Army Corps of Engineers/US Environmental Protection Agency. 2002b. ERDC-WES BSAF database. Available from: http://www.wes.army.mil/el/bsaf/bsaf.html. Accessed 25 Oct 2004.

[USEPA] US Environmental Protection Agency. 1989. Assessing human health risks for chemically contaminated fish and shellfish: A guidance manual. Washington DC: USEPA Office of Water Regulations and Standards. EPA/503/8-89/002.

[USEPA] US Environmental Protection Agency. 1998. Guidelines for ecological risk assessment. Washington DC: USEPA. EPA/630/R-95/002F.

[USEPA] US Environmental Protection Agency. 2000. Appendix to bioaccumulation testing and interpretation for the purpose of sediment quality assessment: Status and needs: Chemical-specific summary tables. Washington DC: USEPA Office of Water. EPA-823-R-00-002.

[USEPA] US Environmental Protection Agency. 2002. Integrated risk information system. Available from: http://www.epa.gov/iris/index.html. Accessed 28 Jul 2004.

[USFDA] US Food and Drug Administration. 2002. Action levels for poisonous or deleterious substances in human food and animal feed. Available from: http://vm.cfsan.fda.bov/~lrd/fdaact.html. Accessed 28 Jul 2004.

Vaes WHJ, Urrestarazu RE, Verhaar HJM, Hermens JLM. 1998. Acute toxicity of nonpolar versus polar narcosis: Is there a difference? *Environ Toxicol Chem* 17: 1380–1384.

Vaes WHJ, Urrestarazu RE, Hamwijk C, Van Holsteijn I, Blaauboer BJ, Seinen W, Verhaar HJM, Hermens JLM. 1997. Solid phase microextraction as a tool to determine membrane/water partition coefficients and bioavailable concentrations in in vitro systems. *Chem Res Toxicol* 10:1067–1072.

Valiela I. 1995. Marine ecological processes. 2nd ed. New York: Springer-Verlag. 686 p.

Vander Zanden MJ, Rasmussen JB. 1996. A trophic position model of pelagic food webs: Impact on contaminant bioaccumulation in lake trout. *Ecol Monogr* 66: 451–477.

Van Hoof PL, Hseih JL, Robinson SD. 2001. Dependency of amphipod biota-sediment accumulation factors (BSAFs) on sediment concentrations of PAHs and PCBs. Abstract PT244, 22nd Annual Meeting of the Society of Environmental Toxicology and Chemistry; 2001 Nov 11–15; Baltimore, MD, USA.

Van Hoof PL, Eadie BJ. 1999. Particle size distributions of PCBs and PAHs in sediments from varying depositional regions of Lake Michigan. Abstract PWA028, 20th Annual Meeting of the Society of Environmental Toxicology and Chemistry; 1999 Nov 14–18; Philadelphia, PA, USA.

Van Wezel AP, Opperhuizen A. 1995. Narcosis due to environmental pollutants in aquatic organisms: residue-based toxicity, mechanisms and membrane burdens. *Crit Rev Toxicol* 25:255–279.

Varanasi U, Reichert WL, Stein JE, Brown DW, Sanborn HR. 1985. Bioavailability and biotransformation of aromatic hydrocarbons in benthic organisms exposed o sediment from an urban estuary. *Environ Sci Technol* 19:836–841.

Verbruggen EMJ, Vaes WHJ, Parkerton TF, Hermens JLM. 2000. Polyacrylate-coated SPME fibers as a tool to simulate body residues and target concentrations of complex organic mixtures for estimation of baseline toxicity. *Environ Sci Technol* 34:324–331.

Voie ØA, Johnsen A, Rossland HK. 2002. Why biota still accumulate high levels of PCB after removal of PCB contaminated sediments in a Norwegian fjord. *Chemosphere* 46:1367–1372.

Von Stackelberg K, Burmistrov D, Linkov I, Cura J, Bridges TS. 2002. The use of spatial modeling in an aquatic food web to estimate power and risk. *Sci Total Environ* 288:97–110.

Wang WX, Fisher NS, Luoma SN. 1996. Kinetic determinations of trace element bioaccumulation in the mussel *Mytilus edulis*. *Mar Ecol Prog Ser* 140:91–113.

Wang WX, Fisher NS. 1999. Assimilation efficiencies of chemical contaminants in aquatic invertebrates: A synthesis. *Environ Toxicol Chem* 18:2034–2045.

Wang WX, Stupakoff I, and Fisher NS. 1999. Bioavailability of dissolved and sediment-bound metals to a marine filter-feeding polychaete. *Mar Ecol Prog Ser* 178:281–293.

Washington State Department of Ecology. 1997. Developing health-based sediment quality criteria for clean-up sites: A case study report. Washington State Department of Ecology Report 97-114. 115 p.

Weeks JM, Rainbow PS. 1993. The relative importance of food and seawater as sources of copper and zinc to talitrid amphipods (Crustacea: Amphipoda; talitridae). *J Appl Ecol* 30:722–735.

Wells JB, Lanno RP. 2001. Passive sampling devices (PSDs) as biological surrogates for estimating the bioavailability of organic chemicals in soil, In: Greenberg BM, Hull RN, Roberts Jr MH, Gensemer RW, editors. Environmental toxicology and risk assessment: Science, policy, and standardization: Implications for environmental decisions: 10th Volume. West Conshohocken (PA): American Soc for Testing and Materials. ASTM STP 1403. p 253–270.

Weston DP, Mayer LM. 1998. Comparison of in vitro digestive fluid extraction and traditional in vivo approaches as measures of polycyclic aromatic hydrocarbon bioavailability from sediments. *Environ Toxicol Chem* 17:830–840.

Weston DP, Penry DL, Gulmann LK. 2000. The role of ingestion as a route of contaminant bioaccumulation in a deposit-feeding polychaete. *Arch Environ Contam Toxicol* 38:446–454.

Weston DP, Maruya KA. 2002. Predicting bioavailability and bioaccumulation using in vitro digestive fluid extraction. *Environ Toxicol Chem* 21:962–971.

Weston DP. 1990. Hydrocarbon bioaccumulation from contaminated sediment by the deposit feeding polychaete *Abarenicola pacifica. Mar Biol* 107:159–169.

Wong CS, Capel PD, Nowell LH. 2001. National field-based evaluation of the biota-sediment accumulation factor model. *Environ Sci Technol* 35:1709–1715.

Zabik ME, Zabik MJ, Booren AM, Nettles M, Song JH, Welch R, Humphrey H. 1995. Pesticides and total polychlorinated biphenyls in chinook salmon and carp harvested from the Great Lakes: Effects of skin-on and skin-off processing and selected cooking methods. *J Agr Food Chem* 43:93–1001.

Ability of SQGs to estimate effects of sediment-associated contaminants in laboratory toxicity tests or in benthic community assessments

12

CHRISTOPHER G INGERSOLL, STEVEN M BAY, JUDY L CRANE, L JAY FIELD, THOMAS H GRIES, JEFFERY L HYLAND, EDWARD R LONG, DONALD D MACDONALD, THOMAS P O'CONNOR

Numerical effects-based sediment quality guidelines (SQGs) have been developed by federal, provincial, or state agencies in the United States (US), Canada, Australia, Italy, the Netherlands, and Hong Kong (Chapters 1, 9, and 10). The approaches that have been selected by different jurisdictions for different management situations depend on the receptors to be protected (e.g., sediment-dwelling organisms, wildlife, or humans), the degree of protection that is to be afforded, the geographic area to which the SQGs are intended to apply, and their intended uses. SQGs for assessing sediment quality relative to the potential for adverse effects on sediment-dwelling organisms have been derived using both

- mechanistic approaches:
  - the equilibrium partitioning (EqP) approach (Di Toro et al. 1991; USEPA 1997, 1999; NYSDEC 1999; Chapter 13) and
- empirical approaches:
  - screening-level concentration (SLC) approach (Persaud et al. 1993),
  - effects range approach (ERA; Long et al. 1995; USEPA 1996),
  - effects level approach (ELA; MacDonald et al. 1996; Smith et al. 1996; USEPA 1996),

Use of Sediment Quality Guidelines and Related Tools for the Assessment of Contaminated Sediments
Wenning RJ, Batley GE, Ingersoll CG, Moore DW, editors.

- apparent effects threshold (AET) approach (Barrick et al. 1988; Ginn and Pastorok 1992; Cubbage et al. 1997),
- consensus-based approach (Swartz 1999; MacDonald, DiPinto, et al. 2000; MacDonald, Ingersoll, Berger 2000), and
- logistic regression modeling approach (Field et al. 1999, 2002).

These SQGs have been used to interpret historical data, identify potential problem chemicals or areas at a site, design monitoring programs, classify hot spots and rank sites, and make decisions for more detailed studies (Long and MacDonald 1998). Additional suggested uses for SQGs include identifying the need for source controls of problem chemicals before release, linking chemical sources to sediment contamination, triggering regulatory action, and establishing target remediation objectives (USEPA 1997). Use of SQGs with other tools such as sediment toxicity tests, bioaccumulation, and benthic community surveys can provide a powerful weight of evidence (WOE) for assessing the hazards associated with contaminated sediments (Ingersoll et al. 1997; Chapter 6).

The underlying supposition in the derivation of effects-based SQGs is that these guidelines can sometimes be substituted for direct measures of the adverse effects that contaminants in sediment have on benthic communities. Thus, in all of the approaches, measures of sediment toxicity in the laboratory or in the field are equated with the concentrations of chemicals in sediments that either cause or are associated with a response. Although large databases have typically been used to derive SQGs, the ability of these guidelines to accurately predict whether sediments are toxic or nontoxic to benthic communities is the cornerstone of their application.

Several concerns have been expressed regarding the application of SQGs in assessments of sediment quality (Chapter 3). One of these concerns is the ability of the SQGs to predict the presence or absence of acute or chronic toxicity to sediment-dwelling organisms in field-collected sediment. A second concern is the ability of SQGs to predict effects resulting from bioaccumulation of sediment-associated contaminants. A third concern deals with the ability of SQGs to establish cause-and-effect relationships. A fourth concern deals with the ability of SQGs developed for 1 endpoint (e.g., amphipod mortality) to be predictive of effects on other organisms in the field.

The primary objective of this chapter is to summarize data from published studies that have evaluated the ability of SQGs to predict effects or no effects, observed in laboratory toxicity tests or in field studies of benthic communities. A second objective of this paper is to summarize published studies that have evaluated the ability of SQGs to establish cause-and-effect relationships in sediment assessments. The ability of mechanistically based (EqP) SQGs to predict the toxicity in laboratory tests is presented in Chapter 13. The ability of SQGs based on bioaccumulation to predict toxicity on sediment-dwelling organisms or to predict toxicity to higher trophic level organisms that consume these contaminated sediment-dwelling organisms is presented in Chapter 11. Approaches that can be used to address potential confounding factors associated with endpoints measured in sediment assessments are presented in Chapter 16. Finally, Chapter 4 summarizes the discussions by participants at the workshop regarding the predictive ability of SQGs in laboratory toxicity tests, in benthic community assessment, and in sediment bioaccumulation assessments.

Summary of Published Studies

This section provides a summary of publications that have evaluated the predictive ability of SQGs in laboratory toxicity tests or in benthic community assessments (based on more than 8000 sediment samples). Information on procedures used to derive the SQGs described in these studies is presented in the literature cited and in Chapter 3. The primary data requirement for including laboratory toxicity studies in this summary was the generation of matching sediment chemistry and effects data (i.e., both measured on a split of the same sample). However, sediment chemistry was typically measured on a composite sample collected from the location where separate replicate samples were collected for benthic community assessments. Data included in this chapter were summarized from peer-reviewed literature, whether published in technical reports or scientific journals, and use of unpublished information has been kept to a minimum. Many of the databases described in this chapter (e.g., Crane et al. 2002; Field et al. 2002; Ingersoll et al. 2001; Long, Field, MacDonald 1998; MacDonald, Ingersoll, Berger, et al. 2000) were developed by using a screening process to include only the highest quality sediment chemistry and toxicity data (e.g., in accordance with ASTM methods; see Appendix 4 in Ingersoll and MacDonald 2002 for additional detail on this screening process).

Identifying a sample as toxic in a laboratory test or as impacted in a benthic community assessment was typically based on the designation in the publication describing the individual studies. For laboratory toxicity tests, several methods have been used to identify samples as toxic. Most commonly, the response of organisms in test sediment is compared to responses in negative control sediment using a variety of statistical procedures (ASTM 2004). Samples in which the observed response of the organism is significantly different from the control are designated as toxic. Similarly, the response in test sediment also may be compared to that in reference sediment, provided that the reference sediment is demonstrated to be appropriate (i.e., nontoxic, chemical concentrations below threshold effect-type SQGs; USEPA 2000a; ASTM 2004). For some toxicity tests (i.e., 10-d marine amphipod survival), power analyses have been used to identify minimum significant differences (MSDs) from the control (Swartz et al. 1985; Mearns et al. 1986; Thursby et al. 1997). For these tests, test sediment may be designated as highly toxic if the response of the organism is significantly different from the control and the response rate exceeds the MSD from the control (e.g., Long, Field, MacDonald 1998). Such MSDs have not been established for the freshwater toxicity tests that are commonly used in sediment quality assessments. Benthic community responses are often compared to responses at reference sites similar in environmental conditions. The comparison to reference conditions may be done within a study if appropriate reference conditions can be established. Alternatively, Reynoldson et al. (1995) recommended that the normal range of benthic invertebrate community metrics be established a priori using the 95% confidence intervals, and impacted benthic communities can then be established as the upper or lower limits of the normal range for each metric. These different approaches to designating samples "toxic" in laboratory or field studies may contribute to the variability in the prediction of sediment toxicity by SQGs.

Laboratory toxicity tests with marine or estuarine sediments

Predictive ability of individual ERL or ERM and TEL or PEL sediment quality guidelines

Matching chemistry and toxicity data were assembled by Long, Field, and MacDonald (1998) from 1068 samples collected in the US by USEPA and NOAA to quantify the predictive ability of the ERLs and ERMs (Long et al. 1995) and TELs and PELs (MacDonald et al. 1996). Results were reported from chemical analyses and 10-d toxicity tests with amphipods (*Ampelisca abdita* in 900 samples, *Rhepoxynius abronius* in 168 samples) and other sublethal tests (microbial bioluminescence, mollusk embryo development, or sea urchin fertilization) in 437 samples. The predictive ability of individual ERMs and PELs was reported as the incidence of significant toxicity in amphipod tests alone and in any of 2 or 3 tests performed for each chemical or chemical class (Table 12-1). In most cases, the incidence of toxicity was >50% when individual ERMs were exceeded by any amount (range of 52% for Cu to 100%

Table 12-1 Incidence of toxicity in either marine amphipod tests alone or any of 2 to 3 tests performed among samples in which individual ERMs or PELs were equaled or exceeded[a]

	ERMs				PELs			
	Amphipod tests (n = 1068)		Any of 2 to 3 tests (n = 437)		Amphipod tests (n = 1068)		Any of 2 to 3 tests (n = 437)	
Chemical	Nr.	% significantly toxic	Nr.	% significantly toxic	Nr.	% significantly toxic	Nr.	% significantly toxic
Metals								
Cd	2	100	0	na	21	81	6	100
Cr	5	100	2	100	41	66	24	92
Cu	25	52	22	82	179	59	146	87
Pb	35	83	20	95	122	63	85	92
Hg	126	66	81	90	127	66	82	89
Ni	21	76	5	100	74	66	37	94
Ag	38	65	22	100	109	59	82	88
Zn	56	66	32	88	126	62	87	86
PAHs								
2-methylnaphthalene	6	100	4	100	47	73	22	95
Dibenz(*a,h*)anthracene	43	72	31	81	80	64	65	85
Acenaphthene	13	77	7	100	84	62	56	95
Acenaphthylene	7	100	6	100	47	77	40	98
Anthracene	25	76	19	89	131	56	100	89
Benz(*a*)anthracene	47	77	30	90	116	61	93	88
Benzo(*a*)pyrene	63	64	46	83	126	59	100	88
Chrysene	38	68	26	92	116	56	93	88
Fluoranthene	32	72	21	95	103	59	80	88
Fluorene	17	71	10	90	74	70	51	94
Naphthalene	4	75	4	100	38	74	23	100
Phenanthrene	40	75	25	96	106	60	77	92
Pyrene	66	67	46	89	117	60	94	89
Sum LPAHs	48	80	31	96	117	64	79	92
Sum HPAHs	76	60	56	84	114	58	90	87
Sum total PAHs	11	91	6	100	56	68	38	89
Chlorinated hydrocarbons								
p,p′-DDE	89	55	70	92	3	33	3	67
p,p′-DDD					144	65	115	92
p,p′-DDT					97	67	68	94
Total DDTs	107	61	82	96	101	64	78	95
Total PCBs	194	51	162	84	191	49	159	83
Dieldrin					41	80	25	96
Lindane					54	19	50	86

[a] Reprinted with permission from Long, Field, MacDonald 1998. Copyright Society of Environmental Toxicology and Chemistry [SETAC].

for 4 other chemicals). With the results from sublethal toxicity tests considered, the incidence of toxicity generally increased by factors of about 20% to 30% (overall effects ranging from 81% to 100%). The predictive ability of individual PELs was similar to those for the ERMs (Table 12-1). In most cases, 60% or more of the samples were toxic in amphipod tests when the individual PELs were exceeded. The exceptions included the results for p,p'-DDE for which there were only 3 samples (33% toxic), total PCBs (49% toxic), and lindane (19% toxic). Generally, the incidence of toxicity increased to >90% (range of 67% to 100%) when data from sublethal tests were considered.

Among the 1068 samples analyzed by Long, Field, and MacDonald (1998), there were 329 in which none of the chemical concentrations exceeded the ERLs. Because the ERLs were calculated as the 10th percentiles of the effects database, it was expected that about 10% of samples would be toxic when all chemical concentrations were less than the ERLs. Examination of the database indicated that 11% of the 329 samples were classified as highly toxic (mean survival significantly different from controls and <80% survival) in amphipod tests (Table 12-2). In the same analysis, 9% of samples were highly toxic when none of the chemical concentrations equaled or exceeded the TELs. Thus, the predicted and observed toxicity were similar with either ERLs or TELs.

O'Connor et al. (1998) evaluated a database with 1508 measures of sediment chemistry and 10-d toxicity tests with marine amphipods obtained from the USEPA Environmental Monitoring and Assessment Program (EMAP) and NOAA Status and Trends Program (NST). A total of 239 of these samples exceeded at least 1 ERM, and 38% of these samples were toxic (<80% survival). A total of 481 of these samples did not exceed ERLs, and 5% of these samples were toxic. O'Connor and Paul (2000) evaluated an expanded database with 2475 records obtained from the EMAP, NST, and Regional EMAP (REMAP) programs. Similarly, among the 453 samples with at least 1 ERM exceeded, 41% were toxic, and of the 730 samples without any ERLs exceeded, only 5% were toxic (Table 12-3). In a database generated in the USEPA National Sediment Quality Survey (NSQS; USEPA 2001), 2761 samples were evaluated with matching sediment chemistry and 10-d amphipod toxicity (Table 12-4). Of the 762 samples with at least 1 ERM exceeded, 48% were toxic, and of the 919 samples without any ERLs exceeded, only 8% were toxic. These data also showed a consistent pattern of increasing incidence of toxicity as the numbers of exceeded ERMs that increased. The percentages of samples classified as toxic exceeded 75% when >5 substances exceeded ERMs. Table 12-3 also contains results from using the EMAP and NST databases to evaluate the

Table 12-2 Incidence of highly toxic samples and average percent survival of marine amphipods in sediment samples classified according to SQGs

Chemical characteristics relative to sediment guidelines	Percent highly toxic[a] samples in amphipod survival tests				Average, control-adjusted amphipod survival (%)			
	National[b] database (n = 1068)	Biscayne Bay (n = 226)	Pearl Harbor (n = 219)	Combined summary (n = 1513)	National[b] database (n = 1068)	Biscayne Bay (n = 226)	Pearl Harbor (n = 219)	Combined summary (n = 1513)
Category 1								
• mean ERM-Q < 0.1	11	3	4	9	93	95	94	93
• mean PEL-Q < 0.1	10	3	6	8	93	95	92	93
• no ERLs exceeded	11	5	14	9	92	94	93	92
• no TELs exceeded	9	6	0	8	92	93	102	92
Category 2								
• mean ERM-Q 0.11–0.5	30	15	5	21	81	85	97	86
• mean PEL-Q 0.11–1.5	25	23	12	21	84	81	93	86
• 1 to 5 ERMs exceeded	32	46	24	32	79	66	86	79
• 1 to 5 PELs exceeded	24	37	2	18	83	70	99	88
Category 3								
• mean ERM-Q 0.51–1.5	46	73	61	49	74	48	68	70
• mean PEL-Q 1.51–2.3	50	67	33	49	66	46	76	68
• 6 to 10 ERMs exceeded	52	67	71	57	63	47	66	59
• 6 to 20 PELs exceeded	47	59	48	48	71	57	73	70
Category 4								
• mean ERM-Q > 1.5	75	75	100[c]	76	43	31	44	41
• mean PEL-Q > 2.3	77	75	33[d]	73	47	31	56	46
• >10 ERMs exceeded	85	75	33[d]	80	41	31	56	41
• >20 PELs exceeded	88	75	50[c]	83	38	10	45	37

[a] Mean survival significantly different from controls and <80% of controls
[b] Data from Long and MacDonald (1998)
[c] 100% (2 of 2) either marginally toxic ($p < 0.05$, mean survival ≥80% of control) or highly toxic
[d] 100% (3 of 3) either marginally toxic ($p < 0.05$, mean survival ≥80% of control) or highly toxic
ERM-Q = Effect range median quotient

Table 12-3 Incidence of toxicity or no toxicity in 10-d marine or estuarine amphipod tests in samples based on exceedances of SQGs (Reprinted from *Marine Pollution Bulletin,* O'Connor and Paul, Misfit between sediment toxicity and chemistry, p 59–64, copyright 2000, with permission from Elsevier)

Samples	Total	Toxic
All	2475	388 (16%)
With 1 chemical concentration > ERM	453	186 (41%)
With 6 or more chemical concentrations > ERM	46	29 (63%)
With mean ERM quotient > 1	59	37 (63%)
With sum PAH_{iw}/LC50) > 1	17	9 (53%)
With Sum PAH_{23} > 18 μMg-OC	12	7 (58%)
With Sum PAH_{23} > 11 μMg-OC	25	12 (48%)
With Sum PAH_{23} > 3.5 μMg-OC	123	54 (44%)
With no chemical concentration > ERL	730	33 (5%)
Samples with Sum total metal[a] <AVS	1256	1117 (89%)

[a]Sum total metal = sum of total Zn, Cu, Ni, Pb, Cd, Hg, and Ag extracted with HF

predictability of SQGs based EqP for PAHs (Swartz et al. 1995; Di Toro and McGrath 2000).

The predictive ability of these mechanistically base SQGs (44% to 53%) was somewhat higher than the predictive ability of SQGs based on single exceedances of ERMs (41%) in the NSQS database.

Predictive ability of ERM-Qs and PEC-Qs on a national and regional basis in North America

During the analyses of the predictive ability of the SQGs, Long, Field, and MacDonald (1998) observed that the incidence of toxicity increased with increasing numbers of chemicals that exceeded the ERMs and PELs. To further clarify this relationship and to provide a quantitative method of estimating probabilities of toxicity, Long and MacDonald (1998) identified the chemical characteristics equivalent to percent frequencies of toxicity of about 10%, 25% to 30%, 50%, and >75%. The incidence of toxicity (based on survival different from control and <80% of control response) in marine amphipod tests increased from <10% with no ERLs or TELs exceeded, to 24% to 32% with 1 to 5 ERMs or PELs exceeded, to 85% to 88% with >10 ERMs or >20 PELs exceeded (Table 12-2). In addition, the incidence of toxicity was evaluated within ranges of mean SQG quotients (SQG-Qs). Individual quotients

Table 12-4 Results for samples with matching chemistry and 10-d marine and estuarine amphipod data in National Sediment Quality Survey Database (USEPA 2001) compiled by Tom O'Connor (unpublished)

	Nr. samples		
Nr. ERM exceedances	Total	Toxic	% toxic
0	1999	289	14
1 or more	762	362	48
2 or more	343	220	64
3 or more	219	152	69
4 or more	181	130	72
5 or more	136	101	74
6 or more	102	79	77
7 or more	82	63	77
8 or more	55	41	75
9 or more	41	33	80
10 or more	31	25	81
11 or more	22	20	91
12 or more	15	14	93
13 or more	10	9	90
14 or more	6	5	83
15 or more	4	4	100
16 or more	3	3	100
17 or more	2	2	100
ERM-Q			
<0/1	1392	120	9
0.1–0.5	1045	330	32
0.5-1	175	99	56
>1	155	94	61

were calculated by dividing the measured concentration of a chemical by its SQG. A mean quotient was then calculated by summing these individual quotients and dividing by the number of quotients. The mean SQG-Q was used to account for the presence and additive toxicity of mixtures of substances in the sediments, presuming these diverse groups of chemicals exert some form of joint toxic action. The incidence of effects increased steadily from about 10% to 11% at mean SQG-Qs <0.1 to about 75% to 77% at higher mean quotients (Table 12-2). In addition, the average control-adjusted survival of the amphipods decreased from about >90% to <50% in the 4 categories of chemical concentrations listed in Table 12-2. The average control-adjusted survival approximated a statistical threshold of 80% in the Category 2 samples and decreased below 80% in Category 3 samples (Table 12-2).

Despite the fact that these analyses were performed on a relatively large database (1068 samples) collected along portions of all the coastlines of the US, concerns remained that these SQGs might not be applicable over a wider variety of estuarine conditions. Thus, the analyses reported by Long and MacDonald (1998) were repeated with 2 additional data sets from Biscayne Bay, Florida, and Pearl Harbor, Hawaii (Long et al. 2000; Jeff Grovhoug, US Navy, San Diego, CA, USA, unpublished data). The data from Biscayne Bay and Pearl Harbor generally indicated patterns of increasing toxicity in association with increasing chemical concentrations similar to those observed in the original "national" database (Table 12-2). Addition of the data from these 2 areas to the combined database had little effect upon either the incidence of toxicity or average survival. The percentages of samples indicating toxicity remained low (<10%), and average survival remained very high (92% to 93%) for samples in which all chemical concentrations were less than the ERLs or TELs or where mean SQG-Qs were <0.1. The incidence of toxicity increased steadily and average survival decreased incrementally as chemical concentrations increased relative to these SQGs. In the category with highest concentrations, the incidence of toxicity reached 73% to 83% and average survival dropped to 37% to 46%. Tables 12-3 and 12-4 contain results of analyses of the NSQS database based on either increasing mean ERM-Qs or multiple exceedances of individual ERMs. Similar to the analyses in Tables 12-1 and 12-2, toxicity increased with increasing contamination based on multiple exceedances of ERMs or based on increasing mean ERM-Qs (Tables 12-3 and 12-4).

Predictive ability of ERM-Qs in San Francisco Bay

Thompson et al. (1999) reported a significant negative correlation between mean ERM-Qs and percent survival in 10-d toxicity tests with *Eohaustorius estuarius* in sediments from San Francisco Bay. No samples were identified as toxic with mean ERM-Qs <0.11. The incidence of toxicity increased to about 50% at mean ERM-Qs of 0.11 to 0.18, increased to 89% with mean ERM-Qs >0.19, and peaked at 100% with mean ERM-Qs >0.22. In contrast, no significant correlation was observed between the bivalve embryo tests conducted with sediment elutriates and mean ERM-Qs. The composition of the mixtures of chemicals that were correlated with amphipod survival often differed among the regions of the bay. The authors concluded that there was evidence that the additive effects of numerous substances in the sediments contributed to toxicity observed in the amphipod tests.

Predictive ability of revised SQG-Qs in a national marine and estuarine database

The mean quotients derived for evaluation in Table 12-2 were based upon the normalization of chemical concentrations with all of the available SQGs (e.g., either all ERMs or all PELs for individual chemicals). It is reasonable to assume that some chemicals were less important contributors to toxicity than others in the mixtures and that the chemical nature of the mixtures would differ among study areas from which data were compiled. As a result, the toxicity would differ among regions. By compiling data from many regions, the exposure–response relationships apparent in individual areas would tend to be masked. If this were true, the agreement between the predicted toxicity and observed toxicity should improve if the data from the nonconcordant chemicals were ignored when the mean quotients were calculated (Fairey et al. 2001). By iteratively adding and eliminating chemicals from consideration and merging SQGs from several approaches, the makeup of the mean quotients that provided the best exposure–response relationship was identified by Fairey et al. (2001). Matching sediment chemistry and 10-d amphipod survival data were compiled from marine and estuarine surveys conducted in California coastal bays (n = 605), the NOAA SEDTOX database (n = 2753) from estuaries and marine bays nationwide, and NOAA data from Biscayne Bay, Florida (n = 226). Samples were classified as toxic when survival was significantly less than in negative controls and less than the respective MSD for each test.

With mean quotients derived with data from 9 chemicals (5 trace metals, 4 organics), toxicity occurred in only 6% of samples in which the mean quotients were <0.1 (Table 12-5). Average survival of amphipods in these samples was 95%. These results are slight improvements over the results summarized in Table 12-2 for the mean ERM-Qs and mean PEL-Qs. The incidence of toxicity increased steadily to 91%, and average survival decreased steadily to 31% over 10 ranges of mean quotients, again demonstrating an improvement in the chemistry and toxicity relationships summarized in Table 12-2 for the mean ERM-Qs and mean PEL-Qs. Results of these analyses indicate that predictive ability of the mean quotients was improved by selecting the SQGs used to derive the mean quotients.

Table 12-5 Incidence of toxicity in marine amphipod tests and survival within 10 ranges of mean SQG-Qs[a]

Ranges in mean SQG quotients	Nr. samples	Significantly toxic (%)	Average survival (%)
0–0.1	804	6	95
0.1–0.25	490	19	85
0.25–0.5	304	32	77
0.5–0.75	139	32	77
0.75–1.0	90	47	74
1.0–1.25	58	59	63
1.25–1.5	26	77	60
1.5–2.0	23	70	62
2.0–2.5	20	80	54
>2.5	33	91	31

[a] Reprinted with permission from Fairey et al. 2001. Copyright Society of Environmental Toxicology and Chemistry (SETAC)

Predictive ability of ERM-Qs in Southern California Bight

The data presented in Table 12-6 were collected during 2 studies: Bight 98, which was a regional assessment of sediment quality in the Southern California Bight (SCB, coastal region between Point Conception and the border with Mexico) and a compilation of dredge material characterization data from the Los Angeles area (Bay et al. 2000). The Bight 98 study measured sediment toxicity at 241 stations using 2 marine tests. These stations included offshore areas, bays, and harbors. Acute sediment toxicity was measured using the standard 10-d amphipod (*Eohaustorius estuarius*) survival test and sublethal toxicity was evaluated by measuring the reduction in phytoplankton (*Gonyaulax polyhedra*) luminescence following a 24-h exposure to a sediment elutriate (Bay et al. 2000). The classification of a sample as toxic was based upon 2 criteria: 1) the presence of a statistically significant reduction in survival or luminescence and 2) a response that exceeded a species-specific threshold (i.e., <80% control survival or <15% control luminescence). In the dredge study, acute toxicity tests with amphipods were conducted, but the species varied among studies. The state of California has not established SQGs for assessment or cleanup activities, although several local regulatory agencies are developing SQGs for specific applications. For the analysis presented in Table 12-6,

Table 12-6 Toxicity in marine sediments within ranges of mean ERM-Qs, based on the results of 10-d amphipod or 24-h dinoflagellate tests[a]

		Incidence of toxicity (%)			
Priority category	Mean ERM-Q[b]	SCB coastal	SCB bay or harbor	SCB all stations	Los Angeles dredge sites
10-d amphipod survival test[c]					
High	>1.5	0 (2)	0 (0)	0 (2)	100 (2)
Medium-high	0.51–1.5	0 (1)	100 (2)	66.7 (3)	75 (4)
Medium-low	0.11–0.5	4.5 (22)	33.8 (71)	26.9 (93)	57.6 (33)
Low	<0.1	10 (100)	11.7 (43)	10.5 (143)	30.9 (55)
24-h dinoflagellate (*Gonyaulax polyedra*) test					
High	>1.5	0 (2)	0 (0)	0 (2)	
Medium-high	0.51–1.5	0 (1)	0 (2)	0 (3)	
Medium-low	0.11–0.5	20 (16)	25.4 (67)	24.1 (83)	
Low	<0.1	4.5 (44)	37.5 (40)	20.2 (84)	

[a] Nr. stations given in parenthesis; Bay et al. 2002.

[b] ERM-Q = Effects range median quotient

[c] *Eohaustorius estuarius* was the test species for the SCB studies. Data for the dredge sites were obtained using *E. estuarius, Ampelisca abdita, Rhepoxynius abronius*, or *Grandidierella japonica.*

the mean ERM-Q was used to evaluate the predictive ability of SQGs (Long and MacDonald 1998). The incidence of toxicity to amphipods or dinoflagellates within the following ranges of mean ERM-Qs was calculated: >1.5, 0.51 to 1.5, 0.11 to 0.5, <0.1. The data are summarized separately for each study and toxicity test method, and also by 2 spatial strata within the Bight 98 study area: offshore coastal areas (10 to 150 m depth) and inshore bays and harbors (>6 m depth).

The results for all SCB stations indicate that the incidence of amphipod toxicity is low (10%) when concentrations of contaminants are below levels of concern (mean ERM-Q <0.1). The incidence of amphipod toxicity increased with increasing contamination (mean ERM-Q of 0.11 to 0.5), as would be expected from the results of others (Long and MacDonald 1998; Long, Field, MacDonald 1998). Higher contamination levels (mean ERM-Q >0.5) were present at only 3 sites, an insufficient number to draw conclusions regarding the relationship with sediment toxicity. A similar trend of increasing toxicity with elevated mean ERM-Q was also present when the subset of data from bay or harbor areas was examined. Offshore coastal areas did not show a similar relationship. However, the lack of a relationship between toxicity and ERM-Q in coastal samples is probably due to the overall low incidence of toxicity present in each category (≤10%). Amphipod toxicity data from Los Angeles dredge material studies also exhibited a pattern of increasing toxicity with elevated mean ERM-Q. The incidence of toxicity ranged from 31% in the lowest contamination group to 100% in the highest contamination group (ERM-Q > 1.5). Changes in mean ERM-Q did not correspond to the incidence of sublethal toxicity, as measured by the dinoflagellate luminescence test. The incidence of sublethal toxicity at slightly elevated contamination levels (mean ERM-Q of 0.11 to 0.5) was similar to or slightly greater than the incidence of toxicity at the least contaminated sites. The lack of correspondence between sublethal toxicity and the mean ERM-Q may be related to 2 factors: changes in contaminant concentration or bioavailability as a result of the elutriate preparation (ERM-Qs were based on whole sediment concentrations) or the sensitivity of the sublethal test to noncontaminant sediment factors (e.g., dissolved organics, grain size, major ion composition).

Predictive ability of apparent effects threshold in Puget Sound

The AET approach was originally developed to help guide remediation of contaminated tidal flats at the Commencement Bay, Washington, Superfund site (Barrick et al. 1985) and was developed further to evaluate dredged material in Puget Sound (Beller et al. 1986). These AETs were subsequently updated and evaluated for their ability to predict sediment samples that exhibit a significant adverse effect ("hit") or lack thereof ("no hit"; Barrick et al. 1988). The AETs derived from measures of amphipod mortality and benthic abundance from synoptic data collected from 8 Puget Sound embayments were used to predict analogous effects observed in sediment samples from 3 separate bays (Barrick et al. 1988). The 2 independent sets of data were then merged, and 4 types of AETs representing different indicators of adverse effects were calculated (AETs for amphipod mortality, abundance of major taxa of benthos, Microtox, and oyster larval development).

The predictive ability of these AETs to identify "hits" and "no hits" in an independent data set was then evaluated using a different method. This approach involved an iterative process of randomly removing data sets from the database, calculating AETs from the remaining data sets, evaluating predictive ability by using the AETs to predict the missing effects data, replacing the missing data, and repeating the process. Barrick et al. (1988) also evaluated whether or not AETs normalized to organic carbon were better predictors of adverse effects than those on the basis of dry weight. Finally, the predictive ability of the lowest AET (LAET) of these 4 AETs was evaluated based on the recommendation by USEPA (1989) to evaluate a battery of effects endpoints.

Results of the independent evaluation of predictive ability, perhaps the first of its kind, were reported in terms of sensitivity, efficiency, and overall reliability. Sensitivity was defined as the ability to detect environmental problems (i.e., correctly predict hit samples). Efficiency was defined as the ability to predict "hits" accurately (i.e., the number of samples correctly predicted to be hits divided by the total number of samples predicted to be "hits"). Overall reliability was defined as the ability to accurately predict hits and no hits.

For the 1986 amphipod AETs, predictive ability averaged 66% for sensitivity, 56% for efficiency, and 72% overall reliability (Table 12-7). Comparable predictive ability for the 1986 benthic AETs averaged 82% for sensitivity, 74% for efficiency, and 80% for overall reliability. The iterative method for independently evaluating the predictive ability of 1988 AETs indicated that AETs for Microtox and oysters were the most sensitive (93% and 88%) but exhibited the lowest efficiency (37% and 61%) and lowest overall reliability (28% and 31%). The 1988 benthic AETs exhibited moderate sensitivity, efficiency, and overall reliability (71% to 75%). The 1988 amphipod AETs were least sensitive (57%) but exhibited moderate efficiency and overall reliability (67% and 74%). Calculating AETs normalized to organic carbon had a minimal effect or lowered predictive ability of the AETs. The LAETs accurately predicted >90% of those samples exhibiting at least 1 type of significant effect (Barrick

Table 12-7 Summary of some early measures of the predictive ability for some groups of Puget Sound AET values (dry weight)[a]

AET group	Sensitivity (%)	Efficiency (%)	Overall reliability (%)
1986 amphipod[b]	66	56	72
1986 benthic[b]	82	74	80
1988 amphipod[c]	57	67	74
1988 benthic[c]	75	72	71
1988 Microtox[c]	93	37	28
1988 oyster[c]	88	61	31
1988 LAETs[d]	>90	NA	NA

[a] Barrick et al. 1988
[b] Ability of AET values derived from data collected from 8 bays in Puget Sound to predict analogous effects observed in 3 separate different bays.
[c] Iterative method to "independently" evaluate predictive ability (see main text).
[d] Group of lowest of 4 AET values used to predict significant adverse effect.
NA = not available or efficiency of 100% (by definition) limits utility of these values.

et al. 1989). This accuracy is most likely due to the LAET approach that addresses many chemicals of potential concern, possible interactions among these chemicals, several indicators of effects, different organism sensitivities, and different test endpoints.

Regional AETs thus appear to be moderately to highly predictive of various direct and surrogate measures of adverse effects in the benthic macroinvertebrate community, depending on several factors.

These factors include

1) the size of the synoptic database used to derive the AETs;
2) whether AETs for a single chemical, class of chemicals (e.g., metals versus organics) or robust list of chemicals of concern are evaluated; and
3) the number of groups of AETs that are evaluated.

Limited synoptic data generally yield AETs with relatively low predictive ability (Barrick et al. 1988). The predictive ability of a single AET for a single indicator of significant adverse effects was expected to be low and was rarely found to exceed 5%. Similarly, the predictive ability of AETs for a single class of chemicals and biological indicator was also not expected to be substantial. For example, the 1988 AET for benthos and trace metals correctly predicted 32% effects on benthos (Figure 12-1). No one chemical class could account for half the total sensitivity attained by using all classes of chemicals (Barrick et al. 1988).

USEPA (1989) concluded that AETs represent a range of effects that are most predictive when developed from a large database with wide ranges of chemical concentrations and a wide diversity of measured contaminants (as demon-

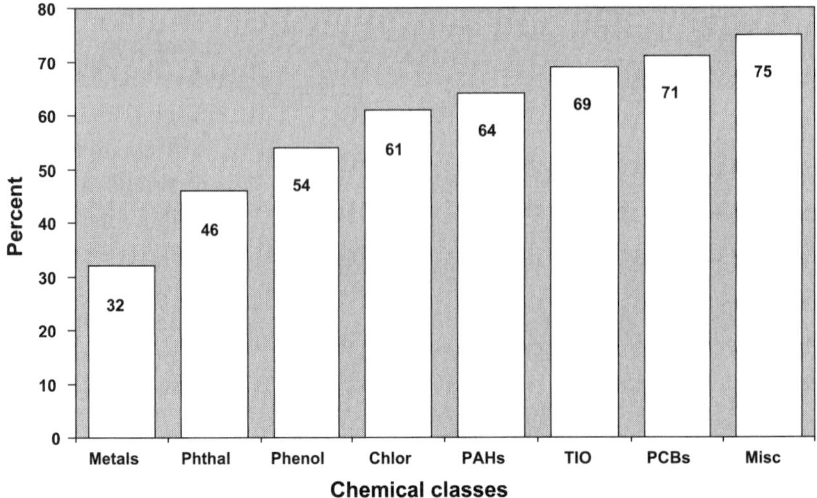

Figure 12-1 AETs based on significant decrease in major taxa of benthic infauna: Cumulative sensitivity and sensitivity by class of chemical AETs. "Chlor" represents chlorinated compounds (i.e., chlorobenzenes), and TIO represents tentatively identified organic compounds (i.e., substitituted PAHs; from Barrick et al. 1988)

strated by the sensitivity of the LAETs). As a result of these analyses, LAETs are now used as a basis for screening-level guidelines (SLs) in the regional dredged material management program (PSDDA 1988). The LAETs are also incorporated into the sediment source control and cleanup standards adopted by the state of Washington (Chapter 173-204 of the Washington State Administrative Code). The latter are considered by the USEPA to be federally enforceable water quality criteria and have been used in Washington to help define cleanup site objectives and boundaries, place water bodies on the Clean Water Act Section 303d list, and establish USEPA-approved sediment total maximum daily loads (TMDLs).

Predictive ability of logistic regression models

Individual chemical logistic regression models were developed for 37 chemicals of potential concern in contaminated sediments to predict the probability of toxicity, based on 10-d survival toxicity tests with the marine amphipods *Ampelisca abdita* and *Rhepoxynius abronius* (Field et al. 1999, 2002). These models were derived from a large database of matching sediment chemistry and toxicity data (n = 3223), which includes contaminant gradients from a variety of habitats in coastal North America. Samples were classified as toxic based on statistical comparison to controls and <90% survival. The individual chemical regression models were combined into a single model to estimate the probability of toxicity for a sample, using the maximum (P-Max model) probability predicted from the chemicals analyzed in a sample (Figure 12-2). The average predicted probability of toxicity within probability quartiles closely matched the incidence of toxicity within the same ranges, demonstrating the overall reliability of the P-Max model for the database that was used to derive the model (Table 12-8). The magnitude of the toxic effect (decreased survival) in the amphipod test increased as the predicted probability of toxicity increased.

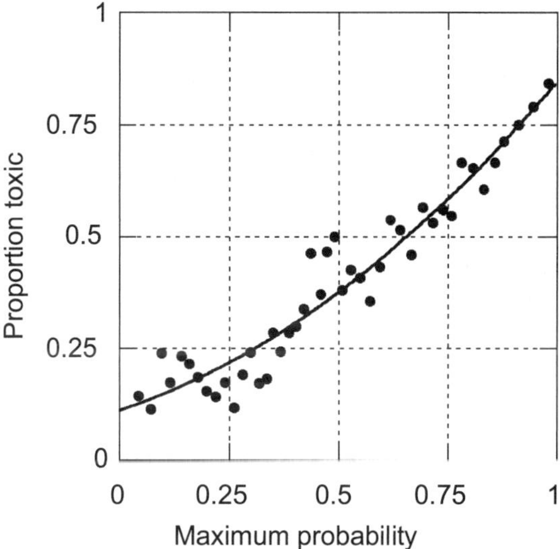

Figure 12-2 Maximum (P_Max) probability model (curve) and probability interval plot (points). Each point represents the median sample probability of a minimum of 50 individual samples within the interval and the proportion of samples toxic within the interval (Field et al. 1999, 2002).

Laboratory toxicity tests with freshwater sediments

Predictive ability of SQG-Qs in laboratory toxicity tests with amphipods (Hyalella azteca) *and chironomids* (Chironomus spp.)

Ingersoll et al. (2001) evaluated the predictive ability of the consensus-based probable effect concentrations (PECs) developed by MacDonald, Ingersoll, and Berger (2000). The database developed by MacDonald, Ingersoll, and Berger (2000) was expanded to include a total of 1657 samples with matching sediment toxicity and chemistry data (USEPA 2000b). The majority of the data from these reports were for 10- to 28-d toxicity tests with the amphipod *Hyalella azteca* or 10- to 14-d toxicity tests with the midges *Chironomus tentans* or *C. riparius*. The endpoints measured in these toxicity tests primarily included survival or growth of test organisms at the end of the sediment exposures. Mean PEC-Qs were calculated to provide an overall measure of chemical contamination and to support an evaluation of the combined ef-

Table 12-8 The average predicted probability toxicity, proportion of samples classified as toxic, and the control-adjusted survival in toxic samples within probability quartiles defined by the P-Max model (Field et al. 1999, 2002).

Probability of toxicity from P-Max Model	Average predicted probability of toxicity	Proportion of samples toxic	Percent control-adjusted survival in toxic samples	Total nr. samples
≤0.25	0.19	0.18	78.4	911
>0.25–0.50	0.36	0.37	71.5	1280
>0.5–0.75	0.61	0.58	61.9	680
>0.75	0.80	0.86	50.1	186

fects of multiple contaminants in sediments. Mean quotients were calculated by equally weighting the contribution of metals, total PCBs, and total PAHs (Ingersoll et al. 2001).

An increase in the incidence of toxicity was observed with an increase in mean quotients in all 3 tests (Figure 12-3). A consistent increase in the toxicity in all 3 tests (short-term midge and amphipod tests and long-term amphipod tests) occurred at a mean quotient >0.5; however, the overall incidence of toxicity was greater in the long-term amphipod test compared to the short-term tests with amphipods or midge. The longer-term tests in which survival and growth are measured tend to be more sensitive than the shorter-term tests, with acute-to-chronic ratios on the order of 6 indicated for *H. azteca* (Ingersoll et al. 2001). A similar relationship has been observed in 10- and 42-d water-only exposures of *H. azteca* to Cd, fluoranthene, or DDD (CG Ingersoll, unpublished data).

Different patterns were observed among the various procedures used to calculate mean quotients. For example, in the long-term test with *H. azteca*, a relatively abrupt increase in toxicity was associated with elevated PCBs alone or with elevated PAHs alone, compared to the pattern of a gradual increase in toxicity observed with quotients calculated using a combination of metals, PAHs, and PCBs (Table 12-9). These analyses indicate that the different patterns in toxicity may be the result of unique chemical signals associated with individual contaminants in samples. While mean quotients can be used to classify samples as toxic or nontoxic, individual quotients might be useful in helping to identify substances that may cause or substantially contribute to the observed toxicity.

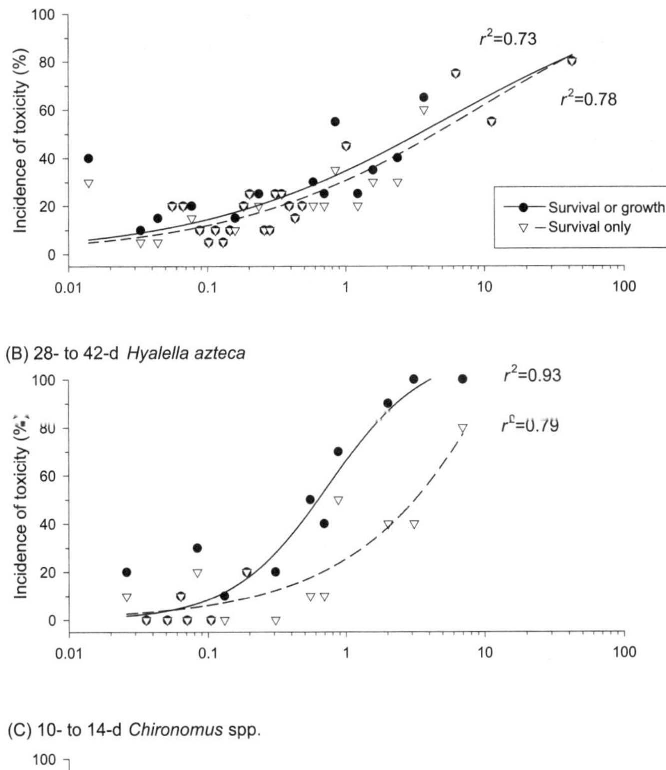

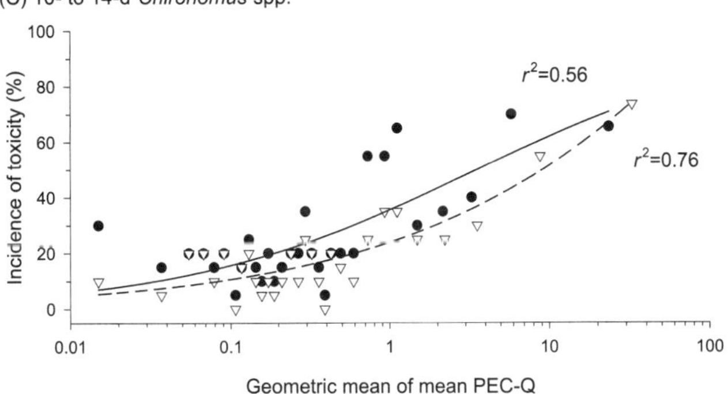

Figure 12-3 Relationship between geometric mean of mean PEC-Qs and incidence of toxicity in 3 tests, based on survival or growth or based on survival alone (USEPA 2000b)

Predictive ability of SQG-Qs at the St. Louis River Area of Concern

Sediment quality targets (SQTs) for the protection of sediment-dwelling organisms have been established for the St. Louis River Area of Concern (AOC) in Minnesota, 1 of 42 current AOCs in the Great Lakes basin (Crane et al.

Table 12-9 Incidence of sediment toxicity within ranges of PEC quotients (calculated using various approaches) for freshwater tests based on survival or growth[a]

PEC quotient	Incidence of toxicity (%) based on mean PEC quotients (nr. samples in parentheses)						Total nr. samples
	<0.1	0.1 to <0.5	0.5 to <1.0	1.0 to <5.0	>1.0	>5.0	
Hyalella azteca 10- to 14-d tests							
Mean: metals	20 (104)	19 (354)	39 (72)	63 (72)	63 (51)	62 (21)	623
Total PAHs	20 (178)	26 (160)	23 (47)	37 (54)	57 (103)	80 (49)	488
Total PCBs	26 (109)	21 (91)	46 (49)	51 (47)	60 (77)	73 (30)	326
Mean PEC-Q	18 (147)	16 (288)	37 (73)	41 (92)	54 (162)	71 (70)	670
H. azteca 28- to 42-d tests							
Mean: metals	8 (50)	20 (51)	62 (37)	NC[b]	86 (22)	NC	160
Total PAHs	17 (98)	61 (46)	56 (9)	NC	86 (7)	NC	160
Total PCBs	4 (26)	6 (35)	17 (12)	NC	97 (36)	NC	109
Mean PEC-Q	10 (63)	13 (39)	56 (27)	NC	97 (31)	NC	160
Chironomus spp. 10- to 14-d tests							
Mean: metals	22 (88)	23 (338)	25 (89)	39 (61)	44 (84)	57 (23)	599
Total PAHs	14 (178)	33 (133)	57 (28)	67 (33)	74 (53)	85 (20)	392
Total PCBs	46 (91)	22 (49)	36 (33)	53 (51)	58 (74)	70 (23)	247
Mean PEC-Q	20 (121)	17 (313)	43 (63)	43 (88)	52 (132)	68 (44)	629

[a] Sources: USEPA 2000b, Ingersoll et al. 2001
[b] NC = not calculated

2000, 2002). Level I SQTs were established primarily as TECs, and Level II SQTs were established primarily as PECs developed by MacDonald, Ingersoll, and Berger (2000). The predictive ability of the Level II SQT quotients (i.e., mean PEC-Qs) was evaluated using the incidence of toxicity information contained in the matching sediment chemistry and toxicity database for the St. Louis River AOC (Crane et al. 2000, 2002). This evaluation involved determination of the incidence of toxicity to amphipod *H. azteca,* and midge *C. tentans,* within 5 ranges of mean PEC-Qs: ≤0.1, >0.1 to ≤0.5, >0.5 to ≤1.0, >1.0 to ≤5.0, and >5.0 (Crane et al. 2000, 2002). Overall toxicity was assigned to a sediment sample if toxicity was observed for 1 or more metrics in 10-d sediment toxicity tests with either amphipods or midges (endpoints of survival and growth for both tests). The incidence of toxicity observed in these tests was also compared to that for other geographic areas in the Great Lakes region and in North America for 10- to 14-d amphipod (*H. azteca*) and 10- to 14-d midge (*C. tentans* or *C. riparius*) toxicity tests (USEPA 2000b; Ingersoll et al. 2001).

The results of the predictive ability evaluation indicate that the incidence of acute toxicity to amphipods (*H. azteca*) and midges (*C. tentans*), measured in 10-d toxicity tests, was only 7% when the concentrations of sediment-associated contaminants in the St. Louis River AOC were low (i.e., as indicated by mean PEC-Qs of ≤0.1; Crane et al. 2000, 2002). Importantly, the incidence of sediment toxicity in the St. Louis River AOC sediments generally increased with increasing contaminant concentrations (Table 12-10). In particular, a high incidence of toxicity (i.e., ≥75%) was observed at mean PEC-Qs >5.0 for 10-d amphipod and midge toxicity tests (Crane et al. 2000, 2002). The predictive ability of the mean PEC-Qs was generally similar across geographic areas (Table 12-10). However, these comparisons were most appropriate for the 2 lowest mean PEC-Q ranges (i.e., ≤0.1 and >0.1 to ≤0.5) because the minimum data requirements (i.e., 20 samples) were met for most geographic areas. The results of these predictive ability evaluations indicate that, collectively, the mean PEC-Qs provide a reliable basis for classifying sediments as toxic or not toxic in the St. Louis River AOC compared to other locations in North America (Ingersoll et al. 2001; Crane et al. 2002).

Predictive ability of SQG-Qs in the Grand Calumet River and Indiana Harbor Canal, IN

Ingersoll et al. (2002) conducted 10-d sediment exposures with the amphipod *H. azteca* using 30 sediment samples collected from the Grand Calumet River and Indiana Harbor Canal located in northwestern Indiana. Toxic effects on amphipod survival were observed in 60% of the samples from the assessment

Table 12-10 Incidence of toxicity in freshwater sediments within ranges of mean PEC-Qs, based on results of 10- to 14-d amphipod or midge tests[a]

Mean PEC-Q range	Incidence of toxicity (%)[b]					
	St. Louis River AOC	Other Great Lakes sites[c]	Other North American sites[c]	Non-Great Lakes sites	All Great Lakes sites	All North American sites
Hyalella azteca 10- to 14-d tests						
≤ 0.1	7% (44)	50% (6)[d]	24% (96)	22% (90)	12% (50)	18% (140)
>0.1 to ≤0.5	11% (80)	25% (56)	19% (198)	16% (142)	17% (136)	16% (278)
>0.5 to ≤1.0	30% (10)	52% (27)	37% (62)	26% (35)	46% (37)	36% (72)
>1.0 to ≤5.0	27% (11)	68% (37)	42% (79)	19% (42)	58% (48)	40% (90)
>5.0	75% (4)	77% (30)	70% (63)	64% (33)	76% (34)	70% (67)
Chironomus tentans 10- to 14-d tests						
≤ 0.1	7% (46)	19% (52)	28% (72)	50% (20)[e]	13% (98)	19% (118)
>0.1 to ≤0.5	12% (74)	23% (135)	18% (222)	12% (87)	19% (209)	17% (296)
>0.5 to ≤1.0	20% (10)	59% (34)	46% (52)	22% (18)	50% (44)	42% (62)
>1.0 to ≤5.0	36% (11)	47% (57)	43% (75)	28% (18)	46% (68)	42% (86)
>5.0	100% (5)	63% (30)	61% (36)	50% (6)	68% (35)	66% (41)

[a] Reprinted with permission from Crane et al. 2002. Copyright Springer-Verlag GmbH.

[b] Excluded 2 sites (Breneman et al. 2000) due to incomplete sediment chemistry data (i.e., PAHs, PCBs) for these known contaminated areas.

[c] Excluded the St. Louis River AOC data.

[d] All toxic samples came from Sheboygan River, WI (USEPA 2000b; Ingersoll et al. 2001). The Sheboygan River samples had high control survival (i.e., 97%), resulting in samples with survival as high as 88% being designated toxic.

[e] Most of toxic samples came from a study of Tennessee portion of lower Mississippi River (USEPA 2000b; Ingersoll et al. 2001); data set was not examined to determine possible factors contributing to sediment toxicity.

area. Relationships between sediment chemistry and toxicity were evaluated by calculating individual PEC-Qs based on metals, PCBs, and PAHs, or by calculating a combined mean PEC-Qs based on concentrations of metals, PCBs, and PAHs.

Samples that were toxic tended to have the highest concentrations of metals, PAHs, and PCBs. Figure 12-4 illustrates the relationship between survival of amphipods and the average PEC-Q for metals (Figure 12-4a), the PEC-Q for total PCBs (Figure 12-4b), and the PEC-Q for total PAHs (Figure 12-4c). With increasing quotients for metals, PCBs, or PAHs, there was an increasing degree of toxicity observed in the amphipod test. For metals, total PCBs, and total PAHs, samples with a PEC-Q within an individual chemical class at or above 1.0 to 2.0 were frequently toxic to amphipods. However, there were some samples with quotients ranging from 0.1 to 1.0 that were also toxic to amphipods (Figure 12-4). In these instances, multiple chemicals may be contributing to the observed toxicity. Therefore, the mean PEC-Q based on

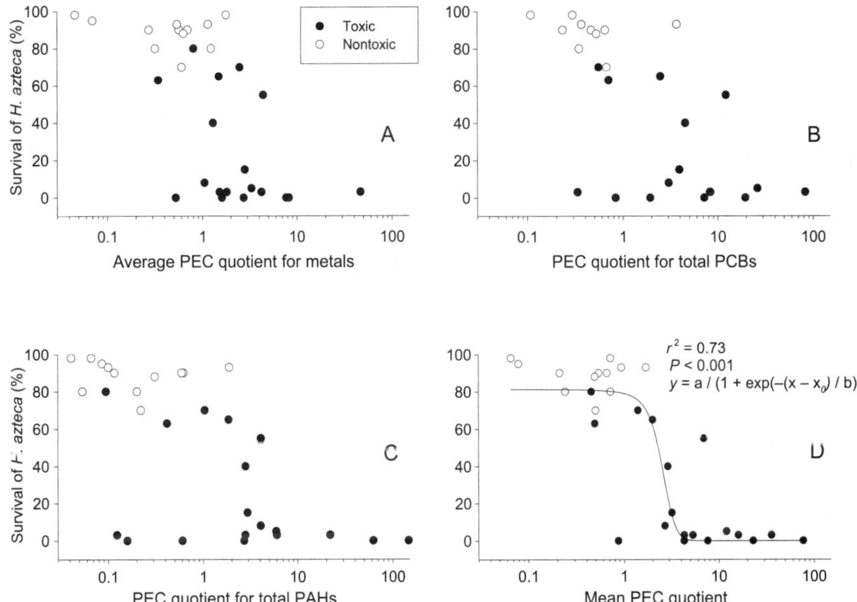

Figure 12-4 Relationship between probable effect concentration (PEC) quotient for metals, PCBs, PAHs, or mean PEC quotient and response of 28- to 48-d *H. azteca* in 10-d sediment toxicity tests (Reprinted with permission from Ingersoll et al. 2002. Copyright Springer-Verlag GmbH.)

metals, total PCBs, and total PAHs was calculated for each sample in order to equally weight the contribution of metals, PCBs, and PAHs in the evaluation of sediment chemistry and toxicity.

Figure 12-4d illustrates the relationship between survival of amphipods and mean PEC-Q in the sediment samples. All of the samples with a mean quotient >2.0 were toxic, none of the samples with a mean quotient <0.3 were toxic, and 38% of the samples between a mean quotient of 0.3 to 2.0 were toxic. Plotting the data based on mean PEC-Q reduced the variability in toxicity observed at quotients <1.0 based on metals, PAHs, or PCBs alone. A 50% reduction in survival of amphipods was estimated at a mean quotient of 2.3. Results of these analyses are similar to the findings reported in Ingersoll et al. (2001) where a 50% incidence in toxicity was observed in a database for sediment tests with *H. azteca* at a mean quotient of 3.4 in 10-d exposures and at a mean quotient of 0.63 in 28-d exposures (Figure 12-3).

Predictive ability of SQGs on a regional basis within North America

Ingersoll et al. (2001) evaluated the predictive ability of mean PEC-Qs developed by MacDonald, Ingersoll, and Berger (2000) on a regional basis within a larger database developed in USEPA (2000b). The goal of this evaluation was to determine if there were differences in the predictive ability of the PECs across the entire database, compared to various geographic areas within the database, such as all of the samples from the Great Lakes or all of the samples from an area within a Great Lake, such as Indiana Harbor or Waukegan Harbor located within Lake Michigan. Evaluations were made for 10- to 14-d exposures with *H. azteca* and midge and 28- to 42-d exposures with *H. azteca* (USEPA 2000b; Ingersoll et al. 2001). Table 12-10 presents a subset of these analyses for the short-term exposures with amphipods and midge. Table 12-11 presents results of these analyses for the long-term exposures with amphipods.

In the 28- to 42-d exposures with *H. azteca*, the incidence of toxicity increased with an increase in the mean quotient within most of the regions, basins, and areas (Table 12-11). The incidence of toxicity for samples from each of the Great Lakes and within areas of each Great Lake was relatively consistent with the overall pattern of toxicity in the entire database. Similarly, the incidence of toxicity for samples from the Great Lakes was consistent with the incidence in toxicity for samples from non-Great Lakes areas. Therefore, no one area seemed to influence the overall incidence of toxicity. The incidence of toxicity for the Great Lakes samples was 13% at mean quotients of <0.5, 71% at mean quotients of 0.5 to <1.0, and 100% at mean quotients of >1.0. This pattern is consistent with incidence of toxicity for the entire database of 17% at mean quotients of <0.5, 56% at mean quotients of 0.5 to <1.0, and 97% at mean quotients of >1.0 (Table 12-11). The results of the analyses in Tables 12-10 and 12-11 indicate that mean PEC-Qs can be used to reliably predict toxicity of sediments on both a regional and national basis.

The predictive ability of mean PEC-Qs was evaluated for 3 separate databases generated for

1) the southeastern US (643 samples; MacDonald et al. 2003; Figure 12-5a),
2) the Calcasieu estuary in Louisiana (128 samples; MacDonald, Ingersoll, Moore, et al. (2002; Figure 12-5b), and
3) the Columbia River Basin (147 samples; MacDonald, Ingersoll, Smorong, Lindskoog 2002; Figure 12-5c).

The incidence of toxicity observed in these 3 separate regions was compared to the incidence of toxicity observed in a national database described in Ingersoll et al. (2001; Figure 12-3). More specifically, the relationship between mean PEC-Qs (concentration) and incidence of toxicity (response) was evaluated by fitting logistic regression models to summarize data for each toxicity test endpoint (Figure 12-5). The regional concentration–response relationships

Table 12-11 Incidence of sediment toxicity within ranges of PEC-Q for *H. azteca* 28- to 42-d test in various geographic areas based on survival or growth[a,b]

Region	Basin	Area	Toxicity (%)	Incidence of toxicity (%) based on mean PEC-Qs					
				<0.1	0.1 to <0.5	0.5 to <1.0	1.0 to <5.0	>1.0	>5.0
North America		All (with controls)	35 (160)	10 (63)	13 (39)	56 (27)	96 (25)	97 (31)	100 (6)
		All (w/o controls)	37 (151)	10 (63)	17 (30)	56 (27)	96 (25)	97 (31)	100 (6)
Great Lakes		All	79 (42)	(0)	13 (8)	71 (7)	100 (22)	100 (27)	100 (5)
	Lake Erie	Buffalo River	80 (5)	(0)	(0)	67 (3)	100 (2)	100 (2)	(0)
	Lake Huron	Saginaw River	30 (10)	(0)	0 (6)	50 (2)	100 (2)	100 (2)	(0)
	Lake Michigan	All	96 (27)	(0)	50 (2)	100 (2)	100 (18)	100 (23)	100 (5)
		Indiana Harbor	100 (4)	(0)	(0)	(0)	100 (1)	100 (4)	100 (3)
		Waukegan Harbor	96 (23)	(0)	50 (2)	100 (2)	100 (17)	100 (19)	100 (2)
Non-Great Lakes		All	21 (109)	10 (63)	18 (22)	50 (20)	67 (3)	75 (4)	100 (1)
Upper Mississippi River		All	4 (53)	4 (45)	0 (8)	(0)	(0)	(0)	(0)
State of Alabama	Gulf of Mexico	Mobile Bay	0 (5)	(0)	0 (4)	(0)	0 (1)	0 (1)	(0)
State of Maine	Kennebec River	Brunswick	25 (4)	0 (2)	50 (2)	(0)	(0)	(0)	(0)
State of Maryland	Chesapeake Bay	Aberdeen	75 (4)	100 (1)	0 (1)	100 (2)	(0)	(0)	(0)
State of Montana	Pend Oreille River	Clark Fork River	62 (13)	50 (2)	50 (2)	57 (7)	100 (1)	100 (2)	100 (1)
State of Texas		All	20 (15)	15 (13)	50 (2)	(0)	(0)	(0)	(0)
	Gulf of Mexico	Galveston Bay	60 (5)	50 (4)	100 (1)	(0)	(0)	(0)	(0)
		Rio Grande	0 (5)	0 (5)	(0)	(0)	(0)	(0)	(0)
		Trinity River	0 (4)	0 (3)	0 (1)	(0)	(0)	(0)	(0)
		Austin	0 (1)	0 (1)	(0)	(0)	(0)	(0)	(0)
Washington DC	Chesapeake Bay	Chesapeake Bay	40 (15)	(0)	33 (3)	36 (11)	100 (1)	100 (1)	(0)

[a] Nr. samples in parentheses
[b] USEPA 2000b

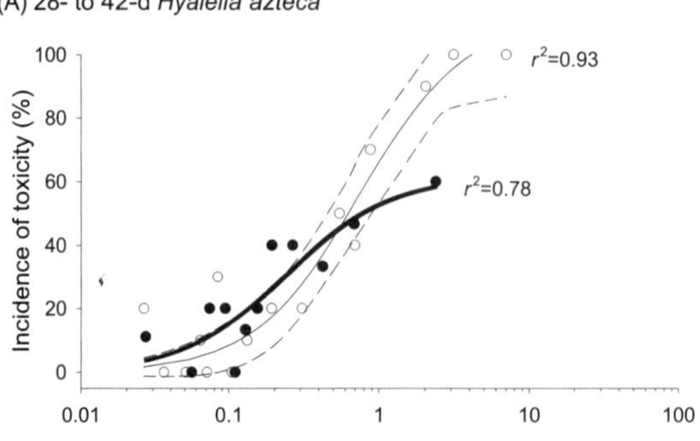

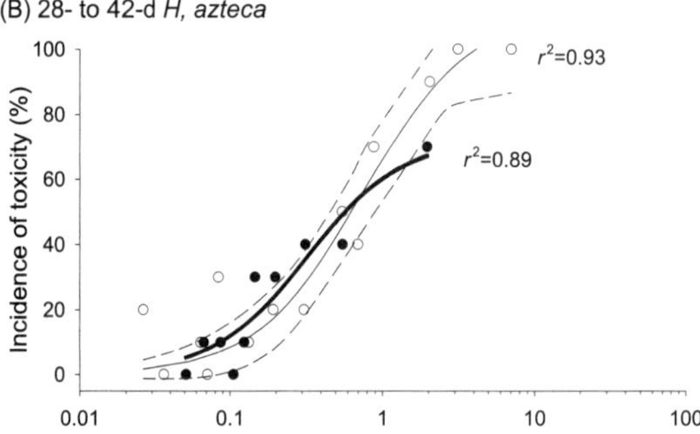

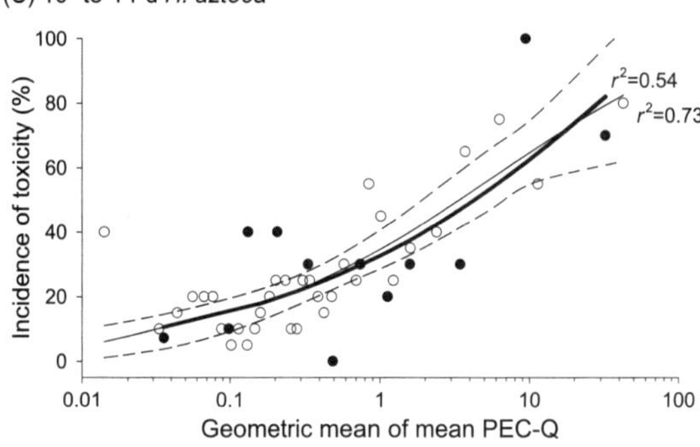

Figure 12-5 Relationship between geometric mean of mean PEC-Qs and incidence of toxicity in (A) southeastern US, (B) Calcasieu estuary, and (C) Columbia River Basin (MacDonald, Ingersoll, Moore, et al. 2002; MacDonald, Ingersoll, Smorong, Lindskoog 2002; MacDonald et al. 2003; bold lines and closed circles) compared to a national database (Ingersoll et al. 2001; open circles and lighter lines; see Figure 12-2).

were compared to the concentration–response curves in the national database using the results of 10- to 14-d or 28- to 42-d toxicity tests with *H. azteca*. The concentration–response curve generated using the matching sediment chemistry and toxicity data from the 3 regional databases generally fell within the 95% prediction intervals for the dose–response relationship that was generated using the information contained in the national database (Figure 12-5). Because systematic differences were not observed in the toxicity of regionally collected or nationally collected samples (Figure 12-5), MacDonald, Ingersoll, Moore, et al. (2002); MacDonald, Ingersoll, Smorong, and Lindskoog (2002); and MacDonald, et al. (2003) concluded that the mean PEC-Qs provide a reliable basis to accurately classify sediments as toxic or not toxic using sediment chemistry data alone.

Benthic communities assessments

Benthic–SQG relationships in southeastern US estuaries

Synoptic data on concentrations of sediment-associated chemical contaminants and benthic macroinfaunal community structure from 231 subtidal stations in southeastern US estuaries (Cape Henry, VA to St. Lucie Inlet, FL) were used to develop an empirical framework for evaluating risks of benthic community-level effects within different chemical-contaminant ranges (Hyland et al. 1999). Sediment contamination was evaluated using mean ERM-Qs or mean PEL-Qs (Long and MacDonald 1998). The computation of mean ERM-Q and PEL-Qs was based on 24 individual ERMs and 29 individual PELs (Hyland et al. 1999). Benthic condition was assessed using a multimetric Benthic Index of Biotic Integrity (BIBI), developed as a tool for classifying benthic samples as degraded versus nondegraded relative to optimum conditions expected under nonpolluted reference conditions by major habitat type (Van Dolah et al. 1999). Such an approach was used as a means to help account for the influence of natural controlling facts that appeared to have the strongest effect on the benthic distributions throughout the sampling area (i.e., salinity and latitudinal gradients). Cumulative percentages of stations with a degraded benthic community were plotted against ascending mean ERM-Q or PEL-Qs. The 10th and 50th percentile points from these cumulative-frequency plots were then used as critical points to define ranges in mean SQG-Qs associated with low, moderate, or high probabilities of detecting adverse benthic conditions.

The incidence of degraded benthos increased with increasing chemical concentration across the 3 mean SQG-Q ranges (Table 12-12). Predictive ability within these ranges was high. Within the low range (mean ERM-Q <0.020 and mean PEL-Qs <0.035), only 5% of the stations had degraded benthic assemblages. In contrast, within the high range (mean ERM-Qs >0.058 and mean PEL-Qs >0.096) the majority of stations (73% to 78%) had degraded benthic assemblages. These upper critical points, marking the beginning of the contaminant range associated with a high incidence of benthic impacts, are well below those required to elicit a similar incidence of sediment toxicity in laboratory survival tests conducted with single species (Table 12-2). For example, a similar high incidence of sediment toxicity in amphipod assays (74% to 76%) is associated with much higher contaminant levels: mean ERM-Qs >1.5 and mean PEL-Qs >2.3 (Table 12-2). The results of the benthic analyses are important for 2 related reasons: 1) in demonstrating that mean quotients provide a reliable tool for evaluating risks within different ranges of sediment contamination from mixtures of multiple chemicals present at varying con-

Table 12-12 Incidence of degraded v. nondegraded benthic condition within 3 ranges of sediment contamination expressed as mean ERM and PEL quotients (SQG-Qs; data from Hyland et al. 1999). Benthic condition was assessed using a multimetric index of biotic integrity (Van Dolah et al. 1999).

SQG type	Risk of benthic impacts	Mean SQG-Q range	Nr. stations	% with degraded benthos	% with healthy benthos
ERM	Low	≤0.020	135	5	95
	Moderate	>0.020 to 0.058	51	55	45
	High	>0.058	45	78	22
PEL	Low	≤0.035	131	5	95
	Moderate	>0.035 to 0.096	52	54	46
	High	>0.096	48	73	27

centrations and 2) in showing that condition of the ambient benthic community provides a reliable and sensitive indicator of the biological significance of sediment-associated stressors.

Similar results were obtained when this initial analysis of benthic-contaminant relationships in southeastern estuaries was extended to additional data sets from mid-Atlantic and Gulf of Mexico estuaries (Hyland et al. 2003). Based on data from all 3 regions combined, the incidence of degraded benthos was 74% at mean ERM-Qs >0.36 and 77% at mean PEL-Qs >0.78. Benthic condition in this expanded analysis was assessed using multimetric benthic indices that had been developed specifically for each region through prior independent research efforts (Hyland et al. 2003). Slightly different approaches were used to develop these indices, and each was based on a different combination of biological attributes. However, the indices were all similar in that multiple benthic attributes were combined into a single measure designed to maximize the ability to distinguish between degraded versus nondegraded benthic condition. Also, all 3 indices were developed with methods included to help account for biological variability attributable to key, natural controlling factors. Efforts to account for such variability focused on factors shown, through multivariate statistical analysis, to have the strongest influence on benthic distributions under relatively nondegraded (reference) conditions within each specific study area (e.g., as mentioned above, latitude and salinity in the case of the southeastern data set).

Benthic–SQG relationships in Gulf of Mexico estuaries

In a study of the benthos in Gulf of Mexico estuaries, Brown et al. (2000) showed that the trophic structure of infauna in samples in which ERLs were exceeded differed from that in randomly selected sites (Table 12-13). The data came from the Louisianian Province EMAP surveys conducted by USEPA. The proportion of the infauna composed of surface deposit feeders was considerably lower (12% to 17%) in samples with chemical concentrations greater than the ERLs than in samples chosen randomly (29%). Subsurface deposit feeders, primarily oligochaetes, showed the opposite pattern with higher abundance (42% to 64%) in sediments with chemical concentrations equal to or greater than the ERLs, compared to randomly selected sites (28%). The abundance of the latter trophic group was positively correlated with the concentrations of some toxicants. These samples had chemical concentrations that approximated or exceeded the ERLs for metals, PAHs, or DDT, but not the ERMs. Some of these samples were toxic in laboratory tests performed with amphipods. Therefore, results of this study are consistent with the observations of Hyland et al. (1999, 2003), indicating that shifts in macrobenthic community structure could be detected in samples with slight to moderate degrees of contamination that were not acutely toxic in 10-d laboratory toxicity tests. For example, Hyland et al. (2003) reported that samples from the Gulf of Mexico estuaries with mean ERM-Qs of 0.036 to 0.062 had an 86% incidence of degraded benthic communities. Similarly, 85% of the samples with mean PEL-Qs of 0.062 to 0.12 had degraded benthic communities. These ranges in mean quotients are usually associated with a lower incidence of sediment toxicity (<10 to 11%) in 10-d laboratory toxicity tests (Table 12-2).

Table 12-13 Macrobenthic trophic structure in estuarine sediments of northern Gulf of Mexico in randomly selected samples versus those in which chemical concentrations were nearly ≥ ERLs[a]

		% total abundance			
Trophic group	Random sites	Metals > ERLs	Metals > ERLs	PAHs > ERLS	DDT > ERLs
Surface deposit	29	15	17	12	14
Subsurface deposit	28	46	56	42	64
Filter	25	27	13	40	15
Carnivore	12	10	11	3	4
Omnivore	2	1	<1	1	1
Others	4	2	2	2	3

[a] Reprinted with permission from Brown et al. 2000. Copyright Estuarine Research Federation.

Again, the benthic community may have been responding to natural controlling factors in addition to contaminants in sediment (see "Benthic-SQG relationships in southeastern US estuaries," p 523), though methods included steps to help account for such influences.

Benthic–SQG relationships in Tampa Bay estuary

MacDonald, et al. (2004) evaluated the relationship between laboratory toxicity tests and benthic community surveys conducted with sediments collected from the Tampa Bay estuary. The toxicity tests included 10-d whole-sediment exposures with the amphipod *A. abdita* and 1-h porewater exposures with the sea urchin *Arbacia punctulata.* Effects on benthos were evaluated using the Shannon-Weiner diversity index. Concentrations of contaminants in sediments were evaluated using mean PEL-Qs (MacDonald et al. 1996; Long and MacDonald 1998). Relationships between effects and mean PEL-Qs were evaluated using logistic regression models. Based on data from 10-d toxicity tests with amphipods, a mean PEL-Q of 0.28 corresponded to a 10% probability of observing acute toxicity to sediment-dwelling organisms (i.e., P10 value); the P20 was 0.54, and the P50 was 1.3 for the 10-d toxicity tests with amphipods (Table 12-14). By comparison, the P10 was 0.05 and P50 was 0.34 for benthic community impairment. Sublethal toxicity responses, based on sea urchin fertilization toxicity tests conducted in 25% pore water, yielded a P10 of 0.10 and a P50 of 0.55, which correspond more closely to the values for benthic community impairment. The site-specific concentration–response relationships that were developed for Tampa Bay sediments were consistent with those that have been developed using other data sets for 10-d tests with amphipods (Table 12-2). The results of this study also indicate that benthic invertebrate community impairment can occur at relatively low levels of sediment contamination (similar to the findings of Hyland et al. 1999, 2003 and Long et al. 2002).

In this study, the P50 for benthic impairment was roughly a factor of 3.8 lower than the P50 for acute toxicity to amphipods. This is not surprising because the duration of exposure to contaminated sediments is likely to be >10 d for in situ benthic invertebrate communities. Ingersoll et al. (2001) compared the P50 values for acute (10 d) and chronic (28 d) toxicity to the amphipod *H. azteca* (measuring survival and growth) and concluded that acute-to-chronic ratios (ACRs) for that species were on the order of 6 ("Predictive ability of SQG-Qs in laboratory toxicity test...," p 513). Therefore, the relationship between acute toxicity and benthic impairment thresholds for marine inver-

Table 12-14 Summary of sediment quality targets generated using matching sediment chemistry and biological effects data from Tampa Bay[a]

		Sediment quality targets (expressed as mean PEL-Qs[b])				
Endpoint measured	n	P_{10}	P_{20}	P_{50}	P_{80}	Logistic model parameters[c]
Acute toxicity to amphipods (endpoint: survival)	96	0.28	0.54	1.3	2.0	a = 15983.8265; b = –1.0589; x_0 = 296.1373 (r^2 = 0.91; p = <0.001)
Chronic toxicity to sea urchins (endpoint: fertilization)						
100% pore water	141	NA[d]	NA[d]	NA[d]	0.38	a = 336.9413; b = –0.2043; x_0 = 116.0202 (r^2 = 0.90; p = <0.05)
50% pore water	141	NA[d]	0.04	0.24	0.83	a = 136.1706; b = –0.7167; x_0 = 0.5095 (r^2 = 0.89; p = <0.05)
25% pore water	141	0.10	0.19	0.55	1.2	a = 126.6869; b = –1.1760; x_0 = 0.7868 (r^2 = 0.90; p = <0.05)
Benthic community impairment (endpoint: Shannon-Weiner Diversity Index)	535	0.05	0.10	0.34	NA[d]	a = 74.2956; b = –1.3634; x_0 = 0.1989 (r^2 = 0.97; p = <0.0001)

[a] Reprinted with permission from MacDonald et al. 2004. Copyright Springer-Verlag GmbH.

[b] PEL-Q = probable effects level quotient from MacDonald et al. (1996)

[c] Logistic Model Equation: $y = a / [1 + (x / x_0)^b]$

[d] NA = not applicable; concentration–response data did not support calculation of *P* value.

tebrates in Tampa Bay sediments is similar to the ACR that was reported for freshwater invertebrates.

MacDonald et al. (2004) noted that benthic invertebrate communities can be adversely affected by both physical or chemical stressors. It was not possible to fully evaluate the effects of physical factors on the structure of benthic communities in Tampa Bay. For example, data on salinity or the levels of dissolved oxygen measured when sediment samples were collected provide only a snapshot of potential stressors on the benthic community. The concentration of total organic carbon in sediment was weakly correlated with benthic community response (r^2 = 0.23; p = <0.001); therefore, the concentration of organic carbon in sediments was not a good predictor of benthic impairment. Nevertheless, it is possible that effects thresholds would be higher if physical stressors on the benthic community could be more directly assessed.

Evaluation of laboratory versus benthic community assessments

The work by Hyland et al. (1999, 2003) has suggested that condition of the ambient benthic fauna can serve as a reliable and sensitive indicator of disturbance from chemical stressors at exposure levels that are well below those associated with a similar incidence of sediment toxicity in laboratory survival tests with single species. Long and MacDonald (1998) reported a relatively high probability of observing sediment toxicity in samples with mean ERM-Qs >1.5 and mean PEL-Qs >2.3 (Table 12-2). The incidence of highly toxic samples (i.e., mean survival in sample <80% of control, and significantly different at p <0.05) in these upper chemical contamination ranges was 75% for mean ERM-Qs and 76% for mean PEL-Qs. In comparison, Hyland et al. (1999) found a high incidence of degraded benthic condition in samples from southeastern estuaries at mean ERM-Qs >0.058 (78% of samples) and at mean PEL-Qs >0.096 (73% of samples). Similar results were obtained when this initial analysis of data from southeastern estuaries was extended to additional data sets from mid-Atlantic and Gulf of Mexico estuaries (Hyland et al. 2003). Based on data from all 3 regions combined, the incidence of degraded benthos was 74% at mean ERM-Qs >0.36 and 77% at mean PEL-Qs >0.78. Hyland et al. (2003) suggested that benthic community-level responses may be observed at lower chemical concentration ranges because these responses reflect the varying sensitivities of multiple species and life stages to longer-term exposures that may persist over several generations. Such measures can provide a sensitive indication of how the ambient benthic ecosystem responds to chemical exposure. Additional data sets are beginning to show consistent response patterns in other regions as well (Table 12-15). Additional analyses are needed, however, to further evaluate the relative influence of abiotic factors (i.e., grain size, total organic carbon, water depth, dissolved oxygen, salinity; Chapter 16) versus chemical gradients in sediment on the response of benthic invertebrates.

As noted by Hyland et al. (2003), such comparisons should not be misinterpreted to imply that laboratory toxicity data are invalid or of lesser value than the benthic assessment data. Rather, these comparisons help to show that the sensitivity to contaminants may vary with different receptors, endpoints, and exposure regimes. If other toxicological indicators (e.g., those based on sublethal endpoints, longer-term chronic exposures, earlier life stages) were evaluated in relation to mean SQG-Qs, one might find toxicity thresholds that are much closer to these benthic community-level responses, as illustrated in the previous section with Tampa Bay data and the similarities shown between benthic community and sublethal laboratory toxicity responses. As an-

Table 12-15 Examples of recent studies showing benthic community-level effects in relation to chemical contamination of sediments

Study region	Mean SQG quotient range	Effect	Reference
SE estuaries (*n* = 231)	ERM-Q > 0.058	78% incidence of degraded benthos	Hyland et al. 1999
SE estuaries (*n* = 290)	ERM-Q > 0.196	100% incidence of degraded benthos	Hyland et al. 2003
Mid-Atlantic estuaries (*n* = 511)	ERM-Q > 0.473	85% incidence of degraded benthos	Hyland et al. 2003
Gulf of Mexico estuaries (*n* = 588)	ERM-Q > 0.062	92% incidence of degraded benthos	Hyland et al. 2003
Puget Sound (*n* = 297)	ERM-Q > 0.1	Reduced H′ and SDI	Hyland et al. 2003
Biscayne Bay	ERM-Q > 0.1	Reduced species and arthropod abundance (amphipod toxicity at ERM-Q > 1.25)	Long et al. 2002
Tampa Bay	PEL-Q of 0.34	Regression model P50 for reduced H′ (amphipod toxicity P50=1.3)	MacDonald et al. 2004
Gulf of Mexico estuaries	Concentrations below ERLs and ERMs	Shift in benthic trophic structure (no amphipod toxicity	Brown et al. 2000
San Francisco Bay	> 100 µg/g-OC DDT	Loss of amphipods in field at DDT concentration 3-fold < concentration causing significant amphipod toxicity in 10-d lab tests	Swartz et al. 1994
Delaware Bay	Mean ERM-Q of about 0.1 to 0.2	Reduction in benthic diversity and densities of ambient amphipod populations.	ER Long, unpublished data

other example, Chung (1999) found 50% mortality in young juvenile clams (*Mercenaria mercenaria*) at a mean ERM-Q of about 0.4, which is similar to the mean ERM-Q, upper-range critical point of 0.36 reported by Hyland et al. (2003) for benthic data from mid-Atlantic, southeastern US, and Gulf of

Mexico estuaries combined. Also, acute–chronic ratios from results of aqueous-phase toxicity tests are typically about a factor of 10 for many chemicals, which is similar to the magnitude of variation that has been observed between benthic community-level responses and results of acute toxicity tests conducted with adult amphipods.

In a study of the toxicity thresholds of total DDT, Swartz et al. (1994) reported that adverse effects in the benthos occurred at concentrations below those that caused 50% mortality in laboratory tests (Table 12-16). The study focused upon DDT-contaminated sediments in an industrialized channel of San Francisco Bay. In tests of Lauritzen Canal sediments performed with *E. estuarius*, the LC50 for total DDT of 2500 µg/g-OC was similar to the LC50 of 2580 µg/g-OC for a site in Hunstville, AL and was similar to the LC50 of 1040 µg/g-OC for a site on the Continental Shelf off Palos Verdes, CA. In the Lauritzen Canal, toxicity to amphipods in 10-d exposures was reported at total DDT concentrations >300 µg/g-OC, which is about 3-fold above the concentrations associated with losses of resident amphipods in the field. Thus, these data indicate that benthos exposed in the field may be more sensitive than 10-d acute tests performed in the laboratory.

Table 12-16 Toxicity thresholds for total DDT in sediments derived from field-collected samples[a]

Biological test	Total DDT (µg/g-OC)
Field-derived 10-d LC50	
Lauritzen Canal, CA *(E. estuarius)*	2500
Huntsville, AL *(H. azteca)*	2580
Palos Verdes, CA *(R. abronius)*	1040
10-d sediment toxicity threshold	
R. abronius	>300
Benthic amphipods missing or rare	
Lauritzen Canal, CA	>100
Palos Verdes, CA	>200

[a] Reprinted with permission from Swartz et al. 1994. Copyright Society of Environmental Toxicology and Chemistry (SETAC).

To compare the degree of response in the laboratory tests to benthic infauna over a gradient in chemical contamination, data developed by NOAA in a survey of Biscayne Bay, FL were compiled by Long et al. (2002). These data from the central region of the bay indicated an inflection point in both amphipod survival (*A. abdita*, n = 83) and in benthic indices (n = 29) at a mean ERM-Q of about 0.25 (Table 12-17). Amphipod survival averaged about 35% in the 18 samples in which the mean quotients exceeded 0.25 and about 95% in the 65 samples in which mean quotients were <0.25, representing a factor of 3

Table 12-17 Percent survival of marine amphipods (n = 83) and numbers of benthic organisms (n = 29) within 2 ranges in mean ERM-Qs for samples from lower Miami River and central Biscayne Bay[a]

Laboratory toxicity tests		Benthic community analyses						
Mean ERM-Qs	Amphipod survival (%)	Mean ERM-Qs	Nr. organisms	Nr. species	Nr. arthropods	Nr. amphipods	Nr. ampeliscids	Nr. capitellids
Range: <0.25								
Average 0.05	95	0.05	512	73	134	61	2	10
Std Dev 0.05	14	0.04	324	32	127	78	5	11
Count 65		22						
Range: >0.25								
Average 0.85	35	0.85	471	6	10	6	0	204
Std Dev 0.48	28	0.61	408	2	14	12	0	279
Count 18		7						

[a] Reprinted with permission from Long et al. 2002. Copyright Estuarine Research Federation.

difference in the averages. Benthic samples were collected at 29 of the 83 stations in which chemistry and toxicity information was generated. Averages in total abundance were similar in both groups of stations. However, differences between the 2 groups of stations in average numbers of species, arthropods, and amphipods were factors of about 10. Ampeliscid amphipods were not found in the samples with mean ERM-Qs >0.25. The abundance of opportunist capitellids was higher in the stations with the higher chemical concentrations. These data suggest, as reported by others (Swartz et al. 1994; Hyland et al. 1999, 2003; Brown et al. 2000; MacDonald, Lindskoog, et al. 2003), that the response of benthic infauna to chemical contamination in sediments may occur at concentrations that are lower than those associated with a similar incidence of toxicity based on short-term laboratory toxicity tests that measure survival of amphipods.

Principal component analysis (PCA) was performed with the data from central Biscayne Bay (Long et al. 2002). The analysis was performed with the toxicity, chemical, benthic, and physical data (22 variables) collected at 29 stations. The results of this analysis indicated that the first PCA component was heavily weighted with data from the amphipod survival tests, concentrations of several classes of substances, 3 indices of benthic community composition, and 2 physical variables. The first PCA component (representing 44% of total variability) indicated that the concentrations of most classes of toxicants increased as concentrations of total organic carbon increased and as salinity decreased. That is, the degree of contamination generally increased progressing upstream from the bay. Along this same contamination gradient, there was a decrease in amphipod survival, numbers of benthic taxa, and abundance of arthropods and amphipods. The second PCA component (representing 12% of total variability) was associated with decreasing Microtox EC50s (i.e., increasing toxicity) along with decreasing total abundance of the benthos and abundance of polychaetes (capitellids). The third PCA component (representing 10% of the total variability) was associated with increasing capitellid abundance, and increasing sea urchin fertilization success, and normal embryological development as the concentrations of unionized ammonia in the pore water increased. The 7 riverine stations clustered relatively near each other along the first component axis, whereas the bay stations indicated a wide degree of variability along the first component axis. Therefore, the stations sampled in the river were different from those sampled in the bay in terms of lower amphipod survival, higher toxicant concentrations, lower benthic diversity and abundance, higher concentrations of total organic carbon in sediment, and lower bottom water salinity. Long et al. (2002) concluded that it was not possible to distinguish the effects of natural factors from the effects of toxic chemicals in the responses observed in the indices of benthic structure. Moreover, organisms in the field may avoid contaminated sediment (Roper et al. 1995). Long et al. (2002) speculated that because the samples with diminished benthic abundance and diversity had the highest level of both chemical contamination and acute toxicity, toxicant-induced toxicity had an important role in the impacts observed on the benthic community.

Canfield et al. (1994,1996,1998) evaluated the composition of benthic invertebrate communities in sediments from a variety of freshwater locations including the upper Clark Fork River in Montana, the Great Lakes, and the upper Mississippi River. Results of these benthic invertebrate community assessments were compared to SQGs (ERMs; Ingersoll et al. 1996) and chronic 28-d sediment toxicity tests with *H. azteca*. Good concordance was evident between measures of laboratory toxicity, SQGs, and benthic invertebrate community composition in extremely contaminated samples. However, in moderately contaminated samples, less concordance was observed between the composition of the benthic community and either laboratory toxicity tests or SQGs. These differences could be due to the influence of other environmental factors such as alteration of habitat on benthic invertebrate communities. The use of chronic laboratory toxicity tests better identified gradients in chemical contamination in sediments, compared to many of the commonly used measures of benthic invertebrate community structure. Therefore, the use of longer-term toxicity tests in combination with SQGs may provide a more sensitive and protective measure of potential toxic effects of sediment contamination on benthic communities, compared to use of 10-d sediment toxicity tests. Colonization studies conducted by Ingersoll et al. (2004) indicate that benthic community responses to sediment spiked with DDD or dilutions of contaminated sediment collected from the field were predicted from results of chronic laboratory toxicity tests conducted with *H. azteca* for 28 to 42 d, but not from acute 10-d tests with *H. azteca*.

Evaluation of Concordance among Sediment Quality Guidelines

The concordance among SQGs was evaluated on a chemical-specific basis for PAHs, PCBs, and metals in the following 3 sections. Concordance among SQGs was also evaluated by comparing the predictive ability among the various SQGs ("Comparison of the predictive ability of various SQGs," p 539). Finally, the use of SQGs to evaluate cause-and-effect relationships is discussed ("Use of SQGs to evaluate cause-and-effect relationships," p 546).

Predictive ability of SQGs for PAHs

Swartz et al. (1995) derived mechanistically based SQGs for concentrations of total PAHs with a theoretical approach involving use of EqP models. To quantify the predictive ability of the SQGs, the incidence of acute toxicity was determined in field-collected samples over 6 ranges in PAH toxic units (TUs; Table 12-18). Data for these analyses were assembled from random sampling (EMAP surveys) and focused studies of sites contaminated with PAHs. At relatively low chemical concentrations, the results differed between the data set in which PAHs were of primary concern and the data set in which other mixtures of contaminants were present. Among samples with high PAH TUs (i.e., >1.0), the correct predictions of mortality ranged from 78% to 100% in samples known to have high PAH concentrations and from 85 %to 100% in samples known to have mixtures of chemicals, including PAHs. Total correct predictions of mortality were 87% and 57% in the 2 data sets. Swartz et al. (1995) also reported that the probabilities of toxicity ranged from 60% to 100% for several different SQGs derived to predict effects and ranged from 5% to 6% for several SQGs derived to identify thresholds of toxicity (i.e., predictive of nontoxicity).

Table 12-18 Percentages of field-collected samples that were correctly predicted to be toxic in marine amphipod tests in relation to predicted sum of PAH toxic units in samples in which PAHs were principal contaminants of concern and samples in which other chemicals were present[a]

Sum of PAH toxic units	Correct predictions where PAHs were of concern		Correct predictions where PAHs were not of concern	
	Nr. samples[b]	% of total	Nr. samples[b]	% of total
<0.10	34	95	185	58
0.10–0.24	25	93	22	44
0.25–0.49	19	72	7	55
0.50–0.99	9	77	6	75
1.00–4.99	6	78	3	85
≥5.00	6	100	1	100
Total correct predictions		87		57

[a] Reprinted with permission from Swartz et al. 1995. Copyright Society of Environmental Toxicology and Chemistry (SETAC).

[b] Sum of the lower number of samples observed or predicted to be in each of 3 mortality classes (<13%, 13 to 24%, and >24% mortality).

In the calculation of consensus-based SQGs for total PAHs, Swartz (1999) reported that many of the SQGs derived with a similar narrative intent, but derived using different empirical or mechanistic approaches, resulted in very similar concentrations, and that this similarity probably was not coincidental. MacDonald, Ingersoll, and Berger (2000) used a similar approach to calculate consensus-based SQGs for a variety of compounds in freshwater sediments, including total PAHs. Swartz (1999) calculated 3 types of SQGs for total PAHs, including a threshold effects concentration (TEC = 290 µg/g-OC), a median effects concentration (MEC = 1800 µg/g-OC), and an extreme effects concentration (EEC = 10,000 µg/g-OC) as the means of clusters of individual SQGs derived for the same narrative intent. Field verifications of the predictive ability of the 3 concentrations were performed with data from EMAP surveys and a compilation of data sets from surveys of PAH-contaminated sites in the US and Canada. In the PAH-contaminated sites, the incidence of toxicity increased steadily from 6% to 43% to 50% and to 100% in the 4 concentration ranges defined by the 3 SQGs (Table 12-19). In these same samples, average mortality of amphipods increased from 8% to 34% to 38% and to 97% within the 4 ranges. The pattern in response with the EMAP samples was similar in the first 2 concentration ranges, but the analyses were limited by too

Table 12-19 Incidence of toxicity in marine amphipod toxicity tests in samples within 4 ranges in total PAH concentrations defined by consensus-based sediment effect concentrations for total PAHs[a]

	Total PAH concentrations			
	<TEC	TEC < MEC	MEC < EEC	>EEC
PAH-contaminated sites				
Nr. samples	36	60	24	12
Toxic (%)	6	43	50	100
Mean mortality (%)	8	34	38	97
EMAP sites				
Nr. samples	633	43	2	no data
Toxic (%)	13	44	100	
Mean mortality (%)	11	28	40	

[a] Reprinted with permission from Swartz 1999. Copyright Society of Environmental Toxicology and Chemistry (SETAC).
TEC = threshold effect concentration
MEC = median effect concentration
EEC = extreme effect concentration

little data in the third and fourth ranges. Swartz (1999) also reported that the numbers of species of crustaceans and mollusks in samples from San Diego Bay (CA) and Elliott Bay (WA) declined as PAH concentrations increased across the 4 ranges.

Additional field validation of various versions of the sum PAH models was reported in a study of a creosote site in Elliott Bay (WA), using data from 2 amphipod survival tests (*Leptocheirus plumulosus, R. abronius*) performed on 30 sediment samples that were analyzed for PAHs (Ferraro and Cole 1997; Ozretich et al. 2000). PAH TUs were determined using a variety of methods, including those for bulk sediment concentrations and those for concentrations dissolved in pore waters. Data from this study were summarized in 4 categories (Table 12-20). There were no samples in which TUs ranged between 5.2 and 10.3. Percent mortality in the 2 toxicity tests averaged 9% to 12% in 8 samples in which TUs were <1.0. The mean of the responses in the 2 tests was about 11%. As PAH concentrations increased in the summarized categories, percent mortality incrementally increased to overall means of about 25% (range in TU: 1.0 to 2.0) and 70% (range in TU: 2.0 to 5.2) and, in the last category, 100%. Amphipod mortality was 100% in all 5 samples in which PAH TUs exceeded 10.0. At 1 TU, one would predict about 50% toxicity. Only 25% toxicity was observed in the range of 1.0 to 2.0 TU, suggesting

Table 12-20 Summary of percent mortality data for two 10-d marine amphipod toxicity tests conducted with samples in Elliott Bay (WA) over ranges in sum PAH toxic units (TUs). TUs computed for sums of 20 parent and 14 alkylated PAH groups in sediments and for sums of 33 individual PAHs in both sediments and pore waters.[a]

	Sum PAH TU					
	20 parent + 14 alkylated	33 individual PAH		Percent mortality		
Categories of TUs[b]	Sediments only	Sediments only	Porewater	*L. plumulosus*	*R. abronius*	Mean
<1.0 (*n* = 8)						
Average	1.4	1.0	0.7	12	9	11
Std. Dev.	0.7	0.6	0.2	13	10	7
Median	1	0.8	0.75	8	5	11
>1.0 to 2.0 (*n* = 8)						
Average	2.8	1.8	1.5	31	18	25
Std. Dev.	1.9	1.2	0.3	23	24	22
Median	2.5	1.65	1.6	25	8	16
>2.0 to 5.2 (*n* = 9)						
Average	6.4	4.2	3.9	74	66	70
Std. Dev.	4.7	3.0	1.2	29	41	33
Median	4	2.7	4.3	70	85	78
>10.0 (*n* = 5)						
Average	20.6	14.3	23.7	100	100	100
Std. Dev.	12.6	12.2	12.5	0	0	0
Median	16	8.7	27.6	100	100	100

[a] Reprinted with permission from Ozretich et al. 2000. Copyright Society of Environmental Toxicity and Chemistry (SETAC).

[b] Categories based upon sum PAH TUs for 33 individual compounds in pore water.

that bioavailability of PAHs may be underestimated using this procedure. The Elliott Bay data largely corroborated the observations reported in previous studies of the predictive ability of the SQGs derived with the sum PAH models.

MacDonald and Ingersoll (2001) compiled published SQGs for total PAHs based on mechanistic-based approaches, empirical-based approaches, or results of spiked-sediment toxicity tests (Table 12-21). These SQGs were then grouped on the basis of narrative intent to estimate chronic, acute, or severe effects in sediments. SQGs derived using these 3 different approaches resulted in similar thresholds for total PAHs, depending on narrative intent.

Table 12-21 Summary of SQGs for PAHs assuming 1% total organic carbon in sediment[a]

Endpoint		Threshold (µg/g dry wt)	Threshold (µg/g-OC)
Site specific: Benthic community		2–10	400–1000
Site specific: *H. azteca* 10-d LC50		4.9	490
Chronic			
Mechanistic:	Swartz (1999)	2.1	210
	USEPA (1999)	8.7	870
	Di Toro and McGrath (2000)	9.9	990
Empirical:	TEC (marine; Swartz 1999)	2.9	290
	TEC (freshwater; MacDonald, Ingersoll, Berger 2000)	1.6	160
Acute			
Spiked sediment:	*R. abronius* 10-d LC50	24	2400
Mechanistic:	Di Toro and McGrath (2000)	50	5000
Empirical:	MEC (marine; Swartz 1999)	18	1800
	PEC (freshwater; MacDonald, Ingersoll, Berger 2000)	23	2300
Severe			
Empirical:	EEC (marine; Swartz 1999)	100	10,000

[a] MacDonald and Ingersoll 2001

MacDonald and Ingersoll (2001) then used these SQGs to evaluate site-specific data for a benthic community assessment or for laboratory exposures with *H. azteca* conducted with sediments collected from a site primarily contaminated with PAHs. Effects observed in the laboratory and in the field ranged between 2 to 10 µg/g (dry weight) or 400 to 1000 µg/g-OC) for total PAHs, which was consistent with effects that would be predicted from the range of chronic SQGs listed in Table 12-21 for total PAHs.

Predictive ability of sediment quality guidelines for PCBs

Using the approach of Swartz (1999), many different SQGs derived with either mechanistic or empirical approaches for freshwater, estuarine, or marine sediments were assembled to calculate consensus-based sediment effect concentrations (SECs) for total PCBs (MacDonald, DiPinto, et al. 2000). In this evaluation, previously published SQGs were assembled with the narrative intent of identifying total PCB concentrations below which adverse effects were rare. The geometric mean of these SECs (the TEC) was calculated to be 0.04 mg/kg (dry weight). The geometric mean of SQGs derived as midrange

SQGs, above which effects were expected more frequently, was calculated to be 0.4 mg/kg (dry weight; MEC). Finally, the geometric mean of SQGs derived to be highly predictive of toxicity was determined to be 1.7 mg/kg (dry weight; EEC).

MacDonald, DiPinto, et al. (2000a) assembled data sets from toxicity assessment with a broad range of PCBs to evaluate the performance of the SQGs for total PCBs. Samples from these surveys had measurable concentrations of substances other than PCBs. The incidence of toxicity in amphipod survival tests increased from 12% with PCB concentrations <TEC to 33% with concentrations between the TEC and MEC, to 50% with concentrations between the MEC and EEC, and to 86% at concentrations >EEC (Table 12-22). Average amphipod survival in these categories was 90% below the TEC, 76% between the TEC and MEC, 66% between the MEC and EEC, 58% above the MEC, and 38% above the EEC. Therefore, the predictive ability (i.e., as measured by percent incidence of toxicity and magnitude of survival) of the consensus-based SEC for total PCBs was similar to that observed in Table 12-2 for all the ERMs and PELs. MacDonald, DiPinto, et al. (2000) concluded that PCBs were causing or substantially contributing to the toxicity of the sediment samples. Also, these data indicated a relationship between PCB concentrations and toxicity almost identical to that reported by Swartz (1999) for the SQGs derived for total PAHs for samples in which PAHs were of primary concern.

Table 12-22 Incidence of toxicity and average percent survival in marine amphipod tests within 5 ranges in total PCB concentrations defined by consensus sediment effects concentrations[a]

Sediment effect concentration[b]	Ranges in total PCB concentrations	Nr. samples	% toxic	Average % survival
< TEC	<0.04 mg/kg, dw	599	12	90
TEC to MEC	>0.04–0.4 mg/kg, dw	391	33	76
> MEC to EEC	>0.4–1.7 mg/kg, dw	133	50	66
> MEC	>0.4 mg/kg, dw	161	56	58
> EEC	>1.7 mg/kg, dw	28	86	38

[a] Reprinted with permission from MacDonald, DiPinto, et al. 2000. Copyright Society of Environmental Toxicology and Chemistry (SETAC)

[b] TEC = threshold effect concentration; MEC = midrange effect concentration; EEC = extreme effect concentration

Predictive ability of sediment quality guidelines for metals

Long, MacDonald, et al. (1998) compared the predictive ability of SQGs for trace metals derived with empirical approaches and mechanistic approaches using a marine data set (n = 77) assembled by Hansen et al. (1996). These data were compiled from samples collected in locations in North America and elsewhere with matching metals concentrations and amphipod survival (*A. abdita*). Samples were selected to represent a range in exposures to metals. The metals SQGs derived with empirical approaches included the AETs, ERLs, and ERMs, which were expressed in units of dry weight. The SQGs derived with mechanistic methods were the SEM–AVS approach (expressed as a molar ratio or as a difference in molar concentration; Hansen et al. 1996). Samples were classified as toxic if percent mortality exceeded 24%.

Among the 52 samples in which SEM–AVS ratios were ≤1.0 (i.e., predicting nontoxicity attributable to metals), 81% were not toxic (Table 12-23). In comparison, 97% of the samples were not toxic when metals concentrations in the same samples were less than all AETs, and 100% of the samples were not toxic when metal concentrations were less than all ERLs. Similarly, 93% of samples were not toxic when mean ERM-Qs were <1.0. With SEM–AVS ratios >1.0 (i.e., when toxic conditions might be expected), 28% of samples were toxic in the amphipod tests. Among the samples in which one or more metals concentrations exceeded an AET or ERM, 35% were toxic. Therefore, the empirically based SQGs were somewhat better at predicting nontoxic conditions compared to mechanistically based SQGs; however, both sets of SQGs were similar in predicting the presence of toxicity. There were 8 samples in which SEM–AVS ratios were >5.0, and 50% of these samples were toxic. Among the 11 samples in which the mean ERM-Qs were >5.0, 64% were toxic. Therefore, the predictive ability of both sets of SQGs increased as metal concentration increased.

Comparison of the predictive ability of various sediment quality guidelines

Table 12-24 was compiled to compare the predictive ability of various SQGs based on 4 ranges of incidence in toxicity of samples:

1) <25%,
2) 25 to 50%,
3) 50 and 75%, and
4) >75%.

Endpoints evaluated include laboratory toxicity tests and estuarine benthic community assessments. In laboratory toxicity tests, the incidence of toxicity

Table 12-23 Percentages of samples that were toxic in marine amphipod survival tests when predicted to be either toxic or nontoxic with different sets of SQGs for metals[a]

Chemical characteristics relative to SQGs	Nr. samples	Correct classification as toxic (%)	Nr. samples	Correct classification as nontoxic (%)
SEM–AVS ratios ≤ 1.00			52	81
SEM – AVS differences < 0.00			52	81
All metals concentrations < AETs			31	97
All metals concentrations < ERLs			6	100
Mean ERM quotients for metals <1.0			40	93
SEM–AVS ratios > 1.00	25	28		
SEM – AVS differences ≥ 0.00	25	28		
SEM–AVS ratios > 5.00	8	50		
One or more metals concentrations > AET	46	35		
One or more metals concentrations > ERM	49	35		
Mean ERM quotients for metals ≥ 1.00	37	38		
Mean ERM quotients for metals ≥ 5.00	11	64		

[a] Reprinted with permission from Long, MacDonald, et al. 1998. Copyright Society of Environmental Toxicology and Chemistry (SETAC).

was <25% when either none of the lower-range empirical SQGs were exceeded or when mean SQG-Q were generally <0.1. In benthic community assessments, the incidence of toxic samples was <25% when mean SQG-Qs ranged from about <0.01 to <0.04 (Table 12-24).

In laboratory toxicity tests, the incidence of toxicity ranged from 25% to 50% as chemical concentrations began to increase relative to those in the <25% category. Generally, these were sediments where several low-range empirical SQGs were exceeded, a few midrange empirical SQGs were exceeded, or mean SQG-Qs ranged from about 0.1 to 1.0. In the benthic community assessments, a 36% to 37% incidence in toxicity was observed at mean SQG-Qs ranging from about 0.01 to 0.08.

In laboratory toxicity tests, the incidence of toxicity ranged between 50% to 75% and >75% as chemical concentrations continued to increase relative to those in the lower categories. Generally, these were sediments where several midrange empirical SQGs were exceeded or when mean SQG-Qs were in excess of about 1.0 to 5.0. In the benthic community assessments, a 54% to

Table 12-24 Summary of analyses of predictive ability of SQGs in laboratory toxicity tests or in benthic community assessments

Study, location	Nr. samples within range	Total nr. samples within study	Indicator	Toxicity endpoint	SQG	Incidence of toxic samples (%)	Reference
Incidence of toxic samples <25%							
Marine/estuarine, metals sites	6	77	*A. abdita*	10-d survival	All metals < ERLs	0	Long. MacDonald, et al. 1998
Marine/estuarine, metals sites	31	77	*A. abdita*	10-d survival	All metals < AETs	3	Long. MacDonald, et al. 1998
Estuarine, southeastern USA	135	231	Benthic IBI	Benthic composition	Mean ERM-Q < 0.020	5	Hyland et al. 1999
Estuarine, southeastern USA	131	231	Benthic IBI	Benthic composition	Mean PEL-Q < 0.035	5	Hyland et al. 1999
Marine, PAH sites	36	132	*A. abdita*	10-d survival	<TEC for PAHs	6	Swartz 1999
Marine/estuarine, various sites	804	1987	*A. abdita, R. abronius*	10-d survival	<mean SQG-Q of 0.1	6	Fairey et al. 2001
Marine/estuarine, various sites	631	1513	*A. abdita, R. abronius*	10-d survival	Mean PEL-Q < 0.1	8	Long et al. 2000
Marine/estuarine, various sites	317	1513	*A. abdita, R. abronius*	10-d survival	No TELs exceeded	8	Long et al. 2000
Marine/estuarine, various sites	861	1513	*A. abdita, R. abronius*	10-d survival	Mean ERM-Q < 0.1	9	Long et al. 2000
Marine/estuarine, various sites	428	1513	*A. abdita, R. abronius*	10-d survival	No ERLs exceeded	9	Long et al. 2000
Freshwater, various sites	63	160	*H. azteca*	28- to 42-d survival, growth	Mean PEC-Q < 0.1	10	Ingersoll et al. 2001
Marine, PCB sites	599	1312	*A. abdita, R. abronius*	10-d survival	<TEC for PCBs	12	MacDonald, DiPinto, et al. 2000
Marine, EMAP sites	633	678	*A. abdita*	10-d survival	<TEC for PAHs	13	Swartz 1999
Marine/estuarine, southern Califorina, USA	198	576	Various amphipods	10-d survival	Mean ERM-Q < 0.1	14	Bay et al. 2002
Freshwater, various sites	147	670	*H. azteca*	10- to 14-d survival, growth	Mean PEC-Q < 0.1	18	Ingersoll et al. 2001; Crane et al. 2002
Marine/estuarine, various sites	354	1513	*A. abdita, R. abronius*	10-d survival	1 to 5 PELs exceeded	18	Long et al. 2000
Estuarine, Atlantic and Gulf of Mexico	165	1389	Benthic IBI	Benthic composition	Mean ERM-Q < 0.007	18	Hyland et al. 2003
Estuarine, Atlantic and Gulf of Mexico	177	1389	Benthic IBI	Benthic composition	Mean PEL-Q < 0.012	18	Hyland et al. 2003
Marine/estuarine, various sites	911	3057	*A. abdita, R. abronius*	10-d survival	P-Max Logistic Model < 0.25	18	Field et al. 2002

Table 12-24 *contd*

Study, location	Nr. samples within range	Total nr. samples within study	Indicator	Toxicity endpoint	SQG	Incidence of toxic samples (%)	Reference
Incidence of toxic samples <25% *continued*							
Marine/estuarine, metals sites	52	77	*A. abdita*	10-d survival	SEM–AVS ratios < 1.0	19	Hansen et al. 1996
Freshwater, various sites	121	629	*Chironomus* spp.	10- to 14-d survival, growth	Mean PEC-Q < 0.1	19	Ingersoll et al. 2001; Crane et al. 2002
Marine/estuarine, various sites	490	1987	*A. abdita, R. abronius*	10-d survival	mean SQG-Q 0.1 to 0.25	19	Fairey et al. 2001
Marine/estuarine, various sites	458	1513	*A. abdita, R. abronius*	10-d survival	mean ERM-Q 0.11 to 0.5	21	Long, Field, MacDonald 1998
Marine/estuarine, various sites	783	1513	*A. abdita, R. abronius*	10-d survival	mean PEL-Q 0.11 to 1.5	21	Long, Field, MacDonald 1998
Incidience of toxic samples 25% to 50%							
Marine/estuarine, metals sites	25	77	*A. abdita*	10-d survival	SEM–AVS ratios >1.0	28	Hansen et al. 1996
Marine/estuarine, various sites	308	1513	*A. abdita, R. abronius*	10-d survival	1 to 5 ERMs exceeded	32	Long et al. 2000
Marine/estuarine, southern Califorina, USA	126	576	Various amphipods	10-d survival	mean ERM-Q 0.11 to 0.5	32	Bay et al. 2002
Marine/estuarine, various sites	139	1987	*A. abdita, R. abronius*	10-d survival	mean SQG-Q 0.25 to 0.75	32	Fairey et al. 2001
Marine, PCB sites	391	1312	*A. abdita, R. abronius*	10-d survival	>TEC to <MEC for PCBs	33	MacDonald, DiPinto, et al. 2000
Marine, various sites	46	77	*A. abdita*	10-d survival	one or more metals >AET	35	Long, MacDonald, et al. 1998
Marine, various sites	49	77	*A. abdita*	10-d survival	one or more metals >ERM	35	Long, MacDonald, et al. 1998
Estuarine, Atlantic and Gulf of Mexico	752	1389	Benthic IBI	Benthic composition	mean ERM-Q 0.007 to 0.044	36	Hyland et al. 2003
Freshwater, various sites	288	670	*H. azteca*	10- to 14-d survival, growth	mean PEC-Q >0.5 to <1.0	37	Ingersoll et al. 2001; Crane et al. 2002
Estuarine, southeastern USA	737	1389	Benthic IBI	Benthic composition	mean PEL-Q 0.012 to 0.076	37	Hyland et al. 2003
Marine/estuarine, various sites	1280	3057	*A. abdita, R. abronius*	10-d survival	P-Max Logistic Model 0.25 to 0.5	37	Field et al. 2002

Table 12-24 *contd*

Study, location	Nr. samples within range	Total nr. samples within study	Indicator	Toxicity endpoint	SQG	Incidence of toxic samples (%)	Reference
Incidence of toxic samples 25% to 50% *continued*							
Freshwater, various sites	313	629	*Chironomus* spp.	10- to 14-d survival, growth	mean PEC-Q >0.5 to <1.0	43	Ingersoll et al. 2001; Crane et al. 2002
Marine, PAH sites	60	132	*A. abdita*	10-d survival	TEC to <MEC for PAHs	43	Swartz 1999
Marine, EMAP sites	43	678	*A. abdita*	10-d survival	TEC to < MEC for PAHs	44	Swartz 1999
Marine/estuarine, various sites	90	1987	*A. abdita, R. abronius*	10-d survival	mean SQG-Q 0.75 to 1.0	47	Fairey et al. 2001
Marine/estuarine, various sites	211	1513	*A. abdita, R. abronius*	10-d survival	6 to 20 PELs exceeded	48	Long et al. 2000
Marine/estuarine, various sites	160	1513	*A. abdita, R. abronius*	10-d survival	mean ERM-Q 0.51 to 1.5	49	Long et al. 2000
Marine/estuarine, various sites	47	1513	*A. abdita, R. abronius*	10-d survival	mean PEL-Q 1.51 to 2.3	49	Long et al. 2000
Incidence of toxic samples 50 to 75%							
Marine, metals sites	8	77	*A. abdita*	10-d survival	SEM–AVS ratios >5.0	50	Hansen et al. 1996
Marine, PAH sites	24	132	*A. abdita*	10-d survival	MEC to <EEC for PAHs	50	Swartz 1999
Marine, PCB sites	133	1312	*A. abdita, R. abronius*	10-d survival	MEC to <EEC for PCBs	50	MacDonald et al. 2000
Freshwater, various sites	44–132	629	*Chironomus* spp.	10- to 14-d survival, growth	mean PEC-Q > 1.0 or > 5.0	52 to 68	Ingersoll et al. 2001; Crane et al. 2002
Freshwater, various sites	70–162	670	*H. azteca*	10- to 14-d survival, growth	mean PEC-Q > 1.0 or > 5.0	54 to 71	Ingersoll et al. 2001, Crane et al. 2002
Estuarine, southeastern USA	52	231	Benthic IBI	Benthic composition	mean PEL-Q 0.035 to 0.096	54	Hyland et al. 1999
Estuarine, southeastern USA	51	231	Benthic IBI	Benthic composition	mean ERM-Q 0.020 to 0.058	55	Hyland et al. 1999
Freshwater, various sites	27	160	*H. azteca*	28- to 42-d survival, growth	mean PEC-Q >0.5 to <1.0	56	Ingersoll et al. 2001
Marine, PCB sites	161	1312	*A. abdita, R. abronius*	10-d survival	>MEC for PCBs	56	MacDonald, DiPinto, et al. 2000
Marine/estuarine, various sites	68	1513	*A. abdita, R. abronius*	10-d survival	6 to 10 ERMs exceeded	57	Long et al. 2000
Marine/estuarine, various sites	680	3057	*A. abdita, R. abronius*	10-d survival	P-Max Logistic Model 0.5–0.75	58	Field et al. 2002
Estuarine, Atlantic and Gulf of Mexico	435	1339	Benthic IBI	Benthic composition	mean PEL-Q 0.076–0.781	62	Hyland et al. 2003

Table 12-24 *contd*

Study, location	Nr. samples within range	Total nr. samples within study	Indicator	Toxicity endpoint	SQG	Incidence of toxic samples (%)	Reference
Incidence of toxic samples 50% to 75% *continued*							
Estuarine, Atlantic and Gulf of Mexico	430	1389	Benthic IBI	Benthic composition	mean ERM-Q 0.044–0.361	63	Hyland et al. 2003
Marine, various sites	8	77	*A. abdita*	10-d survival	mean ERM-Q > 10.0	68	Long, MacDonald, et al. 1998
Marine/estuarine, various sites	52	1513	*A. abdita, R. abronius*	10-d survival	mean PEL-Q > 2.3	73	Long et al. 2000
Estuarine, southeastern USA	48	231	Benthic IBI	Benthic composition	mean PEL-Q > 0.096	73	Hyland et al. 1999
Estuarine, Atlantic and Gulf of Mexico	42	1389	Benthic IBI	Benthic composition	mean ERM-Q > 0.361	74	Hyland et al. 2003
Incidence of toxic samples >75%							
Marine/estuarine, various sites	34	1513	*A. abdita, R. abronius*	10-d survival	mean ERM-Q > 1.5	76	Long et al. 2000
Estuarine, Atlantic and Gulf of Mexico	40	1389	Benthic IBI	Benthic composition	mean PEL-Q > 0.781	77	Hyland et al. 2003
Estuarine, southeastern USA	45	231	Benthic IBI	Benthic composition	mean ERM-Q > 0.058	78	Hyland et al. 1999
Marine/estuarine, various sites	25	1513	*A. abdita, R. abronius*	10-d survival	>10 ERMs exceeded	80	Long et al. 2000
Marine/estuarine, various sites	106	1987	*A. abdita, R. abronius*	10-d survival	mean SQG-Q > 1.25	80	Fairey et al. 2001
Marine/estuarine, various sites	18	1513	*A. abdita, R. abronius*	10-d survival	>20 PELs exceeded	83	Long et al. 2000
Marine, PCB sites	28	1312	*A. abdita, R. abronius*	10-d survival	>EEC for PCBs	86	MacDonald , DiPinto, et al. 2000
Marine/estuarine, various sites	186	3057	*A. abdita, R. abronius*	10-d survival	P-Max Logistic Model > 0.75	86	Field et al. 2002
Freshwater, various sites	31	160	*H. azteca*	28- to 42-d survival, growth	mean PEC-Q > 1.0	97	Ingersoll et al. 2001

78% incidence in toxicity was observed at mean SQG-Qs ranging from about 0.04 to 0.8.

The information summarized in Table 12-24 can be used to estimate the probability of observing adverse effects within ranges of chemical concentrations defined using SQGs and a number of different toxicity endpoints. Sediments in which it is important to protect estuarine benthic communities generally would fall in the first category, and sediments in which there is a modest risk of toxicity or estuarine benthic impacts would fall in the second category. Sediments in which adverse effects would be expected in more than 50% to 75% the cases would be those in the third or fourth categories. As might be expected using different sets of SQGs and different toxicity endpoints at different sites, there was a fair amount of variability among the indicators.

Importantly, effects observed in laboratory toxicity tests were observed at sediment concentrations about 10-fold higher than effects observed in estuarine benthic-community assessments. These analyses suggest that if effects are observed in laboratory tests, then effects would be frequently observed on the estuarine benthic communities. Alternatively, effects may be occurring on estuarine benthic communities at concentrations in sediment that are not observed to be toxic in laboratory tests (particularly in acute 10-d tests). Thus, the condition of the ambient estuarine benthic community appears to be a sensitive indicator of contaminants in sediment (Hyland et al. 2003). This pattern has been confirmed by consistent observations across several recent independent studies (Table 12-15). These benthic community responses may be occurring at lower concentrations of contaminants in sediment as a result of the varying sensitivities of the multiple species and life stages to long-term exposures that may be persisting over several generations. However, as noted by these authors, there is a need for additional research to better differentiate between the contributions of contaminants in sediments from the contributions of other possible stressors that may be co-varying with contaminants (e.g., ammonia, sulfide, grain size, salinity, dissolved oxygen, water depth; Chapter 16). Other unmeasured contaminants may be a factor as well. The incidence of degraded benthic communities at elevated concentrations of contaminants in sediment does not preclude the possibility that the observed effects were due in part to other co-varying stressors. For example, results from multivariate redundancy analyses on 13 environmental variables in a Great Lakes embayment indicated that the majority of the variation in the benthic community was attributed to water depth and distance from the headwaters (Breneman et al. 2000). Only a small portion of the variability in the benthic community structure was explained by sediment chemistry. Where co-occurrence of elevated sediment chemistry and degraded benthic community condition is observed, follow-up studies are recommended to determine the specific causes and stressors contributing to these impacts (Chapter 4).

Use of sediment quality guidelines to evaluate cause-and-effect relationships

While it is important to be able to accurately predict the presence and absence of toxicity in field-collected sediments, it is also important to identify the factors that are causing or substantially contributing to sediment toxicity. Such information enables limited resources to be focused on the highest priority sediment issues. One approach that has been used to identify concentrations of sediment-associated contaminants that cause or substantially contribute to toxicity has been to compare empirically based SQGs to mechanistically based EqP SQGs. Swartz (1999) observed that SQGs for total PAHs with a consistent narrative intent were similar among guidelines derived using mechanistic (EqP) and empirical (SLCs, ERMs, PELs, AETs) approaches. Furthermore, Swartz (1999) concluded that these SQGs reflect causal rather than correlative effects when expressed as total PAHs rather than as SQGs for individual compounds. Similarly, MacDonald, DiPinto, et al. (2000) and MacDonald, Ingersoll, and Berger (2000) compared midrange or upper-range empirical SQGs and mechanistic SQGs derived for total PAHs (Table 12-21) or for total PCBs (0.2 to 0.8 mg/L no-effect concentrations determined in freshwater or marine water-only chronic toxicity tests with individual Aroclors). The results of this evaluation indicate that the consensus-based PECs were generally comparable to the mechanistically based SQGs (i.e., within a factor of 3; MacDonald et al. 1996; Smith et al. 1996; MacDonald, DiPinto, et al. 2000; MacDonald, Ingersoll, Berger 2000). Therefore, MacDonald, DiPinto, et al. (2000) and MacDonald, Ingersoll, and Berger (2000) also concluded that the empirically based SQGs can be used to define concentrations of sediment-associated contaminants that are sufficient to cause or substantially contribute to sediment toxicity in field-collected samples. The fact that multiple applications of mechanistic and empirical approaches yield similar results, with the total PAH results being very similar, suggests that for many of the sediments, it is not the individual compounds but may be the mixtures of PAH and, to a lesser extent, PCBs that may cause or substantially contribute to toxicity observed in many field-collected sediments (Chapter 4).

It has been suggested that the results of spiked-sediment toxicity tests or toxicity identification evaluations (TIEs) can also be used to provide a basis for identifying the concentrations of sediment-associated contaminants that are causing or contributing to sediment toxicity (Ingersoll et al. 1997; MacDonald, DiPinto, et al. 2000; MacDonald, Ingersoll, Berger 2000). Unfortunately, only limited relevant data are available to assess effects of spiked sediments. For example, the available data from spiked sediment toxicity tests for freshwater sediments are limited to just a few of the chemical substances for which reliable PECs are available (primarily Cu and fluoranthene; MacDonald, Ingersoll, Berger 2000). Additionally, differences in spiking procedures, equilibration time, and lighting conditions during exposures confound the interpretation of the results of sediment spiking studies, especially for PAHs (DeWitt et al. 1999; ASTM 2004). Moreover, many sediment spiking studies have been conducted to evaluate bioaccumulation using relatively insensitive test organisms or in sediments containing mixtures of chemical substances. TIEs might also be useful in helping to identify major groups of chemicals that contribute to the observed toxicity. However, TIE methods may lack the specificity needed to identify specific chemicals of concern (Besser et al. 1998). Future studies are needed to better evaluate relationships between either spiked-sediment toxicity tests or sediment TIEs and SQGs derived using either empirical or mechanistic approaches.

Conclusions

Considerable effort has been expended to publish studies that quantify the predictive ability of various SQGs. Thus far, the majority of studies have been focused on evaluating the predictive ability of SQGs derived with empirical approaches. Numerous studies have been published in which matching sediment chemistry and either laboratory toxicity data or benthic structure data were compared. These studies have been completed in freshwater, estuarine, and marine systems to document predictive ability of SQGs. Conditions along all of the US coastlines and Hawaii were represented in estuarine and marine studies. Conditions in the Great Lakes, midwestern lakes, and many large rivers and harbors across North America were represented in freshwater studies. The level of effort put forth to quantify the predictive ability of SQGs is indicative of the importance of this issue.

Several important themes emerge from a review of the predictive ability of SQGs. Perhaps most importantly, studies were seldom designed specifically to quantify the predictive ability of SQGs. Most studies were performed with data acquired from monitoring or research programs conducted for other purposes. The predictive ability of SQGs can vary as a function of the endpoints measured and the statistical approach used to classify samples as either toxic in laboratory tests or as impacted in benthic community assessments.

The underlying hypothesis and supposition in studies of predictive ability is that the percentages of samples indicating toxicity or adverse benthic impacts would be similar to the narrative intent of the SQG being used in an evaluation. That is, the lowest incidence of effects would be expected at concentrations less than low-range empirical SQGs, and the incidence would increase incrementally in chemical concentration ranges defined by midrange and upper-range empirical SQGs.

This may be reasonable, given that data compiled from many independent laboratory toxicity studies have demonstrated a relatively low incidence of false positive errors when low-range empirical SQGs were not exceeded (i.e., samples indicating toxic conditions at low chemical concentrations). The narrative intent of low-range empirical SQGs is to be indicative of chemical concentrations below which effects would be infrequent; however, low-range empirical SQGs were not intended to be highly predictive of adverse effects. Although SQGs exist for a relatively limited list of chemicals and for a limited number of toxicity tests, these chemicals frequently co-vary with each other and, therefore, serve as surrogates for other chemicals that were either not quantified in analyses or for which there are no SQGs. If SQGs were poor surrogates for other chemicals, the rates of false positive errors would be expected to be higher than the observed range of 0% to 21%.

Further, data compiled from 10 studies have demonstrated that the percentage of samples identified as adversely affected was highest (76% to 97% indicating laboratory toxicity or benthic impacts) when mean SQG-Qs were highest or when many chemical concentrations exceeded individual SQGs (Table 12-15). Thus, indices based upon mixtures of chemicals in Table 12-24 can be derived and used to accurately classify sediments as toxic using the midrange or upper-range empirical SQGs.

Another important theme is the recognition of the influence that a variety of environmental factors can have on chemical toxicity in sediments. While it is important to be able to accurately predict the presence and absence of toxicity in field-collected sediments, it is also important to identify the factors that cause or substantially contribute to sediment toxicity. Studies have been conducted to quantify the predictive abilities of SQGs derived for specific chemical classes, such as the PAHs or PCBs, by using data only from areas in which these respective chemical classes are the predominant chemicals of concern. Data from such focused studies demonstrate strong exposure–response relationships whether chemical concentrations were expressed in units of dry weight or organic carbon. Concordance among mechanistically based SQGs, spiked-sediment toxicity tests, or empirically based SQGs has been used to provide evidence of SQGs that identify chemicals that cause or substantially contribute to sediment toxicity. These evaluations of concordance have focused primarily on SQGs that have been developed for total PAHs and total PCBs.

A growing body of evidence from several recent independent studies suggests that impacts on estuarine benthic communities may occur at chemical concentrations below those associated with a similar incidence of toxicity in 10-d laboratory survival tests with amphipods. A possible explanation for this difference is that the in situ benthic community-level effects may be occurring at lower contaminant thresholds because of the varying sensitivities of multiple species and life stages to longer-term exposures that may be persisting over several generations. However, the relative contribution of natural confounding factors versus chemical-induced toxicity as a cause of the observed effects is not well understood and should be given further consideration. Some recent studies have applied multimetric benthic indices and reference-range approaches to help account for the influence of such factors in the evaluation of benthic–contaminant relationships from field data. While this is a reasonable approach, the complexity of nature makes it difficult to account for all such factors.

Because conclusions about potential contaminant-induced effects in benthic community assessments are usually drawn from associations between the chemical and biological data, there is always some uncertainty over potential contributions from other environmental factors (including unmeasured chemical stressors) that have not been accounted for in a study. The use of additional chronic laboratory toxicity tests, simulated field studies (mesocosms), and manipulative field experiments (e.g., sediment-tray colonization studies under varying exposure scenarios) is also encouraged in the future as a means to further validate the ability of SQGs to predict impacts on benthic communities accurately under controlled conditions (Chapter 4).

And finally, whenever possible, decisions regarding the management of contaminated sediments should be made using a weight-of-evidence (WOE) approach, which includes sediment chemistry and other relevant data on sediment quality (Ingersoll and MacDonald 2002; Chapter 6). Nevertheless, the results of numerous evaluations of the predictive ability of SQGs indicate that sediment chemistry data can be used to accurately classify sediments as toxic or not toxic. Therefore, it is appropriate to make sediment management decisions using sediment chemistry data alone (i.e., with SQGs) at sites where the costs of further sediment assessments are likely to approach or exceed the costs of sediment remediation. At sites where multiple indicators of sediment quality conditions are to be measured, sampling strategies should be developed and implemented that facilitate the collection of matching sediment chemistry and effects data (i.e., by preparing split samples for toxicity, bioaccumulation, chemistry, and benthos evaluations).

References

[ASTM] American Society for Testing and Materials. 2004. Standard test methods for measuring the toxicity of sediment-associated contaminants with freshwater invertebrates. E1706-00. In: ASTM annual book of standards, Vol. 11.05. West Conshohocken (PA): ASTM.

Barrick RC, Becker DS, Watson DP, Ginn TC. 1985. Commencement Bay nearshore/ tideflats remedial investigation, final report. Prepared for the Washington State Department of Ecology and U.S. Environmental Protection Agency. EPA-910/9-85-134b. Bellevue (WA): Tetra Tech.

Barrick R, Becker S, Brown L, Beller H, Pastorok, R. 1988. Sediment quality values refinement: 1988 update and evaluation of Puget Sound AET, Vol. I. Bellevue (WA): PTI Environmental Services. PTI Contract C717-01.

Barrick RB, Beller H, Becker S, Ginn T. 1989. Use of the apparent effects threshold approach (AET) for classifying contaminated sediments. Proceedings of Oceans 1989 Conference; 1989 Sep 18–21; Seattle, WA, USA.

Bay SM, Lapota D, Anderson J, Armstrong J, Mikel T, Jirik AW, Asato S. 2000. Southern California Bight 1998 Regional Monitoring Program: IV. Sediment toxicity. Westminster (CA): Southern California Coastal Water Research Project.

Beller HR, Barrick RC, Becker DS. 1986. Development of sediment quality values for Puget Sound, final report. Prepared for the Puget Sound Dredged Disposal Analysis and Puget Sound Estuary Programs. Bellevue (WA): Tetra Tech.

Besser JM, Ingersoll CG, Leonard E, Mount DR. 1998. Effect of zeolite on toxicity of ammonia in freshwater sediments: Implications for sediment toxicity identification evaluation procedures. *Environ Toxicol Chem* 17:2310–2317.

Breneman D, Richards C, Lonzao S. 2000. Environmental influences on the benthic community structure in a Great Lakes embayment. *J Gt Lakes Res* 26:287–304.

Brown SS, Gaston GR, Rakocinski CF, Heard RW. 2000. Effects of sediment contaminants and environmental gradients on macrobenthic community trophic structure in Gulf of Mexico estuaries. *Estuaries* 23:411–424.

Canfield TJ, Kemble NE, Brumbaugh WG, Dwyer FJ, Ingersoll CG, Fairchild JF. 1994. Use of benthic invertebrate community structure and the sediment quality triad to evaluate metal-contaminated sediment in the Upper Clark Fork River, Montana. *Environ Toxicol Chem* 13:1999–2012.

Canfield TJ, Dwyer FJ, Fairchild JF, Haverland PS, Ingersoll CG, Kemble NE, Mount DR, La Point TW, Burton GA, Swift MC. 1996. Assessing contamination in Great Lakes sediments using benthic invertebrate communities and the sediment quality triad approach. *J Gt Lakes Res* 22:565–583.

Canfield TJ, Brunson EL, Dwyer FJ, Ingersoll CG, Kemble NE. 1998. Assessing sediments from the upper Mississippi river navigational pools using a benthic community invertebrate evaluations and the sediment quality triad approach. *Arch Environ Contam Toxicol* 35:202–212.

Chung, K. 1999. Toxicity of cadmium, DDT, and fluoranthene to juvenile *Mercenaria mercenaria* in aqueous and sediment bioassays [Masters thesis]. Charleston: Univ Charleston South Carolina. 123 p.

Crane JL, MacDonald DD, Ingersoll CG, Smorong DE, Lindskoog RA, Severn CG, Berger TA, Field LJ. 2000. Development of a framework for evaluating numerical sediment quality targets and sediment contamination in the St. Louis River Area of Concern. Chicago: USEPA Great Lakes National Program Office. EPA 905-R-00-008.

Crane JL, MacDonald DD, Ingersoll CG, Smorong DE, Lindskoog RA, Severn CG, Berger TA, Field LJ. 2002. Evaluation of numerical sediment quality targets for the St. Louis River Area of Concern. *Arch Environ Contam Toxicol* 43:1–10.

Cubbage J, Batts D, Briedenbach S. 1997. Creation and analysis of freshwater sediment quality values in Washington State. Olympia (WA): Environmental Investigations and Laboratory Services Program, Washington Department of Ecology.

DeWitt TH, Hickey CW, Morrisey DJ, Nipper MG, Roper DS, Williamson RB, Van Dam L, Williams EK. 1999. Do amphipods have the same concentration-response to contaminated sediment in situ as in vitro? *Environ Toxicol Chem* 18: 1026–1037.

Di Toro DM, McGrath JA. 2000. Technical basis for narcotic chemicals and polycyclic aromatic hydrocarbon criteria. II. Mixtures and sediments. *Environ Toxicol Chem* 19:1971–1982.

Di Toro DM, Zarba CS, Hansen DJ, Berry WJ, Swartz RC, Cowan CE, Pavlou SP, Allen HE, Thomas NA, Paquin PR. 1991. Technical basis for establishing sediment quality criteria for non-ionic organic chemicals using equilibrium partitioning. *Environ Toxicol Chem* 10:1541–1583.

Fairey R, Long ER, Roberts CA, Anderson BS, Phillips BM, Hunt JW, Puckett HR, Wilson CJ. 2001. An evaluation of methods for calculating mean sediment quality guideline quotients as indicators of contamination and acute toxicity to amphipods by chemical mixtures. *Environ Toxicol Chem* 20:2236–2286.

Ferraro S, Cole F. 1997. Field verification of sediment-amphipod toxicity tests and calibration of PAH effects on the macrobenthos. Abstracts from the 18th Annual

Meeting of the Society of Environmental Toxicology and Chemistry (SETAC); 1997 Nov 16–20; San Francisco, CA, USA. Pensacola (FL): SETAC.

Field LJ, MacDonald DD, Norton SB, Severn CG, Ingersoll CG. 1999. Evaluating sediment chemistry and toxicity data using logistic regression modeling. *Environ Toxicol Chem* 18:1311–1322.

Field LJ, MacDonald DD, Norton SB, Ingersoll CG, Severn CG, Smorong D, Lindskoog R. 2002. Predicting amphipod toxicity from sediment chemistry using logistic regression models. *Environ Toxicol Chem* 21:1993–2005.

Ginn TC, Pastorok RA. 1992. Assessment and management of contaminated sediments in Puget Sound. In: Burton Jr GA, editor. Sediment toxicity assessment. Boca Raton (FL): Lewis. p 371–397.

Hansen DJ, Berry WJ, Mahony JD, Boothman WS, Di Toro DM, Robson DL, Ankley GT, Ma D, Yan Q, Pesch CE. 1996. Predicting the toxicity of metal-contaminated field sediments using interstitial concentrations of metals and acid-volatile sulfide normalizations. *Environ Toxicol Chem* 15:2080–2094.

Hyland JL, Van Dolah RF, Snoots TR. 1999. Predicting stress in benthic communities of southeastern U.S. estuaries in relation to chemical contamination of sediments. *Environ Toxicol Chem* 18:2557–2564.

Hyland JL, Balthis WL, Engle VD, Long ER, Paul JF, Summers JK, Van Dolah RF. 2003. Incidence of stress in benthic communities along the U.S. Atlantic and Gulf of Mexico coasts within different ranges of sediment contamination from chemical mixtures. *Environ Monitor Assess* 81:149–161.

Ingersoll CG, Dillon T, Biddinger RG, editors. 1997. Ecological risk assessment of contaminated sediment. Pensacola (FL): Society of Environmental Toxicology and Chemistry (SETAC).

Ingersoll CG, Haverland PS, Brunson EL, Canfield TJ, Dwyer FJ, Henke CE, Kemble NE, Mount DR, Fox RG. 1996. Calculation and evaluation of sediment effect concentrations for the amphipod *Hyalella azteca* and the midge *Chironomus riparius*. *J Gt Lakes Res* 22:602–623.

Ingersoll CG, Wang N, Hayward JMR, Jones JR. 2004. A field assessment of long-term laboratory sediment toxicity tests with the amphipod *Hyalella azteca* and the midge *Chironomus dilutus*. Chicago: USEPA. Forthcoming.

Ingersoll CG, MacDonald DD. 2002. Guidance manual to support the assessment of contaminated sediments in freshwater ecosystems. Volume III: Interpretation of the results of sediment quality investigations. Chicago: USEPA Great Lakes National Program Office. EPA-905-B02-001-C.

Ingersoll CG, MacDonald DD, Brumbaugh WG, Ireland SD, Johnson BT, Kemble NE, Kunz JL, May TW. 2002. Toxicity assessment of sediments from the Grand Calumet River and Indiana Harbor Canal in northwestern Indiana. *Arch Environ Contam Toxicol* 43:153–167.

Ingersoll CG, MacDonald DD, Wang N, Crane JL, Field LJ, Haverland PS, Kemble NE, Lindskoog RA, Severn C, Smorong DE. 2001. Predictions of sediment toxicity using consensus-based freshwater sediment quality guidelines. *Arch Environ Contam Toxicol* 41:8–21.

Long ER, Field LJ, MacDonald DD. 1998. Predicting toxicity in marine sediments with numerical sediment quality guidelines. *Environ Toxicol Chem* 17:714–727.

Long ER, Hameedi MJ, Sloane GM, Read L. 2002. Chemical contamination, toxicity, and benthic community indices in sediments of the lower Miami River and adjoining portions of Biscayne Bay, Florida. *Estuaries* 25:622–637.

Long ER, MacDonald DD. 1998. Recommended uses of empirically derived, sediment quality guidelines for marine and estuarine ecosystems. *Hum Ecol Risk Assess* 4:1019–1039.

Long ER, MacDonald DD, Smith SL, Calder FD. 1995. Incidence of adverse biological effects within ranges of chemical concentrations in marine and estuarine sediments. *Environ Manag* 19:81–97.

Long ER, MacDonald DD, Cubbage JC, Ingersoll CG. 1998. Predicting the toxicity of sediment-associated trace metals with simultaneously-extracted trace metal: acid-volatile sulfide concentrations and dry weight-normalized concentrations: A critical comparison. *Environ Toxicol Chem* 17:972–974.

Long ER, MacDonald DD, Severn CG, Hong CB. 2000. Classifying the probabilities of acute toxicity in marine sediments with empirically-derived sediment quality guidelines. *Environ Toxicol Chem* 19:2598–2601.

MacDonald DD, Carr RS, Calder FD, Long ER, Ingersoll CG. 1996. Development and evaluation of sediment quality guidelines for Florida coastal waters. *Ecotoxicology* 5:253–278.

MacDonald DD, DiPinto LM, Field J, Ingersoll CG, Long ER, Swartz RC. 2000. Development and evaluation of consensus-based sediment effect concentrations for polychlorinated biphenyls. *Environ Toxicol Chem* 19:1403–1413.

MacDonald DD, Ingersoll CG. 2001. An overview of the toxic effects of sediment-associated polycyclic aromatic hydrocarbons (PAHs), with special reference to the 8335 Meadow Avenue site in Burnaby, British Columbia. Prepared for Legal Services Branch, BC. Vancouver (BC): Ministry of the Attorney General.

MacDonald DD, Ingersoll CG, Berger TA. 2000. Development and evaluation of consensus-based sediment quality guidelines for freshwater ecosystems. *Arch Environ Contam Toxicol* 39:20–31.

MacDonald DD, Ingersoll CG, Moore DRJ, Bonnell M, Brenton RL, Lindskoog RA, MacDonald DB, Muirhead YK, Pawlitz AV, Sims DE, Smorong DE, Teed RS, Thompson RP, Wang N. 2002. Calcasieu estuary remedial investigation/feasibility study (RI/FS): Baseline ecological risk assessment (BERA). Prepared for USEPA, Region 6, Dallas TX by MacDonald Environmental Sciences, Ltd., Nanaimo (BC), September 2002. Document control nr 3282-941-RTZ-RISKZ-14858.

MacDonald DD, Ingersoll CG, Smorong DE, Lindskoog RA. 2002. Development and evaluation of sediment quality standards for the waters of the Colville Indian

Reservation including Lake Roosevelt and Okanogan River. Prepared for the Confederated Tribes of the Colville Reservation, Nespelem (WA).

MacDonald DD, Ingersoll CG, Smorong DE, Lindskoog RA, Sloane G. 2003. Development and evaluation of numerical sediment quality assessment guidelines for Florida inland waters. Prepared for the Florida Department of Environmental Protection, Tallahassee (FL), January 2003.

MacDonald DD, Carr RS, Eckenrod D, Greening H, Grabe S, Ingersoll CG, Janicki S, Janicki T, Lindskoog RA, Long ER, Pribble R, Sloane G, Smorong DE. 2004. Development of an ecosystem-based framework for assessing and managing sediment quality conditions in Tampa Bay, Florida. *Arch Environ Contam Toxicol*: 46:147–161.

Mearns A, Swartz RC, Cummins J, Dinnell D, Plesha P, Chapman P. 1986. Interlaboratory comparison of a sediment toxicity test using the marine amphipod *Rhepoxynius abronius*. *Mar Environ Res* 19:13–37.

[NYSDEC] New York State Department of Environmental Conservation. 1999. Technical guidance for screening contaminated sediments. Albany (NY): Division of Fish and Wildlife, Division of Marine Resources. 45 p.

O'Connor TP, Daskalakis KD, Hyland JL, Paul JF, Summers JK. 1998. Comparisons of measured sediment toxicity with predictions based on chemical guidelines. *Environ Toxicol Chem* 17:468–471.

O'Connor TP, Paul JF. 2000. Misfit between sediment toxicity and chemistry. *Mar Pollut Bull* 40:59–64.

Ozretich RJ, Ferraro SP, Lamberson JL, Cole FA. 2000. Test of Sigma polycyclic hydrocarbon model at a creosote-contaminated site, Elliott Bay, Washington, USA. *Environ Toxicol Chem* 19:2378–2389.

Persaud D, Jaagumagi R, Hayton A. 1993. Guidelines for the protection and management of aquatic sediment quality in Ontario. Toronto (ON): Water Resources Branch, Ontario Ministry of the Environment. 27 p.

Puget Sound Dredged Disposal Analysis (PSDDA). 1988. Final environmental impact statement and management plan reports (Phase I and II). Prepared by the Puget Sound Dredged Disposal Analysis agencies, Seattle (WA).

Reynoldson TB, Day KE, Bailey RC, Norris RH. 1995. Methods for establishing biologically based sediment guidelines for freshwater quality management using benthic assessment of sediment. *Aust J Ecol* 20:198–219.

Roper DS, Nipper MG, Hickey CW, Martin ML, Weatherhead MA. 1995. Burial, crawling and drifting behaviour of the bivalve Macomona liliana in response to common sediment contaminants. *Mar Pollut Bull* 31:4–12.

Smith SL, MacDonald DD, Keenleyside KA, Ingersoll CG, Field J. 1996. A preliminary evaluation of sediment quality assessment values for freshwater ecosystems. *J Gt Lakes Res* 22:624–638.

Swartz RC. 1999. Consensus sediment quality guidelines for polycyclic aromatic hydrocarbon mixtures. *Environ Toxicol Chem* 18:780–787.

Swartz RC, Cole FA, Lamberson JO, Ferraro SP, Schults DW, DeBen WA, Lee H, Ozretich JR. 1994. Sediment toxicity, contamination and amphipod abundance at a DDT- and dieldrin-contaminated site in San Francisco Bay. *Environ Toxicol Chem* 13:949–962.

Swartz RC, DeBen WA, Jones JKP, Lamberson JO, Cole FA. 1985. Phoxocephalid amphipod bioassay for marine sediment toxicity. In: Cardwell RD, Purdy R, Bahner RC, editors. Aquatic toxicology and hazard assessment: 7th Symposium. Philadelphia: American Soc for Testing and Materials. ASTM STP 854. p 284–307.

Swartz RC, Schults DW, Ozretich RJ, Lamberson JO, Cole FA, DeWitt TH, Redmond MS, Ferraro SP. 1995. Sigma PAH: A model to predict the toxicity of polynuclear aromatic hydrocarbon mixtures in field-collected sediments. *Environ Toxicol Chem* 14:1977–1987.

Thompson B, Anderson B, Hunt J, Taberski K, Phillips B. 1999. Relationships between sediment contamination and toxicity in San Francisco Bay. *Mar Environ Res* 48:285–309.

Thursby GB, Heltshe J, Scott KJ. 1997. Revised approach to toxicity test acceptability criteria using a statistical performance assessment. *Environ Toxicol Chem* 16: 1322–1329.

[USEPA] US Environmental Protection Agency. 1989. Report by the USEPA Science Advisory Board, Sediment Criteria Subcommittee. Evaluation of the apparent effects threshold (AET) approach for assessing sediment quality. Final report. Washington DC: USEPA. EPA SAB-EETFC-89-027.

[USEPA] US Environmental Protection Agency.1996. Calculation and evaluation of sediment effect concentrations for the amphipod *Hyalella azteca* and the midge *Chironomus riparius*. Chicago: USEPA. EPA 905/R-96/008.

[USEPA] US Environmental Protection Agency. 1997. The incidence and severity of sediment contamination in surface waters of the United States. Volume 1: National sediment quality survey. Washington DC: USEPA Office of Science and Technology. EPA 823-R-97-006.

[USEPA] US Environmental Protection Agency. 1999. Equilibrium-partitioning sediment guideline (ESG) for the protection of benthic organisms: PAH mixtures. Draft report. Washington DC: USEPA. (As cited in Di Toro and McGrath 2000.)

[USEPA] US Environmental Protection Agency. 2000a. Methods for measuring the toxicity and bioaccumulation of sediment-associated contaminants with freshwater invertebrates, second edition, Washington DC: USEPA. EPA 600/R-99/064.

[USEPA] US Environmental Protection Agency. 2000b. Prediction of sediment toxicity using consensus-based freshwater sediment quality guidelines. Chicago: USEPA. EPA 905/R-00/007.

[USEPA] US Environmental Protection Agency. 2001. National sediment quality survey database. Washington DC: USEPA. EPA-823-C-01-001.

Van Dolah RF, Hyland JL, Holland AF, Rosen JS, Snoots TR. 1999. A benthic index of biological integrity for assessing habitat quality in estuaries of the southeastern USA. *Mar Environ Res* 48:1–15.

13

Predictive ability of sediment quality guidelines derived using equilibrium partitioning

DOMINIC M DI TORO,
WALTER J BERRY,
ROBERT M BURGESS,
DAVID R MOUNT,
THOMAS P O'CONNOR,
RICK C SWARTZ

The principal difficulty in establishing the relationship between chemical concentrations in sediments and biological effects is the varying extent to which the chemicals in the sediment are available to the sediment biota. This is the problem of bioavailability. It is especially acute for sediments since there are large variations in the physical and chemical properties of sediments that determine the extent of biological interaction with potentially toxic chemicals. These interactions can result in chemical–sediment associations that are biologically inert, thereby complicating any simple relationship between chemical concentration and biological effects.

The equilibrium partitioning (EqP) model of sediment toxicity is an attempt to solve this problem. The method was first suggested by Pavlou and Weston (1983). The first critical data set that demonstrated the principle was presented by Adams et al. (1985). A systematic presentation of the EqP model for nonionic hydrophobic chemicals, with larger and more diverse sets of supporting data, was presented subsequently by Di Toro et al. (1991). This paper summarizes the available data that relate to the predictive ability of the EqP model. Data for organic chemicals and metals are presented.

Use of Sediment Quality Guidelines and Related Tools for the Assessment of Contaminated Sediments
Wenning RJ, Batley GE, Ingersoll CG, Moore DW, editors.
 ISBN 1-880611-71-6

Equilibrium Partitioning Sediment Quality Guidelines

EqP SQGs are derived from effects concentrations measured in water-only exposures. In sediment exposures, the effect is predicted to occur at the same concentration—more precisely, at the same chemical activity or fugacity—in the pore water of the sediment. The equivalent sediment concentrations are derived from partitioning theory applied to the pore water–sediment particle system, which is assumed to be at equilibrium, hence the name "equilibrium partitioning." Figure 13-1 presents the schematic diagram that illustrates these ideas. Note that the organisms are not assumed to be at equilibrium: the arrows to the biota are one-way. Only the pore water and sediment particles are assumed to be at equilibrium. Further it is assumed that the exposure can be from both pore water and sediment ingestion. The reason is that the chemical activity of the pore water–sediment particle system is the same at equilibrium, and therefore exposure from either or both media exerts the same "chemical pressure" or fugacity on the organism. An issue with deposit-feeding organisms is whether conditions in the gut of the organism modify the chemistry sufficiently so that ingested sediment cannot be assumed to be in equilibrium with pore water. This issue can be addressed only by examining the experimental data presented below.

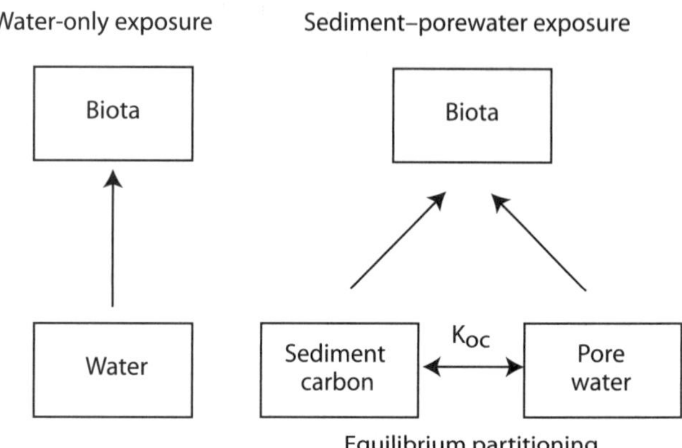

Figure 13-1 Organism exposure routes for a water-only exposure (left) and a sediment exposure (right). "Equilibrium partitioning" assumes that an equilibrium exists between the chemicals sorbed to particulate sediment organic carbon and pore water. The partition coefficient is K_{OC} (Reprinted with permission from Di Toro et al. 1991. Copyright Society of Environmental Toxicology and Chemistry [SETAC]).

The basic premise of the EqP SQGs is that if chemical concentrations (strictly speaking, thermodynamic activities) in pore water are not at toxic levels, then the sediment will not be toxic. The pore waters of sediments are often anoxic, for example, those containing acid volatile sulfide (AVS), which is formed by sulfur-reducing bacteria under anoxic conditions. Most benthic organisms must avoid immersion with such environments, either by respiring and eating in burrows they ventilate with overlying water or directly at the sediment–water interface. These organisms are exposed to waters that may be intermediate between pore water and overlying water. Nonetheless, so long as overlying water is not toxic, sediments with nontoxic pore water will be not be toxic to test organisms.

For nonionic hydrophobic organic chemicals, the partitioning is assumed to be to sediment organic carbon, although studies performed over the last decade have proposed the existence of more than 1 "type" of carbon in sediments (Luthy et al. 1997). These other types of carbon include soot carbon formed by the combustion of biomass and fossil fuels as well as "glassy" and "rubbery" carbon domains. For metals, the AVS and, more recently, sediment organic carbon are the phases considered.

The EqP model predicts the results of a sediment toxicity test on the basis of a known response to a water-only exposure by the same organism. The toxicity prediction is made using either observed porewater concentrations, or porewater concentrations calculated with the EqP assumptions from measured sediment concentrations. This is the reason EqP SQGs are referred to as mechanistic SQGs (Chapter 3). Their predictive ability can be determined directly from laboratory experiments that measure each of the chemical parameters required to calculate porewater concentrations. This is in contrast to empirically derived SQGs, which rely on correlations between chemical concentrations in field-collected sediment and observed biological effects (Chapter 12).

The validity of the EqP model rests first on the observation that toxicity can be predicted from water-only LC50s and porewater chemistry. Figure 13-2 presents data from experiments with 6 individual organic chemicals, each tested on 3 sediments in which pore water concentrations were measured directly (Adams et al. 1985; Nebeker et al. 1989; Schuytema et al. 1989; Swartz et al. 1990; Di Toro et al. 1991; USEPA 2002). EqP predicts that 50% mortality should occur at a porewater concentration equal to the water-only LC50. There is about a factor of 2 variation in the observed results, relative to the predicted ratio of 1.

Similar data established that the same principle applied to metals (Di Toro et al. 1990; Ankley et al. 1993). Figure 13-3 presents the most recent compilations of metal toxicity data for laboratory-spiked sediments (Berry et al. 1996), where interstitial toxic units are defined as the ratio of porewater concentrations to water-only LC50s. The organisms tested include both de-

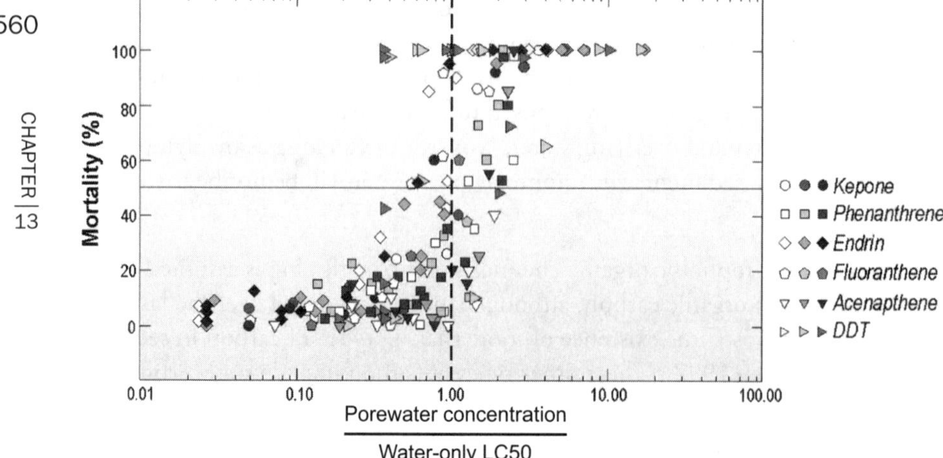

Figure 13-2 Mortality versus porewater concentration normalized by the water-only LC50, measured independently. Six chemicals and 3 sediments per chemical. Shading indicates increasing fraction organic carbon in the sediment. The EqP model predicts that 50% mortality should occur at a ratio of 1 (vertical dashed line) (Di Toro et al. 1991; USEPA 2002).

posit and suspension feeders. The results clearly demonstrate that no toxicity is observed if the porewater concentrations are below the water-only LC50s. This is one test of the predictive ability of the EqP model. It is also clear that, unlike the organic chemical data in Figure 13-2, exceeding the water-only LC50 does not necessarily cause toxicity. The reason is that metal speciation in the pore water, for example, forming complexes to dissolved organic carbon, also influences the bioavailability of metals. A proposal for establishing SQGs for metals (Ankley, Mattson, et al. 1996) discusses this issue in more detail. Progress has been made in determining the speciation of metals in water-only exposures and in predicting the extent of toxicity. The biotic ligand model (BLM; Di Toro et al. 2001; Paquin et al. 2002) is one such attempt. The use of this type of model within the framework of EqP is an obvious next step in the development of pore water–based, mechanistic SQGs for metals. This point is addressed more completely below (p 565), where an initial application of the BLM to sediment is presented.

Sediment Data: Individual Organic Chemicals

Given that porewater concentrations predict toxicity, the second test of EqP SQGs is their ability to predict porewater concentrations and, therefore, observed toxicity solely from sediment chemical data. This type of comparison

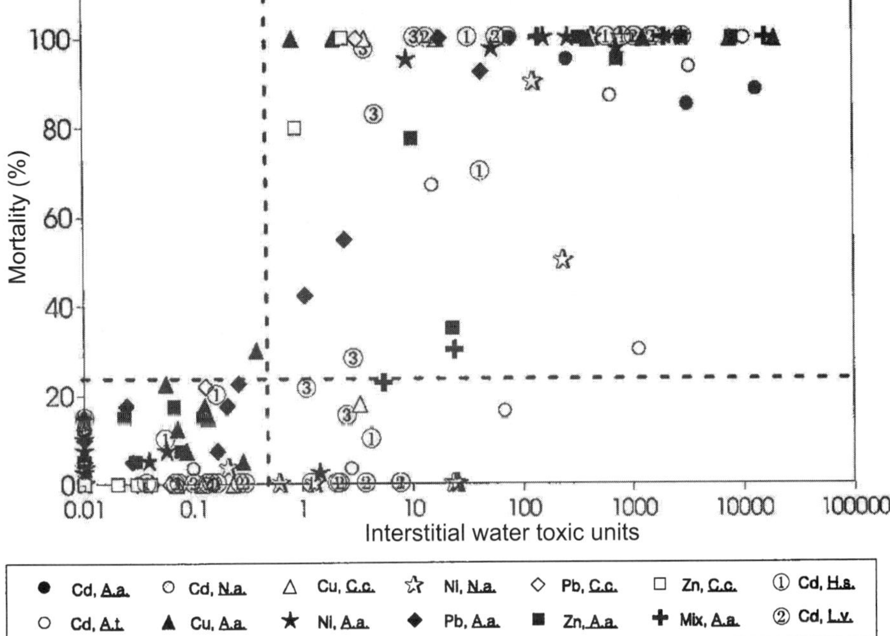

Figure 13-3 Percentage mortality of saltwater and freshwater benthic species versus interstitial water toxic units (IWTU = porewater concentration / water-only LC50). Species: oligochaetes (*L. variegatus*, L.v.), polychaetes (*C. capitata*, C.c., and *N. arenaceodentata*, N.a.), amphipods (*A. abdita*, A.a.), harpacticoids (*Amphiascus tenuiremis*, A.t.), and snails (*Helisoma* sp., H.s.) exposed to sediments spiked with Cd (A.t.; A.a.; N.a.; H.s.; L.v.). Cu (A.a.; C.c.), Pb (A.a.; C.c.), nickel (A.a.; N.a.), Zn (A.a.; C.c.) or a mixture of metals (A.a.) as a function of IWTUs. Data below the detection limit are plotted at IWTU = 0.01 (Reprinted with permission from Berry et al. 1996. Copyright Society of Environmental Toxicology and Chemistry [SETAC]).

tests both the exposure equivalence (water-only versus porewater sediment equilibrium) and the partitioning model used in EqP. Figure 13-4 presents the same mortality data as in Figure 13-2, but using the sediment concentrations, suitably normalized using sediment organic carbon concentration (f_{OC}) and the sediment organic carbon–porewater partition coefficient (K_{OC}) computed from the chemical's octanol–water partition coefficient (K_{OW}) (see Di Toro et al. 1991 and USEPA 2002 for details). EqP predicts that 50% mortality should occur at a ratio of 1. There is slightly more variation in the observed results than with the porewater comparison. This is not unexpected because the porewater concentrations are being predicted from a partitioning model. However, the 95% confidence limits still agree with the EqP prediction within about a factor of 2.

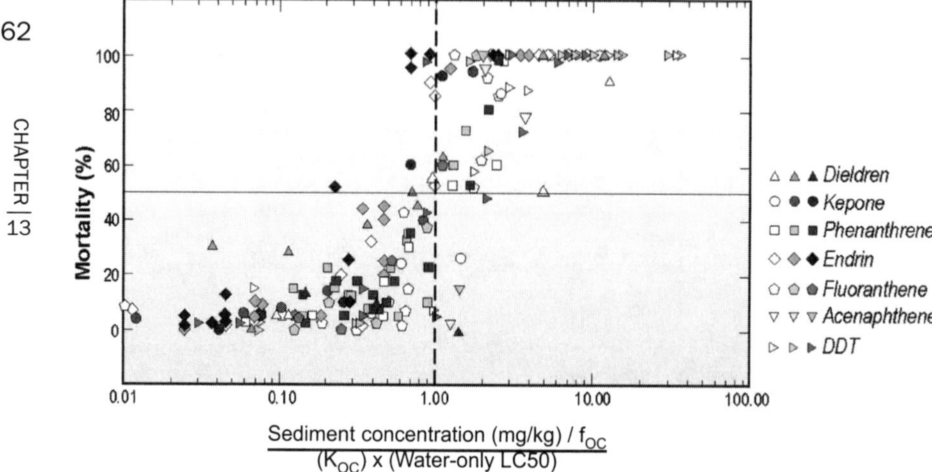

Figure 13-4 Mortality versus predicted porewater concentration (Sediment concentration/f_{OC}/K_{OC}) normalized by the water-only LC50, measured independently. Seven chemicals and 3 sediments per chemical. Shading indicates increasing fraction organic carbon in the sediment. The EqP model predicts that 50% mortality should occur at a ratio of 1 (vertical dashed line) (Di Toro et al. 1991; USEPA 2002)

A comparison of predicted and observed LC50s for selected nonionic hydrophobic chemicals is shown in Figure 13-5. The variation is about a factor of 2 to 3, as shown by the dotted lines in Figure 13-5. Thus, EqP SQGs can predict the LC50 for a single chemical in spiked laboratory sediment toxicity tests to within a factor of 2 to 3.

Sediment Data: PAH Mixtures

In order to use the EqP model to derive SQGs for a single chemical, there are 2 requirements: 1) It is necessary to know or predict the LC50s of the chemical in water-only exposures, and 2) it is necessary to predict the partitioning between sediment particles and pore water. For mixtures of chemicals, it is necessary to predict the water-only toxicity of the mixture as well. A model of this type has been proposed for mixtures of PAHs (Swartz et al. 1995). The toxicity of each PAH in water-only exposures is predicted using K_{OW}. An additive toxic unit (TU) model is used to predict the toxicity of the mixture. Figure 13-6 illustrates the success of the model in predicting the probability of observing toxicity in field-collected sediments.

A more recent model of this type has been proposed on the basis of the target lipid model of narcosis toxicity (Di Toro et al. 2000). PAHs have been

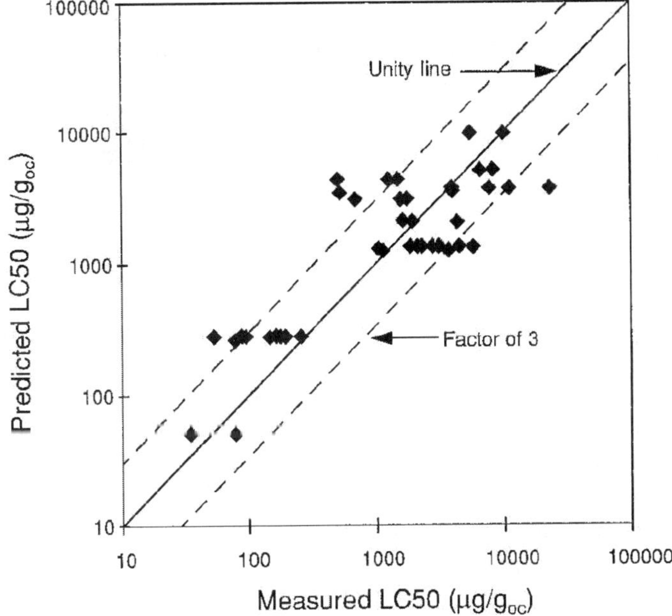

Figure 13-5 Sediment LC50s predicted using EqP versus measured LC50 from EqP "check" tests (Reprinted with permission from Ankley, Berry, et al. 1996. Copyright Society of Environmental Toxicology and Chemistry [SETAC].)

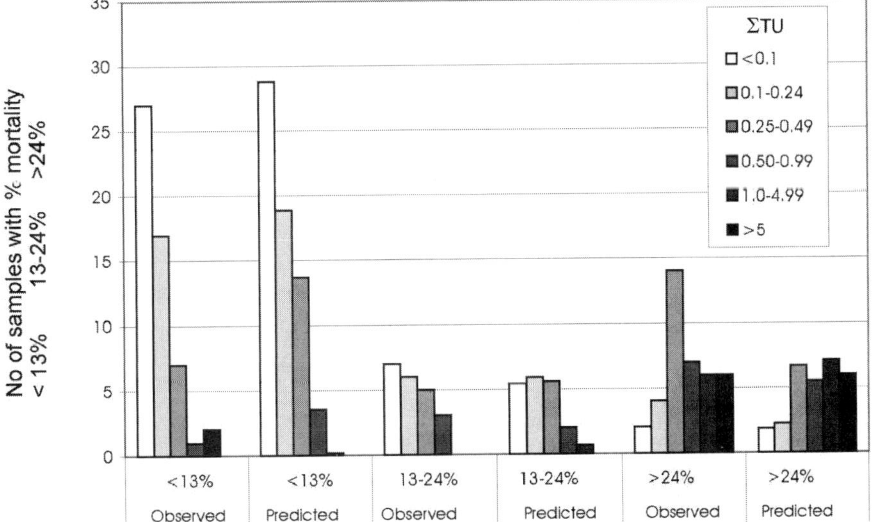

Figure 13-6 Predicted and observed amphipod mortality in relation to predicted sum of PAH toxic units (Sum TUi) in field-collected sediment samples (Reprinted with permission from Swartz et al. 1995. Copyright Society of Environmental Toxicology and Chemistry [SETAC].)

demonstrated to exhibit narcotic toxicity (Di Toro and McGrath 2000), as do hydrocarbons in general. The second requirement, a model for the toxicity of mixtures, is also available. Narcotic chemicals are strictly additive on a TU basis (see Hermens 1989 for a review). Therefore, it is possible to derive SQGs for individual PAHs and petroleum components. But more important, it is possible to analyze mixtures of these chemicals.

An interesting result of the target lipid model–derived SQGs (see Di Toro and McGrath 2000 for details) indicates that, to a reasonable level of accuracy, the molar sum of the concentrations of the individual PAHs and other narcotic chemicals, normalized to organic carbon, can be used as an SQG for the mixture. The reason is that the target for narcosis is the lipid compartment in the organisms. And it happens that the partition coefficients for target lipid–water and sediment organic carbon–water are almost the same. This would not be the case if the sediment contained significant quantities of soot carbon that more strongly binds PAHs (Burgess et al. 2003). Thus, adding carbon–normalized sediment molar concentrations (e.g., μmol/g OC) is equivalent to adding the lipid-normalized molar concentrations in the organism (e.g. μmol/g lipid). Because narcotic toxicity is thought to be caused by the chemical's physical disruption of the nerve cell lipid membranes, the specific identity of the chemical in the lipid is less important than the number of molecules (i.e., the moles) per unit volume of lipid. This is the mechanistic explanation for the success of the total PAH empirical criteria discussed in Chapter 4, which is based on the sum of the PAH dry weight normalized concentrations in the sediment. Although the empirical criteria are not organic carbon–normalized and use the mass rather than the moles of chemical (e.g., mg chemical/kg dry wt sediment versus mmol chemical/kg OC), the majority of the sediments in the database have a relatively narrow range of organic carbon concentrations, and the PAHs have a relatively narrow range of molecular weights, so that the use of moles and organic carbon normalization, although it should improve the predictive ability, would not affect the empirical correlation. The empirical SQG for the sum of PAH concentrations in mg chemical/kg dry wt sediment is equivalent to SQG EqP target lipid mechanistic guideline using f_{OC} = 1% and MW = 173 g/mol to convert the units. It should be pointed out that this is not true for the individual PAH empirical SQGs that are 1 to 2 orders of magnitude lower than the corresponding mechanistic EqP SQGs. The reason is that PAHs in field-contaminated sediment always occur as mixtures, and attributing the toxicity to only 1 component of the mixture, as is the case for empirical SQGs, leads to a much lower apparent effect concentration.

Figure 13-7 illustrates the ability of the target lipid model to predict the observed mortality of *Rhepoxynius abronius* to 5 PAHs exposed individually and an equitoxic mixture. The percent mortality is presented versus the organic carbon–normalized PAH concentration. For total PAHs the use of the organic carbon–normalized, total molar concentration is equivalent to using the total TU concentration. The vertical solid line is the predicted toxicity for *R. abronius* using EqP (SQG = 18.3 μmol/g OC; Di Toro and McGrath 2000). The dotted lines encompass a factor of 2 uncertainty. The results for the single chemical exposure are equivalent to the equitoxic mixture as predicted by the target lipid model. Again, a factor of 2 appears to be the accuracy associated with the predictions.

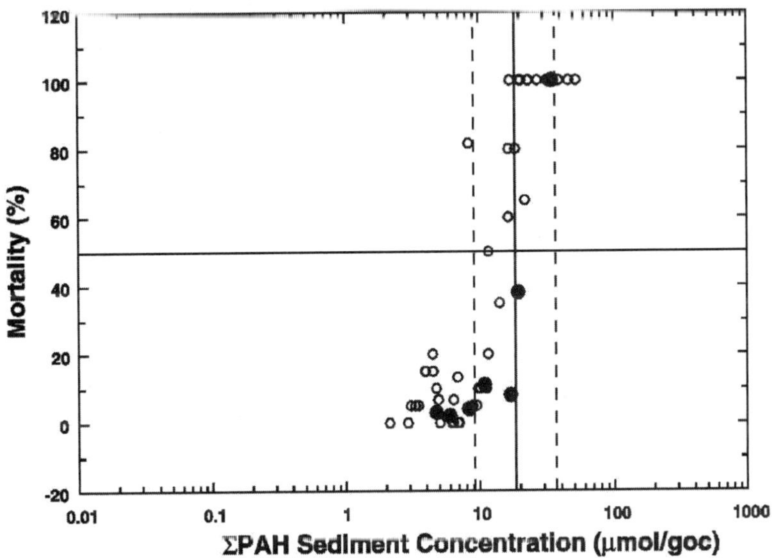

Figure 13-7 Comparison of observed morality in *Rhepoxynius abronius* 10-d sediment toxicity tests. Data from the spiked mixture experiment of Swartz et al. (1997) for either the individual PAHs (o) or the sum of the PAH concentratios (ΣPAH) (•).

Sediment Data: Metals

The partitioning of metals in sediments is more complicated than nonionic organic chemicals because metals can form complexes with other dissolved and particulate species, in addition to particulate organic carbon (POC). In particular, metals can form essentially insoluble metal sulfides. Thus, the

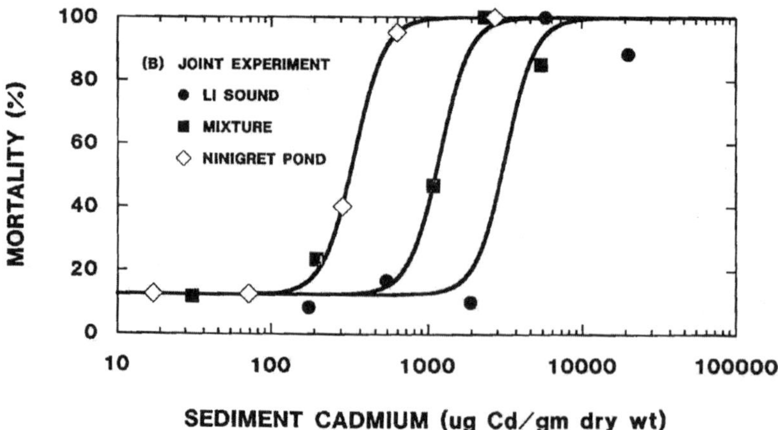

Figure 13-8 Observed percent mortality for Ninigret Pond sediment (*Rhepoxynius hudsoni*), Long Island (LI) Sound, and a 50/50% (v/v) mixture of the 2 sediments (*Ampelisca abdita*) versus sediment Cd concentration. Lines are fitted to the data to aid data interpretation (Reprinted with permission from Di Toro et al. 1990. Copyright Society of Environmental Toxicology and Chemistry [SETAC]).

presence of sulfides in sediments was the basis for the initial EqP guideline for metals (Di Toro et al. 1990). It was observed that the toxicity of Cd in 3 sediments was quite different, varying by more than an order of magnitude (Figure 13-8). However, the concentration of cold acid-extractable sulfide in sediments, quantified as acid-volatile sulfide (AVS), predicted the toxicity of Cd in these sediments quite nicely (Figure 13-9). If the Cd exceeded the binding capacity of the sulfide, that is if the molar concentration of Cd exceeded the molar concentration of AVS, then free Cd could exist in the pore water, and toxicity was observed.

It was fortunate that Cd was the first metal tested in the development of mechanistic or bioavailability-based SQGs because it is completely extracted in the AVS extraction. For other metals, Cu and Ni, for example, the extraction is not complete. As a consequence, it is necessary to measure the metal concentration in the same extraction as the AVS. This metal concentration is the SEM (Di Toro et al. 1992). Thus, if the metal sulfide is not extracted, it appears neither in the sulfide nor in the metal analysis and is therefore not considered in the evaluation of the quantity of bioavailable metal present in the sediment. In fact, because there are 5 commonly present metals that form sulfides more insoluble than iron sulfide, it is necessary to consider them together. ΣSEM is the sum of molar concentrations of the cold 1N HCl-extracted metals (Cd, Cu, Ni, Pb, and Zn). Although not usually present in large concentrations, and therefore not usually included in ΣSEM, Ag toxic-

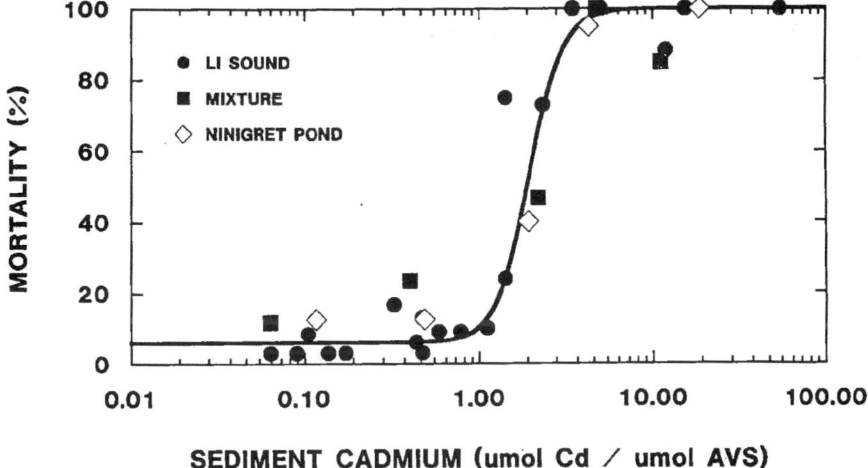

Figure 13-9 Observed percent mortality for Ninigret Pond sediment (*R. hudsoni*), Long Island (LI) Sound, and a 50/50% (v/v) mixture of the 2 sediments (*A. abdita*) versus sediment Cd: AVS ratio. Line is fitted to the data to aid data interpretation (Reprinted with permission from Di Toro et al. 1990. Copyright Society of Environmental Toxicology and Chemistry [SETAC]).

ity has also been demonstrated to be predictable using SEM_{Ag} = [Ag] / 2 and AVS (Berry et al. 1999). Thus, the ΣSEM/AVS method was suggested as the method to predict the lack of toxicity in sediments. If ΣSEM/AVS < 1, it can be shown that the concentration of free metal ion in the pore water is less than the ratio of the solubilities of the metal sulfide and iron sulfide (Di Toro et al. 1992). This result is true, independent of the details of the porewater chemistry, for example, the pH and concentration of other complexing ligands. This is the chemical reason that the condition ΣSEM/AVS < 1 is successful in predicting the lack of toxicity. If this condition is satisfied, then the porewater free metal concentrations are well below the water-only toxicity of any organism, and no toxicity results. Figures 13-10 and 13-11 demonstrate the predictive ability of the SQG based on the condition ΣSEM/AVS < 1. The lack of toxicity is predicted almost perfectly. However, for the case when ΣSEM/AVS > 1, no prediction is possible because at this point, the other phases that influence metal partitioning and speciation become important.

It should be recognized that application of EqP to metals is somewhat different from its application to neutral organic compounds. In the latter case, calculated porewater concentrations increase smoothly as the bulk sediment concentration increases and eventually can reach toxic levels because the 2 are related by a linear partition coefficient. For metals, however, the presence of AVS limits solubility to below toxic levels if ΣSEM/AVS < 1, regardless of the

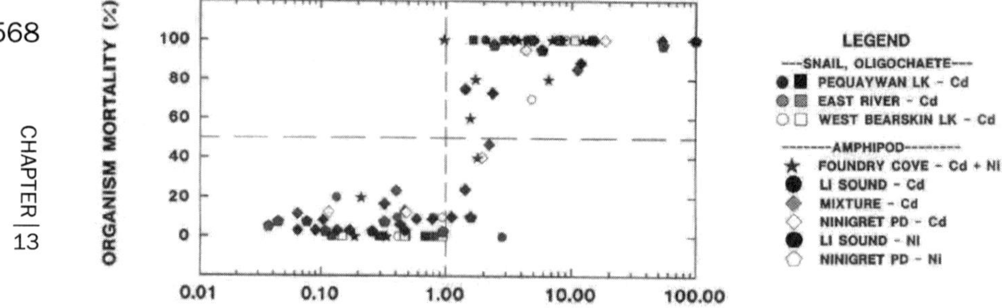

Figure 13-10 Organism mortality vs SEM/AVS ratio. Legend identifies the sediment, metals, and the organism tested. For the Foundry Cove sediments, the metal concentration is the molar sum of the Cd and Ni SEM (Reproduced with permission from Di Toro et al. 1992, *Environ Sci Technol* 26:96–101. Copyright 1992 Am Chem Soc).

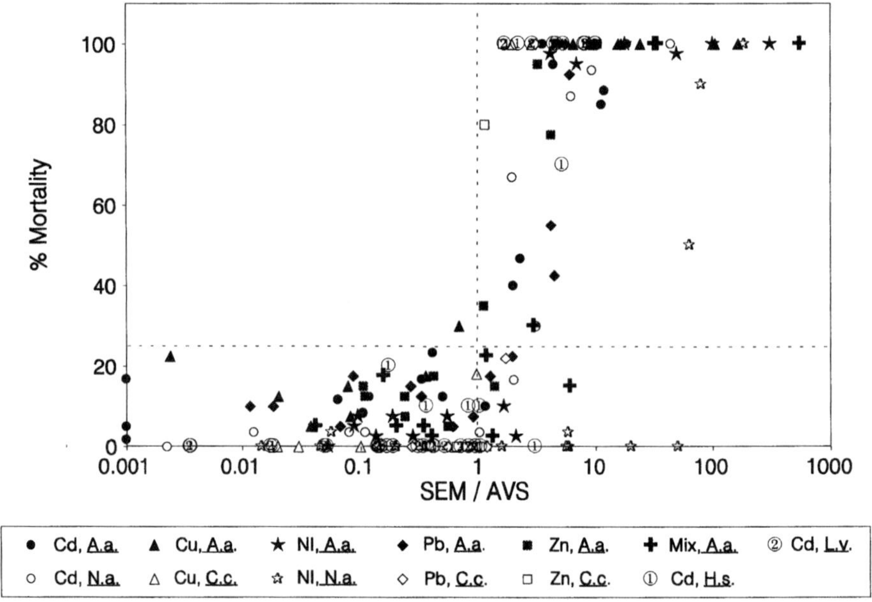

Figure 13-11 Percentage mortality of saltwater and freshwater benthic species versus SEM/AVS ratio. Species: oligochaetes (*L. variegatus,* L.v.), polychaetes (*C. capitata*, C.c., and *N. arenaceodentata*, N.a.), amphipods (*A. abdita*, A.a.), harpacticoids (*Amphiascus tenuiremis*, A.t.), and snails (*Helisoma* sp., H.s.) exposed to sediments spiked with cadmium (A.t.; A.a.; N.a.; H.s.; L.v.). copper (A.a.; C.c.), lead (A.a.; C.c.), nickel (A.a.; N.a.), zinc (A.a.; C.c.) or a mixture of metals (A.a.) as a function of SEM/AVS ratio. Data below the detection limit are plotted at SEM/AVS = 0.001 (Reprinted with permission from Berry et al. 1996. Copyright Society of Environmental Toxicology and Chemistry [SETAC]).

sediment metal concentration. The strongest prediction for metals is for a lack of toxicity when an excess of sulfide is present. The reason is that this prediction is based solely on the formation of highly insoluble metal sulfides that limit porewater concentrations to well below toxic levels and not on validity of a partitioning model. When ΣSEM > AVS, then partitioning comes into play and the EqP model for metals and organic chemicals are similar in this regard.

Recently, a modification of the ΣSEM/AVS method has been suggested that can be applied to the situation in which ΣSEM > AVS. The idea is to include the effect of organic carbon partitioning. The SQG is based on organic carbon–normalized excess SEM, that is, $\Sigma SEM / f_{OC} = (\Sigma SEM - AVS) / f_{OC}$ (Di Toro et al. 2005; USEPA 2002). Figure 13-12 presents the results. Data for Cd, Cu, Ni, Pb, and Zn are included. Remarkably, a metal-specific analysis reveals no differences. This puzzling finding is discussed below (p 568). The vertical lines are the 95% uncertainty range. If $\Sigma SEM / f_{OC} < 130$ μmol/g OC, no toxicity is predicted. If $\Sigma SEM / f_{OC} > 3400$ μmol/g OC, then toxicity is predicted. About 25% of the data are in the ambiguous range where no prediction is possible. Note that this figure includes both laboratory-spiked

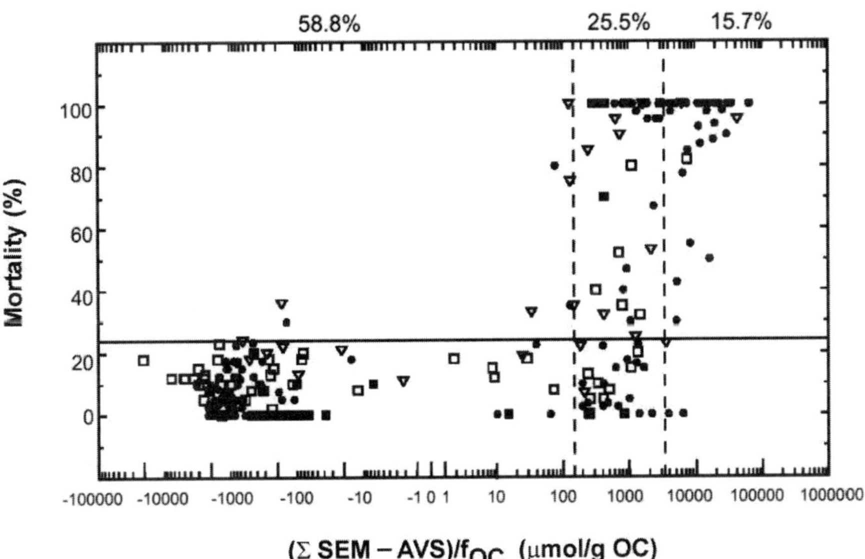

Figure 13-12 Percentage mortality versus organic carbon–normalized excess SEM. Spiked and field-collected sediment. The data within the vertical dotted lines enclose 90% of the misclassified data. The percentages above the figure are the percent of the total number of data points in the 3 regions (no toxicity, mortality < 24%; uncertain toxicity; and toxic, mortality = 100%). Data are saltwater field data without Bear Creek and Jinzhou Bay (❐), freshwater field data (■), freshwater spiked sediment data (❍), and saltwater spiked data (●) (USEPA 2002).

and field-collected sediments. The use of organic carbon–normalized excess SEM makes it possible to make a more definitive prediction of toxicity for those samples with metal in excess of AVS.

Applicability of Sediment Quality Guidelines to Field-Collected Sediments

Mechanistic SQGs, in particular those derived from EqP, were initially developed using laboratory-spiked sediments. The immediate question is this: How predictive are EqP SQGs when applied to field-collected sediments? Validation of SQGs is complicated by the fact that the chemicals that are causing the toxicity in field-collected sediments are not known. This is the reason that empirical SQGs cannot establish causal links between chemical concentrations and the observed toxicity. Nevertheless, mechanistic SQGs have been tested using field-collected sediments. The usual strategy is to use sediments that are heavily contaminated with a particular class of compounds.

As outlined in the following truth table, the critical failing of an SQG occurs when sediments that are predicted to be toxic are observed not to be toxic. All other outcomes are inconclusive with regard to validating the SQG.

		EqP predicts sample will be	
		Toxic	Not toxic
Toxicity test finds sample	Toxic	Correct prediction but does not prove guideline is correct, because other chemicals may be causing toxicity	No conclusion, because uncertain whether toxicity is caused by chemicals having EqP guideline
	Not toxic	Suggests EqP guideline is incorrect, assuming exceedance is beyond uncertainty in the model	Correct prediction, but does not prove guideline is correct

Organic Chemicals

Hoke et al. (1997) tested DDT-contaminated sediments from a field site in a freshwater stream. Using laboratory water-only tests with *Chironomus tentans* and the K_{OC}s for the DDT compounds, predicted sediment toxic units (PSTUs) and porewater toxic units (PWTUs) were calculated for each sediment. When *C. tentans* mortality is plotted against PSTU, most sediments with less than 3.0 PSTU elicited less than 50% mortality, while most sediments with greater than 3.0 PSTU elicited greater than 50% mortality (Figure

13-13). This is slightly higher than the theoretical prediction, which would predict 50% mortality at 1.0 PSTU.

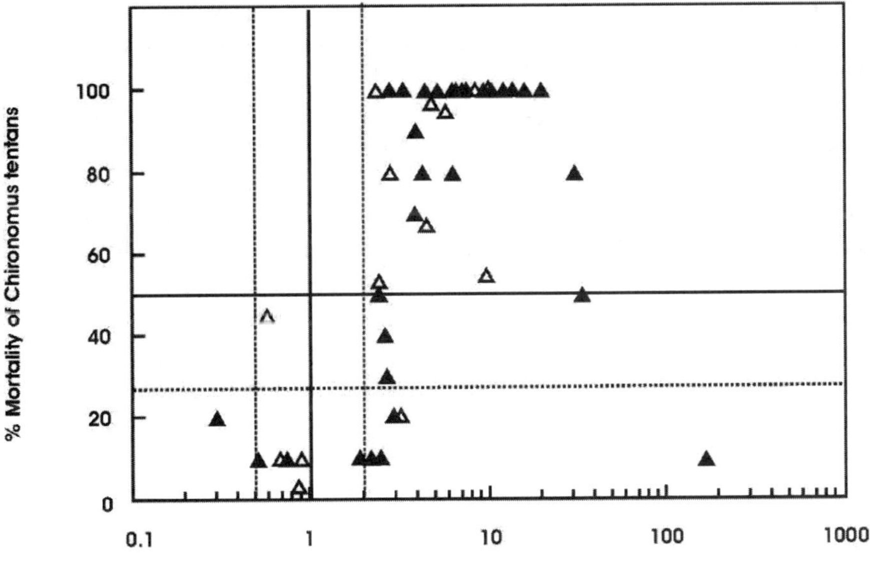

Figure 13-13 Mortality of *C. tentans* from 10-d laboratory sediment tests of grab (open symbol) and core horizon (filled symbol) samples of study site sediments versus predicted sediment toxic units (PSTUs) for total DDTR = DDT + DDD + DDE in the sediments. Vertical line at TU = 1, dotted lines at 0.5 and 2.0 TU, the range of uncertainty (Data from Hoke et al. 1997).

While EqP theory allows predictions of the extent of toxicity in terms of percent of a response, such as 50% mortality, it is helpful to define sediments as toxic when there is more than 24% mortality of amphipods during 10-d exposures under laboratory conditions. The amphipod test, introduced by Swartz et al. (1979), and the 24% mortality endpoint (Mearns et al. 1986) have become routine for testing and classifying sediments in large monitoring programs (Long et al. 1996; Maccauley et al. 1996). All toxicity results discussed below are from 10-d exposures of amphipods, and sediments are deemed toxic when there is more than 24% mortality. The validity of EqP predictions can be judged by examining how frequently sediments predicted to cause more than 24% mortality actually do so.

Figure 13-14 compares the PAH mixture SQG prediction for sediments in which PAHs are the predominant contaminant. The vertical solid line is the prediction. The complicating feature in these data sets is the number of PAHs actually measured in the data set. The target lipid narcosis model suggests that all of the PAHs present in the sediment contribute to the observed toxicity. This issue is discussed at some length in the USEPA's EqP sediment guidelines (ESGs; USEPA 2003). The most complete data sets quantify 34 PAHs that include the commonly measured parent (e.g., the unsubstituted PAHs such as naphthalene), as well as some alkylated homologous series that are less frequently measured (e.g., methyl-substituted naphthalenes). The toxicity of sediment-associated PAHs can be severely underestimated if these alkylated PAHs are not considered. A method for estimating the total of the 34 PAH concentrations, using either a subset of 13 or 23 commonly measured PAHs, is provided. The median ratio of the concentration of 34 PAHs to the concentration of the 13 PAHs is 2.75 (USEPA 2003). However, this ratio varies considerably, especially when petrogenic and pyrogenic mixtures are compared. Therefore, the portion of the data for which the total concentration is estimated from a measured subset is identified with a different plotting symbol in Figure 13-14.

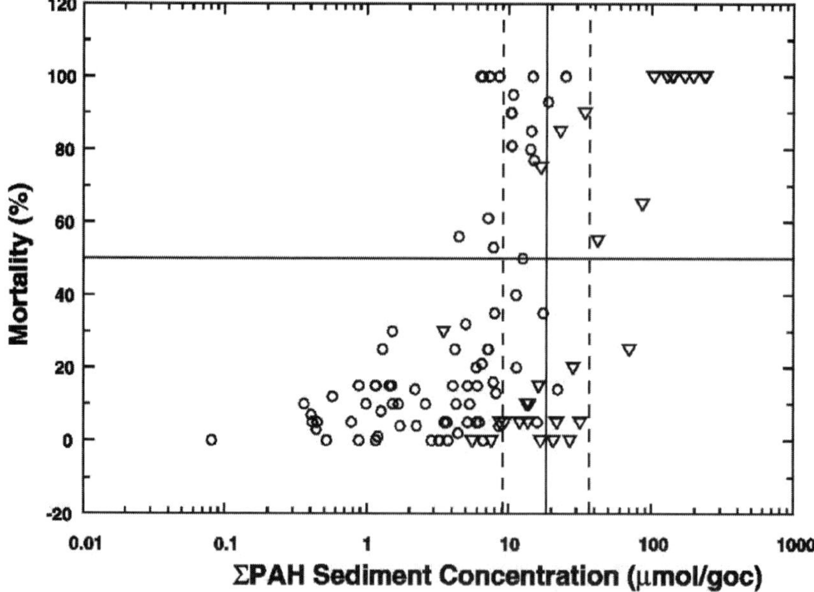

Figure 13-14 Comparison of observed mortality in *R. abronius* 10-d sediment toxicity tests. Data from Swartz et al. (1997) using the estimated total of 34 PAHs from the measured 13 PAHs (o). Data from Ozretich et al. (1997) from Elliot Bay (∇).

A set of data from a creosote-contaminated site (Ozretich et al. 2000) can be compared to the target lipid model–derived SQGs for total PAHs. In this case, 33 parent and alkylated PAHs were measured, so there is no need to estimate the total PAH concentrations in the sediment. Figure 13-15 presents the comparison. The vertical solid line is the predicted EqP SQG for total PAHs. The vertical dotted lines represent a factor of 2 uncertainty in the predicted extent of toxicity. The line at 24% mortality separates toxic from nontoxic samples. All of the 13 samples predicted to be toxic were toxic (the upper right-hand quadrant). This is further verification from field data of the EqP target lipid model for total PAHs.

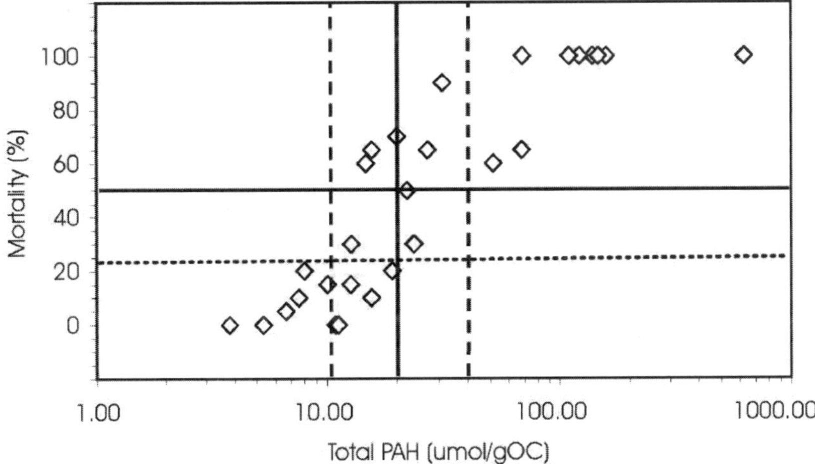

Figure 13-15 Comparison of observed mortality in *R. abronius* 10-d sediment toxicity tests to total PAH concentration from a creosote-contaminated site (adapted from Ozretich et al. 2000). The vertical solid line is the predicted EqP SQG for total PAHs. The vertical dotted lines represent a factor of 2 uncertainty in the predicted extent of toxicity. The line at 24% mortality separates toxic from nontoxic samples.

Sediments contaminated with fuel oil from a spill (Ho et al. 1999) provide another data set that can be examined. Samples were collected at 2 locations (HR1, HR2) for a period of 9 months following the spill. The 33 PAHs (parent and alkylated compounds) were measured (29 of the compounds measured were PAHs, and 4 compounds were heterocycles containing sulfur). The predicted mortality is compared to the EqP prediction in Figure 13-16. The observed mortality is predicted quite accurately. Only 4 samples were predicted to be toxic, and amphipod survival in all 4 was very low.

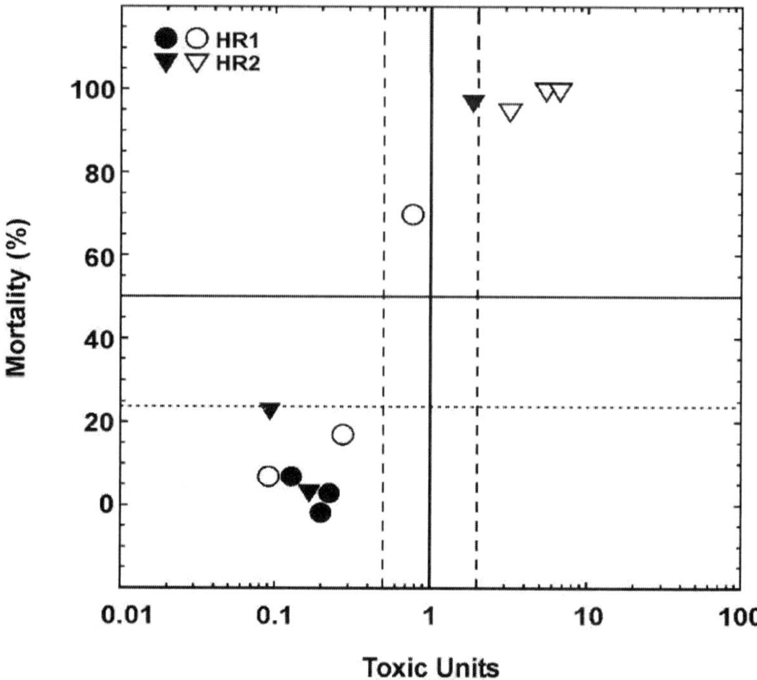

Figure 13-16 Mortality of *A. abdita* versus total PAH toxic units. Sediments from 2 locations (HR1, HR2) contaminated with fuel oil (data from Ho et al. 1999). Samples were collected at 2 locations (HR1, HR2) for a period of 9 months following the spill. Open symbols <100 d post spill; closed symbols >100 d post spill. The 33 PAHs (parent and alkylated compounds) were measured.

Another test of the EqP SQGs is observations of the abundance of sensitive organisms in field-collected samples. Figure 13-17 presents the observed *Ampelisca abdita* abundance versus estimated total PAH from sediments collected as part of the Virginian and Louisianian province EMAP and the New York–New Jersey Harbor REMAP sediment sampling programs. While amphipods may avoid sediments for many reasons independent of PAH concentration, the results are very encouraging. All the sediments that exceed the chronic EqP SQG (5.7 μmol/g OC) (USEPA 2003) have markedly lower amphipod abundance, consistent with prediction that if this concentration is exceeded, detrimental effects should be seen.

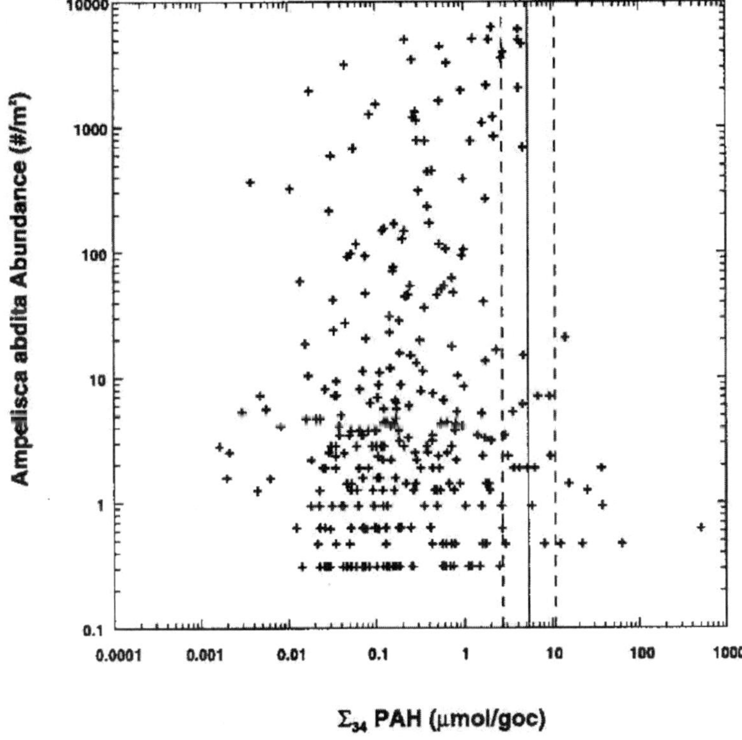

Figure 13-17 *A. abdita* abundance versus estimated total of the 34 PAHs using the measured 23 PAHs. Data from Virginian and Louisianian province EMAP and the New York–New Jersey Harbor REMAP. Vertical lines are average final chronic value (FCV) SQG a factor of 2 (Reprinted with permission from Di Toro and McGrath 2000. Copyright Society of Environmental Toxicology and Chemistry [SETAC]).

Metals

Field-contaminated sediments with high metals concentrations also have been tested. The sediments were chosen so that chemicals whose bioavailability is not affected by sulfides, for example, toxic organic chemicals, were at low concentrations and not likely sources of toxicity. Figure 13-18 presents the results. The ΣSEM/AVS method correctly predicts the lack of toxicity in field-collected sediments. It was shown previously that the organic carbon–normalized excess SEM SQG predicts the occurrence of toxicity in field-collected sediments as well outside of the uncertain range (Figure 13-12).

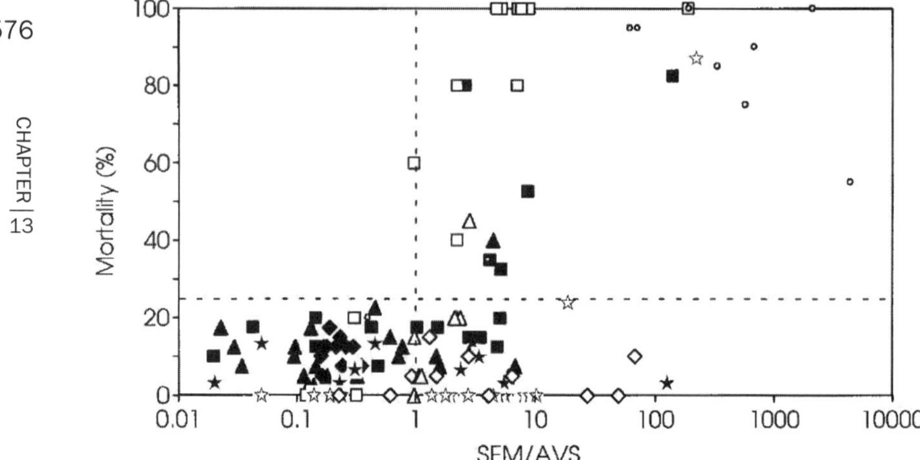

Figure 13-18 Percent mortality of the amphipods *A. abdita* and *Hyalella azteca*); oligochaete *Lumbriculus variegates* and polychaete *Neanthes arenaceodentata* exposed to sediments from freshwater (open symbols) and saltwater (solid symbols) locations as a function of the SEM/AVS ratio (Reprinted with permission from Hansen et al. 1996. Copyright Society of Environmental Toxicology and Chemistry [SETAC]).

SQGs for metals have also been tested in chronic exposures in both laboratory and field settings. The data in Figure 13-12 from which the effect concentrations were developed are mostly from 10-d exposures. The results of much longer-term chronic toxicity tests with metals-spiked sediments (Hare et al. 1994; DeWitt et al. 1996; Hansen, Mahony, et al. 1996; Liber et al. 1996; Sibley et al. 1996; Boothman et al. 2001) can also be compared to organic carbon–normalized excess SEM. Some of these tests are field deployments of spiked sediments and rely on the integrity of the colonizing benthic fauna as the measure of toxicity. Therefore, they are sensitive and unambiguous tests of the method. They expose the entire benthic community to the metal-contaminated sediment, across the entire spectrum of exposure routes. The data are presented in Figure 13-19 (USEPA 2002). The metals used in the experiment are labeled. The multiple metals are from a field-deployed mixture experiment (Boothman et al. 2001). The chronic data indicate that the lower bound for organic carbon–normalized excess $\Sigma SEM/f_{OC}$ derived from the 10-d toxicity tests also is predictive for chronic exposure because all but 1 sediment exhibited no effects below $\Sigma SEM / f_{OC} < 130$ μmol/g OC.

The fact that organic carbon–normalized excess SEM can apparently predict the onset of toxicity of Cd, Cu, Ni, Pb, and Zn without regard to their specif-

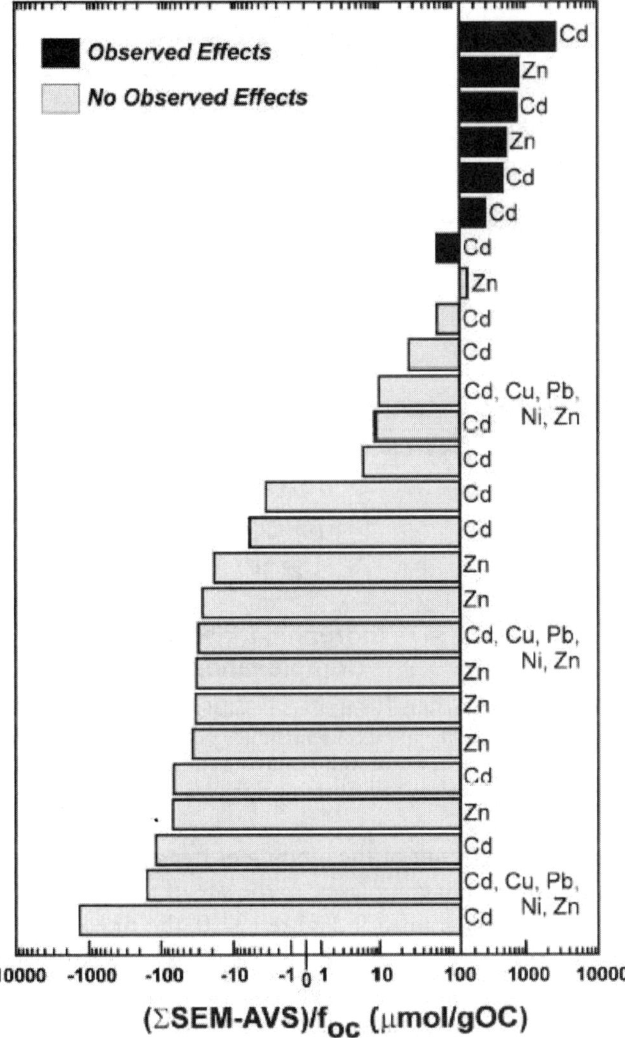

Figure 13-19 Results of chronic toxicity tests with metal-spiked sediments compared to organic carbon–normalized excess SEM. Data from Hare et al. 1994; DeWitt et al. 1996; Hansen, Mahony, et al. 1996; Liber et al. 1996; Sibley et al. 1996; Boothman et al. 2001. All but 1 sediment exhibited no effects below $\Sigma SEM / f_{OC} < 130$ mmol/g OC (USEPA 2002).

ic characteristics is remarkable, and an explanation has been sought for some time. Recently, a computation has been done using the BLM (Di Toro et al. 2001; Paquin et al. 2003) that is diagrammed in Figure 13-20. The BLM, (HydroQual 2003) is based on the assumption that the toxicity of metals to

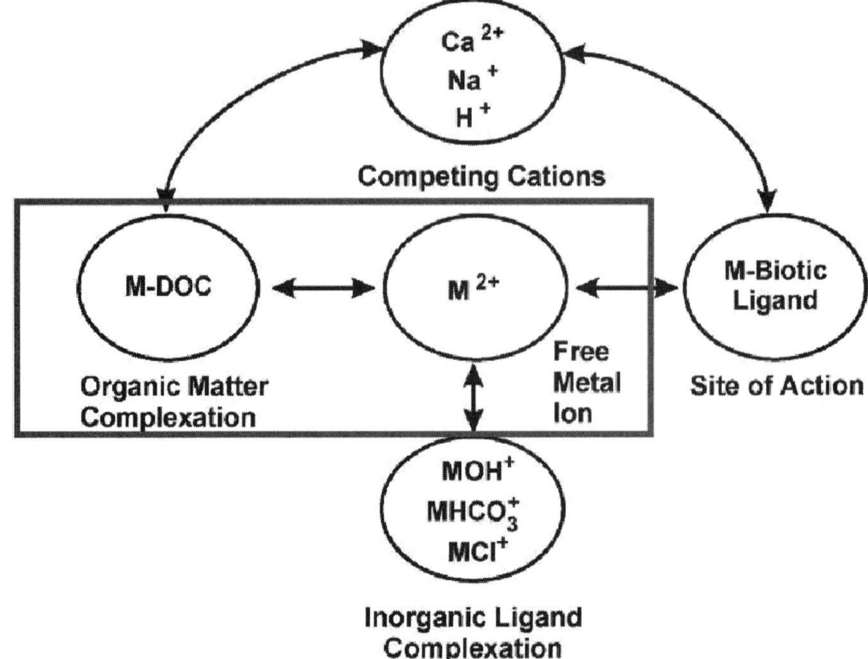

Figure 13-20 Schematic diagram of the biotic ligand model (BLM). The box indicates the interaction used for computing the metal concentration on organic carbon corresponding to the LC50 for *Daphnia magna* (Reprinted with permission from Di Toro et al. 2001. Copyright Society of Environmental Toxicology and Chemistry [SETAC]).

aquatic organisms is the result of the binding of free metal cation, and in some cases metal-inorganic anion complexes, to the site of toxic action that is called the "biotic ligand." As diagramed in Figure 13-20, the BLM determines the binding of metal cation (M^{2+}) to the biotic ligand, to inorganic ligands and DOC, and the competition by other cations (Ca^{2+}, Mg^{2+}, etc) for the biotic ligand. The BLM is the subject of intensive research and development (see the papers accompanying Paquin et al. 2003) and is currently being incorporated into the USEPA freshwater Cu criteria.

The BLM was applied to the problem of determining the effects concentration for metals on sediment organic carbon as follows. Using the BLM parameters that characterize *Daphnia magna* acute toxicity, the concentration of metal on organic carbon was calculated when the aqueous metal concentration was at the LC50 for *D. magna.* The organic carbon was assumed to be entirely humic acid. This is the assumption that is made in applying the speciation model used in the BLM (WHAM V, Tipping 1994) to soils (Lofts

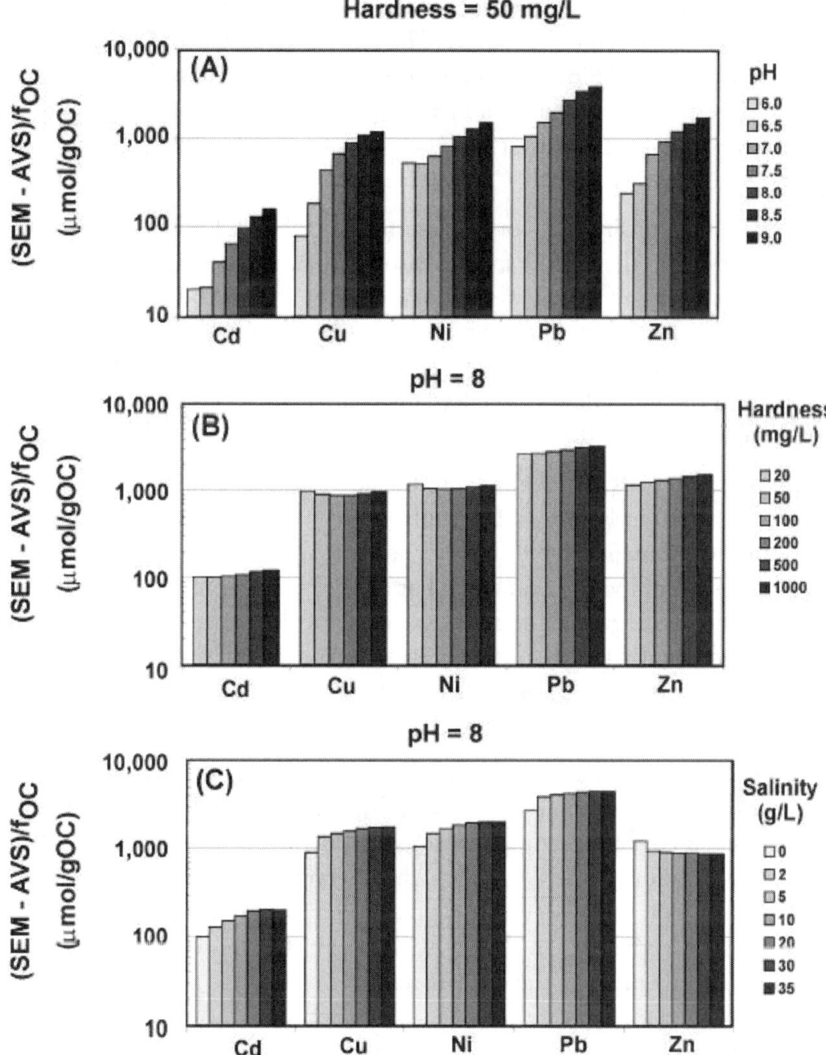

Figure 13-21 Metal concentrations on organic carbon corresponding to the LC50 for *D. magna:* (A) pH varying (B) hardness varying (C) salinity varying (Di Toro et al. 2005)

and Tipping 1998). The results are shown in Figure 13-21. For all the metals except Cd, the resulting LC50 metal concentration on organic carbon ranges from approximately 100 to 3000 µmol/g OC, the values found previously from simply analyzing the mortality data (Figure 13-12). The primary variable influencing the predicted LC50 is interstitial water pH (Figure 13-21A).

Remarkably, hardness variations do not substantially change the LC50 on an organic carbon basis (Figure 13-21B). The reason is that Ca and Mg compete with the metals on the biotic ligand (this is the mechanism in the BLM for the protective effect of hardness) and also on organic carbon, apparently to the same extent. The effect of salinity variations is also small for the same reason (Figure 13-21C).

It is important to realize that complexation of metal by DOC in the pore water has no effect on these results. The reason is that the partitioning to organic carbon is due only to the free metal ion and is not affected by the formation of metal–DOC complexes. Further, because the quantity of particulate organic carbon (POC) greatly exceeds the DOC, the small amount of metal that desorbs from the POC does not materially change the metal concentration on the sediment organic carbon.

The computed LC50s are compared to the acute amphipod toxicity data (Berry et al. 1996) in Figure 13-22. Only the data with excess metal (SEM > AVS) in the sediment are presented. The vertical dotted lines are the computed LC50s for pH = 6, 7, 8, and 9. For all the metals except Cd, the BLM predictions are quite good. The BLM-computed Cd concentration on organic matter is significantly less than the observed effect concentrations. Because these are saltwater tests, it is possible that the cadmium chloride complex is sorbing to organic matter but not to the biotic ligand.

Figure 13-23 presents the single metal chronic data comparison to the BLM computations. For both Zn and Cd in freshwater, the comparison is quite good. The predicted effect concentrations straddle the observed no-observed-effects concentration (NOEC) and lowest-observed-effects concentration (LOEC). Note that the observed and predicted freshwater Cd LOEC is about 100 μmol/g OC, whereas the Zn LOEC is about 500 μmol/g OC, and that the BLM correctly predicts the difference. It is only the Cd seawater results that are not consistent. Thus the BLM, which accounts for metal speciation and competitive cation interaction, appears to be able to predict the onset of chronic toxicity with the exception of Cd in seawater, for which it is suspected that the metal partitioning model (WHAM V) is not applicable. These results are described in more detail elsewhere (Di Toro et al. 2005).

Application to field-collected sediments

With the exception of the amphipod abundance example (Figure 13-17) the validations described above employed either laboratory-spiked sediments or field-collected sediment for which there is ample evidence which class of con-

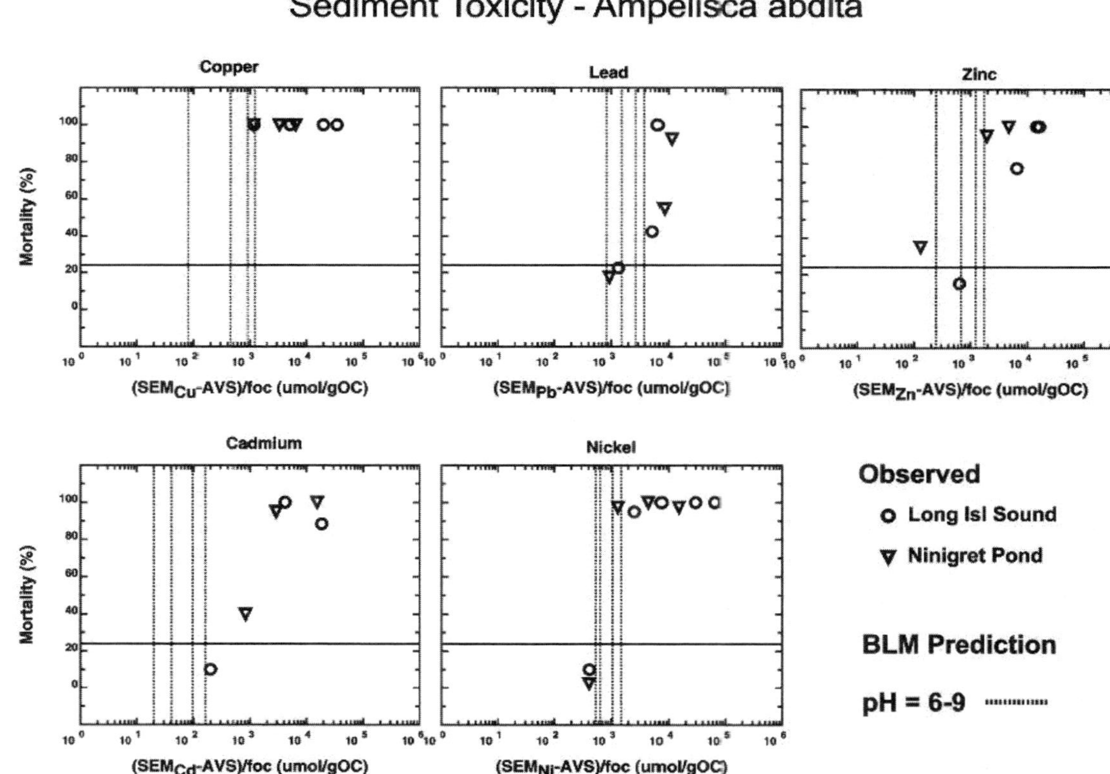

Figure 13-22 Comparison of metal concentrations on organic carbon corresponding to the LC50 for *D. magna* to observed *A. abdita* mortality (Berry et al. 1996). The vertical dotted lines correspond to pH = 6, 7, 8, and 9 (Di Toro et al. 2005).

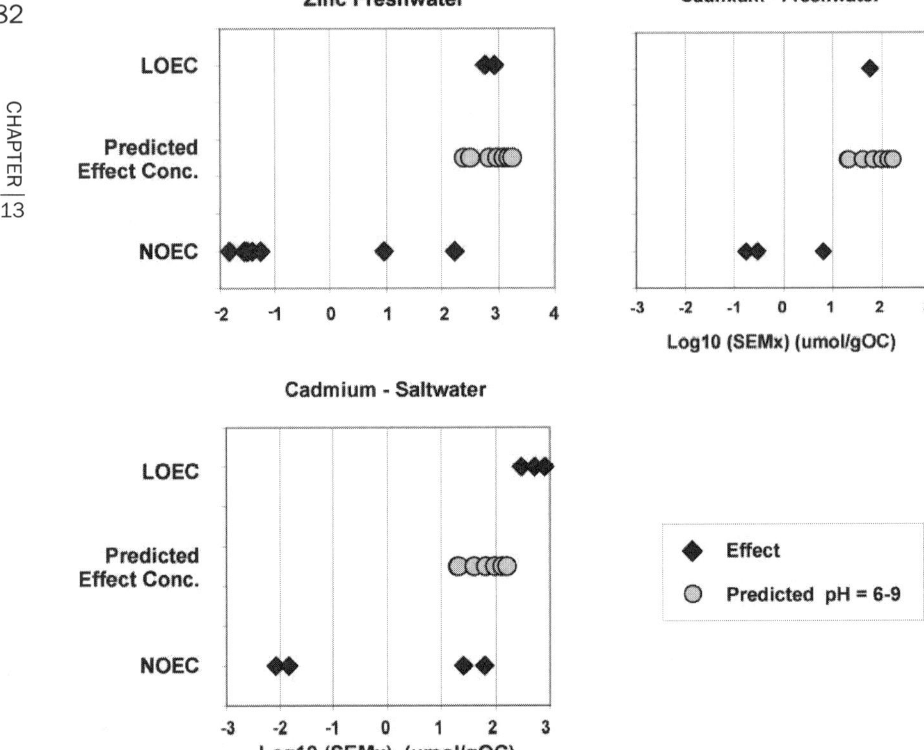

Figure 13-23 Comparison of metal concentrations on organic carbon corresponding to the LC50 for *D. magna* to observed NOEC and LOEC concentrations from chronic sediment toxicity tests (Hare et al. 1994; DeWitt et al. 1996; Hansen, Mahony, et al. 1996; Liber et al. 1996; Sibley et al. 1996). The diamonds are the experimental NOECs and LOECs. The circles correspond to pH = 6, 7, 8, and 9. The vertical spacing is chosen for visual convenience.

taminant is responsible for the toxicity. The situation is quite different when the larger data sets that are used to derive the empirical SQGs are examined. Attempting to validate EqP guidelines using this approach is difficult because of the uncertainties introduced by the presence of mixtures of chemicals in field samples. Because EqP SQGs are designed to address only specific chemicals (or groups of chemicals, such as cationic metals or nonionic narcotic toxicants), they cannot be expected to correctly predict the existence of toxicity when other chemicals are present in toxic amounts. For these reasons, the conclusions that can be drawn from this exercise are limited.

In a data set of 2475 field-collected samples with paired measurements of 23 PAHs and 10-d survival of amphipods, O'Connor (2002) found that carbon–normalized total PAH concentrations for 12, 25, and 123 samples were predicted to be acutely toxic based on exceeding the final acute value (FAV) from USEPA water quality criteria, the 96-h LC50, and chronically toxic based on exceeding the final chronic value (FCV), as shown in Table 13-1.

Table 13-1 Predicted and observed toxicity of PAHs to marine amphipods based on FAV, LC50, and FCV effect concentrations

Biological basis	Guideline	Predicted	Observed	Correct %
FAV	Σ_{23}PAH > 18 µmol/gTOC	12	7	58
LC50	Σ_{23}PAH > 11 µmol/gTOC	25	12	48
FCV	Σ_{23}PAH > 3.5 µmol/TOC	123	54	44

Whichever of those concentrations is chosen to predict less than 80% survival of amphipods over 10-d exposures, the predicted toxicity was observed in less than half of the cases, putting more than half the observations in the lower left-hand box of the truth table. O'Connor attributed this evidence of the failure of the EqP SQG to the likely presence of soot in high-PAH samples. Except in cases of creosote or spilled oil, the majority of PAHs in sediments are from combustion sources and thus may be associated with soot carbon and other combusted forms of organic carbon (Boehm and Farrington 1984). However, in general, the distribution of pyrogenic PAHs in the environment is low in terms of concentration but high in regards to areal extent (Burgess et al. 2003). For example, McGroddy and Farrington (1995) showed that PAHs in sediments from Boston Harbor would not partition into an aqueous phase except to a very small extent. Thus, predictions of porewater concentrations (and thus toxicity) that use partition coefficients derived from octanol–water coefficients and total organic carbon concentrations in sediment are overestimates. Conversely, a unique strength of the EqP approach is that the model can be amended, if deemed necessary, to include new partitioning phases like soot carbon. If there is evidence suggesting that bioavailable concentrations of PAHs are being overpredicted, the EqP model can have terms added that address the effects of soot carbon. Depending on the availability of these terms, the EqP model can be adjusted to make better predictions because of the presence of soot carbon. To date, few studies have evaluated the importance of pyrogenic PAHs potentially associated with soot carbon on the toxicity of sediment samples.

Summary of Predictive Capability

The data presented above suggest that, for neutral organic chemicals and narcotic mixtures, a factor of 2 (0.5× to 2.0× the predicted value) is the accuracy that can be expected for predicting the outcome of a single-species toxicity test. That encompasses the field data and the majority of the laboratory data. The data sets for metal prediction indicate that for ΣSEM < AVS, there is no case for which the prediction of lack of toxicity, either acute or chronic, and from either laboratory-spiked or field colonization experiments, has failed. For the case that ΣSEM > AVS, the organic carbon–normalized excess SEM has an uncertain range that encompasses 25% of the data but is 95% accurate outside of that range. For chronic data, 1 sample is misclassified. However, there are too few data for even a rough statistical evaluation. The application of the BLM to sediments promises to reduce this uncertainty because it provides a method for directly calculating the SQG for a specific sediment. The initial results (Figures 13-22 and 13-23) are quite encouraging.

It should be pointed out that EqP predictions are bound to fail if the cause of toxicity is not among the measured chemicals. This can be the case in situations where many chemicals are present and only a few are quantified, for example, in oil-contaminated sediment where only 13 PAHs are measured. Additionally, the partitioning theory can fail. The EqP model requires a partition coefficient to predict the sediment concentration corresponding to the toxic concentrations established in water-only exposures. The presence of significant quantities of unusual partitioning phases, for example, soot or wood particles that are not properly taken into account, would invalidate the predictions. However, the data presented in this chapter suggest that these are relatively rare occurrences. For the many sediment samples tested in laboratory-spiked experiments and for the field-collected sediments that are heavily contaminated, the EqP predictions are consistently within the error bounds presented in the preceding paragraph. The exception to this appears to be the results from the large data sets, for which the EqP SQGs are overprotective in approximately 50% of the cases.

One final comment on the relationship between mechanistic and empirical SQGs is worth mentioning. The use of mechanistically based SQGs derived from the EqP method is warranted for situations in which the chemical cause of the toxicity is in question: that is, what is the concentration *C* of chemical *X* in sediment *Y* that will cause an adverse effect to organism *Z*. The empirical SQGs do not answer that question. Rather they make a probability estimate of whether a field collected sediment with concentration *C* of chemical *X* will cause an adverse effect to organism *Z*. That is, they would predict that there is a *P*% chance that a particular sediment will cause an adverse effect if tested. But the identity of the chemical or chemicals causing the adverse effect is unknown. In fact, it may be an unmeasured chemical that covaries with the measured chemicals. Thus, the 2 methods are complementary. Mechanistic EqP SQGs seek to establish cause and effect using partitioning models. Empirical SQGs seek to make predictions based on empirical correlations. Each has its uses, and each has its limitations. Perhaps the most useful outcome of this Pellston Workshop was this clarification and the resulting cessation of hostilities between the proponents of mechanistic and empirical SQGs.

References

Adams WJ, Kimerle RA, Mosher RG. 1985. Aquatic safety assessment of chemicals sorbed to sediments. In: Cardwell RD, Purdy R, Bahner RC, editors. Aquatic toxicology and hazard assessment: Seventh Symposium. STP 854. p 429–453. Philadelphia: American Soc for Testing and Materials.

Ankley GT, Berry WJ, Di Toro DM, Hansen DJ, Hoke RA, Mount DR, Reiley MC, Swartz RC, Zarba CS. 1996. Use of equilibrium partitioning to establish sediment quality criteria for nonionic chemicals: A reply to Iannuzi et al. *Environ Toxicol Chem* 15:1019–1024.

Ankley GT, Di Toro DM, Hansen DJ, Berry WJ. 1996. Technical basis and proposal for deriving sediment quality criteria for metals. *Environ Toxicol Chem* 15:2056–2066.

Ankley GT, Mattson VR, Leonard EN, West CW, Bennett JL. 1993. Predicting the acute toxicity of copper in freshwater sediments: evaluation of the role of acid volatile sulfide. *Environ Toxicol Chem* 12:312–320.

Berry WJ, Hansen DJ, Mahony JD, Robson DL, Di Toro DM, Shipley BP, Rogers B, Corbin JM, Boothman WS. 1996. Predicting the toxicity of metals-spiked laboratory sediments using acid volatile sulfide and interstitial water normalizations. *Environ Toxicol Chem* 15:2067–2079.

Berry W, Cantwell M, Edwards P, Serbst JR, Hansen DJ. 1999. Predicting the toxicity of sediments spiked with silver. *Environ Toxicol Chem* 18:40–48.

Boehm PD, Farrington JW. 1984. Aspects of the polycyclic aromatic hydrocarbon geochemistry of recent sediments in the Georges Bank region. *Environ Sci Technol* 18:840–845.

Boothman W, Hansen D, Berry W, Robson D, Helmstetter A, Corbin J, Pratt S. 2001. Biological response to variation of acid volatile sulfides and metals in field-exposed spiked sediments. *Environ Toxicol Chem* 20:264–272.

Burgess RM, Ahrens MJ, Hickey CW. 2003. Geochemistry of PAHs in aquatic environments: A synthesis of source, distribution and persistence. In: Douben PET, editor. PAHs: An ecotoxicological perspective. London: J Wiley.

Burgess RM, Ahrens MJ, Hickey CW, den Besten PJ, ten Hulscher D, van Hattum B, Meador JP, Douben PET. 2003. An overview of the partitioning and bioavailability of PAHs in sediments and soils. In: Douben PET, editor. PAHs: An ecotoxicological perspective. Chichester: Wiley. p 90–126.

DeWitt T, Swartz R, Hansen D, McGovern D, Berry W. 1996. Bioavailability and chronic toxicity of cadmium in sediment to the estuarine amphipod *Leptocheirus plumulosus. Environ Toxicol Chem* 15:2095–2101.

Di Toro DM, Allen HE, Bergman HL, Meyer JS, Paquin PR, Santore RC. 2001. A biotic ligand model of the acute toxicity of metals. I. Technical basis. *Environ Toxicol Chem* 20:2383–2396.

Di Toro DM, Mahony JD, Hansen DJ, Scott KJ, Carlson AR, Ankley GT. 1992. Acid volatile sulfide predicts the acute toxicity of cadmium and nickel in sediments. *Environ Sci Technol* 26:96–101.

Di Toro DM, Mahony JD, Hansen DJ, Scott KJ, Hicks MB, Mayr SM, Redmond MS. 1990. Toxicity of cadmium in sediments: The role of acid volatile sulfide. *Environ Toxicol Chem* 9:1487–1502.

Di Toro DM, McGrath JA. 2000. Technical basis for narcotic chemicals and polycyclic aromatic hydrocarbon criteria. II. Mixtures and sediments. *Environ Toxicol Chem* 19:1971–1982.

Di Toro DM, McGrath JA, Hansen DJ. 2000. Technical basis for narcotic chemicals and polycyclic aromatic hydrocarbon criteria. I. Water and tissue. *Environ Toxicol Chem* 19:1951–1970.

Di Toro DM, McGrath JM, Hansen DJ, Berry WJ, Paquin PR, Mathew R, Wu KB, Santore RC. 2005. Predicting sediment metal toxicity using a sediment biotic ligand model: Methodology and initial application. *Environ Toxicol Chem* Forthcoming.

Di Toro DM, Zarba CS, Hansen DJ, Berry WJ, Swartz RC, Cowan CE, Pavlou SP, Allen HE, Thomas NA, Paquin PR. 1991. Technical basis for establishing sediment quality criteria for nonionic organic chemicals using equilibrium partitioning. *Environ Toxicol Chem* 10:1541–1583.

Hansen DJ, Berry WJ, Mahony JD, Boothman WS, Robson DL, Ankley GT. 1996. Predicting the toxicity of metals-contaminated field sediments using interstitial concentration of metal and acid volatile sulfide normalizations. *Environ Toxicol Chem* 15:2080–2094.

Hansen D, Mahony J, Berry W, Benyi S, Corbin J, Pratt S, Di Toro D, Able M. 1996. Chronic effect of cadmium in sediments on colonization by benthic marine organisms: An evaluation of the role of interstitial cadmium and acid volatile sulfide in biological availability. *Environ Toxicol Chem* 15:2126–2137.

Hare L, Carignan R, Huerta-Diaz M. 1994. A field study of metal toxicity and accumulation by benthic invertebrates; implication for the acid-volatile sulfide (AVS) model. *Limnol Oceanogr* 39:1653–1668.

Hermens JLM. Quantitative structure-activity relationships of environmental pollutants. 1989. In: Hutzinger O, editor. Handbook of environmental chemistry. Volume 2E, Reactions and processes. Berlin: Springer Verlag. p 111–162.

Ho KT, Patton L, Latimer JS, Pruell RJ, Pelletier M, McKinney R, Jayaraman S. 1999. The chemistry and toxicity of sediment affected by oil from the North Cape spilled into Rhode Island Sound. *Mar Pollut Bull* 38:314–323.

Hoke RA, Ankley GT, Kosian PA, Cotter AM, Vandermeiden FM, Balcer M, Phipps GL, West C, Cox JS. 1997. Equilibrium partitioning as the basis for an integrated laboratory and field assessment of the impacts of DDT, DDE and DDD in sediments. *Ecotoxicology* 6:101–125.

HydroQual Inc. 2003. The Biotic Ligand home page. Available from: http://www.hydroqual.com/blm/.

Liber K, Call D, Markee T, Schmude K, Balcer M, Whiteman F, Ankley G. 1996. Effects of acid volatile sulfide on zinc bioavailability and toxicity to benthic macroinvertebrates: A spiked-sediment field experiments. *Environ Toxicol Chem* 15:2113–2125.

Lofts S, Tipping E. 1998. An assemblage model for cation binding by natural particulate matter. *Geochim Cosmochim Acta* 62:2609–2625.

Luthy RG, Aiken GR, Brusseau ML, Cunningham SD, Gschwend PM, Pignatello JJ, Reinhard M, Traina SJ, Weber WJ, Westall JC. 1997. Sequestration of hydrophobic organic contaminants by geosorbents. *Environ Sci Technol* 31: 3341–3347.

McGroddy SE, Farrington JW. 1995. Sediment porewater partitioning of polycyclic aromatic hydrocarbons in three cores from Boston Harbor, Massachusetts. *Environ Sci Technol* 29:1542–1550.

Nebeker AV, Schuytema GS, Griffis WL, Barbitta JA, Carey LA. 1989. Effect of sediment organic carbon on survival of *Hyalella azteca* exposed to DDT and endrin. *Environ Toxicol Chem* 8:705–718.

O'Connor TP. Empirical and theoretical shortcomings of sediment quality guidelines. 2002. In: Whittemore R, editor. A treatise on sediment quality. Alexandria (VA): Water Environment Federation.

Ozretich R, Swartz R, Lamberson J, Ferraro S. 1997. Toxicity of alkylated aromatic hydrocarbons in sediment to a marine amphipod. Test of the ΣPAH model quantitative structure activity relationship. Proceedings, 18th Annual Meeting; Society of Environmental Toxicology and Chemistry (SETAC); San Francisco, CA, USA. p 289.

Ozretich RJ, Ferraro SP, Lamberson JA, Cole FA. 2000. Test of Σ polycyclic aromatic hydrocarbon model at a creosote-contaminated site, Elliott Bay, Washington, USA. *Environ Toxicol Chem* 19:2378–2389.

Paquin PR, Gorsuch JW, Apte S, Batley GE, Bowles KC, Campbell PGC, Delos CG, Di Toro DM, Dwyer RL, Galvez F, Gensemer RW, Goss GG, Hogstrand C, Janssen CR, McGeer JC, Naddy RB, Playle RC, Santore RC, Schneider U, Stubblefield WA, Wood CM. 2003. The biotic ligand model: a historical overview. *Comp Biochem Physiol Part C* 133:3–35.

Schuytema GS, Nebeker AV, Griffis WL, Miller CE. 1989. Effects of freezing on toxicity of sediment contaminated with DDT and endrin. *Environ Toxicol Chem* 8:883–891.

Sibley P, Ankley G, Cotter A, Leonard E. 1996. Predicting chronic toxicity of sediments spiked with zinc: an evaluation of the acid volatile sulfide model using a life cycle test with the midge *Chironomus tentans. Environ Toxicol Chem* 15: 2102–2112.

Swartz RC, DeBen WA, Cole FA. 1979. A bioassay for the toxicity of sediment to the marine macrobenthos. *J Water Pollut Control Fed* 51:944–950.

Swartz RC, Schults DW, Ozretich RJ, Lamberson JO, Cole FA, DeWitt TH, Redmond MS, Ferraro SP. 1995. PAH: A model to predict the toxicity of polynuclear aromatic hydrocarbon mixtures in field-collected sediments. *Environ Toxicol Chem* 14:1977–1987.

Swartz RC, Schults DW, Dewitt TH, Distort GR, Lamberson JO. 1990. Toxicity of fluoranthene in sediment to marine amphipods: A test of the equilibrium partitioning approach to sediment quality criteria. *Environ Toxicol Chem* 9: 1071–1080.

Tipping E. 1994. WHAM: A computer equilibrium model and computer code for waters, sediments, and soils incorporating a discrete site/electrostatic model of ion-binding by humic substances. *Comput Geosci* 20:973–1023.

[USEPA] US Environmental Protection Agency. 2000. Procedures for the derivation of equilibrium partitioning sediment benchmarks (ESBs) for the protection of benthic organisms: Metal mixtures (cadmium, copper, lead, nickel, silver, and zinc) [draft]. Washington DC: USEPA Office of Research and Development. EPA-600-R-02-011.

[USEPA] US Environmental Protection Agency. 2002. Technical basis for the derivation of equilibrium partitioning sediment benchmarks (ESBs) for the protection of benthic organisms: Nonionic organics [draft]. Washington DC: USEPA Office of Research and Development. EPA-600-R-02-014.

[USEPA] US Environmental Protection Agency. 2003. Procedures for the derivation of equilibrium partitioning sediment benchmarks (ESBs) for the protection of benthic organisms: PAH mixtures. Washington DC: USEPA Office of Research and Development. EPA-600-R-02-013.

Use of sediment quality guidelines in damage assessment and restoration at contaminated sites in the United States

14

RON GOUGUET

Today we live in modern societies that rely on complex manufacturing and transportation networks to support our large populations and burgeoning growth. We move tremendous quantities of fuels and toxic chemicals to support our lifestyles. Sometimes these commodities are accidentally released into the environment. We typically react to stop the release and mitigate its impacts; these first immediate steps often evolve to long-term remediation to clean up the spill. Chronic toxic chemical releases or historical persistent chemical releases can also have adverse effects on natural resources. Generally, the release has had some impact on natural resources. If the contaminants have the potential to cause harm to the environment, active measures (i.e., cleanup) may have to be employed. The additional impact of cleanup activities can be substantial in the medium term. This "collateral damage" should be taken into consideration during spill response. This chapter describes an approach that uses sediment quality guidelines (SQGs) and other lines of evidence (LOEs) in the cooperative evaluation of benthic injury due to sediment contamination.

Natural Resource Damage Assessment in the United States

Most societies value natural resources: their birds, crabs, fish, turtles, water, air, etc. The United States broadly categorizes natural resources such as land,

Use of Sediment Quality Guidelines and Related Tools for the Assessment of Contaminated Sediments
Wenning RJ, Batley GE, Ingersoll CG, Moore DW, editors.
 ISBN 1-880611-71-6

fish, wildlife, biota, air, water, groundwater, drinking water supplies, and other such resources into 5 groups:

1) surface water resources,
2) groundwater resources,
3) air resources,
4) geologic resources, and
5) biological resources (43 CFR §11.14 (z)).

Society places value on the resources or their uses; that is, the societal value of the resource. "Natural resource services" is a concept used to measure the societal value of the physical and biological functions performed by the natural resources, including human uses of those services and services to other resources and ecosystems.

In the US, federal and state laws provide a scheme to compensate the public for the loss of natural resources and their services using natural resource damage assessment (NRDA) and compensatory restoration. The purpose of an NRDA is to determine the quantity of the losses and create habitat, increase stocks, or otherwise provide additional natural resources and services to compensate for the lost resources and services that result from a chemical or oil release to the environment. Rules for conducting NRDAs were developed by the US Department of the Interior (43 CFR Part 11) and the National Oceanic and Atmospheric Administration (NOAA; 15 CFR Part 990). European, South American, and Southeast Asian governments also have begun to consider losses of their natural resources that occur when oil or hazardous chemicals are accidentally released into the environment.

Sometimes we are lucky and a release dissipates quickly (Figure 14-1, scenario 1), so there are few impacts to natural resources and little loss of natural resources or their services. Sometimes our cleanup activities are quite successful and have little long-term impact, themselves, on natural resources (Figure 14-1, scenario 2). In each of these cases, while some organisms and habitats are briefly affected, few medium- to long-term impacts remain in the environment. (Figure 14-1, scenario 2). In another situation, a more persistent, more toxic, more difficult to remove release would generally result in many organisms and habitats being affected on a medium- to long-term basis (Figure 14-1, scenario 3).

Under the US Comprehensive Environmental Response, Compensation, and Liability Act (CERCLA) 122j, trustees have the authority to release potentially responsible parties (PRPs) from natural resource damages liability if PRPs take appropriate measures to protect and restore injured natural resources. Under

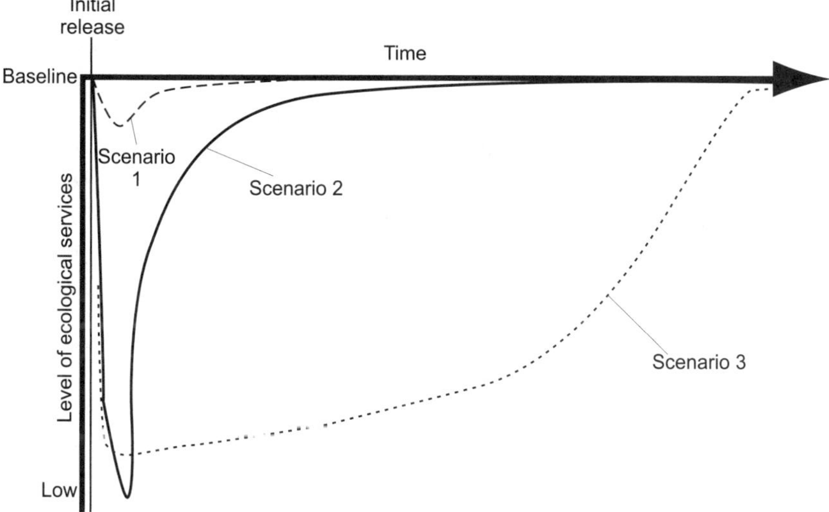

Figure 14-1 Simplified natural resource service loss and recovery curves associated with oil or hazardous substance release scenarios

CERCLA, trustees authorized to act on behalf of the public include the federal and state government and Native American tribes (40 CFR §300.600). The approach described here protects and restores coastal natural resources by collaboratively resolving natural resource liability as part of or in parallel to the cleanup process. By working with the responsible parties and trustees during the cleanup process, trustees can quickly and efficiently plan for and cooperatively implement restoration actions that compensate the public for natural resource injuries, including interim service losses, caused by contamination.

Achieving restoration of injured resources requires that the following questions be answered:

- What measures must be taken to protect natural resources from existing and future threats?
- What resources have been injured, and what is the loss to the public?
- How can the resources be restored, and what type and amount of restoration are appropriate to make the public and the environment whole?

Integration of the trustees into the site characterization, risk assessment, and cleanup planning process ensures that the first question above is answered at the appropriate point in the cleanup process. The damage assessment and restoration planning process addresses the last 2 questions and has 3 primary phases: injury assessment, restoration planning (including identification, evaluation and selection from among restoration alternatives and scaling of restoration), and restoration implementation. "Injury" means a measurable adverse change, either long or short term, in the chemical or physical quality or the viability of a natural resource, resulting either directly or indirectly from exposure to a discharge of oil or release of a hazardous substance or from exposure to a product of reactions resulting from the discharge of oil or release of a hazardous substance [43 CFR Part 11]. This chapter explains a way of cooperatively scaling natural resource injuries and restoration. Regarding estuaries of coastal Louisiana and Texas, the trustees and cooperating parties have evaluated injury to benthos and loss of benthic services (LOBS) using SQGs and SQG indices (Texas Trustees 2001).

Barnthouse and Stahl (2002) provide views on specific approaches drawn from ecological risk assessment (ERA) practices that could improve NRDA. Their recommendations refer to the conduct of rigorous and costly investigations that could be required of trustees to prove NRDA claims. They state that, "Perhaps the greatest challenge facing Trustees and PRPs involved in NRDA proceedings is the development of an assessment process that leads to cost-effective and properly scaled restoration of natural resource services." The approach outlined herein has been successfully applied to meet this challenge.

Ecological Risk Assessment and Risk Management: Trustee Integration

ERA is used to determine the probability that an ecological receptor (resource) will be harmed through exposure to a particular hazardous substance in its environment (USEPA 1997). Some form of ERA is usually performed as part of risk-based corrective action investigations under various environmental statutes (e.g., US CERCLA, Texas Risk Reduction Rule, US Federal Water Pollution Control Act [FWPCA]). The US Superfund ERA process consists of several steps, ranging from screening to risk characterization (USEPA 1997). A properly conducted risk assessment will provide useful information to guide the party responsible for cleaning up the release. Ideally, trustees' concerns are integrated into work plans for the remedial investigations (which determine the nature and extent and risk) to reduce potentially duplicative studies and efficiently fill any data gaps without additional studies.

Typically, conservative SQGs are used in an initial screening step when ecological risk to benthos from oil or chemical releases is assessed (USEPA 1997). If the environmental concentration of the particular toxic substance is higher than its respective, conservative SQG (e.g., effects range low [ERL]), then that contaminant is carried forward in the ERA. Further investigation during the

ERA must generally be conducted to determine if that contaminant significantly contributes to risk to benthos at the site. For many contaminants, no-observed-adverse-effect levels (NOAELs) and lowest-observed-adverse-effects levels (LOAELs) define the "risk range" within which benthos risk can safely be managed (USEPA 1997). The "risk range" defines contamination levels identified as posing no ecological risk and the lowest contamination levels identified as likely to produce adverse ecological effects on the benthos assessment endpoint (USEPA 1997).

Agreement on conservative ecological risk management at screening levels (or NOAEL) can be the most appropriate strategy, provided the costs in collateral damage and dollars are acceptable. This option should be considered when small areas are to be addressed and can save the time and expense of a "full-blown" ERA (Chapter 4). USEPA defines this as a "removal action" to screening concentrations (USEPA 1992). More complex sites require more in-depth analysis and comparison of more complex cleanup alternatives (Chapter 6).

Cooperative Injury Evaluation and Restoration Scaling

For the past decade, the Texas and Louisiana Trustees have been cooperatively engaged with PRPs in a cooperative approach to NRDA. Reasonably Conservative Injury Evaluation (RCIE) uses existing information from remedial investigations and other data sources, along with information from the scientific literature and other sites, to develop restoration packages that compensate the public for natural resource injuries at a site. In these states, the trustees' concerns for the chemical characterization investigations, risk assessments, etc., are often integrated into state or federal required remedial work plans. Additional injury assessment studies rarely have been required.

The Texas and Louisiana Trustees' RCIE approach to NRDA recognizes that it is sometimes better to make reasonable, conservative estimates of natural resource injuries or losses, using information obtained for other purposes, than to spend additional time and money on injury assessment studies. In the RCIE approach, the parties seek to err on the side of conservatism in favor of finding "resource injury" for an exposure level which at least one data or information source indicated was reasonably likely to result in an adverse effect. Integration of the trustees' unique concerns and perspective into the site characterization and risk assessment processes nearly eliminates the need for additional NRDA data in areas of overlapping concern.

Conservatism versus uncertainty

An underlying principle of RCIE is that there is a relationship between the need for conservatism, the uncertainty in the assessment, and the cost of reducing the need for conservatism and uncertainty (Figure 14-2).

Uncertainty is high when little is known about the nature or scale of an injury; in this case, a higher estimate of injury is assumed to be needed to ensure that enough compensation was provided. Additional information is needed to reduce uncertainty and the need for conservatism. However, additional information comes with costs in time and money. Early in the RCIE process when there is a great deal of uncertainty, acquiring some additional information can efficiently reduce the need for a conservative injury estimate. However, as an example of the law of diminishing returns, reducing uncertainty in the estimate can become less cost efficient, especially when continuing transaction costs are taken into account. At some point, the additional costs to refine conservative injury estimates do not justify further investment considered against the costs of providing additional habitat as compensation (this assumes that resources are protected from future or ongoing harm). In this case, the trustees and PRPs often can agree that sufficient information has been obtained to reach an agreeable level of certainty and that the issue at hand can be settled.

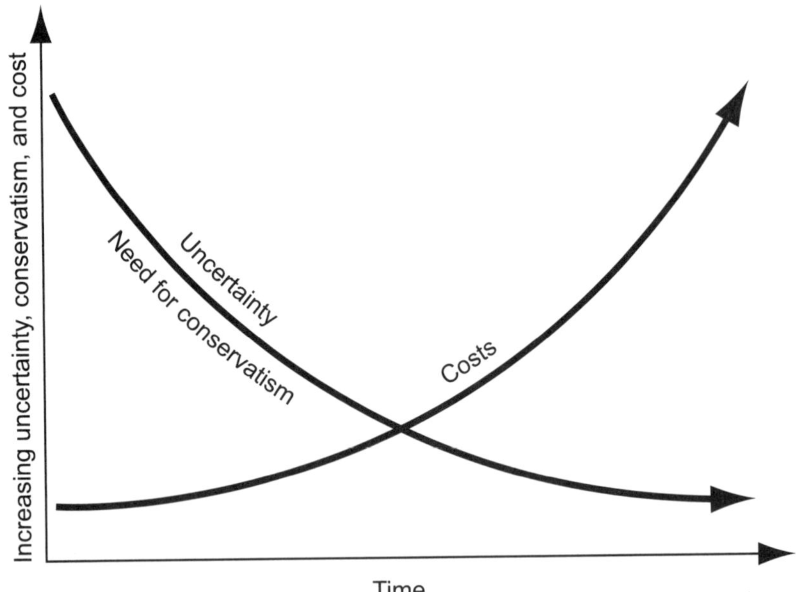

Figure 14-2 Idealized representation of the relationship between the need for conservatism, uncertainty, and the cost of reducing the need for conservatism and uncertainty in the RCIE process

RCIE scaling approach

As an early step at any site where RCIE is being considered, the trustees and PRP should assemble all relevant data and information bearing on the injury to each resource, preferably in an electronic database. Reliable historical data, data collected as part of any remedial process, and the results of prior relevant scientific studies or literature reviews should be located and included in the database. These data and any relevant historical data ideally should be managed in a relational database linked to a geographical information system (GIS) and freely shared among the trustees and the PRP. The RCIE approach attempts to match the weight of information with the weight of the decision to be made.

Generally, RCIE scales restoration of spatial units (habitat acres) using the appropriate SQG or some other benchmark or index (See "Use of Sediment Quality Guidelines in Benthic Community RCIE," p 597), GIS, and habitat equivalency analysis (HEA). Under US CERCLA, the public is considered to have been made whole for ecological losses when the scale of restoration needed to offset losses of resources and services is achieved. HEA establishes the discounted service acre year (DSAY) as the "common currency" for comparison of the public's value of past injury and future restoration in a common time frame (Julius 1998). One service acre year is defined as the ecological service provided by 1 acre in 1 year. Economic discounting is used to express past injury and future restoration units in a common time (Julius 1998). So, 1 DSAY is the service provided by 1 acre in 1 year "discounted" to net present value. Area of injured habitat, percent loss of ecological services, duration of injury, etc. are considered in HEA to determine DSAYs. The public is made whole when replaced resources and services equal lost resources and services, that is, when the areas under the curves are equal (Figure 14-3).

Screening-level RCIE

Under the screening-level RCIE approach, before proceeding to plan and implement any specific studies to further investigate and/or quantify any resource injury or loss, the trustees and PRP should try to reach agreement on resource injury determinations on the basis of the available data and scientific information, using conservative scientific assumptions. Where sufficient information exists to support technically sound and reasoned analyses, injury determinations can be based on that information, by agreement of the parties.

For the Texas RCIE, a database built by the trustees or provided by the cooperating party is used by the trustees to make a conservative, screening-level estimate of injury and scale it to a preferred restoration option. The trustee con-

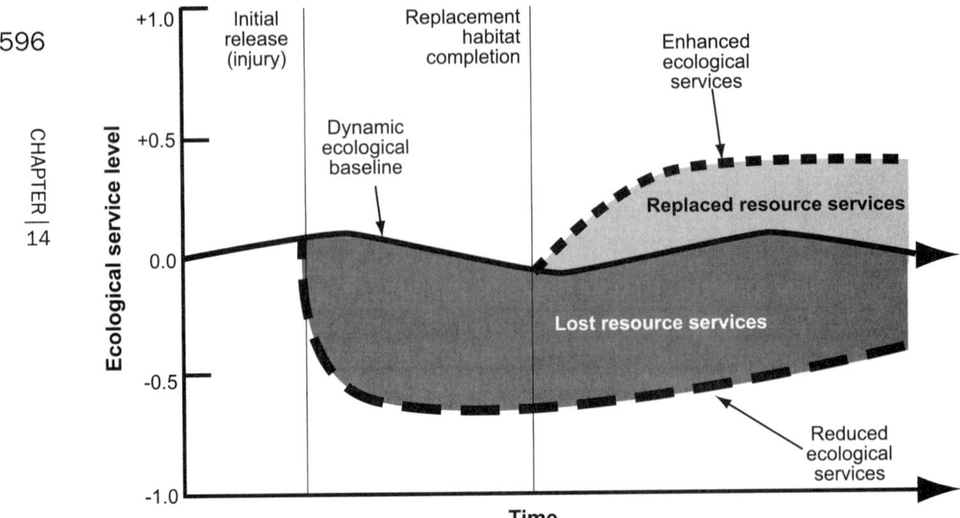

Figure 14-3 Timeline for a habitat equivalency analysis (HEA) used to determine the scale of restoration needed to just offset losses of resources and services

siders the remedy, any construction damage to the environment, any natural recovery period, and conservative, habitat-specific injury thresholds to develop an estimate of the lost DSAYs of the preferred habitat type. This value is discounted to determine a finite number of habitat units required by the trustees (NOAA 2000). If the PRP wishes to apply an additional level of rigor to the assessment, the trustees prefer that a memorandum of agreement (MOA) be developed to conduct a cooperative assessment.

Collaborative RCIE

Using a database built by the trustees or provided by the cooperating PRP, both parties can agree upon reasonable, conservative estimates of injury and scale the extent of the injury to a preferred restoration option. Both parties can consider the remedy, any construction damage to the environment, any natural recovery period (USEPA 2002), and conservative, habitat-specific injury thresholds to develop a estimate of the lost DSAYs of the preferred habitat type. This value is discounted (NOAA 2000) to determine a finite number of habitat units required by the Trustees, with agreement from the cooperating party.

If the estimate of injury is too uncertain and the parties cannot agree on the scale of the injury or if sufficient information is not available to make a reasonable estimate, then specific NRDA injury studies should be jointly developed to address the data gaps critical to determining or quantifying that resource injury, and the assessment process should be continued for that resource. The trustee and the PRP, through their participation in the cleanup process, strive to ensure that any additional data or samples would be taken simultaneously with response samples to answer injury-specific questions. Uncertainty and conservatism about various parameter estimates can usually be revised on the basis of additional, high-quality, site-specific data.

Where agreements can be reached, the resulting injury determination can then be used to quantify the injuries to that resource using tools such as GIS and sediment recovery models as input to HEA (NOAA 2000).

If a conservative review of the available data and literature indicate that there is little or no reasonable likelihood of an injury occurring, then the assessment process should conclude with a finding that no further consideration of that resource injury category is required.

Each resource injury determination agreed to by the trustees and the PRP using the RCIE approach should be memorialized in a jointly developed document that describes the methods and information on which the agreed assessment of injury is based.

Use of Sediment Quality Guidelines in Benthic Community RCIE

When injury to the benthic community is assessed, agreement on injury thresholds based on SQGs can lead to rapid agreement with trustee agencies when the RCIE approach to injury assessment is followed. SQGs can be used as reference points during the negotiation of injury values. Generally, marginally exceeding a single conservative SQG provides less evidence of potential injury than exceeding multiple SQGs or greatly exceeding an individual SQG. Generally, a higher level of resource injury would be assumed as the number of less conservative SQGs increased (Long et al. 1995, 1998). SQG-based sediment injury determinations have been conducted using SQG exceedances and SQG quotients elsewhere in the US (Ingersoll et al. 2002; MacDonald, Moore, et al. 2002).

The trustee and the cooperating PRP develop the scale that describes level of injury (LOBS) associated with the rate of increase in the contaminant gradient. A reasonably conservative threshold, near the LOAEL, is typically selected for the onset of injury. In some cases, ERL and ERM values are thought of as NOAEL and LOAEL values (Long et al. 1998). Similarly, threshold effects level–probable effects level (TEL–PEL; MacDonald et al. 1996), threshold effects concentration–probable effects concentration (TEC–PEC; MacDonald et al. 2000) pairs can be used, depending on ecosystem characteristics; the PEL, PEC, ERM most often are used. Less conservative SQGs (e.g., individual apparent effects threshold [AET]) can be agreed upon to represent an increased level of severity of injury, allowing construction of a "scale of potential benthos injury."

The trustees and cooperating PRP can agree on LOBS based on comparison of site chemistry and injury thresholds based on SQG. GIS allows quantification of spatially explicit injury maps to quantify injury areas and tally LOBS. Applying these agreed-upon injury thresholds using GIS allows quantification of consequences of exposures that affect only part of a population or the spatial distribution of organisms.

Severity of endpoints and scale of benthos injury

Mortality (loss of biomass) or lost reproductive capacity are on the upper end of the injury scale and would have a greater LOBS value; biomarkers without a clear, scalable connection would have a lower LOBS value; behavioral endpoints (e.g., avoidance, reburial, and aggressiveness) are usually considered intermediate LOBS values. More sensitive endpoints are considered to contribute less to the LOBS fraction than are less sensitive endpoints. The assumption is that if you observe a response with a less sensitive endpoint, then more sensitive endpoints are accounted for and a greater portion of the population potentially is affected, thus LOBS increases. There are some indications that, when the SQG quotient for a particular contaminant is low, only highly sensitive endpoints show effects (Chapter 4).

Sediment quality guideline indices

A higher probability of injury to the benthic community can be shown using multiple contaminant models such as mean ERM quotient, mean PEC quotient, or logistic regression P-Max model (Long et al. 1995, 1998; Field et al. 2002). As part of a comprehensive investigation of hazardous substance releases, the USEPA conducted the Calcasieu Estuary Ecological Risk Assessment (MacDonald, Moore, et al. 2002). In the benthic community risk portion of the investigation, site mean PEC quotients were compared to the results of amphipod acute and chronic toxicity tests (Figures 14-4 and 14-5). The

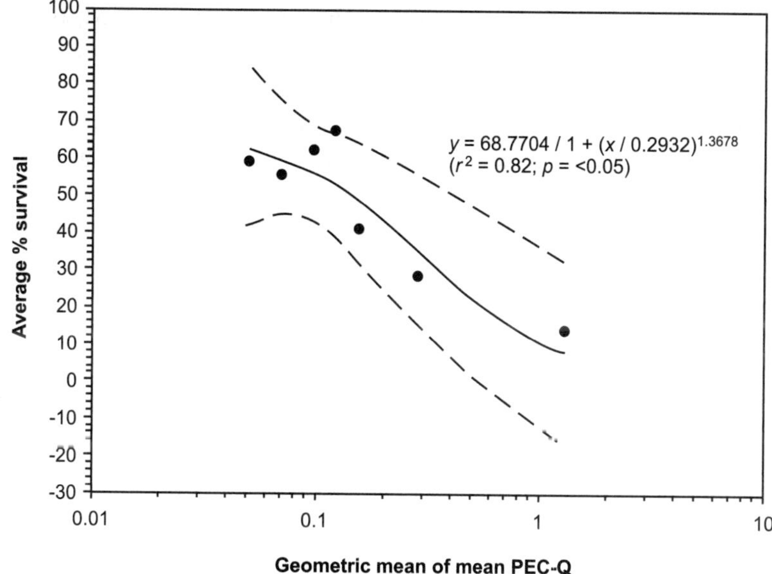

Figure 14-4 Relationship between geometric mean of the PEC-Q and survival of freshwater amphipod *A. abdita*, in 10-d whole sediment toxicity tests (MacDonald, Moore, et al. 2002)

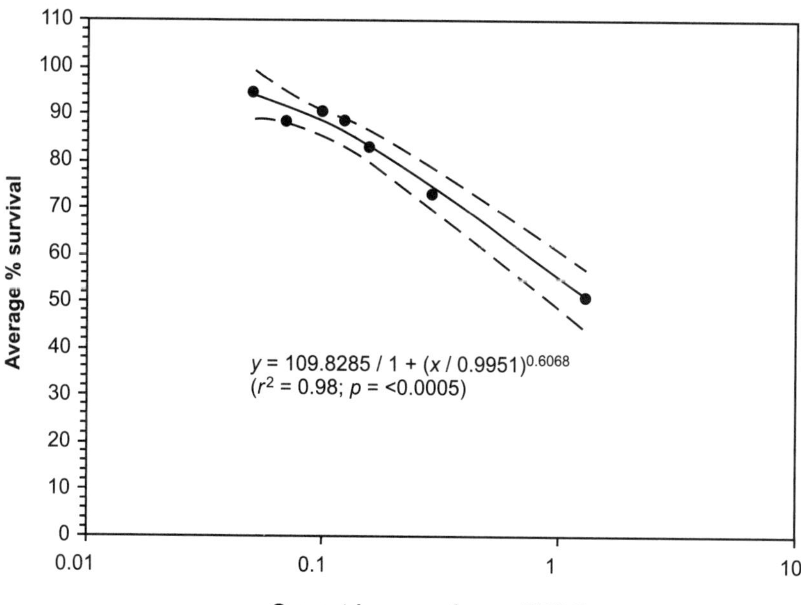

Figure 14-5 Relationship between geometric mean of the PEC-Q and survival of freshwater amphipod *H. azteca*, in 28-d whole sediment toxicity tests (MacDonald, Moore, et al. 2002)

logistic regression modeling technique, maximum probability (P-Max) value, has been shown to correlate well with observed toxicity in paired chemistry or toxicity test data (Figure 14-6). This approach shows promise at sites where only a few hazardous substances are problematic but may overpredict toxicity at marine metals-only sites (L. Jay Field, 2002, personal communication). These comparisons showed a high level of correlation with increased SQG quotients and reduced survival in the toxicity tests (MacDonald, Moore, et al. 2002). The high degree of correlation between the increase of an SQG index value and LOBS is being used to facilitate rapid settlement in several ongoing assessments.

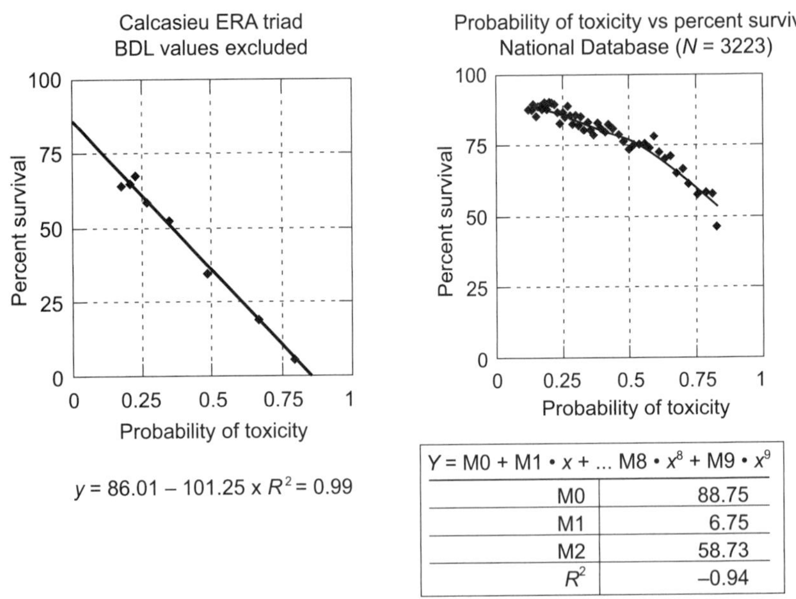

Figure 14-6 Relationship between logistic regression P-Max model and survival of estuarine amphipod *A. abdita,* in 10-d whole sediment toxicity tests using the Calcasieu ERA and the Biological Effects Databases (L. Jay Field, Seattle, Washington, USA, unpublished data)

Example: Benthic Community Assessment, Lavaca Bay, Texas

The conservative and reasonable estimate of injury thresholds (RCIE) approach derives from one that was developed during the negotiation of a settlement for injuries related to the Lavaca Bay CERCLA case in Texas. Multiple natural resource injury categories were assessed and quantified using the RCIE approach (then called "Reasonable Worst Case"), including the benthic community, birds, fish, and terrestrial habitats. Additionally, injury to ground water and surface water was assessed, but no compensation was sought. Mercury and polynuclear aromatic hydrocarbons (PAHs) were determined to significantly contribute to injury to the benthic community (Texas Trustees 2001).

The trustee participated in most facets of the remedial investigation, risk assessment, and feasibility phases on the Superfund process. All data created or considered was assembled into a GIS database system provided by the PRP. To assess benthic community loss of services, the following sources of information were used:

- analytical chemistry RI/FS data (extent of contamination),
- sediment quality triad (SQT) study, and
- ERA and literature survey for
  - Hg and PAH growth effects,
  - Hg and PAH survival effects, and
  - Hg and PAH reproduction effects.

Estimates of percent loss of ecological services were based on a weight-of-evidence (WOE) approach, multiple (LOEs), and consensus judgment (of the trustees and the cooperating PRP) using the results of available studies reported in the scientific literature, results of SQT, and knowledge of Texas estuarine ecosystems (Table 14-1). For Hg, the percent loss of ecological services identified and evidence used were as follows:

- The SQT found no observed decrease in survival for the amphipod *Leptocheirus* spp., no apparent reduced growth for the polychaete *Neanthes,* and no difference in the macroinvertebrate assemblages when compared to the reference area. The SQT was conducted over a Hg concentration range from 0.3 to 4.6 mg/kg, which encompasses the range of both the ERM and AET. However, sublethal toxicological endpoints such as behavioral changes and loss of reproductive capacity were not measured in SQT. To be consistent with the reasonable worst-case approach, a 10% loss of services was estimated for sediment Hg concentrations > the ERM (0.71 mg/kg) but < the benthic AET, in open water, intertidal mudflats, and fringe marsh habitats.
- A significant decrease in activity behavior of *Pontoporeia affinis* was noted at concentrations exceeding 2.15 to 3.35 mg/kg from sediment bioassays, as cited by Long and Morgan (1990). A 25% loss of services was calculated for concentrations > benthic AET (2.1 mg/kg).

Table 14-1 Loss of benthic community services values used at Lavaca Bay for open water and fringe marsh habitats (Texas Trustees 2001)

Hg in sediments	PAHs in sediments	Service (%) losses
Hg < ERM	ERL < PAHs < ERM	10%
Hg < ERM	PAHs > ERM	25%
ERM < Hg <AET	PAHs < ERL	10%
ERM < Hg <AET	ERL < PAHs < ERM	20%
Hg > AET	PAHs < ERL	25%
ERM < Hg <AET	PAHs > ERM	35%
Hg > AET	ERL < PAHs < ERM	35%
Hg > AET	PAHs > ERM	50%

- A 25% loss of services was estimated for Hg concentrations above the oyster larvae AET on oyster reefs. This injury value was consistent with the RWC approach and was based on best professional judgment.

For PAHs, the percent loss of services and identified evidence are as follows:

- In open water and intertidal mudflats or fringe marshes, a 10% percent loss of services was determined for high molecular weight PAH (HPAH) concentrations > ERL and for HPAH concentrations < HPAH ERM. This level of severity was established because of the following:
  - The NOAA ERL is based on sites with co-occurring chemicals. Additivity between Hg and PAHs was considered separately in this analysis.
  - The NOAA ERL is lower than reported apparent AETs for PAHs, including the benthic AET (Puget Sound: low molecular weight PAHs [LPAHs] > 13 mg/kg and HPAHs > 69 mg/kg), amphipod AETs (Puget Sound: LPAHs > 24 mg/kg and HPAHs > 69 mg/kg; San Francisco Bay: total PAHs >15 mg/kg; Mississippi Sound total PAHs > 205 mg/kg), and mysid AETs (Mississippi Sound total PAHs > 99 mg/kg).
- A 25% percent loss of services was estimated for sediment concentrations > HPAH ERM. PAHs detected in Lavaca Bay sediments are predominately the higher-molecular-weight PAHs. The NOAA HPAH ERM (9.6 mg/kg) is lower than reported effects thresholds (AETs) for PAHs, including the benthic AET (Puget Sound: HPAHs > 69 mg/kg), amphipod AETs (HPAHs > 69 mg/kg; Mississippi Sound total PAHs

> 205 mg/kg), and mysid AETs (Mississippi Sound total PAHs > 99.4 mg/kg).

- A 25% loss of services was estimated for total LPAHs (AET: 5.2 mg/kg) or total HPAH (AET: 17 mg/kg) concentrations above the oyster larvae AETs for sediments on oyster reefs.

For co-occurring PAHs and Hg,

- estimates of percent loss of services for Hg and the individual PAH constituents were assumed to be additive.

Based on available site-specific data, a greater uncertainty exists for estimating direct injury from PAHs because no site-specific studies were conducted to assess PAH toxicity. Therefore, it was conservatively determined that injury would begin where sediment concentrations exceeded the lowest reported sediment benchmark criterion (i.e., the ERL).

For locations where there were co-occurring elevated concentrations of Hg and PAHs, estimates for the percent loss of services were assigned to areas exceeding chemical benchmark criteria, assuming additivity of effects for Hg and PAHs. Additivity was assumed because no data were available to suggest either synergistic or antagonistic toxicity between these chemicals to benthic invertebrates. Therefore, where co-occurring chemicals were present, ranges were used to assess the percent loss of services, as indicated in Table 14-1.

GIS spatial analysis tools were used to describe and quantify zones where each combination occurred in Lavaca Bay. A spatially explicit map of benthic community injury was developed and served to sufficiently document the result (Figure 14-7). HEA was used to develop the number of acres of habitat required to compensate for the injury. When this and several other categories, like marsh and oyster reef, were similarly assessed, the restoration of 70 acres of estuarine marsh would be required by the trustees (Texas Trustees 2001).

Summary

The cooperative NRDA process is advantageous for all parties. The cooperating PRP saves time and money, and the public can receive compensation earlier for its losses. SQGs and indices are being used successfully at dozens of sites in Delaware, South Carolina, Texas, and Louisiana, involving federal and state oversight. In the US, courts prefer that potential litigants settle issues without resorting to litigation. This process, along with other tools, has been used to settle cases such as Bailey Waste, Orange County, TX; Conoco EDC,

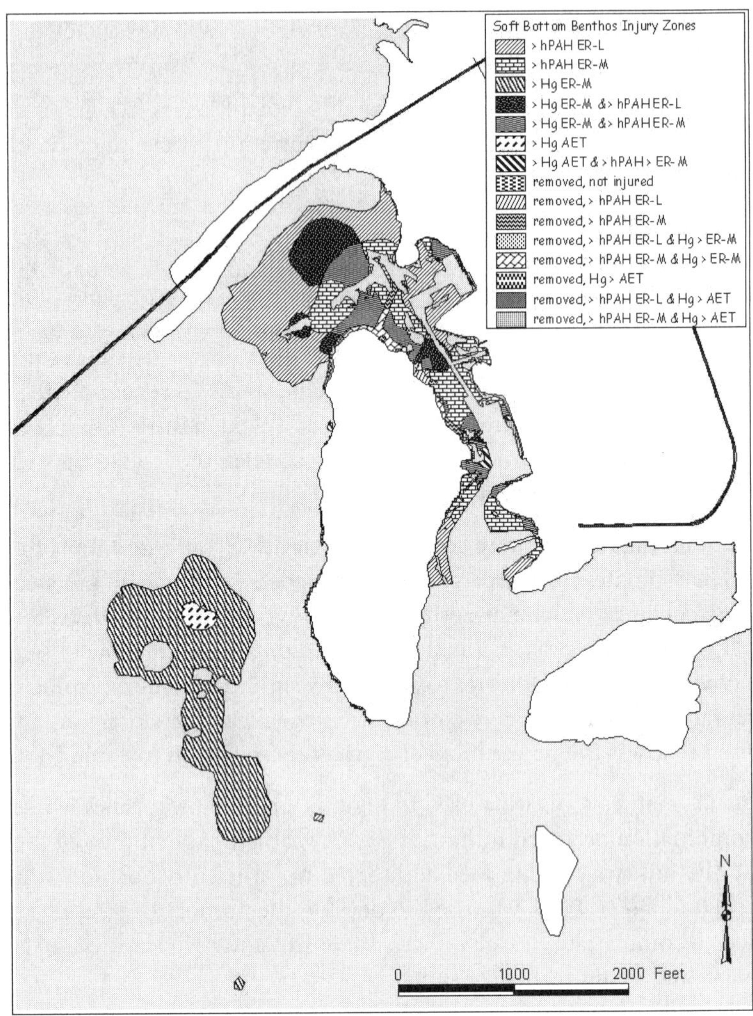

Figure 14-7 Spatially explicit distribution of injury to soft-bottom benthic community, Lavaca Bay CERCLA site (Note: Dredged 38- and 12-foot channels were excluded from the assessment.)

Calcasieu Parish, LA; Tex-Tin, Galveston County, TX; Brio NPL, Harris County, TX. In these cases, RCIE or a similar approach was used, and each was settled with the State of Texas or Louisiana and the United States with a federal consent decree. Many more cases are in the process of settling. This approach allows PRPs to settle their liability issues cost effectively and to restore damaged habitats.

Disclaimer—The information in this paper reflects the views of the author, and does not necessarily reflect the official positions or policies of NOAA or the Department of Commerce.

References

Barnthouse LW, Stahl RG. 2002. Quantifying natural resource injuries and ecological service reductions: Challenges and opportunities. *Environ Manag* 30:1–12.

Field LJ, MacDonald DD, Norton SB, Ingersoll CG, Severn CG, Smorong D, Lindskoog R. 2002. Predicting amphipod toxicity from sediment chemistry using logistic regression models. *Environ Toxicol Chem* 21:1993–2005.

Ingersoll CG, MacDonald DD, Brumbaugh WG, Johnson BT, Kemble NE, Kunz JL, May TW, Wang N, Smith JR, Sparks DW, Ireland SD. 2002. Toxicity assessment of sediments from the Grand Calumet River and Indiana Harbor Canal in northwestern Indiana. *Arch Environ Contam Toxicol* 43:153–167.

Long ER, Field LJ, MacDonald DD. 1998. Predicting toxicity in marine sediments with numerical sediment quality guidelines. *Environ Toxicol Chem* 17:714–727.

Long ER, MacDonald DD, Smith SL, Calder FD. 1995. Incidence of adverse biological effects within ranges of chemical concentrations in marine and estuarine sediments. *Environ Manag* 19:81–97.

Long ER, Morgan LG. 1990. The potential for biological effects of sediment-sorbed contaminants tested in the National Status and Trends Program. Rockville (MD): National Oceanic and Atmospheric Agency. Technical Memorandum NOS OMA 5.

MacDonald DD, Moore DR, Pawlisz A, Smorong DE, Breton RL, MacDonald DB, Thompson R, Lindskoog RA, Hanacek MA. 2002. Calcasieu Estuary Remedial Investigation/Feasibility Study (RI/FS): Baseline Ecological Risk Assessment (BERA). Volume 2. Baseline problem formulation: Appendices. Prepared for: CDM Federal Programs Corporation. Dallas, Texas. Available from: http://www.epa.gov/earth1r6/6sf/pdffiles/bera_report_text.pdf. Accessed 21 Aug 2004.

MacDonald DD, Ingersoll CG, Smorong DE, Lindskoog RA, Sparks DW, Smith JR, Simon TP, Hanacek MA. 2002. Assessment of injury to sediments and sediment-dwelling organisms in the Grand Calumet River and Indiana Harbor Area of Concern, USA. *Arch Environ Contam Toxicol* 43:141–155.

[NOAA] National Oceanic and Atmospheric Agency. 2000. Habitat equivalency analysis: an overview, damage assessment and restoration program, March 21, 1995 (Revised October 4, 2000). Silver Spring (MD): NOAA.

Texas Trustees. 2001. Damage assessment and restoration plan and environmental assessment for the Point Comfort/Lavaca Bay NPL site, ecological injuries and service losses, final. Trustees for the State of Texas (TX Trustees); General Land Office, Texas Parks and Wildlife Department, Texas Natural Resource Conservation Commission, National Oceanic and Atmospheric Administration,

US Fish and Wildlife Service, US Department of the Interior. Available from: http://www.darp.noaa.gov/library/pdf/lavdarpe.pdf. Accessed 2 Nov 2004.

[USEPA] US Environmental Protection Agency. 1993. Guidance on conducting non-time-critical removal actions under CERCLA. Washington DC: USEPA, Office of Emergency Response and Solid Waste. OSWER Dir. 9360.0-32.

[USEPA] US Environmental Protection Agency. 1997. Ecological risk assessment guidance for Superfund: Process for designing and conducting ecological risk assessments. Washington DC: USEPA, Office of Water. Report EPA 540-R-97-006.

Evaluation of sediment quality guidelines for management of contaminated sediments in New York–New Jersey Harbor

15

W Scott Douglas,
Mark Reiss,
James Lodge

The New York–New Jersey Harbor ("the Harbor") complex is situated in the metropolitan center of the Hudson–Raritan estuary (Figure 15-1). The Hudson–Raritan estuary is the terminus of a 33 700 km^2 watershed, with land use ranging from wilderness park to urban industrial. The estuary itself covers some 770 km^2 and contains 80 000 hectares of tidal wetlands and more than 600 km of shoreline. The tidal range of the estuary is as much as 2 m, with salinity in the Harbor averaging 25 g/L.

The New York–New Jersey (NY–NJ) metropolitan area is the second largest metropolitan area in the United States, with a population approaching 30 million people (10% of the population of the US). The Port of New York and New Jersey ("the Port"), located within the Harbor, is the largest container port on the East Coast and the largest petroleum distribution point in the US. The Port is the main conduit for goods to and from the region, and its activity generates more than $29 billion for the regional economy.

The Harbor has been contaminated by a variety of chemicals as a result of its historical and ongoing use as an industrial port and urban center. Sediment quality is degraded throughout most of the system and is particularly poor in the inner Harbor complex. Levels of contaminants are markedly elevated in sediments in the Passaic River–Newark Bay–Arthur Kill complex in the western portion of the system. Sediment concentrations of metals, chlorinated organics (particularly pesticides), polycyclic aromatic hydrocarbons (PAHs), and dioxin or furans are all higher in these areas than in the rest of

Use of Sediment Quality Guidelines and Related Tools for the Assessment of Contaminated Sediments
Wenning RJ, Batley GE, Ingersoll CG, Moore DW, editors.

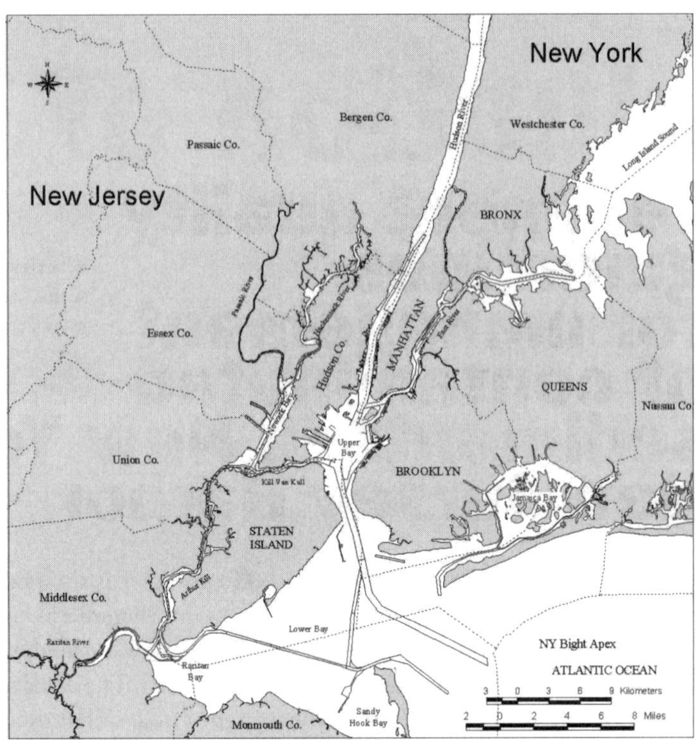

Figure 15-1 Hudson–Raritan estuary complex and New York–New Jersey Harbor

the Harbor. Sediments with high contaminant levels are also sparsely distributed throughout Upper New York Bay and the East River. Substantially lower levels are present in sediments taken from Lower New York Bay and Raritan Bay. Sediment contaminant levels co-vary across the Harbor (Table 15-1). Typical distributions of sediment contaminants in and around the Harbor are presented in Figure 15-2.

Maintaining safe navigation depths in the 400 km of engineered waterways serving the Port requires the dredging and management of some 3 to 5 million m^3 of sediment each year. Prior to 1992, about 95% of Harbor dredged materials were determined to be suitable for open water disposal in the Atlantic Ocean. In 1991, the US Environmental Protection Agency (USEPA) and the US Army Corps of Engineers (USACE) revised their national guidance for conducting evaluations of dredged materials proposed for open water disposal in ocean waters (USEPA, USACE 1991). The change having the most significant impact in the Harbor was the introduction of the 10-d whole sediment

Table 15-1 Correlations of selected contaminants in Harbor sediments (USEPA 1998) (Spearman's rank, $n = 168$, all correlations significant [$p < 0.001$])

Contaminant	Total metals	Hg	Total PAH	Total chlordane	Total DDT	Dieldrin
Hg	0.88	—	—	—	—	—
Total PAH	0.82	0.85	—	—	—	—
Total chlordane	0.90	0.92	0.88	—	—	—
Total DDT	0.90	0.93	0.89	0.97	—	—
Dieldrin	0.85	0.81	0.82	0.90	0.89	—
Total PCB	0.88	0.90	0.88	0.96	0.96	0.88

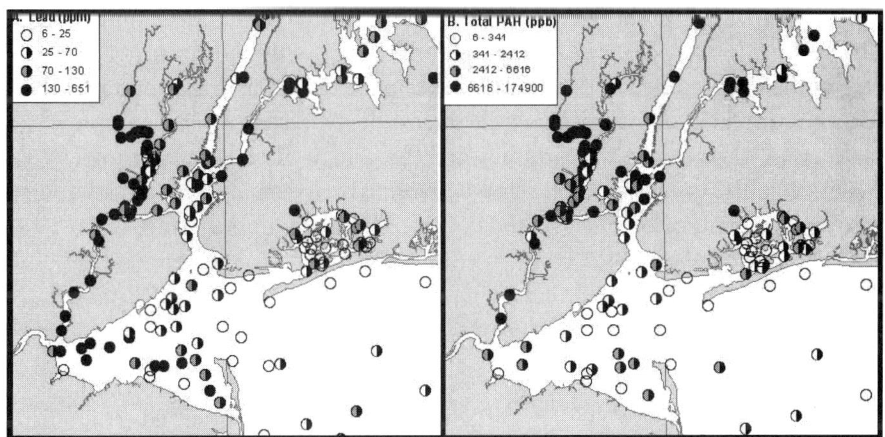

Figure 15-2 Distribution of total PAH and Pb contamination in surficial Harbor sediments by quartiles

amphipod acute toxicity test. About 70% of Harbor dredged material that has been evaluated using these testing methods has been determined to be unsuitable for ocean disposal because of the acute toxicity of sediments to amphipods. Stricter guidelines for interpreting bioaccumulation test results have further reduced the amount of Harbor dredged material that can be placed in the ocean by 5% to 10%. With no readily available and cost-effective alternatives to ocean disposal, the cost of dredging and disposal of this dredged material increased from approximately \$7 per m^3 to more than \$150 per m^3. Despite regional efforts to identify and develop dredged material management alternatives, assessment and management of contaminated sediments continues to be a critical priority for Harbor managers (McDonough et al. 1999; Douglas et al. 2003).

Nearly the entire Harbor complex is severely impacted by sediment contamination. Therefore, the cost and scope of any remedial efforts must be justified by a direct link to benefits generated. Navigational dredging is this link. The Contaminant Assessment and Reduction Project (CARP) was established in 1994 under the NY–NJ Harbor Estuary Program (HEP) to study Harbor sediment contamination and to identify and prioritize management alternatives to reduce it. The CARP working group was specifically tasked with estimating how long it would take before newly deposited sediments in navigation channels would be clean enough to meet requirements for disposition in the ocean and to identify what, if any, actions could be taken to decrease that time. To achieve this, the CARP commissioned a contaminant fate and transport model for the Harbor. The CARP model, building off existing models for the Harbor (Blumberg et al. 1999; Farley et al. 1999; Landeck-Miller et al. in press) will incorporate estuarine hydrodynamics, sediment transport, and organic carbon models into a framework capable of predicting contaminant concentrations in the various compartments of the system (water, sediment, and tissue). The model will not, however, be able to directly predict toxicity to amphipods. Given the direct link between toxicity in amphipod tests and the management of dredged material, it is particularly important that target concentrations selected for Harbor sediments be specifically protective of this endpoint. Once developed, these benchmarks will enable the CARP model to be used to identify, prioritize, and set goals for remedial and regulatory actions.

It is the purpose of this chapter to examine the ability of published sediment quality guidelines (SQGs) to predict acute sediment toxicity to amphipods in and around the Harbor, to evaluate the utility of SQGs in identifying priorities and goals for Harbor remediation, and to identify the research and SQG development needs for the region.

Evaluation of Sediment Quality Guideline Performance in Predicting Harbor Sediment Toxicity

Data used to evaluate SQGs

The nature and extent of sediment contamination and toxicity in the Harbor estuary has been assessed in a variety of studies sponsored by federal and state agencies as well as private industry. Data generated in the following studies provide synoptic measurements of sediment chemistry and 10-d amphipod toxicity test data for 230 sediment samples from throughout the

Hudson–Raritan estuary: National Oceanic and Atmospheric Administration (NOAA) National Status and Trends (NS&T) Hudson–Raritan Phase I and Phase II (Long, Wolfe, et al. 1995), USEPA Environmental Monitoring and Assessment Program (EMAP) 90-92 (Weisberg et al. 1993), and USEPA Regional Environmental Monitoring and Assessment Program (R-EMAP) 93-94 (USEPA 1998). Of these, 167 sediment samples were considered to be in the Harbor proper (excluding New York Bight and Long Island Sound) and were used in this analysis.

The 167 Harbor sediment samples with matching chemistry and *Ampelisca abdita* toxicity test data were used to evaluate the ability of various published SQGs to predict results of toxicity tests of Harbor sediments. Sediment chemistry data for the Harbor samples (in quartile ranges) and published empirical guidelines for the contaminants are summarized in Table 15-2. Using the SQG approaches described below, a prediction of toxicity was made for each of the 167 samples and compared to the toxicity test results. Sediments observed to cause more than 20% amphipod mortality were considered toxic for this evaluation (67 [40%] of the 167 samples in the database were toxic by this criterion). The majority of samples (about 70%) collected from the Passaic–Arthur Kill–Newark Bay complex and about 25% to 30% of the samples collected in Upper New York and Jamaica Bays were acutely toxic to amphipods (Figure 15-3a).

Sediment quality guideline approaches evaluated in this study

The SQG approaches evaluated in this study include empirical approaches (effects ranges [MacDonald 1994; Long and Morgan 1991] and apparent effects thresholds [AETs; Barrick et al. 1998]), mechanistic approaches (equilibrium partitioning [EqP; Hansen et al. 1996; Di Toro et al. 2000a, 2000b;]), and approaches that combine aspects of both empirical and mechanistic methodologies (consensus-based approaches [Swartz 1999; MacDonald et al. 2000; Fairey et al. 2001]). A more detailed description of these approaches is provided in Chapter 3.

Empirical sediment quality guidelines

Some SQGs have been used as threshold values for predicting toxicity based on levels of individual contaminants; other SQGs require compilation and consideration of data for multiple contaminants (Chapters 4 and 12). The empirical SQGs evaluated in this chapter represent both of these approaches. The State of Washington developed an AET approach, using various indicators of biological effects, that has resulted in SQGs for individual contami-

Table 15-2 Distribution of bulk contaminant levels (by quartile) in Harbor sediments and corresponding empirical SQGs

	Surficial sediment characteristics						Empirical SQGs		
Contaminant	*n*	Minimum	25th quartile	Median	75th quartile	Maximum	PEL[a]	ERM[b]	PSSDA ML[c]
Acenaphthene (µg/kg)	167	4	12	41	110	56300	88.9	500	2000
Acenaphthylene (µg/kg)	167	2.4	22	66	162	12900	128	640	1300
Anthracene (µg/kg)	167	4.4	56	200	440	89400	245	1100	13000
Benzo(*a*)anthracene (µg/kg)	167	5.2	140	410	954	59300	693	1600	5100
Benzo(a)pyrene (µg/kg)	167	4	123	454	944	54900	763	1600	3600
Benzo(*b*)fluoranthene (µg/kg)	51	88	499	833	2450	39200	NA	NA	NA
Benzo(*ghi*)-perylene (µg/kg)	167	6	84	320	725	31700	NA	NA	3200
Benzo(*k*)fluoranthene (µg/kg)	55	54	217	486	1780	35000	NA	NA	NA
Chrysene (µg/kg)	167	4.4	146	490	1240	60300	864	2800	21000
Dibenz(*ah*)anthracene (µg/kg)	163	2	12.2	67	165	4530	135	260	1900
Fluoranthene (µg/kg)	167	2.8	190	680	1550	108000	1490	5100	30000
Fluorene (µg/kg)	167	0.8	21.5	71	130	54200	144	540	3600
Indeno(123*cd*)-perylene (µg/kg)	167	6	83	335	665	22800	NA	NA	4400
Naphthalene (µg/kg)	167	5.6	51	160	250	17400	391	2100	2400
Phenanthrene (µg/kg)	167	3.5	100	312	615	194000	544	1500	21000
Pyrene (µg/kg)	167	3.2	200	659	1500	143000	1400	2600	16000
LPAH (µg/kg)	167	6	337	961	2040	552000	1400	3160	29000
HPAH (µg/kg)	167	6	1100	3640	8490	462000	6680	9600	69000
Total PAH (µg/kg)	140	6	1100	4600	10700	175000	16800	44800	NA
Total chlordane (µg/kg)	31	0.1	2.2	3.6	6.6	19	4.8	6	NA

Table 15-2 *continued*

	Surficial sediment characteristics						Empirical SQGs		
Contaminant	*n*	Minimum	25th quartile	Median	75th quartile	Maximum	PEL[a]	ERM[b]	PSSDA ML[c]
Total DDT (μg/kg)	140	0.4	3.2	20.4	66	4630	52	46.1	69
Dieldrin (μg/kg)	167	0.01	0.5	1.3	3.4	21.2	4.3	8	NA
Total PCB (μg/kg)	167	1	33	163	325	2480	189	180	3100
As (mg/kg)	163	0.12	5	10	15	114	42	70	700
Cd (mg/kg)	165	0.009	0.4	1.1	2.1	12.9	4.2	9.6	14
Cr (mg/kg)	165	1	50	99	142	420	160	370	NA
Cu (mg/kg)	165	0.2	25	87	143	1030	108	270	1300
Pb (mg/kg)	165	6.7	35	98	172	651	112	218	1200
Hg (mg/kg)	167	0.02	0.3	1.1	2.3	15	0.7	0.7	2.3
Ni (mg/kg)	165	2.03	15.8	30	44	181	43	51.6	370
Ag (mg/kg)	161	0.01	0.6	2.1	2.9	5.7	1.8	3.7	8.4
Zn (mg/kg)	165	11	91	202	342	1400	271	410	3800

[a] MacDonald et al. 1996; [b] Long, MacDonald, et al. 1995; [c] USACE 2000

nants or contaminant classes. The AETs are used in combination for a given sediment, where exceedence of any one AET is predictive of a high likelihood of a deleterious ecological effect (Barrick et al. 1988). The Puget Sound Dredged Disposal Analysis (PSDDA) program uses an AET approach to screen sediments proposed for dredging (USACE 2000). The maximum contaminant level (ML) values are based on the highest of 4 biological indicators, usually the one set by results of a 10-d amphipod toxicity test, and therefore, MLs were deemed to be the most relevant AET for the objectives of this study.

Long et al. (1998, 2000) suggested that effects ranges (specifically, effects range median [ERM] and probable effects level [PEL]) can be used to estimate the probability of amphipod toxicity. In this approach, an effects range quotient is calculated for each contaminant in the mixture by dividing the bulk sediment concentration by its respective effects range concentration. As the mean of the individual quotients for a given sediment sample increases, the likelihood of acute toxicity is predicted to increase correspondingly (Chapter 4). Long et al. (1998) defined 4 categories of probability based on threshold mean effects range quotients. If the mean of the individual quotients (using ei-

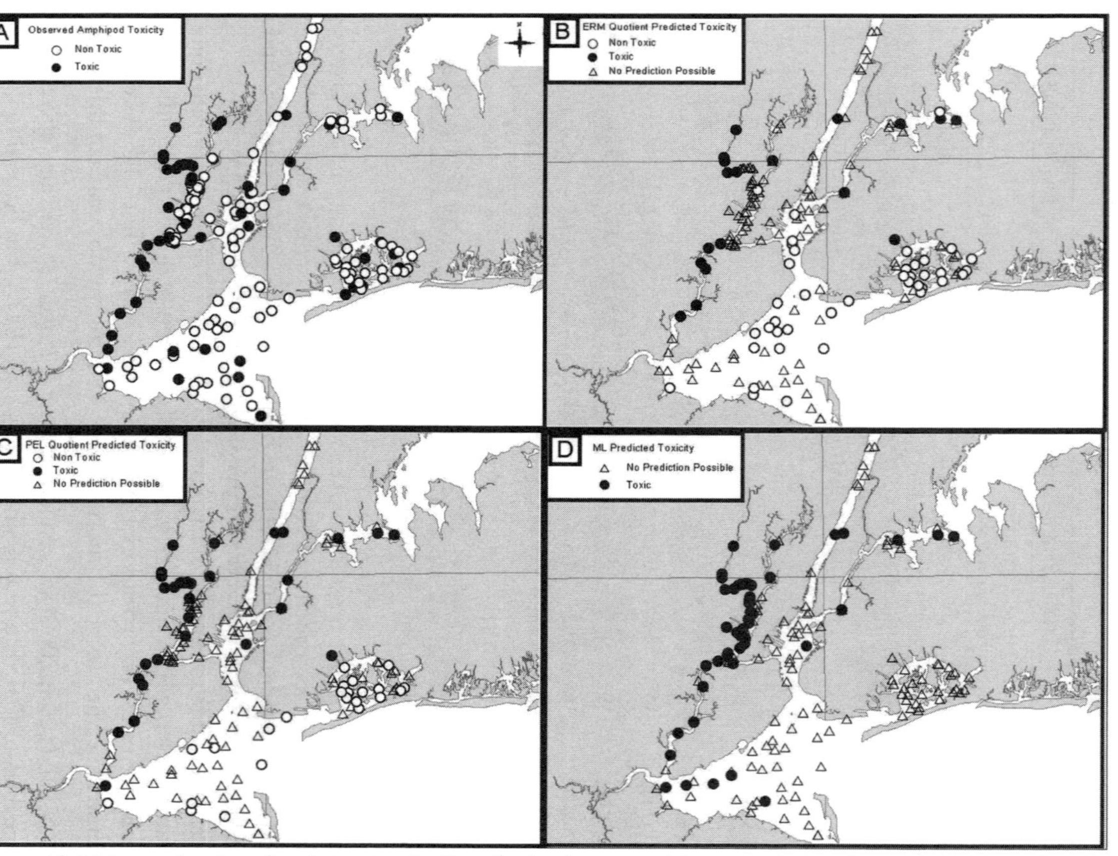

Figure 15-3 Measured and predicted toxicity of 167 surficial sediments in NY–NJ Harbor to the amphipod, *Ampelisca abdita* in 10-d sediment toxicity tests

ther ERM-Q or PEL-Q) for any given sediment sample is less than 0.1, there is a <10% probability for acute sediment toxicity to amphipods for the sediment. If the mean of the individual quotients for any given sediment sample exceeds 1.5 (if ERM-Q is used) or 2.3 (if PEL-Q is used), there is a high probability for acute toxicity to amphipods. Confident predictions cannot be made for sediments with mean quotients between these thresholds. These threshold quotients were used to make toxicity predictions for individual Harbor sediments.

Mechanistic sediment quality guidelines

Sediment levels of individual toxicants or classes of toxicants can also be evaluated for their potential to cause toxicity to amphipods using mechanistic (i.e., EqP-based) SQGs. EqP sediment benchmarks (ESBs) have been published for metals mixtures including Cd, Cu, Ni, Pb, Ag, and Zn (called "simultaneously extracted metals" [SEM]; USEPA 2003a), polycyclic aromatic hydrocarbon (PAH) mixtures (USEPA 2003b), and dieldrin (USEPA 2003c). EqP theory posits that the bioavailability of contaminants in sediment is governed by the partitioning of the contaminant within the sediment matrix or pore water, particularly either the organic carbon or sulfide phases. ESBs are designed to be protective of benthic species (Chapter 13).

EqP theory suggests that if molar concentrations of SEM are lower than molar concentrations of acid volatile sulfide (AVS) in sediment pore water, the metals are not bioavailable to organisms and therefore cannot cause toxicity (Hansen et al. 1996; only 132 samples in the synoptic dataset had AVS and SEM concentrations reported). ESBs for the organic contaminants dieldrin and PAH mixtures are based on the organic carbon–normalized sediment concentrations of those compounds. If PAH mixtures and dieldrin sediment concentrations are below their respective ESBs, then the sediment is not predicted to contain enough of this particular contaminant to be toxic to benthic organisms. The ESB for dieldrin, which is based on the USEPA water quality criteria (WQC) value, is 28 µg/g OC (USEPA 2003a). The PAH mixtures ESB is derived based on the sum of the toxic units (TUs) of 34 bioavailable PAHs present in sediments. The TU contribution of each PAH is based on the critical body burden. Because EqP theory is mechanistic, ESBs can be calibrated to be specifically predictive of selected endpoints (USEPA 2003a). For example, for individual PAHs, the relationship between critical body burden in a specific organism and K_{ow} can be used to calculate an organic carbon–normalized LC50 ESB. This relationship is described by the equation below:

$$\text{Log}\,(C^*_{S,OC}) = \log\,(C^*_{L}) + 0.038 \log\,(K_{ow}),$$

where $C^*_{S,OC}$ is equivalent to the ESB, and C^*_L is the critical body burden for the species of interest (Di Toro et al. 2000b). The critical body burden reported for *A. abdita* is 30.8 µmol/g octanol (USEPA 2003b). An ESB quotient (TU) can be calculated for each PAH on the basis of its concentration in the sediment. TUs are then summed for all PAHs in the sample. A sum of TUs greater than 1 indicates a prediction of toxicity. The original sum of TUs approach was based on contributions of narcotic dose from 34 PAHs (USEPA 2003b), including 21 alkylated compounds. The sum of the TUs for 13 key unsubstituted PAHs can be multiplied by 2.75 to provide an estimate with 50% confidence for the contribution of the 21 PAH compounds that were not quantified in sediments (Di Toro et al. 2000b). Data that have been collected for 9 samples of Harbor dredged material show that this correction factor is appropriate for estimating the concentration of total (parent and alkylated) PAHs from the total concentration of the key 13 PAHs (M. Reiss, unpublished data). Higher correction factors can be used to improve the confidence in the estimate; however, the potential for false positives increases proportionally (USEPA 2003b). For example, the correction factor for 95% confidence is 11.5.

Consensus-based SQGs

Swartz (1999) and MacDonald et al. (2000) have proposed consensus-based SQGs for PAHs and polychlorinated biphenyls (PCBs), respectively. These approaches combine available SQGs for those contaminants into 3 categories on the basis of their original predictive intent (e.g., screening, midrange, high). The central tendency of SQGs in each category is then calculated and presented as a consensus sediment effect concentration (SEC) to be used for that predictive function. SECs are evaluated by direct comparison to sediment chemistry results for these contaminants, with or without organic carbon normalization. The threshold effects concentration (TEC) is a concentration below which effects are unlikely. The median effects concentration (MEC) represents a concentration above which effects frequently occur, and the extreme effects concentration (EEC) represents the concentration at which effects are expected. The consensus TEC–MEC–EECs are 40–400–1700 µg PCBs/kg and 290–1800–10000 µg PAH/g OC.

Another approach has been suggested by Fairey et al. (2001). Fairey et al. analyzed a national database to determine which contaminants or groups of contaminants were most useful for predicting toxicity to amphipods. The resulting SQGQ1 model calculates a mean SQG quotient for 9 compounds: Cd, Cu, Pb, Ag, Zn, total chlordane, dieldrin, total PAH_{OC}, and total PCB. The

SQGs used are PEL for Cd, Ag, and Pb; ERM for Cu, Zn, total chlordane, and dieldrin; and MEC for total PAH_{OC} and total PCB (Chapter 12). The resulting mean quotient is then evaluated using the following regression equation (Fairey et al. 2001) relating the SQGQ1 to amphipod mortality, where

$$\text{Fraction response} = -0.1774(\text{SQGQ1}) + 0.8277.$$

Fairey et al. (2001) suggest that this approach serve as a method for evaluating sediment toxicity using mean SQG quotients. Total chlordane concentrations were available for only 38 of the synoptic Harbor samples; therefore, evaluation of the approach was limited to these samples. Twenty-one of these 38 samples were toxic to amphipods in toxicity tests.

Sediment Quality Guideline Performance in New York–New Jersey Harbor Region

Predictions of toxicity made for the 167 samples using the various SQG approaches and observed toxicities in those samples are shown in Figure 15-3. Performance of the various approaches in predicting toxicity is summarized in Table 15-3.

Empirical sediment quality guidelines

Overall, predictions of toxicity made with empirical SQGs agreed with results of toxicity tests for most Harbor samples (Table 15-2). The 3 empirical approaches predicted widespread toxicity in the Passaic–Arthur Kill–Newark Bay complex (Figures 15-3b–d). Agreement between predicted and observed toxicity was high (92%) using mean effect range quotients. However, because most Harbor sediments had mean quotients between the upper and lower threshold effect range quotients, predictions could be made only for relatively few samples (56 samples using mean PEL-Qs; 59 samples using mean ERM-Qs). PEL-Qs overpredicted toxicity in 5 samples (3 in the Newark Bay complex and 1 each in the East and Hudson Rivers). ERM-Qs failed to identify 2 toxic samples in Jamaica Bay and overpredicted toxicity in 3 samples (1 sample each in the Hackensack, Hudson, and East Rivers). ML values failed to predict toxicity in Jamaica Bay (missed all 8 toxic samples) and most Upper New York Bay–East River–Hudson River sites (missed 7 of 10 toxic samples). ML values also predicted toxicity in 4 nontoxic sites in Lower New York Bay, while failing to identify 5 of the toxic Lower Bay samples where toxicity was observed.

Table 15-3 Results of evaluation of various SQG approaches using Harbor-wide synoptic database of 167 surficial sediment samples (67 toxic, 100 not toxic in *A. abdita* toxicity tests)

	Empirical			Consensus-based[d]			Mechanistic[d]		
	ERM-Q	PEL-Q	ML[e]	total PAHoc[e]	tPCB[e]	SQGQ[a]	ΣPAHs[e]	Dieldrin[e]	SEM/AVS
A Not predicted to be toxic[a,c]	19	28	54	5	3	35	5	0	(23)
B Predicted toxic when toxic	16	23	38	3	3	21	3	NA	ND
C Predicted toxic when nontoxic	3	5	16	2	0	14	2	NA	ND
D Not predicted to be nontoxic[b]	40	28	—	(42)	(47)	3	(162)	(167)	(109)
E Predicted nontoxic when nontoxic	38	28	—	(39)	(42)	3	(98)	(100)	(69)
F Predicted nontoxic when toxic	2	0	—	(3)	(5)	0	(64)	(67)	(40)
% Overall agreement (B + E / A + D)[c]	92	91	70	(89)	(90)	63	60	(70)	(57)
% Underprediction (F / D)	5	0	—	—	—	0	—	—	—
% Overprediction (C/A)	16	18	30	40	0	14	40	0	—

[a] ERM-Q > 1.2; PEL-Q > 2.3; Any ML exceeded; [PAHoc], [PCB] > EEC, ΣPAH TUs >1, SEM/AVS > 1
[b] ERM-Q < 0.1; PEL-Q < 0.1; [PAHoc], [PCB] < TEC, ΣPAH TUs < 1, SEM/AVS < 1
[c] For SEM/AVS, predictions are of metals bioavailability, not toxicity. Agreement is based on E / D.
[d] Values in parentheses denote toxicity predictions based on specific contaminants, not overall sediment toxicity.
[e] SQGs can be confidently used only for positive prediction of mixture toxicity.
NA = not available; ND = not detected

Mechanistic sediment quality guidelines

EqP approaches identified few samples as having levels of PAHs, dieldrin, or metals in forms or concentrations that are predicted to be toxic. Total PAHs exceeded the ESB in 5 samples; 3 of these samples were shown to be acutely toxic to amphipods in laboratory toxicity tests. These high PAH samples were located in areas of Upper New York Bay and the Newark Bay–Passaic River–Arthur Kill complex. Twenty-three samples were predicted to have porewater metals present in bioavailable forms (i.e., $[SEM_{molar}] > [AVS_{molar}]$). Most (15) of these samples were located in Raritan Bay. Predictions of toxicity based on metals were not made for these samples. Dieldrin concentrations did not exceed the ESB in any sample.

Consensus-based sediment quality guidelines

Predictions of toxicity made with consensus-based SQGs were limited to those samples with PAH and PCB concentrations below TECs (no toxicity predicted) or above the EEC (toxicity predicted). Forty-two samples were below the TEC for PAHs; 47 samples were below the TEC for PCBs. Twenty-eight samples were below TECs for both compounds. A total of 8 samples were predicted to be toxic on either PAH (5 samples) or PCB (3 samples) concentrations above EECs. Forty-two (89%) of the predictions made on the basis of PAH concentrations, and 45 (90%) of the predictions made on the basis of PCB concentrations agreed with toxicity tests. The 8 stations predicted to be toxic were located in the Upper New York Bay or in the East and Passaic Rivers. All samples predicted to be toxic based on PCB concentrations agreed with laboratory results; 2 of the 5 samples predicted to be toxic on the basis of PAHs were not toxic in the laboratory.

The SQGQ1 model was used to predict toxicity to amphipods in the 38 samples that had chlordane data available. The model predicted >20% mortality of amphipods in 35 of the 38 samples. The 3 samples that were predicted to be nontoxic agreed with the results of toxicity tests using those samples. Overall agreement between predicted and observed toxicity was 63% using this model.

Discussion of Published Sediment Quality Guideline Performance in New York–New Jersey Harbor

In recent years, much has been written in the literature emphasizing the weakness of some of the relationships used to establish the empirically based SQGs and arguing against overreliance on these numbers for regulatory decision making (Long, MacDonald, et al. 1995; Chapman et al. 1997; Luoma and Fisher 1997; Long et al. 1998). The results of the evaluation presented in Tables 15-1 and 15-2 suggest that such cautions are indeed warranted.

The degree of contamination and toxicity is an important factor to consider when evaluating the predictive ability of SQGs. In this study, sediments in the inner Harbor were highly to moderately contaminated and sediments in the lower New York Bay were predominately sand with lower levels of contamination. As a result, about 60% of these samples were not toxic to amphipods in laboratory tests. It is difficult to use nontoxic and relatively uncontaminated samples to evaluate the ability of SQGs to predict toxicity in contaminated samples. Therefore, SQG performance was examined separately in the inner Harbor and the other Harbor areas.

Empirical sediment quality guidelines

The overall rate of agreement of toxicities predicted using mean effect range quotients with observed toxicities was quite high (i.e., >90%) for the samples in this database. Despite the high degree of agreement between observed toxicity and toxicity predicted using mean quotients, the utility of the approach for directing Harbor management efforts is tempered by the fact that predictions could be made only for those sediments having contaminant levels at the tails of the distribution (i.e., <60 of the 167 samples). In addition, the use of mean quotients for the entire contaminants mixture present in the sediments makes it difficult to determine which compound or group of compounds is most directly responsible for predicted toxicity. Given the high covariance of contaminants, this is especially problematic for Harbor sediments. Therefore, the utility of mean quotients as a method for selecting sediment remediation benchmarks for reducing toxicity to amphipods in the Harbor is limited.

Predictions could be made for only 54 samples using ML values (i.e., those having sediment concentrations exceeding respective MLs). Thirty-eight of these sediments were toxic to amphipods in laboratory tests; therefore, the overall rate of agreement of predictions with laboratory toxicity tests was 70%. The ML approach failed to predict toxicity in 29 (43%) of 67 toxic samples. Agreement of predictions made using ML values with laboratory toxicity test results was higher (78% agreement) in the more contaminated portions of the Harbor (i.e., in the Newark Bay, Passaic, and Hackensack Rivers and in the Kills complex) than elsewhere in the Harbor (46% agreement). Had the evaluation of ML performance been restricted to predicting toxicity for the 69 samples taken from Lower New York and Jamaica Bays (see Figure 15-3a), the values would identify only 1 of the 15 samples that were toxic to amphipods in Jamaica Bay and would incorrectly predict toxicity in 4 lower New York Bay samples.

The lack of agreement between ML-predicted and laboratory toxicity in less heavily contaminated areas of the Harbor suggests that ML values derived using empirical relationships between sediment contaminants and effects in one locale (i.e., in this case Puget Sound) cannot reliably be used to predict the outcome of acute toxicity tests with amphipods in the NY–NJ Harbor. The values reflect the contaminant correlations that are present within the system from which the supporting toxicity data were obtained; ML values that are based on matched NY–NJ Harbor toxicity and chemistry data may perform better.

Several of the Harbor studies included in the dataset were performed before protocols to control ammonia toxicity were widely used in toxicity testing with amphipods (as per USACE et al. 1992). Static exposures were used in these studies; original documentation did not describe any methods used to reduce or control ammonia. Therefore, it is also possible that some of the toxicity observed in Harbor samples may have been caused by ammonia. Samples collected by USEPA R-EMAP surveys (USEPA 1998) were taken from the uppermost 2 cm of surficial sediments and tests were conducted without procedures to control ammonia. Several authors have demonstrated that porewater ammonia concentrations are significantly lower near the sediment–water interface than in deeper sediment strata (Blackburn and Henriksen 1983; Chambers et al. 1992; Frazier et al. 1996). All data in our dataset from samples in Jamaica Bay (an area where MLs did not predict any acute toxicity) were generated during R-EMAP (USEPA 1998). Although ammonia toxicity cannot be ruled out as a factor contributing to the lack of agreement between ML-predicted and observed toxicity in Jamaica Bay, we believe the lack of agreement is more likely to reflect the site-specificity of the ML values used to predict the toxicity.

Mechanistic and consensus-based sediment quality guidelines

Performance of the EqP and consensus-based SQGs is more difficult to interpret than that of the empirical SQGs because they predict toxicities on the basis of concentrations of individual contaminants (e.g., dieldrin) or groupings of contaminants (e.g., PAH mixtures, PCBs, SEM), rather than the entire mixture. Incidence of toxicity at concentrations below the ESB or TEC or in sediments where SEM/AVS predicts metals to be unavailable to biota may be associated with toxicity of other compounds in the mixture. As a result, only affirmative predictions of toxicity made for samples exceeding ESBs and EECs (metals toxicity was not evaluated) can be confidently used to evaluate their predictive ability in the Harbor region.

Only 5 sediments in the Harbor dataset had PAH levels exceeding the ESB. No sediment samples in the Harbor dataset exceeded the dieldrin ESB. The evaluation may be incomplete in characterizing the ability of the PAH mixture ESB to predict toxicity to amphipods in the Harbor. The PAH mixture ESB is based on narcotic effects to exposed organisms, but PAHs are not the only contaminants present in sediment that contribute to the narcotic dose. All nonionic organics in the mixture will contribute equally (on a molar basis) and additively to this endpoint (Chaisuksant et al. 1999). Therefore, restricting the evaluation of narcosis to PAHs may underestimate sediment toxicity because it does not consider the narcotic contributions of these other contaminants. An additional factor that may have contributed to the PAH mixture (and dieldrin) ESB identifying relatively few toxic samples is that ESBs are calculated using concentrations that cause 50% mortality in exposed organisms (i.e., LC50s), but an acutely toxic sediment was defined for this study as one that caused 20% mortality in exposed *A. abdita* (i.e., LC20). The ESB would be correspondingly lower had it been based on LC20s and would presumably have predicted more Harbor sediments to be toxic.

Few sediments in the dataset were contaminated with PAHs or PCBs to levels exceeding EECs. Although the 3 sediments with PCBs in excess of the EEC were toxic to amphipods in toxicity tests and there was approximately 90% agreement of predictions made using comparisons to TECs and EECs with toxicity test results, the results do not necessarily validate the consensus-based SQGs as causal because of the high degree of covariance between contaminants in the dataset. Furthermore, 2 of the 5 samples predicted to be toxic based on PAHs above the EEC (the same samples were identified by EqP; this agreement is not surprising because EqP approaches for establishing SQGs for individual contaminants [USEPA 1993a–1993c] were included in the calculation of the EEC) were actually not toxic to *A. abdita* in the laboratory. This result may be explained by the presence of PAHs in the samples in a matrix that was not bioavailable (e.g., soot or coal particles; Chapter 4) but highlights the need for a cautionary approach to using SQGs, such as mechanistic and consensus-based SQGs, that infer causality to screen sediments.

The overprediction of toxicity by the SQGQ1 model is likely to have resulted from the models lack of consideration of metals availability. SEM and AVS data for the Harbor samples showed that a small minority of the samples actually had SEM concentrations exceeding AVS, and therefore, metals are not generally expected to be available to cause toxicity in the Harbor.

Sediment quality guidelines for evaluation of dredged material

Because of the significance of the results of amphipod toxicity tests to management of Harbor dredged material, special attention must be given to the performance of SQGs in making predictions for these sediments. An evaluation of the predictive ability of the published SQG approaches was performed on a sample-by-sample basis for a limited data set specific to the dredging program. Data from 4 composited samples of Harbor dredged material were used to make a preliminary assessment of the ability of SQGs to predict toxicity to amphipods in Harbor dredged sediments. Samples of sediment proposed for dredging in the Arthur Kill (AK), Raritan Bay (RB), Upper New York Bay (UB), and the Hudson River (HR) were evaluated. Each of these composites is representative not only of the material from that portion of the Harbor but, taken together, are representative of the diversity in Harbor contaminant profiles. Only the AK composite was shown to be acutely toxic to amphipods in the laboratory. The bulk sediment chemistry and toxicity test results and predictions of toxicity made using the various SQGs for the 4 samples are presented in Table 15-4.

The AK sample was correctly predicted to be toxic by the 3 empirical SQG approaches. Predictions made using PEL-Qs incorrectly predicted the UB composite to be toxic; toxicity of this sample could not be predicted using ERM-Qs. Predictions of toxicity could not be made for the HR or RB samples using either mean quotient approach. Toxicity predictions could not be made for the HR, RB, or UB samples using ML values. No samples exceeded the dieldrin or PAH ESBs. Therefore, the predictive ability of the ESBs could not be further evaluated. Predictions could not be made using the consensus-based PCB guidelines because sediment concentrations in the 4 sediments were between the TEC and EEC. Two of the nontoxic samples (RB and HR) were below the TEC for PAHs; predictions could not be made for the other compounds. The Fairey et al. (2001) model overpredicted toxicity in the 3 nontoxic samples but correctly identified the toxic AK sample.

It is clear from the foregoing analysis that SQGs alone can lead to erroneous conclusions regarding sediment toxicity if applied to NY–NJ Harbor dredged materials. Although only 4 samples were evaluated, these samples represent more than a half million cubic meters of sediment, and the economic implication of regulatory decisions on these sediments, with regard to management options, is in excess of $10 million. Implications of erroneous conclusions with respect to the overall management of the Harbor sediment contamination (e.g., costs associated with remedial actions) are much greater.

Conclusions and Next Steps

The Dredged Material Management Plan for the Port assumes that aggressive contaminant reduction and remediation can achieve an 80% decrease in the amount of dredged material unsuitable for ocean disposition by the year 2040 (USACE 1999). USEPA has specified reduction of sediment toxicity as a major component of total maximum daily loads (TMDLs) of individual contaminants that are to be implemented for the Harbor by 2006. In addition

Table 15-4 Results of SQG screening for applicability to predict toxicity in recent amphipod toxicity tests of 4 navigational dredged material samples from NY–NJ Harbor

Contaminant	Arthur Kill (AK)	Hudson River (HR)	Raritan Bay (RB)	Upper Bay (UB)
Naphthalene (μg/kg)	580	72	53	1650
Acenaphthylene (μg/kg)	593	180	18	154
Acenaphthene (μg/kg)	500	58	16	1120
Fluorene (μg/kg)	521	120	30	1200
Phenanthrene (μg/kg)	2490	470	148	3560
Anthracene (μg/kg)	934	490	68	1650
Fluoranthene (μg/kg)	4190	1000	423	3670
Pyrene (μg/kg)	4000	580	421	3360
Benzo(*a*)anthracene (μg/kg)	1270	450	195	2600
Benzo(*b*)fluoranthene (μg/kg)	5000	420	234	3180
Benzo(*k*)fluoranthene (μg/kg)	500	360	228	1750
Benzo(*a*)pyrene (μg/kg)	2100	480	239	3100
Indeno(123*cd*)perylene (μg/kg)	1500	200	183	1740
Chrysene (μg/kg)	2240	520	263	2510
Dibenz(*ah*)anthracene (μg/kg)	500	75	45	484
Benzo(*ghi*)perylene (μg/kg)	1030	180	174	1520
Low MW PAH (μg/kg)	5120	1390	333	9330
High MW PAH (μg/kg)	17500	3630	2000	20200
Total PAH (μg/kg)	22600	5020	2330	29500
Total DDT (μg/kg)	773	11.1	38.1	22.4
Total chlordane (μg/kg)	1	13.7	2.97	3.02
Dieldrin (μg/kg)	1	1.2	1.46	1.51
Total PCB (μg/kg)	394	123	228	234
TOC (%)	2.2	2.8	3.36	4.8
As (mg/kg)	24.7	11.6	33.6	11
Cd (mg/kg)	4.85	1.26	2.23	1.29
Cu (mg/kg)	365	104	244	97.1
Cr (mg/kg)	159	95.5	111	96.9
Pb (mg/kg)	292	104	172	137
Hg (mg/kg)	6.04	1.28	2.1	1.32

Table 15-4 *continued*

Contaminant	Arthur Kill (AK)	Hudson River (HR)	Raritan Bay (RB)	Upper Bay (UB)
Ni (mg/kg)	49.2	29.1	40.6	34.4
Ag (mg/kg)	5.37	3.99	4.64	3.97
Zn (mg/kg)	457	173	380	175
Acutely toxic?	Yes	No	No	No
Mortality observed	41%	17%	2%	10%
Mean PEL quotient[a]	3.17	0.85	0.73	2.91
Predicted toxicity	Yes	No prediction	No prediction	Yes
Mean ERM quotient[b]	1.86	0.44	0.46	1.12
Predicted toxicity	Yes	No prediction	No prediction	No prediction
% chemicals exceeding MCL[c]	3	0	0	0
Predicted toxicity	Yes	No prediction	No prediction	No prediction
Sum toxic units from PAH ESB[d]	0.40	0.07	0.02	0.23
Predicted toxicity	No	No	No	No
% Chemicals exceeding EEC[e]	0	0	0	0
Predicted toxicity	No	No	No	No
SQGQ1[f]	1.23	0.79	0.87	0.69
Predicted mortality	39%	31%	33%	29%

[a] MacDonald 1994
[b] Long, MacDonald, et al. 1995
[c] USACE 2000
[d] USEPA 2003b
[e] Swartz 1999 (PAH); MacDonald et al. 2000 (PCB)
[f] Fairey et al. 2001

to reducing sediment toxicity, the TMDLs will also be set to achieve other Harbor environmental quality objectives (e.g., harbor-wide compliance with WQC, acceptable levels of bioaccumulation in aquatic organisms, benthic community integrity).

The ability and performance of some published SQG values to predict amphipod toxicity in the NY–NJ Harbor system has been shown to be limited. Overall, the published SQG values from various SQG approaches may be useful for broadly characterizing the probability of encountering sediment toxicity in different areas of the harbor system. However, predictions of toxicity made using published values derived from different SQG approaches for individual samples are often inconsistent. Some level of screening of toxic (e.g., using the PCB EEC) and nontoxic samples (e.g., using the PAH and PCB TECs) may also be possible. However, because of the high covariance of contaminants in Harbor sediments this is just as likely to be an indication of correlation rather than indicating causality. The limitations of all of the SQGs evaluated were, unfortunately, evident for sediments in Newark Bay and Upper New York Bay where the Port's facilities (and dredging) are centered and the need for accurate predictions is most urgent. It is possible that further examination and regional calibration (where possible) using published SQG approaches might provide more useful SQGs for the NY–NJ Harbor.

SQGs are only intended to specifically predict impacts to aquatic organisms that are directly exposed to sediments; they do not provide information regarding risks associated with bioaccumulation (Chapter 11) or sufficient information about the specific agents of toxicity in the Harbor to direct remediation efforts. Additional work is needed before a preferred approach can be selected to derive numerical guidelines that will be suitable for addressing all of the Harbor's assessment and management needs. Ongoing efforts to obtain the necessary information to complement published SQGs include the projects outlined below.

Regional calibration of published sediment quality guideline approaches

The CARP working group has recently begun calculating empirical SQGs (TECs, ERLs–ERMs, PELs–TELs, and AETs) using the Harbor dataset described above and synoptic chemistry and toxicity datasets for Harbor dredged materials. At least in theory, these regional SQGs are expected to perform better in the Harbor than published empirical SQGs because they should better reflect the bioavailability and covariance of Harbor contaminants (MacDonald 1994; MacDonald et al. 1996; Hyland et al. 1999) and will be based solely on the specific endpoint of interest (acute toxicity to amphipods).

Toxicity identification evaluation

There is a critical need for a better understanding of which contaminants and concentrations are actually causing sediment toxicity in the Harbor. Because of the high degree of covariance of Harbor contamination, existing SQGs cannot be confidently used to identify toxic agents. A regional workshop to be hosted by the Hudson River Foundation in Manhattan is being planned to evaluate the state of the science with regard to sediment toxicity identification evaluation (TIE).

Regional biota–sediment accumulation factors

A regional database of synoptic field observations of sediment contaminant concentration and organic carbon content and of polychaete tissue and lipid concentrations is being assembled to calculate site-specific biota–sediment accumulation factors (BSAFs) for organic contaminants in the Harbor system. These regionally validated factors will allow managers to set sediment quality goals that protect upper trophic-level receptors (including humans) from the effects of bioaccumulation. These new sediment quality goals can then be used in tandem with sediment quality goals based on toxicity, allowing managers to select a final concentration that will be protective of both direct toxicity and bioaccumulation pathways.

Critical body residue

USACE and USEPA have proposed using the target lipid model of Di Toro et al. (2002a, 2000b) to set acceptable levels for bioaccumulation of nonpolar organic contaminants (i.e., narcotics) from dredged materials proposed for ocean disposition. This approach is also being considered for organic contaminants that have specific modes of toxic action. The USACE and USEPA are also investigating setting acceptable levels based on effect–residue relationships for individual contaminants or contaminant groups. These benchmarks could then be used to identify sediment quality targets using the regional BSAFs discussed above.

Because only a small number of narcotic chemicals are routinely measured and tracked in sediment assessments, predictions of narcotic toxicity may suffer from their exclusion from toxicity models (Di Toro DM, Hydroqual Inc., Mahwah, NJ, personal communication). A research plan for measuring and evaluating the narcotic contribution of an extensive list of the hydrocarbons in Harbor sediments is being developed. It is anticipated that this effort will improve the reliability of predictions of sediment toxicity that results from narcosis.

Benthic community analysis

Long et al. (2001) noted that the ERL and ERM methodology was not as effective at predicting benthic effects in NY–NJ Harbor as in other parts of the nation. There are ongoing efforts to better understand the highly modified benthic community in NY Harbor. The New Jersey Marine Sciences Consortium has developed a preliminary index of biotic integrity (IBI) for Barnegat Bay, NJ (Weinstein MP, New Jersey Marine Sciences Consortium,

Sandy Hook, NJ, personal communication). It is anticipated that this index may aid in calibrating site-specific SQGs for community effects.

References

Barrick R, Becker S, Brown L, Beller H, Pastorok R. 1988. Volume 1, Sediment quality values refinement: 1988 update and evaluation of Puget Sound AET. Bellevue (WA): PTI Environmental Services. EPA Contract No. 68-01-4341. PTI Contract No. C717-01. 144 p.

Blackburn TH, Henriksen K. 1983. Nitrogen cycling in different types of sediments from Danish waters. *Limnol Oceanogr* 28:477–493.

Blumberg AF, Khan LA, St. John JP. 1999. Three-dimensional hydrodynamic model of New York Harbor Region. *J Hydraul Engineer* 125:799–816.

Chaisuksant Y, Yu Q, Connell, D.W. 1999. The internal critical level concept of nonspecific toxicity. *Rev Environ Contam Toxicol* 162:1–41.

Chambers PA, Prepas EE, Gibson K. 1992. Temporal and spatial dynamics in riverbed chemistry: The influence of flow and sediment composition. *Can J Fish Aquat Sci* 49:2128–2140.

Chapman PM, Cano M, Fritz AT, Gaudet C, Menzie CA, Sprenger M, Stubblefield WA. 1997. Workgroup summary report on contaminated site cleanup decisions. . In: Ingersoll CG, Dillon T, Biddinger GR, editors. Ecological Risk Assessment of Contaminated Sediments. Pensacola (FL): Society of Environmental Toxicology and Chemistry (SETAC). p 83–114.

Crawford DW, Bonnevie NL, Gillis CA, Wenning RJ. 1994. Historical changes in ecological health of the Newark Bay Estuary, New Jersey. *Ecotoxicol Environ Saf* 29:276–303.

Cubbage J, Batts D, Breidenbach S. 1997. Creation and analysis of freshwater sediment quality values in Washington State. Olympia (WA): Washington State Department of Ecology. Publication Nr. 97-323a.

Di Toro DM, McGrath JA, Hansen DJ. 2000a. Technical basis for narcotic chemicals and polycyclic aromatic hydrocarbon criteria. I. Water and tissue. *Environ Toxicol Chem* 19:1951–1970.

Di Toro DM, McGrath JA, Hansen DJ. 2000b. Technical basis for narcotic chemicals and polycyclic aromatic hydrocarbon criteria. II. Mixtures and sediment. *Environ Toxicol Chem* 19:1971–1982.

Douglas WS, Baier L, Gimello RJ, Lodge J. 2003. A comprehensive strategy for managing contaminated dredged materials in the Port of New York and New Jersey. *J Dredging Eng* 5:1–12.

Fairey R, Long ER, Roberts CA, Anderson BS, Phillips BM, Hunt FW, Puckett HR, Wilson CJ. 2001. An evaluation of methods for calculating mean sediment quality guideline quotients as indicators of contamination and acute toxicity to amphipods by chemical mixtures. *Environ Toxicol Chem* 20:2276–2286.

Farley KJ, Thomann RV, Cooney TF, Damiani DR, Wands JR. 1999. An integrated model of organic chemical fate and bioaccumulation in the Hudson River estuary. Report prepared for the Hudson River Foundation, March, 1999. New York: Hudson River Foundation.

Frazier BE, Naimo TJ, Sandheinrich MB. 1996. Temporal and vertical distribution of total ammonia nitrogen and un-ionized ammonia nitrogen in sediment pore water from the Mississippi River. *Environ Toxicol Chem* 15:92–99.

Hansen DJ, Berry WJ, Mahony JD, Boothman WS, Di Toro DM, Robson DL, Ankley AT, Ma D, Yan Q, Pesch CE. Predicting the toxicity of metal-contaminated sediments using interstitial concentration of metals and acid-volatile sulfide normalizations. *Environ Toxicol Chem* 15:2080–2094.

Hyland JL, Van Dolah RF, Snoots TR. 1999. Predicting stress in benthic communities of souteastern U.S. estuaries in relation to chemical contamination of sediments. *Environ Toxicol Chem* 18:2557–2564.

Landeck-Miller RE, St. John JP. *In Press*. Modeling production in the Lower Hudson River estuary. In: Levinton J, editor. The Hudson River ecosystem. New York: Oxford Univ Pr.

Long ER, Field LJ, MacDonald DD. 1998. Predicting toxicity in marine sediments with numerical sediment quality guidelines. *Environ Toxicol Chem* 17:714–727.

Long ER, Hong CB, Severn CG. 2001. Relationships between acute sediment toxicity in laboratory tests and abundance and diversity of benthic infauna in marine sediments: A review. *Environ Toxicol Chem* 20:46–60.

Long ER, MacDonald DD, Severn CG, Hong CB. 2000. Classifying probabilities of acute toxicity in marine sediments with empirically derived sediment quality guidelines. *Environ Toxicol Chem* 19:2598–2601.

Long ER, MacDonald DD, Smith SL, Calder FD. 1995. Incidence of adverse biological effects within ranges of chemical concentrations in marine and estuarine sediments. *Environ Manag* 19:81–97.

Long ER, Morgan LG. 1991. The potential for biological effects of sediment-sorbed contaminants tested in the National Status and Trends Program. Seattle (WA): US National Oceanic and Atmospheric Administration. NOAA Tech. Mem. NOS OMA 52. 175 p.

Long ER, Wolfe DA, Scott KJ, Thursby GB, Stern EA, Peven C, Schwartz T. 1995. Magnitude and extent of sediment toxicity in the Hudson–Raritan Estuary. Silver Spring (MD): US National Oceanic and Atmospheric Administration. NOAA Technical Memorandum NOS ORCA 88. August 1995. 230 p.

Luoma SN, Fisher N. 1997. Uncertainties in assessing contaminant exposure from sediments. In: Ingersoll CG, Dillon T, Biddinger GR, editors. Ecological Risk Assessment of Contaminated Sediments. Pensacola (FL): Society of Environmental Toxicology and Chemistry (SETAC). p 211–238.

MacDonald DD. 1994. Approach to the assessment of sediment quality in Florida coastal waters. Volume 1, Development and evaluation of sediment quality guidelines. Tallahassee (FL): Florida Department of Environmental Protection.

MacDonald DD, Carr RS, Calder FD, Long ER, Ingersoll CG, 1996. Development and evaluation of sediment quality guidelines for Florida coastal waters. *Ecotoxicology* 5:253–278.

MacDonald DD, DiPinto LM, Field J, Ingersoll CG, Long ER, Swartz RC. 2000. Development and evaluation of consensus-based sediment effect concentrations for polychlorinated biphenyls. *Environ Toxicol Chem* 19:1403–1413.

McDonough FM, Boehm GA, Douglas WS. 1999. Dredged material management in New Jersey: A multifaceted approach for meeting statewide dredging needs in the 21st century. Presented at WEDA XIX 31st Annual Dredging Seminar; 1999 June 16–18.

Swartz RC. 1999. Consensus sediment quality guidelines for polycyclic aromatic hydrocarbon mixtures. *Environ Toxicol Chem* 18:780–787.

[USACE] US Army Corps of Engineers. 1999. Dredged material management plan for the port of New York and New Jersey, Draft implementation report. New York: USACE.

[USACE] US Army Corps of Engineers, US Environmental Protection Agency, Washington Department of Natural Resources, Washington Department of Ecology. 2000. Dredged material evaluation and disposal procedures: A users manual for the Puget Sound Dredged Disposal Analysis (PSSDA) Program. Seattle (WA): US Army Corps of Engineers.

[USACE/USEPA] US Army Corps of Engineers, US Environmental Protection Agency. 1992. Draft guidance for performing tests on dredged material proposed for ocean disposal. New York: USACE NY District, USEPA Region 2.

[USEPA] US Environmental Protection Agency. 1993a. Sediment quality criteria for the protection of benthic organisms: Fluoranthene. Washington DC: USEPA. EPA 822/R/93/012.

[USEPA] US Environmental Protection Agency. 1993b. Sediment quality criteria for the protection of benthic organisms: Acenaphthene. Washington DC: USEPA. EPA 822/R/93/013.

[USEPA] US Environmental Protection Agency. 1993c. Sediment quality criteria for the protection of benthic organisms: Phenanthrene. Washington DC: USEPA. EPA 822/R/93/014.

[USEPA] US Environmental Protection Agency. 1998. Sediment quality of the NY/NJ Harbor system: an investigation under the regional environmental monitoring and assessment program (R-EMAP). Final Report. March 1998. New York: USEPA. EPA/902-R-98-001.

[USEPA] US Environmental Protection Agency. 2003a. Procedures for the derivation of equilibrium partitioning sediment benchmarks (ESBs) for the protection of benthic organisms: metals mixtures (cadmium, copper, lead, nickel, silver, zinc).

Washington DC: USEPA Office of Research and Development. EPA-600-R-02-011. (draft)

[USEPA] US Environmental Protection Agency. 2003b. Procedures for the derivation of equilibrium partitioning sediment benchmarks (ESBs) for the protection of benthic organisms: PAH mixtures. Washington DC: USEPA Office of Research and Development. EPA-600-R-02-013. (draft)

[USEPA] US Environmental Protection Agency. 2003c. Procedures for the derivation of equilibrium partitioning sediment benchmarks (ESBs) for the protection of benthic organisms: Dieldrin. Washington DC: USEPA Office of Research and Development. EPA-600-R-02-010. (draft)

[USEPA/USACE] US Environmental Protection Agency, US Army Corps of Engineers. 1991. Evaluation of dredged material proposed for ocean disposal: testing manual. Washington DC: USEPA. EPA 503/8-91/001.

Weisberg, SB, Frithsen JB, Holland AF, Paul JF, Scott KJ, Summers JK, Wilson HT, Valente R, Heimbuch DG, Gerritsen J, Schimmel SC, Latimer RW. 1993. EMAP-estuaries Virginian Province 1990 demonstration project report. Narragansett (RI): US Environmental Protection Agency NHEERL-AED. EPA 620/R-93/006.

Influence of confounding factors on SQGs and their application to estuarine and marine sediment evaluations

16

JACK Q WORD, WILLIAM W GARDINER, DAVID W MOORE

Empirical sediment quality guidelines (SQGs) have been established on the basis of the co-occurrence of statistically significant toxicological or benthic community changes and the presence of elevated sediment chemistry (Chapter 12). While concordant responses do not necessarily indicate causal relationships, the current group of empirical SQG relationships has proven to be an effective method for screening chemical data to identify sediments in which there may be adverse impacts (Long and Chapman 1985; Chapman 1996; Chapter 7). However, sediment toxicity or benthic community effects can occur independently of chemical contaminants of potential concern (COPCs) and can be misinterpreted as being related to the COPCs. The physical, chemical, and biological factors that can result in observed biological effects, potentially implicating COPCs, are termed "confounding factors" (CFs).

However, a significant proportion of sediments for which effects are predicted by SQGs do not actually elicit significant biological effects (Chapter 12). One possible reason for lack of predicted effects may be the influence of CFs on tests used in the original development of the SQGs. Conversely, the presence of adverse biological effects where they would not be predicted by SQGs may be due to CFs or the presence of unmeasured contaminants that were present in sediments associated with the development of the original SQGs. Incorporating an evaluation of CFs in empirical SQG development may improve the predictive abilities of SQGs. In addition, confirmatory analyses following an SQG exceedance should incorporate CF evaluations to

Use of Sediment Quality Guidelines and Related Tools for the Assessment of Contaminated Sediments
Wenning RJ, Batley GE, Ingersoll CG, Moore DW, editors.
 ISBN 1-880611-71-6

better link COPCs with biological effects and to strengthen subsequent regulatory decisions.

The objective of this chapter is to present CFs that have been shown to interfere with the identification of contaminant-related effects and to suggest one proposed method for addressing CFs. It is not the objective of this chapter to demonstrate that SQGs are inherently flawed because of CFs, but rather that improvements in the predictive ability of SQGs can be realized if CFs are incorporated in sediment evaluations.

For the purposes of this chapter, CFs will be divided into 4 categories on the basis of their source or the related analytical process (Table 16-1):

1) persistent physical sediment characteristics,

2) persistent chemical characteristics,

3) nonpersistent chemical characteristics, and

4) nonmatrix characteristics.

Therefore, this list is subject to change as our understanding of CFs sources and effects grows. It is important to note that some factors identified as CFs can also be important components of toxicity, depending on the type of sediment evaluation that is being performed or the source of the CF (i.e., natural vs. anthropogenic). In such cases, the factors should not be simply dismissed as "confounding." This chapter addresses approaches that can be used to help identify and address CFs in estuarine and marine systems. Many of these approaches may also be applicable in addressing CFs in freshwater systems.

The London Dumping Convention, which is the basis for many of the dredged material evaluation methods, identifies persistent chemicals of concern as those that are beyond the "capacity of the sea to assimilate and render them harmless" because "its ability to regenerate natural resources is not unlimited" (LDC 1972; Chapter 6). While this definition is useful for characterizing chemicals in dredged material evaluations, the definition of which chemicals are persistent may differ for various site evaluations, especially where in-place toxicity is evaluated. For the purposes of this discussion, persistence is used to help define the tools that may be available for evaluating the effects of a particular CF. Persistent characteristics are those that cannot be easily manipulated during the sediment evaluation process. Nonpersistent characteristics are those that may change during normal evaluative manipulations or may be manipulated using specialized laboratory techniques.

Table 16-1 Categories of confounding factors

Categories of confounding factors	Confounding factors
Persistent physical characteristics	Sediment grain size Shell debris Sediment grain angularity Sediment consolidation Sediment water content Water depth Sediment depth Physical alteration of habitat
Persistent chemical characteristics	Total organic carbon Total organic nitrogen Mineral matrix–associated metals Residual weathered compounds Poorly degradable compounds Woody debris
Non- or less-persistent chemical characteristics	Organic carbon and nitrogen quality Ammonia Salinity pH Sediment biochemical oxygen demand Sediment acclimation Dissolved organic carbon and nitrogen Labile sulfides
Nonmatrix characteristics	Test organism selection Test organism acclimation and handling Test organism, population of species sensitivity Inter- and intralaboratory differences Aeration induced turbulence Sampling and homogenization methods Sediment sieving Test exposure selection Habitat differences Diet Recruitment variation Organism behavior Identification of taxa levels Experience of analysts

Confounding Factors

The purpose of this section is to introduce a number of CFs, discuss their potential implications for SQGs and sediment evaluations, and suggest possible evaluative tools for understanding and/or reducing their influence during sediment assessments. CFs can influence each leg of the sediment quality triad:

chemistry, toxicity testing, and benthic assessments. The principal issue associated with CFs and physical and chemical analyses is whether the method of measuring sediment attributes is equivalent or proportional to the manner in which an organism interacts with that attribute. Many CFs influence the biological responses (toxicity and benthic community changes). Sediment-dwelling organisms have adapted to their environment, developing both behavioral and physiological mechanisms that permit them to live in various types of environments. Indeed, it is the combination of the physical, chemical, and biological interactions that determines infaunal distribution and abundance in the natural environment. Marine and estuarine toxicity tests rely heavily on the use of benthic infaunal organisms. As a consequence, they are also susceptible to the physical and chemical characteristics of test sediment.

Because benthic communities are so sensitive to the naturally occurring features of benthic environments, as well as to contaminants, determining cause and effect in the benthic community is a complex task. While benthic community analysis is a well-established process, determining the significance of change and the related causes is not. Species distribution and abundance are closely tied to sediment grain size, salinity, hydrology, and feeding strategies, as well as to resource competition and predation. There are numerous established metrics for community structure, including species diversity, relative abundance, spatial structure or pattern size structure, functional group characterizations, and the use of indicator species. However, the application of these metrics to determine management decisions is problematic. The application of benthic metrics for sediment evaluation is outside the scope of this discussion but are reviewed in Diaz (1992), Hyland et al. (1999) and Chapter 12. However, some of the sediment characteristics that introduce increased variability or lead to erroneous conclusions regarding benthic community response will be discussed.

Persistent physical characteristics

Persistent physical characteristics are those associated with physical features of sediment, such as consolidation, sediment grain size, or sediment angularity. Physical properties of sediment are generally persistent, and the associated biological effects are difficult to remove or alter. For this reason, CFs associated with physical sediment characteristics are generally accounted for by comparison to appropriate reference sediments.

Confounding influence of sediment grain-size

Sediment particles are comprised of mineral grains, organic-detrital materials, organic materials coating the mineral grains, fecal pellets, and other biogenic structures, each with varying shapes and densities (Shepard 1973). The composition of these different particle types for a given grain size can influence chemical availability (Tada and Suzuki 1982; Literathy et al. 1987), toxicological responses, and biological communities (Parker 1975). The relative importance of sediment grain size and benthic organisms has long been recognized (Petersen 1913; Thorson 1971). In fact, an early theory regarding community structure (trophic group amensalism) suggests that suspension-feeding and deposit-feeding communities cannot live together, separating themselves into sandier and siltier environments on the basis of sediment preference and mutually exclusive behavioral responses (Sanders 1958). Although this theory has been modified through time, the main precept that communities of organisms prefer different environmental grain sizes still holds true. These grain-size preferences also apply to toxicity tests that are conducted with benthic infauna.

Despite the apparent importance of sediment composition, grain-size determinations are typically based on the size of individual mineral particles, also called the "true" grain size (PSEP 1986). Determining the true size of sediment is based on removal of organic materials that may bind sediment grains together or occur as a component part (fecal pellets, plant debris, etc.) of the sediment (PSEP 1986). Organic material is removed from the sample using hydrogen peroxide, and a dispersant is used to disassociate particles within the settling column (Plumb 1981). The grain size for these samples is then based on the dry weight size distribution of the "cleaned" minerals; however, the grain size apparent to the organism may be quite different.

A secondary effect related to grain-size analyses in areas of high organic loads is related to the dry weight basis for the analysis. The effect that occurs is associated with the lower dry mass per unit volume of organic materials versus the higher dry mass per unit volume of quartz sediment grains. Because sediment grain size is based on the dry mass of material trapped on various screens, it requires a larger volume of organic material to influence the grain-size determinations. Sediment evaluations based on dry weight mass of true grain size can affect reference site selections, which can in turn affect the performance standards for chemical analyses, toxicity tests, and benthic community analyses.

Grain size can also be determined using "apparent" grain-size analysis. The apparent sediment-grain size is based on the same measurements (weight and settlement) as true grain size, without using the hydrogen peroxide to remove the organic matrix. While the true grain-size analysis is the common analytical method used for most sediment characterization studies (PSEP 1986; USEPA/USACE 1991, 1998), apparent grain size is more representative of the sediment that benthic organisms are exposed to. The differences in apparent grain size and true grain size are most likely to affect sediment evaluations in areas with high deposition rates of aquatic or terrigenous organic material, such as the head of submarine canyons; within estuaries (especially in well-vegetated zones); in depositional sites near the mouths of rivers, deeper channels; in locations with surfaced or buried peat layers; in areas dominated by gravel, shell hash, or other debris; and in locations adjacent to eddy-producing objects.

For benthic community assessments, the potential for erroneous relationships based on the apparent versus the true grain-size analysis is especially true when comparisons are made between sites or when temporal shifts are determined in areas with high organic or different distributions of gravel, shell, or other types of debris. Extreme examples of benthic communities that are responding to the organic loads in sediment rather than the size of individual sediment grains can be seen in areas of woody deposits surrounding lumber processing yards or accumulation of fish or shellfish processing wastes (Rasnake and Gruber 1998; USEPA/USACE 1998). These debris piles overlay the original sediment, and the organisms living in the debris respond to the size of materials in the debris and the related spaces rather than to the sediment grains underneath or mixed within them.

Another example includes the presence of fecal pellets as a dominant component of sediment. Fecal pellets have been noted as being extremely common in certain environments, accounting for 70% to 90% of the "sediment." The apparent size of the sediment grains (fecal pellets) is much different than the individual components of the fecal pellets (true grain size). Destroying the organic matrix containing the fecal pellets as individual particles to produce the true rather than the apparent grain size would provide a grain-size distribution that was dominated by finer sediments than is actually observed by the organisms.

A third example would include areas where fine-grained sediment materials are tightly bound to each other, similar to the submerged dry clays found in glacially compacted sediment environments. These very fine-grained materials were ground to fine glacial flour that was compressed by the glacier mass, and although they were submerged for >10,000 years, their water content is less than 5%. The organisms that live in these environments (80% to 90% clay by true assessment) are more likely to be boring clams that are suspension feeders than to be the deposit-feeding species that would be predicted to occur based on the true grain-size determinations. Grain size, and preferably apparent grain size, is an important component of selecting an appropriate reference site.

The Puget Sound Estuary Program (PSEP) has a number of approved reference sites, which represent a range of sediment grain sizes (PSEP 1995). Each testing program must use appropriate grain-size references. Gardiner, Word, and Swan (1996) used cluster analysis and triangle plots to determine appropriate reference sites from an array of suitable reference sites based on grain size. The objective of the grain-size reference sites is to match the grain size of the test sediments as closely as possible and account for the added influence of the varying grain size on the population of test organisms. The American Society for Testing and Materials (ASTM) and the US Environmental Protection Agency (USEPA) methods also provide guidance on tolerance limits for laboratory testing of organisms in the US and Canada (USEPA 2001; ASTM 2004a). However, if apparent grain size is altered by the presence of organic agglutinations, shell debris, large gravel and rocks, or heavy organic debris, a reference selected on the basis of true grain-size analysis may result in an inappropriate comparison.

Benthic communities respond to seasonal changes in grain-size distributions, such as spring and summer periods when seasonal storm patterns in coastal areas alter the distribution of sediment grain size. Changes in the abundance and dominance hierarchy of species at deep sites off the Columbia River have been observed to coincide with changes in grain size and total organic carbon (TOC) observed during 2 sampling events separated by 3 months (Word et al. 2003). This study demonstrated that benthic communities responded to these seasonal changes in sediment characteristics through increased abundance, decreased species richness, changes in dominance hierarchy, and changes in dominant mobility patterns of infaunal species when sediment grain size decreased in coarseness and TOC increased in concentration. Those observed changes in the benthic community indicate that seasonal cycles in sediment characteristics influence the benthic communities, and those cycles in sediment characteristics need to be taken into account in establishing reference locations.

The confounding influences of sediment grain size on toxicity tests are generally ameliorated by selecting test species that are either tolerant of the sediment grain size in the test sediments or by including appropriate sediment grain-size controls or references to account for the added biological effects associated with varying grain sizes. Factors that need to be considered for the selection of appropriate reference sites relative to grain size are the apparent versus true grain-size distributions, water content of sediment, depth of water over the sediments surface, appropriate faunal region, season of sampling, and presence of different types of debris. Use of these screens to provide data from comparable types of benthic environments will then permit a better estimate of the effects of chemical contaminants of concern on benthic species from comparable environments. Shepard (1973) recommends a procedure for determining the relative mass of various constituents within each size fraction and a graphical and pie chart presentation that portrays these differences, especially in the coarser size fractions. This level of detail, combined with the apparent grain-size analysis, may be a useful addition to the standard analyses of sediment grain size for the assessment of toxicity and alterations to benthic communities.

Confounding influence of sediment grain angularity

Another factor associated with sediment grain size is the shape of the sediment grains themselves. The shape of sand-sized particles can range within 6 classes, from very angular to well rounded (Shepherd 1973). The significance of sediment angularity to organism survival in test sediments has not generally been addressed for toxicity assessments. Anecdotal observations have documented the effect that highly angular sediment particles appeared to have on the polychaete *Nephtys caecoides.* These organisms were observed on the surface of the sediment with small pools of blood adjacent to everted proboscises. Examination revealed that the surfaces of the proboscis integument were lacerated, that the sediment was uncharacteristically angular, and that the concentrations of contaminants in sediment from the Merritt Sand formation in Oakland Harbor was low. The conclusion was that the adverse impacts on survival (<42% survival) of this test organism on these uncontaminated sediments over a 10-d exposure period was apparently due to the damage received during burrowing and/or feeding on these angular sediments (Word 1990). While this test uncovered the presence of a CF associated with sediment grain angularity, the possible occurrence of this CF in other tests has not been examined.

The potential implication to SQGs is that sediment with relatively low contaminant concentrations may be paired with unacceptable toxicity, although the toxicity was due to sediment angularity. Furthermore, confirmatory testing of similar sediments would result in a similar response. For coarse-grained samples, especially those from areas with recently fractured slate and shale surfaces (land slides, drilling or mining activities) observations on sediment angularity and test organism health should be recorded and the potential for effect of this factor addressed.

Confounding influence of sediment consolidation and sediment water content

In general, the water content of coastal marine sediment is closely associated with the sediment grain size (Trask 1932). As the average grain size decreases, the percentage of water content increases. However, certain environments depart from this association, with the percent water content being lower and in some cases much lower than would be predicted based on the average grain size. These low water content sediments include those sediments that have been consolidated because of ancient overburdens (e.g., clays that have been consolidated by glaciers), sediment that is deeper within cores (e.g., consolidation by sediment overburden), sediments that have been desiccated as a result of being on the surface and then resubmerged (e.g., Merritt and Posey sand formations in San Francisco Bay; Trask and Rolston 1951), or those that have been compacted by human activity. The potential problems associated with these drier sediments include

1) the lack of biogenic reworking by microbial communities,
2) the inability of benthic organisms to penetrate the surface of sediment or to use small particles for construction of tubes and burrows,
3) the decrease in potentially interactive surface area because individual particles are compacted together and act as a paved surface, and
4) the unavailability of organic materials locked into the sediment.

Each of these attributes of consolidation can influence organism survival in the sediment, either directly or indirectly.

The water content and degree of compaction of sediment with the same grain size can vary widely. Fine-grained clays can be compressed by the weight of overlying sediment layers squeezing submerged layers and reducing the water content to less than 1% (sediment overburden). Examples of overburden can be seen in areas with a past history of glacial activity (blue clays in Puget Sound, WA) or consolidated at depth within a sediment column (Posey Formations in San Francisco Bay, CA). In each case, the sediment has been compressed to such an extent that little water is present within the formations. The effect that the lack of water and consolidation has on test organisms or benthic communities can be substantial. Tightly compacted clays are impenetrable by burrowing organisms; consequently, obligate burrowers (e.g., *Nephtys caecoides, Rhepoxynius abronius,* or *Eohaustorius estuarius* generally) will perform poorly in such sediments (Lamberson et al. 1992).

In contrast, highly flocculent, newly settled fine-grain clays can have water content of more than 70%. These clays are likely to be found in quiescent waters with high, fine-grained sediment loads. In test chambers, flocculent clays are more easily disturbed, can remain in suspension for prolonged periods, and are more likely to clog gills, entrain larval forms, provide unstable sediments for burrowing and tube formation, and potentially reduce the overlying water oxygen content through increased biochemical oxygen demand.

Inclusion of data based on consolidated sediments in SQG databases can result in deceptively low COPC levels paired with toxic effects. Confirmatory tests may also indicate unacceptable toxicity in sediments with low COPC concentrations. An introduced bias may occur when ancient clays and sands predating anthropogenic input are used. Therefore, when reduced water content and increased consolidation are observed, it is necessary to account for these sediment characteristics by using specific reference sediments or sites that are free of contamination but exhibit similar sediment characteristics, or to modify the conditions of the sediment to provide conditions for successful toxicity tests or to evaluate the responses of the benthic community. Measurement of water content is a relatively straightforward process, and it helps account for consolidation effects that are not generally measured for our current sediment investigations. Test organism selection can ameliorate the effects of extremely high or low water-content sediments. Some limited sediment manipulation can also be considered, such as breaking up compacted or cemented clays; however, the possible effects caused by altering COPC availability must also be considered.

Water depth

Benthic communities are sensitive to changes in water depth, as shown by species distributions and abundance patterns (Barnes and Hughes 1999). Benthic community responses that are observed in the presence of increased chemical contamination (shifts in dominance, decreased richness, abundance, losses of specific taxa) are similar to those observed with changing depth (Pearson and Rosenburg 1978). Comparisons of structural indices of benthic communities must control for depth. Large-scale environmental monitoring programs or evaluations based on a reference-site comparison are particularly at risk for attributing depth-related community differences to chemical contaminants.

Confounding influence of water and sediment depth

The depth at which sediment is collected can affect the physical, chemical, and biological characteristics of the sample. In depositional zones, the age of sediments and the associated contaminants are directly related to depth. This is especially true at sites that have experienced temporal shifts in contaminant sources. Steep vertical gradients in chemical concentrations can result in vastly different chemical concentrations between the operational depths of samplers or sampling programs. When such sites are evaluated, the depth of chemical analyses should match those of the biological endpoints. If the depths of sediment chemistry are not paired with those of the biological endpoints, there may be little basis for comparison, and SQGs may be higher or lower, depending upon the nature of the chemical stratification.

Benthic community analysis endpoints can also be affected by sampling depth. In most cases, the biologically active zone is generally considered to be 10 to 20 cm, with approximately 90% of the organisms living in the top 10 cm (Oliver et al. 1980; Weston 1990). However, the depth of the biologically active zone can be altered by geographic location, by the dominant sediment grain size at a site, or by the presence or absence of certain burrowing species at the site. For example, certain species of burrowing shrimp are exposed to sediment up to 1 m deep. If contaminants are associated with buried layers and chemical data are collected from 2 to 10 cm, benthic effects may be linked to low COPC concentrations. For such cases, it is necessary to understand the zone of exposure for the collected benthos and match depths for sediment chemistry.

Confounding influence of physical alterations

Changes in hydrology, either seasonal or long term, or the introduction of materials that change the physical structure can alter the benthic community. Seasonal or long-term changes in sedimentation can lead to high suspended-sediment loads and flocculent sediment or can lead to a hard erosional, "paved" surface. Also, the addition of an organic matrix, such as algal mats or woody debris, can alter the physical nature of the sediment surface. Such shifts in the physical nature of sediment surface can affect the interactions of benthic infauna and epifauna with the sediment surface. Corresponding shifts in the benthic community can be inappropriately attributed to COPCs.

Finding a suitable reference site for such CFs may be difficult. One strategy may be to select several reference sites, some of which are on or close to the study site but at the edge or outside the contaminant gradient. At the very least, detailed observations regarding the nature and extent of such physical alterations should be recorded.

Persistent chemical characteristics

Persistent chemical characteristics are chemical constituents of sediment that are not readily changed and have effects that are difficult to remove or alter. Many contaminants of concern fit into this category, as do CFs such as metals that comprise the sediment mineral matrix, organic carbon, and organic nitrogen. As with physical features, the use of reference values or reference sediments can be used to evaluate this group of CFs.

Confounding factors associated with total organic carbon quantity and quality

In toxicity tests, reduced growth or survival of test organisms may sometimes indicate an inappropriate supply of sufficient quality food, even if the quantity of TOC or total organic nitrogen (TON) is relatively high. Pinza et al. (1997) reported high levels of mortality in sediment that was buried at depths that predate human activities, with no measurable contamination by priority pollutant organics (pesticides, polychlorinated biphenyls [PCBs], polycyclic aromatic hydrocarbons [PAHs]), reference concentrations of heavy metals, and 0.5 to 1.0 mg/L TOC. Typically, these sediments should have been acceptable to the test organism *Nephtys caecoides*. However, the observed high mortality suggested that some other noncontaminant or unmeasured contaminant factor might have been a cause of the toxicity. A series of sediment manipulations demonstrated that survival increased to greater than 90% with a 0.1% increase in the concentration of "high-quality" TOC (produced from dried and pulverized *Enteromorpha* algae [Pinza et al. 1997]). It should also be recognized that the addition of a small amount of high quality food to a test container may alter toxicity and uptake of contaminants into tissues during laboratory exposures. The addition of these materials to a laboratory exposure should be evaluated in terms of the purpose of the testing.

The quantity of organic carbon can also affect test organism response. Marine sediment typically ranges between 0.5% and 10% TOC. Benthic organisms generally have a range of TOC tolerance, above and below which population densities decrease (Word and Rosman 1996). Riverine sediment or sediment that is organically enriched with woody debris or sewage effluent may not be suitable for some marine test species. Measuring TOC, understanding the nature and source of the TOC and selecting appropriate test species may be the best approach for addressing highly TOC-enriched sediments.

In benthic communities, it is generally expected that increased concentrations of TOC in sediment will result in increased abundance and productivity, with decreased species richness and diversity resulting in changes to benthic community structure and possibly to function (Pearson and Rosenburg 1978). However, not all of the organic carbon in a sediment sample is easily oxidized or used by indigenous marine benthic organisms, and it is presumed that the quality of the organic carbon or its availability to infaunal organisms may be equally important in determining the type and structural richness of a benthic community. Coal particles, petroleum hydrocarbons, recalcitrant wood, or plant fibers are examples of organic carbon that can produce an elevated estimate of potential productivity and yet not be fully utilized by benthic infauna (Gray 1981; Word 1990). Sources of highly usable organic carbon are contained in benthic organisms themselves and bacterially aged waste materials (Levinton 1972; Gray 1981). The quality of the available organic carbon can influence potential productivity and community structure.

TOC also sequesters chemical contaminants by preferentially binding nonionic organic chemicals or creating anaerobic environments that produce sulfide materials that can bind certain metals (Chapter 13). This binding of contaminants can decrease the availability of these contaminants to the benthos, resulting in less toxicity than if the same concentrations of contaminants had been present in a more available form (Adams et al. 1985). Thus, the relationship between sediment contaminant concentrations and biological effects can vary from one location to another on the basis of contaminant availability.

Reduced diversity and species richness can occur with increased TOC quantity, while reduced production (biomass, abundance) can occur with reduced TOC quality. These same changes in benthic community can be observed with increased concentrations of available chemical contaminants. The increase in abundance and productivity observed in organically stimulated areas can occur even in the presence of elevated chemical contaminants. That increase in productivity can be a result of rapid responder reproductive activity (r-strategy) taking advantage of the increased organic materials in the sediment, with the organisms becoming more resistant to the contaminants that are present (Gray 1981). It can also be a result of the sediment contaminants adsorption or binding to other sediment characteristics where they are not available to interfere with the lives of the organisms living at these locations. Accurately ascribing the reduction in species numbers and reduced diversity of benthic communities along gradients from point sources of contamination to organic enrichment or chemical contaminants can be problematic.

Several methods have been developed to address the quality of organic matter in sediments, including extraction of amino acids (Buchanan and Longbottom 1970), bioavailable amino acids (Mayer et al. 1995), use of the biochemical oxygen demand of sediments relative to the TOC as a normalizing estimate (Word 1990), and protein extraction of sediments (Mayer et al. 1986). Reduction in availability of organic carbon to infaunal organisms can also occur when organic materials are buried beyond the sediment depth at which they can be used, when there is reduced bioavailability of materials at depths within a sediment, or when placement in a particular part of the environment excludes contact with the species that reside in these sediments (Gray 1981; Rhoads and Germano 1982; Mayer et al. 1995).

Species composition and abundance in soft-bottom communities can also be closely related to food availability and organism feeding strategies. Soft-bottom benthic communities have been classified into 4 feeding groups based on the principal location of food resources above, on, and within sediment as well as the predominant size of food (< or >50 μm in diameter. The categories under one set of groupings included suspended detritus feeders, surface detritus feeders, surface deposit feeders, and subsurface deposit feeders (Rhoads and Germano 1982; Word 1990). The species in these categories are similar to the distribution to taxa placed into feeding groups by other authors (Enequist 1950; Turpaeva 1953; Savilov 1961; Kuznetov 1964; Zatespin 1970; Walker and Bambach 1974; Dauer et al. 1981). Thus, differences in both the size and source of food can help determine the benthic community structure.

Benthic communities are sensitive to the amount of TOC as well. Community shifts have been observed with changes in percentage organic carbon as small as 0.5% (Word et al. 2003; Word 1978). Sediment amended with as little as 0.1% of high-quality TOC became suitable to *Nepthys caecoides* under laboratory evaluations. These changes in TOC are within the seasonal variability that has been observed at some long-term monitoring sites in California, USA (Southern California Coastal Water Research Project 1994) and that has been observed over relatively small areas (over several meters) (Hakanson 1992).

As with grain size, suitable reference-site selection is the best tool for accounting for the influence of TOC quantity and quality. This can be problematic for sites in which organic enhancement has been a component of anthropogenic influence (i.e., sewage effluent or woody debris). Understanding the feeding groups and incorporating feeding strategies into the benthic community assessment, such as with the infaunal trophic index (Word 1978), may also provide a useful tool to distinguish the influence of sediment contaminants on communities of organisms that are also responding to organic stimulation.

Confounding influence of contaminant bioavailability: Matrix-associated metals and weathered organic compounds

Determination of the bioavailable fraction of sediment contaminants has been approached in a variety of ways (Samilov and Wells 1984; Baudo et al. 1990). Bioavailable fractions of COPCs may occur in the water column, in the pore waters, or at the sediment–water interface (Lamberson et al. 1992). The bioavailable fraction may also be associated with various sediment compartments, some that may be used as food (e.g., total or some fraction of the organic carbon, dissolved organic carbon), or others that influence the various chemical states that can reduce the availability of the contaminant (acid volatile sulfides [AVS], or mineral matrices of sediment), or in chemical states or species that accentuate the availability of the chemical to organisms (e.g., methyl mercury, hexavalent versus trivalent chromium; distinctions of selenate, selenite, and organic forms; Samilov and Wells 1984; Baudo et al. 1990; Burgess and Scott 1992). Measuring chemical concentrations in whole sediment rather than on the bioavailable fraction incorporates variation that can bias the predictive ability of SQGs.

Processing of sediment for analysis of chemical contamination is designed to obtain repeatable measurements of those chemicals from a variety of sources (water, sediment, and tissue). The best way to obtain repeatable and consistent measurements of chemicals in these matrices is to extract or digest all of the contaminants that are contained or associated with the matrix. However, complete digestion or extraction from sediment matrices is unlikely to be accomplished by organisms. As a result, the standard methods are not designed, nor are they the most effective means, to measure the bioavailable fractions of the analytes. Examples of sediment digestion methods that introduce this type of error into metals analyses include

1) total dissolution using hydrofluoric and mixtures of other acids,
2) strong acid digestion using hot concentrated acid (e.g., nitric), and
3) mild acid leach with dilute acid (0.025M HCL) (PSEP 1989).

The total dissolution using stronger acids completely dissolves the silicate minerals so that the total concentration of metals in sediment analyzed with this method is determined. The strong acid digestion dissolves nearly all of the heavy metals that may become environmentally available in fine-grained sediments (Cd, Cu, Pb, Hg, Ag, and Zn analysis; Method 3050B) (USEPA 1986) but does not completely remove all metals that are in the matrices of the minerals (Fe, Al, Mn, Cr, and Ni). Use of a mild acid leach intentionally provides a lower recovery of metals from the sediment; thus, a correlation between metal concentrations in sediment, the toxic fraction of those metals, and the concentration of metals that may be incorporated into tissues may be more representative of the fraction that impacts benthic infauna (PSEP 1989).

Despite the acknowledgement that mild acid leach is more closely correlated with uptake, some established guidelines state that "methods that do not extract 100% of the metals from the sediment are not recommended for use in sediment analyses within Puget Sound" (PSEP 1989). Similarly, methods promulgated for San Francisco Bay require the use of either USEPA Method 3010 or 3051 for digestion of metals from sediment, which are equivalent to the total or strong acid digestion methods used in Puget Sound (USEPA 1986). These digestion methods do not result in metals extractions that are well correlated to the fraction of metals that are available for uptake resulting in observed toxicity. Metals simultaneously extracted with the AVSs are better at mimicking available fractions in sediments. Indeed Casas and Crecelius (1994) showed that metals exceeding, on a molar basis, the concentration of AVSs in a sample were those more closely associated with observations of toxicity (Chapters 4 and 13).

Similarly, the extraction of organic contaminants from sediment provides total concentrations that can be removed with an organic solvent. However, nonionic organic contaminants reach an equilibrium between that fraction associated with the sediment-bound TOC (generally less available) and that fraction that is dissolved (more available and probable toxic fraction) into the surrounding waters. Thus, the organic contaminant measured during the chemical analysis is not necessarily present in a directly toxic form. Some organic materials, as they become weathered, sequester themselves into less biologically available and, consequently, less toxic forms (e.g., tar balls and granules or mineralization of toxic forms of contaminants into less toxic degradation products). Also, some contaminants can be present in a form that may be less available, such as tributyltin (TBT) or Cu in paint chips. Only a fraction of the total TBT or Cu are actually in a dissolved or more available state. The total concentration of these contaminants can provide overestimates of potential toxicity because they measure both the bioavailable and nonbioavailable forms of the contaminants.

Through the application of equilibrium partitioning (EqP) models and testing associated with EqP, some investigators have concluded that the bulk concentration of a contaminant is not the best estimate of the bioavailable fraction (Adams et al. 1985). Other investigators have reported that dry weight normalization is more predictive of toxicity (Barrick et al. 1988; Ingersoll et al. 1996; Long et al. 1998). The TOC (and possibly the dissolved organic carbon) adsorbs nonionic organic contaminants, leaving a small fraction of the total concentration in an available form. EqP methods permit an estimation of that concentration of a contaminant that is released from a sediment matrix, becoming available to organisms. The biological effect associated with this water-soluble fraction provides a better estimate of acute toxicity. Some SQGs, such as ERLs and ERMs, incorporate EqP, addressing, in part, availability. However, the relative contribution of contaminants associated with particles consumed by organisms is poorly understood (Chapter 3).

In summary, CFs associated with changes in bioavailability that can confound interpretation of test results include metal speciation, organometallic complexes created by biological processes, weathering and decreasing availability of organic compounds, different levels of association with the mineral matrix of a sample, and biological sequestration of organisms from contact with the chemicals of concern. The toxic form of the contaminant may be a small fraction of the amount of the chemical measured through standard chemical extraction procedures. SQGs should account for the bioavailable fraction of a contaminant, not the total amount that can be measured. If assessments do not analyze bioavailable fractions, then the correlation between the measured contaminant concentrations in the sediment and the biological effects will be less significant.

Nonpersistent chemical characteristics

Nonpersistent chemical characteristics are constituents with concentrations that can be readily changed during sampling or storage and also by application of various test modifications. In some cases, these CFs are a natural occurrence in the sample, while in other cases, the concentrations may be a result of sampling, storing or handling processes, or an anthropogenic source. Interstitial water salinity sulfides and ammonia are examples of nonpersistent chemical characteristics. Reference values and/or specialized laboratory methods are important tools to determine the related effects of these 3 CFs.

Confounding influence of ammonia

The natural degradation of organic material in sediment can result in the formation of ammonia. This is especially true of sediments with high amounts of organic material, such as river mouths or tidal estuaries where organic materials can accumulate and where the efficiency of nitrification processes can be reduced (Rysgaard et al. 1999). Sediment porewater ammonia concentrations can range from <1 mg/L to >400 mg/L total ammonia (ammonium and unionized ammonia; Jack Word, William Gardiner, and David Moore, personal observations). Ammonia usually does not persist as a sediment contaminant in areas where it will either be used by plants or undergo bacterial mineralization to its component parts (N, H, and O). This is true if the sediment organic load is in balance with the biological communities that are present. However, if the sediment organic load is not in balance, the organic materials can accumulate, and the amount of time required to undergo mineralization becomes longer. Under these conditions, the ammonia would be considered persistent, and its influence on benthic animals would be adverse.

Total ammonia is measured either with specific ion probes or spectophotometrically after all ammonia-containing compounds are converted to a measurable ionized form. The ionized and unionized fractions of ammonia are then determined using a relationship between the percentage of the ionized compound (NH_4^+) and the unionized portion (NH_3) as a function of the pH, temperature, and salinity (hardness in freshwater environments). Over normal ranges of pH, the fraction of NH_3 is relatively small (about 1% to 5%). Many assume that the toxic fraction is the unionized portion of the mixture. However, there is evidence that NH_4^+ is also able to diffuse across the gill in some crustaceans (Borgmann 1994), and others have demonstrated that NH_4^+ interferes with the sodium pump (Regnault 1987). While there will continue to be arguments in favor of presenting total and/or unionized concentrations of ammonia, it seems more reasonable to be sure that each is presented in combination with pH, temperature, and salinity (hardness or conductivity data). Spotte (1992) stated that, "…the empirical argument favoring the greater toxicity of NH_3 has not been confirmed, despite the belief by most that it has. Most investigators have interpreted any increase in toxicity to the greater concentration of NH_3, while ignoring the possible additive or synergistic effects of pH and ammonia." Erickson (1985) noted that any toxic effects of NH_3 and pH may be inseparable. The most recent update of water quality criteria for ammonia in freshwaters (USEPA 1999) is based on total ammonia.

Ammonia is considered to be very toxic to marine and estuarine animals, while it can also be considered an essential nutrient to marine and estuarine plants. Ammonia is toxic to test organisms at concentrations on the order of 3 mg/L to 100 mg/L total ammonia as nitrogen (TAN). Larval forms can be among the most sensitive to ammonia, with LC50s of 3 to 5 mg/L total ammonia for larval bivalves and echinoderms (Gardiner et al. 1995). Ammonia can cause decreased feeding and respiratory distress, gill hyperplasia, reduced growth, abnormal development, and increased mortality in test treatments. In fact, recommendations for extended holding of fish in marine systems indicate that concentrations of total ammonia should be less than 0.01 mg/L (Spotte

1992). The class and onset of these effects are pH dependent, with higher pH resulting in more rapid and severe effects.

When ammonia concentrations persist in sediment at concentrations known to cause toxicity, the benthic community is likely to reflect that of a degraded site. The site would have reduced species richness, diversity, and possibly abundance as well as an alteration in the functional roles of the surviving infaunal invertebrates, while the more balanced systems might show increases in productivity associated with the nutrient enrichment (Pearson and Rosenburg 1978).

In urban waterways, a substantial number of sediment samples have ammonia concentrations that are equal to or exceed probable effects levels. Moore et al. (1997) reported that the probability of toxicity from ammonia alone in sediment toxicity tests conducted with the marine amphipod *Leptocheirus plumulosus* may run as high as 18% in dredged material evaluations. Additionally, ammonia concentrations were observed to influence the response of marine amphipods in more than 60% of the dredged materials evaluated at a specific site. Moore et al. (1997) also reported that porewater ammonia concentrations in dredged material were higher in dredged materials than in surface sediment by a factor of 4, with the porewater ammonia concentrations in dredged material averaging 40 mg/L TAN. This average concentration in dredged material exceeds the threshold of 20 mg/L TAN established by USEPA Region II and USACE NY District dredging projects by a factor of 2 (Ferretti et al. 2000). Frazier et al. (1996) also reported that porewater ammonia concentration increases with sediment depth at all study sites, and as such, deeper cores would be expected to have more toxicity associated with this factor. Additionally, ammonia concentrations were observed to influence the response of marine amphipods in all dredged materials evaluated at a single site (Barrows et al. 1996).

Whether or not ammonia is considered to be a COPC at the study site, porewater ammonia concentrations should be measured to determine its potential impact to the site relative to the other COPCs. Sediment samples that have produced inordinate amounts of ammonia during laboratory toxicity tests include organically enriched sediment, sediment that has been stored or altered to maintain or increase its reduced state (e.g., maintenance of high nitrogen head space), and sediment that comes from marine locations that are also associated with freshwater and estuaries (Rysgaard et al. 1999). Overlying water quality measurements of ammonia during toxicity tests (in conjunction with pH, salinity, and temperature) should also be measured, preferably at multiple points throughout the duration of the test, to determine any changes in ammonia that may have occurred during the testing period. Further actions, such as concurrent ammonia reference-toxicant tests or removal of ammonia from the test system, are program-specific considerations. Removal of ammonia using water exchanges has been incorporated into some dredged material programs (USEPA 1994; USEPA/USACE 1998; Barton 2002). In all cases, it is recommended that the concentration of total ammonia, pH, and salinity or conductivity (in freshwater) be measured in pore waters in addition to the overlying water. Measurements of ammonia in both compartments (overlying and pore waters) are necessary assessments because the behavior of a variety of test organisms can influence their exposure to either compartment.

Confounding influence of sulfides

Sulfur chemistry is highly complex, with oxidation states ranging from –2 to +6. Sulfur is an important constituent of many organic molecules, and it complexes with many metals and occurs in various simplified chemical states, ranging from sulfates, sulfides, and various oxides to elemental sulfur (Woollins 1996). The influence of sulfur on the availability of selected metals in aquatic sediment has been examined with AVS methods (Di Toro et al. 1991). Sulfides are highly toxic to marine and estuarine invertebrates and the effects of sulfides have been an important field of study for many years (Knezovich et al. 1996; Wang and Chapman 1999).

The influence of sulfides on sediment evaluations is not, as yet clearly understood. It is clear that metals availability and toxicity are influenced by sulfides and that SQGs could be altered by the effects of sulfides. Some sediment evaluation programs include dissolved and total sulfides as a standard measurement required in aquatic testing of sediment (PSEP 1989).

Confounding influence of salinity

The basic definition of salinity includes those ions that make up a significant contribution of a water sample and generally includes those ions at concentrations in excess of 1 mg/L. Salinity determinations can be made with a variety of methods, including conductivity or chloride measurement, optical or electrochemical determinations, and specific gravity determinations (Wilson 1975). Ideally, multiple techniques should be used to compare the salinity values obtained in order to determine potential interferences with any of the methods. However, using multiple methods to measure salinity is not always possible, and measurement generally is based on a single technique. A refractometer is one of the most common techniques for measurement of salinity when water quality of bioassay testing is being monitored and the salinity of pore water is being determined. This instrument compares the refractive index of a small sample of water to known refractive indices with various salinity waters. While this instrument is simple and convenient, there are potential problems with its use, especially when samples of pore water are examined. The issue is the presence of pigments and dissolved or particulate solids in pore water that also refract light. As a result, these materials can falsely indicate higher salinity values than actually exist within porewater samples. The same holds for suspensions of sediment that have elevated suspended solids concentrations (Zaneveld 1974). Annotations on the clarity of the water sample, colors that are present in the sample, any observations of particles, and decreases in water quality salinity through time should be made; these annotations can indicate that the salinity measurements may be erroneous.

Salinity measurements can confound toxicity evaluations in 2 ways. First, marine tests with freshwater sediment can be influenced by the sediment pore water driving down overlying water salinity. Marine animals may be stressed when placed into sediment with low-salinity pore waters. Refractometer readings can be misleading because of the presence of light-refracting pigments and particulate solids. Thus, necessary adjustments may not be made. Freshwater sediments can also be problematic in marine evaluations because of dramatic shifts in ionic and chemical balances, causing releases of ammonia and sulfides as well as some metals. Such changes could significantly alter the relationship between sediment COPC concentrations and the resulting toxicity. Multiple measurements of salinity and alternatives to refractometer may help to more accurately measure salinity and interpret test results. Furthermore, salinity measurements are important in defining benthic infaunal distributions (Hyland et al. 1999; Chapter 12). Incorporating salinity in reference site selection criteria is critical to establishing a valid reference site.

Confounding influence of sediment acclimation

The majority of toxicity testing is required to be conducted within short periods of time after sediment collection. However, acclimation of sediments in test containers is a necessary and complex process in which microbial communities attain sufficient populations to allow sediments to become suitable environments for aquatic organisms (Spotte 1992). This is especially true when terrestrial soils, sediment from deep within a core, or sediment from freshwater environments are tested with marine and estuarine benthic organisms. The lack of appropriate microbial communities can result in the creation of sediment and overlying water conditions that are inappropriate for the survival of test organisms. Sediment acclimation time, that amount of time required to attain an adequate microbial community that will provide sediment conditions conducive to the survival of test organisms, is generally at least 1 to

3 weeks (Spotte 1992). This acclimation period is typically not acknowledged, nor is it provided for in testing programs.

Experiments on acclimation have been performed using the amphipod *Eohaustorius estuaries,* the polychaete *Neanthes arenaceodentata* (Word et al. 2001), and the amphipod *Rhepoxynius abronius* (Pinza et al. 1997). Both experiments revealed high mortality in unacclimated sediment. However, after periods of acclimation, toxicity was reduced, in most cases dramatically. In each case, test organism survival increased significantly after sediment acclimation in tests using aliquots of well-mixed sediment from 3 locations for each study (Table 16-2).

Table 16-2 Percent survival of 3 test organisms in a sediment acclimation study

	Sediment 1		Sediment 2		Sediment 3	
Species	Unacclimated	Acclimated	Unacclimated	Acclimated	Unacclimated	Acclimated
Rhepoxynius abronius[a]	13	94	1	95	65	70
Eohaustorius estuarius[b]	36	80	51	84	77	88
Neanthes arencaceodentata[b]	0	80	0	84	0	88

[a] Stations and data (Pinza et al. 1997).
[b] Stations and data (Word et al. 2001).

Nonmatrix characteristics

Nonmatrix characteristics include any CF that is not directly related to a matrix characteristic and can be related to field or laboratory activities or to organism behavior and sensitivity. Behavior can affect how an organism is exposed to a contaminant and the potential influence of that contaminant, and these CFs can generally be addressed by standardizing and improving methodology (such as organism handling procedures), establishing appropriate species for evaluating different exposure scenarios or matrices, and selecting appropriate reference sites. Accounting for behavioral affects also requires a better understanding of organism behavior to improve selection of appropriate test organisms or to better understand the responses that are observed in the field.

Confounding influence of organism acclimation

Standard practices with toxicity testing use either field-collected or laboratory-raised test organisms. These organisms can come from local suppliers or can be cultured and obtained by laboratory staff for testing. Organisms received from field collections or laboratory cultures should be collected, handled, and transported to minimize stress from these procedures (USEPA 1994; ASTM 2004a). Additionally, the rate of change in temperature and salinity from the collection to testing conditions is designed to be slow (multiple days of gradual change) to permit acclimation of test organisms without causing additional unavoidable stress to test organisms (PSEP 1995; USEPA 1994; ASTM 2004a, 2004b). However, the influence of rapid acclimation of test organisms, even those following the industry standards recommended by PSEP, USEPA, and ASTM, has not been critically evaluated.

A study evaluated the influence of rapid and slow acclimation on the responses of *E. estuarius,* an estuarine amphipod with a wide salinity tolerance. In its natural habitat, this species is exposed to ranges of salinity from nearly fresh water to full seawater (DeWitt et al. 1989) over relatively short time periods. Similarly, survival of test organisms exposed to native control sediment after rapid acclimation to test salinities is generally high. As a result of its ability to survive in wide ranges of salinity and under rapid acclimation, it was concluded that slow acclimation to test salinities was not required for this species (DeWitt et al. 1989).

However, reduced survival and increased variation between replicates occurred in tests with sediments from San Francisco. The test organisms for this evaluation were obtained during the winter from a less saline environment with little to no periods of acclimation before their use in a toxicity test. It was also noted that the longer the test organisms were held prior to use during these tests, the lower the toxicity and the less variable the test results. To directly evaluate the impact of acclimation on survival of test organisms, sediment was collected from 16 stations and tested with organisms that were either acclimated immediately to test conditions or acclimated slowly at a rate of 5‰ per day. Both sets of test organisms were tested within 4 d of the completion of the acclimation process. Survival was greater and the variability in survival rates lower in the slow-acclimation treatments, relative to those organisms that were acclimated rapidly. The correlation of survival under each condition for these stations was statistically significant and represented about 18% increased mortality with the rapidly acclimated *E. estuarius* (Figures 16-1 and 16-2).

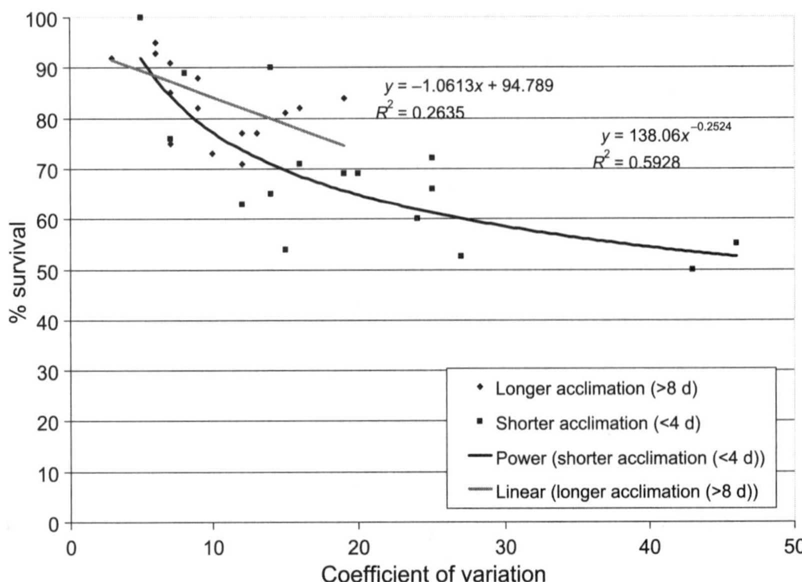

Figure 16-1 Coefficient of variation for survival in replicate test exposures is greater for shorter holding times (Word 1999)

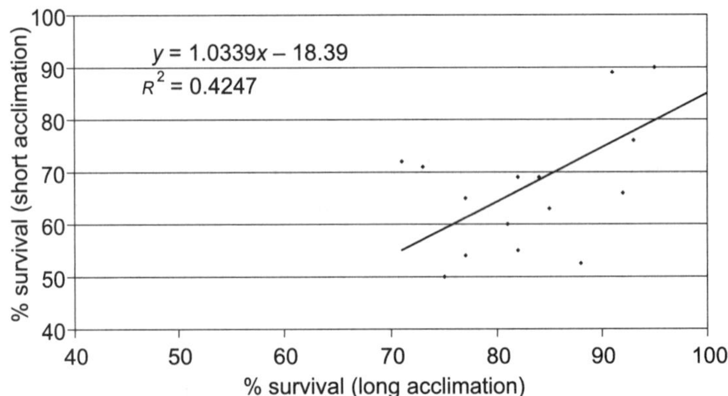

Figure 16-2 Comparison of survival after short and long test-organism acclimation periods (data show an average increase in survival of about 18%; Word 1999)

This increase in mortality under rapid acclimation is similar to the increased mortality in the response of rapidly acclimated organisms to PAH contamination noted by DeWitt et al. (1989). Again, under both the rapid- and slow-acclimation treatments, the survival and variation in survival for organisms exposed to native sediment was acceptable (DeWitt et al. 1989).

Confounding influence of exposure and organism sensitivity

Benthic invertebrates have adapted many different life styles that influence their exposure to sediment contaminants. Some species live on the sediment surface and feed in the water column; others live on the sediment surface in tubes and feed on detrital materials that are at the surface of the sediment or suspended in the water column. Other organisms live within mucous-lined burrows within the sediment and obtain their food supplies from the surface of the sediment adjacent to the burrows. Finally, other benthic infauna live freely within the sediment. Each of these strategies can alter the degree of exposure to chemicals that may be contained in sediment. Species also show varying sensitivity to chemical contaminants. The combination of sensitivity to chemicals and the presence or absence of behavior that enhances or excludes exposure to contaminants combine to produce the observed effect.

The varying sensitivity of benthic marine and estuarine amphipods is exemplified by comparing results from sediment versus water-only exposures. Marine or estuarine amphipods used in toxicity testing either burrow freely through the sediment (*R. abronius* or *E. estuarius*) or construct tubes or burrows and feed primarily in the water column above the sediment–water interface or just off the surface of the sediment (*Ampelisca abdita, Leptocheirus plumulosus,* and *Grandidierella japonica*). These different behavioral interactions with the sediment provide different levels of exposure to overlying water and pore water, altering the observed response to contaminants. However, the relative sensitivity of these organisms tested in a water-only exposure may differ. The relative sensitivity of these species to ammonia or Cd is provided in Table 16-3. In water-only exposures, the most sensitive of these species to either chemical is *A. abdita,* while *E. estuarius* (Cd) and *G. japonica* (ammonia) were the least sensitive. In general, *R. abronius* and *L. plumulosus* are relatively similar in sensitivity.

With this type of sensitivity, it might be expected that the responsiveness to sediment contamination by ammonia or Cd would show a similar relative pattern, if the organisms received the same types of exposure. A unique experiment was performed that addressed the apparent sensitivity of two of these species to overlying water and pore water ammonia (Barrows et al. 1994; Kohn et al. 1994). In these experiments, the concentration of porewater ammonia was spiked or measured in sediment to attain a concentration series that encompassed concentrations reported from the water-only tests with overlying water equal to or much less than the porewater concentrations. *R. abronius* toxicity was directly related to pore water and not to the overlying water ammonia concentrations, whereas *A. abdita* survival was directly related

Table 16-3 Comparison of amphipod sensitivity to sediment characteristics

	Amphipods typically used for toxicity testing of marine sediment and characteristics that allow survival comparable to "native" controls				
Parameter	*Rhepoxynius abronius*[a]	*Eohaustorius estuarius*[b]	*Ampelisca abdita*[b]	*Leptocheirus plumulosus*[b,c]	*Grandidierella japonica*[f]
Grain size (% silt clay)	<93	No apparent grain size effect on survival of test organisms[b] although some decreased performance with >90% fines	>90	No apparent grain size effect on survival of test organisms[b]	Coarse sand to clay environments are occupied but no relationships between toxicity and grain size have been established[f]
Grain size (% clay)	0 to <53.5		Not apparently influenced by high % of clay		
Grain size (% sand)	>10 to 100		>10 but <95		
Salinity	>22‰	1–25‰ in pore water	0–34‰ if overlying water is >28‰[b]; naturally in waters from 10‰ to full marine[g]	5–30‰[e]	>10‰[f]
Feeding behavior	Meiofaunal predator[a]; sedimentary organic material[a,d]	Deposit feeders	Filter feeder on particles that are suspended or on sediment surface	Filter feeder on particles that are suspended or on sediment surface	Assumed to be a filter feeder on particles that are suspended or on sediment surface
Life habits	Free burrowing	Free burrowing	Tube dweller	Tube dweller	Tube dweller
LC50 Total ammonia (mg/L)[h]	78.7	125.5	49.8	80.6	148.3
LC50 Cd (mg/L)[i]	0.70	7.25	0.52	4.1	2.2

[a]Swartz et al. 1985
[b]USEPA 1994
[c]DeWitt et al. 1988, 1992
[d]Oakden 1984
[e]Schlekat et al. 1992
[f]Greenstein and Tiefenthaler 1997
[g]USEPA 1994
[h]Kohn et al. 1994
[i]MEC Analytical Systems, Carlsbad, CA, unpublished reference toxicant information (published with permission).
Note: Color blocks indicate similar range of response by amphipod species to sediment characteristics.

to overlying and not to the porewater concentrations. Furthermore, *A. abdita* was able to survive at porewater concentrations >200 mg/L, as long as the overlying water was lower than 20 mg/L. This indicates that the relative sensitivity of test organisms based on water-only exposures is not a good predictor for routes of exposure that are based on sediment and pore water, and lifestyle must be taken into account when an appropriate test species is selected.

Other important aspects of organism behavior are based on their life habits relative to sediment grain size, salinity changes, and feeding behaviors that occur in their native habitat. Some prominent characteristics of marine amphipods typically used in marine sediment bioassays are presented in Table 16-3. These characteristics are used in many regulatory programs to guide species selection.

Confounding influence of biological interactions

Benthic communities not only respond to the physical and chemical environment but to the biological environment as well. Indeed, the benthic environment can be a hostile and competitive place. Predator–prey relationships, inter- and intraspecies competition for food or space, and seasonal migrations or transformations can drastically alter community structure. In addition, colonization, successional shifts, or genetic shifts can mimic contaminant affects.

While there is no one tool that can be used to ameliorate these CFs, an understanding of spatial–temporal shifts at a site, the use of multiple indices, and careful selection of reference sites can greatly increase the usefulness of benthic community endpoints.

The STARR Process: Proposed Method for Evaluating Potential Confounding Factors

The persistent and nonpersistent sediment characteristics that are described in this chapter may affect the perceived relationship between COPCs and the biological impacts that are observed at a test site. As a result, they can potentially alter sediment management decisions and the empirical or mechanistic relationships that have been used in developing SQGs. The procedures that are used to determine how to best address the influence of CFs are dependent upon the type of sediment assessment that is being made, the question that is being asked, and the nature of the biological response to the CF.

Three general types of sediment assessments are used for sediment management decisions (Figure 16-3):

1) assessing the risks of leaving sediment and contaminants in place (in situ evaluation),
2) assessing the risks associated with the process of removing sediment from its current location, and
3) assessing the risks associated with placing the sediment at another location.

	In-situ	Removal	Placement
Organism exposure		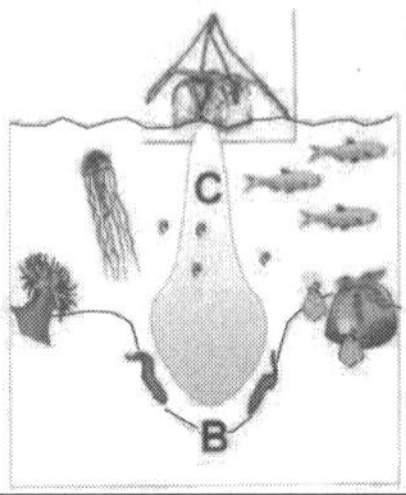	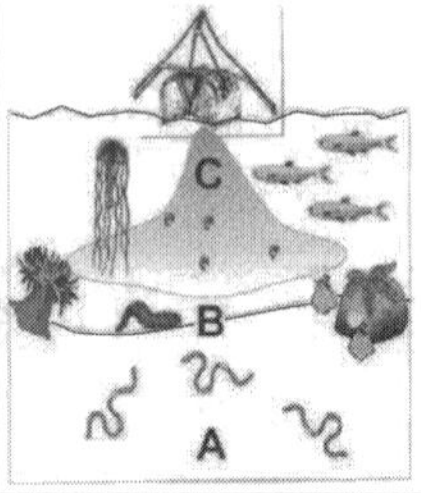
Sediment disturbance	Minimize	Maximize	Maximize
Storage	Minimize	Minimize	Minimize

A Infaunal sediment dweller
B Sediment water interface or epifaunal dweller
C Water column dweller

Figure 16-3 Testing options to account for factors under 3 assessment types of contaminant availability. The nature of organism exposure differs between sediment management types. The amount of disturbance that should be allowed during sediment assessment should mimic the amount of disturbance associated with the management types. In each case, storage times should be minimized.

The nature and extent of sediment disturbance and the resulting contaminant mobility differ for each assessment type. Likewise, the extent to which certain CFs may affect biological–chemical relationships differs with each assessment type. Some CFs remain an important component of the test sediment beyond the management action and should not be ameliorated, while others may not persist once the management action is taken and their removal from the evaluation may be warranted. A third class of CFs may be created as laboratory artifacts and should be addressed regardless of the assessment type.

Organism response is another important component in developing a strategy to address CFs. There are basically 4 different ways that test organisms respond to CFs (Table 16-4). First, some sediment features (e.g., sediment grain size, TOC, and salinity) have optimum concentrations or quantities above or below which adverse responses will be observed. Second, some sediment features have threshold concentrations, above which a dose–response will be observed (e.g., ammonia and sulfides). Third, some sediment features result in a change in the quality of the testing environment (TOC quality, labile nitrogenous compounds). Increases can provide increased positive benefit to the organisms, which may plateau at a certain concentration (e.g., growth rates and quality of the diet, not quantity). The fourth class is not related to sediment features but to poor laboratory practices that result in organisms responding in a highly variable and unpredictable fashion. This is particularly true when an organism has been handled poorly or is held under inappropriate conditions that may alter an organism's response to other factors within the test system, such as COPCs. Understanding the effects of these sediment features requires an understanding of sediment–biota interactions and the goals of each particular sediment evaluation and may require methods of analyses beyond methods described in standardized toxicity tests.

The proposed method for understanding, and in some cases ameliorating, the effects of CFs in sediment evaluations has been termed the "STARR Test Process" (select, test, account, remove, replace). The STARR process incorporates previously developed methods into a stepwise approach for assessing the impact of CFs on a sediment evaluation. The goal of this evolving process is to more clearly identify the source of toxicity and more accurately assess the toxicity related to COPCs. In this way, the SQGs may become more refined and may more accurately predict toxicity in test sediments. This method applies the following steps:

- Select the appropriate test species for the specific sediment assessment that is being addressed.
- Test under the appropriate conditions provided for the test species and this sediment toxicity assessment type.
- Account for the influence of the CF by direct comparison to established references or compare the test results with reference toxicant tests conducted at the same time to evaluate the CF being addressed.
- Remove the CF and test.
- Replace the CF and test to determine if the same response is obtained relative to the presence or absence of the CF.

Although developed for toxicity testing, this approach may also be useful in benthic community studies and methods of measuring various sediment char-

Table 16-4 Organism responses to confounding factors with bioassays or benthic communities

Confounding factor	Suspended particulate phase (SPP)	Solid phase (SP)	Bioaccumulation
Persistent sediment properties			
Sediment grain size	Fine grain suspensions increase TSS, increasing gill fouling and potentially decreasing contaminant bioavailability	Optimum grain size: Increasing response with grain size above or below the optimum	Change in bioavailability of sediment contaminants
Sediment angularity	None	Lacerations on soft bodied organisms; decreased survival	Lacerations on soft bodied organisms; decreased survival
Sediment compaction	Highly compacted sediment does not produce appropriate SPP	Highly compact sediment does not permit burrowing; loosely compacted sediment potentially "drowns" small epifaunal tube dwellers	Highly compact sediment does not permit burrowing; loosely compacted sediment usually not an issue
Total organic carbon (TOC) quantity Total organic nitrogen (TON)	Should have little impact on most test species	Optimum TOC and TON: Increasing response with TOC or TON greater or lesser than optimum	Increasing TOC or TON may alter bioavailability of contaminants, potentially increasing transfer of contaminants from sediment into tissues
Metals content of minerals	No affect expected from metals in mineral matrix		
Sediment water content	Lower sediment water content reduces the quantity of releasable soluble contaminants into SPP water.	Lower sediment water content reduces the mass of releasable contaminants into pore water.	Lower sediment water content reduces the mass of releasable contaminants into pore water.
Hardness of water	Optimum hardness values for each species and hardness influences bioavailability of ionizable contaminants released into SPP.	Optimum hardness values for each species and hardness influences bioavailability of ionizable contaminants in pore water.	Optimum hardness values for each species and hardness influences bioavailability of ionizable contaminants for uptake into tissues of test organisms.

Table 16-4 *continued*

Confounding factor	Suspended particulate phase (SPP)	Solid phase (SP)	Bioaccumulation
Nonpersistent sediment properties			
Ammonia			
Sulfides	No-effects concentrations plus toxicity associated with ammonia or sulfide concentrations		
Salinity	Optimum salinity range: Increasing toxicity associated with salinity values that are higher or lower than the acceptable range		
TOC quality	Unlikely to influence survival of organisms exposed to short-term SPP elutriates	As quality of organic increases the survival, growth and reproductive potential will increase	As quality increases, survival and growth will increase. Net effect on contaminant uptake is based on kinetics of contaminant.
Labile nitrogenous compounds			
pH	Optimum pH: Increasing response associated with pH ranges outside of acceptable range for each species. pH can also alter bioavailability of ionizable contaminants.		
Temperature	Generally controlled in toxicity testing. No impact expected on test organisms.		
Suspended solids	Increases can increase effects on species that filter water through gills. Can also alter the availability of chemical contaminants.	No expected affect	No expected effect
Nonmatrix parameters including laboratory methods			
Acclimation and handling	Increased sensitivity resulting from stress related to handling and organism application.		
Population or species sensitivity	Variation in sensitivity of EC50 responses is probably not the best criterion for looking at these effects because they become statistically and biologically significant at a response of about 10%. We should probably be looking at LOEC or 10% effects to ascertain the influence of handling on toxicity testing.		
Inter- and intralaboratory			

acteristics. A similar process is followed for benthic assessments: One must select the appropriate benthic measure, test the appropriate sites and reference sites, account for the influence of the CF by comparison with an appropriate reference site, and research other relevant seasonal or long-term changes that are likely to have altered the benthic community. This process is further detailed in Table 16-5 and discussed in detail in the following section.

Select: Were appropriate sampling methods and test species selected to address the specific assessment type?

Tests species should be selected on the basis of the type of sediment assessment in question and the behavior of the test organism and its relationship to the available fraction of the COPCs in the appropriate environmental compartment (Figure 16-3). Three basic sediment assessments are considered:

1) "no action" or in-place option,
2) disturbance or removal process, and
3) disposal evaluation (ocean, inland waters, or upland).

Three basic environmental compartments (water column, sediment–water interface, and bedded sediment) address potential exposure mechanisms. A selection of test species that might be used in North America is presented in Table 16-6. This information can be used to select appropriate species to address the specific assessment type. Improper selection of species reduces the value of the test in predicting the consequences of leaving sediment in place, of removing the sediment, or of disposing of the sediment and may create a need for selection of a new and more appropriate assessment endpoint.

No-action evaluation

The no-action alternative evaluates the availability and effects of contaminants related to the relative risks associated with leaving contaminated sediment in place or the existing effects of contaminants at a hazardous waste site. If sediment is to be left in place, the testing scenario for both toxicity and benthic comparisons should attempt to represent the in situ conditions of the site as much as possible. This means that sampling and handling procedures for toxicity testing need to minimize disturbance of the chemical–physical relationships that have been established in the environment as much as possible. Therefore, collection methods should disrupt sediment as little as possible. Sediment mixing and compositing of sediment treatments is not appropriate for this assessment objective. Benthic community assessments should be based on reference condition comparisons that mirror these site conditions as closely as possible without the presence of the contaminants.

Table 16-5 STARR process used in the evaluation of confounding factors

Step	Evaluation
Select Were the appropriate sampling methods and test species selected to address the specific assessment type?	1) Was the assessment evaluating the effects of COPCs in place? 2) Was the assessment evaluating the effects of COPCs during the removal of sediment? 3) Was the assessment evaluating the effects of COPCs during sediment placement?
Test Was the toxicity test performed correctly?	1) What were the acclimation conditions and rates of acclimation for test organisms? 2) What was the mortality to test organisms prior to test initiation? 3) Were water quality parameters within appropriate limits throughout the test? 4) How variable were the test result replicates? In excess of 30% CV? 5) How consistent were the reference-toxicant test results with past information? Inter and intralaboratory comparisons? 6) Did the performance of the test meet established acceptability criteria?
Account Were the sediments free from potential confounding factors (CFs)?	1) Was the assessment evaluating the effects of COPCs in place, during removal or sediment placement or disposal? 2) Was the sediment near a source of fresh water? Surface or ground water? 3) What is the porewater salinity concentration? (Before sediment compositing and mixing, after compositing and mixing and before the test initiation, during and at the end of the test)? 4) What is the porewater ammonia concentration? (Before sediment compositing and mixing, after compositing and mixing and before the test initiation, during and at the end of the test)? 5) What is the porewater sulfide concentration? (Before sediment compositing and mixing, after compositing and mixing and before the test initiation, during and at the end of the test)? 6) Were the sediments collected from sediment depths in excess of 10 cm? 7) Was there a source of recent organic enrichment? 8) Were the sediments collected from sediment depths not at the sediments surface during man's residence in the area? 9) Were the sediments being evaluated highly compact? 10) What is the sediment grain size? 11) Were there sharp angles on sediment grains? 12) Was the sediment elutriate sufficiently cloudy to preclude observations of test organisms during or after the tests?

Table 16-5 *continued*

Step	Evaluation
Remove, Replace, Research Was the chemistry of sediment or water conducted for this test based on "bioavailable" fractions of contaminants? Were there significant correlations between the biological effects and the concentrations of any contaminants or CFs?	1) Did the digestion process for the heavy metal concentrations include the metals tied into the mineral fraction of the sediment? 2) Did the extraction process for the organic contaminants include all forms of the contaminant? 3) Were there correlations between the measured biological effects and selected chemicals of potential ecological concern? Did all concentrations exceed the effects level? If so, the regression relationships may not demonstrate this as a cause. 4) Did these correlations or sediment concentrations for the samples reflect the expected relationships of the contaminant to known responses of these organisms? 5) Were there correlations between the measured biological effects and CFs? 6) Did these correlations reflect the expected relationships of the CF response? Did all concentrations exceed the effects level? If so, the regression relationships may not demonstrate this as a cause.

An evaluation of the no-action alternative also requires that the test species be exposed to test sediments in an appropriate manner. Test species should be surrogate organisms for those living under physical and chemical conditions of the site (i.e., sediment modifications should be minimized, and test organisms used should have characteristics that permit them to live without sediment adjustments being made). They should represent the exposure pathways that are present under in situ conditions. Marine and estuarine organisms typically interact with sediment in 3 basic ways. First, there are those organisms that live within the sediments and have direct exposure to contaminants within the sediment pore water and sorbed to sediment grains. Second, there are organisms that live at the sediment–water interface and are exposed to COPCs near the sediment–water interface that are released from the sediment or that drift in the near bottom-water currents. The final group of test organisms are the ones that live in the water column and are only exposed to COPCs in the water. The actual number of these exposure pathways at a particular site may vary, depending on the characteristics of the site environment.

In all cases, organisms should be selected on the basis of their tolerance to the conditions of the site and the testing environment should not be modified to accommodate an inappropriate species. Modification of the testing regime to accommodate inappropriate species selection negates the value of this assessment type in 2 ways. Sediment disruption occurs, maximizing the release of

Table 16-6 List of species and representative attributes

Species	Attribute
Amphipod crustacean	
Eohaustorius estuarius	Infaunal sediment dweller, deposit feeder, 2–28‰, 90% fine or coarser, low TOC
Rhepoxynius abronius	Infaunal sediment dweller, detrital and predatory feeder, >22‰, 95% fine or coarser, low TOC
Ampelisca abdita	Epifaunal tube dweller, detrital feeder, 20–35‰, 95% coarse or finer, higher TOC
Grandidierella japonica	Epifaunal tube dweller, detrital feeder, 30–35‰, 95% coarse or finer, higher TOC
Leptocheirus plumulosus	Epifaunal tube dweller, detrital feeder, 5–30‰, <90% coarse, <75% clay, tolerant to TOC ranges.
Mysid	
Mysidopsis bahia	Epifaunal or water column, detrital feeder and carnivore, 25-34‰, wide tolerance to grain size and TOC
Holmesimysis costata	Epifaunal on kelp, detrital feeder and carnivore, 32–36‰, sediment grain size not applicable to test organism use
Neomysis mercedis	Epifaunal, detrital feeder and carnivore, low salinity environment, sediment grain size not applicable to test organism
Larval species	
Mytilus galloprovincialis	Larval mussel test. 20–32‰, sediment grain size not applicable to test organism use elutriate testing organism. Adult attaches to rocks or fouling communities and is suspension feeder.
Crassostrea gigas	Larval oyster test. 15–35‰, sediment grain size not applicable to test organisms. Adult attaches to shell or rock within sand environment, filter feeder.
Dendraster excentricus	Larval echinoderm test (sanddollar), 26–30‰, sediment grain size not applicable to test organism use. Adult is sand dweller, detrital feeder.
Strongylocentrotus purpuratus	Larval echinoderm test (sea urchin), 26–30‰, sediment grain size not applicable to test organism use. Adult lives on rock habitat, episammic grazer.
Fish	
Citharichthys stigmaeus	Epifaunal or water column, carnivore, 28-34‰, tolerant to wide range of grain size.
Platichthys stellatus	Epifaunal or water column, carnivore, <10-34‰, tolerant to wide range of grain size.
Menidia berylina	Water column, carnivore, <10–35‰. Water-column toxicity test.
Clam	
Macoma nasuta	Infaunal sediment dweller, epifaunal and water column or detrital browser, 25–32‰, tolerant to wide range of sediment grain size.

Table 16-6 *continued*

Species	Attribute
Polychaeta / Oligochaeta	
Lumbriculus spp	Infaunal or epifaunal dweller, detrital feeder (B and G), 0–5‰, tolerant to wide range of sediment grain size, bioaccumulation test organism
Nephtys caecoides	Infaunal sediment dweller, detrital feeder (A and B), 28–32‰, tolerant to wide range of sediment grain size, toxicity and bioaccumulation test organism
Nereis virens	Infaunal sediment dweller, sediment engulfer and carnivore (B), 20–35‰, tolerant to wide range of sediment grain size, bioaccumulation test organism.
Neanthes arenaceodentata	Infaunal or epifaunal nestler, tube dweller, sediment engulfer and carnivore (A, B, and G), 20–35‰, tolerant to wide range of sediment grain size, bioaccumulation and acute and chronic toxicity test organism.

potentially available COPCs through interference with chemical equilibria that may have been established between the sediment compartments (sediment, porewater, organic carbon, and overlying water). Modification of the sediment regime to accommodate the selected species (e.g., changes in salinity) alters the potential availability of the COPCs between the sediment compartments. Both attributes strongly influence the outcome of the tests and can result in inaccurate assessments of the risk of leaving sediment in place.

Sediment removal evaluation

The second type of sediment assessment evaluates the availability and effects of COPCs that are released during the removal process. If sediment is to be removed, then disturbance of the chemical–physical relationships that have been established at the site is reasonable and acceptable. Therefore, collection methods can be more disruptive of the sediment. Mixing and compositing of sediments are appropriate under this assessment type. The test species to be exposed to these sediments should satisfy the following criteria:

1) The test species should be likely surrogate organisms for living under the environmental and sediment conditions of the removal site (i.e., sediment disruption and mixing can occur, and test organisms should have characteristics that permit them to live under the conditions of the removal site).
2) Test species should also represent the exposure pathways that are present during removal and most likely influenced by the removal process. These exposure pathways do not normally include the organisms that

live within the sediments because they will generally be removed from the site and lost. Organisms that live at the sediment–water interface in areas surrounding the removal site likely are exposed to COPCs in both the dissolved phase (COPCs released to the water column during removal) and in the particulate-bound phase (transport of suspended particulates off site). Water column organisms potentially are exposed to COPCs in both phases as water and particulates move away from the removal site. Test organisms should represent these exposure pathways and should be tolerant of the conditions of the removal site.

Again, inappropriate selection of species and modification of the test environment to accommodate those species can reduce the value of the estimation of the risk of removing COPC-containing sediment from a particular location as a result of modifying the availability of the COPC.

Sediment disposal evaluation

The third type of sediment assessment evaluates the availability and effects of COPCs when sediment is disposed of at an ocean, inland aquatic site, or upland location. If sediment is removed and placed at a new location, the sediment will be highly disturbed, and testing scenarios should attempt to maximize disturbance of the chemical–physical relationships that have been established in the original location. Therefore, collection methods can be disruptive. Again, sediment mixing and compositing is appropriate for this assessment objective. The test species that will be exposed to these sediments should satisfy the following criteria:

1) Test species should be surrogate organisms living under the salinity and sediment conditions of the disposal site, not the removal site. Sediment modifications should occur, and the test organisms should have characteristics that permit them to live at the disposal site. Sediment adjustments must be made to accommodate environmental conditions at the disposal site. An assessment should also be made regarding the need to acclimate the sediment to the conditions of the disposal site. Upland soils placed in water or freshwater sediments placed in marine waters will not be appropriate habitats for surrogate organisms at these disposal sites unless the sediment is acclimated to those conditions.

2) Test species should represent the exposure pathways at the disposal site. These exposure pathways include organisms that live within the sediments and have direct exposure to contaminants in the pore water and adsorbed onto sediments, as well as those that live in or on the sediment surface and feed at the sediment–water interface. The test organisms

in ocean disposal and inland aquatic site evaluations also include those that live in the water column and are only exposed to COPCs during sediment disposal.

In each case, organisms should be selected on the basis of their tolerance to the conditions of the disposal site. The testing environment should be modified to accommodate the selected disposal site surrogate species, including sediment acclimation. Inappropriate selection of species to accommodate the conditions of the removal site, rather than the disposal site, may provide inaccurate estimations of the risk of placing sediment in a new location. Under upland conditions, the types of biological tests would need to be defined on the basis of the specific exposure pathways at the confined disposal site.

Test: Was the toxicity testing performed correctly?

The test phase of the investigation relates to the health and sensitivity of the test population as well as to the test methods that are followed during the evaluation. This discussion will, however, focus on organism handling and changes in sensitivity that result from variation in handling. It is important to note that any data set should also be reviewed against the appropriate protocols, should ensure that water quality ranges were not affecting test responses, and should ensure that performance criteria were met. Datasets used in establishing SQGs typically follow these evaluative steps (Chapter 12), and they will not be discussed further here.

Organisms that are captured in the field and brought to the laboratory are introduced to a variety of stresses that they do not normally encounter. Stressed populations generally show increased sensitivity to contaminants, resulting in greater adverse biological effects than those observed for unstressed populations. To limit the effects of these stressors, standardized handling protocols and test population assessment methods have been established (e.g., ASTM 2004a, 2004b). These include standard methods for collecting, handling, and shipping test organisms, which vary depending upon species. The protocols also include acclimation schedules that limit the degree of change that test organisms are exposed to during preparation for testing conditions. The health of the test population generally is assessed just before test initiation to ensure a higher probability of test success.

Tests that do not follow these protocols run the risk of increased population sensitivity with increased mortality and/or other test anomalies appearing in test treatments. Lack of adverse biological effects resulting from compromised collecting, handling, and acclimation procedures does not necessarily negate the test results. In fact, the lack of adverse biological effects occurring in stressed organisms indicates an even lower risk associated with COPCs in sediment. If adverse biological effects are identified in samples for which collecting, handling, or acclimation procedures are compromised, the value of these results in evaluating risk is also compromised, and resampling and analyses should be considered. In one study, it was observed that, although there was acceptable control survival, there was an increase in the range of survival for test sediments (>30% range in replicate mortality) and an average increase of 18% mortality for amphipods that were acclimated too rapidly. Experimental manipulation of the acclimation rates reproduced these observations (Word 1999).

Test organism collection and handling before use in a toxicity test is difficult to evaluate retrospectively. Quality of test organisms should be based on estimating the mortality or behavior shortly before the start of a test. This is a part of the art of toxicity testing and relies on the laboratory staff to notice aberrant behavior or excess mortality in holding tanks. Experienced personnel may more easily notice aberrant behavior than increased mortality, but it is still a nonquantitative assessment of the health of the initial population of test organisms. Because toxicology labs often deal with variable numbers of test organisms for particular test setups, the estimation of premortality conditions under holding conditions is difficult and subject to error during the estimation process. Often the pretest mortality is not completely known until after a test has been set up and most of the organisms in the holding tanks are used or accounted for. Making a decision not to initiate a test on a specific date because of the apparent health of the test organisms necessitates delaying a program that is often on a tight time schedule and budget. The resulting cost and time delays are difficult to support with quantitative data. Therefore, it is rare that a test will be delayed because of the perceived health of test organisms, but investigators may see increased variability of response as well as greater than expected mortality in the assessment of experimental results.

Appropriate acclimation of test organisms is an easier procedure to implement. In general, standard practices are that temperature should not change by more that 3 °C/d and salinity changes must be less than 5‰/d. Acclimated organisms should then be held for a minimum of 2 d under these conditions prior to their use. A laboratory that reduces the time for these acclimation schedules runs the risk of biasing test results by testing more-sensitive organisms. Survival of test organisms in native control sediment and measurement of the sensitivity of a test species to a reference toxicant provide an indication of the relative sensitivity of the test population. Successful survival in the native control sediment is a necessity, but it does not always indicate that the population was completely healthy. Survival simply means that, under the most ideal laboratory conditions, the test organisms chosen for testing will survive at a high level (e.g., <10% mortality over the test period).

Comparison of the control chart limits for reference toxicant exposure within a laboratory indicates whether that laboratory consistently handles the test organisms (narrow control limits; USEPA 1994; ASTM 2004a, 2004b), but it does not indicate whether Lab A would have a different result than Lab B. Multiplying the standard deviation of the response by 2 and adding and subtracting that amount from the mean creates the control chart ranges within a laboratory. Typical control chart ranges for reference toxicant responses are generally about 3-fold, indicating a standard deviation with a 25% coefficient of variation. Ranges of 10-fold or more should be viewed with concern that the populations being tested are either naturally more variable (e.g., seasonal trends in sensitivity) or not being handled in a consistent manner. These larger test ranges also do not indicate whether an increased sensitivity brought about by faster acclimation rates caused marginal responses to become greater during the testing of sediment COPCs or CFs. A challenge toxicity test is often used in drug efficacy studies to determine the ability of a drug to increase the susceptibility of an organism under stress. If acclimation schedules are not performed in accordance with standard protocols and increased range of replicate responses (range in excess of 30%) are noted, then one of the expected CFs in these types of data is increased sensitivity from internal (seasonal effects) or external stresses that can result from the handling of test organisms.

Account: Were the sediments tested free from potential confounding factors?

The account step of the investigation uses nonmanipulative tools to assess the potential influence of CFs on toxicity test results (Table 16-5). This is accomplished by comparison to

1) reference site responses,
2) literature values, or
3) measured responses observed in specialized concurrent reference-toxicant tests.

Three water quality properties (e.g., salinity, ammonia, and sulfides) can individually or in combination cause adverse biological effects if the parameters exceed critical no-observed effects concentrations (NOECs). These characteristics may be natural components of the sediment or may be produced in the laboratory during handling of the sediment.

If the sediment is near a source of fresh water (e.g., rivers and streams, effluent discharges, or underwater seeps from submerged aquifers), the porewater salinity concentrations may be very low. Proximity to known sources of fresh water cannot always be determined accurately, especially with underwater seeps. Therefore, the only sure way of determining the salinity of pore water is to measure that characteristic when the sediment is collected. If the porewater salinity differs from the proposed test conditions, then additional measurements should be made

1) after appropriate compositing and mixing,
2) just before starting a test,
3) during the test, and
4) at the end of the test.

This measurement schedule will document changes to the porewater environment during handling of sediment and may indicate potential causes for increases in ammonia or sulfides, such as disruption in the indigenous populations of microbes that normally handle the processing of these materials to less toxic forms. Additionally, a salinity-response profile should be used (or created) to properly evaluate the sensitivity of the selected test organisms to salinity. If the selected test organisms are not tolerant of the existing porewater salinity, alternative species may need to be considered. In some cases, such as with benthic assessments, a suitable reference site may need to be evaluated and test results compared to reference-site responses.

As mentioned in "Confounding influence of ammonia" (p 649), in a well-balanced, acclimated aquatic system, the direct release of ammonia or the production of ammonia are parts of the nitrogen cycle that regenerate nitrogen for plant uptake from biological waste products. In this type of system, the bacterial populations are capable of altering the available ammonia as it is being produced into relatively less toxic nitrites and nitrates. Unbalanced systems are not as capable of enabling this process, and as a result, ammonia may increase in concentration, often to very high levels that are toxic to many different species. This imbalance can occur naturally, or it may result from laboratory manipulations of environmental samples. In the simplest case, freshwater microbial communities in freshwater sediment samples will die when placed in marine waters for testing. As a result, ammonia concentrations will increase. Ammonia increases will then continue until the marine microbial community has replaced the freshwater-based community that was disrupted.

This sediment acclimation process occurs over extended periods of time and can occur in less than one week, but it can also take up to 6 to 8 weeks to complete. As a result, typical bioassays that are performed on sediment or elutriates of sediment that have undergone modified salinity conditions that may have unbalanced bacterial communities, and elevated ammonia concentrations may occur. These artifacts of laboratory processes need to be evaluated during the data review. Further evaluation can include acclimating sediments in a manner similar to that described in Word et al. (2001) concurrent to the standard test procedure.

Regardless of the source of ammonia and sulfides, the potential for toxic effects needs to be accounted for by measuring the ammonia and sulfide concentrations in the sediment pore water, and measuring the ammonia and sulfide concentrations in the test chambers, both in the pore water and the overlying water. The potential for ammonia- and sulfide-related toxicity can then be assessed by comparing the measured concentrations in the appropriate exposure compartment for each test organism to document sensitivity, either from literature values or from concurrent reference-toxicant test values. If the ammonia and/or sulfide concentrations exceed the NOEC and lowest observed effects concentration (LOEC) values, additional manipulative steps may be necessary to fully understand the role of these factors in the observed sediment toxicity (see the following section).

The available quantity and quality of organic materials contained in sediment and the vertical distribution of those organic materials are factors that should be accounted for, especially in deeply buried sediment and those with high peat content. Most sediment generally contains sufficient quantity of detrital food materials to support the test organisms used in short-term laboratory testing (e.g., ≤10 d). However, in some sediment, the amount of usable detritus is insufficient to maintain certain test organisms, especially over longer testing periods. Low survival in sediment with low TOC (e.g., <0.5%) and low concentration of COPCs should be viewed with suspicion, especially in tests that are greater than 10 d in length. If TOC amounts are exceptionally low, or if test sediments are ancient, deeply buried, or have low-value carbon sources, observations regarding the nature of the sediment should be evaluated, and further manipulations may be required (See the following section).

Sediment grain size, compaction, water content, and angularity are persistent sediment properties that can produce adverse biological effects, unrelated to COPCs. These CFs are best evaluated using reference sediments selected for these attributes. Reference site selection is a critical component to this evaluative process. It may not be possible to find 1 reference site to fulfill each requirement in a given sediment, and multiple sites may be evaluated. Response at the reference is then compared to that of the test sediment. A decision to modify the sediment to meet the needs of test organisms or to select a test organism that is not affected by the sediment feature should also be considered. This type of assessment goal can establish the likely biological impact that might occur as a result of exposure to COPCs in this material at a disposal site or during the process of sediment removal.

Remove and replace, research: Were there significant correlations between the biological effects and the concentrations of any contaminants or CFs?

The final steps of the STARR process feature sediment manipulations designed to further identify the role of CFs in sediment toxicity. These typically involve removing and sometimes replacing the CF and are similar to toxicity identification/reduction evaluation (TI/RE) methods. These manipulations are appropriate for all types of sediment investigations; however, the regulatory or interpretative use of the resultant data may differ, depending on the purpose of testing.

CF removal has proven to be an effective evaluative tool for nonpersistent characteristics, such as ammonia, salinity, and sulfides. For marine and estuarine tests, removal has been most commonly used with ammonia. As mentioned in "Confounding influence of ammonia" (p 649), ammonia can occur in test chambers at concentrations that are toxic to test organisms. Ammonia can be successfully removed from test chambers by reducing the ammonia concentrations in the overlying water and thereby also reducing the porewater concentrations. These reductions can be achieved by renewing overlying water prior to test initiation (USEPA 1994; USEPA/USACE 1998), the placement of algae, such as *Ulva* sp. in the test chamber, or the use of acclimated sediments (Pinza et al. 1997; Word et al. 2001; Ho et al. 2002). Sulfide concentrations can also be reduced through overlying water removals.

Removal of lower-salinity pore water to accommodate selected test species may be necessary, especially when potential effects of freshwater sediments at a marine disposal site are evaluated. This evaluation should be done carefully while addressing other CFs that may be created as a result of modifying the salinity of the porewater environment. Salinity should not be modified to assess the effects of COPCs that may be present in sediment if the proposed option is to leave that sediment in place, nor should salinity be modified when evaluating the potential COPC-related effects of sediment removal. Instead, the appropriate species that can accommodate the salinity in the particular environment should be selected. Selection of inappropriate test species or modification of the sediment to accommodate those species will compromise the potential value of these data.

Removal of sediment compaction or TOC-related effects requires more specialized manipulations. Compacted sediment can be simply broken apart to the extent that animals may be able to burrow or construct tubes. It is important to minimize the extent to which these sediments are disturbed because this disturbance can also affect contaminant equilibria in the test container. TOC-depleted sediments can be supplemented with small amounts of high-quality TOC, such as *Enteromorpha* sp. Care should be taken to minimize the extent to which TOC is added because food supplementation can interfere with contaminant availability through the ingestion pathway in the test container or can create a BOD or ammonia issue.

Replacement is another TIE procedure that can be applied to CF evaluations. Following CF removal, the same concentration of a CF is spiked back into the test sediment, and toxicity tests are repeated to determine if the response observed in the unmanipulated sediment returns (USEPA 1989). This method has been used with ammonia in suspended particulate–phase (SPP) test with *Mytilus* larvae (Gardiner, Barrows, Word 1996). The original sediment–water preparation had high toxicity and ammonia concentrations exceeding the toxicity threshold for *Mytilus*. Removal of ammonia from sediment using the USEPA-approved overlying water exchanges before making the SPP resulted in reduced toxicity. In a second set of test treatments, the ammonia was spiked back into the ammonia-reduced SPP, and the same level of toxicity observed in the original, unmanipulated SPP was observed. It is advisable to combine replacement tests with other confirmatory tools to ensure that other constituents are not contributing to toxicity.

The question of bioavailability requires a different set of evaluative tools. As noted in "Confounding influence of contaminant bioavailability" (p 647), the concentration of heavy metals in the mineral fraction of sediments is not a direct assessment of the bioavailable fraction of metals in sediment. Data used for the establishment of SQGs (e.g., ERL, ERM, screening level, maximum level, AET; Chapter 3) are obtained using a variety of extraction techniques. These techniques, while generally consistent within a particular set of data used for SQG development, range from full acid digestion of the internal matrices of sediment to weaker acid digestions of sediment that do not digest the mineral matrices of the sediment. These are not equivalent digestion techniques, and as such different SQGs values based on different methods of extraction have marginal value in establishing NOECs or LOECs. Observations of exceedances of SQGs should be set aside when these levels are not coupled with adverse biological effects.

There are 3 steps in determining the availability of a particular COPC in sediment:

1) laboratory testing during which sediment is exposed to (spiked with) various concentrations of a particular contaminant and the effects related to that concentration directly established,
2) determination of the correlation or concordance of contaminants to biological responses and direct assessment of the relationship, and
3) comparison of the concentration of a contaminant to a known effects-based concentration in terms of toxic units (TUs).

Each of these processes has strengths and weaknesses in determining the availability of a particular contaminant and the influence of complex mixtures on increasing or decreasing the level of effect of a particular chemical. Of particular value for determining available fractions is the TU approach. For this approach, the observed COPC concentration in the sediment is divided by known effects levels, such as the LC50, LOEC, or NOEC. In this way, a proportional estimate of each COPCs, contribution to toxicity can be evaluated. Replacement can also be applied to CFs to determine their relative contribution. It is important to note that TUs are estimates and are subject to synergistic or antagonistic interactions.

Refinements and Validation of SQGs

SQGs have been shown to be effective screens for the prediction of adverse biological effects (Chapters 4 and 12). If the co-occurrence of elevated sediment contaminant concentrations predicted the presence or absence and extent of adverse biological effects (toxicity or benthic community changes) 100% of the time under all conditions, then this workshop would have been unnecessary. However, the empirical nature of SQGs makes them subject to the influences, not only of CFs (overprediction of contaminant related effects) but also of unmeasured contaminants that are not normally evaluated (under prediction of contaminant-related effects) or testing under inappropriate conditions (both under- and overprediction of effects).

As indicated in Chapter 4, the current group of SQGs provides concordant responses of biological effects associated with mixtures of contaminants and was never meant necessarily to implicate a particular COPC. A generally accepted approach to defining the relative toxicity of COPCs is the use of composite values, such as the ERM values or ERM quotients (Long and Chapman 1985; Long et al. 1998; Chapter 12). The ERM value is the median concentration associated with effects measured at many stations throughout North America. ERM quotients are calculated by dividing the concentration of the chemical (when detected) by the ERM value (when available) and determining the average quotient for all of the detected contaminants. Composite values attempt to overcome the variability introduced by multiple ERM exceedances as well as the CFs noted above. This approach permits an evaluation of the presence or absence of a chemical or mixture of chemicals at levels that have been previously related to biological effects (e.g., Long and Chapman 1985; Long et al. 1998; Chapter 12).

An incorporation of CFs in the refinement of the current SQGs or in the formation of new SQGs would help reduce uncertainty associated with individual values and would increase the predictive ability of individual SQGs. For some datasets, it may be possible to review the supporting data for the existing SQGs to determine those values that may be driving a particular SQG but are influenced by CFs. The goal of such an analysis is to refine and strengthen the link between environmental concentration of a particular contaminant and resulting biological effects.

It is important to note that the inappropriate use of SQGs has a major impact on the predictive ability of the SQGs. Often the nature of an investigation and the applicability of a certain set of SQGs do not match. An important factor for toxicity testing in SQG comparisons was the depth of sediment sampled (Chapter 4). The sediment depth used to develop the guidelines was typically the upper few centimeters of sediment surface, whereas the depth sampled by programs applying those guidelines is often deeper (e.g., dredged material assessment programs with sediment obtained predominantly below the biogenic zone). The effects of sediment depth on SQG concordance are compounded by CFs because many of the CFs identified in this chapter are observed in sediment collected below the biologically active zone. For benthic assessments and SQG comparisons, an important factor identified was the selection of the appropriate measurement endpoints (e.g., number of species, abundance, diversity measures, fecundity) as well as the selection of appropriate reference comparison conditions (e.g., grain size, water depth, salinity; Chapter 4).

A variety of unintended applications are presented in Chapter 4, including comparison of SQG concentrations to acid-leached mining wastes, peat materials, birds' eggs, etc. These inappropriate applications of SQG predictions are not advisable and should be discouraged. The dredged material testing program also recognized that the relationship between sediment contaminant concentrations and biological effects from dredged materials may not show the same relationship as the SQGs. As a result, the implementation manuals prepared by the USEPA and the USACE use an effects-based approach rather than a chemical concentration–based approach, similar to the Jensen Criteria that had been used previously (Chapter 8). The USEPA/USACE effects-based approach includes Tier 4 evaluations that permit applicants to investigate unexpected effects and to determine whether those observed effects were related to persistent chemical contaminants requiring additional programmatic controls, or whether the effects were due to other factors that were more transitory in nature and were not targeted for control by the dredged material disposal programs (e.g., CFs, less persistent sediment attributes like ammonia or salinity, lack of food).

The influence of CFs on empirically derived SQGs should be expected and addressed where necessary, especially when the testing is being performed under conditions that are not comparable to the original purpose or design of the testing procedure. Application of the STARR process for toxicity testing and the complementary STARR process for benthic community analysis permits the dissection of potential causes of observed effects under various testing programs and the determination of whether alternative management decisions are required.

Future development of program-specific SQGs, especially for application to different assessment programs (e.g., dredged material testing, beneficial use, wetland application, upland assessment of soil contamination, development of cleanup criteria) will need to evaluate the influence of CFs in order to properly ascertain COPC-related effects. The ultimate predictor of adverse biological effects is the measurement endpoint, and the objective of many efforts is to develop measurement endpoints that are more sensitive to contaminants. As more sensitive measurement endpoints are developed, it is expected that the influence of CFs will increase.

References

Adams WJ, Kimerle RA, Mosher RG. 1985. Aquatic safety assessment of chemicals sorbed to sediments. In: Cardwell RD, Purdy R, Bahmer RC, editors. Aquatic toxicology and hazard assessment: 7th Symposium. Philadelphia: American Soc for Testing and Materials. ASTM STP 854.p 429–453.

[ASTM] American Society for Testing and Materials. 2004a. Conducting acute toxicity test on aqueous ambient samples and effluents with fishes, macroinvertebrates, and amphibians. Philadelphia: ASTM. #E1192 (11.05). p 440–452.

[ASTM] American Society for Testing and Materials. 2004b. Standard test method for measuring the toxicity of sediment-associated contaminants with estuarine and marine invertebrates. Philadelphia: ASTM. E1367-02. p 711–737.

Barnes RSK, Hughes RN. 1999. An introduction to marine ecology. Malden (MA): Blackwell Science. 286 p.

Barrick R, Becker S, Brown L, Beller H, Pastorok R. 1988. Sediment quality values refinement: 1988 update and evaluation of Puget Sound AET. Volume I. Bellevue (WA): PTI Environmental Services. PTI Contract C717-01.

Barrows ES, Word JQ, DeWitt TH, Pinza MR, Gardiner WW. 1994. Test animal mortality in dredged material toxicity tests: Contributions of ammonia, salinity, and grain size. Presented at the Dredging '94, American Society of Civil Engineers Conference; 1994 Nov 13–16; Lake Buena Vista, FL, USA.

Barrows, ES, Mayhew HL, Word JQ, Tokos JJS. 1996. Evaluation of dredged material proposed for ocean disposal from Port Chester, New York. Prepared for the U.S. Army Corps of Engineers, New York District, by Battelle Marine Sciences Laboratory, Sequim, WA. Richland (WA): Pacific Northwest National Laboratory. PNNL-11286.

Barton J. 2002. Ammonia and amphipod toxicity testing. DMMP Clarification Paper; presented to Sediment Management Annual Review Meeting (SMARM); 2002 April 29; Seattle, WA, USA.

Baudo R, Giesy J, Muntau H. 1990. Sediment: Chemistry and toxicity of in-place pollutants. Boca Raton (FL): Lewis. 405 p.

Borgmann U. 1994. Chronic toxicity of ammonia to the amphipod *Hyalella azteca*: Importance of ammonium ion and water hardness. *Environ Poll* 86:329–335.

Buchanan JB, Longbottom MR. 1970. The determination of organic matter in marine muds: the effect of the presence of coal and the routine determination of protein. *J Exp Mar Biol Ecol* 5:158–169.

Burgess RM, Scott KJ. 1992. The significance of in-place contaminated marine sediments on the water column: processes and effects. In: GA Burton, editor. Sediment toxicity assessment. Boca Raton (FL): Lewis. p 129–166.

Casas AT, Crecelius EA. 1994. Relationship between acid volatile sulfide and the toxicity of zinc, lead, and copper in marine sediments. *Environ Toxicol Chem* 13: 529–536.

Chapman PM. 1996. Presentation and interpretation of sediment quality triad data. *Ecotoxicology* 5:327–339.

Dauer DM, Maybury CA, Ewing RM. 1981. Feeding behavior and general ecology of several spionid polychaetes from the Chesapeake Bay. *J Exp Mar Biol Ecol* 54: 21–38.

DeWitt TH, Ditsworth GR, Swartz RC. 1988. Effects of natural sediment features on survival of the Phoxocephalid amphipod, *Rhepoxynius abronius*. *Mar Environ Res* 25:99–124.

DeWitt TH, Redmond MS, Sewall JE, Swartz RC. 1992. Development of a chronic sediment toxicity test for marine benthic invertebrates. CBP/TRS/89/93. Annapolis (MD): US Environmental Protection Agency, Chesapeake Bay Program.

Dewitt TH, Swartz RC, Lamberson JO. 1989. Measuring the acute toxicity of estuarine sediments. *Environ Toxicol Chem* 8:1035–1048.

Diaz RJ. 1992. Ecosystem assessment using estuarine and marine benthic community structure. In: Burton Jr GA, editor. Sediment toxicity assessment. Ann Arbor (MI): Lewis. p 67–86.

Di Toro DM, Mahony JD, Hansen DJ, Scott KJ, Carlson AR, Ankley GT. 1991. Acid volatile sulfide predicts the acute toxicity of cadmium and nickel in sediments. *Environ Sci Technol* 26:96–101.

Emery VL, Moore DW, Gibson AB, Gray BR, Duke BM, Wright RB, Farrar JD. 1997. Development of a chronic sublethal sediment bioassay using the estuarine amphipod *Leptocheirus plumulosus* (Shoemaker). *Environ Toxicol Chem* 16:1912–1920.

Enequist P. 1950. Studies on the soft-bottom amphipods of the Skagerrak. *Zool Bidrag fran Uppsala* 28:297–492.

Erickson RJ. 1985. An evaluation of mathematical models for the effects of pH and temperature on ammonia toxicity to aquatic organisms. *Water Res* 19:1047–1058.

Ferretti JA, Calesso DF, Hermon TR. 2000. Evaluation of methods to remove ammonia interference in marine sediment toxicity tests. *Environ Toxicol Chem* 19: 1935–1941.

Frazier BE, Naimo TJ, Sandheinrich MB. 1996. Temporal and vertical distribution of total ammonia nitrogen and un-ionized ammonia nitrogen in sediment pore water from the upper Mississippi River. *Environ Toxicol Chem* 15:92–99.

Gardiner WW, Barrows ES, Word JQ. 1996. Ecological evaluation of proposed reference sites in the New York Bight, Great South Bay, and Ambrose Light, New York. Report submitted to the U.S. Army Corps of Engineers, New York District. Sequim and Richland (WA): Battelle Marine Sciences Laboratory, Pacific Northwest Laboratory. PNNL-11361.

Gardiner WW, Word JQ, Swan BK. 1996. Ammonia as a confounding factor in suspended-particulate-phase sediment tests. Presented at the SETAC 17th Annual Meeting; Washington DC, USA.

Gardiner, WW, Antrim LD, Word JQ. 1995. An evaluation of ammonia toxicity in suspended-particulate phase testing. Battelle Marine Sciences Laboratory, Sequim, WA, and Battelle, Pacific Northwest Laboratories, Richland, WA. Paper presented at the Second SETAC World Congress, sponsored by the Society of Environmental Toxicology and Chemistry; Vancouver, BC, Canada.

Gray JS. 1981. The ecology of marine sediments: An introduction to the structure and function of benthic communities. Cambridge Studies in Modern Biology 2. New York: Cambridge Univ Pr. 185 p.

Greenstein DJ, Tiefenthaler LL. 1997. Reproduction and population dynamics of a population of *Grandidierella japonica* (Stephensen) (Crustacea:Amphipoda) in upper Newport Bay, California. *Bull So Cal Acad Sci* 96:34–42.

Hakanson L. 1992. Sediment variability. In: Burton Jr GA, editor. Sediment toxicity assessment. Ann Arbor (MI): Lewis. p 19–36.

Ho KT, Burgess RM, Pelletier MC, Serbst JR, Ryba SA, Cantwell MG, Kuhn A, Hyland JL, Van Dolah RF, Snoots TR. 2002. An overview of toxicant identification in sediments and dredged 17 materials. *Mar Pollut Bull* 44:286–293.

Ingersoll CG, Haverland PS, Brunson EL, Canfield TJ, Dwyer FJ, Henke CE, Kemble NE, Mount DP, Fox RG. 1996. Calculation and evaluation of sediment effect concentrations for the amphipod *Hyalella azteca* and the midge *Chironomus riparius. J Gt Lakes Res* 22:602–623.

Knezovich, JP, Steichen DJ, Jelinski JA, Anderson SL. 1996. Sulfide tolerance of four marine species used to evaluate sediment and pore-water toxicity. *Bull Environ Contam Toxicol* 57:450–457.

Kohn NP, Word JQ, Niyogi DK, Ross LT, Dillon T, Moore DW. 1994. Acute toxicity of ammonia to four species of marine amphipod. *Mar Environ Res* 38:1–15.

Kuznetov AP. 1964. Distribution of the sea bottom fauna in the western part of the Bering Sea and its trophical zonation. *Akademiia Nauk SSR Inst Okeano* (Summary in English, p 177).

Lamberson JO, DeWitt TH, Swartz RC. 1992. Assessment of sediment toxicity to marine benthos. In: Burton Jr GA, editor. Sediment toxicity assessment. Ann Arbor (MI): Lewis. p 183–211.

[LDC] London Dumping Convention. 1972. The Convention of the Prevention of Marine Pollution by Dumping Wastes and Other Matter. London: Office for the London Convention 1972, International Maritime Organization.

Levinton JS. 1972. Stability and trophic structure in deposit-feeding and suspension-feeding communities. *Am Naturalist* 106:472–486.

Literathy P, Nasser AL, Zarba MA, Ali MA. 1987. The role and problems of monitoring bottom sediment for pollution assessment in the coastal marine environment. *Environ Sci Technol* 19:781–792.

Long ER, Chapman PM. 1985. A sediment quality triad: Measures of sediment contamination, toxicity and infaunal community composition in Puget Sound. *Mar Pollut Bull* 16:405–415.

Long ER, Field LJ, MacDonald DD. 1998. Predicting toxicity in marine sediments with numerical sediment quality guidelines. *Environ Toxicol Chem* 17:714–727.

Mayer LM, Schick LL, Setchell F. 1986. Measurement of protein in nearshore marine sediments. *Mar Ecol Progr Ser* 30:159–165.

Mayer LM, Schick LL, Sawyer T, Plante CJ. 1995. Bioavailable amino acids in sediments: A biomimetic, kinetics-based approach. *Limnol Oceanogr* 40:511–520.

Moore DW, Bridges TS, Gray BR, and Duke BM. 1997. Risk of ammonia toxicity during sediment bioassays with the estuarine amphipod *Leptocheirus plumulosus*. *Environ Toxicol Chem* 16:1020–1027.

Oakden JM. 1984. Feeding and sediment selection in five species of Central California Phoxocephalid amphipods. *J. Crustacean Biol* 4:233–247.

Oliver JS Slattery PN, Hulberg LW, Nybakken JW. 1980. Relationships between wave disturbance and zonation of benthic invertebrate communities along a subtidal high energy beach in Monterey Bay, California. *Fishery Bulletin* 78:437–454.

Parker RH. 1975. The study of benthic communities: A model and a review. Elsevier Oceanography Series, 9. New York: Elsevier. 279 p.

Pearson TH, Rosenburg R. 1978. Macrobenthic succession in relation to organic enrichment and pollution of the marine environment. *Oceanogr Mar Biol Ann Rev* 16:229–311.

Petersen CG. 1913. Valuation of the sea. II. The animal communities of the sea bottom and their importance for marine zoogeography. *Rep Dan Biol Stat* 21:44.

Pinza MR, Kohn NP, Ohlrogge SL, Ferguson CJ, Word JQ. 1997. Fate and effects of ammonia sediment toxicity tests with the amphipod, *Rhepoxynius abronius*. Proceedings, 5th Symposium on Environmental Toxicology and Risk Assessment: Biomarkers. Sponsored by ASTM. Sequim and Richland (WA): Battelle/Marine Sciences Laboratory, Pacific Northwest Laboratories. PNL-SA-25992A.

Plumb RH. 1981. Procedures for handling and chemical analysis of sediment and water samples. Vicksburg (MS): US Army Corps of Engineers. Technical Report EPA/CE-81-1.

[PSEP] Puget Sound Estuary Program. 1986. Recommended protocols for measuring conventional sediment variables in Puget Sound. Prepared by Washington Department of Ecology, Olympia, WA. Seattle (WA): US Environmental Protection Agency Region 10.

[PSEP] Puget Sound Estuary Program. 1989. Recommended protocols for measuring metals in puget sound water, sediment and tissue samples. Prepared by Washington Department of Ecology, Olympia, WA. Seattle (WA): US Environmental Protection Agency Region 10.

[PSEP] Puget Sound Estuary Program. 1995. Recommended guidelines for conducting bioassays on Puget Sound sediments. Prepared by Washington Department of Ecology, Olympia, WA. Seattle (WA): US Environmental Protection Agency Region 10.

Rasnake WJ, Gruber D. 1998. Minimizing ecological impact from a seafood processing facility. Presented at SETAC 19th Annual Meeting; The Natural

Connection: Environmental Integrity and Human Health; 1998 Nov 15–19; Charlotte, NC. USA.

Regnault M. 1987. Nitrogen excretion in marine and fresh-water crustacea. *Biol Rev* 62:1–24.

Rhoads DC, Germano JD. 1982. Characterization of benthic processes using sediment profile imaging: an efficient method of remote ecological monitoring of the seafloor (REMOTS System). *Mar Ecol Progr Ser* 8:115–128.

Rysgaard S, Thastum P, Dalsgaard T, Christensen PB, Sloth NP. 1999. Effects of salinity on NH4+ adsorption capacity, nitrification, and denitrification in Danish estuarine sediments. *Estuaries* 22:21–30.

Samilov MR, Wells PG. 1984. Future trends in marine ecotoxicology. In: Ecotoxicological testing for the marine environment. Bredene (BE): State Univ Ghent, Belgium, Inst Marine Scientific Research. Volume 1 of 2.

Sanders HL. 1958. Benthic studies in Buzzards Bay: Animal-sediment relationships. *Limnol Oceangr* 3:245–258.

Savilov SI. 1961. Ecological characteristics of bottom communities. *Tr Inst Okeanol Akad Nauk SSSR* 11:3–84.

Schlekat CE, McGee BL, Reinharz E. 1992. Testing sediment toxicity in Chesapeake Bay with the amphipod *Leptocheirus plumulosus*: An evaluation. *Environ Toxicol Chem* 11:225–236.

Shepard FP. 1973. Submarine geology. 3rd edition. New York: Harper & Row. 517 p.

Southern California Coastal Water Research Project. 1994. Sediment grain size: Results of an interlaboratory intercalibration experiment. In: Annual Report 1993–94. Westminster (CA): So Calif Coastal Water Res Project.

Spotte S. 1992. Captive seawater fishes: Science and technology. New York: Wiley-Interscience. 942 p.

Swartz RC, DeBen WA, Jones JKP, Lamberson JO, Cole FA. 1985. Phoxocephalid amphipod bioassay for marine sediment toxicity. In: Cardwell RD, Purdy R, Bahner RC, editors.Aquatic toxicology and hazard assessment: 7th Symposium. ASTM, Philadelphia: American Soc for Testing and Materials. ASTM STP 854.

Tada F, Suzuki S. 1982. Adsorption and desorption of heavy metals in bottom mud of urban rivers. *Water Res* 16:14890–1494.

Thorson G. 1971. Life in the sea. New York: McGraw-Hill. 256 p.

Trask PD. 1932. Origin and environment of source sediments of petroleum. Houston: American Petroleum Inst, Gulf Publ Co. 323 p.

Trask PD, Rolston JW. 1951. Engineering geology of San Francisco Bay, California. *Geol Soc Am Bull* 62:1079–1109.

Turpaeva EP. 1953. Feeding habits and food groups of marine bottom invertebrates. *Tr Inst Okeanol Akad Nauk SSSR* 11:3–84.

[USEPA] US Environmental Protection Agency. 1986 (Revised 1990). Test methods for evaluating solid wastes: Physical/chemical methods, 3rd ed. Washington DC: USEPA Office of Solid Waste and Emergency Response. SW-846.

[USEPA] US Environmental Protection Agency. 1989. Methods for aquatic toxicity identification evaluations: Phase III toxicity confirmation procedures. Duluth (MN): USEPA Environmental Research Laboratory. EPA/600/388/036.

[USEPA] US Environmental Protection Agency. 1994. Methods for assessing the toxicity of sediment-associated contaminants with estuarine and marine amphipods. Washington DC: USEPA. EPA/600/R-94/025.

[USEPA] US Environmental Protection Agency. 1999. Update of ambiwent water quality criteria for ammonia. Washington DC: USEAP Office of Water 4304. EPA-822-99-014 December 1999.

[USEPA] US Environmental Protection Agency. 2001. Methods for assessing the chronic toxicity of marine and estuarine sediment-associated contaminants with the amphipod *Leptocheirus plumulosus*. Washington DC: USEPA. EPA/600/R01/0.

[USEPA/USACE] US Environmental Protection Agency, US Army Corps of Engineers. 1991. Evaluation of dredged material proposed for ocean disposal, testing manual. Washington DC: USEPA, Office of Water. EPA/503/891/001.

[USEPA/USACE] US Environmental Protection Agency, US Army Corps of Engineers. 1998. Evaluation of dredged material proposed for discharge in waters of the U.S.: Inland testing manual. Washington DC: USEPA. EPA/823/B98/004.

Walker KR, Bambach RK. 1974. Feeding by benthic invertebrates: Classification and terminology for paleoecological analysis. *Lethaia* 7:67–78.

Wang F, Chapman PM. 1999. Biological interactions of sulfide in sediment: A review focusing on sediment toxicity. *Environ Toxicol Chem* 18:2526–2532.

Weston DP. 1990. Quantitative examination of macrobenthic community changes along an organic enrichment gradient. *Mar Ecol Progr Ser* 61:233–244.

Wilson TRS. 1975. Salinity and major elements of seawater. In: Riley JP, Skirrow G, editors. Chemical oceanography, Volume 1. New York: Academic Pr. p 365–413.

Woollins JD. 1996. Chemistry of sulfur. In: Mitchell S, editor. Biological interactions of sulfur compounds. London: Taylor & Francis. p 1–19.

Word JQ. 1978. The infaunal trophic index. Annual Report Southern California Coastal Water Research Project. Long Beach (CA): So Calif Coastal Water Res Project. p 19–40.

Word JQ. 1990. The infaunal trophic index: A functional approach to benthic community analyses [PhD dissertation]. Seattle: Univ Washington. 297 p.

Word JQ. 1999. Pretest to evaluate confounding factors observed in past toxicity tests. A presentation of Sediment Work Group for U.S. Navy BRAC Program; 1999 Dec 7; San Francisco, CA, USA.

Word JQ, Bodensteiner S, Word L. 2001. Results from the biological and chemical evaluation of sediment testing in Bahia Lagoon, CA. Report to the Bahia Homeowners Association, 4 May 2001. Sequim (WA): MEC Analytical Systems. 41 p + appendices.

Word JQ, Gardiner WW, Hammermeister T, Hunt C. Environmental studies at proposed ocean disposal sites off the mouth of the Columbia River. Report to USACE Portland District, Portland, OR. Sequim (WA): MEC Analytical Systems. 106 p.

Word JQ, Rosman LB. 1996. Evaluation of dredged material proposed for ocean disposal from federal projects in New York and New Jersey and the Military Ocean Terminal (MOTBY). PNNL-11280. Prepared for the U.S. Army Corps of Engineers, New York District. Sequim and Richland (WA): Battelle Marine Sciences Laboratory, Pacific Northwest National Laboratory.

Word JQ, Word LS. 2001. Hazardous contaminants in marine sediments. In: Lehr J, Hyman M, Gass TE, Seevers WJ, editors. Handbook of complex environmental remediation problems. New York: McGraw-Hill. Chapter 5.

Zaneveld JRV. 1974. Spatial distribution of the index of refraction of suspended matter in the ocean. In: Gibbs RJ, editor. Suspended solids in water. *Mar Sci* 4: 87–100.

Zatespin VI. 1970. On the significance of various ecological groups of animals in the bottom communities of the Greenland, Norwegian, and Barents Sea. In: Stell JH, editor. Marine food chains. Berkeley (CA): Univ Calif Pr. 552 p.

Uncertainties in assessments of complex sediment systems 17

PETER M CHAPMAN, WESLEY J BIRGE, ROBERT M BURGESS, WILLIAM H CLEMENTS, W SCOTT DOUGLAS, MICHAEL C HARRASS, CHRISTER HOGSTRAND, DANNY D REIBLE, AMY H RINGWOOD

Sediment assessment is challenging for 3 reasons. First, sediments are heterogeneous, sometimes at the scale of millimeters, possessing very different physicochemical characteristics both horizontally and with depth (e.g., grain size, organic carbon content). Second, organisms are exposed to sediment-associated contaminants through a variety of exposure routes, including the sediment–water interface, pore water, or direct contact or ingestion of the sediments. In addition, species-specific differences in physiology, biochemistry, behavior, and tolerance can mitigate adverse effects of contaminants. Third, sediments tend to be contaminated by mixtures of chemicals whose potential interactions are not well characterized and whose bioavailability can be variable and challenging to predict (Swartz and Di Toro 1997), particularly when compounded by variability in human and wildlife habitat use patterns and estuarine, hydrological, and/or seasonal cycling. This variability will also change in nature and magnitude depending on the situation and/or habitat being assessed and the pool of receptors.

Both scale and time need to be considered in any sediment assessment. For example, a dredging or other significant anthropogenic disturbance event may be relatively brief compared to seasonal or multi-year climatic cycles, but during the event, the sediment system is generally affected far more than an average tidal cycle. Further, it is challenging to separate physical, chemical, and biological variables from each other. For example, a system experiencing low amounts of energy will most likely become depositional. Depositional systems tend to accumulate organic carbon and fine-grained

Use of Sediment Quality Guidelines and Related Tools for the Assessment of Contaminated Sediments
Wenning RJ, Batley GE, Ingersoll CG, Moore DW, editors.
 ISBN 1-880611-71-6

sediments more readily than do high-energy systems in which low organic carbon and large particles are most common. The dramatically different physical and chemical conditions in these 2 systems will attract different fauna and flora. Finally, because of the elevated presence of, for example, organic carbon, many contaminants will have greater affinity for a depositional environment than a high-energy erosional system.

Further, it is difficult to determine whether or not some observed effects are significant when compared with existing variability. Often, natural variability in a system "cloaks" important trends and behaviors. To combat this problem, statistical mathematics emphasizes designs seeking to reduce variability (Dowdy and Wearden 1983). For example, reference sites might be selected to mimic exposed sites in as many ways as possible except for the presence of contaminants, such that significant adverse effects will be more apparent. Another approach is to composite sediments, thus removing the uniqueness of the individual components.

This background chapter outlines some of the uncertainties faced by sediment assessments in different aquatic systems, in freshwater (lotic and lentic), estuarine, and marine ecosystems, and provides information to support the conclusions reached in Chapter 7. Although the focus is on sediment quality guidelines (SQGs), and the uncertainties outlined herein apply to SQGs, these uncertainties are often relevant to other sediment assessment tools (e.g., sediment bioassays, community structure evaluations).

Sediment Assessment Tools

A variety of tools in addition to SQGs are employed to assess the level of sediment contamination in aquatic habitats (Chapter 5). The historical emphasis on physical and chemical measures of contaminant effects that dominated the field of pollution biology until the 1970s has been replaced by an understanding that biological indicators of integrity are equally important. Empirical approaches for assessment of biological integrity in aquatic ecosystems generally involve combinations of analytical chemistry, toxicity testing, and in situ measures of community structure. The strongest evidence for a causal relationship between the level of contamination and important ecological impacts is obtained when these 3 approaches are used simultaneously. For example, the Sediment Quality Triad (SQT; Long and Chapman 1985; Chapman 1990, 1996) combines chemical measures of contaminants, sediment toxicity tests, and field assessments to characterize the degree of sediment impairment. Addition of bioaccumulation measurements to the SQT can assist in determining causality (Borgmann et al. 2001), as can the use of toxicity identification evaluation (TIE) procedures (Chapman 2000). Addition of biomarkers can provide valuable data on sublethal effects and can be used to identify causation (Ringwood et al. 1999). The strength of the SQT lies in the weight of evidence (WOE) approach and in its ability to discern direct toxicological effects from natural variation in habitat characteristics.

Field assessments of community structure are the most common in situ measure of sediment contaminant effects on benthic organisms. These assessments are based on the important assumption that composition and organization of communities reflect local environmental conditions and respond predictably to sediment contaminants. Typical community-level endpoints in sediment quality assessments include species richness, abundance of sensitive and tolerant species, and functional organization. One of the key challenges in relating levels of sediment contamination to changes in community composition is to quantify the direct effects of sediment characteristics on these measures of community structure. Many of the same features that influence contaminant bioavailability and toxicity in sediments (e.g., substrate composition, grain size, percent organic carbon) also influence community composition. Experimental approaches in which these sediment characteristics are manipulated can provide information on their relative importance. In situ experimental approaches can also be employed to quantify the relative importance of water quality and sediment quality in systems subjected to multiple stressors. Measuring colonization of contaminated sediment or conducting community-level experiments with macroinvertebrates allows researchers to separate the direct effects of sediment contaminants from those associated with poor habitat or degraded water quality (Chapter 4). Natural stressor gradients and regression analyses can assist in determining thresholds of effects related to both sediment and water column stressors (Hickey and Clements 1998; Hickey and Golding 2002).

The second major tool used to assess sediment quality in marine and freshwater ecosystems is the sediment toxicity test. Standard protocols for conducting sediment toxicity tests have been developed to reduce uncertainty, provide consistent procedures for assessing bioavailability of contaminants in sediment, and improve reliability. These protocols specify appropriate test conditions, organisms, test duration, and other features appropriate for assessing the quality of marine, estuarine, and freshwater sediments. The uncertainty associated with sediment toxicity tests may be reduced by conducting experiments with ecologically relevant organisms and by measuring ecologically meaningful responses.

Ultimately, test endpoints measured in routine sediment toxicity tests (e.g., growth, mortality) need to be related to patterns of community structure observed at contaminated sites in the field. Similarly, concentrations of contaminants measured in benthic macroinvertebrates from laboratory bioaccumulation tests may correlate with levels observed in the field (Chapter 11). Although sediment toxicity and bioaccumulation tests are generally conducted with surrogate organisms, evidence for a causal relationship between the presence of sediment contaminants and community responses in the field is strengthened if tests are conducted with indigenous organisms from those communities.

Complete sediment assessments must consider the temporal framework in which the potential exposure occurs (i.e., not just present conditions but also future expectations). Chapman et al. (1997) referred to this as retrospective versus prospective assessment. Retrospective assessment refers to an evaluation of the risk of contamination given current conditions and resource use patterns. Prospective assessments, on the other hand, consider future conditions and resource use patterns. The predictive nature of this type of approach requires input from stakeholders and the public. While the proper assessment tool or framework may remain constant regardless of the temporal frame chosen, the appropriate interpretation of the data generated may change dramatically. Obviously, the potential impacts on either remedial or regulatory programs can be substantial, but this type of forecasting is the only way historical contamination can be properly managed.

SQGs and other sediment assessment tools typically provide a reference to some desired situation. Earlier Society of Environmental Toxicology and Chemistry (SETAC) workshops on sediment contamination noted the critical role that SQGs, reference sites, reference ecotoxicity data, reference benthic communities, and reference sediments play in decision making (Dickson et al. 1987; Ingersoll et al. 1997). A critical determination is which type of reference and reference comparison is most useful and appropriate, including the need to determine (to the extent possible) what benthic community should be present, given factors unrelated to contamination. The identification of a reference condition is particularly complicated in assessments of benthic systems that have been highly modified by human activity, such as ports and harbors, channelized streams, and hardened shorelines, or where development has essentially eliminated upland habitat (Chapter 7).

Unfortunately, very little is known about the nature of the natural benthic communities in the systems most prone to contamination, including natural variability. The most widely applied assessment protocols are for small streams and rivers (Plafkin et al. 1989; Barbour et al. 1999), not the large, slow-moving rivers, estuaries, lakes, and wetlands where industry is likely to have produced widespread sediment contamination (USEPA 1997). Field studies and assessments are limited by inadequate reference sites, incompatible habitat types, and sampling error; in addition, many metrics are based more on number theory than on ecology (Washington 1984). The potential for error is even higher when one attempts to apply bioassessment protocols to large or diverse systems where there is known variability in abiotic factors such as salinity, temperature, vegetation, elevation, or geomorphology. Guidance on addressing these uncertainties is provided in Chapter 7.

Physical and Chemical Factors

In addition to uncertainties related to the nature of contaminant mixture interactions, there are 3 major physical and chemical sources of uncertainty associated with current SQGs:

1) effects of dynamic conditions leading to disequilibrium,
2) variability in critical sediment-normalizing constituents (e.g., acid volatile sulfide [AVS], particulate organic carbon [POC]), and
3) desorption resistance and limited availability of organic contaminants partitioned to organic carbon phases. Each of these uncertainties is discussed in the following sections.

Dynamic conditions

Sediment assessments employing SQGs, especially equilibrium partitioning (EqP)-based SQGs, and other tools assume that both sediment and contaminant conditions are time independent and at equilibrium or near equilibrium (Chapter 3). Such assessments are inherently incapable of addressing problems governed by dynamics of a time scale comparable to the times scales implicit in SQG measurements. In addition, sediment assessments typically indicate conditions at a particular location and time. They do not indicate how conditions might change in the future. Thus, for example, comparison of sediment conditions to SQGs may not provide an indication of the future response of the system to remedial efforts. However, because SQGs and other assessment tools typically represent a snapshot in time, it is a failure of the experimental design, not of the tools, if dynamic conditions and temporal changes (i.e., disturbances) are not addressed.

Sediment fate and transport

Sediment contaminants of interest are generally strongly associated with sediment particles. Hence, a first-order approximation to the dynamics and fate of contaminants is the dynamics and fate of the sediment particles. Movement of sediment that results from currents, tidal fluctuations, waves, and storm events will also move contaminants into the water column where exposure to filter-feeders, fish, or other organisms may occur directly or through release to the dissolved phase. These types of dynamic conditions are not considered under the assumptions of many SQGs. Consequently, the dynamics and spatial variability of these processes may lead to wide variability and ambiguity in interpreting assessments by SQGs and other sediment assessment tools.

Given the potential uncertainty introduced by dynamic conditions, it is critical to have means for measuring the magnitude of their effects on sediment conditions. Crude assessment of sediment erosion over time can be accomplished using bathymetric measurements, but accuracy is limited because of the lack of reproducibility of location measurement. Radionuclide tracers such as ^{137}Cs, ^{7}Be, and ^{210}Pb can be useful indicators of sediment movement and depositional processes, but significant sediment reworking that results from storms or other events may complicate interpretation. Sediment traps can also be used to infer rates of deposition of sediment.

Prediction of erosion and deposition in noncohesive sediment systems is relatively well understood (Reible et al. 2002). Unfortunately, contaminants are normally more strongly associated with cohesive, fine-grained sediments. For cohesive sediments, the friction and energy required to produce particle motion is significantly larger for a particular particle size than would be expected from noncohesive sandy sediments. Because of the very fine particle size in cohesive sediments, however, the energy required may represent relatively low velocities. The property of cohesiveness is a complicated function of particle size, bulk density, mineralogy, organic content, and salinity. These properties vary significantly with position and time. Often, because of the lack of sufficient data on the deposit properties with position and time, these variations are not fully incorporated into sediment transport models.

Erosion in cohesive sediment beds depends on shear stress and particularly on water velocity. The strong dependence on stream velocity emphasizes the critical need for high-resolution modeling of stream hydraulics when sediment transport is being predicted. A necessary precursor to sediment transport modeling and understanding is a detailed description of the hydraulics of the system. Even given our understanding of stream hydraulics, it is not yet possible to predict erosion without bed measurements. Deposition varies significantly with time and space because of the effects of bed shear stress (inversely related to deposition), cohesiveness of the depositing sediment, and flocculation processes.

Although a priori prediction is not yet possible, erodibility measurements with shear stress and sediment density can be obtained using experimental techniques such as Sedflume (McNeill et al. 1996) or particle entrainment simulator (PES; Tsai and Lick 1986). In the Sedflume apparatus, the sample to be evaluated is raised continuously to maintain a constant-level bottom surface as the sediment erodes. The fundamental constitutive relationships developed from such measurements can be used in realistic sediment transport models. Extensive measurements are needed to support high-resolution models of sediment transport that account for variability in the physical properties of the sediment beds with space and time. In addition, the association of contaminants with particles can be used to predict likely depositional areas in estuaries and increases in contaminant concentrations with time (Morrisey et al. 2000; Williamson and Morrisey 2000).

Contaminant fate and transport

Sediment transport only partly describes contaminant migration. Contaminant release from resuspended sediment, migration, and fate mechanisms from stable bed sediment are also critical processes for defining exposure in the real world. Diffusion, advection, and bioturbation, as well as the distribution of contaminant sources, result in the heterogeneous distribution of sediment contaminants (Valette-Silver 1993) also contributing to nonequilibrium conditions. Because of the slower dynamics of contaminant transport and fate processes in stable bed sediments, SQGs and other tools are inherently more valid in such systems. Dynamic conditions, however, may preclude effective use of SQGs and other tools. Further, there remain problems such as prediction of the long-term response of the system to remedial efforts. The following discussion emphasizes real-world processes affecting hydrophobic organic contaminants, but many of the processes are also relevant to inorganic stressors, including metals, ammonia, and hydrogen sulfide.

Under conditions of minimal sediment transport, contaminant migration to the overlying water is likely controlled by the presence of benthic organisms that process surficial sediments (Reible et al. 1991). Sediment processing by animals residing in the upper layers includes burrowing, ingestion and defecation, tube building, and biodeposition. Taken together, these processes are termed "bioturbation." The net result of bioturbation is the vertical and horizontal movement of sediment particles and pore water. Bioturbation influences the variability of sediment characteristics such as texture, organic carbon, redox potential, and AVS. Here, we focus on the direct effects of bioturbation on encouraging nonequilibrium conditions of contaminants. Contaminants on the particles or in the pore spaces are transported in the bioturbation process. Especially important is particle transport for hydrophobic contaminants that are heavily retarded by porewater processes (i.e., under ordinary circumstances, very slowly transported from pore waters). Worm tubes and other macroscopic animal burrows can also significantly enhance contaminant transport by advection across the sediment–water interface. In addition, direct ingestion of sediment deposits can lead to rapid transport of sediment and associated contaminants to the surface (Reible et al. 1996).

The presence of seasonal variations in biomass density and organism activity suggests that the dynamic changes caused by organisms in the surficial sediment layers will change dramatically over the course of the year. Thus, the depth of oxygenated sediments and the reducing conditions influencing parameters such as AVS may also change with seasonal conditions. In addition, horizontal variability in biomass density and organism activity will ensure horizontal variations in these same parameters.

Effective biodiffusion or exchange coefficients that describe bioturbation are crude approximations at best because they do not separate the various modes of particle movement exhibited by an organism. In addition, because they are normally measured with strongly sorbing contaminants, porewater pumping, or the effect of bioturbation on other diffusive and advective processes (i.e., creation of secondary porosity, changes in texture or permeability) are not included. Effective biodiffusion coefficients estimated by radionuclide tracers of particle mixing rates collected by Thoms et al. (1995) indicate that particle diffusion rates were typically in the range of 0.3 to 30 cm^2/y and limited to surficial sediments. More than 90% of the 240 observations of bioturbation mixing depths in both freshwater and saltwater reported by Thoms et al. (1995) were 15 cm or less, and more than 80% were 10 cm or less. The vast majority of these measurements of effective bioturbation diffusivities were made by estimating particle reworking rates using strongly sorbed radionuclides. The time and amounts of particular radionuclide release and their current distribution within sediments are an indirect measurement of the reworking rates of those sediments. A variety of radionuclides can be used to assess particle reworking rates on various time scales. Long-lived species such as ^{137}Cs and ^{210}Pb can be used to estimate long-term average reworking rates on the timescale of years, while isotopes such as ^{7}Be can be useful for estimating short-term rates (i.e., days) of transport from the overlying water (Aller and Cochran 1976; Nittrouer et al. 1979; Rice 1986; Bentley and Nittrouer 1997, 1999).

Advective processes such as porewater seepage or gas bubble migration create the best potential for contaminant migration in the deeper layers of sediment and in the upper layers when neither bioturbation nor sediment erosion are significant. Advection can also explain the flux of contaminants from sediments into the overlying water. Groundwater seepage can change porewater geochemical properties. Seasonal variability or heterogeneity in groundwater flows can thus also cause changes in conditions such as redox potential and sulfide reduction of metals. Groundwater seepage, however, is very difficult to measure and subject to significant variability. Hydraulic gradients between the surface and groundwaters below or adjacent to the body of water are relatively easily measured with piezometers. Variations in the depth and texture of sediments, however, can dramatically affect permeability and give rise to significant variations in response to those gradients. More difficult to assess is the advective flow in high-energy systems associated with pressure variations over an uneven bottom surface (Savant et al. 1987; Elliot and Brooks 1997; Packman and Benacala 2000; Work et al. 2002). Local pressure variations of the order of 100 to 1000 Newtons per square meter (N/m^2) can be observed between the upstream and downstream faces of the typically triangular-shaped, dune-like sediment structures that form at the sediment–water interface. Because advective processes are operative only in the pore water, contaminants are subject to retardation by sorption onto the adjacent immobile sediments. For strongly sorbing contaminants (i.e., $K_{ow} > 5.0$), this limits the significance of advective processes relative to bioturbation or sediment erosion. The use of seepage meters (i.e., containers covering a portion of the sediment bed that collect water that seeps through the sediment) is common. Net seepage measurements, however, do not fully describe contaminant transport in that the bulk of any observed seepage can be through sandy portions of the bed that contain little or no contamination. A means of detecting slow vertical transport by groundwater flow is through tracers as described by Cornett et al. (1989). Chemical tracers in the overlying water have also been used to estimate the net effect of seepage on hyporheic zone contaminant transport (Bencala 1984).

The final real-world transport process that occurs in contaminated sediments and that may introduce dynamic conditions and uncertainty into the application of SQGs and other tools is diffusion. Diffusion of contaminants is migration associated with random molecular motion. Diffusive processes are ubiquitous but extremely slow. In areas not influenced by significant advective processes or particle resuspension and bioturbation, however, diffusion may be the most important contaminant migration process. As with advection, diffusion is retarded by sorption of contaminants onto adjacent sediment particles. Unlike other sediment transport processes, it is possible to describe diffusion with a relatively high degree of confidence, limited only by the knowledge of the chemical concentrations and local concentration gradients and the sorption–desorption characteristics that contribute to retardation. Tools for the measurement of various sediment and chemical fate and transport processes are summarized in Table 17-1.

Table 17-1 Field measurement and analytical tools for assessing sediment and contaminant transport (adapted from Apitz et al. 2002)

Process	Tools
Sediment transport	• Measure in situ sediment depths over time (e.g., bathymetric surveys) • Evaluate historical sedimentation using ^{210}Pb, ^{137}Cs, ^{7}Be, or similar dating of sediment cores; pollen or chemical layers (Hume et al. 1989) can also be used • Develop a hydraulic sedimentation model to characterize sedimentation
Sediment transport under storm conditions	• Conduct storm event sediment and water column sampling (e.g., grab samples, sediment traps) • Measure sedimentation and sediment scouring processes under differing flow conditions (e.g., bathymetric surveys, sediment traps) • Measure critical shear stress with in situ or laboratory flumes
Vertical diffusive flux	• Measure diffusive flux for surface sediments using laboratory column studies • Measure diffusive flux for exposed and buried sediments using in situ benthic flux chambers
Vertical groundwater advection and hyporheic zone advective transport	Perform hydraulic studies: • dye studies • Piezometer studies • seepage flux meters • multilevel pressure transducers • radon isotope studies
Sediment mixing through bioturbation	• Use sediment age dating analyses (e.g., ^{210}Pb) to assess bioturbation • Assess macroorganisms and their reported mixing depths • Use underwater photography to assess surface benthic activity

Variability in critical normalizing constituents

The above dynamic, real-world, physical and chemical, nonequilibrium conditions result in the transport of sediments and contaminants. Simultaneously, these same processes contribute to the heterogeneous distribution of critical normalizing constituents in space and time. While numerous sediment constituents that affect bioavailability, this discussion focuses on 2 of the most critical: AVS and POC as used in the derivation of EqP-based estimates of metal and organic contaminant bioavailability, respectively (Di Toro et al. 1991; Ankley et al. 1996; Chapter 13). Diffusion, advection, and bioturbation all contribute to the heterogeneous distribution of sediment contaminants. Contaminant source will also contribute to the heterogeneous distribution of contaminants in sediments.

The important issue raised by the variation in normalizing parameters is as follows: In general, sediment contaminant concentrations, such as Cd and polycyclic aromatic hydrocarbons (PAHs), and normalizing parameters, such as AVS and organic carbon, are collected simultaneously in the field. In the laboratory, contaminant concentrations and constituent levels in these samples are measured and analyzed to make estimates of contaminant bioavailability. What has been observed on some occasions is that the distributions of organic carbon and, in particular, AVS can vary significantly more than contaminant concentrations. At best, this variability creates an impression of limited usefulness for the normalizing constituent. At worst, it suggests a causal disconnect between the constituent and the hypothetical mechanism that explains bioavailability (Luoma et al. 2001).

Distributions of AVS and POC as well as total organic carbon (TOC) are known to vary with depth in contaminated sediments. For example, studies by Leonard et al. (1993), Van den Berg et al. (1998), Boothman et al. (2001), and Wijsman et al. (2001) demonstrated the variability of AVS in sediment cores from freshwater and marine systems (Figure 17-1A). POC also varies with depth, although freshwater POC concentrations vary to a lesser degree than POC or AVS concentrations in either freshwater or marine sediments (Figure 17-1B; Burgess et al. 1996; Tenzer et al. 1999). The vertical variability in both AVS and POC raises the following question: At what depth should sediments be sampled to accurately assess metal and organic contaminant bioavailability? The answer to this question depends on the purpose of the sediment assessment and the depth to which organisms of interest might interact with sediments.

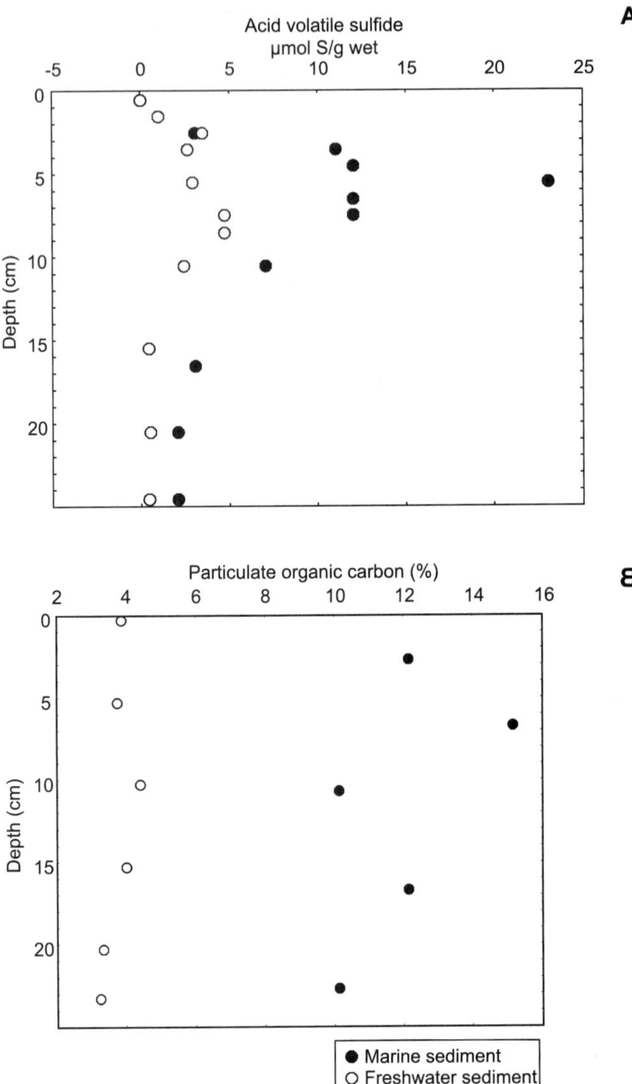

Figure 17-1 Concentration of (A) acid volatile sulfide (AVS) and (B) particulate organic carbon (POC) with depth in fine-grained marine and freshwater sediment cores (adapted from Howard and Evans 1993; Burgess et al. 1996; Tenzer et al. 1999; Wijsman et al. 2001)

More significant than sediment depth, with regard to sampling AVS, is seasonality. Several studies have shown that AVS concentrations at the same sites vary significantly between seasons (Howard and Evans 1993; Leonard et al. 1993; van den Hoop et al. 1997; Grabowski et al. 2001). Generally, because of the biogeochemical processes that form AVS in sediments, AVS concentrations are elevated during the summer and are lowest during the winter. For example, Grabowski et al. (2001) reported coefficients of variation (CVs) for AVS and simultaneously extracted metals (SEM) concentrations measured in the spring and summer at 6 sites along the Mississippi River. The CVs for AVS ranged from 115% to 140% between seasons, while SEM CVs ranged from only 5% to 37% (Figure 17-2). Similarly, Ringwood and Keppler (unpublished data, USEPA STAR, Grant No. R826201) measured the concentrations of AVS and TOC in sediments from 3 stations over 3 years in Rathall Creek, Charleston Harbor, South Carolina (USA)(Table 17-2). Over time, they observed concentrations of TOC from the same site ("A" in Table 17-2) had a CV of 66%, while AVS had a CV of 139%. Spatially, TOC varied by 83%, and AVS varied by 146%. These levels of variability between seasons and at the same site and stations over time indicate a lack of constancy in AVS and TOC concentrations that, without proper experimental design (i.e., more

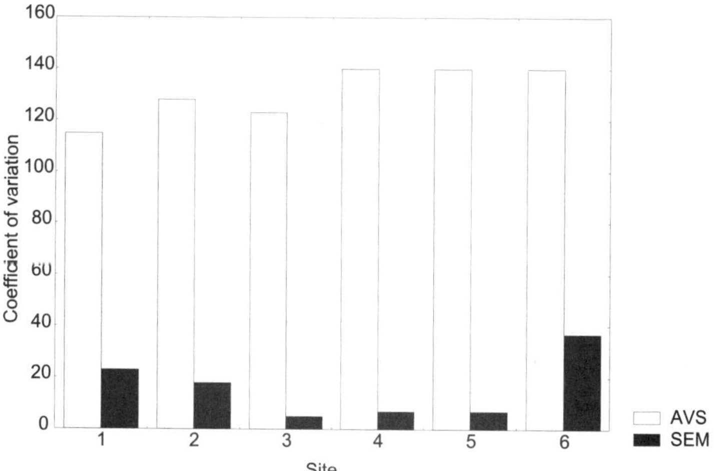

Figure 17-2 Coefficient of variation (CV) for sediment AVSs and simultaneously extracted metals (SEMs) in fine-grained sediment core samples collected in spring and summer at several freshwater sites along the Mississippi River (adapted from Grabowski et al. 2001)

Table 17-2 Concentrations of total organic carbon and acid volatile sulfides in sediments from 3 stations collected over 3 y from Rathall Creek, Charlestown Harbor, South Carolina (USA) (Ringwood and Keppler, unpublished data)

Year	Station	Total organic carbon (TOC) (%)	Acid volatile sulfide (AVS) (μmol/g)
1998	A	0.44	0.37
1999	A	1.76	11.7
2000	A	0.85	1.43
2000	B	0.90	1.12
2000	C	3.30	21.5

than just a "snapshot in time"), could contribute ultimately to poor predictions of bioavailability and inaccurate derivation of SQGs. This is especially true when, as shown by Grabowski et al. (2001), the levels of the contaminants of interest do not vary to the same degree. Other authors have also reported concerns with horizontal variability in AVS concentrations (Mackay and Mackay 1996; van den Hoop et al. 1997), and De Witt et al. (1999) found marked changes in AVS concentrations with manipulation and during a 10-d sediment toxicity test.

Despite the observed variability in AVS and TOC with depth, location, and season, these sediment-normalizing constituents can be useful for predicting the bioavailability of some contaminants during sediment assessments. In a sediment assessment where the distributions of metals and organic contaminants with depth and in space are being determined, AVS and TOC should also be measured. However, as noted in Chapter 7, it is critical that the measurements of AVS and TOC are performed such that the variability in their distributions is well characterized (e.g., using power analyses to estimate the appropriate number of replicates or compositing to reduce overall variability). Well-characterized distributions of both constituents will greatly improve predictions of contaminant bioavailability. For example, the measurement of AVS is often recommended during the coldest times of the year in a given aquatic setting in order to capture the likely lowest concentrations and thus provide an estimate of metal bioavailability in the worst case.

Desorption resistance and limited availability of organic contaminants

Many of the current SQGs estimate the bioavailability of organic contaminants on the basis of a 2-phase model. Other sediment assessment techniques, including toxicity tests and bioaccumulation studies, may also implicitly be based on a 2-phase model. This conventional model defines the 2 phases as the organic coating surrounding sediment particles and the dissolved phase (Chiou et al. 1979; Karickhoff et al. 1979). Based on this model, organic contaminants in the dissolved phase (i.e., interstitial water) are considered bioavailable, while contaminants in the particulate phase are considered non-bioavailable (Di Toro et al. 1991). In the last decade or so, several studies have shown that hydrophobic contaminants exist in a system composed of more than 2 relatively simple phases. Rather, the aquatic environment consists of multiple phases, including colloidal carbon, glass-like (hard) and rubbery (soft) particulate carbon, and soot carbon (Luthy et al. 1997) (Figure 17-3). At this time, the magnitude of the effect of these other partitioning phases on estimates of bioavailability, and consequently SQGs, is not fully understood. Some studies have shown that conventional estimates of hydrophobic contaminant partitioning and bioavailability do not appear to be impacted to a large degree by these phases (DeWitt et al. 1992; Burgess et al. 2000). Despite these studies, there is sufficient evidence to indicate that, in order for SQGs to be effective in as many applications as possible, it is critical for them to be as accurate in their depiction of partitioning and bioavailability as possible. Consequently, where it is likely that more than 2 phases are present, SQGs and other sediment assessment tools should incorporate the other phases.

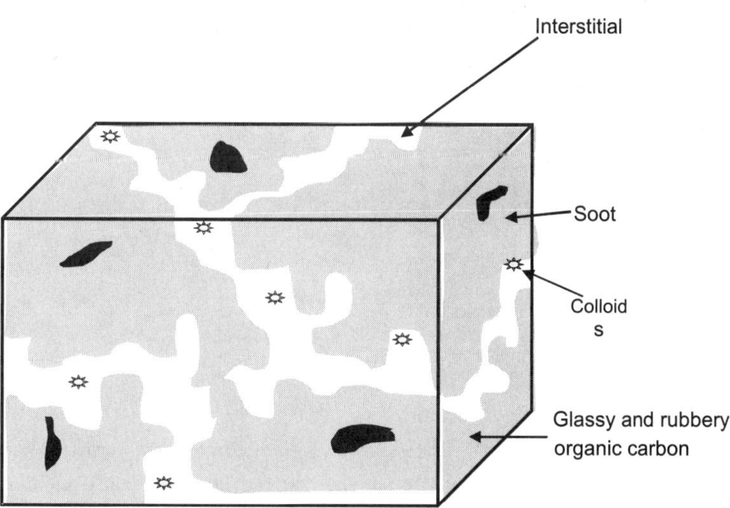

Figure 17-3 Conceptual illustration of recognized sediment phases affecting the partitioning of organic contaminants (adapted from Burgess et al. 2004)

Specific effects of these additional phases vary, but arguably the most critical with regard to the derivation of SQGs and other sediment assessment tools is the effect on hydrophobic contaminant desorption. Desorption from the particulate phase is a critical process because it regulates how much contaminant will be present in the dissolved phase and thus bioavailable (but note that the dissolved phase is not the only bioavailable phase in sediment). The results of many laboratory and field observations indicate that a significant fraction of sediment-bound organic contaminants do not desorb linearly and reversibly as suggested by conventional 2-phase partitioning models, are not biodegraded, and are difficult to remove by extraction with surfactants or solvents. For example, Pereira et al. (1988) found that the concentrations of halogenated hydrophobic contaminants in native water, suspended sediments, and biota were far below the values predicted with respect to concentrations in the contaminated sediments collected from Bayou d'Inde, Louisiana. Similarly, McGroddy and Farrington (1995) and Readman and Mantoura (1987) observed that a fraction of PAHs in harbor and river sediments was not available for partitioning compared to levels predicted by 2-phase partitioning models. For PAHs, Ghosh et al. (2000) related highly sorbed concentrations and high partition coefficients to the presence of soot carbon in sediments. Similar findings have been reported along with mechanisms for incorporation of both PAH and polychlorinated biphenyls (PCBs) into soot-like particles by Jonker and Koelmans (2002). In addition, it is common for hydrophobic contaminants to persist in sediments even when conditions are appropriate for biodegradation to occur (Connaughton et al. 1993; Kan et al. 1994; Hatzinger and Alexander 1996). For most of these sediments, contaminant input had ceased for periods of many years, yet the sediment-bound contaminants persisted over decades without significant concentration reduction or hydrocarbon fingerprint loss.

Desorption, as well as sorption, of hydrophobic contaminants to soils and sediments is a complex process, given the diversity, magnitude, and activity of chemical species, phases, and interfaces. In addition to the hypothesis that desorption resistance is associated with the different character of the sediment organic carbon (Luthy et al. 1997; Ghosh et al. 2000), it has been postulated that the observed phenomenon is due to the occlusion of contaminants from desorption by cooperative conformational changes of the organic phase during the sorption process (Kan et al. 1994, 1997, 1998; Hunter et al. 1996). The conformational rearrangement of the solid organic matter in the presence of sorbed chemicals could cause the chemical environment of the adsorbate to be different and, hence, be the source of desorption resistance and hysteresis (i.e., the relationship between contaminant sorption to and desorption from the sediment are different when, in theory, they would be expected to be similar). At least one hypothesis has stated that the quality of organic carbon determines the degree of hysteresis in sorption desorption experiments (Huang and Weber 1997), suggesting that the magnitude of the desorption resistance can be related to organic carbon age. An alternative hypothesis is that "apparent" hysteresis is only the result of slow desorption (Lick and Rapaka 1996; Pignatello and Xing 1996). Field sediments may have been contaminated for decades, allowing time for sorption into phases or portions of sediments that allow for only slow desorption. Because, during a toxicity test or bioaccumulation study, the period of desorption can be very short compared to the period of sorption, the net effect is observation of an apparent hysteresis that does not reflect true equilibrium.

Regardless of the mechanism of desorption resistance and whether it is rate or extent (equilibrium) limited, recent work has shown that the interstitial water exposure paradigm remains a good model for prediction of bioavailability (Lu et al. 2003), that is, the amount of hydrophobic contaminants that can accumulate in plants and animals in intimate contact with the sediment and interstitial water is controlled by partitioning into the interstitial water. Lu et al. (2003) have shown that reductions in steady-state biota–sediment accumulation factors (BSAFs) in a freshwater oligochaete resulting from desorption resistance can be related to interstitial water concentration reductions resulting from the higher-than-expected sediment–water partition coefficient in these systems. The result is consistent with a 2-stage process for the uptake into lipid of an organism: first, partitioning from a solid phase into either interstitial water or worms' gut juices, the extent of which is hindered by desorption resistance; second, partitioning into the lipid from interstitial water or worms' gut juices, which is not hindered by desorption resistance.

Thus, the apparent increase in partition coefficients associated with the desorption-resistant hydrophobic contaminants reduces the normalized uptake into an organism. This is a direct consequence of the reduction in interstitial water concentrations relative to the sediment loading that is employed in the definition of the BSAF. A comparison of this model of reduced bioavailability to selected laboratory and literature data for 3 PAHs can be found in Figure 17-4 (Lu et al. 2003). Gomez-Hermosillo et al. (2004) have shown similar behavior in wetland plants (Figure 17-5). These results suggest that the EqP approach used in deriving SQGs can be a useful tool, but that in desorption-resistant systems, it is important to recognize that the effective partition coefficient is not that given in many tabulated sources that assume a 2-phase model. In such cases, either revised models or actual measurements of interstitial

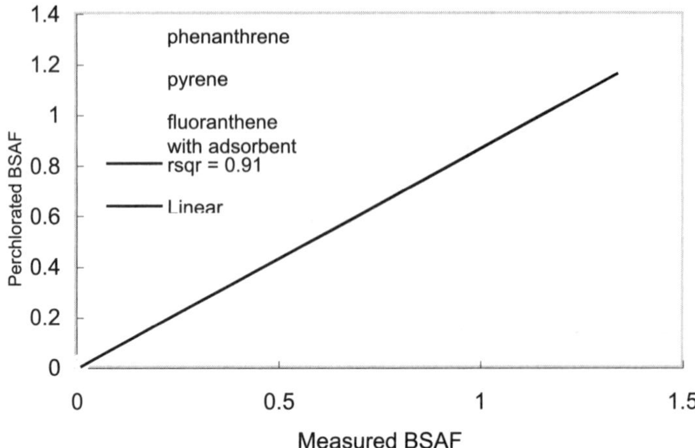

Figure 17-4 BSAFs for 3 PAHs predicted from the partition coefficient versus measured BSAFs (adapted from Lu et al. 2003)

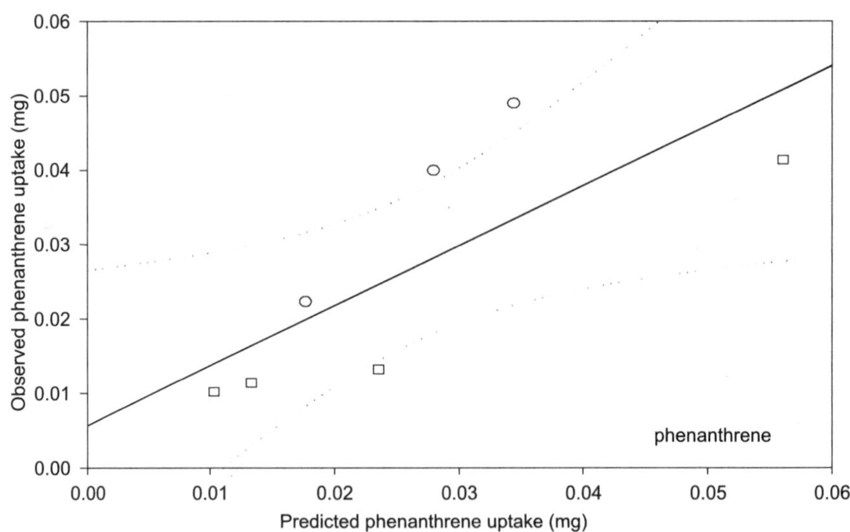

Figure 17-5 Comparison of wetland plants (*Salix n.* and *Scirpus o.*) accumulation from desorption resistant sediments to a model prediction based upon interstitial water concentrations of phenanthrene (adapted from Gomez-Hermosillo et al. (2004)

water concentrations or the partition coefficient must be employed to assess bioavailability and accumulation in benthic organisms. Site-specific BSAFs are one possible way to account for site-specific availability concerns.

Biological Issues

Protecting populations and communities

Protecting populations and communities requires appropriate measurement endpoints relative to assessment endpoints, understanding natural variability relative to anthropogenic and other stressors and knowledge of exposure routes at the individual and population or community levels.

Appropriate measurement endpoints

Typically, sediment toxicity tests are conducted with single species and the results extrapolated to populations and communities. Single-species tests are clearly not ideal. Microcosms and mesocosms are more similar to the environment but have their own problems (Ingersoll et al. 1997). The endpoints of death, growth, and reproduction in single species tests can provide information on the ability of populations or communities of similarly sensitive organisms to survive, persist, and prosper in the presence of the tested chemicals. This conservative approach is based on the presumption of accumulated effects but needs to be based on at least the potential for population or community-level impacts (Chapman, Ho, et al. 2002).

However, the utility of such information depends on testing "similarly sensitive organisms" to those present in the environment. There are 2 components to such similarity: same taxon and same tolerance. The latter component is not necessarily taxon specific. Laboratory test organisms tend to be naïve, whereas the same or similar organisms in the real environment can be more tolerant to some chemicals. Tolerance or resistance to the toxicity of chemical contaminants may be due to 4 mechanisms, which in the case of metabolism and genetic tolerance may not have a metabolic cost (Table 17-3). Because it is highly unlikely that laboratory testing, whether involving

Table 17-3 Tolerance mechanisms in biota

Tolerance mechanism	Metabolic cost?	Representative references
Acclimation	Yes	Millward and Klerks (2002); Weis (2002)
Nongenetic adaptation	Yes	
Genetic adaptation	Possibly not	Barata et al. (2002)
Metabolism	Possibly not	Lovell et al. (2002)

single species or multiple species, can duplicate field conditions, measurement endpoints derived from such testing, or even from field testing, comprise only one component of a WOE evaluation (Burton et al. 2002; Chapman, MacDonald, Lawrence 2002; MacDonald and Ingersoll 2002). And predictive linkages are required "to extrapolate between what is being measured in toxicity tests and the ecological responses of assessment endpoints in the environment" (Chapman, Ho, et al. 2002).

Understanding natural variability

Biological communities are naturally variable. Multiple, alternate steady states that are not readily predictable can occur in nature (May 1977), and some communities are naturally unstable (Mills 1969). Natural variability can be regular in pattern or irregular and unpredictable, and can occur because of biotic and abiotic disturbances (Paine and Levin 1981) including but not restricted to periodic anoxia (Seliger et al. 1985), variations in sediment type (Poore and Rainer 1979), food pulses (Josefson et al. 1993), temperature (Levings 1975), salinity (Holland 1985), bioturbation (Eagle 1975), storms, waves, and currents (Hall 1994). Regular patterns are generally seasonal, often a result of faunal depletion due to abiotic factors such as salinity (Holland et al. 1977; Chapman and Brinkhurst 1981) or biological cycles (Beukema 1974). Patterns of variation can vary from abrupt to gradual and may be relatively subtle. For instance, Lie and Evans (1973), in a study of the long-term variability (over 7 y) of subtidal marine benthos in Puget Sound (WA, USA), found that although species composition remained relatively constant, relative dominance varied. Similarly, Buchanan et al. (1974) found that subtidal marine benthos off the Northumberland Coast (UK) had approximately the same number of species over a 4-y period, but the number of individuals more than doubled. Seasonal recruitment of juvenile biota is a major variable in many environments.

Regular seasonal changes are predictable and need to be recognized as occurring. "Snapshot" assessments provide no information on such changes and should be used only where previous studies have established the level of natural variability. For instance, mudflat communities in San Francisco Bay (CA, USA) demonstrate seasonal changes but persist because of recurrence of minor abiotic disturbances (sedimentation, salinity) on time scales comparable to life cycles, life history strategies that are opportunistic and permit continued colonization, and/or rapid recolonization (Nichols and Thompson 1985).

Other changes are not predictable, and the reason for their occurrence is not easily determined. For instance, van de Koppel et al. (2001) demonstrated a positive feedback mechanism between primary producers (diatoms) and their environment (silt erosion) in an intertidal environment and noted that such feedback mechanisms can result in "dramatic and unexpected reactions of ecosystems to small disturbances," thus reducing our ability to predict the behavior of specific ecosystems on the basis of physical or environmental parameters. Researchers must be aware of the possibility of such changes when conducting their investigations, tailoring their experimental design to address natural variability and/or ensuring that any snapshots in time are representative.

Exposure routes

Exposure routes to sediment contaminants comprise whole sediments, pore water, and overlying water. Overlying water may be a more important exposure route than pore water for some contaminants (Warren et al. 1998), depending upon sediment binding, which can vary with depth in the sediments (Borgmann and Norwood 2002). For example, water column species such as salmonids are influenced by sediment contaminants via this exposure route (Stehr et al. 2000). Exposure routes from sediment to overlying water can be facilitated by bioirrigation and bioturbation. Such processes can also "detoxify" sediments by decreasing sediment sulfides concentrations and increasing oxygen concentrations, thus preparing sediments for further colonization (Modig and Ólafsson 2001). The effects of contaminated sediments on benthic colonization or recruitment can be assessed by small-scale manipulative field experiments (Roach et al. 2001; Chapter 4).

In other cases, dietary uptake may be more important than water (pore or overlying) for some species. For instance, sediments are a direct source of metals to sipunculid worms (Yan and Wang 2002) and to at least some amphipods (Wiklund and Sundeline 2002).

However, contaminant uptake routes can change. For example, Yunker et al. (2002) demonstrated a shift in chlorinated dibenzo-*p*-dioxin and dibenzofuran uptake by marine biota near pulp mills, from pelagic to benthic food webs. Such shifts emphasize the importance of realistic assessments involving initial and evolving conceptual diagrams that document exposure routes for key taxa.

Physical habitat and biological heterogeneity

Physical characteristics that are highly variable include the rate and extent of deposition and the resulting texture, porosity, and permeability of the sediments. Sediments will accumulate in depositional areas, for example, in pools of a pool-riffle environment and at the mouth of a river or estuary.

Physical tides and to some extent, wind-induced waves create a regular flow of water across the sediment–water interface, causing the resuspension of sediments at various levels of intensity. Storm events or floods cause unpredictable motions of water across sediment surfaces, also resuspending sediments; the magnitude of resuspension events is a direct result of the size of the storm or flood (Kennet 1982). Scouring in high-flow areas will limit accumulation of sediment and sediment-associated contaminants.

Anthropogenic activities such as navigational and remedial dredging contribute to the further potential for sediment resuspension. In principle, uncertainty caused by physical variables such as wind, water, and tides increases in proportion to the energy added to the system.

Physical characteristics also provide an indication of, and are partly responsible for, biological heterogeneity. Fine-grained sediments will tend to contain larger amounts of organic material, which not only control the extent of sorption of hydrophobic organic contaminants but will also attract deposit-feeding benthos seeking the organic material. Coarse-grained sediments may also attract specific species of benthos that will exhibit different modes, intensity, and depths of sediment interaction.

Organisms do not distribute themselves homogeneously across or within the sediment. Further, the presence of organisms creates several potential effects that increase heterogeneity in sediment systems (Parsons et al. 1984). For instance, a surface that may have remained relatively flat or slightly furrowed by wave action is instead permeated with burrows, comprises piles of sediments and pseudofeces, and may also contain structures rising above the surface. All of these alter the flow of water across the sediment surface, creating microhabitats and high energy and deposition zones (Valiela 1984). Through reworking of the sediment (e.g., irrigation, bioturbation), organisms change sediment density, water saturation, and cohesion (e.g., porosity and permeability), which further increase overall heterogeneity (Rhoads et al. 1978; Rhoads and Boyer 1982). Organism life histories contribute to the ways in which the benthos are affected: Infaunal, epibenthic, and tube-dweller life styles as well as level of mobility all contribute to sediment system variability.

The above organism-induced physical changes to the sediments lead to chemical alterations including variations in the redox boundary, which tend to move down into the sediment related to the presence of macrobenthic organisms (Lee and Swartz 1980). A deeper oxidation–reduction zone indicates the presence of more oxygen at depth and results in more microbial activity by aerobic microbes as well as release of some contaminants such as metals that may be bound under anaerobic conditions (e.g., AVS) in the sediments. Active populations of aerobic microbes enhance the mineralization of organic carbon, which further alters the sediment composition.

Organic contaminants may be metabolized by microbial communities in sediments, and such modification or removal constitutes a source of uncertainty. Contaminants vary in their susceptibility to aerobic or anaerobic metabolism and the affiliated microbial consortia. Chlorinated organics susceptible to anaerobic microbes include chloroethenes (Drzyzga et al. 2002), chlorinated dioxins (Beurskens et al. 1995; Vargas et al. 2001), and simple chlorinated aromatics (Myers et al. 1994). Not only is microbial activity likely to modify the mass of contaminants present, it may also alter the chemicals present. Methylation of inorganic mercury to organic methylmercury is one well-known example. Excessive microbial activity is likely to result in secondary issues, such as oxygen depletion.

The most dramatic effects resulting from the presence of organisms are bioturbation and bioirrigation (Levinton 1982; Valiela 1984). Bioturbating organisms transport large quantities of sediment into the sediment–water interface and surrounding sediments, further altering the sediment surface. Bioirrigating organisms tend to push material (water, sediment, contaminants) above the sediment–water interface. Both processes allow aerated surface water to penetrate deeper into the sediment mass, but diffusion outside of burrows occurs only where these are not lined and impermeable. Both bioturbation and bioirrigation further increase sediment heterogeneity as well as transport contaminants as described previously.

In summary, physical and related biological factors contribute to the lack of homogeneity in sediment systems. The resulting heterogeneity leads to variability that complicates measurements and predictions of the fate and effects of sediment contaminants. Thus, any attempt to assess sediment risks to human health and the environment must recognize the different environments in which contaminated sediments are found and address site-specific issues of heterogeneity. Risk is largely defined by the balance between transport processes that give rise to exposure and the fate and transport processes that attenuate that exposure. The relative importance of the natural processes leading to exposure and risk differ significantly between lacustrine, riverine, estuarine, and marine environments (Chapter 7). The range and significance of these natural processes are also heavily influenced by site-specific characteristics within these general environments. These characteristics may also vary significantly with a specific site.

Spatial and temporal scaling

Spatial and temporal scaling issues introduce significant uncertainty into sediment quality assessments and complicate our ability to relate contaminant levels to observed ecological effects. Understanding the influence of spatial and temporal scale on ecological processes is critical for predicting how communities will respond to contaminants in natural systems (Ellis et al. 2000; Hewitt et al. 2001; Thrush et al. 2000). Such understanding is particularly important both for establishing clear linkages between contaminants and effects (if such exist) and for extrapolating up to community and ecological effects. Investigators should consistently question if the physical, chemical, and biological processes they study are at the most relevant spatial and temporal scales. Although we generally assume that results of small-scale, short-term experiments conducted in the laboratory are relevant to natural systems, these assumptions are rarely tested (Cairns 1983, 1986).

Despite concerns over scaling issues, few studies have tested the hypothesis that experiments conducted at one spatiotemporal scale are appropriate for predicting responses at a different scale. Scale-dependent responses of aquatic organisms to contaminants in microcosms, mesocosms, and field recolonization experiments will complicate our ability to predict effects in natural systems. Perez et al. (1991) provide one of the few detailed analyses of the effects of spatial scale on community responses to contaminants in water and sediments. Results of microcosm experiments in which communities were exposed to kepone in 9, 35, and 140 L containers showed that fate and effects were size dependent. For example, the concentration of kepone in surface sediments and the potential exposure to benthic organisms increased with microcosm size because of greater mixing and bioturbation in larger test chambers. Based on these results, Perez et al. (1991) concluded that small microcosms would underestimate the effects of kepone on aquatic communities. The dependency of community responses on container size has obvious implications for ecological assessments and is central to the debate over the usefulness of laboratory test systems for assessing sediment contaminants.

The issue of spatiotemporal scale is equally important in field investigations. The spatial and temporal scales of anthropogenic stressors should coincide with the endpoints being measured in ecological assessments (Suter 1993), but often they do not. Samples collected at discrete points reflect conditions at those points, and information between locations is either smoothed over or considered background variation. For example, attempts to relate the incidence of liver disease to mortality in fish collected from PAH-contaminated sites will be difficult for widely distributed and highly mobile populations. Similarly, recruitment of benthic organisms to contaminated sites may occur at a broader spatial scale and may not reflect local conditions where samples are collected. The issue of spatial scale must be recognized and considered in experimental designs; it is especially relevant for analyses of physical, chemical, and biological conditions along longitudinal gradients in rivers. Recent studies of riverine landscapes suggest that our relatively poor understanding of the structure and function of lotic ecosystems is a result of conducting studies at the inappropriate spatial and temporal scale (Fausch et al. 2002). Similar concerns have been expressed by Chapman and Wang (2001) with regard to estuarine ecosystems. In contrast, seasonality in contamination and toxicity, though present, may be less common in littoral marine ecosystems (DelValls et al. 2002).

Regulatory Issues

Use of sediment quality guidelines as regulatory criteria

Originally, SQGs were envisioned as a means of protecting water quality, similar to water quality criteria (WQC), (i.e., as a set of readily available and widely applicable values that could be used to make management decisions and set goals for aquatic systems). For example, it was envisioned that SQGs could be translated into end-of-pipe effluent limits for wastewater discharges. Chemical-specific guidelines and sediment bioassays have, in fact, been incorporated into offshore oil and gas well wastewater permit requirements for coastal waters in Texas, USA (USEPA 2001a, 2001b).

Under the US Clean Water Act, USEPA has mandated that total maximum daily load (TMDL) calculations be made on waterbodies designated as impaired or threatened (e.g., that fail to meet designated uses or WQC on a consistent basis). For sediment contaminants, derivation of a TMDL can be problematic. SQGs exist for a limited number of contaminants and are generally based on empirical approaches. Some SQGs have been shown to accurately predict the point at which sediment will be either toxic or nontoxic in certain situations (Long et al. 2000, 2001). However, they were not designed to translate between the mass of a contaminant entering a water body and the precise concentration to which a particular contaminant must be reduced in order to ensure that the sediments do not result in adverse effects at a particular location. This is particularly true when there are many contaminants at varying concentrations across a particular site and when there are multiple diffuse sources throughout the system. Consequently, SQGs may fail to provide good predictions of what waste load reductions will achieve. Given the expense of the TMDL process and any resulting waste load reductions, the uncertainty of achieving satisfactory results is problematic.

Nor do empirical SQGs reflect the dynamics of water–sediment exchanges of contaminants. If one assumes that contaminants are in equilibrium with the sediments, then decreasing the concentration of contaminant in the water column may result in contaminant release from the sediments. This could counteract the benefit of a reduced waste load in an effluent discharge. Burial of historical sediment contamination by cleaner materials, resuspension during severe flows or tidal flux, or changes in bioavailability either temporally or spatially are also not built into SQGs but require special consideration in any TMDL process.

An important aspect of using SQGs as regulatory criteria is the ability to determine SQGs for new contaminants. However, development of new SQGs is very resource intensive (e.g., spiked sediment toxicity tests) and probably must involve the private sector, such as recently occurred with development of the methyl tert-butyl ether (MTBE) WQC (Mancini et al. 2002). No set procedure for derivation of numerical SQGs has been agreed, unlike the procedures established for new ambient WQC (Stephan 1985). Such considerations are recommended during the development and promulgation of any site-specific or regional SQGs.

Use of sediment quality guidelines as surrogates for ecological status and benthic performance

Simple measurements are very appealing as tools to report the condition of aquatic systems. SQGs have been proposed to distinguish areas where contaminants are not believed to be affecting ecological conditions or the performance of the benthic component of aquatic systems, where there may be contaminant impacts, and where impacts are highly likely. Analytical chemistry can provide extensive data at reasonable cost, has high replicability, and has good public credibility. The lure of the "bright line" SQG is that it appears to remove the manager from the evaluative process, making the outcome more certain for the affected party and more defensible for the regulator. Consequently, numeric SQGs based on chemical concentrations have appeal as a driving metric in pollution assessment and control programs. However, this is not appropriate in all situations, as described in Chapters 5 and 6 and below. It is apparent that in many cases, SQGs are best used in a WOE approach (Chapman 1989; Burton et al. 2002; Chapman, MacDonald, Lawrence 2002; Suter et al. 2002; Chapters 5 and 6).

Use of sediment quality guidelines for sediment management decisions

SQGs may have a significant role in 2 common types of sediment management decisions involving dredging: dredging as a transportation necessity (navigational dredging) or for pollution abatement (remedial dredging). Dredging of sediments has long been controversial, regardless of the management reason. The recent increase in public awareness of the role that sediments play in sequestration of contaminants has occasionally resulted in delays on the order of months or even years for navigational dredging or sediment removal actions deemed to be in the public interest.

The potential impact of sediment-bound contaminants on the aquatic environment has been evaluated in dredging projects since the early 1970s (USEPA/USACE 1977). The evolution and current status of the evaluative tests used in navigational dredging are discussed in detail in Ingersoll et al. (1997). Because of the real or perceived expense of tools such as sediment toxicity testing, assessment of resident communities, and bioaccumulation, SQGs have been suggested as a potential way to screen sediments during the evaluation process (USEPA/USACE 1998), but to date not all US Regions or Districts of USEPA have adopted them (USACE/USEPA 1994).

Before deciding on the appropriateness of an SQG in this process, it is important to clearly identify the decisions that will be made using the information obtained. Decisions are made using the results of these tests to evaluate the risk of both the dredging operation itself and the proposed management of the dredged material. Dredging contaminated materials will resuspend some of the material in the water, and freshly exposed historical sediments may have more contamination than recently deposited materials. Managers need to determine the level of control needed during the dredging operation itself and for the appropriate management of the dredged materials. SQGs can play many different roles, and the investigator and manager need to understand the technical ramifications of the SQGs used and the system or action being evaluated. It may not always be appropriate to use the same set of numerical guidelines to evaluate the dredging site and the management (disposal) site.

Many investigators have suggested that SQGs are particularly useful for screening (Chapman 1989; MacDonald 1994; Long et al. 2001; USEPA

2002). Screening in a tiered evaluation framework eliminates from further consideration those samples that are believed to represent such a small chance of being problematic that the cost of further assessment is not warranted. The use of numerical guidelines simplifies the screening process, providing time and resources for areas of higher concern. For example, SQGs can be used to screen out clean from contaminated sediments, obviating the need for further testing or stringent controls on the disposal of the dredged material. SQGs can also be used to identify the contaminants of concern in dredged (or other) sediments. Sediments deemed to be unacceptably contaminated may require special removal or disposal procedures.

This tiered approach has considerable appeal because it focuses resources as necessary. Clean sediments may be left in place or, if removed for navigational needs, may be disposed with reasonable certainty that they will not contaminate a disposal site. Contaminated materials may be further evaluated to determine what limitations in handling and disposal are needed to provide environmental protection.

Perhaps one of the most controversial topics today involves pollution abatement via the remediation of contaminated sediments. Historical contamination in urban areas can be so great as to act as a continual source of contamination to a particular system, resulting in loss of resources and an inability to achieve water quality goals. Unfortunately, not only are the potential costs of remedial dredging extremely high, but there are risks from both disturbing the sediments and leaving them in place. In many cases, sediments that have been historically contaminated are buried under layers of cleaner sediments. Removal of sediments will clearly destroy or severely disturb the existing benthic communities. If the system is stable, historically contaminated sediments may be entombed, but storm events or tidal action pose a risk that the contaminated sediments will be re-exposed or resuspended. In such cases, the manager needs to balance the risks and costs of removal and management against the risks of leaving the contaminants in place. In some cases, extensive analyses are not necessary, because it may be more cost effective to simply remove small amounts of contaminated sediments to a given numerical SQG than to perform the necessary additional tests to support a different remedy. This will primarily be a business decision for the responsible party. For more complex evaluations, tools that are necessary include sediment stability and human and/or ecological risk assessment, including both present and future risk (Ingersoll et al. 1997; Chapman, MacDonald, Lawrence 2002).

Once the decision to dredge contaminated sediments has been made, the practitioner will need to determine the extent of removal required and define a concentration that will serve as a guide for proper engineering and selection of equipment. Decisions regarding the presence of pollution (where to start dredging) may use different information than decisions regarding the absence of pollution (where to stop dredging). While SQGs may provide a convenient target, many SQGs are not designed to predict which chemicals cause toxicity, and none of the current SQGs can determine which contaminant or group of contaminants needs to be reduced and how much it needs to be reduced to achieve acceptable benthic conditions. In addition, sites in urban areas may be subject to significant recontamination from diffuse sources. Remedies that do not account for ambient sediment contamination may result in significant waste of resources without achieving environmental goals. Targeted action (i.e., hot spot removal) may be an option that reduces overall site risks to acceptable levels without removing the entire contaminant load, while at the same time reducing time and cost and reducing the risks associated with dredging and management.

Use of sediment quality guidelines in natural resource damage assessments

Natural resource damages refer here to a US law that entitles the public to recover compensation from responsible parties for the injuries to natural resources resulting from the unpermitted release of a hazardous substance or oil. Natural resource damage assessments (NRDAs) have more frequently included contaminated sediments in recent years. However, how best to evaluate injury is complicated in even the most straightforward situations. The NRDA should determine the condition of the natural resource but for the unpermitted release (i.e., establish a baseline), quantify the injury, and identify the responsible party or parties. SQGs may provide a convenient way to screen datasets in these situations. They may also provide a convenient point of departure for negotiations. In systems where, historically, multiple sources of various contaminants have occurred, lack of historical information regarding the natural condition of the system may make determining a baseline problematic. It is probably also unrealistic to expect that a highly urbanized system can be restored to pre-industrial conditions (Ianuzzi et al. 2002). The likelihood for recontamination from diffuse ambient and other point sources must be considered while remedial goals in these situations are being developed. SQGs developed from national guidelines will be unlikely to provide the necessary information to properly develop a remedy. The proper assessment tools in these cases should include a combination of site-specific SQGs, conventional assessment tools, cost negotiations, or frameworks developed for realistic future use scenarios.

Water Bodies: Similarities and Differences

Perhaps the only thing in common among complex sediment systems is the fact that they are all covered by water, either constantly or periodically. This section describes how different water bodies affect their underlying sediments, the driving forces and their limiting factors, and how these affect both biota and contaminants. Table 17-4 summarizes differences between sediment environments related to key fate and transport processes. Chapter 7 provides discussion and guidance regarding the utility and relevance of SQGs and other sediment assessment tools in each setting.

Marine

Sediment assessments in offshore or coastal marine environments use a variety of techniques. The remoteness and depth of sampling stations distinguish such assessment programs from most nearshore and freshwater sampling. Sampling gear is operated remotely, and unless marked by fixed objects on the seabed, sampling does not occur at exactly the same location. Procedures usually involve removal of sediment, water, or biota for testing and analysis, rather than in situ experimentation or manipulation.

The use of reference stations can be additional to or a replacement for numeric chemical analytical values (i.e., SQGs). Collection of biota and interpretation of species abundance data appear to be a suitable approach for evaluating impacts, provided that adequate natural diversity and abundance are present at reference sites. Monitoring programs at a regional scale provide information about trends. Uncertainties about the system are thus addressed by consideration of sampling program design, selection of appropriate metrics, and, in many instances, multivariate comparison.

The marine sediment environment reflects a few major physical factors: water velocity and direction (e.g., currents, tides, wave action, storms), topography, particularly bottom slope, and particulate properties (e.g., grain size). Temperature, salinity, and oxygen availability typically show less variability with increasing depth. Of most importance for aquatic life is the Continental Shelf, regions ranging to depths of 100 to 200 m and extending, on average, for about 65 km from coastal margins.

Sediment types range from silts to clays to sand. Deposition rates vary with particle size and density, interacting with water velocity and turbulence. Because smaller particles tend to travel farther, offshore reaches tend to have finer grain substrates. Subsurface contours may be minimal (i.e., resulting in a broad Continental Shelf, such as off the northeastern US) or may reflect

Table 17-4 Sediment processes and relationship to various sediment environments

Environment	Environmental characteristics	Key fate and transport processes
Lacustrine	Low-energy environment Generally depositional environment Groundwater interaction decreasing away from shore Organic matter decreasing with distance from shore Often fine-grained sediment	Sediment deposition Water-side mass transfer limitations Groundwater advection in near-shore area Bioturbation and bioirrigation (especially in near-shore area) Diffusion in quiescent settings Metal sorption Aerobic and anaerobic biotransformation Biotransformation of organic matter (e.g., gas formation)
Riverine	Low- to high-energy environment Depositional or erosional environment Potential for significant groundwater interaction Significant variability in flow and sediment characteristics within and between rivers	Local and generalized groundwater advection Sediment deposition and resuspension Aerobic biotransformation processes in surficial sediments (potentially anaerobic at depth) Bioturbation and bioirrigation
Estuarine	Generally low or moderate energy environment Generally depositional environment Generally fine-grained sediment Grading to coarse sediment at ocean boundary	Bioturbation and bioirrigation Sediment deposition Water-side mass transfer limitations Aerobic and anaerobic biotransformation of contaminants Biotransformation of organic matter (e.g., gas formation)
Coastal marine	Relatively high-energy environment, decreasing with depth and distance from shore Often coarse sediments	Bioturbation and bioirrigation Sediment erosion and deposition Localized advection processes
Offshore marine	Generally low-energy environment Generally depositional environment Generally fine-grained sediment	Bioturbation and bioirrigation Sediment deposition Anaerobic biotransformation of contaminants

the steep topography seen above the water, such as the cliffs and fjords along northwestern Europe and western Canada.

The complexity of depositional processes interacts with a variety of contaminant properties and sources. Anthropogenic contaminants may be introduced from terrestrial activities, such as dissolved or particulate materials carried from land by rivers and estuaries, from direct discharges of sewage and wastewaters, from airborne materials deposited to surface waters, or by activities in the marine environment. Examples of marine activities that may directly impact sediments include marine transportation, oil and gas exploration and production, disposal of wastes or dredge spoils, and bottom fishing.

Benthic communities in marine environments are typically below the photosynthetic zone, other than along the coastal margins. Consequently, benthic food chains are typically built on organic materials carried into the system; thus, the food chain is primarily allochthonous. Materials such as phytoplankton may be filtered from the water, or deposits may provide organic material for bacterial growth, which can then be harvested by filtering or grazing organisms.

Limiting factors in nearshore benthic communities are nutrients, light, and habitat. Patterns of currents, thermoclines, and salinity gradients affect nutrient supply, enriching some areas continually, seasonally, or periodically, for example, El Niño. Habitat limitations often take the form of competition for substrate. Construction of a fixed structure, such as an oil and gas platform, provides a surface for colonization by immobile organisms and a habitat for mobile organisms. This leads to a number of changes in the local sediment community, which can be interpreted as an ecological improvement. For example, Texas (USA) has established a program to leave oil rigs in place after use as artificial reefs (http://www.tpwd.state.tx.us/fish/reef/artreef.htm).

Because the exploration, production, and transportation of oil and gas from offshore sites is a particular activity of concern, many related sediment assessments have been conducted in the marine environment. Such sites may be fixed to the ocean floor or float in deeper water, tethered in place. Because these activities are high profile, they have been the subject of numerous studies and provide case studies of how marine sediments may be studied and impacts evaluated and managed.

Baseline studies are conducted before activities begin. Their purpose can be to describe conditions in the absence of the proposed activity or to address particularly sensitive issues that may require specific management or preventive measures. Baseline studies are typically required as part of the licensing or permitting process. Data likely to be obtained in baseline studies include benthic organism data (number of individuals distinguished by species), physical sediment characteristics (granulometry, type of material), chemical sediment profile (organic matter, organic compounds of interest, such as petroleum hydrocarbons and metals), water column data (dissolved oxygen, pH, salinity, temperature, nutrient concentrations, suspended particulates, chlorophyll, selected organic compounds and metals), and photographic observations.

Baseline studies may include specific investigations to evaluate concerns to protect the environment. For example, a baseline assessment associated with an oil and gas development in Alaska collected representative samples of the material proposed to be excavated for construction of a subsea pipeline, to support a model predicting the suspended sediment transport (URS 2001).

Monitoring studies include regular programs to track conditions and impacts during an active operation or following cessation (or remediation). Typically, monitoring programs will use reference sites for comparison. For example, Norwegian authorities have established requirements for environmental monitoring of petroleum activity on the Norwegian Continental Shelf (SFT 2002). One aspect of this program is a shift from monitoring at specific fields (platforms) to a regional program, and from sole focus on sediments to an integrated program assessing both the sediments and the water column. Regional data are collected every 3 years. Condition monitoring includes measurement of the tissue levels of selected contaminants in fish. Impact monitoring includes measures of biota distribution and abundance in conjunction with chemical analyses of sediments and water. Because a large number of stations are surveyed, analyses are presented at the level of the individual station, the field, and the region. Biological information includes community indices, number of species and individuals per 0.5 m^2 sediment sample, dominant species, distributional gradients, similarity calculations and correlations between biota, sediment characteristics, and contaminant concentrations.

Special investigations include studies that may represent research or retrospective evaluations of the extent and status of historic activities. In onshore or nearshore sites, these investigations might be analogous to major terrestrial contaminated-site investigations. An illustrative example comes from investigations of seabeds contaminated by oily cuttings in the North Sea. Projects to evaluate present conditions and management options have been commissioned by both the Netherlands and UK authorities. Dulfer (1999) evaluated parameters that can be used to assess pollution of the seabed in the Netherlands Continental Shelf by residues from past discharges of oil-based mud (OBM) cuttings (OBM were used prior to 1993 to carry drill cuttings up a well shaft. Although reused to the extent possible, OBM residues remained with the rock and were discharged to the sea floor, creating cuttings piles around the bases of oil and gas production platforms. More recent drilling has used water-based muds, which are discharged near the ocean surface and are thus dispersed before reaching the sea floor). Dulfer (1999) suggested that oil concentration in sediment was an effective parameter and noted several types of criteria equivalent to SQGs: target values (representing negligible risk), natural background levels, limit values (representing objectives that can be attained in the short term), test or content values (representing criteria for accepting dredged materials for disposal in the environment), and intervention values (representing significant risk and probably requiring immediate action).

Ecotoxicological risk values, based on ecotoxicity tests, have been used to set different risk levels (serious, maximum permissible, and negligible). Other biological data suggested by Dulfer (1999) include microbial activity (measured as turnover rate), change in morphology (such as loss of ventral spines in the sea urchin *Echinocardium cordatum*), change in physiology (measured in bioassays), and changes in ecosystem structure (population density, biodiversity, and similarity). A review of conditions around former OBM drilling locations (TNO 1999) reported that the sea urchin *E. cordatum* has been identified as a universal and sensitive species, so its distribution and abundance provide an indication of affected seafloor. Echinoderms have long been recognized as sensitive species in marine benthic surveys at contaminated sites, which Chen et al. (2002) suggest may be due to their high sensitivity to digestive inhibition by contaminants. Other parameters included species richness and similarity. TNO (1999) noted that no field data on microbial activity, morphology, or physiology were available.

A more recent project on UK and Norwegian OBM drill cuttings sites examined some additional parameters in an effort to evaluate current status and future management options, including cuttings pile removal (UKOOA 2002). Investigations using cell assay techniques to look at endocrine disruption were not conclusive, finding effects to be weak and variable. Toxicokinetic studies used the amphipod *Corophium volutator* and the phytoplankton *Skeletonema costatum*. Bioaccumulation patterns were evaluated in the amphipod and in mesocosms containing mussels (*Mytilus edulis*), the ragworm *Nereis virens*, and turbot *Scophthalmus maximus*. Toxicity tests used *C. volutator*, *E. cordatum,* and the copepod *Acartia tonsa*.

Estuarine

Estuaries are dynamic, complex, and unique systems that can have strong physical–chemical gradients, particularly salinity, dissolved oxygen (DO), pH, nutrients, sediment grain size, and organic content (e.g., TOC), all of which may affect the bioavailability of sediment contaminants. Estuarine sediments should not be assessed without understanding both seasonal and spatial variability, particularly for these strong gradients. Overlying water and porewater characteristics will affect the flux of sediment-associated contaminants. The shear forces at the freshwater–saltwater interfaces cause turbulent mixing that serves to trap sediments as well as nutrients. Shallow systems tend to be less stratified vertically, but tidal regimes result in more horizontal salinity gradients so the turbulent zones are dictated more by tidal cycles. While the nutrient and sediment trap effects are essential for maintaining the high productivity of estuaries, these also cause estuaries to function as major contaminant sinks. Periodic shifts in salinity associated with rain events can contribute to the increased flux of contaminants on spatial as well as temporal scales, as the turbulent zone shifts along the modified salinity gradient. Because of the dynamic nature of sediments in estuaries with strong flows or currents, knowledge of sediment stability is necessary to fully assess contaminant flux and effects on bioavailability via porewater and overlying water exposure routes. Therefore, initial sediment assessments, particularly in salt wedge estuaries, should be conducted during both low and high freshwater flow periods.

Estuarine systems can be divided into 3 general classes based on salinity. Estuarine areas where salinity is >30 g/L can be considered marine, and those where salinity is <5 g/L can be considered freshwater for the purposes of sediment assessments. The estuarine areas with intermediate salinities (e.g., >5 to 30 g/L) are the most challenging and may require tools different from those used for freshwater or marine systems; and subdivisions within this broad range may be needed. Furthermore, because many of these systems are tidally driven, salinities can vary substantially over a tidal cycle. Therefore, salinity-based divisions must be based on measurements of overlying bottom waters, taken periodically over at least a full tidal cycle, and ideally with knowledge of seasonal variability in both overlying and porewater salinities.

In large, well-mixed systems (e.g., bays, harbors), daily variations in dissolved oxygen concentrations and pH levels may be minimal, as illustrated by the stable water chemistry profiles observed in Charleston Harbor, USA (Figure 17-6A). In deep, large estuarine systems, stratification (associated with temperature as well as salinity oscillations) can cause extended periods (weeks to months) of hypoxia or anoxia (Officer et al. 1984; Seliger and Boggs 1988; Rosenberg 1990; Rabalais et al. 1994). In shallow or more stratified systems, low DO regimes tend to occur periodically rather than for extended periods and are controlled by diurnal and tidal factors (Borsuk et al. 2001; Ringwood and Keppler 2002). These patterns change in dendritic systems (Figure 17-7), moving from a large seaward body of water where there is little diurnal variation in DO or pH (Figure 17-6A) to the upper reaches, into a tidal river

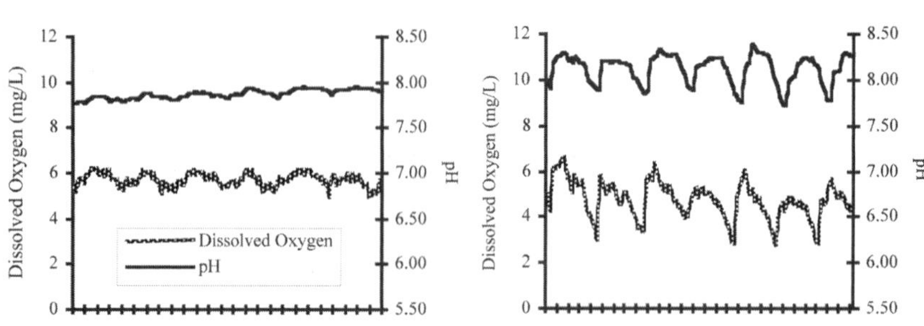

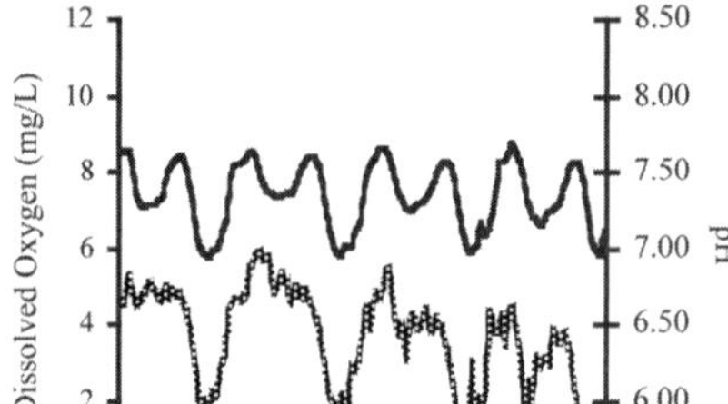

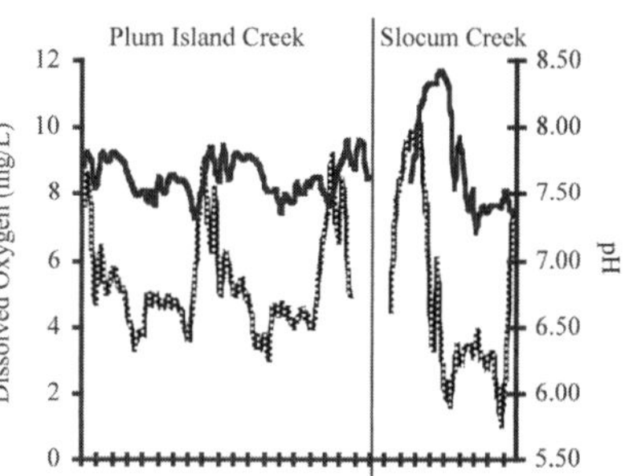

Figure 17-6 Examples of water quality profiles generated from in situ data loggers (Hydrolab, Datasondes), programmed to record dissolved oxygen (DO) and pH at 30-min intervals for >24 h periods during the summer (Jul to Aug). Boldface values of 8 mg/L on the DO axes and of 7.5 on the pH axes indicate 100% saturation. PH values are lower than full-strength seawater because of freshwater inflows. (Unpublished data collected and validated by A. Ringwood and colleagues as part of the Environmental Monitoring and Assessment Program [EMAP] in the Carolinian Province or the Bioavailability Program [USEPA STAR])

system, and then into tidal creeks (Figures 17-6B, 17-6C), where substantial variations can occur on a daily basis. These systems are subject to a relatively high level of natural fluctuations; for example, DO can be highly variable on a daily basis, ranging from hypoxic during the late night or early morning to normoxic during the day. The magnitude and frequency of these daily cycles are exacerbated by nutrients and eutrophication (e.g., the DO cycles from sites near sewage treatment plants range from severe hypoxia or anoxia to supersaturation [i.e., >8 mg/L] on a daily basis) (Figure 17-6D).

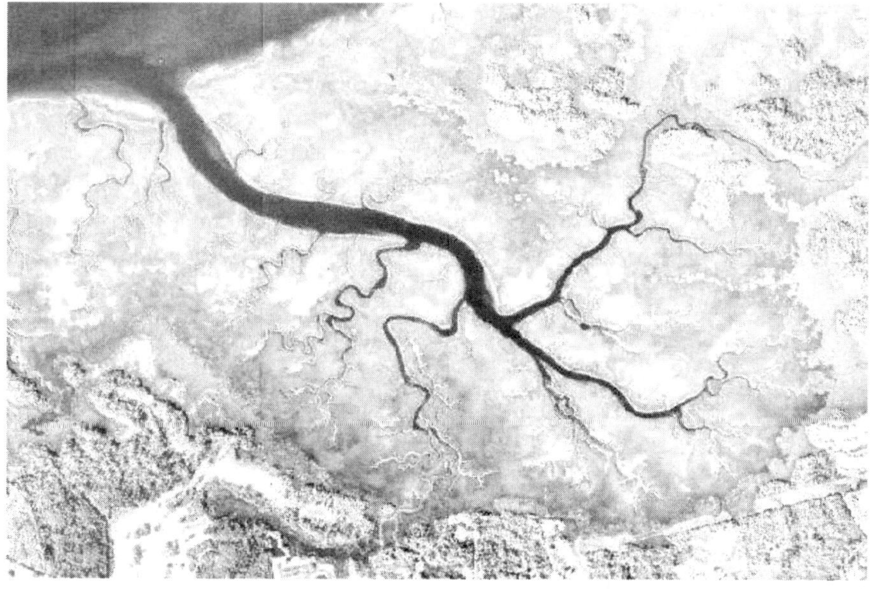

Figure 17-7 Example of dendritic nature of tidal creek and tidal river estuarine systems. The tidal creek (Foster's Creek) and associated branches can be further categorized into multiple secondary creeks that drain from the associated watersheds into the primary creek; other primary creeks, like Foster's Creek, are connected to the Wando River, which flows into Charleston Harbor, South Carolina, USA. Aerial photograph by George S. Steele, SCDNR.

Low DO is accompanied by high CO_2 (i.e., hypercapnia) that drives pH down, so pH is an important covariate. Because pH levels are tightly linked to DO conditions, pH will also cycle, with lower values during hypoxic periods than during normoxic periods (Ringwood and Keppler 2002; Figure 17-6). Salinity also affects pH, which decreases as salinity decreases. Low pH conditions frequently lead to an increased flux of metal contaminants (Allen 1995). The pH minimum occurs at the base of the aerobic zone and can facilitate increased flux of metals to the overlying waters (Van Cappellen and Wang 1995). As with DO cycles, the magnitude of the pH cycles is affected by eutrophication, salinity shifts, and iron sulfides. Like Eh, pH reflects fundamental changes in redox status that perpetuate contaminant flux; thus, estuarine sediments can function as contaminant sources as well as contaminant sinks.

Depending on the freshwater sediment load and the estuarine circulation patterns, estuarine sediments can come from inland and/or from the sea. Deposition and resuspension are ongoing processes. Estuarine sediments can range from sandy (coarse particles) to muddy (fine particles). Generally, there is a tendency for an upstream–downstream gradient from mud to sand (the opposite of many rivers, where the gradient is coarse to fine); in tidal habitats, there can be a gradient of fine particles in subtidal areas to coarser particles in higher intertidal zones (Knox 1986). However, these trends may be profoundly affected by the shape of the system, the tidal effects, and the activities of biological components. Tidal river and tidal creek systems that have curves and winding turns or are dendritic are characterized by depositional zones similar to river systems (Figure 17-7), so the upstream–downstream gradient model does not always apply. Similarly, variations in depth profiles characteristic of fjords and bays will affect sediment distribution so that finer-grain materials will collect in deeper zones, leaving coarser materials in the shallow regions.

Sediment TOC concentration is a commonly measured variable that provides a good overall index of organic loading and composition. It integrates carbon enrichment from multiple sources, including land-based inputs, detritus, and algal and microbial metabolism. These diverse organics can affect the bioavailability of both organic and inorganic contaminants. Typically, TOC co-varies with fine sediment particles.

Rivers, streams and other erosional environments

Unique features

High-gradient streams and other erosional environments offer unique challenges for assessing sediment contamination. Such lotic systems differ significantly from lentic systems in terms of major physical processes, factors that limit primary production, nutrient dynamics, types of primary producers, and the relative importance of allochthonous versus allochthonous energy sources (Table 17-5). These differences can influence the toxicity of sediment contaminants and their potential effects on aquatic organisms. The major physical characteristic and indeed the defining feature of lotic environments is the unidirectional flow of water. Many aquatic organisms are dependent on this downstream transport of biotic and abiotic materials. The unidirectional flow of water is also responsible for the redistribution of sediments and associated contaminants and of biota (e.g., downstream recolonization after disturbance, drift). Although sediments in lentic environments may be resuspended by a variety of physical and biological processes (e.g., lake turnover, wave action, bioturbation), the potential for movement of sediments in lotic environments is much greater. For instance, temporal changes in stream discharge associated with storm events, spring runoff from snowmelt, or seasonal rains (e.g., monsoons) will resuspend sediments and may transport these materials significant distances downstream.

Table 17-5 Ecological and physical differences between lentic, lotic depositional, and lotic erosional ecosystems

Characteristics	Lentic	Lotic (depositional)	Lotic (erosional)
Major primary producers	Phytoplankton	Phytoplankton, macrophytes	Periphyton
Ecosystem energetics	Primarily autochthonous (P > R)	Primarily allochthonous (P < R)	Allochthonous (P < R), autochthonous (P > R) in unshaded reaches
Factors limiting primary production	Nutrients	Light (+ nutrients)	Light (+ nutrients)
Nutrient dynamics	Cycling	Spiraling	Spiraling
Physical factors responsible for sediment movement	Lake turnover and wave action	Unidirectional flow	Unidirectional flow
Sediment transport	Deposition and accumulation	Deposition, accumulation, and resuspension	Deposition, accumulation at stream edges, and resuspension
Transport of organisms	Negligible; primarily by wind and waves	Downstream drift	Downstream drift

P = photosynthesis; R = respiration

Other differences between lentic and lotic environments must be considered when effects of sediment contaminants are characterized. Lakes are generally considered autotrophic systems in which the major primary producers consist of phytoplankton. There is relatively little interaction between sediments and phytoplankton until these organisms settle to the bottom of the lake. In contrast, major primary producers in streams consist of attached algae and periphyton, which are an integral component of most benthic habitats. These primary producers are an important route of exposure to benthic organisms and may influence contaminant fate and effects in streams (Stewart and Hill 1993). The relative importance of autochthonous production by periphyton in streams varies longitudinally and is generally dependent on the amount of available light. Open reaches of shallow, mid-order streams that receive sufficient levels of light may be autotrophic (photosynthesis > respiration), but most streams are heterotrophic and depend on allochthonous input of energy.

Examples of high-energy, erosional habitats include high-gradient streams and rocky intertidal habitats. Many of the physical, chemical, and biological characteristics that differ between depositional and erosional habitats are likely to influence the exposure and toxicity of sediment contaminants to benthic organisms (Table 17-5). Because of greater energy levels and greater potential for sediment transport, grain size is larger and organic carbon content is generally lower in erosional environments. These differences are likely to increase the bioavailability of sediment-associated contaminants in erosional environments. Unlike depositional habitats, fine-grain sediments in erosional environments are highly mobile. For example, the unidirectional flow of streams and rivers transports nutrients, sediments, and contaminants downstream. These materials may be deposited in slower-moving sections of the river and then resuspended during periods of high discharge. Thus, unlike depositional environments, sediments and associated contaminants in lotic ecosystems are characterized by a cycle of deposition and resuspension along the river continuum (Figure 17-8). The amounts and relative size of particles transported downstream are dependent on the magnitude of stream discharge, which may show considerable temporal variation. In environments dominated by snowmelt, maximum discharge generally occurs during spring runoff, flushing materials to depositional areas downstream. Episodic events such as summer convection storms may rapidly increase stream discharge and transport large volumes of contaminated sediments to downstream reaches.

The unidirectional movement of water in stream environments continuously redistributes sediment particles from upstream to downstream. These materials are sorted, resulting in a variable size distribution of particles, ranging from clay and sand to large cobble and boulder. Sediments in depositional zones of larger rivers are generally dominated by fine silts and clays. Traditional sediment assessment methods employed in these systems will be similar to those developed for many lentic and marine environments. Standard sediment toxicity tests conducted with freshwater benthic and epibenthic test organisms are appropriate. However, because these depositional zones are in a constant state of flux, sediment characteristics measured at one point in time may be quite different from those on another sampling occasion. Particle size distributions, the amount of organic material, level of contamination, and oxic conditions will vary temporally.

In contrast to the situation for large, low-gradient environments, fine sediments often comprise a relatively small component of the benthic habitat in high-gradient streams and other erosional habitats. It is unlikely that fine sediments are an important source of contamination in streams where the

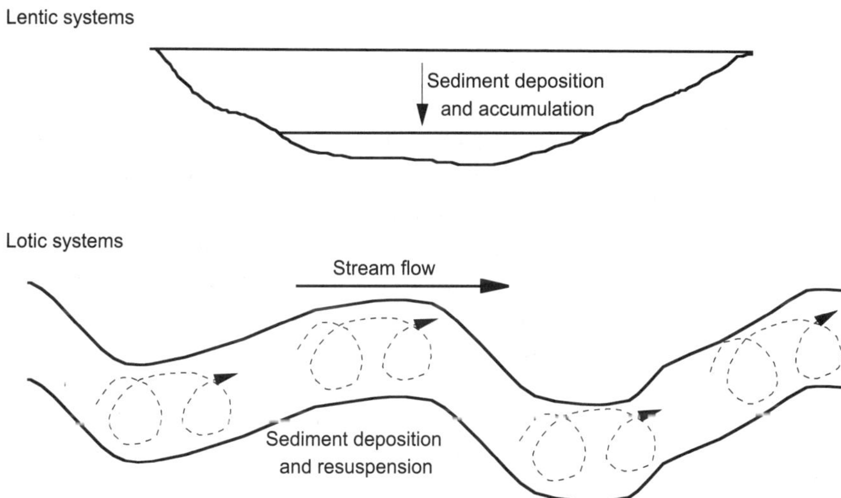

Figure 17-8 Comparison of sediment transport in lentic and lotic ecosystems. Although physical processes may resuspend lake sediments, these systems are generally characterized by deposition and accumulation. In contrast, sediments in lotic ecosystems are continuously deposited, resuspended, and transported downstream.

substrate is dominated by large cobble and boulder. Therefore, the traditional approaches used to assess sediment contamination in many lentic and estuarine environments will be inappropriate in high-energy, erosional habitats. For instance, in cobble and gravel substrates, porosity and permeability will be high, such that the composition and contaminant concentrations in pore and overlying waters will be similar (Chapman 1981). Protocols for assessing sediment contamination in cobble bottom streams need to be developed to reflect these differences in routes of exposure and accompanying different benthic communities from soft-bottom water bodies.

Importance of longitudinal changes

The most important distinguishing feature of lotic ecosystems is the unidirectional transport of materials from upstream to downstream. The river continuum concept (RCC) predicts important changes in the structure and function of stream communities along a longitudinal gradient from upstream to downstream (Vannote et al. 1980). Although the details of these changes will vary among geographic locations (e.g., deciduous, mountain, desert), differences in community composition along natural longitudinal stream gradients have been reported in numerous systems. The most consistent findings include an increase in species richness in mid-order streams and a shift in the relative abundance of functional feeding groups, corresponding with changes in the primary sources of energy (e.g., allochthonous versus autochthonous) and light levels.

The RCC is widely regarded as one of the most significant developments in stream ecology and has provided an important framework for relating physical changes along longitudinal gradients to predictable ecological characteristics. This paradigm has important implications for the design and implementation of sampling programs in lotic ecosystems. The traditional approach in most stream monitoring programs is to compare upstream reference sites to downstream impacted and recovery sites. Natural variation in physicochemical characteristics, community structure, and ecosystem function along longitudinal gradients is superimposed on contaminant-induced variation and will complicate assessment of the effects of anthropogenic disturbance (Clements and Kiffney 1995).

For example, if species richness increases naturally from upstream to downstream as predicted by the RCC, our ability to assess the effects of a stressor is compromised (Figure 17-9). If a perturbation occurs in the lower reaches of a watershed, quantifying the effects of a stressor by making comparisons to upstream reference sites is not appropriate because of the natural decrease in species richness. Similarly, natural shifts in the abundance of dominant species along a longitudinal gradient also limit the usefulness of community composition data for assessing perturbation. Clements and Kiffney (1995) reported that distinguishing effects of contaminants on stream benthic communities required that natural longitudinal changes in community composition be considered. Natural variation in community composition also confounded the effects of contaminants on diatom communities in polluted streams (Medley and Clements 1998).

Various approaches may be used to account for the influence of natural longitudinal variation in community composition along stream gradients. The most obvious approach is to select reference and impacted sites with similar elevations, stream sizes, watershed areas, and other habitat characteristics that influence community structure and function. In areas where locating reference sites in different watersheds is not possible, an alternative is to adjust the number of species expected at a site by accounting for stream size or watershed area. This is essentially the approach used by Karr and others in the development of the index of biotic integrity (IBI) (Karr and Chu 1997). Several of the metrics used in the IBI are highly sensitive to watershed area (e.g., abundance, species richness), and therefore, expected values for a particular metric are

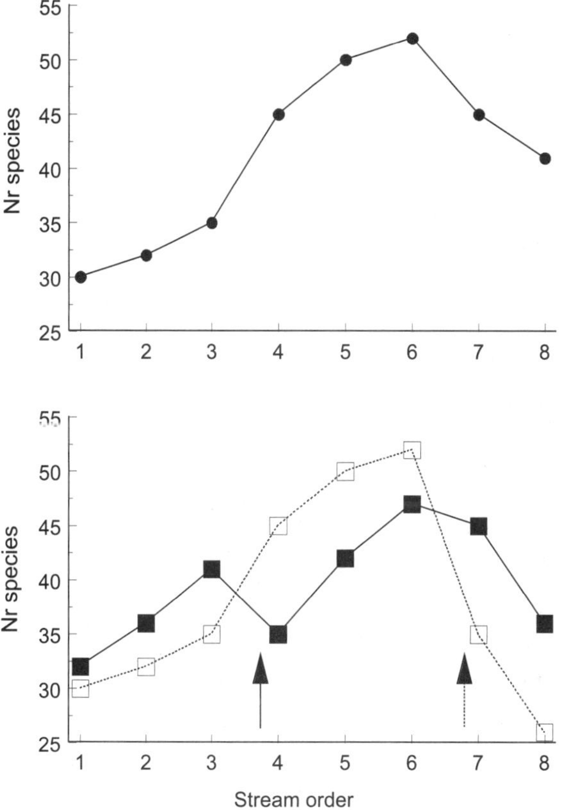

Figure 17-9 Influence of natural longitudinal changes in community composition on responses to anthropogenic stressors. Upper panel shows predicted change in number of species from upstream to downstream as predicted by the river continuum concept (Vanotte et al. 1980). Lower panel shows hypothetical influence of a perturbation that occurs either in the upper reaches of a stream as species richness is naturally increasing (solid line) or in the downstream reaches as species richness is naturally declining (dashed line) (cf. Clements and Kiffney 1995; Medley and Clements 1998)

adjusted to reflect differences among streams.

Other physical changes that occur along longitudinal gradients in streams will influence toxicity and bioavailability of contaminants in sediments. Because of lower velocity and stream power, mean particle size will generally decrease and the amounts of organic carbon will increase from headwater streams to larger rivers (Figure 17-10). As a consequence, bioavailability of many contaminants will increase from upstream to downstream. These natural changes must be considered when sampling programs for contaminated sediments are designed.

A more problematic source of longitudinal variation associated with lotic environments is the difference in sensitivity to stressors among communities from different locations. Several researchers have speculated that some communities may be inherently more sensitive to anthropogenic stressors than other communities (Rapport et al. 1985; Poff and Ward 1990; Howarth 1991). For example, Rapport et al. (1985) suggested that communities in unstable environments may be

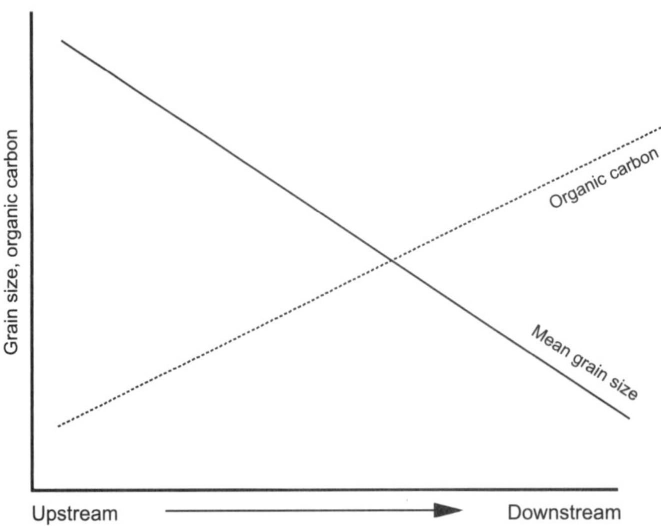

Figure 17-10 Predicted longitudinal changes in organic carbon content and grain size of sediments in lotic ecosystems. These natural changes in grain size and organic carbon content occur because of reduced stream gradient, power, and current velocity (Minshall et al. 1985); as a result, it is expected that contaminant bioavailability in sediments will decrease from upstream to downstream.

preadapted to moderate levels of anthropogenic stress. Kiffney and Clements (1996) used microcosm experiments and a field study to show that benthic communities in small headwater streams were more sensitive to metals than communities from low elevation streams. Increased tolerance in communities from lower elevations most likely resulted from the greater levels of natural variability these populations experienced. In contrast to these findings, Medley and Clements (1998) observed reduced effects of metals on diatoms from headwater communities, compared to those from lower elevations.

Although the precise mechanism explaining variation in sensitivity to contaminants among communities is uncertain, these differences greatly complicate our ability to assess the effects of sediment contaminants in lotic ecosystems. Differences in sensitivity between locations suggest that WQC and SQGs protective of some communities alone may not be protective of others (Kiffney and Clements 1996), thus requiring site-specific evaluation of sediment contamination.

Discussion

This paper provides a discussion of some of the uncertainties involved in sediment assessments. As such, it serves as detailed technical background information in support of the conclusions reached in Chapter 7.

Investigators embarking on a sediment quality assessment need to make decisions regarding the appropriate approach and tools to use to support the goals of the assessment. Each available tool has its inherent strengths and weaknesses and intended applicability. SQGs are but one tool in the investigators' toolbox. The tools that may be appropriate in one ecosystem or situation may not be appropriate in another, for either technical or logistical reasons. Neither do all sediment quality assessments require nor will they all obtain the same level of certainty. The level of certainty required will depend in no small part on the potential consequences of management decisions to be made with the data generated. Simply put, a screening-level assessment does not require the same level of certainty as a remedial design. It is the responsibility of the investigators to choose the appropriate tools, given the particular details of the assessment. The investigators need to familiarize themselves with the underlying scientific principles of the tools to be used, including their strengths and weaknesses. Cost, timing, and consequences of erroneous decisions are all important factors in this decision making process, and all parties involved need to take the time to discuss these factors prior to the initiation of the project. Choosing the wrong tools can result in exceedingly expensive and slow data generation or, worse, data generated rapidly and inexpensively that are too general to provide defensible decisions. The investigators need, a priori, to clearly state the objectives of the study, the inherent strengths and weaknesses of the suggested approach, and the anticipated uncertainty in the data to be generated. Once the data have been generated, these same points need to be revisited as the results are interpreted and decisions made.

As outlined by ANZECC/ARMCANZ (2000), SQGs are guidelines, not goals. They comprise part of the first line of defense of the environment but should not be used alone for management decision making (e.g., remediation) except when a simple situation exists and the costs of additional investigations to resolve uncertainties would exceed the cost of remediation and when the responsible parties waive their rights to conduct further investigations (Chapter 7). Neither should they be used as "bright lines." Promulgation of SQGs into "criteria" or other regulatory instruments should be avoided to ensure that there is flexibility to incorporate new information as it is brought to light.

Successfully addressing uncertainty in sediment assessments requires the following 3 principles for managing contaminated sediments developed by the USEPA (Horinko 2002):

- Develop and refine a conceptual site model that considers sediment stability. "A conceptual site model should identify all known and suspected sources of contamination, the types of contaminants and affected media, existing and potential exposure pathways, and the known or potential human and ecological receptors that may be threatened."
- Use an iterative approach in a risk-based framework ("risk assessment should play a critical role in evaluating options"). "An iterative approach is defined broadly to include approaches which incorporate testing of hypotheses and conclusions and foster re-evaluation of site assumptions as new information is gathered." The need for an iterative, or multi-tiered approach, has also been recommended by European investigators (Sijm et al. 2001).
- Carefully evaluate the assumptions and uncertainties associated with site characterization data and site models. "The amount of site-specific data required and the complexity of models used to support site decisions should depend on the complexity of the site and the significance of the decision (e.g., level of risk, response cost, community interest)."

References

Allen HE, editor. 1995. Metal contaminated sediments. Ann Arbor (MI): Ann Arbor.

Aller RC, Cochran JK. 1976. $^{234}Th^{238}U$ disequilibrium and diagenetic time scales. *Earth Planet Sci Lett* 29:37–50.

Ankley GT, Di Toro DM, Hansen DJ, Berry WJ. 1996. Technical basis and proposal for deriving sediment quality criteria for metals. *Environ Toxicol Chem* 15:2056–2066.

[ANZECC/ARMCANZ] Australian and New Zealand Environment and Conservation Council and Agriculture and Resource Management Council of Australia and New Zealand. 2000. Australian and New Zealand guidelines for fresh and marine water quality. Canberra: ANZECC/ARMCANZ.

Apitz SE, Kirtay VJ, Davis JW, Finkelstein K, Hohreiter DL, Hoke R, Jensen RH, Mack EE, Stahl R, Jersak J, Magar V, Moore D, Reible D. 2002. Critical issues for contaminated sediment management. Washington DC: US Environmental Protection Agency, Marine Environmental Support Office. Report # MESO-02-TM-01. Available from http://www.hsrc-ssw.org/sedmgt.pdf. Accessed 21 Aug 2004.

Barata MD, Baird DJ, Mitchell SE, Soares AMVM. 2002. Among- and within-population variability in tolerance to cadmium stress in natural populations of *Daphnia magna*: Implications for ecological risk assessment. *Environ Toxicol Chem* 15:2056–2066.

Barbour MT, Gerritsen J, Snyder BD, Stribling JB. 1999. Rapid bioassessment protocols for use in streams and wadeable rivers: Periphyton, benthic macroinvertebrates and fish, 2nd ed. Washington DC: US Environmental Protection Agency, Office of Water. EPA 841-B-99-002.

Bencala KE. 1984. Interactions of solutes and streambed sediment: 2. A dynamic analysis of coupled hydrologic and chemical processes that determine solute transport. *Water Resour Res* 20:1804–1814.

Bentley SJ, Nittrouer CA. 1997. Environmental influences on the formation of sedimentary fabric in a fine-grained carbonate-shelf environment: Dry Tortugas, Florida Keys. *Geo-Mar Lett* 17:268–275.

Bentley SJ, Nittrouer CA. 1999. Physical and biological influences on the formation of sedimentary fabric in an oxygen-restricted depositional environment: Eckernförde Bay, southwestern Baltic Sea. *Palaios* 14:585–600.

Beukema JJ. 1974. Seasonal changes in the biomass of the macro-benthos of a tidal flat area in the Dutch Wadden Sea. *Neth J Sea Res* 8:94–107.

Beurskens JEM, Toussaint M, De Wolf J, Van der Steen JMD, Slot PC, Commandeur LCM, Parsons JR. 1995. Dehalogenation of chlorinated dioxins by an anaerobic microbial consortium from sediment. *Environ Toxicol Chem* 14:939–943.

Boothman WS, Hansen DJ, Berry WJ, Robson DL, Helmstetter A, Corbin JM, Pratt SD. 2001. Biological response to variation of acid-volatile sulfides and metals in field-exposed spiked sediments. *Environ Toxicol Chem* 20:264–272.

Borgmann U, Norwood WP. 2002. Metal bioavailability and toxicity through a sediment core. *Environ Pollut* 116:159–168.

Borgmann U, Norwood WP, Reynoldson TB, Rosa F. 2001. Identifying cause in sediment assessments: bioavailability and the sediment quality triad. *Can J Fish Aquat Sci* 58:950–960.

Borsuk ME, Stow CA, Luettich RA Jr, Paerl HW, Pinckney JL. 2001. Modeling oxygen dynamics in an intermittently stratified estuary: Estimation of process rates using field data. *Estuar Coast Shelf Sci* 52:33–49.

Buchanan JB, Kingston PF, Sheader M. 1974. Long-term population trends of the benthic macrofauna in the offshore mud of the Northumberland coast. *J Mar Biol Assoc UK* 54:785–795.

Burgess RM, Ahrens MJ, Hickey CW, den Besten PJ, ten Hulscher D, van Hattum B, Meador JP, Achazi RK, Van Gestel CAM, Douben PET. 2004. An overview of the partitioning and bioavailability of PAHs in sediments and soils. In: Douben PET, editor, PAHs: An ecotoxicological perspective. London: J Wiley. p 99–126.

Burgess RM, McKinney RA, Brown WA. 1996. Enrichment of marine sediment colloids with polychlorinated biphenyls (PCBs): Trends resulting from PCB solubility and chlorination. *Environ Sci Technol* 30:2556–2566.

Burgess RM, Ryba SA, Cantwell MG. 2000. Importance of organic carbon quantity on the variation of Koc in marine sediments. *Toxicol Environ Chem* 77:9–29.

Burton Jr GA, Batley GE, Chapman PM, Forbes VE, Schlekat CE, Smith PE, den Besten PJ, Bailer J, Reynoldson T, Green AS, Dwyer RL, Berti WR. 2002. A weight-of-evidence framework for assessing sediment (or other) contamination: Improving certainty in the decision-making process. *Hum Ecol Risk Assess* 8: 1675–1696.

Cairns Jr J. 1983. Are single species toxicity tests alone adequate for estimating environmental hazard? *Hydrobiologia* 100:47–57.

Cairns Jr J. 1986. The myth of the most sensitive species. *BioScience* 36:670–672.

Chapman PM. 1981. Measurements of the short-term stability of interstitial salinities in subtidal estuarine sediments. *Estuar Coastal Shelf Sci* 12:67–81.

Chapman PM. 1989. Current approaches to developing sediment quality criteria. *Environ Toxicol Chem* 8:589–599.

Chapman PM. 1990. The Sediment Quality Triad approach to determining pollution-induced degradation. *Sci Total Environ* 97/98:815–825.

Chapman PM. 1996. Presentation and interpretation of Sediment Quality Triad data. *Ecotoxicology* 5:327–339.

Chapman PM. 2000. The Sediment Quality Triad (SQT) – then, now and tomorrow. *Int J Environ Pollut* 13:351–356.

Chapman PM, Brinkhurst RO. 1981. Seasonal changes in interstitial salinities and seasonal movements of subtidal benthic invertebrates in the Fraser River estuary, BC. *Estuar Coastal Shelf Sci* 12:49–66.

Chapman PM, Cano M, Fritz AT, Gaudet C, Menzie CA, Sprenger M, Stubblefield WA. 1997. Workgroup summary report on contaminated site cleanup decisions. In: Ingersoll CG, Dillon T, Biddinger GR, editors. Ecological risk assessment of contaminated sediments. Pensacola (FL): Society of Environmental Toxicology and Chemistry (SETAC). p 83–114.

Chapman PM, Ho K, Munns W, Solomon K, Weinstein MP. 2002. Issues in sediment toxicity and ecological risk assessment. *Mar Pollut Bull* 44:271–278.

Chapman PM, McDonald B, Lawrence GS. 2002. Weight of evidence issues and frameworks for sediment quality (and other) assessments. *Hum Ecol Risk Assess* 8: 1489–1515.

Chapman PM, Wang F. 2001. Assessing sediment contamination in estuaries. *Environ Toxicol Chem* 20:3–22.

Chen Z, Mayer LM, Weston DP, Bock MJ, Jumars PA. 2002. Inhibition of digestive enzyme activities by copper in the guts of various marine benthic invertebrates. *Environ Toxicol Chem* 21:1243–1248.

Chiou CT, Peters LJ, Freed VH. 1979. A physical concept of soil-water equilibria for nonionic organic compounds. *Science* 206:831–832.

Clements WH, Kiffney PM. 1995. The influence of elevation on benthic community responses to heavy metals in Rocky Mountain streams. *Can J Fish Aquat Sci* 52: 1966–1977.

Connaughton DF, Stedlinger JR, Lion LW, Shuler ML. 1993. Description of time-varying desorption kinetics: Release of naphthalene from contaminated soils. *Environ Sci Technol* 30:1145–1151.

Cornett RJ, Risto BA, Lee DR. 1989. Measuring groundwater transport through lake sediments by advection and diffusion. *Water Resour Res* 25:1815–1823.

DelValls TÁ, Forja JM, Gómez-Parra A. 2002. Seasonality of contamination, toxicity, and quality values in sediments from littoral ecosystems in the Gulf of Cádiz (SW Spain). *Chemosphere* 46:1033–1043.

DeWitt TH, Hickey CW, Morrisey DJ, Nipper MG, Roper DS, Williamson RB, Van Dam L, Williams EK. 1999. Do amphipods have the same concentration-response to contaminated sediment *in situ* as *in vitro*? *Environ Toxicol Chem* 18: 1026–1037.

DeWitt TH, Ozretich RJ, Swartz RC, Lamberson JO, Schults DW, Ditsworth GR, Jones JKP, Hoselton L, Smith LM. 1992. The influence of organic matter quality on the toxicity and partitioning of sediment-associated fluoranthene. *Environ Toxicol Chem* 11:197–208.

Dickson KL, Maki AW, Brungs WA. 1987. Fate and effects of sediment-bound chemicals in aquatic systems. New York: Pergamon.

Di Toro DM, Zarba CS, Hansen DJ, Berry WJ, Swartz RC, Cowan CE, Pavlou SP, Allen HE, Thomas NA, Paquin PR. 1991. Technical basis for establishing sediment quality criteria for nonionic organic chemicals using equilibrium partitioning. *Environ Toxicol Chem* 10:1541–1583.

Dowdy SM, Wearden S. 1983. Statistics for research. New York: J Wiley.

Drzyzga O, El Mamouni R, Agathos SN, Gottschal JC. 2002. Dehalogenation of chlorinated ethene and immobilization of nickel in anaerobic sediment columns under sulfidogenic conditions. *Environ Sci Technol* 36:2630–2635.

Dulfer JW. 1999. OBM drill cuttings discharges: Assessment criteria. Report RIKZ-99.018. Den Haag: National Institute for Coastal and Marine Management/ RIKZ (Netherlands). Available from: http://www.ukooa.co.uk/issues/ drillcuttings/pdfs/OBMDrill.pdf. Accessed 21 Aug 2004.

Eagle RA. 1975. Natural fluctuations in a soft bottom benthic community. *J Mar Biol Assoc UK* 55:865–878.

Elliot AH, Brooks NH. 1997. Transfer of nonsorbing solutes to a streambed with bed forms: theory. *Water Resour Res* 33:123–136.

Ellis JI, Schneider DC, Thrush SF. 2000. Detecting anthropogenic disturbance in an environment with multiple gradients of physical disturbance, Manukau Harbour, New Zealand. *Hydrobiologia* 440:379–391.

Fausch KD, Torgersen CE, Baxter CV, Li HW. 2002. Landscapes to riverscapes: bridging the gap between research and conservation of stream fishes. *BioScience* 52:483–498.

Ghosh U, Luthy RG, Clemett SJ, Zare RN. 2000. Microscale location, characterization, and association of polycyclic aromatic hydrocarbons on harbor sediment particles. *Environ Sci Technol* 34:1729–1736.

Gomez-Hermosillo C, Pardue J, Reible D. 2004. Uptake of desorption resistant organic chemicals from sediments. *Environ Sci Technol.* Forthcoming.

Grabowski LA, Houpis JLJ, Woods WI, Johnson KA. 2001. Seasonal bioavailability of sediment-associated heavy metals along the Mississippi river floodplain. *Chemosphere* 45:643–651.

Hall SJ. 1994. Physical disturbance and marine benthic communities: Life in unconsolidated sediments. *Oceanogr Mar Biol Ann Rev* 32:179–239.

Hatzinger PB, Alexander M. 1996. Biodegradation of organic compounds sequestered in organic solids or in nanopores within silica particles. *Environ Toxicol Chem* 15: 2215–2221.

Hewitt JE, Thrush SE, Cummings VJ. 2001. Assessing environmental impacts: effects of spatial and temporal variability at likely impact scales. *Ecol Appl* 11: 1502–1516.

Hickey CW, Clements WH. 1998. Effects of heavy metals on benthic macroinvertebrate communities in New Zealand streams. *Environ Toxicol* Chem 17:2338–2346.

Hickey CW, Golding LA 2002. Response of macroinvertebrates to copper and zinc in a stream mesocosm. *Environ Toxicol Chem* 21:1854–1863.

Holland AF. 1985. Long-term variation of macrobenthos in a mesohaline region of Chesapeake Bay. *Estuaries* 8:93–113.

Holland AF, Mountford NK, Mihursky JA. 1977. Temporal variation in upper bay mesohaline benthic communities: 1. The 9-m mud habitat. *Chesapeake Sci* 18: 370–378.

Horinko ML. 2002. Principles for managing contaminated sediment risk at hazardous waste sites. Washington DC: US Environmental Protection Agency. OSWER Directive 9285.6-08.

Howard DE, Evans RD. 1993. Acid-volatile sulfide (AVS) in a seasonally anoxic mesotrophic lake: Seasonal and spatial changes in sediment AVS. *Environ Toxicol Chem* 12:1051–1057.

Howarth RW. 1991. Comparative responses of aquatic ecosystems to toxic chemical stress. In: Cole J, Lovett G, Findlay S, editors. Comparative analyses of ecosystems: Patterns, mechanisms, and theories. New York: Springer-Verlag. p 169–195.

Huang W, Weber WJ. 1997. A distributed reactivity model for sorption by soils and sediments 10. Relationships between desorption, hysteresis, the chemical characteristics of organic domains. *Environ Sci Technol* 31:2562–2569.

Hume TM, Fox ME, Wilcock RJ. 1989. Use of organochlorine contaminants to measure sedimentation rates in estuaries: a case study from Manukau Harbour. *J Roy Soc New Zealand* 19:305–317.

Hunter MA, Kan AT, Tomson MB. 1996. Development of a surrogate sediment to study the mechanisms responsible for adsorption/desorption hysteresis. *Environ Sci Technol* 30:2278–2285.

Ianuzzi TJ, Ludwig DF, Kinnell JC, Wallin JM, Desvousges WH, Dunford RW. 2002. A common tragedy: history of an urban river. Amherst (MA): Amherst Scientific. 200 p.

Ingersoll CG, Dillon T, Biddinger GR, editors. 1997. Ecological risk assessment of contaminated sediments. Pensacola (FL): Society of Environmental Toxicology and Chemistry (SETAC).

Jonker MTO, Koelmans AA. 2002. Sorption of polycyclic aromatic hydrocarbons and polychlorinated biphenyls to soot and soot-like materials in the aqueous environment: Mechanistic considerations. *Environ Sci Technol* 36:3725–3734.

Josefson AB, Jensen JN, Ærtebjerg G. 1993. The benthos community structure anomaly in the late 1970s and early 1980s – a result of a major food pulse? *J Exp Mar Biol Ecol* 172:31–45.

Kan AT, Fu G, Hunter MA, Chen W, Ward CH, Tomson MB. 1998. Irreversible sorption of neutral hydrocarbons to sediments: Experimental observations and model predictions. *Environ Sci Technol* 32:892–902.

Kan AT, Fu G, Hunter MA, Tomson MB. 1997. Irreversible adsorption of naphthalene and tetrachlorobiphenyl to lula and surrogate sediments. *Environ Sci Technol* 31:2176–2185.

Kan AT, Fu G, Tomson MB. 1994. Adsorption desorption hysteresis in organic pollutant and soil-sediment interactions. *Environ Sci Technol* 28:859–867.

Karickhoff SW, Brown DS, Scott TA. 1979. Sorption of hydrophobic pollutants on natural sediments. *Water Res* 13:241–248.

Karr JR, Chu EW. 1997. Biological monitoring: essential foundation for ecological risk assessment. *Hum Ecol Risk Assess* 3:993–1004.

Kennett JP. 1982. Marine geology. Englewood Cliffs (NJ): Prentice-Hall.

Kiffney PM, Clements WH. 1996. Effects of metals on stream macroinvertebrate assemblages from different altitudes. *Ecol Appl* 6:472–481.

Knox GA. 1986. Estuarine ecosystems: A systems approach. Boca Raton (FL): CRC.

Lee H, Swartz RC. 1980. Biological processes affecting the distribution of pollutants in marine sediments. Part II. Biodeposition and bioturbation. In: Baker RA, editor. Contaminants and sediments: Volume 2, Analysis, chemistry, biology. Ann Arbor (MI): Ann Arbor Science. p 533–553.

Leonard DE, Mattson VR, Benoit DA, Hoke RA, Ankley GT. 1993. Seasonal variation of acid-volatile sulphide concentration in sediment cores from three north eastern Minnesota lakes. *Hydrobiologia* 271:87–95.

Levings CD. 1975. Analyses of temporal variation in the structure of a shallow-water benthic community in Nova Scotia. *Int Revue ges Hydrobiol* 60:449–470.

Levinton JS. 1982. Marine ecology. Englewood Cliffs (NJ): Prentice-Hall.

Lick W, Rapaka V. 1996. A quantitative analysis of the dynamics of the sorption of hydrophobic organic chemicals to suspended sediments. *Environ Toxicol Chem* 15:1038–1048.

Lie U, Evans RA. 1973. Long-term variability in the structure of subtidal benthic communities in Puget Sound, Washington, USA. *Mar Biol* 21:122–126.

Long ER, Chapman PM. 1985. A sediment quality triad: Measures of sediment contamination, toxicity and infaunal community composition in Puget Sound. *Mar Pollut Bull* 16:405–415.

Long ER, Hong CB, Severn CG. 2001. Relationships between acute sediment toxicity in laboratory tests and abundance and diversity of benthic infauna in marine sediments: A review. *Environ Toxicol Chem* 20:46–60.

Long ER, MacDonald DD, Severn CG, Hong CB. 2000. Classifying probabilities of acute toxicity in marine sediments with empirically derived sediment quality guidelines. *Environ Toxicol Chem* 19:2598–2601.

Lovell CR, Eriksen NT, Lewitus AJ, Chen YP. 2002. Resistance of the marine diatom *Thalassiosira* sp. to toxicity of phenolic compounds. *Mar Ecol Progr Ser* 229: 11–18.

Lu XX, Reible DD, Fleeger JW, Chai YZ. 2003. Bioavailability of desorption-resistant phenanthrene to the oligochaete, *Ilyodrilus templetoni*. *Environ Toxicol Chem* 22: 153–160.

Luoma SN, Clements WH, DeWitt T, Gerritsen J, Hatch A, Jepson P, Reynoldson T, Thom RM. 2001. Role of environmental variability in evaluating stressor effects. In: Baird DJ, Burton Jr GA. Ecological variability: Separating natural from anthropogenic cause of ecosystem impairment. Pensacola (FL): Society of environmental Toxicology and Chemistry (SETAC). p 141–177.

Luthy RG, Aiken GR, Brusseau ML, Cunningham SD, Gschwend PM, Pignatello JJ, Reinhard M, Traina SJ, Weber WJ, Westall JC. 1997. Sequestration of hydrophobic organic contaminants by geosorbents. *Environ Sci Technol* 31: 3341–3347.

MacDonald DD. 1994. Approach to the assessment of sediment quality in Florida coastal waters. Volume 1, Development and evaluation of sediment quality guidelines. Final Report. Tallahassee (FL): Florida Department of Environmental Protection.

MacDonald DD, Ingersoll CG. 2002. Guidance manual to support the assessment of contaminated sediments in the Great Lakes Basin. Chicago: US Environmental Protection Agency Great Lakes National Program Office. USEPA series number pending: In press.

Mackay AP, Mackay S. 1996. Spatial distribution of acid-volatile sulphide concentration and metal bioavailability in mangrove sediment from the Brisbane River, Australia. *Environ Pollut* 93:205–209.

Mancini ER, Steen A, Rausina GA, Wong DCL, Arnold WR, Gostomski GE, Davies T, Hockett JR, Stubblefield WA, Drottar KR, Springer TA, Errico P. 2002. MTBE ambient water quality criteria development: A public/private partnership. *Environ Sci Technol* 36:125–129.

May RM. 1977. Threshold and breakpoints in ecosystems with a multiplicity of stable states. *Nature* 269:471–477.

McGroddy SE, Farrington JW. 1995. Sediment porewater partitioning of polycyclic aromatic hydrocarbons in three cores from Boston Harbor, Massachusetts. *Environ Sci Technol* 29:1542–1550.

McNeill J, Taylor C, Lick W. 1996. Measurements of erosion of undisturbed bottom sediments with depth. *J Hydrol Res* 12:361–369.

Medley CN, Clements WH. 1998. Responses of diatom communities to heavy metals in streams: the influence of longitudinal variation. *Ecol Appl* 8:631–644.

Mills EL. 1969. The community concept in marine zoology, with comments on continua and instability in some marine communities: A review. *J Fish Res Bd Canada* 26:1415–1428.

Millward RN, Klerks PL. 2002. Contaminant adaptation and community tolerance in ecological risk assessment: Introduction. *Hum Ecol Risk Assess* 8:921–932.

Minshall GW, Clements KW, Petersen RC, Cushing CE, Bruns DA, Sedell JR, Vanotte RL. 1985. Developments in stream ecosystem theory. *Can J Fish Aquat Sci* 42:1045–1052.

Modig H, Ólafsson E. 2001. Survival and bioturbation of the amphipod *Monoporeia affinis* in sulphides-rich sediments. *Mar Biol* 138:87–92.

Morrisey DJ, Williamson RB, Van Dam L, Lee DJ. 2000. Stormwater contamination of urban estuaries. 2. Testing a predictive model of the build-up of heavy metals in sediments. *Estuaries* 23:67–79.

Myers CR, Alatalo LJ, Myers JM. 1994. Microbial potential for the anaerobic degradation of simple aromatic compounds in sediments of the Milwaukee Harbor, Green Bay, and Lake Erie. *Environ Toxicol Chem* 13:461–471.

Nichols FH, Thompson JK. 1985. Persistence of an introduced mudflat community in South San Francisco Bay, California. *Mar Ecol Prog Ser* 24:83–97.

Nittrouer CA, Sternberg RW, Carpenter R, Bennett JT. 1979. The use of ^{210}Pb geochronology as a sedimentological tool: Application to the Washington continental shelf. *Mar Geol* 31:297–316.

Officer CB, Biggs RB, Taft JL, Cronin LE, Tyler MA, Boynton WR. 1984. Chesapeake Bay anoxia: origin, development, and significance. *Science* 223: 22–25.

Packman AI, Bencala KE. 2000. Chapter 2: Modeling surface-subsurface hydrological interactions. In: Jones JB, Mulholland PJ, editors. Streams and ground waters. San Diego: Academic. p 46–80.

Paine RT, Levin SA. 1981. Intertidal landscapes: Disturbance and the dynamics of pattern. *Ecol Monogr* 51:145–178.

Parsons TR, Takahashi M, Hargrave B. 1984. Biological oceanographic processes. New York: Pergamon.

Pereira WE, Rostad CE, Chiou CT, Brinton TI, Barbar LBII. 1988. Contamination of estuarine water, biota, and sediment by halogenated organic compounds: A field study. *Environ Sci Technol* 22:772–778.

Perez KT, Morrison GE, Davey EW, Lackie NF, Soper AE, Blasco RJ, Winslow DL, Johnson RL, Murphy PG, Heltshe JF. 1991. Influence of size on fate and ecological effects of kepone in physical models. *Ecol Appl* 1:237–248.

Pignatello JJ, Xing B. 1996. Mechanisms of slow sorption of organic chemicals to natural particles. *Environ Sci Technol* 30:1–11.

Plafkin JL, Barbour MT, Porter KD, Gross SK, Hughes RM. 1989. Rapid bioassessment protocols for use in streams and rivers: Benthic macroinvertebrates and fish. Washington DC: US Environmental Protection Agency, Office of Water. EPA 440-4-89-001.

Poff NL, Ward JV. 1990. Physical habitat template of lotic systems: recovery in the context of historical pattern of spatiotemporal heterogeneity. *Environ Manag* 14: 629–645.

Poore GCB, Rainer S. 1979. A three-year study of benthos of muddy environments in Port Phillip Bay, Victoria. *Estuar Coastal Mar Sci* 9:477–497.

Rabalais NN, Wiseman WJ, Turner RE. 1994. Comparison of continuous records of near-bottom dissolved oxygen from the hypoxia zone along the Louisiana coast. *Estuaries* 17:850–861.

Rapport DJ, Regier HA, Hutchinson TC. 1985. Ecosystem behavior under stress. *Am Nat* 125:617–640.

Readman JW, Mantoura RFC. 1987. A record of polycyclic aromatic hydrocarbon (PAH) pollution obtained from accreting sediments of the Tamar Estuary U.K.: Evidence for nonequilibrium behaviour of PAH. *Total Environ* 66:73–94.

Reible DD, Jensen RH, Bentley SJ, Dannel MB, DePinto JV, Dyer JA, Farley KJ, Garcia MH, Glaser D, Hamrick JM, Lick WJ, Pastorok RA, Schwer RF, Ziegler CK. 2002. The role of modeling in managing contaminated sediments. In: Chien CC, Medina MA Jr, Pinder GF, Reible DR, Sleep BE, Zheng C, editors. Environmental modeling and management: Theory, practice, and future directions. Wilmington (DE): Today Media. p 63–110.

Reible DD, Popov V, Valsaraj KT, Thibodeaux LJ, Lin F, Dikshit M, Todaro MA, Fleeger JW. 1996. Contaminant fluxes from sediment due to tubificid oligochaete bioturbation. *Water Res* 30:704–714.

Reible DD, Valsaraj KT, Thibodeaux LJ. 1991. Chemodynamic models for transport of contaminants from sediment beds. In: Hutzinger O, editor. Handbook of environmental chemistry. Heidelberg: Springer-Verlag. p 187–228.

Rhoads DC, Boyer LF. 1982. The effects of marine benthos on physical properties of sediments: A successional perspective. In: McCall P, Tevesz M, editors. Animal-sediment relations. New York: Plenum. p 3–52.

Rhoads DC, Yingst JY, Ullman WJ. 1978. Seafloor stability in Central Long Island Sound: Part I. Temporal changes in erodibility of fine-grained sediment. In: Wiley ML, editor. Estuarine interactions. New York: Academic. p 221–244.

Rice DL. 1986. Early diagenesis in bioadvective sediments: Relationships between the diagenesis of ^{7}Be, sediment reworking, and the abundance of conveyor-belt deposit feeders. *J Mar Res* 44:149–184.

Ringwood AH, Hameedi MJ, Lee RF, Brouwer M, Peters EC, Scott GI, Luoma SN, DiGiulio RT. 1999. Bivalve biomarker workshop: overview and discussion group summaries. *Biomarkers* 4:391–399.

Ringwood AH, Keppler CJ. 2002. Water quality variation and clam growth: Is pH really a non-issue in estuaries? *Estuaries* 25:901–907.

Roach AC, Jones AR, Murray A. 2001. Using benthic recruitment to assess the significance of contaminated sediments: The influence of taxonomic resolution. *Environ Pollut* 112:131–143.

Rosenberg R. 1990. Negative oxygen trends in Swedish coastal bottom waters. *Mar Pollut Bull* 21:335–339.

Savant SA, Reible DD, Thibodeaux LJ. 1987. Modeling convective transport in stable river sediments. *Water Resour Res* 23:1763–1768.

Seliger HH, Boggs JA. 1988. Long term pattern of anoxia in the Chesapeake Bay. In: Lynch MP, Krome EC, editors. Understanding the estuary: Advances in Chesapeake Bay research. Gloucester Point (VA): Chesapeake Research Consortium. p 87–106.

Seliger HH, Boggs JA, Biggley WH. 1985. Catastrophic anoxia in the Chesapeake Bay in 1984. *Science* 228:70–73.

[SFT] Norwegian Pollution Control Authority. 2002. Appendix 1 to the Activity Regulations: Requirements for environmental monitoring of the petroleum activity on the Norwegian Continental Shelf. Available from: http://www.ukooa.co.uk/issues/drillcuttings/rdintro.htm.

Sijm D, de Bruijn J, Crommentuijn T, van Leeuwen K. 2001. Environmental quality standards: Endpoints or triggers for a tiered ecological effect assessment approach? *Environ Toxicol Chem* 20:2644–2648.

Stehr CM, Brown DW, Hom T, Anulacion BF. 2000. Exposure of juvenile chinook and chum salmon to chemical contaminants in the Hylebos Waterway of Commencement Bay, Tacoma, Washington. *J Aquat Ecosyst Stress Recov* 7:215–227.

Stephan CE. 1985. Guidelines for deriving numerical national water quality criteria for the protection of aquatic organisms and their uses. PB85-227049. Washington DC: US Environmental Protection Agency, Office of Water Regulations and Standards.

Stewart AJ, Hill WR. 1993. Grazers, periphyton and toxicant movement in streams. *Environ Toxicol Chem* 12:955–957.

Suter II GW. 1993. Ecological risk assessment. Chelsea (MI): Lewis.

Suter II GW, Norton SB, Cormier SM. 2002. A methodology for inferring the causes of observed impairments in aquatic ecosystems. *Environ Toxicol Chem* 21:1101–1111.

Swartz RC, Di Toro DM. 1997. Sediments as complex mixtures: An overview of methods to assess ecotoxicological significance. In: Ingersoll CG, Dillon T, Biddinger GR, editors. Ecological risk assessment of contaminated sediments. Pensacola (FL): Society of Environmental Toxicology and Chemistry (SETAC). p 255–270.

Tenzer GE, Meyers PA, Robbins JA, Eadie BJ, Morehead NR, Lansing MB. 1999. Sedimentary organic matter record of recent environmental changes in the St. Marys River ecosystem, Michigan-Ontario border. *Org Geochem* 30:133–146.

Thoms SR, Matisoff G, McCall PL, Wang X. 1995. Models for alteration of sediments by benthic organisms. Final Report Project 92-NPS-2. Alexandria (VA): Water Environment Research Foundation.

Thrush SF, Hewitt JE, Cummings VJ, Green MO, Funnell GA, Wilkinson MR. 2000. The generality of field experiments: Interactions between local and broad-scale processes. *Ecology* 81:399–415.

[TNO] TNO Institute of Environmental Sciences, Energy Research and Process Innovation. 1999. Assessment of sediment contamination and biological effects around former OBM drilling locations on the Dutch Continental Shelf. Delft (NL): TNO. Report R-98/515. TNO-MEP.

Tsai C, Lick W. 1986. A portable device for measuring sediment resuspension. *J Gt Lakes Res* 12:314–321.

[UKOOA] United Kingdom Offshore Operators Association. 2002. UKOOA drill cuttings initiative final report. Available from: http://www.ukooa.co.uk/issues/drillcutti ngs/rdintro.htm. Accessed 21 Aug 2004.

URS. 2001. Liberty Development 2001 sediment quality study. Final report. Anchorage: BP Exploration (Alaska) Inc.

[USACE/USEPA] US Army Corps of Engineers NY District/US Environmental Protection Agency. 1994. Draft guidance for performing tests on dredged material proposed for ocean disposal. New York.

[USEPA] US Environmental Protection Agency. 1997. The incidence and severity of sediment contamination in surface waters of the United States, Volume 1: National Sediment Quality Survey. Washington DC: USEPA, Office of Water. EPA 823-R-97-006.

[USEPA] US Environmental Protection Agency. 2001a. Notice of proposed NPDES general permit for discharges from the coastal subcategory of the oil and gas extraction point source category in Texas (TXG330000). *Federal Register* 66: 6607–6610.

[USEPA] US Environmental Protection Agency. 2001b. Effluent guidelines: synthetic-based drilling fluids in the oil and gas extraction point source category – effluent limitations guidelines and new source performance standards. Available from: www.epa.gov/waterscience/guide/sbf/. Accessed 21 Aug 2004.

[USEPA] US Environmental Protection Agency. 2002. Technical basis for the derivation of equilibrium partitioning sediment benchmarks (ESBs) for the protection of benthic organisms: nonionic organics. Technical Report. Washington DC: USEPA Office of Research and Development. EPA-600-R-02-014.

[USEPA/USACE] US Environmental Protection Agency/US Army Corps of Engineers. 1977. Ecological evaluation of proposed discharge of dredged material into ocean waters. Vicksburg (MI): US Army Corps of Engineers, Waterways Experiment Station.

[USEPA/USACE] US Environmental Protection Agency/US Army Corps of Engineers. 1998. Evaluation of dredged material proposed for discharge in waters of the U.S. Testing manual. Washington DC: USEPA/USACE. EPA-823-B-98-004.

Valette-Silver NJ. 1993. The use of sediment cores to reconstruct historical trends in contamination of estuarine and coastal sediments. *Estuaries* 16:577–588.

Valiela I. 1984. Marine ecological processes. New York: Springer-Verlag.

Van Cappellen P, Wang Y. 1995. Metal cycling in surface sediments: Modeling the interplay of transport and reaction. In: Allen HE, editor. Metal contaminated sediments. Champaign (MI): Ann Arbor. p 21–64.

Van den Berg GA, Loch JPG, van der Heijdt LM, Zwolsman JJG. 1998. Vertical distribution of acid-volatile sulfide and simultaneously extracted metals in a recent sedimentation area of the River Meuse in the Netherlands. *Environ Toxicol Chem* 17:758–763.

Van den Hoop MAGT, den Hollander HA, Kerdijk HN. 1997. Spatial and seasonal variations of acid volatile sulphide (AVS) and simultaneously extracted metals (SEM) in Dutch marine and freshwater sediments. *Chemosphere* 35:2307–2316.

Van de Koppel J, Herman PMJ, Thoolen P, Heip CHR. 2001. Do alternate stable states occur in natural ecosystems? Evidence from a tidal flat. *Ecology* 82:3449–3461.

Vannote RL, Minshall GW, Cummins KW, Sedell JR, Cushing CE. 1980. The river continuum concept. *Can J Fish Aquat Sci* 30:130–137.

Vargas C, Fennell DE, Haeggblom MM. 2001. Anaerobic reductive dechlorination of chlorinated dioxins. *Appl Microbiol Biotech* 57:786–790.

Warren LA, Tessier A, Hare L. 1998. Modeling cadmium accumulation by benthic invertebrates in situ: The relative contributions of sediment and overlying water reservoirs to organism cadmium concentrations. *Limnol Oceanogr* 43:1442–1454.

Washington HG. 1984. Diversity, biotic and similarity indices: A review with special relevance to aquatic ecosystems. *Water Res* 18:653–694.

Weis JL. 2002. Tolerance to environmental contaminants in the mummichog, *Fundulus heteroclitus*. *Hum Ecol Risk Assess* 8:933–954.

Wijsman JWM, Middelburg JJ, Herman PMJ, Bottcher ME, Heip CHR. 2001. Sulfur and iron speciation in surface sediments along the northwestern margin of the Black Sea. *Mar Chem* 74:261–278.

Wiklund A-KE, Sundelin B. 2002. Bioavailability of metals to the amphipod *Monoporeia affinis*: interactions with authigenic sulfides in urban brackish-water and freshwater sediments. *Environ Toxicol Chem* 21:1219–1228.

Williamson RB, Morrisey DJ. 2000. Stormwater contamination of urban estuaries. 1. Predicting the build-up of heavy metals in sediments. *Estuaries* 23:56–66.

Work PA, Moore PR, Reible DD. 2002. Bioturbation, advection and diffusion of a conserved contaminant in a laboratory flume. *Water Resour Res* 38:241–249.

Yan Q-L, Wang W-X. 2002. Metal exposure and bioavailability to a marine deposit-feeding Sipuncula, *Sipunculus nudus*. *Environ Sci Technol* 36:40–47.

Yunker MB, Cretney WJ, Ikonomou MG. 2002. Assessment of chlorinated dibenzo-*p*-dioxin and dibenzofuran trends in sediment and crab hepatopancreas from pulp mill and harbor sites using multivariate- and index-based approaches. *Environ Sci Technol* 36:1869–1878.

Abbreviations

AET	apparent effects threshold
ANOVA	analysis of variance
ANZECC	Australian and New Zealand Environment and Conservation Council
AOC	area of concern
AQEM	Assessment System for the Ecological Quality of Streams and Rivers throughout Europe using Benthic Macroinvertebrates
ARCS	Assessment and Remediation of Contaminated Sediments (US)
ARGE Elbe	Arbeitsgemeinschaft für die Reinhaltung der elbe (Germany)
ARMCANZ	Agriculture and Resource Management Council of Australia and New Zealand
ASTM	American Society for Testing and Materials
AT	averaging time
ATV	German Association for Water Pollution Control
AVS	acid volatile sulfide
AWQC	ambient water quality criteria
BACI	before and after, control and impact
BC	background concentration
BCF	bioconcentration factor
BCL	base chemical level
BEAST	Benthic Assessment of Sediment Toxicity
BEDS	biological effects database for sediments
BF	bioaccumulation factor
BfG	Federal Institute for Hydrology (Germany)
BIBI	Benthic Index of Biotic Integrity
BLM	Biotic Ligand Model
BMV	Federal Ministry of Transport (Germany)
BSAF	biota–sediment accumulation factor
BW	body weight
CARP	Contaminant Assessment and Reduction Project
CBR	critical body residue
CCME	Canadian Council of Ministers of the Environment
CDF	confined disposal facility
CE	critical element
CEPA	Canadian Environmental Protection Act
CERCLA	Comprehensive Environmental Response, Compensation, and Liability Act of 1980 ("Superfund" US)
CF	concentration of contaminant in fish
CF	confounding factors
CFR	Code of Federal Regulations (US)
CH	chlorinated hydrocarbon
CI	control and impact
C/I	control/impact
CIW	Commissie Integraal Waterbeheer (The Netherlands)
CLPP	community-level physiological profileling
COC	contaminant of concern
COD	chemical oxygen demand
COPC	chemical potential concern
CTT	chemistry toxicity test
CUWVO	Dutch Commission for the Implementation of the Act on Pollution of Surface Waters
CV	coefficient of variation
CWA	Clean Water Act (US)
DCE	Director of Civil Engineering (Hong Kong)

DEP	Director of Environmental Protection (Hong Kong)
DGGE	denaturing gradient gel electrophoresis
DGT	diffusive gradients in thin film
DM	dredged material
DMAF	dredged material assessment framework
DMRP	Dredged Material Research Program (US)
DMSO	dimethyl sulfoxide
DNA	deoxyribonucleic acid
DO	dissolved oxygen
DOC	dissolved organic carbon
DSAY	discounted service acre year
EAC	ecotoxicological assessment criteria
EBI	extended biotic index
ED	exposure duration
EEC	extreme effects concentration
Eh	redox potential
ELA	effects level approach
EMAP	Environmental Monitoring and Assessment Program (US)
EOX	extractable organohalogen
EPIWIN	Estimations Programs Interface for Windows
EqP	equilibrium partitioning
EQS	environmental quality stgandard
ER50	effective residue
ERA	ecological risk assessment, effects range approach
ERED	Environmental Residue Effects Database (US)
ERL	effects range low
ERM	effects range median
ERM*i*	individual ERM
ERM-Q	effects range median quotient
ESB	EqP sediment benchmark
ESG	equilibrium partitioning sediment guideline
ETWB	Environment Transport and Works Bureau (Hong Kong)
EU	European Union
FCV	final chronic value
FMC	Fill Management Committee (Hong Kong)
FWPCA	Federal Water Pollution Control Act (US)
FWQA	Federal Water Quality Administration
GC	gas chromatography
GIPME	Global Investigation of Pollution in the Marine Environment
GIS	geographic information system
GLNPO	Great Lakes National Program Office (US)
HAET	highest apparent effects threshold
HARS	Historic Area Remediation Site
HCB	hexachlorobenzene
HEA	habitat equivalency analysis
HELCOM	Helsinki Commission (Finland)
HEP	Harbor Estuary Program (NY–NJ)
HPAH	high-molecular-weight PAH
HPLC	high-performance liquid chromatography
HRGS	Human Reporter Gene System
HRL	high risk level
IBE	Indice Biotico Esteso
ICPR	International Commission for the Protection of the Rhine
ICRAM	Istituto Centrale per la Ricerca Scientifica e Tecnologica Applicata al Mare
ICRCL	Interdepartmental Committee on the Redevelopment of Land
IJC	International Joint Commission
IMO	International Maritime Organization
IR	ingestion rate
IRIS	Integrated Risk Information System (US)
ISO	International Organization for Standardization
ISQV	interim sediment quality value
ITM	Inland Testing Manual
IWTU	interstitial water toxic unit
LADD	lifetime average daily dose
LAET	lowest apparent effects threshold
LC	London Convention
LC50	lethal concentration for 50% of test organisms
LCEL	lower chemical exceedance level
LCSG	London Convention Scientific Group

LDC London Dumping Convention
LEL lowest effect level
LOBS loss of benthic services
LOAEL lowest observed adverse effect level
LOE line of evidence
LOEC lowest-observed-effect concentration
LOER lowest observed effect residue
LPAH low-molecular-weight PAH
LPDE low-density polyethylene
LR50 median lethal residue
LRM logistic regression modeling

MANOVA multiple analysis of variance
MC/I multiple control/impact
ME method error
ME midrange effect, median effect
MEC median effect concentration
MNR monitored natural recovery
MOA memorandum of agreement
MPA maximum permissible addition
MPC maximum permissible concentration
MPRSA Marine Protection, Research and Sanctuaries Act (US)
MTBE methyl tert-butyl ether

NAPL nonaqueous-phase liquid
NEC no-effect concentration
NEL no-effect level
NOAA National Oceanic and Atmospheric Administration (US)
NOEC no-observed-effects concentration
NOAEL no-observed adverse effect level
NPHC nonpolar hydrocarbon or mineral oil
NRC National Research Council (US)
NRDA natural resource damage assessment
NSI National Sediment Inventory (US)
NSQS National Sediment Quality Survey (US)
NST National Status and Trends (US)
NY–NJ New York–New Jersey
NYSDEC New York State Department of Environmental Conservation

OBM oil-based mud
OC organic carbon
OCP organochlorine pesticide
OECD Organization for Economic Cooperation and Development
OM organic matter
ORD Office of Research and Development (USEPA)
OSPAR Convention for the Protection of the Marine Environment of the North-East Atlantic
OTM Ocean Testing Manual
OW Office of Water (USEPA)

PA polyacrylate
PAET probable apparent effects threshold
PAH polycyclic aromatic hydrocarbon
PCA principal component analysis
PCB polychlorinated biphenyl
PCR polymerase chain reaction
PDMS polydimethylsiloxane
PE probable effect
PEC probable effects concentration
PEL probable effects level
PERA probabilistic ecological risk assessment
PES particle entrainment simulation
PNEC predicted no-effect concentration
POC particulate organic carbon
POM particulate organic matter
PRP potentially responsible party
PSDDA Puget Sound Dredged Disposal Analysis Program (US)
PSEP Puget Sound Estuary Program
PSTU predicted sediment toxic unit
PWTU porewater toxic unit

QA quality assurance
QC quality control
Q-PCR quantitative polymerase chain reaction
QSAR quantitative structure–activity relationship

RBCA risk-based corrective action
RCA reference condition approach
RCC river continuum concept
RCIE Reasonably Conservative Injury Evaluation
REMAP Regional Environmental Monitoring and Assessment Program (US)

RMG radial or multiple gradient
RNA ribonucleic acid
ROC receptor of concern
RSS root sum of squares
RV reference value

SAB Science Advisory Board
SARA Superfund Amendments and Reauthorization Act of 1986 (US)
SCB Southern California Bight
SEC sediment effect concentration
SEL severe effect level
SEM simultaneously extracted metal
SEQ-Eau Système d'Evaluation de la Qualité de l'Eau
SETAC Society of Environmental Toxicology and Chemistry
SFT Norwegian Pollution control Authority
SG simple gradient
SLC screening-level concentration
SMDC scientific/management decision point
SOPC sum of pesticides
SP solid phase
SPMD semipermeable membrane devices
SPME solid-phase micro-extraction
SQG sediment quality guideline
SQG-Q sediment quality guideline quotient
SQT sediment quality target
SQT Sediment Quality Triad
SSD species sensitivity distribution
SSLC species screening-level concentration
SSP suspended particulate pahse
STARR selection, testing, accounting, removal, replacement

TAN total ammonia as nitrogen
TBP theoretical bioaccumulation potential
TBT tributyl tin
TCA trichloroaniline
TCL threshold chemical level
TE threshold effect, total error
TEC threshold effects concentration
TEL threshold effects level
TIE toxicity identification evaluation
TIO tentatively identified organic compound
TI/RE toxicity identification/reduction evaluation
TKN total Kjeldahl notrogen
TMDL total maximum daily load
TNO TNO Institute of Environmental Sciences, Energy Research and Process Innovation (The Netherlands)
TOC total organic carbon
TON total organic nitrogen
TSS total suspended solid
TTP trophic transfer potential
TU toxic unit

UCEL upper chemical exceedance level
UK United Kingdom
UKOOA United Kingdom Offshore Operators Association
UNEP United Nations Environment Programme
US United States
USACE United States Army Corps of Engineers
USEPA United States Environmental Protection Agency
USFDA United States Food and Drug Administration
USGS United States Geological Survey

WAC Washington Adminisrative Code
WDOE Washington Department of Ecology
WEF Water Environment Federation
WOE weight of evidence
WQC water quality criteria
WQS water quality standard
WRDA Water Resources Development Act (US)

Index

C

D

E

I

M

N

Q

S

T

SETAC

A Professional Society for Environmental Scientists and Engineers and Related Disciplines Concerned with Environmental Quality

The Society of Environmental Toxicology and Chemistry (SETAC), with offices currently in North America and Europe, is a nonprofit, professional society established to provide a forum for individuals and institutions engaged in the study of environmental problems, management and regulation of natural resources, education, research and development, and manufacturing and distribution.

Specific goals of the society are:

- Promote research, education, and training in the environmental sciences.
- Promote the systematic application of all relevant scientific disciplines to the evaluation of chemical hazards.
- Participate in the scientific interpretation of issues concerned with hazard assessment and risk analysis.
- Support the development of ecologically acceptable practices and principles.
- Provide a forum (meetings and publications) for communication among professionals in government, business, academia, and other segments of society involved in the use, protection, and management of our environment.

These goals are pursued through the conduct of numerous activities, which include:

- Hold annual meetings with study and workshop sessions, platform and poster papers, and achievement and merit awards.
- Sponsor monthly and quarterly scientific journals, a newsletter, and special technical publications.
- Provide funds for education and training through the SETAC Scholarship/Fellowship Program.
- Organize and sponsor chapters to provide a forum for the presentation of scientific data and for the interchange and study of information about local concerns.
- Provide advice and counsel to technical and nontechnical persons through a number of standing and ad hoc committees.

SETAC membership currently is composed of more than 5,000 individuals from government, academia, business, and public-interest groups with technical backgrounds in chemistry, toxicology, biology, ecology, atmospheric sciences, health sciences, earth sciences, and engineering. If you have training in these or related disciplines and are engaged in the study, use, or management of environmental resources, SETAC can fulfill your professional affiliation needs.

All members receive a newsletter highlighting environmental topics and SETAC activities, and reduced fees for the Annual Meeting and SETAC special publications. All members except Students and Senior Active Members receive monthly issues of Environmental Toxicology and Chemistry (ET&C), a peer-reviewed journal of the Society. Student and Senior Active Members may subscribe to the journal. Members may hold office and, with the Emeritus Members, constitute the voting membership.

If you desire further information, contact the appropriate SETAC Office.

1010 North 12th Avenue	Avenue de la Toison d'Or 67
Pensacola, Florida 32501-3367 USA	B-1060 Brussels, Belgium
T 850 469 1500 F 850 469 9778	T 32 2 772 72 81 F 32 2 770 53 83
E setac@setac.org	E setac@setaceu.org

www.setac.org
Environmental Quality Through Science®